AF881165

Biomarkers in Medicine

Edited by

Pınar Atukeren
Department of Medical Biochemistry,
Cerrahpasa Faculty of Medicine,
Istanbul University-Cerrahpasa,
Istanbul, Turkey

Faculty of Pharmacy,
Istanbul University-Cerrahpasa,
Istanbul, Turkey

&

Hafize Uzun
Department of Biochemistry,
Faculty of Medicine
Istanbul Atlas University,
Istanbul, Turkey

Biomarkers in Medicine

Editors: Pınar Atukeren & Hafize Uzun

ISBN (Online): 978-981-5040-46-3

ISBN (Print): 978-981-5040-47-0

ISBN (Paperback): 978-981-5040-48-7

First published in 2022.

Bentham Science Publishers Pte. Ltd.
80 Robinson Road #02-00
Singapore 068898
Singapore
Email: subscriptions@benthamscience.net

BENTHAM SCIENCE

CONTENTS

PREFACE

"Basically, there are two things in life, science and personal thoughts. The first leads to knowledge, the second to ignorance (Hippocrates)."

This detailed book will provide a vital guide to further understanding the current and future biomarkers, one of the most important trends in healthcare today. This book will facilitate the combination of therapeutics with diagnostics and will consequently feature the key correlations between diseases and classifications of biomarkers. The book has chapters on topics such as cancer, neuroscience, cardiology, immunology/immunotherapy, metabolism, pharmacology, haematology, obstetrics and gynecology, hepatology, aging, obesity, urology, nephrology, microbiology, gastroenterology, pediatrics, surgery, pulmonary diseases, pathology and also emphasizes on the preclinical and clinical manifestation of the injury and disease process. This book will supply the unique insight of an expert with extensive experience in diagnostics and clinical laboratory, offering case studies and practical examples from different classes of biomarkers on different platforms, including new data for biomarkers in different therapeutic indications. Clinicians managing patients or clinical trials, clinical researchers, clinical laboratories, diagnostic companies, regulatory agencies, medical school graduate students, academic students, and the general public will all benefit from this book. We would like to thank all my colleagues, Bentham publishing house, and the staff for their support in the preparation of this book. A good book is a real treasure.

Pınar Atukeren
Department of Biochemistry
Cerrahpasa Faculty of Medicine
Istanbul University-Cerrahpasa
Istanbul, Turkey
Faculty of Pharmacy
Istanbul University-Cerrahpasa
Istanbul, Turkey

&

Hafize Uzun
Department of Biochemistry
Faculty of Medicine, Istanbul Atlas University
Istanbul, Turkey

List of Contributors

Abdullah Sonsuz	Department of Internal Medicine-Gastroenterology, Cerrahpasa Medical Faculty, Istanbul University – Cerrahpasa, Istanbul, Turkey
Abdurrahman Yigit	Department of Internal Medicine and Medical Oncology, Canakkale 18 Mart University Medical Faculty, Canakkale, Turkey
Ahmet Gurbuz	Department of Urology, Istanbul University- Cerrahpasa, Cerrahpasa School of Medicine, Istanbul, Turkey
Ahmet Emre Eşkazan	Division of Hematology, Department of Internal Medicine, Cerrahpasa Faculty of Medicine, Istanbul University-Cerrahpasa, Istanbul, Turkey
Ali Kirik	Department of Internal Medicine, Balikesir University Medical Faculty, Balıkesir, Turkey
Ali Uğur Soysal	Department of Cardiology, Istanbul University-Cerrahpaşa, Cerrahpaşa Faculty of Medicine, Istanbul, Turkey
Anıl Orhan	Department of General Surgery, University of Health Sciences, Haseki Training and Research Hospital, Istanbul, Turkey
Aykut Oruc	Department of Physiology, Istanbul University-Cerrahpaşa, Cerrahpasa Faculty of Medicine Istanbul, Istanbul, Turkey
Ayse Pelin Yigider	Istanbul Research and Training Hospital Otorhinolaryngology, Istanbul, Turkey
Aysun Tekeli Taskomur	Department of Obstetrics and Gynecology, Amasya University, Faculty of Medicine, Amasya, Turkey
Berin Selcuk	Department of Urology, Istanbul University- Cerrahpasa, Cerrahpasa School of Medicine, Istanbul, Turkey
Berrin Papila Kundaktepe	Department of General Surgery, Istanbul University-Cerrahpasa, Cerrahpasa Medical Faculty, İstanbul, Turkey
Bulent Onal	Department of Urology, Istanbul University- Cerrahpasa, Cerrahpasa School of Medicine, Istanbul, Turkey
Burcu Yesilkaya	Departments of Nutrition and Dietetics, İstanbul Okan University, İstanbul, Turkey
Cigdem Usul Afsar	Department of Internal Medicine and Medical Oncology, Istinye University Medical Faculty,, İstinye Üniversitesi Topkapı Kampüsü, Maltepe Mah., Teyyareci Sami Sk., No. 3 Zeytinburnu, Istanbul 34010, Turkey
Deniz Özen	Department of Medical Pharmacology, Istanbul University-Cerrahpasa, Cerrahpasa Medical Faculty, Istanbul, Turkey
Fatih Orkun Kundaktepe	Department of Internal Medicine, Istanbul Taksim Research and Training Hospital, Istanbul, Turkey
Fatma Köksal Çakırlar	Department of Medical Microbiology, Faculty of Cerrahpaşa Medicine, University of İstanbul-Cerrahpaşa, Istanbul, Turkey
Gonul Simsek	Department of Physiology, Istanbul University-Cerrahpaşa, Cerrahpasa Faculty of Medicine Istanbul, Istanbul, Turkey
Hafize Uzun	Department of Biochemistry, Faculty of Medicine, Istanbul Atlas University, Istanbul, Turkey

Halit Eren Taskin	Department of General Surgery, Cerrahpasa Medical Faculty, Istanbul University-Cerrahpasa, Istanbul, Turkey
Ismail BIYIK	Department of Obstetrics and Gynecology, Kutahya Health Sciences University, Kutahya, Turkey
Kivanc Yalin	Department of Cardiology, Istanbul University-Cerrahpaşa, Cerrahpaşa Faculty of Medicine, Istanbul, Turkey
Lebriz Uslu-Beşli	Department of Nuclear Medicine, Istanbul University-Cerrahpasa, Cerrahpasa Medical Faculty, Istanbul, Turkey
Mehmet Hamza Gultekin	Department of Urology, Istanbul University- Cerrahpasa, Cerrahpasa School of Medicine, Istanbul, Turkey
Muhammed Fatih Simsekoglu	Department of Urology, Istanbul University- Cerrahpasa, Cerrahpasa School of Medicine, Istanbul, Turkey
Mustafa Albayrak	Department of Obstetrics and Gynecology, Sisli Florence Nightingale Hospital, Istanbul, Turkey
Nuran Dariyerli	Faculty of Medicine, Istanbul University-Cerrahpaşa, Cerrahpaşa, Istanbul, Turkey
Nurhan Seyahi	Department of Nephrology, Cerrahpasa Medical Faculty, Istanbul University, Cerrahpasa, Turkey
Nurgül Özgür Yurttaş	Division of Hematology, Department of Internal Medicine, Cerrahpasa Faculty of Medicine, Istanbul University-Cerrahpasa, Istanbul, Turkey
Oguz Kagan Bakkaloglu	Department of Internal Medicine-Gastroenterology, Cerrahpasa Medical Faculty, Istanbul University – Cerrahpasa, Istanbul, Turkey
Ozgur Yigit	Istanbul Research and Training Hospital Otorhinolaryngology, Istanbul, Turkey
Ozlem Erten	Department of Obstetrics and Gynecology, Kutahya Health Sciences University, School of Medicine, Kutahya, Turkey
Pelin Uysal	Department of Chest Diseases, Faculty of Medicine, Mehmet Ali Aydınlar University, Atakent Hospital, Istanbul, Turkey
Pınar Atukeren	Department of Medical Biochemistry, Cerrahpaşa Faculty of Medicine, Istanbul University-Cerrahpasa, Istanbul, Turkey Faculty of Pharmacy, Istanbul University-Cerrahpasa, Istanbul, Turkey
Ravindri Jayasinghe	Department of Surgery, Faculty of Medicine, University of Colombo, Colombo, Sri Lanka
Remise Gelisgen	Department of Biochemistry, Cerrahpasa Faculty of Medicine, Istanbul University-Cerrahpasa, Istanbul, Turkey
Ruhsen Ozcaglayan	Department of Internal Medicine, Balikesir University Medical Faculty, Balıkesir, Turkey
Sanjeewa Seneviratne	Department of Surgery, Faculty of Medicine, University of Colombo, Colombo, Sri Lanka
Selim Afsar	Department of Obstetrics and Gynecology, Balıkesir University Medical Faculty, Balıkesir, Turkey
Seyda Gul Ozcan	Department of Internal Medicine, Cerrahpasa Medical Faculty, Istanbul University, Cerrahpasa, Turkey

Sibel Özyazgan	Department of Medical Pharmacology, Istanbul University-Cerrahpasa, Cerrahpasa Medical Faculty, Istanbul, Turkey
Sila H. Bozdogan Polat	Faculty of Medicine, Istanbul University-Cerrahpaşa, Cerrahpaşa, Istanbul, Turkey
Sinem Durmus	Department of Biochemistry, Cerrahpasa Faculty of Medicine, Istanbul University-Cerrahpasa, Istanbul, Turkey
Suleyman Demiryas	Department of General Surgery, Istanbul University – Cerrahpaşa, Cerrahpaşa School of Medicine, Istanbul, Turkey
Umesh Jayarajah	Department of Surgery, Faculty of Medicine, University of Colombo, Colombo, Sri Lanka
Yusuf Elgormus	Department of Pediatric Health and Diseases, Medicine Hospital, Istanbul, Turkey
Zeynep Gizem Todurga Seven	Department of Medical Pharmacology, Istanbul University-Cerrahpasa, Cerrahpasa Medical Faculty, Istanbul, Turkey

CHAPTER 1

New Biomarkers and Immunotherapy Decision

Abdurrahman Yigit[1], **Berkay Kuscu**[2], **Ali Kirik**[2], **Ruhsen Ozcaglayan**[2] and **Cigdem Usul Afsar**[3,*]

[1] *Department of Internal Medicine and Medical Oncology, Canakkale 18 Mart University Medical Faculty, Canakkale, Turkey*

[2] *Department of Internal Medicine, Balikesir University Medical Faculty, Balıkesir, Turkey*

[3] *Department of Internal Medicine and Medical Oncology, Istinye University Medical Faculty, İstinye Üniversitesi Topkapı Kampüsü, Maltepe Mah., Teyyareci Sami Sk., No.3 Zeytinburnu, Istanbul 34010, Turkey*

Abstract: As immune checkpoint blockade and other immune-based therapy approaches lead to broad treatment advances among patients with advanced cancer, an important consideration is how to best select patients whose tumors will respond to these therapies. As a consequence predictive and prognostic markers are needed. There are genomic features, such as tumour mutation burden (TMB), microsatellite instability (MSI), and immune phenotype features, such as programmed death-ligand 1 (PD-L1), CTLA-4 and tumour infiltrating lymphocytes (TILs), to predict response to immunotherapies (ITs). Several studies show the correlation between TMB and predicted neoantigen load across multiple cancer types. Response to immune checkpoint inhibitors is higher in tumours with high TMB. The candidate biomarker that has been studied mostly other than TMB is PD-L1 expression in trials utilizing programmed cell death-1 (PD-1) blockade. PD-L1 and PD-1 expression are dynamic markers that change in relation to local cytokines and other factors, and the thresholds that separate "positive" and "negative" PD-L1 expressions remain under debate. PD-L1 expression is now a routine diagnostic marker for patients with newly diagnosed NSCLC. The potential applicability of PD-L1 in other disease settings is still uncertain. Microsatellite instability is characterised by high rates of alterations to repetitive DNA sequences caused by impaired mismatch repair (MMR); MSI was the biomarker was approved according to tumor's initial location. Combining TMB with specific genomic alterations is crucial. Moreover, new biomarkers are being investigated.

Keywords: Checkpoint inhibitor, Immunotherapy, MSI, PD-L1, PD1, Predictive , Prognostic , TIL, TMB.

* **Corresponding author Cigdem Usul Afsar:** Department of Internal Medicine and Medical Oncology, Istinye University Medical Faculty, İstinye Üniversitesi Topkapı Kampüsü, Maltepe Mah., Teyyareci Sami Sk., No.3 Zeytinburnu, Istanbul 34010, Turkey; Tel: +90 (530) 016 42 58; E-mails: cigdemusul@yahoo.com, cigdem.afsar@istinye.edu.tr

Pınar Atukeren and Hafize Uzun (Eds.)

NEW BIOMARKERS AND IMMUNOTHERAPY DECISION

Principles of Cancer İmmunotherapy

Immunotherapy (IT) is a type of biological therapy. Biological therapy is a type of treatment that uses cytokines made from living organisms to treat cancer. What is expected is that the immune system detects and destroys the abnormal cells and likely inhibits the growth of many cancers. For example, immune cells are sometimes found in and around tumors. These cells, called tumor-infiltrating lymphocytes or TILs, are a sign that the immune system is responding to the tumor. People whose tumors contain TIL generally do better than people who do not have tumors [1].

Even though the immune system can prevent or slow cancer growth, cancer cells have ways to avoid destruction mediated by the immune system. Cancer cells may have genetic changes that make them less visible to the immune system; they also have proteins on their surface that turn off immune cells and change the normal cells around the tumor, thus interfering with how the immune system responds to the cancer cells [2]. Immunotherapy helps the immune system to act better against cancer.

Tumor Immunology

Cell types involved in tumor recognition and rejection — An effective and specific cytotoxic immune response against a tumor requires a complex, serially evolving interaction between various immune cell types in the adaptive and innate immune systems. The Th1/Th2 subclasses of CD8+ lymphocytes and CD4+ T lymphocytes are conventionally referred to as cytotoxic T cells and helper T cells. CD8+ and CD4+ lymphocytes initiate the distinction between self and non-self antigens through recognition at the "immune synapse." Natural killer (NK) cells do not require antigen presentation by the major histocompatibility complex (MHC) for cytotoxic activity. In fact, NK cells target cells with low MHC class 1 expression for degradation. Like T cells, NK cells express numerous inhibitory molecules, notably the very different lethal immunoglobulin-like receptor (KIR) subtypes [3].

Additional cell types, such as FoxP3+, CD25+, CD4+, T regulatory (Treg), and myeloid-derived suppressor cells (MDSCs), largely inhibit cytotoxic T lymphocyte activity [4, 5]. Th17 cells, subsets of CD4+ T cells that secrete interleukin (IL)-17, are implicated in autoimmunity and cancer [6]. Macrophages differentiate into at least two different phenotypes: M1 macrophages, which release interferon (IFN) gamma and are responsible for phagocytosis, and M2 macrophages, which release cytokines, such as IL-4, IL-10, transforming growth factor-beta (TGF-beta), and dampen inflammatory responses and foster tolerance [7]. The "immune synapse," the most widely studied phenomenon in immunologic surveillance, is the ability of T lymphocytes to distinguish self *versus* non-self antigens, which are presented by antigen-presenting cells (APCs) such as dendritic cells. Overall, the cytotoxic activity of a CD8+ T cell is regulated by the presence and spatial orientation of a set of stimulatory and inhibitory receptors whose expression is regulated by a myriad of cytokines. Together, this configuration is often referred to as the "immune synapse."

Therapeutic Approaches

A number of therapeutic approaches are being studied to unleash the immune system and control malignancy. These approaches include cytokines, T cells (checkpoint inhibitors, agonism of costimulatory receptors), manipulation of T cells, oncolytic viruses, therapies directed at other cell types, and vaccines.

Cytokines — Early approaches to immunotherapy exploited the numerous diverse effects of cytokines and other substances that affect immune cell activity. Examples include Interleukin (IL)-2, which was originally discovered as a T cell growth factor. IL-2 has pleiotropic effects on both cytotoxic T cell function and T-regulatory (Treg) cell maintenance. Effects vary in part with the dose and timing of IL-2 administration [8]. Although IL-2 use has been largely supplanted by the use of checkpoint inhibitors, bolus high-dose IL-2 achieved durable objective responses in a minority of patients with melanoma and renal cell carcinoma (RCC), serving as proof of principle that the immune system could eliminate cancer cells [9, 10].

Immune Checkpoint Inhibitors (ICI)

1) PD-1 and PD Ligand 1/2: Programmed cell death 1 (PD-1) is a transmembrane protein expressed on T cells, B cells, and NK cells. It is an inhibitory molecule that binds to the PD-1 ligand (PD-L1; also known as B7-H1) and PD-L2 (B7-H2). PD-L1 is expressed on the surface of multiple tissue types, including many tumor cells, as well as hematopoietic cells; PD-L2 is more restricted to hematopoietic cells. The PD-1:PD-L1/2 interaction directly inhibits apoptosis of the tumor cell, promotes peripheral T effector cell exhaustion, and promotes the conversion of T effector cells to Treg cells [11, 12]. Based upon prolonged overall survival in phase III trials and durable responses in phase II studies, antibodies inhibiting PD-1 (pembrolizumab, nivolumab) and PD-L1 (atezolizumab, avelumab, durvalumab) have been approved for a number of clinical indications and are being evaluated in multiple other malignancies; these are discussed in specific disease-related topics.

2) CTLA-4: CTLA-4 exerts its effect when it is present on the cell surface of CD4+ and CD8+ T lymphocytes, where it has a higher affinity for the costimulatory receptors CD80 and CD86 (B7-1 and B7-2) on antigen-presenting cells (APCs) than the T cell costimulatory receptor CD28 [13]. The expression of CTLA-4 is upregulated by the degree of T cell receptor (TCR) activation and cytokines such as IL-12 and IFN gamma, forming a feedback inhibition loop on activated T effector cells. As a result, CTLA-4 can be broadly considered a physiologic "brake" on the CD4+ and CD8+ T cell activation that is triggered by APCs. The anti-CTLA-4 antibody ipilimumab was the first immune checkpoint inhibitor to be approved based upon its ability to prolong survival in patients with metastatic melanoma [14]. Ipilimumab has also been approved as adjuvant therapy for high-risk melanoma as an alternative to IFN.

Vaccines — There is a long history of using adaptive immune recognition of a cancer-associated antigen to influence antitumor responses. Vaccination methods are very diverse, and a full review is beyond the scope of this article. The only vaccine-based therapy currently approved for advanced cancer is sipuleucel-T, an autologous dendritic cell preparation targeted to prostatic acid phosphatase (PAP), which has shown an overall survival benefit in men with castrate-resistant prostate adenocarcinoma [15].

THE PREDICTIVE AND PROGNOSTIC MARKERS OF IMMUNOTHERAPY

As immune checkpoint blockade and other immune-based therapy approaches lead to broad treatment advances among patients with advanced cancer, an important consideration is how to best select patients whose tumors will respond to these therapies. There are needs for predictive and prognostic markers. There are genomic features such as tumour mutation burden (TMB) and microsatellite instability (MSI) and immune phenotype features such as programmed death ligand 1 (PD-L1), CTLA-4 and tumour infiltrating lymphocytes (TILs) to predict response to immunotherapies (ITs). The most important factors of response prediction to ICIs are summarized in Table **1**.

Table 1. Response prediction to ICIs.

Factor	Related to Favourable Outcome	Predictive or Prognostic	Cancer Type
TMB	Yes	Predictive	Multiple
PD-L1	Yes	Predictive	Multiple
Different gut microbial species	Yes or No	Predictive	Melanoma

(Table 1) cont.....

Factor	Related to Favourable Outcome	Predictive or Prognostic	Cancer Type
Copy number variation	No	Predictive, prognostic or both	Multiple
T-cell inflamed microenvironment	Yes	Predictive, prognostic or both	Multiple
HLA class I diversity	Yes	Predictive	Melanoma and NSCLC
T cell repertoire	Yes	Predictive	Melanoma
Different mutations	Yes or No	Mostly predictive	Melanoma and NSCLC
TGF-β expression	No	Predictive	Colon and urothelial cancer

(TMB: tumor mutation burden; PD-L1: programmed death ligand 1; HLA: human leucocyte antigen; TGF: transforming growth factor; ICI: immune checkpoint inhibitor).

TMB: Tumours with a high number of mutations are more likely to produce new, abnormal proteins, and present them as so-called neoantigens, which can potentially be recognised and targeted by the immune system [16, 17].

Across many cancers, higher TMB is associated with higher levels of predicted neoantigens [16, 18]. Several studies show the correlation between TMB and predicted neoantigen load across multiple cancer types [19 - 22]. Response to immune checkpoint inhibitors is higher in tumours with high TMB. In other words, tumors with high levels of TMB (*e.g.*, sun-exposed cutaneous melanoma, NSCLC, bladder cancer, and microsatellite-unstable colorectal carcinomas) can benefit more from immune checkpoint blockade [16, 17, 20, 23, 24]. Higher TMB across cancer types was associated with increased mutagen exposure (eg, lung cancer, melanoma), alterations in specific genes, including those involved in DNA mismatch repair/replication and increased age [25, 26].

Approximately 16.4% of cancers had a TMB greater than 10 mutations/Mb, and 7.3% had a TMB greater than 20 mutations/Mb [27]. TMB are comparable to prior estimates of TMB in other cancer cohorts, including whole-exome studies [27]. TMB stratification lacks standardisation [28, 29] and cut-off points determining high and low TMB vary between cancer types [30] and between studies of the same cancer type [29]. Variability in TMB cut-offs places importance on clinical validation for all TMB assays in each indication [28].

There is an evolution of TMB as an immunotherapy biomarker over the last several years. In CheckMate 026, 032, 038, 275, 227 trials and Keynote 012, 028 trials, it has been shown that higher TMB is associated with a better response to

IT [26]. In a study evaluating the association between atezolizumab efficacy and TMB in NSCLC patients for 2nd line therapy, enroled in POPLAR, BIRCH or FIR trials showed that treatment benefit of atezolizumab (as measured by OS, PFS and ORR) was associated with high TMB [31]. This association occurred in both unselected and PD-L1-selected patients. PD-L1 expression and TMB were both independent predictors of treatment response [31].

In a retrospective study assessing association of TMB and response to various immunotherapies in 151 patients representing 21 different cancer types which were subgrouped into melanoma, NSCLC (58%) or other cancers (42%) revealed that high TMB was independently associated with better outcome parameters (PFS and RR), regardless of subgroup. Of patients with melanoma or NSCLC treated with anti-PD-1 / PD-L1 monotherapy, 44% with high TMB, compared to 5% with low TMB, experienced a complete or partial response to treatment ($p = 0.0023$) [32].

PD-L1: The candidate biomarker that has been studied mostly other than TMB is PD-L1 expression in trials utilizing programmed cell death-1 (PD-1) blockade. PD-L1 and PD-1 expression are dynamic markers that change in relation to local cytokines and other factors, and the thresholds that separate "positive" and "negative" PD-L1 expression remain under debate. Still, most trials with either retrospective or prospective assessments of PD-L1 status have shown trends for increased response rates to PD-1 blockade in PD-L1 "positive" tumors [33 - 36]. Most notably, in patients with newly diagnosed advanced non-small cell lung cancers (NSCLCs) with ≥50 percent PD-L1 expression who were randomized to pembrolizumab or chemotherapy, those randomized to pembrolizumab had a significantly improved objective response rate, progression-free survival, and overall survival [37]. On the basis of this trial, PD-L1 expression is now a routine diagnostic marker for patients with newly diagnosed NSCLC. The potential applicability of PD-L1 in other disease settings is still uncertain.

MSI: Microsatellite instability (MSI) is characterised by high rates of alterations to repetitive DNA sequences caused by impaired mismatch repair (MMR) and MSI was the first instance of the FDA approving a drug 'based on a tumour's biomarker without regard to the tumour's original location [38, 39]. Thirty-one % of MSI-high mCRC patients who progressed on prior therapy achieved an objective response to nivolumab [40]. TMB measurement captures most MSI-high patients. For example; the vast majority (83%) of MSI-high samples were also TMB-high however the opposite was not true, with only 16% of TMB-high specimens also MSI-high [25, 41].

Therefore, combining PD-L1 with emerging biomarkers to address current limitations is spatial. PD-L1 protein expression by immunohistochemistry (IHC) can predict response to PD-(L)1 inhibitors in a variety of tumour types, but is not absolute. Factors potentially limiting sensitivity and specificity of available IHC assays are; assay performance and positivity threshold, spatial and temporal expression heterogeneity, adaptive vs constitutive expression, limited sampling and alternate PD-1 / PD-L1 ligand / receptor activity [42]. So, the emerging biomarkers to potentially address limitations are TILs; high TMB and gene expression signatures (GES) [43].

TMB correlates only weakly with PD-L1 tumour expression in patients with NSCLC [44]. PD-L1 status and TMB showed only a weak correlation, with 60% of PD-L1 negative samples exhibiting high TMB [44].

Studies support TMB and PD-L1 as independent biomarkers for response to immunotherapy in NSCLC [45, 46]. However, PD-L1 expression and TMB are not significantly correlated within most cancer subtypes. PD-L1 expression and TMB may each inform the use of immune checkpoint inhibitor therapy but treatment responsiveness may be regulated by different mechanisms between these biomarkers [27].

Combining all these data; we can conclude that TMB-high identifies a population independent of PD-L1 status and additionally TMB and PD-L1 are complementary biomarkers carrying independent information.

TMB can be integrated with TIL measurement to predict pre-treatment prognosis. TMB-high is associated with immune cell infiltration in many cancers [47]. Only in TMB-high patients, TILs are a prognostic marker in breast cancer. 100% of TMB-high/TIL-high patients survived 10 years, regardless of tumour subtype [48]. For urothelial carcinoma, TMB and TILs are biomarkers which correlate with disease outcome at least partially independently. Therefore, integration of TMB and TILs could aid in predicting diagnosis [49]. TILs can indicate an immunologically 'hot' tumour which may respond better to immunotherapy.

It is important that no patient with an advanced cancer and an established clinical rationale for use of an immune therapy agent should be refused immune therapy on the basis of lack of PD-L1 expression or any other investigational biomarker. Additional gene-expression-based signatures for immune response are also under active investigation and combining TMB with either TIL or PD-L1 measurement enriches for patients who may respond to immune checkpoint inhibitors [50, 51].

Combining TMB with specific genomic alterations is crucial. Development of tailored immunotherapy approaches may be facilitated by combining TMB and/or

PD-L1 measurements with genomic alterations associated with response or resistance to immunotherapy. For instance; **STK11** is associated with poor response to immunotherapies [46, 52, 53]. **EGFR** is a potential marker of resistance in immunotherapy; **MET** is linked with increased immune infiltration and activation phenotype [53] and **BRAF** is associated with prolonged time on immunotherapies [53]. STK11/LKB1 co-mutations are associated with inferior ORR with PD-1 blockade in KRAS-mutant, TMB-high lung cancer [52].

TMB is relevant for identifying patients to combine immune checkpoint inhibitors. High TMB has shown to correlate with efficacy of PD-1 plus CTLA-4 inhibition in lung cancers and their distinct mechanisms of action explains their synergistic effect [54 - 56]. TMB and PD-L1 are independent variables. Additionally benefit with combination immune checkpoint inhibitors was independent of PD-L1 expression in the high TMB patient population and TMB associated strongly with efficacy of immunotherapies [54, 55]. These data are further supported in a phase 3 non-small cell lung cancer trial (Checkmate 227) comparing first-line nivolumab plus ipilimumab with chemotherapy in NSCLC [57]. In the low TMB (< 10 mutations / Mb) patient population PFS does not differ significantly between treatment arms (low vs high PD-L1) [57].

In CheckMate-032 trial which evaluated combination immune checkpoint inhibition in small-cell lung cancer (SCLC) patients with high TMB; 1-year survival almost doubled in TMB-high patients treated with combination *vs* monotherapy [55]. PD-L1 was positive among 10-12% of patients and not predictive of response rate. However, TMB has a role as a biomarker in SCLC and may help identify patients most likely to benefit from combination immunotherapy [55]. In phase 3 CASPIAN and Keynote 604 studies which evaluated the addition of IT to chemotherapy in SCLC patients, the benefit of IT on survival was independent of the PD-L1 status [58, 59].

Immune checkpoint blockade (ICB) provides clinical benefits to a subset of patients with cancer. ICB using inhibitors of programmed death-1 or its ligand (anti-PD-1/PD-L1) has emerged as an effective therapy for many cancer types. However, a minority of patients (<20%) respond to treatment or have durable clinical benefits.

Circulating tumor DNA (ctDNA) within peripheral blood plasma provides noninvasive access to cancer-specific somatic mutations [60]. ctDNA as a biomarker, can be used to manage patients with advanced cancer by replacing tissue biopsy for genotyping of specific mutations which are linked with therapeutic response in a noninvasive manner. However, ctDNA has not yet been clinically implemented for patients treated with ICB [61].

There is a study that evaluated the performance of an amplicon-based bespoke ctDNA detection platform for prognostication and response monitoring in patients treated with the anti-PD-1 monoclonal antibody, pembrolizumab. They hypothesized that the baseline ctDNA levels would be prognostic and early changes in ctDNA levels would head the radiographic response. These findings could help advance the implementation of ctDNA-based testing in the context of ICB treatment across cancer types. Biomarker-directed use of ICB is an important frontier in precision medicine. Despite the dramatic improvement in clinical outcomes in some cancer types, only a minority of patients with solid cancer will derive sustained response or meaningful clinical benefit.

Hypothetical biomarkers could provide clinical utility in this setting by identifying patients most likely to benefit from ICB before or shortly after treatment initiation. Tumor characteristics, including TMB and PD-L1 have shown variable predictive value depending on cancer type. MSI-H phenotype identifies patients with favorable outcomes to ICB treatment but is an uncommon feature in many cancer types, including those enrolled in the INSPIRE trial [62]. In this study, the clinical validity of ctDNA before and during ICB treatment as a prognostic, predictive and pharmacodynamic tool was investigated. Although low baseline ctDNA levels were associated with favorable clinical outcomes in this multi-cohort study, the effect size was modest and partly confounded by tumor type. Patients with malignant melanoma, who may have a higher response rate to pembrolizumab than the other patients enrolled in INSPIRE, displayed the lowest average ctDNA levels. Of note, patients with TNBC were among those with the highest baseline ctDNA levels, which is consistent with other findings that ctDNA levels are higher in TNBC (triple negatif breast cancer) than other breast cancer subtypes [63].

Immunotherapy Toxicity

There are predictive markers of IT toxicity. Immunologic biomarkers are being studied as a way to predict the risk of immunotherapy-related adverse events (irAEs) and as an aid in the early identification of such complications. Examples include interleukin 17 (IL-17) [64], eosinophilia [65, 66], and combined toxicity scores based on gene expression profiling of immunologically predictive cytokines [67, 68]. The optimal predictive biomarker remains to be defined. Some autoantibodies as predictors for survival and immune-related adverse events in checkpoint-inhibition therapy of metastasized melanoma have been discovered recently [69]. Machine learning and artificial intelligence strategies are under investigation to predict the benefit of IT and prognosis.

CONCLUSION

As the principles of cancer, IT and tumor immunology is better understood, there is an urgent need for prognostic and predictive markers. Genomic features such as TMB and MSI and immune features such as PD-L1, CTLA-4 and TILs are predictive markers. Several studies show the correlation between TMB and predicted neoantigen load across multiple cancer types. Response to immune checkpoint inhibitors is higher in tumours with high TMB. Another candidate biomarker that has been studied mostly other than TMB is PD-L1 expression in trials utilizing programmed cell death-1 (PD-1) blockade. PD-L1 protein expression is determined by using Combined Positive Score (CPS) in some types of cancers. PD-L1 expression or CPS is now a routine diagnostic marker for patients with newly diagnosed NSCLC, bladder cancer, and gastrointestinal cancer. MSI was the first instance of the FDA approving a drug 'based on a tumour's biomarker without regard to the tumour's original location. Combining TMB with specific genomic alterations is crucial. Immunologic biomarkers are being studied as a way to predict the risk of irAEs and as an aid in the early identification of such complications. Morever, new biomarkers are being investigated for both therapy benefit and toxicity prediction.

CONSENT FOR PUBLICATION

Not applicable.

CONFLICT OF INTEREST

The authors declare no conflict of interest, financial or otherwise.

ACKNOWLEDGEMENT

Declared none.

REFERENCES

[1] Badalamenti G, Fanale D, Incorvaia L, *et al.* Role of tumor-infiltrating lymphocytes in patients with solid tumors: Can a drop dig a stone? Cell Immunol 2019; 343: 103753. [http://dx.doi.org/10.1016/j.cellimm.2018.01.013] [PMID: 29395859]

[2] Capobianco A, Rovere-Querini P, Rugarli C, Manfredi AA. Melanoma cells interfere with the interaction of dendritic cells with NK/LAK cells. Int J Cancer 2006; 119(12): 2861-9. [http://dx.doi.org/10.1002/ijc.22279] [PMID: 16998790]

[3] Gras Navarro A, Björklund AT, Chekenya M. Therapeutic potential and challenges of natural killer cells in treatment of solid tumors. Front Immunol 2015; 6: 202. [http://dx.doi.org/10.3389/fimmu.2015.00202] [PMID: 25972872]

[4] Savage PA, Leventhal DS, Malchow S. Shaping the repertoire of tumor-infiltrating effector and regulatory T cells. Immunol Rev 2014; 259(1): 245-58. [http://dx.doi.org/10.1111/imr.12166] [PMID: 24712470]

[5] Marvel D, Gabrilovich DI. Myeloid-derived suppressor cells in the tumor microenvironment: expect the unexpected. J Clin Invest 2015; 125(9): 3356-64. [http://dx.doi.org/10.1172/JCI80005] [PMID: 26168215]

[6] Bailey SR, Nelson MH, Himes RA, Li Z, Mehrotra S, Paulos CM. Th17 cells in cancer: the ultimate identity crisis. Front Immunol 2014; 5: 276. [http://dx.doi.org/10.3389/fimmu.2014.00276] [PMID: 24987392]

[7] Laoui D, Van Overmeire E, De Baetselier P, Van Ginderachter JA, Raes G. Functional relationship between tumor-associated macrophages and macrophage colony-stimulating factor as contributors to cancer progression. Front Immunol 2014; 5: 489. [http://dx.doi.org/10.3389/fimmu.2014.00489] [PMID: 25339957]

[8] Boyman O, Sprent J. The role of interleukin-2 during homeostasis and activation of the immune system. Nat Rev Immunol 2012; 12(3): 180-90. [http://dx.doi.org/10.1038/nri3156] [PMID: 22343569]

[9] Rosenberg SA, Yang JC, Topalian SL, *et al.* Treatment of 283 consecutive patients with metastatic melanoma or renal cell cancer using high-dose bolus interleukin 2. JAMA 1994; 271(12): 907-13. [http://dx.doi.org/10.1001/jama.1994.03510360033032] [PMID: 8120958]

[10] Atkins MB, Lotze MT, Dutcher JP, *et al.* High-dose recombinant interleukin 2 therapy for patients with metastatic melanoma: analysis of 270 patients treated between 1985 and 1993. J Clin Oncol 1999; 17(7): 2105-16. [http://dx.doi.org/10.1200/JCO.1999.17.7.2105] [PMID: 10561265]

[11] Francisco LM, Salinas VH, Brown KE, *et al.* PD-L1 regulates the development, maintenance, and function of induced regulatory T cells. J Exp Med 2009; 206(13): 3015-29. [http://dx.doi.org/10.1084/jem.20090847] [PMID: 20008522]

[12] Amarnath S, Mangus CW, Wang JC, *et al.* The PDL1-PD1 axis converts human TH1 cells into regulatory T cells. Sci Transl Med 2011; 3(111): 111ra120. [http://dx.doi.org/10.1126/scitranslmed.3003130] [PMID: 22133721]

[13] Walker LS, Sansom DM. The emerging role of CTLA4 as a cell-extrinsic regulator of T cell responses. Nat Rev Immunol 2011; 11(12): 852-63. [http://dx.doi.org/10.1038/nri3108] [PMID: 22116087]

[14] Schadendorf D, Hodi FS, Robert C, *et al.* Pooled analysis of long-term survival data from phase II and phase III trials of ipilimumab in unresectable or metastatic melanoma. J Clin Oncol 2015; 33(17): 1889-94. [http://dx.doi.org/10.1200/JCO.2014.56.2736] [PMID: 25667295]

[15] Kantoff PW, Higano CS, Shore ND, *et al.* Sipuleucel-T immunotherapy for castration-resistant prostate cancer. N Engl J Med 2010; 363(5): 411-22. [http://dx.doi.org/10.1056/NEJMoa1001294] [PMID: 20818862]

[16] Schumacher TN, Schreiber RD. Neoantigens in cancer immunotherapy. Science 2015; 348(6230): 69-74. [http://dx.doi.org/10.1126/science.aaa4971] [PMID: 25838375]

[17] Lu YC, Robbins PF. Cancer immunotherapy targeting neoantigens. In Seminars in immunology 2016; 28(1): 22-7. [http://dx.doi.org/10.1016/j.smim.2015.11.002]

[18] Alexandrov LB, Nik-Zainal S, Wedge DC, *et al.* Signatures of mutational processes in human cancer. Nature 2013; 500(7463): 415-21. [http://dx.doi.org/10.1038/nature12477] [PMID: 23945592]

[19] Miller A, Asmann Y, Cattaneo L, *et al.* High somatic mutation and neoantigen burden are correlated with decreased progression-free survival in multiple myeloma. Blood Cancer J 2017; 7(9): e612. [http://dx.doi.org/10.1038/bcj.2017.94] [PMID: 28937974]

[20] Rizvi NA, Hellmann MD, Snyder A, *et al.* Cancer immunology. Mutational landscape determines sensitivity to PD-1 blockade in non-small cell lung cancer. Science 2015; 348(6230): 124-8. [http://dx.doi.org/10.1126/science.aaa1348] [PMID: 25765070]

[21] Rooney MS, Shukla SA, Wu CJ, Getz G, Hacohen N. Molecular and genetic properties of tumors associated with local immune cytolytic activity. Cell 2015; 160(1-2): 48-61. [http://dx.doi.org/10.1016/j.cell.2014.12.033] [PMID: 25594174]

[22] Chae YK, Arya A, Iams W, *et al.* Current landscape and future of dual anti-CTLA4 and PD-1/PD-L1 blockade immunotherapy in cancer; lessons learned from clinical trials with melanoma and non-small cell lung cancer (NSCLC). J Immunother Cancer 2018; 6(1): 39. [http://dx.doi.org/10.1186/s40425-018-0349-3] [PMID: 29769148]

[23] Rozenblum AB, Ilouze M, Dudnik E, *et al.* Clinical impact of hybrid capture–based next-generation sequencing on changes in treatment decisions in lung cancer. J Thorac Oncol 2017; 12(2): 258-68. [http://dx.doi.org/10.1016/j.jtho.2016.10.021] [PMID: 27865871]

[24] Rosenberg JE, Hoffman-Censits J, Powles T, *et al.* Atezolizumab in patients with locally advanced and metastatic urothelial carcinoma who have progressed following treatment with platinum-based chemotherapy: a single-arm, multicentre, phase 2 trial. Lancet 2016; 387(10031): 1909-20. [http://dx.doi.org/10.1016/S0140-6736(16)00561-4] [PMID: 26952546]

[25] Chalmers ZR, Connelly CF, Fabrizio D, *et al.* Analysis of 100,000 human cancer genomes reveals the landscape of tumor mutational burden. Genome Med 2017; 9(1): 34. [http://dx.doi.org/10.1186/s13073-017-0424-2] [PMID: 28420421]

[26] Chan TA, Yarchoan M, Jaffee E, *et al.* Development of tumor mutation burden as an immunotherapy biomarker: utility for the oncology clinic. Ann Oncol 2019; 30(1): 44-56. [http://dx.doi.org/10.1093/annonc/mdy495] [PMID: 30395155]

[27] Yarchoan M, Albacker LA, Hopkins AC, *et al.* PD-L1 expression and tumor mutational burden are independent biomarkers in most cancers. JCI Insight 2019; 4(6): 126908. [http://dx.doi.org/10.1172/jci.insight.126908] [PMID: 30895946]

[28] Meléndez B, Van Campenhout C, Rorive S, Remmelink M, Salmon I, D'Haene N. Methods of measurement for tumor mutational burden in tumor tissue. Transl Lung Cancer Res 2018; 7(6): 661-7. [http://dx.doi.org/10.21037/tlcr.2018.08.02] [PMID: 30505710]

[29] Stenzinger A, Allen JD, Maas J, *et al.* Tumor mutational burden standardization initiatives: Recommendations for consistent tumor mutational burden assessment in clinical samples to guide immunotherapy treatment decisions. Genes Chromosomes Cancer 2019; 58(8): 578-88. [http://dx.doi.org/10.1002/gcc.22733] [PMID: 30664300]

[30] Samstein RM, Lee CH, Shoushtari AN, *et al.* Tumor mutational load predicts survival after immunotherapy across multiple cancer types. Nat Genet 2019; 51(2): 202-6. [http://dx.doi.org/10.1038/s41588-018-0312-8] [PMID: 30643254]

[31] Kowanetz M, Zou W, Shames D, *et al.* OA20. 01 tumor mutation burden (TMB) is associated with improved efficacy of atezolizumab in 1L and 2L+ NSCLC patients. J Thorac Oncol 2017; 12(1): S321-2. [http://dx.doi.org/10.1016/j.jtho.2016.11.343]

[32] Goodman AM, Kato S, Bazhenova L, *et al.* Tumor mutational burden as an independent predictor of response to immunotherapy in diverse cancers. Mol Cancer Ther 2017; 16(11): 2598-608. [http://dx.doi.org/10.1158/1535-7163.MCT-17-0386] [PMID: 28835386]

[33] Larkin J, Chiarion-Sileni V, Gonzalez R, *et al.* Combined nivolumab and ipilimumab or monotherapy in untreated melanoma. N Engl J Med 2015; 373(1): 23-34. [http://dx.doi.org/10.1056/NEJMoa1504030] [PMID: 26027431]

[34] Robert C, Long GV, Brady B, *et al.* Nivolumab in previously untreated melanoma without BRAF mutation. N Engl J Med 2015; 372(4): 320-30.

[http://dx.doi.org/10.1056/NEJMoa1412082] [PMID: 25399552]

[35] Weber JS, D'Angelo SP, Minor D, *et al.* Nivolumab *versus* chemotherapy in patients with advanced melanoma who progressed after anti-CTLA-4 treatment (CheckMate 037): a randomised, controlled, open-label, phase 3 trial. Lancet Oncol 2015; 16(4): 375-84. [http://dx.doi.org/10.1016/S1470-2045(15)70076-8] [PMID: 25795410]

[36] Weber JS, Kudchadkar RR, Yu B, *et al.* Safety, efficacy, and biomarkers of nivolumab with vaccine in ipilimumab-refractory or -naive melanoma. J Clin Oncol 2013; 31(34): 4311-8. [http://dx.doi.org/10.1200/JCO.2013.51.4802] [PMID: 24145345]

[37] Reck M, Rodríguez-Abreu D, Robinson AG, *et al.* Pembrolizumab *versus* chemotherapy for PD-L1–positive non–small-cell lung cancer. N Engl J Med 2016; 375(19): 1823-33. [http://dx.doi.org/10.1056/NEJMoa1606774] [PMID: 27718847]

[38] Boland CR, Goel A. Microsatellite instability in colorectal cancer. Gastroenterology 2010; 138(6): 2073-2087.e3. [http://dx.doi.org/10.1053/j.gastro.2009.12.064] [PMID: 20420947]

[39] https://www.fda.gov/NewsEvents/Newsroom/PressAnnouncements/ucm560167.htm

[40] Overman MJ, McDermott R, Leach JL, *et al.* Nivolumab in patients with metastatic DNA mismatch repair-deficient or microsatellite instability-high colorectal cancer (CheckMate 142): an open-label, multicentre, phase 2 study. Lancet Oncol 2017; 18(9): 1182-91. [http://dx.doi.org/10.1016/S1470-2045(17)30422-9] [PMID: 28734759]

[41] Fabrizio DA, George TJ Jr, Dunne RF, *et al.* Beyond microsatellite testing: assessment of tumor mutational burden identifies subsets of colorectal cancer who may respond to immune checkpoint inhibition. J Gastrointest Oncol 2018; 9(4): 610-7. [http://dx.doi.org/10.21037/jgo.2018.05.06] [PMID: 30151257]

[42] Cottrell T, Taube JM. PD-L1 and emerging biomarkers in PD-1/PD-L1 blockade therapy. Cancer J 2018; 24(1): 41. [http://dx.doi.org/10.1097/PPO.0000000000000301] [PMID: 29360727]

[43] Yuan J, Hegde PS, Clynes R, *et al.* Novel technologies and emerging biomarkers for personalized cancer immunotherapy. J Immunother Cancer 2016; 4(1): 3. [http://dx.doi.org/10.1186/s40425-016-0107-3] [PMID: 26788324]

[44] Cohen EE, Hong DS, Wise Draper T, *et al.* 1135OPhase 1b/2 Study (SCORES) assessing safety, tolerability, and preliminary anti-tumor activity of durvalumab plus AZD9150 or AZD5069 in patients with advanced solid malignancies and squamous cell carcinoma of the head and neck (SCCHN). Annals of Oncology 2017; 28(suppl_5).

[45] Gandara DR, Paul SM, Kowanetz M, *et al.* Blood-based tumor mutational burden as a predictor of clinical benefit in non-small-cell lung cancer patients treated with atezolizumab. Nat Med 2018; 24(9): 1441-8. [http://dx.doi.org/10.1038/s41591-018-0134-3] [PMID: 30082870]

[46] Rizvi H, Sanchez-Vega F, La K, *et al.* Molecular determinants of response to anti–programmed cell death (PD)-1 and anti–programmed death-ligand 1 (PD-L1) blockade in patients with non–small-cell lung cancer profiled with targeted next-generation sequencing. J Clin Oncol 2018; 36(7): 633-41. [http://dx.doi.org/10.1200/JCO.2017.75.3384] [PMID: 29337640]

[47] Budczies J, Seidel A, Christopoulos P, *et al.* Integrated analysis of the immunological and genetic status in and across cancer types: impact of mutational signatures beyond tumor mutational burden. OncoImmunology 2018; 7(12): e1526613. [http://dx.doi.org/10.1080/2162402X.2018.1526613] [PMID: 30524909]

[48] Thomas A, Routh ED, Pullikuth A, *et al.* Tumor mutational burden is a determinant of immune-mediated survival in breast cancer. OncoImmunology 2018; 7(10): e1490854. [http://dx.doi.org/10.1080/2162402X.2018.1490854] [PMID: 30386679]

[49] Mariathasan S, Turley SJ, Nickles D, *et al.* TGFβ attenuates tumour response to PD-L1 blockade by contributing to exclusion of T cells. Nature 2018; 554(7693): 544-8. [http://dx.doi.org/10.1038/nature25501] [PMID: 29443960]

[50] Cristescu R, Mogg R, Ayers M, *et al.* Pan-tumor genomic biomarkers for PD-1 checkpoint blockade-based immunotherapy. Science 2018; 362(6411): eaar3593. [http://dx.doi.org/10.1126/science.aar3593] [PMID: 30309915]

[51] Lu S, Stein JE, Rimm DL, *et al.* Comparison of biomarker modalities for predicting response to PD-1/PD-L1 checkpoint blockade: a systematic review and meta-analysis. JAMA Oncol 2019; 5(8): 1195-204. [http://dx.doi.org/10.1001/jamaoncol.2019.1549] [PMID: 31318407]

[52] Skoulidis F, Goldberg ME, Greenawalt DM, *et al.* STK11/LKB1 mutations and PD-1 inhibitor resistance in KRAS-mutant lung adenocarcinoma. Cancer Discov 2018; 8(7): 822-35. [http://dx.doi.org/10.1158/2159-8290.CD-18-0099] [PMID: 29773717]

[53] Ross JS, Goldberg ME, Albacker LA, *et al.* 1138PDImmune checkpoint inhibitor (ICPI) efficacy and resistance detected by comprehensive genomic profiling (CGP) in non-small cell lung cancer (NSCLC). Annals of Oncology 2017; 1;28(suppl_5).

[54] Hellmann MD, Nathanson T, Rizvi H, *et al.* Genomic features of response to combination immunotherapy in patients with advanced non-small-cell lung cancer. Cancer Cell 2018; 33(5): 843-852.e4. [http://dx.doi.org/10.1016/j.ccell.2018.03.018] [PMID: 29657128]

[55] Hellmann MD, Callahan MK, Awad MM, *et al.* Tumor mutational burden and efficacy of nivolumab monotherapy and in combination with ipilimumab in small-cell lung cancer. Cancer Cell 2018; 33(5): 853-861.e4. [http://dx.doi.org/10.1016/j.ccell.2018.04.001] [PMID: 29731394]

[56] Boumber Y. Tumor mutational burden (TMB) as a biomarker of response to immunotherapy in small cell lung cancer. J Thorac Dis 2018; 10(8): 4689-93. [http://dx.doi.org/10.21037/jtd.2018.07.120] [PMID: 30233840]

[57] Hellmann MD, Ciuleanu TE, Pluzanski A, *et al.* Nivolumab plus ipilimumab in lung cancer with a high tumor mutational burden. N Engl J Med 2018; 378(22): 2093-104. [http://dx.doi.org/10.1056/NEJMoa1801946] [PMID: 29658845]

[58] Paz-Ares L, Dvorkin M, Chen Y, *et al.* Durvalumab plus platinum-etoposide *versus* platinum-etoposide in first-line treatment of extensive-stage small-cell lung cancer (CASPIAN): a randomised, controlled, open-label, phase 3 trial. Lancet 2019; 394(10212): 1929-39. [http://dx.doi.org/10.1016/S0140-6736(19)32222-6] [PMID: 31590988]

[59] Corcoran RB, Chabner BA. Application of cell-free DNA analysis to cancer treatment. N Engl J Med 2018; 379(18): 1754-65. [http://dx.doi.org/10.1056/NEJMra1706174] [PMID: 30380390]

[60] Cabel L, Proudhon C, Romano E, *et al.* Clinical potential of circulating tumour DNA in patients receiving anticancer immunotherapy. Nat Rev Clin Oncol 2018; 15(10): 639-50. [http://dx.doi.org/10.1038/s41571-018-0074-3] [PMID: 30050094]

[61] Anagnostou V, Forde PM, White JR, *et al.* Dynamics of tumor and immune responses during immune checkpoint blockade in non–small cell lung cancer. Cancer Res 2019; 79(6): 1214-25. [http://dx.doi.org/10.1158/0008-5472.CAN-18-1127] [PMID: 30541742]

[62] Clouthier DL, Lien SC, Yang SYC, *et al.* An interim report on the investigator-initiated phase 2 study of pembrolizumab immunological response evaluation (INSPIRE). J Immunother Cancer 2019; 7(1): 72. [http://dx.doi.org/10.1186/s40425-019-0541-0] [PMID: 30867072]

[63] Magbanua MJ, Brown-Swigart L, Wu HT, *et al.* Circulating tumor DNA in neoadjuvant treated breast

cancer reflects response and survival. Ann Oncol 2021; 32(2): 229-39.
[http://dx.doi.org/10.1101/2020.02.03.20019760]

[64] Tarhini AA, Zahoor H, Lin Y, *et al.* Baseline circulating IL-17 predicts toxicity while TGF-β1 and IL-10 are prognostic of relapse in ipilimumab neoadjuvant therapy of melanoma. J Immunother Cancer 2015; 3(1): 39.
[http://dx.doi.org/10.1186/s40425-015-0081-1] [PMID: 26380086]

[65] Jaber SH, Cowen EW, Haworth LR, *et al.* Skin reactions in a subset of patients with stage IV melanoma treated with anti-cytotoxic T-lymphocyte antigen 4 monoclonal antibody as a single agent. Arch Dermatol 2006; 142(2): 166-72.
[http://dx.doi.org/10.1001/archderm.142.2.166] [PMID: 16490844]

[66] Nakamura Y, Tanaka R, Maruyama H, *et al.* Correlation between blood cell count and outcome of melanoma patients treated with anti-PD-1 antibodies. Jpn J Clin Oncol 2019; 49(5): 431-7.
[http://dx.doi.org/10.1093/jjco/hyy201] [PMID: 30753621]

[67] Shahabi V, Berman D, Chasalow SD, *et al.* Gene expression profiling of whole blood in ipilimumab-treated patients for identification of potential biomarkers of immune-related gastrointestinal adverse events. J Transl Med 2013; 11(1): 75.
[http://dx.doi.org/10.1186/1479-5876-11-75] [PMID: 23521917]

[68] Lim SY, Lee JH, Gide TN, *et al.* Circulating cytokines predict immune-related toxicity in melanoma patients receiving anti-PD-1–based immunotherapy. Clin Cancer Res 2019; 25(5): 1557-63.
[http://dx.doi.org/10.1158/1078-0432.CCR-18-2795] [PMID: 30409824]

[69] Hassel JC, Zucht HD, Mangana J, *et al.* Autoantibodies as predictors for survival and immune-related adverse events in checkpoint inhibition therapy of metastasized melanoma. J Clin Oncology 2020; 38(15_Suppl): 10011-1.
[http://dx.doi.org/10.1200/JCO.2020.38.15_suppl.10011]

CHAPTER 2

Biomarkers in Gynecologic Tumors

Selim Afsar[1,*]

[1] *Department of Obstetrics and Gynecology, Balıkesir University Medical Faculty, Balikesir, Turkey*

Abstract: Gynecologic malignancies are one of the most frequent cancers amongst women. Biomarkers are crucial for the differential diagnosis of adnexal masses; however, their potential for diagnosis is limited. In the era of difficulty in ovarian cancer screening, novel biomarkers are defined, but CA125 still remains the most valuable one. Circulating tumor DNAs, DNA hypermethylation, metabolites, microRNAs, and kallikreins have recently turned out as ovarian cancer biomarkers and are being applied to clinical practice. For uterine cancer, genomic classification has now been described, it will be used as a prognostic tool. In this chapter, we describe ovarian, endometrial, and cervical cancer biomarkers in detail.

Keywords: Biomarker, CA125, Cervical cancer, cfDNA, HE4, High-copy number, miRNA, MSI, MSS, Ovarian cancer, POLE-mutated, ROMA, SCC-Ag, Uterine cancer.

INTRODUCTION

The lifetime risk of developing ovarian cancer (OC) is approximately 1.3% in women, and OC is the most common cause of gynecological cancer deaths [1]. OC is a generalized term for tumors that involve the ovary, and it can be classified into three different subgroups: epithelial, germ cell, and specialized stromal cell tumors. The most common form of ovarian cancer is epithelial ovarian cancer (EOC), which consists of four main histologic subtypes: serous, endometrioid, clear cell, and mucinous cancer. Serous ovarian carcinoma is the most common form [2].

Ovarian and Fallopian Tube Cancers

Carbohydrate Antigen 125 (CA125)

CA125 is a mucin-type glycoprotein that is associated with the cell membrane,

* **Corresponding author Selim Afsar:** Department of Obstetrics and Gynecology, Balıkesir University Medical Faculty, Balikesir, Turkey; Tel: +90 (266) 460 40 00; E-mail: selimafsar@yahoo.com

Pınar Atukeren and Hafize Uzun (Eds.)

and it is recognized by the OC125 murine monoclonal antibody. CA125 is not only expressed in the tubal, endometrial, and endocervical epithelium but also mesothelial cells of the pleura, pericardium, and peritoneum [3 - 5]. Hence, CA125 is not a specific biomarker for OC.

The cutoff value of 35 U/mL is used in routine clinical practice in premenopausal women, but levels of CA125 tend to be lower in postmenopausal women, so a cutoff value of 26 U/mL has been suggested [4, 6]. Elevated serum levels of CA125 can be detected in approximately 85% of patients with EOC, especially in serous type, but also in benign and physiological states such as fibroids, endometriosis, menstruation, and pregnancy as well as in other malignancies such as pancreas, liver, colorectal and breast cancer (Table **1**) [7 - 9].

Screening

Screening modalities for OC are primarily focused on CA125 and the use of transvaginal ultrasonography (TVU). Early detection of OC that requires a screening test with high sensitivity (>75%) and ultrahigh specificity (99.7%) could improve overall survival [10]. False positivity for CA125 has been shown approximately in 1% of the healthy population and 5% of patients with benign disease that limits its use as a stand-alone test [4, 11].

Table 1. CA125 elevations in gynecologic diseases and non-gynecologic malignancies.

Disease Category	Type of Disease	CA125 > 35 u/ML (%)
Benign gynecologic	Ovarian Cysts[a]	14
	Abscess/Hydrosalpinx/PID[a]	37
	Endometriosis/Endometriomas[a]	67
	Fibroid (Leiomyomas)[a]	26
	Cystadenoma, Adenofibroma, Cystadenofibroma[a]	20
	Serous Epithelial Tumors[a]	20
	Mucinous Epithelial Tumors[a]	18
	Germ Cell Tumors (Mature Teratoma)[a]	21
	Sex Cord Stromal Tumors (Thecoma, Fibrothecoma)[a]	52
Ovarian cancer[b]	Serous[b]	80
	Mucinous[b]	69
	Endometriod[b]	75
	Clear cell[b]	78

(Table 1) cont.....

Disease Category	Type of Disease	CA125 > 35 u/ML (%)
Gynecologic cancers[b,c]	Endometrial[b]	25
	Cervical[c]	39
Non-gynecologic cancers[d]	Pancreas[d]	52.6
	Liver[d]	49.0
	Biliary Tract[d]	45.8
	Colorectal[d]	15.1
	Breast[d]	17.6

[a]Moore RG, Miller MC, Steinhoff MM, *et al.* Serum HE4 levels are less frequently elevated than CA125 in women with benign gynecologic disorders. Am J Obstet Gynecol 2012; 206(4):351. [b]Jacobs I, Bast RC Jr. The CA 125 tumour-associated antigen: a review of the literature. Hum Reprod 1989; 4:1-12. [c]Muyldermans M, Cornillie FJ, Koninckx PR. CA125 and endometriosis. Hum Reprod Update 1999; 1(2):173-87. [d]Bast RC Jr, Klug TL, St John E, *et al.* A radioimmunoassay using a monoclonal antibody to monitor the course of epithelial ovarian cancer. N Engl J Med 1983; 309:883-7.

The Prostate, Lung, Colorectal and Ovarian (PLCO) Cancer Screening Trial enrolled 78,216 women aged 55-74 years, with 39,105 women randomized annual screening. Women were screened annually with CA-125 for 6 years and TVU for 4 years. Neither at a median follow-up of 12.4 years nor at an extended follow-up of 14,7 years, the PLCO trial reported reduction in mortality rates [12, 13]. On the contrary, the false positivity in the PLCO trial was approximately 5%, resulting in complicated surgeries [12].

In the UK Collaborative Trial of Ovarian Cancer Screening (UKCTOCS), 202,638 postmenopausal women aged 50-74 were randomized to either control or annually multimodal screening group (MMS) or ultrasonography screening (USS) group in a 2:1:1 ratio [14]. Although the trial reported an insignificant mortality reduction on primary analysis (15% in the MMS group, 11% in the USS group), further analysis revealed that a declined mortality was seen after seven years of follow-up (23% in the MMS group, 21% in the USS group) [14].

Hereditary breast-ovarian cancer (HBOC) syndrome, which is related to mutations in BRCA1 and BRCA2 genes, accounts for 14% of EOC. Other identified genes (Lynch syndrome, RAD51C, RAD51D, and BRIP1) account for a minor percentage of EOC [15]. The average risk of developing OC is around 44-45% for BRCA1-mutated women and 12-17% for BRCA2-mutated women [16, 17]. Serous ovarian cancer is the most common histologic subtype for both BRCA mutation carriers [18].

The lifetime risk of endometrial cancer in Lynch syndrome (LS) carriers is 40% due to mutations in mismatch repair genes (MLH1, MSH2, MSH6, and PMS2)

[19]. LS is also associated with an increased risk of OC that is generally diagnosed at 45 years of age [20]. Furthermore mutations in BRIP1, RAD51C, and RAD51D have been associated with moderate or high risks of OC [21].

Even if routine OC screening with TVU or CA125 is not recommended in women with BRCA mutations, it may be considered for short-term surveillance in women at greater risk of OC. OC screening should start at age 30–35 years until the time of risk-reducing bilateral salpingo-oophorectomy (RRSO) [22]. In the United Kingdom Familial Ovarian Cancer Screening Study (UK-FOCSS), women were screened with CA125 every 4-months and with TVU annually. RRSO was encouraged throughout the study that resulted in the diagnosis of a significant number of early staged OC [23].

Prediction Models

A variety of combined prediction models have been proposed to estimate the risk of OC in adnexal masses prior to surgery [24]. The risk of malignancy index (RMI) is a scoring system based on menopausal status, ultrasound findings, and the serum CA125 level. The last version of RMI is called RMI4, which includes tumor size score additionally [25]. In a meta-analysis, the diagnostic performance of RMI was reported with a sensitivity of 75% and specificity of 92% [26].

Subjective evaluation of adnexal masses prior to surgery by an experienced ultrasonographer is generally considered the best strategy for differential diagnosis. The handicap of this approach is it requires an expert ultrasonographer [26]. To overcome this obstacle, the International Ovarian Tumor Analysis (IOTA) group created different prediction models based on patients' characteristics and TVU findings using a logistic regression (LR) analysis (LR1 and LR2) [27]. LR1, the main IOTA logistic regression model, had the best performance with an area under the receiver operating characteristic curve (AUC) of 0.96, according to an external validation study [28]. The **A**ssessment of **D**ifferent **N**eoplasias in the adne**X**a (ADNEX) model combines 3 clinical variables with 6 TVU features. The ADNEX model was validated with the AUC of 0.94 for the basic discrimination between benign and malignant tumors [29]. LR2 (AUC:0.92) had a better diagnostic performance compared to RMI (AUC:0.88), and all IOTA strategies had a superior test performance compared with RMI [30].

Prognosis

Stage I EOC patients with higher preoperative CA125 levels (≥65U/mL) had a significantly poorer 5-year survival compared to lower CA125 levels (<65U/mL) [31]. In accordance with this report, an Australian multicenter study revealed that stage I EOC patients with CA125 levels ≤30U/mL had an outstanding 5-year overall survival (OS), and this was an independent prognostic factor [32]. On the contrary, CA125 is not an independent prognostic factor in advanced stage OC patients [33].

It was reported that a decline in serum CA125 levels during the initial 2 cycles of platinum-based chemotherapy was a strong independent prognostic factor for survival [34]. Patients in complete clinical remission after standard primary chemotherapy who had a very low CA125 level (less than 5 kU/L) had a significantly longer progression-free survival (PFS) and OS than patients with higher CA125 levels [35]. In a French multicentric study, it was concluded that a prolonged half-life of CA125 was an indicator of both persistent CA125 production and poor response to chemotherapy [36].

Recurrence

Among patients with EOC who are on complete clinical remission, a progressive low-level increase in serum CA-125 levels, even if within the normal range, is strongly related to disease recurrence, but it should be always supported by CT or PET scan imaging [37]. The MRC OVO5/EORTC55955 trial registered 1442 patients with OC in complete remission after first-line platinum-based chemotherapy and a normal CA125 concentration. Clinical examinations and CA125 measurements were done every 3 months. If CA125 concentration exceeded twice the upper limit of normal, patients were randomly allocated (1:1) to the early treatment group or delayed treatment group to whom CA125 measurements were continued. This trial showed no evidence of survival benefit with early treatment of relapse on the basis of a raised CA125 concentration alone [38]. Notwithstanding that, the European Society of Gynecologic Oncologists (ESGO) announced against the withdrawal of CA125 monitoring during routine follow-up. Consequently, CA125 monitoring should be advised in specific patients after who (i) are eligible for future clinical trials on second-line chemotherapy or already are treated as part of a clinical trial after complete response to first-line chemotherapy (iii) are not going routine follow-up (iv) are eligible for secondary surgery at recurrence [39].

Human Epididymis Protein 4 (HE4)

HE4, is a novel serum biomarker and has been reported as the most promising one in EOC. HE4 is primarily expressed from the reproductive and respiratory tracts, and increased serum levels can be detected in OC [40]. The HE4 serum levels in healthy women have been reported to range from 60 pmol/L to 150 pmol/L due to the relationship between increasing HE4 serum levels with age [41]. The serum concentration of HE4 shows no significant variations during the menstrual cycle or during hormonal treatment, but HE4 levels are lower in premenopausal women than in postmenopausal women [42]. HE4 levels are decreased in pregnancy and women with endometriosis [8].

It has been proposed that HE4 improves the diagnostic specificity of CA-125 with a similar sensitivity [43]. There is no clear evidence that HE4 is a better biomarker than CA125 in terms of screening.

Diagnosis

In a metaanalysis it has been shown that the pooled sensitivity and specificity for HE4 among patients with adnexal masses were 74% (95% CI, 69-78%) and 90% (95% CI, 87-92%), respectively [44]. It has been reported that HE4 was not superior to CA125 for differential diagnosis, on the contrary, another meta-analysis revealed that HE4 was better than CA125 for the diagnosis of ovarian cancer in terms of sensitivity and specificity [45, 46]. Consequently, there is no agreement on the diagnostic performance of HE4 compared to CA125.

Prognosis

Elevated HE4 levels are a powerful and independent factor of poorer prognosis in EOC compared to CA125 [47] and are also significantly correlated with higher tumor grade, serous histology, peritoneal involvement, nodal invasion tumor stage, operation time, and residual tumor size [48].

Risk of Ovarian Malignancy Algorithm (ROMA)

The Risk of Ovarian Malignancy Algorithm (ROMA) is a logistic regression algorithm that uses serum HE4 and CA125 levels with menopausal status to assess the risk of OC in women with adnexal masses [49]. The sensitivity and specificity of ROMA algorithm were detected between 76-86% and 74-95%, respectively, using different types of measurement methods [50]. In a meta-analysis, the ROMA algorithm was demonstrated as having a lesser specificity than HE4 (84% *vs* 94%), but a higher correlation than CA125 (84% *vs* 78%) [45]. It was reported that the ROMA algorithm was more sensitive than HE4 (96.7% *vs* 73.3%) but less specific than HE4 (80% *vs* 98.6%), in addition to similar AUC (0.97 vs 0.96) [51]. The dual measurement of CA125 and HE4 is apparently the best diagnostic tool over and above ROMA algorithms [50].

Circulating Cell-Free DNA (cfDNA)

Plenty of studies investigated cfDNA as a potential biomarker in the clinical management of ovarian cancer patients in terms of diagnosis and prognosis. It was concluded that the preoperative plasma total cfDNA levels are significantly elevated in patients with EOC, and this is an independent prognostic factor for death from OC [52].

It was reported that the sensitivity and specificity cfDNA (88.9% and 89.5%) was higher than CA125 (75% and 52.6%) and HE4 (80.6% and 68.4%) in OC detection [53].

The most common mutation in OC was TP53 which accounts for approximately 96% of all somatic mutations [54]. The mutations of TP53 in circulating DNA of advanced OC patients were identified at allele frequencies as low as 2%, with sensitivity and specificity of >97% by means of a new method called tagged-amplicon deep sequencing (TAm-Seq) [55].

It was demonstrated that a significant correlation between serum DNA levels and residual tumor load of >1 cm after primary surgery (p=0.0001), and both parameters were associated with a higher risk of relapse (p=0.0001 and p=0.020, respectively) and a poorer OS (p=0.021 and p=0.010, respectively) [56].

DNA Methylation

It was reported that the 6-gene DNA hypermethylation analysis showed 82% sensitivity and 100% specificity in the serum of OC patients [57]. It was concluded that the 7-gene DNA hypermethylation analysis showed higher sensitivity (85.3% *vs* 56.1%) and specificity (90.5% *vs* 64.15%) for cfDNA compared to CA125 in stage I OC [58]. The 3-gene DNA hypermethylation analysis revealed that cfDNA could detect OC much earlier than the exact diagnosis [59].

Metabolites

Metabolites are the end products of cellular processes and their alterations can be regarded as a response to genetic or environmental changes. Using gas chromatography/time-of-flight mass spectrometry (GC-TOF MS), 66 invasive OC and 9 borderline tumors of the ovary were analysed, and this analysis found that 51 metabolites were significantly different between borderline tumors and OC ($p< 0.01$) [60].

Metabolite profiling of 35-paired plasma samples from 35 EOC patients before and after surgery using rapid resolution liquid chromatography-mass spectrometry was revealed that identified 10 common biomarkers that were predictive for EOC recurrence. The AUC values in pre- and post-operative plasma were 0.815 and 0.909, respectively; the combined-AUC value reached 0.964 [61].

MicroRNAs

MicroRNAs (miRNAs) are small non-coding RNAs that regulate gene expression within cells [62]. In EOC tumors compared to normal ovarian tissues, upregulation of miR-181a, miR-616 and miR-590-3p, and downregulation of miR-125b, miR-148a-3p, and miR-375 levels have been reported [63 - 68]. It was found that some miRNAs were up- or down-regulated among different subtypes of EOCs using microarray analyses to compare serous, endometrioid, and clear cell tumors with normal ovarian tissues [69]. For example; miR-510 expression was higher in low-grade serous carcinoma and clear cell carcinoma subtypes but lower in high-grade serous carcinoma compared to normal ovarian tissues [70]. The potential role of miR-93, miR-325, and miR-492 was reported in the malign transformation of endometriosis to EOC [71].

There was a correlation between upregulation of miR-520h with increased ascites volume and poorer survival of EOC patients. In contrast, downregulation of miR-26b was inversely correlated with tumour stage, grade, and ascites volume [72, 73]. It was demonstrated that there was a correlation between the downregulation of miR-216 with lymphovascular invasion, upregulation of miR-133a-2, miR-145, and miR-126 with uterine invasion, and upregulation of miR-302c with pelvic peritoneum invasion [69].

Carcinoembryonic Antigen (CEA)

CEA is still the widely used tumor marker in the management of colorectal cancer [74]. In a Danish ovarian cancer study, it was revealed that 18% of borderline tumors and 4% of OCs were positive for CEA expression, which was especially higher in mucinous subtype compared with other histological subtypes ($p<0.00001$). In a Cox survival analysis, the CEA expression was found to be a prognostic factor (HR = 2.12, 95%CI 1.11-4.05) [75].

Alpha-Fetoprotein (AFP)

Alpha-Fetoprotein (AFP) is an oncofetal protein produced by the fetal yolk sac, liver, and upper gastrointestinal tract. AFP rarely increases in OC patients but is increased in patients with the germ cell tumors (yolk sac tumor, immature teratoma, dysgerminoma) presented in young women and adolescents [76]. AFP can be used as a reliable biomarker for the prediction of tumor recurrence during the follow-up period of women with endodermal sinus tumors [77].

Inhibin B

Inhibin B is released from granulosa cells and its serum levels are higher in the early follicular phase [78]. Circulating inhibin B concentrations are undetectable after menopause. Inhibin B is superior to inhibin A, with reported sensitivities between 88-100% for inhibin B and 67-77% for inhibin A, for adult-type granulosa cell tumor (AGCT) patients [79 - 81]. Inhibin B levels fluctuate during the menstrual cycle, and it can also be raised in EOC, especially in the mucinous subtype [80]. Additionally, inhibin B concentrations are in normal range in 10-15% of AGCTs [79, 81].

Human Chorionic Gonadotropin (hCG)

hCG is a glycoprotein hormone that is released at very high concentrations by the placental trophoblasts. Placental and other trophoblastic tumors ordinally release hCG and it is a very sensitive marker. In addition, many patients with nontrophoblastic tumors release the free β-subunit of hCG (hCGβ), which is a powerful marker of poorer prognosis [82].

Dysgerminomas generally release hCG and cause precocious puberty in young adults [83, 84]. Other germ cell tumors, especially embryonal carcinomas, may also release hCG. Yolk sac tumors principally produce alpha-fetoprotein and occasionally hCG [85, 86].

It was reported that hCGβ was the only independent prognostic factor correlated with stage and grade in OC [87]. The combined use of urine hCGβcf and serum CA125 was found to improve the diagnostic power in OC [88].

Kallikreins

The major members of the kallikrein-related peptidases (KLKs) have been shown to be promising biomarkers for OC [89]. It was found that OC patients with higher tumor tissue KLK7 ELISA-levels had a two-fold decreased risk of mortality or recurrence compared to lower levels [90]. It has been reported that KLK13 expression in OC tumor tissue correlated with early-stage disease and favorable OS [91]. The OVSCORE, is an algorithm to predict surgical outcome in OC patients based on tumor grade and volume of ascites with KLKs, KLK6 and KLK13. In the multivariate Cox regression analysis for OS, only KLK7 was a significant marker of OS [92].

Other Markers

Ova1 is a biomarker panel assay that consists of 5-serum protein biomarkers (CA125-II, transferrin, beta-2 microglobulin, apolipoprotein A-1, and transthyretin) and it is used for the triage of patients with adnexal masses with 96% sensitivity and 35% specificity [93, 94]. The next generation of Ova1, displayed 91% sensitivity and 69% specificity that combined CA125-II, HE4, apolipoprotein A-1, follicle stimulating hormone and transferrin [94, 95].

The Copenhagen Index (CPH–I) is similar to ROMA and RMI in terms of the combination of HE4, CA125 and age; however, without considering TVU and menopausal status [96].

The concentrations of CA125, HE4, prolactin, IL-2R, CA 15-3, CA19-9, CA72-4, MIF, Cyfra 21-1, TNFR1, TNFR2, IL-6, IL-7, IL-10, IGFBP1, TSH, TNF-α, GH, TIMP-1, ACTH, and osteopontin were reported to be significantly ($P < 0.001$) higher in the serum of patients with early-stage ovarian cancer compared to healthy women, whilst serum levels of HE4, IL-2R, prolactin, CA15-3, CA19-9, CA72-4, Cyfra 21-1, TNFR1, TNFR2, IL-6, IL-7, IL-10, TNF-α, TSH, IGFBP1, MMP-7, VCAM-1, eotaxin-1, FSH, LH, ErbB2, ApoA1,TTR, adiponectin, and CD40L differed significantly ($P < 0.01$) between patients with early-stage (stage I and II) and late-stage (stage III and IV) OC [97]. However, serum biomarkers other than CA125 are not currently used as a detection tool for the early stage detection due to their lower sensitivity or specificity [98].

Uterine Cancer

Endometrial cancer (EC) is the most common gynecologic tumor and the fourth most frequent cancer in women worldwide [99]. The incidence and mortality of EC have increased in the last few decades due to the higher overall prevalence of obesity and metabolic syndrome [100]. To date, there is no established serum marker in the role of the clinical management of EC.

The most important prognostic factors of EC patients include tumor grade, histological subtype, depth of myometrial invasion, cervical involvement, tumor size, lymphovascular space invasion (LVSI) and lymph node status [101].

CA125

It has been found that CA125 level was elevated in 24.6% of EC patients, and only 10% of EC patients with stage I and II disease have elevated CA125 levels [102, 103]. It was reported that CA125 displayed 20.8% sensitivity and 95% specificity for patients with stage I disease and 32.9% sensitivity and 95% specificity for patients with stage II–IV disease [102].

HE4

HE4 expression is significantly increased in EC tissue and the serum of EC patients so that increased HE4 levels can be detected in approximately 88% of EC patients [104]. HE4 has a higher sensitivity and specificity compared to CA125, especially in the early stages of EC. It was shown that serum HE4 had a relatively lower sensitivity (0.65, 95%CI: 0.63–0.67) and a higher specificity (0.93, 95%CI: 0.92–0.95) with the AUC value of 0.7499, proposing a moderate performance in the diagnosis of EC. HE4 is a promising biomarker for the diagnosis of EC, although its sensitivity requires improvement [105].

Genomic Classification

According to The Cancer Genome Atlas (TCGA) Research Network, EC is classified into four categories: POLE ultramutated, microsatellite stability unstable (MSI) hypermutated, copy-number low (microsatellite stable, MSS), and copy-number high (serous-like) [106]. The POLE ultramutated group contains endometrioid EC (EEC) with a few serous EC (SEC) and has an excellent prognosis. The MSI hypermutated and MSS groups are composed of endometrioid histology and they have a moderate prognosis. The copy-number high (serous-like) group includes SEC and high-grade EEC and has the worst prognosis [107].

This classification of EC has been confirmed with the ProMisE (Proactive Molecular Risk Classifier for Endometrial Cancer) study using less costly methods [108]. The TCGA molecular classification has improved the current risk stratification system by providing additional prognostic information [108 - 110]. It has been reported that copy-number high EEC-G1 patients would be ranked as a poor prognostic group, on the contrary all EC patients with POLE-hypermutated would be accepted as an excellent prognostic group even if they are in the higher histological grade or advanced FIGO stage [106].

Other Markers

It has been shown that upregulation of miR-944 and miR-301 were associated with more aggressive disease and lower survival rates [111, 112]. On the contrary, higher levels of miR-205 are related to <50% myometrial invasion and early-stage EECs with improved OS [113].

Three hyper-methylated genes (BHLHE22, CDO1, and CELF4) were identified in cervical scrapings and showed 83-96% sensitivity and 78-96% specificity for EC. The combination of any two genes increased sensitivity to 91.8%, specificity to 95.5% [114].

It has been shown that higher levels of estrogen receptor (ER) or progesterone receptor (PR) were related to an improved OS. In contrast, higher HER2 levels were associated with decreased OS. Especially, HER2 was a powerful predictor of survival in serous EC [115].

Cervical Cancer

The screening of cervical cancer is based on liquid-based cytology and high risk human papilloma virus (HR-HPV) DNA detection in cervical specimens. As yet, there has been no identified biomarker available for screening purposes even though a vast number of serum markers have been inquired in cervical cancer.

Squamous Cell Carcinoma Antigen (SCC-Ag)

Elevated serum levels of SCC-Ag have been detected in 28-88% of cervical squamous cell carcinomas that is correlated with the stage and size of cancer, and depth of the stromal invasion [116, 117]. It was shown that increased serum levels of SCC-Ag were predictive of either prognosis or response to treatment and recurrence [118 - 120].

Other Markers

Increased levels of serum fragments of cytokeratin (CYFRA) have been detected in 42-52% of patients with cervical squamous cell carcinoma [121]. It was reported that elevated CYFRA levels were not related to prognosis [118]. Serum CA125 levels increased in %75 of patients with cervical adenocarcinoma however, only %26 of patients with squamous cell carcinoma of the cervix have high CA 125 levels [122]. Immunosuppressive acidic protein (IAP) levels are elevated in 43-51% of cervical carcinoma that was associated with the stage of the disease and lymph node involvement therefore it may be used for prognosis and evaluation of treatment efficacy [123]. COX-2 was associated with a poorer response to chemoradiotherapy and cancer-related death in patients with stage IIB cervical adenocarcinoma [124].

CONCLUSION

It has utmost importance to differentiate between the benign and malignant adnexal masses prior to the surgery. For that purpose, various prognostic models have been developed and validated. The genomic classification of uterine cancers would be put into use as a risk stratification method for adjuvant treatment decisions.

CONSENT FOR PUBLICATION

Not applicable.

CONFLICT OF INTEREST

The authors declare no conflict of interest, financial or otherwise

ACKNOWLEDGEMENT

I gratefully thank Prof. Dr. Hafize UZUN for giving me the opportunity to be a part of this successful book "Biomarkers in Medicine".

REFERENCES

[1] Torre LA, Trabert B, DeSantis CE, *et al.* Ovarian cancer statistics, 2018. CA Cancer J Clin 2018; 68(4): 284-96.
[http://dx.doi.org/10.3322/caac.21456] [PMID: 29809280]

[2] Kroeger PT Jr, Drapkin R. Pathogenesis and heterogeneity of ovarian cancer. Curr Opin Obstet Gynecol 2017; 29(1): 26-34.
[http://dx.doi.org/10.1097/GCO.0000000000000340] [PMID: 27898521]

[3] Bast RC Jr, Feeney M, Lazarus H, Nadler LM, Colvin RB, Knapp RC. Reactivity of a monoclonal antibody with human ovarian carcinoma. J Clin Invest 1981; 68(5): 1331-7.
[http://dx.doi.org/10.1172/JCI110380] [PMID: 7028788]

[4] Bast RC Jr, Klug TL, St John E, *et al.* A radioimmunoassay using a monoclonal antibody to monitor the course of epithelial ovarian cancer. N Engl J Med 1983; 309(15): 883-7.
[http://dx.doi.org/10.1056/NEJM198310133091503] [PMID: 6310399]

[5] Dochez V, Caillon H, Vaucel E, Dimet J, Winer N, Ducarme G. Biomarkers and algorithms for diagnosis of ovarian cancer: CA125, HE4, RMI and ROMA, a review. J Ovarian Res 2019; 12(1): 28.
[http://dx.doi.org/10.1186/s13048-019-0503-7] [PMID: 30917847]

[6] Bon GG, Kenemans P, Verstraeten R, van Kamp GJ, Hilgers J. Serum tumor marker immunoassays in gynecologic oncology: establishment of reference values. Am J Obstet Gynecol 1996; 174(1 Pt 1): 107-14.
[http://dx.doi.org/10.1016/S0002-9378(96)70381-2] [PMID: 8571992]

[7] Jacobs I, Bast RC Jr. The CA 125 tumour-associated antigen: a review of the literature. Hum Reprod 1989; 4(1): 1-12.
[http://dx.doi.org/10.1093/oxfordjournals.humrep.a136832] [PMID: 2651469]

[8] Moore RG, Miller MC, Steinhoff MM, *et al.* Serum HE4 levels are less frequently elevated than CA125 in women with benign gynecologic disorders. Am J Obstet Gynecol 2012; 206(4): 351.e1-8.
[http://dx.doi.org/10.1016/j.ajog.2011.12.029] [PMID: 22284961]

[9] Chen K, Gentry-Maharaj A, Burnell M, *et al.* Microarray Glycoprofiling of CA125 improves differential diagnosis of ovarian cancer. J Proteome Res 2013; 12(3): 1408-18.
[http://dx.doi.org/10.1021/pr3010474] [PMID: 23360124]

[10] Elias KM, Guo J, Bast RC Jr. Early Detection of Ovarian Cancer. Hematol Oncol Clin North Am 2018; 32(6): 903-14.
[http://dx.doi.org/10.1016/j.hoc.2018.07.003] [PMID: 30390764]

[11] Bonfrer JM, Korse CM, Verstraeten RA, van Kamp GJ, Hart GA, Kenemans P. Clinical evaluation of the Byk LIA-mat CA125 II assay: discussion of a reference value. Clin Chem 1997; 43(3): 491-7.
[http://dx.doi.org/10.1093/clinchem/43.3.491] [PMID: 9068593]

[12] Buys SS, Partridge E, Black A, *et al.* PLCO Project Team. Effect of screening on ovarian cancer mortality: the Prostate, Lung, Colorectal and Ovarian (PLCO) Cancer Screening Randomized Controlled Trial. JAMA 2011; 305(22): 2295-303.
[http://dx.doi.org/10.1001/jama.2011.766] [PMID: 21642681]

[13] Pinsky PF, Yu K, Kramer BS, *et al.* Extended mortality results for ovarian cancer screening in the PLCO trial with median 15years follow-up. Gynecol Oncol 2016; 143(2): 270-5.
[http://dx.doi.org/10.1016/j.ygyno.2016.08.334] [PMID: 27615399]

[14] Jacobs IJ, Menon U, Ryan A, *et al.* Ovarian cancer screening and mortality in the UK Collaborative Trial of Ovarian Cancer Screening (UKCTOCS): a randomised controlled trial. Lancet 2016; 387(10022): 945-56.
[http://dx.doi.org/10.1016/S0140-6736(15)01224-6] [PMID: 26707054]

[15] Alsop K, Fereday S, Meldrum C, *et al.* BRCA mutation frequency and patterns of treatment response in BRCA mutation-positive women with ovarian cancer: a report from the Australian Ovarian Cancer Study Group. J Clin Oncol 2012; 30(21): 2654-63.
[http://dx.doi.org/10.1200/JCO.2011.39.8545] [PMID: 22711857]

[16] Easton DF, Pharoah PD, Antoniou AC, *et al.* Gene-panel sequencing and the prediction of breast-cancer risk. N Engl J Med 2015; 372(23): 2243-57.
[http://dx.doi.org/10.1056/NEJMsr1501341] [PMID: 26014596]

[17] Couch FJ, Shimelis H, Hu C, *et al.* Associations between cancer predisposition testing panel genes and breast cancer. JAMA Oncol 2017; 3(9): 1190-6.
[http://dx.doi.org/10.1001/jamaoncol.2017.0424] [PMID: 28418444]

[18] Mavaddat N, Barrowdale D, Andrulis IL, *et al.* Pathology of breast and ovarian cancers among BRCA1 and BRCA2 mutation carriers: results from the Consortium of Investigators of Modifiers of BRCA1/2 (CIMBA). Cancer Epidemiol Biomarkers Prev 2012; 21(1): 134-47.
[http://dx.doi.org/10.1158/1055-9965.EPI-11-0775] [PMID: 22144499]

[19] Lu KH, Broaddus RR. Gynecologic Cancers in Lynch Syndrome/HNPCC. Fam Cancer 2005; 4(3): 249-54.
[http://dx.doi.org/10.1007/s10689-005-1838-3] [PMID: 16136386]

[20] Crijnen TE, Janssen-Heijnen ML, Gelderblom H, *et al.* Survival of patients with ovarian cancer due to a mismatch repair defect. Fam Cancer 2005; 4(4): 301-5.
[http://dx.doi.org/10.1007/s10689-005-6573-2] [PMID: 16341807]

[21] Lilyquist J, LaDuca H, Polley E, *et al.* Frequency of mutations in a large series of clinically ascertained ovarian cancer cases tested on multi-gene panels compared to reference controls. Gynecol Oncol 2017; 147(2): 375-80.
[http://dx.doi.org/10.1016/j.ygyno.2017.08.030] [PMID: 28888541]

[22] National Comprehensive Cancer Network. Genetic/familial high risk assessment: breast and ovarian Version 22017 NCCN Clinical Practice Guidelines in Oncology. Fort Washington, PA: NCCN 2016.

[23] Rosenthal AN, Fraser LSM, Philpott S, *et al.* Evidence of stage shift in women diagnosed with ovarian cancer during phase II of the United Kingdom Familial Ovarian Cancer Screening Study. J Clin Oncol 2017; 35(13): 1411-20.
[http://dx.doi.org/10.1200/JCO.2016.69.9330] [PMID: 28240969]

[24] Kaijser J, Sayasneh A, Van Hoorde K, *et al.* Presurgical diagnosis of adnexal tumours using mathematical models and scoring systems: a systematic review and meta-analysis. Hum Reprod Update 2014; 20(3): 449-62.
[http://dx.doi.org/10.1093/humupd/dmt059] [PMID: 24327552]

[25] Yamamoto Y, Yamada R, Oguri H, Maeda N, Fukaya T. Comparison of four malignancy risk indices in the preoperative evaluation of patients with pelvic masses. Eur J Obstet Gynecol Reprod Biol 2009; 144(2): 163-7.
[http://dx.doi.org/10.1016/j.ejogrb.2009.02.048] [PMID: 19327881]

[26] Meys EM, Kaijser J, Kruitwagen RF, *et al.* Subjective assessment versus ultrasound models to diagnose ovarian cancer: A systematic review and meta-analysis. Eur J Cancer 2016; 58: 17-29.
[http://dx.doi.org/10.1016/j.ejca.2016.01.007] [PMID: 26922169]

[27] Timmerman D, Testa AC, Bourne T, *et al.* International Ovarian Tumor Analysis Group. Logistic regression model to distinguish between the benign and malignant adnexal mass before surgery: a multicenter study by the International Ovarian Tumor Analysis Group. J Clin Oncol 2005; 23(34):

8794-801.
[http://dx.doi.org/10.1200/JCO.2005.01.7632] [PMID: 16314639]

[28] Van Holsbeke C, Van Calster B, Bourne T, *et al.* External validation of diagnostic models to estimate the risk of malignancy in adnexal masses. Clin Cancer Res 2012; 18(3): 815-25.
[http://dx.doi.org/10.1158/1078-0432.CCR-11-0879] [PMID: 22114135]

[29] Van Calster B, Van Hoorde K, Valentin L, *et al.* International Ovarian Tumour Analysis Group. Evaluating the risk of ovarian cancer before surgery using the ADNEX model to differentiate between benign, borderline, early and advanced stage invasive, and secondary metastatic tumours: prospective multicentre diagnostic study. BMJ 2014; 349(oct07 3): g5920.
[http://dx.doi.org/10.1136/bmj.g5920] [PMID: 25320247]

[30] Testa A, Kaijser J, Wynants L, *et al.* Strategies to diagnose ovarian cancer: new evidence from phase 3 of the multicentre international IOTA study. Br J Cancer 2014; 111(4): 680-8.
[http://dx.doi.org/10.1038/bjc.2014.333] [PMID: 24937676]

[31] Meyer T, Rustin GJ. Role of tumour markers in monitoring epithelial ovarian cancer. Br J Cancer 2000; 82(9): 1535-8.
[PMID: 10789720]

[32] Paramasivam S, Tripcony L, Crandon A, *et al.* Prognostic importance of preoperative CA125 in FIGO stage I epithelial ovarian cancer: an Australian multicenter study. J Clin Oncol 2005; 23(25): 5938-42.
[http://dx.doi.org/10.1200/JCO.2005.08.151] [PMID: 16087942]

[33] Høgdall CK, Nørgaard-Pedersen B, Mogensen O. The prognostic value of pre-operative serum tetranectin, CA-125 and a combined index in women with primary ovarian cancer. Anticancer Res 2002; 22(3): 1765-8.
[PMID: 12168866]

[34] Markman M, Federico M, Liu PY, Hannigan E, Alberts D. Significance of early changes in the serum CA-125 antigen level on overall survival in advanced ovarian cancer. Gynecol Oncol 2006; 103(1): 195-8.
[http://dx.doi.org/10.1016/j.ygyno.2006.02.024] [PMID: 16595148]

[35] van Altena AM, Kolwijck E, Spanjer MJ, Hendriks JC, Massuger LF, de Hullu JA. CA125 nadir concentration is an independent predictor of tumor recurrence in patients with ovarian cancer: a population-based study. Gynecol Oncol 2010; 119(2): 265-9.
[http://dx.doi.org/10.1016/j.ygyno.2010.07.025] [PMID: 20797777]

[36] Riedinger JM, Eche N, Basuyau JP, Dalifard I, Hacene K, Pichon MF. Prognostic value of serum CA 125 bi-exponential decrease during first line paclitaxel/platinum chemotherapy: a French multicentric study. Gynecol Oncol 2008; 109(2): 194-8.
[http://dx.doi.org/10.1016/j.ygyno.2008.01.035] [PMID: 18329083]

[37] Santillan A, Garg R, Zahurak ML, *et al.* Risk of epithelial ovarian cancer recurrence in patients with rising serum CA-125 levels within the normal range. J Clin Oncol 2005; 23(36): 9338-43.
[http://dx.doi.org/10.1200/JCO.2005.02.2582] [PMID: 16361633]

[38] Rustin GJ, van der Burg ME, Griffin CL, *et al.* Early versus delayed treatment of relapsed ovarian cancer (MRC OV05/EORTC 55955): a randomised trial. Lancet 2010; 376(9747): 1155-63.
[http://dx.doi.org/10.1016/S0140-6736(10)61268-8] [PMID: 20888993]

[39] Sölétormos G, Duffy MJ, Othman Abu Hassan S, *et al.* Clinical use of cancer biomarkers in epithelial ovarian cancer: updated guidelines from the European group on tumor markers. Int J Gynecol Cancer 2016; 26(1): 43-51.
[http://dx.doi.org/10.1097/IGC.0000000000000586] [PMID: 26588231]

[40] Anastasi E, Marchei GG, Viggiani V, Gennarini G, Frati L, Reale MG. HE4: a new potential early biomarker for the recurrence of ovarian cancer. Tumour Biol 2010; 31(2): 113-9.
[http://dx.doi.org/10.1007/s13277-009-0015-y] [PMID: 20358424]

[41] Molina R, Escudero JM, Augé JM, *et al.* HE4 a novel tumour marker for ovarian cancer: comparison with CA 125 and ROMA algorithm in patients with gynaecological diseases. Tumour Biol 2011; 32(6): 1087-95.
[http://dx.doi.org/10.1007/s13277-011-0204-3] [PMID: 21863264]

[42] Hallamaa M, Suvitie P, Huhtinen K, Matomäki J, Poutanen M, Perheentupa A. Serum HE4 concentration is not dependent on menstrual cycle or hormonal treatment among endometriosis patients and healthy premenopausal women. Gynecol Oncol 2012; 125(3): 667-72.
[http://dx.doi.org/10.1016/j.ygyno.2012.03.011] [PMID: 22426487]

[43] Havrilesky LJ, Whitehead CM, Rubatt JM, *et al.* Evaluation of biomarker panels for early stage ovarian cancer detection and monitoring for disease recurrence. Gynecol Oncol 2008; 110(3): 374-82.
[http://dx.doi.org/10.1016/j.ygyno.2008.04.041] [PMID: 18584856]

[44] Wu L, Dai ZY, Qian YH, Shi Y, Liu FJ, Yang C. Diagnostic value of serum human epididymis protein 4 (HE4) in ovarian carcinoma: a systematic review and meta-analysis. Int J Gynecol Cancer 2012; 22(7): 1106-12.
[http://dx.doi.org/10.1097/IGC.0b013e318263efa2] [PMID: 22854652]

[45] Li F, Tie R, Chang K, *et al.* Does risk for ovarian malignancy algorithm excel human epididymis protein 4 and CA125 in predicting epithelial ovarian cancer: a meta-analysis. BMC Cancer 2012; 12(1): 258.
[http://dx.doi.org/10.1186/1471-2407-12-258] [PMID: 22712526]

[46] Yu S, Yang HJ, Xie SQ, Bao YX. Diagnostic value of HE4 for ovarian cancer: a meta-analysis. Clin Chem Lab Med 2012; 50(8): 1439-46.
[http://dx.doi.org/10.1515/cclm-2011-0477] [PMID: 22868811]

[47] Steffensen KD, Waldstrøm M, Brandslund I, Jakobsen A. Prognostic impact of prechemotherapy serum levels of HER2, CA125, and HE4 in ovarian cancer patients. Int J Gynecol Cancer 2011; 21(6): 1040-7.
[http://dx.doi.org/10.1097/IGC.0b013e31821e052e] [PMID: 21738039]

[48] Midulla C, Manganaro L, Longo F, *et al.* HE4 combined with MDCT imaging is a good marker in the evaluation of disease extension in advanced epithelial ovarian carcinoma. Tumour Biol 2012; 33(5): 1291-8.
[http://dx.doi.org/10.1007/s13277-012-0376-5] [PMID: 22430259]

[49] Moore RG, McMeekin DS, Brown AK, *et al.* A novel multiple marker bioassay utilizing HE4 and CA125 for the prediction of ovarian cancer in patients with a pelvic mass. Gynecol Oncol 2009; 112(1): 40-6.
[http://dx.doi.org/10.1016/j.ygyno.2008.08.031] [PMID: 18851871]

[50] Kaijser J, Van Belle V, Van Gorp T, *et al.* Prognostic value of serum HE4 levels and risk of ovarian malignancy algorithm scores at the time of ovarian cancer diagnosis. Int J Gynecol Cancer 2014; 24(7): 1173-80.
[http://dx.doi.org/10.1097/IGC.0000000000000181] [PMID: 24987915]

[51] Chen X, Zhou H, Chen R, *et al.* Development of a multimarker assay for differential diagnosis of benign and malignant pelvic masses. Clin Chim Acta 2015; 440: 57-63.
[http://dx.doi.org/10.1016/j.cca.2014.11.013] [PMID: 25447698]

[52] Kamat AA, Baldwin M, Urbauer D, *et al.* Plasma cell-free DNA in ovarian cancer: an independent prognostic biomarker. Cancer 2010; 116(8): 1918-25.
[http://dx.doi.org/10.1002/cncr.24997] [PMID: 20166213]

[53] Shao X, He Y, Ji M, *et al.* Quantitative analysis of cell-free DNA in ovarian cancer. Oncol Lett 2015; 10(6): 3478-82.
[http://dx.doi.org/10.3892/ol.2015.3771] [PMID: 26788153]

[54] Rivlin N, Brosh R, Oren M, Rotter V. Mutations in the p53 tumor suppressor gene: important

milestones at the various steps of tumorigenesis. Genes Cancer 2011; 2(4): 466-74. [http://dx.doi.org/10.1177/1947601911408889] [PMID: 21779514]

[55] Forshew T, Murtaza M, Parkinson C, *et al.* Noninvasive identification and monitoring of cancer mutations by targeted deep sequencing of plasma DNA. Sci Transl Med 2012; 4(136): 136ra68. [http://dx.doi.org/10.1126/scitranslmed.3003726] [PMID: 22649089]

[56] Wimberger P, Roth C, Pantel K, Kasimir-Bauer S, Kimmig R, Schwarzenbach H. Impact of platinum-based chemotherapy on circulating nucleic acid levels, protease activities in blood and disseminated tumor cells in bone marrow of ovarian cancer patients. Int J Cancer 2011; 128(11): 2572-80. [http://dx.doi.org/10.1002/ijc.25602] [PMID: 20715113]

[57] Ibanez de Caceres I, Battagli C, Esteller M, *et al.* Tumor cell-specific BRCA1 and RASSF1A hypermethylation in serum, plasma, and peritoneal fluid from ovarian cancer patients. Cancer Res 2004; 64(18): 6476-81. [http://dx.doi.org/10.1158/0008-5472.CAN-04-1529] [PMID: 15374957]

[58] Zhang Q, Hu G, Yang Q, *et al.* A multiplex methylation-specific PCR assay for the detection of early-stage ovarian cancer using cell-free serum DNA. Gynecol Oncol 2013; 130(1): 132-9. [http://dx.doi.org/10.1016/j.ygyno.2013.04.048] [PMID: 23623832]

[59] Widschwendter M, Zikan M, Wahl B, *et al.* The potential of circulating tumor DNA methylation analysis for the early detection and management of ovarian cancer. Genome Med 2017; 9(1): 116. [http://dx.doi.org/10.1186/s13073-017-0500-7] [PMID: 29268796]

[60] Denkert C, Budczies J, Kind T, *et al.* Mass spectrometry-based metabolic profiling reveals different metabolite patterns in invasive ovarian carcinomas and ovarian borderline tumors. Cancer Res 2006; 66(22): 10795-804. [http://dx.doi.org/10.1158/0008-5472.CAN-06-0755] [PMID: 17108116]

[61] Zhang F, Zhang Y, Ke C, *et al.* Predicting ovarian cancer recurrence by plasma metabolic profiles before and after surgery. Metabolomics 2018; 14(5): 65. [http://dx.doi.org/10.1007/s11306-018-1354-8]

[62] Ambros V. microRNAs: tiny regulators with great potential. Cell 2001; 107(7): 823-6. [http://dx.doi.org/10.1016/S0092-8674(01)00616-X] [PMID: 11779458]

[63] He M, Liu Y, Wang X, Zhang MQ, Hannon GJ, Huang ZJ. Cell-type-based analysis of microRNA profiles in the mouse brain. Neuron 2012; 73(1): 35-48. [http://dx.doi.org/10.1016/j.neuron.2011.11.010] [PMID: 22243745]

[64] Parikh A, Lee C, Joseph P, *et al.* microRNA-181a has a critical role in ovarian cancer progression through the regulation of the epithelial-mesenchymal transition. Nat Commun 2014; 5(1): 2977. [http://dx.doi.org/10.1038/ncomms3977] [PMID: 24394555]

[65] Chen Z, Zhu J, Zhu Y, Wang J. MicroRNA-616 promotes the progression of ovarian cancer by targeting TIMP2. Oncol Rep 2018; 39(6): 2960-8. [http://dx.doi.org/10.3892/or.2018.6368] [PMID: 29658596]

[66] Salem M, O'Brien JA, Bernaudo S, *et al.* miR-590-3p Promotes Ovarian Cancer Growth and Metastasis *via* a Novel FOXA2-Versican Pathway. Cancer Res 2018; 78(15): 4175-90. [http://dx.doi.org/10.1158/0008-5472.CAN-17-3014] [PMID: 29748371]

[67] Ying X, Wei K, Lin Z, *et al.* MicroRNA-125b Suppresses Ovarian Cancer Progression *via* Suppression of the Epithelial-Mesenchymal Transition Pathway by Targeting the SET Protein. Cell Physiol Biochem 2016; 39(2): 501-10. [http://dx.doi.org/10.1159/000445642] [PMID: 27383536]

[68] Yan H, Li H, Li P, *et al.* Long noncoding RNA MLK7-AS1 promotes ovarian cancer cells progression by modulating miR-375/YAP1 axis. J Exp Clin Cancer Res 2018; 37(1): 237. [http://dx.doi.org/10.1186/s13046-018-0910-4] [PMID: 30249278]

[69] Iorio MV, Visone R, Di Leva G, *et al.* MicroRNA signatures in human ovarian cancer. Cancer Res

2007; 67(18): 8699-707.
[http://dx.doi.org/10.1158/0008-5472.CAN-07-1936] [PMID: 17875710]

[70] Zhang X, Guo G, Wang G, *et al.* Profile of differentially expressed miRNAs in high-grade serous carcinoma and clear cell ovarian carcinoma, and the expression of miR-510 in ovarian carcinoma. Mol Med Rep 2015; 12(6): 8021-31.
[http://dx.doi.org/10.3892/mmr.2015.4485] [PMID: 26497752]

[71] Braicu OL, Budisan L, Buiga R, *et al.* miRNA expression profiling in formalin-fixed paraffin-embedded endometriosis and ovarian cancer samples. OncoTargets Ther 2017; 10: 4225-38.
[http://dx.doi.org/10.2147/OTT.S137107] [PMID: 28894379]

[72] Lin J, Zhang L, Huang H, *et al.* MiR-26b/KPNA2 axis inhibits epithelial ovarian carcinoma proliferation and metastasis through downregulating OCT4. Oncotarget 2015; 6(27): 23793-806.
[http://dx.doi.org/10.18632/oncotarget.4363] [PMID: 26204489]

[73] Lu J, Zhang W, Ding Y, Li X, Song J. Expression of miR-26b in ovarian carcinoma tissues and its correlation with clinicopathology. Oncol Lett 2019; 17(5): 4417-22.
[http://dx.doi.org/10.3892/ol.2019.10117] [PMID: 30944634]

[74] Takagawa R, Fujii S, Ohta M, *et al.* Preoperative serum carcinoembryonic antigen level as a predictive factor of recurrence after curative resection of colorectal cancer. Ann Surg Oncol 2008; 15(12): 3433-9.
[http://dx.doi.org/10.1245/s10434-008-0168-8] [PMID: 18846401]

[75] Høgdall EV, Christensen L, Kjaer SK, *et al.* Protein expression levels of carcinoembryonic antigen (CEA) in Danish ovarian cancer patients: from the Danish 'MALOVA'ovarian cancer study. Pathology 2008; 40(5): 487-92.
[http://dx.doi.org/10.1080/00313020802197889] [PMID: 18604735]

[76] Kawai M, Kano T, Kikkawa F, *et al.* Seven tumor markers in benign and malignant germ cell tumors of the ovary. Gynecol Oncol 1992; 45(3): 248-53.
[http://dx.doi.org/10.1016/0090-8258(92)90299-X] [PMID: 1319383]

[77] Chow SN, Yang JH, Lin YH, *et al.* Malignant ovarian germ cell tumors. Int J Gynaecol Obstet 1996; 53(2): 151-8.
[http://dx.doi.org/10.1016/0020-7292(96)02657-4] [PMID: 8735296]

[78] Groome NP, Illingworth PJ, O'Brien M, *et al.* Measurement of dimeric inhibin B throughout the human menstrual cycle. J Clin Endocrinol Metab 1996; 81(4): 1401-5.
[PMID: 8636341]

[79] Petraglia F, Luisi S, Pautier P, *et al.* Inhibin B is the major form of inhibin/activin family secreted by granulosa cell tumors. J Clin Endocrinol Metab 1998; 83(3): 1029-32.
[http://dx.doi.org/10.1210/jcem.83.3.4800] [PMID: 9506769]

[80] Robertson DM, Cahir N, Burger HG, Mamers P, Groome N. Inhibin forms in serum from postmenopausal women with ovarian cancers. Clin Endocrinol (Oxf) 1999; 50(3): 381-6.
[http://dx.doi.org/10.1046/j.1365-2265.1999.00656.x] [PMID: 10435065]

[81] Mom CH, Engelen MJ, Willemse PH, *et al.* Granulosa cell tumors of the ovary: the clinical value of serum inhibin A and B levels in a large single center cohort. Gynecol Oncol 2007; 105(2): 365-72.
[http://dx.doi.org/10.1016/j.ygyno.2006.12.034] [PMID: 17306349]

[82] Stenman UH, Alfthan H, Hotakainen K. Human chorionic gonadotropin in cancer. Clin Biochem 2004; 37(7): 549-61.
[http://dx.doi.org/10.1016/j.clinbiochem.2004.05.008] [PMID: 15234236]

[83] Kurman RJ, Norris HJ. Embryonal carcinoma of the ovary: a clinicopathologic entity distinct from endodermal sinus tumor resembling embryonal carcinoma of the adult testis. Cancer 1976; 38(6): 2420-33.
[http://dx.doi.org/10.1002/1097-0142(197612)38:6<2420::AID-CNCR2820380630>3.0.CO;2-2]

[PMID: 63319]

[84] Roger M, Chaussain JL, Blacker C, Feinstein MC. Tumors secreting choriogonadotropin in children. Ann Med Interne (Paris) 1984; 135(5): 381-4.
[PMID: 6486617]

[85] Stenman UH, Alfthan H. Markers for testicular cancer. In: Diamandis E, Fritsche H, Lilja H, Chan D, Schwartz M, Eds. Tumor markers Physiology, pathobiology, technology, and clinical applications. Washington: AACC Press 2002; pp. 351-9.

[86] Bosl GJ, Motzer RJ. Testicular germ-cell cancer. N Engl J Med 1997; 337(4): 242-53.
[http://dx.doi.org/10.1056/NEJM199707243370406] [PMID: 9227931]

[87] Vartiainen J, Alfthan H, Lehtovirta P, Stenman UH. Identification of choriocarcinoma by the hCG beta-to-hCG proportion in patients with delayed diagnosis caused by contraceptive use. Contraception 1998; 57(4): 257-60.
[http://dx.doi.org/10.1016/S0010-7824(98)00028-6] [PMID: 9649918]

[88] Kinugasa M, Nishimura R, Koizumi T, *et al.* Combination assay of urinary beta-core fragment of human chorionic gonadotropin with serum tumor markers in gynecologic cancers. Jpn J Cancer Res 1995; 86(8): 783-9.
[http://dx.doi.org/10.1111/j.1349-7006.1995.tb02469.x] [PMID: 7559103]

[89] Dorn J, Bayani J, Yousef GM, *et al.* Clinical utility of kallikrein-related peptidases (KLK) in urogenital malignancies. Thromb Haemost 2013; 110(3): 408-22.
[http://dx.doi.org/10.1160/TH13-03-0206] [PMID: 23765253]

[90] Dorn J, Gkazepis A, Kotzsch M, *et al.* Clinical value of protein expression of kallikrein-related peptidase 7 (KLK7) in ovarian cancer. Biol Chem 2014; 395(1): 95-107.
[http://dx.doi.org/10.1515/hsz-2013-0172] [PMID: 23999494]

[91] Scorilas A, Borgoño CA, Harbeck N, *et al.* Human kallikrein 13 protein in ovarian cancer cytosols: a new favorable prognostic marker. J Clin Oncol 2004; 22(4): 678-85.
[http://dx.doi.org/10.1200/JCO.2004.05.144] [PMID: 14966091]

[92] Dorn J, Bronger H, Kates R, *et al.* OVSCORE - a validated score to identify ovarian cancer patients not suitable for primary surgery. Oncol Lett 2015; 9(1): 418-24.
[http://dx.doi.org/10.3892/ol.2014.2630] [PMID: 25436002]

[93] Zhang Z, Chan DW. The road from discovery to clinical diagnostics: lessons learned from the first FDA-cleared *in vitro* diagnostic multivariate index assay of proteomic biomarkers. Cancer Epidemiol Biomarkers Prev 201; 19(12): 2995-9.

[94] Ueland FR, Desimone CP, Seamon LG, *et al.* Effectiveness of a multivariate index assay in the preoperative assessment of ovarian tumors. Obstet Gynecol 2011; 117(6): 1289-97.
[http://dx.doi.org/10.1097/AOG.0b013e31821b5118] [PMID: 21606739]

[95] Coleman RL, Herzog TJ, Chan DW, *et al.* Validation of a second-generation multivariate index assay for malignancy risk of adnexal masses. Am J Obstet Gynecol 2016; 215(1): 82.e1-82.e11.
[http://dx.doi.org/10.1016/j.ajog.2016.03.003] [PMID: 26970494]

[96] Karlsen MA, Høgdall EV, Christensen IJ, *et al.* A novel diagnostic index combining HE4, CA125 and age may improve triage of women with suspected ovarian cancer - An international multicenter study in women with an ovarian mass. Gynecol Oncol 2015; 138(3): 640-6.
[http://dx.doi.org/10.1016/j.ygyno.2015.06.021] [PMID: 26086566]

[97] Yurkovetsky Z, Skates S, Lomakin A, *et al.* Development of a multimarker assay for early detection of ovarian cancer. J Clin Oncol 2010; 28(13): 2159-66.
[http://dx.doi.org/10.1200/JCO.2008.19.2484] [PMID: 20368574]

[98] Muinao T, Deka Boruah HP, Pal M. Multi-biomarker panel signature as the key to diagnosis of ovarian cancer. Heliyon 2019; 5(12): e02826.
[http://dx.doi.org/10.1016/j.heliyon.2019.e02826] [PMID: 31867451]

[99] Bray F, Ferlay J, Soerjomataram I, Siegel RL, Torre LA, Jemal A. Global cancer statistics 2018: GLOBOCAN estimates of incidence and mortality worldwide for 36 cancers in 185 countries. CA Cancer J Clin 2018; 68(6): 394-424.
[http://dx.doi.org/10.3322/caac.21492] [PMID: 30207593]

[100] American Cancer Society. Cancer Facts & Figures 2018. Atlanta, GA, USA: American Cancer Society 2017.

[101] Colombo N, Creutzberg C, Amant F, *et al.* ESMO-ESGO-ESTRO Consensus Conference on Endometrial Cancer: diagnosis, treatment and follow-up. Ann Oncol 2016; 27(1): 16-41.
[http://dx.doi.org/10.1093/annonc/mdv484] [PMID: 26634381]

[102] Moore RG, Brown AK, Miller MC, *et al.* Utility of a novel serum tumor biomarker HE4 in patients with endometrioid adenocarcinoma of the uterus. Gynecol Oncol 2008; 110(2): 196-201.
[http://dx.doi.org/10.1016/j.ygyno.2008.04.002] [PMID: 18495222]

[103] Scambia G, Gadducci A, Panici PB, *et al.* Combined use of CA 125 and CA 15-3 in patients with endometrial carcinoma. Gynecol Oncol 1994; 54(3): 292-7.
[http://dx.doi.org/10.1006/gyno.1994.1213] [PMID: 8088605]

[104] Galgano MT, Hampton GM, Frierson HF Jr. Comprehensive analysis of HE4 expression in normal and malignant human tissues. Mod Pathol 2006; 19(6): 847-53.
[http://dx.doi.org/10.1038/modpathol.3800612] [PMID: 16607372]

[105] Li J, Wang X, Qu W, Wang J, Jiang SW. Comparison of serum human epididymis protein 4 and CA125 on endometrial cancer detection: A meta-analysis. Clin Chim Acta 2019; 488: 215-20.
[http://dx.doi.org/10.1016/j.cca.2018.11.011] [PMID: 30414437]

[106] Kandoth C, Schultz N, Cherniack AD, *et al.* Integrated genomic characterization of endometrial carcinoma. Nature 2013; 497(7447): 67-73.
[http://dx.doi.org/10.1038/nature12113] [PMID: 23636398]

[107] León-Castillo A, Gilvazquez E, Nout R, *et al.* Clinicopathological and molecular characterisation of 'multiple-classifier' endometrial carcinomas. J Pathol 2020; 250(3): 312-22.
[http://dx.doi.org/10.1002/path.5373] [PMID: 31829447]

[108] Talhouk A, McConechy MK, Leung S, *et al.* Confirmation of ProMisE: A simple, genomics-based clinical classifier for endometrial cancer. Cancer 2017; 123(5): 802-13.
[http://dx.doi.org/10.1002/cncr.30496] [PMID: 28061006]

[109] Stelloo E, Nout RA, Osse EM, *et al.* Improved Risk Assessment by Integrating Molecular and Clinicopathological Factors in Early-stage Endometrial Cancer-Combined Analysis of the PORTEC Cohorts. Clin Cancer Res 2016; 22(16): 4215-24.
[http://dx.doi.org/10.1158/1078-0432.CCR-15-2878] [PMID: 27006490]

[110] León-Castillo A, de Boer SM, Powell ME, *et al.* TransPORTEC consortium. Molecular Classification of the PORTEC-3 Trial for High-Risk Endometrial Cancer: Impact on Prognosis and Benefit From Adjuvant Therapy. J Clin Oncol 2020; 38(29): 3388-97.
[http://dx.doi.org/10.1200/JCO.20.00549] [PMID: 32749941]

[111] He Z, Xu H, Meng Y, Kuang Y. miR-944 acts as a prognostic marker and promotes the tumor progression in endometrial cancer. Biomed Pharmacother 2017; 88: 902-10.
[http://dx.doi.org/10.1016/j.biopha.2017.01.117] [PMID: 28178620]

[112] Lu L, Shen Y, Tseng K, *et al.* Oncogenic function of miR-301 to predicts poor prognosis of endometrial cancer. Int J Clin Exp Med 2016; 9: 12937-42.

[113] Wilczynski M, Danielska J, Dzieniecka M, Szymanska B, Wojciechowski M, Malinowski A. Prognostic and clinical significance of mirna-205 in endometrioid endometrial cancer. PLoS One 2016; 11(10): e0164687.
[http://dx.doi.org/10.1371/journal.pone.0164687] [PMID: 27737015]

[114] Huang RL, Su PH, Liao YP, *et al.* Integrated epigenomics analysis reveals a DNA methylation panel for endometrial cancer detection using cervical scrapings. Clin Cancer Res 2017; 23(1): 263-72. [http://dx.doi.org/10.1158/1078-0432.CCR-16-0863] [PMID: 27507616]

[115] Zhang Y, Zhao D, Gong C, *et al.* Prognostic role of hormone receptors in endometrial cancer: a systematic review and meta-analysis. World J Surg Oncol 2015; 13(1): 208. [http://dx.doi.org/10.1186/s12957-015-0619-1] [PMID: 26108802]

[116] Ohara K, Tanaka Y, Tsunoda H, Nishida M, Sugahara S, Itai Y. Assessment of cervical cancer radioresponse by serum squamous cell carcinoma antigen and magnetic resonance imaging. Obstet Gynecol 2002; 100(4): 781-7. [PMID: 12383549]

[117] Takeda M, Sakuragi N, Okamoto K, *et al.* Preoperative serum SCC, CA125, and CA19-9 levels and lymph node status in squamous cell carcinoma of the uterine cervix. Acta Obstet Gynecol Scand 2002; 81(5): 451-7. [http://dx.doi.org/10.1034/j.1600-0412.2002.810513.x] [PMID: 12027820]

[118] Gaarenstroom KN, Kenter GG, Bonfrer JM, *et al.* Can initial serum cyfra 21-1, SCC antigen, and TPA levels in squamous cell cervical cancer predict lymph node metastases or prognosis? Gynecol Oncol 2000; 77(1): 164-70. [http://dx.doi.org/10.1006/gyno.2000.5732] [PMID: 10739706]

[119] Chou CY, Wang ST, Kuo HC, Tzeng CC, Yao BL. Serum level of squamous cell carcinoma antigen and tumor size are useful to identify preoperatively patients at high risk of cervical cancer. Cancer 1994; 74(9): 2497-501. [http://dx.doi.org/10.1002/1097-0142(19941101)74:9<2497::AID-CNCR2820740917>3.0.CO;2-L] [PMID: 7923006]

[120] Bolli JA, Doering DL, Bosscher JR, *et al.* Squamous cell carcinoma antigen: clinical utility in squamous cell carcinoma of the uterine cervix. Gynecol Oncol 1994; 55(2): 169-73. [http://dx.doi.org/10.1006/gyno.1994.1272] [PMID: 7959279]

[121] Suzuki Y, Nakano T, Ohno T, Abe A, Morita S, Tsujii H. Serum CYFRA 21-1 in cervical cancer patients treated with radiation therapy. J Cancer Res Clin Oncol 2000; 126(6): 332-6. [http://dx.doi.org/10.1007/s004320050352] [PMID: 10870643]

[122] Lehtovirta P, Viinikka L, Ylikorkala O. Comparison between squamous cell carcinoma-associated antigen and CA-125 in patients with carcinoma of the cervix. Gynecol Oncol 1990; 37(2): 276-8. [http://dx.doi.org/10.1016/0090-8258(90)90347-N] [PMID: 2344974]

[123] Battaglia F, Scambia G, Panici PB, *et al.* Immunosuppressive acidic protein (IAP) and squamous cell carcinoma antigen (SCC) in patients with cervical cancer. Gynecol Oncol 1994; 53(2): 176-82. [http://dx.doi.org/10.1006/gyno.1994.1112] [PMID: 8188076]

[124] Jung YW, Kim SW, Kim S, *et al.* Prevalence and clinical relevance of cyclooxygenase-1 and -2 expression in stage IIB cervical adenocarcinoma. Eur J Obstet Gynecol Reprod Biol 2010; 148(1): 62-6. [http://dx.doi.org/10.1016/j.ejogrb.2009.09.011] [PMID: 19836124]

CHAPTER 3

Biomarkers in Urological Cancers

Bulent Onal[1,*], Mehmet Hamza Gultekin[1], Muhammed Fatih Simsekoglu[1], Berin Selcuk[1] and **Ahmet Gurbuz[1]**

[1] *Department of Urology, Cerrahpasa School of Medicine, Istanbul University - Cerrahpasa, Istanbul, Turkey*

Abstract: Urological tumours have become one of the most common cancers in the last decade. It is important to apply an approach that evaluates many factors related to the patient and the disease carefully to minimize cancer-associated morbidity and mortality. The clinical use of cancer biomarkers is a valuable part of the clinical management of urological cancers. These biomarkers may lead to optimized detection, treatment, and follow-up of urological cancers. With the development of molecular research, newly developed biomarkers and next-generation sequencing have also contributed to patient management. In this chapter, we will present biomarkers in the most common urological cancers under subheadings of bladder cancer, prostate cancer, kidney cancer, and testicular cancer. Additionally, due to the development that occurred in the next-generation sequencing (NGS), all the above-mentioned malignancies are evaluated with regard to NGS.

Keywords: Biomarkers, Clinical management, Molecular research, Next-generation sequencing, Patient management, Urological cancers.

BIOMARKERS IN BLADDER CANCER

Bladder cancer (BC) is the 11th most commonly diagnosed cancer worldwide [1]. The most common symptom of BC is hematuria. Eighty-five percent of patients newly diagnosed with BC have painless macroscopic hematuria, and almost all present microscopic hematuria [2].

The gold standard test in the diagnosis of BC is cystoscopy. Currently, white light cystoscopy is the standard of care for the diagnosis of BC, and it is considered to be highly sensitive and specific for this type of cancer. However, it remains costly and somewhat invasive. As a result of the search for less invasive methods in the diagnosis and follow-up of bladder tumors, tissue-, blood- and urine-based biomarkers have also been developed.

[*] **Corresponding author Bulent Onal:** Department of Urology, Cerrahpasa School of Medicine, Istanbul University - Cerrahpasa, Istanbul, Turkey; Tel: +90212 2304300; E-mail: bulent.onal@istanbul.edu.tr

Pınar Atukeren and Hafize Uzun (Eds.)

Today, only urine-based biomarkers are used in clinical practice, and although many studies have been conducted on tissue- and blood-based biomarkers, none have found a place in clinical use. In this article, biomarkers are briefly discussed under separate sub-headings, and examples are given.

Blood-based Biomarkers

Tumor Cells

The presence of tumor cells in the blood has been associated with advanced-stage disease in solid organ malignancies. In a study investigating the prognostic effects of circulating tumor cells captured with the CellSerch system, the rates of recurrence, cancer-specific mortality, and overall survival (OS) were significantly lower in those with circulating tumor cells [3].

CD8 Count and Immune Cells

The CD8 count is a measure of the presence of cytotoxic T cells that may be involved in the immune response against tumor cells. One study reported that a low CD8 count was associated with recurrence-free survival after the transurethral resection of bladder tumor (TURT) [4]. However, the literature on the neutrophil-to-lymphocyte ratio (NLR), a marker of acute phase inflammation, is controversial. In a meta-analysis of 21 studies analyzing the prognostic role of NLR in BC, the authors correlated elevated pre-treatment NLR with OS, recurrence-free survival, and disease-free survival in BC [5]. In contrast, a secondary analysis of the SWOG 8710 trial reported that NLR was not a prognostic or predictive biomarker for OS in muscle-invasive bladder cancer, and it was not able to demonstrate OS benefit from neoadjuvant chemotherapy [6].

Matrix Metalloproteinase

Matrix metalloproteinases (MMPs) are calcium-dependent zinc-containing endopeptidases that are capable of degrading all types of extracellular matrix proteins. In the literature, several MMP isoforms have been simultaneously evaluated, but only MMP-7 levels have been associated with cancer-specific mortality [7].

Epigenetic Markers

Epigenetics investigate molecular DNA changes, including methylation and histone alterations. These DNA changes are dynamic processes that determine which parts of DNA are more frequently transcribed and which regions are silenced. In one study, the promoter methylation of 17 genes was investigated, and that of five gene loci (RASSF1, E-cadherin, TNFSR25, EDNRB, and APC) was defined as a marker for tumor progression [8]. In another study, methylation at the RASSF1A and DAPK loci was identified as an independent predictor for tumor progression [9].

Tissue-based Biomarkers

Cell Cycle-based Markers

These biomarkers consist of p53, pRB, Ki67, cyclins, p21, and p27. Cell cycle regulators are among the most common mutated markers in BC and have been shown to be predictors of aggression, metastases, and cancer-specific mortality in many studies [10].

In a retrospective study with a large sample size, the authors examined four cell-cycle regulators (p53, pRB, p21, and p27) and found that a combination of the four markers was able to significantly improve the prediction of recurrence and cancer-specific mortality [11]. Prospective studies have also revealed that cell cycle-based markers can predict the upstage condition after TURT and the risk of recurrence after cystectomy [12, 13].

Transmembrane- and Signaling-based Markers

PDL/PDL1

In recent years, the programmed-death ligand (PDL) system has become an attractive alternative to treat urothelial carcinoma. PDL is a transmembrane protein pair that helps keep CD8 (cytotoxic) T cells from attacking normal cells. High levels of PD-L1 expression have been shown to be associated with advanced BC and poorer survival outcomes [14].

mTOR

The mammalian target of the rapamycin pathway regulates cell growth, proliferation, motility, and survival. These processes are dysregulated in the presence of multiple malignancies and generally downregulated in muscle-invasive bladder cancer [15].

MDR1

This is a transmembrane that plays an important role in multidrug resistance to chemotherapeutic agents. In a prospective trial conducted in 2010, the authors used quantitative PCR on tissue in locally advanced bladder cancer patients and revealed that high MDR1 expression levels were associated with poor clinical responses to adjuvant platinum-based chemotherapy [16].

Human Epidermal Growth Factor Receptor 2 (HER2)

HER2 is a type-I transmembrane growth factor receptor. Its mutations have been noted frequently in patients with variant histology urothelial carcinomas. A meta-analysis revealed that HER2 expression in BC was associated with high histological grade and lymph node metastasis, as well as poor prognosis [17].

Apoptosis- and DNA Repair-Based Markers

Apoptosis is a form of programmed cell death that results from a combination of intrinsic and extrinsic pathways. Disruptions in apoptosis will facilitate the survival of cancer cells; thus, they are important factors in carcinogenesis. Some genes, such as p53, promote apoptosis while others (*e.g.*, Bcl-2) inhibit it [18]. Survivin is another inhibitor of apoptosis, and a recent meta-analysis revealed that survivin RNA had great potential as a biomarker for BC diagnosis [19]. In a study that investigated the expression of apoptosis markers Bcl-2, caspase-3, P53, and survivin, the use of combined apoptosis markers was shown to provide better prognostic data to predict post-cystectomy recurrence and mortality [20].

DNA repair genes have also been studied to predict the course of the disease. Excision repair cross-complementing 1 (ERCC1) is the key enzyme of the nucleotide excision repair pathway, which maintains genomic stability. In a study conducted on patients who underwent radical cystectomy, ERCC1 expression was determined as an independent predictor for longer survival [21].

MRE11, a protein product of the MRN complex, is involved in the repair of double-stranded DNA breaks. In a study assessing DNA damage-signaling proteins as prognostic factors in a radiotherapy cohort, low tumor MRE11 expression was associated with worse cancer-specific survival compared with the high expression of this product [22].

BRCA1 is an oncogene known to affect cancers with known familial bases, including BC and prostate cancer. In patients receiving neoadjuvant chemotherapy, low BRCA1 mRNA expression levels have been found to be an indicator of a high response to treatment [23].

In a multinational study, Cathepsin E, maspin, Plk-1, and survivin were assessed as protein biomarkers for progression in non-muscle invasive BC, and the results revealed that cathepsin E, Plk-1, maspin, and survivin were significant independent prognosticators for disease progression to the muscle-invasive stage [24].

Multiplexed Markers

Twenty-Gene Panel, GenomeDX Gene-Expression Assay

Performing multiple gene assays can provide prognostic information to facilitate clinical decision-making. In this context, Smith *et al.* developed a 20-gene expression model to predict nodal metastases in patients with clinically localized disease [25].

GenomeDX Biosciences is a commercial company offering a 22 Gene expression assay for the prognosis of men with prostate cancer who are planning to undergo prostatectomy. The efficacy of this assay in predicting the risk of metastasis over a five-year period in patients undergoing radical prostatectomy was investigated, and it was shown to have good performance in distinguishing men with metastases from those without metastasis within five years after surgery [26].

There are three ongoing studies (GenomeDX molecular assay, COXEN Trial-S1314, and NCI-MATCH Trial) that aim to determine the most effective method in the treatment of bladder cancer, and the research community has shown a close interest in the results.

Urine-based Biomarkers

Urine-based biomarkers have been developed in addition to standard methods for the diagnosis and follow-up of BC. The desire is to create a non-invasive test that can be an alternative to cystoscopy and is more sensitive than urine cytology [27]. Current guidelines do not encourage clinicians to use urine-based biomarkers for bladder cancer screening and follow-up [28, 29]. To date, many urine-based biomarkers have been identified, and the markers that have been most evaluated in clinical studies are discussed below.

Fluorescence In situ Hybridization (FISH)

UroVysion, a commercially available cell-based FISH assay, detects the aneuploidy of chromosomes 3, 7 and 17, and the homozygous loss of the 9p21 locus in exfoliated urine cells.

The presence of five or more urinary cells with gains of two or more chromosomes aneuploidies, 10 or more cells with gain of a single chromosome aneuploidy, or homozygous deletion of 9p21 in more than 20% of exfoliated cells is defined as a positive test [27]. This test has relatively low sensitivity in detecting low-grade BC, and there continues to be a lack of consensus regarding criteria to be used for the evaluation of abnormal cells [27].

A meta-analysis found that UroVysion had a pooled sensitivity of 0.63 (95% CI 0.50–0.75) and specificity of 0.87 (95% CI 0.79–0.93) in the detection of bladder cancer [30]. The use of this kit in the re-evaluation of patients with normal cystoscopy findings and suspicious or positive cytology findings has been commonly reported. FISH has been reported to detect most tumors and overlook only a small number of high-grade tumors in suspicious cytology cases [31].

In the midterm follow-up of patients with atypical cytology and negative cystoscopy, a positive FISH test predicted the recurrence and progression of BC. However, a positive FISH test was not associated with tumor recurrence when compared with the negative test [32].

In a study that determined the cost-effectiveness of the FISH assay, the authors noted that in patients with atypical cytology and equivocal or negative cystoscopy, the result of the assay was associated with a significant decrease in BC-associated costs [33].

Bladder Tumor Antigen (BTA) Tests

The BTA assay is a protein-based test that has been approved by the United States Food and Drug Administration (FDA) for the diagnosis and follow-up of BC. It detects two basement membrane antigens that are present within the urine of bladder cancer patients: human complement factor H-related protein and complement factor H [34].

The original form of BTA was abandoned due to its low sensitivity and specificity. BTA stat and BTA TRAK are the currently available forms. BTA stat is a point of care test that uses a qualitative dipstick-based immunochromatographic method to show the result in 5 min and does not require specialized training [34]. A meta-analysis of 22 studies on BTA stat reported the pooled sensitivity and specificity of this assay as 0.64 (95% CI 0.58–0.69) and 0.77 (95% CI 0.73–0.81), respectively [30].

BTA TRAK is a quantitative ELISA-based assay, but it is a time-consuming method that requires laboratory assessment and special training [35]. A meta-analysis assessing the accuracy of urinary biomarkers for the diagnosis of BC noted that BTA TRAK had a sensitivity of 0.65 (95% CI 0.54-0.75) and specificity of 0.74 (95% CI 0.64-0.82) [78]. In cases with hematuria, BTA provides inaccurate results due to the cross-reaction of complement factor H, the target of BTA, with red blood cells [36].

Nuclear Matrix Protein-22 (NMP-22) Test

NMP-22, part of a family of proteins that are the building blocks of the cell nucleus, is 20 times more common in malignant urothelial cells than in normal cells [27]. The NMP-22 test is an immunoassay that identifies an NMP involved in the mitotic apparatus, which is presented at a higher concentration within tumor cells [37]. There are currently two NMP-22 tests that have been approved by FDA for use in BC surveillance: the original NMP22 laboratory-based quantitative test and a point-of-care qualitative test [27].

A meta-analysis was performed to reveal the sensitivity and specificity of quantitative and qualitative tests, and the sensitivity and specificity for the qualitative method were determined as 0.58 (95% CI 0.39–0.75) and 0.88 (95% CI 0.78 to 0.94), respectively, and those for the quantitative method as 0.69 (95% CI 0.62–0.75) and 0.77 (95% CI 0.70–0.83), respectively [30].

Many factors affect the safety of the test, and it should not be undertaken in certain cases, such as in the presence of a urinary tract infection, urinary calculi, urinary instrumentation, and dehydration.

ImmunoCyt Test

This is an FDA-approved cell-based test developed in the late 1990s. It consists of fluorescent-labelled antibodies that specifically bind to three cell surface glycoproteins (a glycosylated from of carcinoembryonic antigen and two mucins) found on the membrane of BC cells, making these cells visible. According to a meta-analysis of 14 studies, ImmunoCyt has a sensitivity of 0.78 (95% CI 0.68–0.85) and specificity of 0.78 (95% CI 0.72–0.82) [30]. The sensitivity of this method increases with BC grade and stage. ImmunoCyt has widely used İn combination with urine cytology to improve the sensitivity of urine cytology. In one study, improvement was shown as an increase of sensitivity from 50% to 90%; however, specificity was reduced (from 80% to 72%) when compared with urine cytology alone [38].

Similar to the case in urine cytology, the greatest limitation of ImmunoCyt is its operator-dependent nature with high interobserver variability and poor agreement. However, according to a previous study, these limitations can be overcome with adequate training [39].

CxBladder Test

This is a cell-based urine assay that comprises five mRNA fragments (MDK, HOXA13, CDC2, IGFBP5, and CXCR2) expressed at high levels in BC [40]. In a study conducted with 152 patients, the specificity of the test was found to be 85%, and its sensitivity increased according to the T stage and grade of the tumor. The sensitivity values for Ta, T1, and > T1 were reported to be 48, 90, and 100, respectively [40].

Although clinical studies on CxBladder are limited, its sensitivity and specificity both reach >80% in the current literature [41]. In a study that compared four diagnostic tests (cytology, NMP22, FISH, and CxBladder), it was found that Cxbladder had higher sensitivity and specificity in comparison with urine cytology and other protein-based assays [42].

SUMMARY

In conclusion, urine biomarkers readily available to detect bladder cancer cannot identify some patients with bladder cancer and give false positive results in others. None of these markers are recommended for diagnostic or follow-up use in routine practice by clinical guidelines [28]. Although clinical studies are promising, there is still no biomarker with high sensitivity and specificity values that can be used as a standard. An ideal marker should be safe, easy to use, rapid, and inexpensive, and have high sensitivity and specificity in the target population.

BIOMARKERS IN PROSTATE CANCER

Prostate cancer (PCa) is an important health problem that affects middle-aged and elderly men. It is the second most commonly diagnosed cancer in men [43]. The frequency of PCa (59% by age> 79 years) in autopsy series is also high [44, 45].

PCa biomarkers play an important role in the management of cancer. These biomarkers have led to serious changes in the approach to the treatment of patients with PCa. In recent years, the development of PCa biomarkers has continued rapidly. Among these biomarkers, a breakthrough is a prostate-specific antigen (PSA), also known as human kallikrein 3 (hK3), whose function was clearly determined by Stamey *et al.* in 1987 [46]. With the routine use of PSA and other biomarkers, it has become possible to diagnose primary and recurrent PCa at an earlier stage, and differentiate this cancer from benign processes more effectively [46, 47].

In this section, PCa biomarkers are presented under three sub-sections as urine-based, tissue-based and blood-based biomarkers.

Blood-Based Biomarkers

PSA or hK3

PSA, the first developed prostate biomarker, is a serine protease member of the human kallikrein gene family [48]. PSA is a 34 KD glycoprotein of 237 amino acids. hK2 and hK3 (PSA) are secreted in the zymogen form from the prostatic epithelium. Therefore, they are found in seminal fluid, as well as serum [47].

PSA is one of the commonly used biomarkers in PCa, approved by the United States Food and Drug Administration (FDA) in 1994. It is produced by normal prostatic luminal epithelial cells; therefore, it is highly organ-specific and not cancer-specific [49]. Traditionally, a serum PSA level above 4 ng/mL is considered to predict PCa. However, the normal levels of PSA vary according to age (Table **1**) [50].

Table 1. Age-related cut-off levels of blood PSA [50].

Age (years)	PSA (ng/mL)
<50	0 - 2.5 ng/mL
50 - 59	0 - 3.5 ng/mL
60 - 69	0 - 4.5 ng/mL
70 - 79	0 - 6.5 ng/mL

An elevated serum PSA level is an indicator of damage to the basal membrane architecture within the prostate gland epithelium [51]. PCa is the most serious cause of PSA elevation. Some other conditions, such as inflammation, prostate biopsy, physical pressure on the prostate or benign prostatic hyperplasia also cause an increase in serum PSA levels [46, 52]. In these cases, PSA derivatives (PSA density, velocity and doubling time, and age-adjusted values) and molecular derivatives can be used to improve clinical management.

In recent years, PSA has become the most valuable tool for detecting, staging and monitoring PCa. The clinical use of PSA is considered necessary for diagnosis and follow-up in PCa patients. With the widespread use of PSA, the incidence of PCa detection in patients younger than 50 years has increased by five-fold within the last few decades [53]. On the other hand, the literature data report a nearly 20% reduction in mortality among the patients that have undergone PCa screening with PSA [54].

Free PSA (fPSA)

Of the total PSA, 5% to 45% circulate in the fPSA form in the intravascular area [55]. fPSA is a useful option for providing better prediction of PCa in patients with a PSA between 2 and 10 ng/mL [55]. Benign prostatic tissue tends to release free PSA; therefore, at the free/total PSA cut-off value of 0.18 (18% fPSA), this parameter has significantly improved the ability to detect PCa, which has led to a decrease in unnecessary biopsies [56].

Prostate Health Index (PHI)

PHI incorporates the measurements of PSA, fPSA, and [-2] pro-PSA to determine a probability score *via* a mathematical algorithm for the risk determination of PCa [57]. PHI received FDA approval in 2012. Studies have shown that PHI has increased accuracy in predicting cancer compared with PSA alone. Catalona *et al.* reported that increased PHI was associated with a 4.7-fold increase in the PCa risk [58].

4Kscore Test

This is a novel blood test that contains a panel of four kallikrein protein biomarkers (total PSA, fPSA, intact PSA, and human kallikrein-related peptidase 2) [59]. It provides a predictive value for high-grade (Gleason score $\geq$ 7) cancer on prostate biopsy. The 4Kscore test can thus have a positive effect on reducing unnecessary prostate biopsies [60]. Recent publications report that the 4Kscore test can also predict surgical pathologies and the occurrence of aggressive, metastatic disease.

Circulating Tumor Cells (CTCs) and Circulating Tumor DNA (ctDNA)

CTCs are cancerous cells isolated from intravascular components [61] and they constitute a popular area of research in recent years. Genomic structures can be analyzed to detect and characterize ctDNA Studies on CTCs in PCa have used a wide range of different methods that assess these cells in the blood [61]. However, CellSearch is the only FDA-approved CTC isolation system. This method uses antibodies against EpCAM for the CTC assessment [62].

Prostate-Specific Membrane Antigen (PSMA)

This is a type II transmembrane protein found within the prostatic epithelial cell membrane [63]. PSMA has been identified in the central nervous system, intestine, and prostate [63]. It demonstrates limited expression in benign prostate tissue, but it can be found up to 100 to 1,000 times more in PCa [64]. This has led to PSMA becoming a useful biomarker for radiopharmaceuticals in the diagnosis and treatment of PCa. In this context, ^{68}Gallium prostate-specific membrane antigen positron emission tomography (^{68}Ga-PSMA PET) has become an increasingly used diagnostic tool in the assessment of biochemical recurrence after primary therapy. In addition, PSMA-targeted radionuclide treatments (beta and alpha) can be used effectively in metastatic prostate cancer patients.

Urine-Based Biomarkers

PCA3

Progensa PCA3 is a noncoding large-chain RNA [65], which is the first FDA-approved RNA-based urinary marker used in the diagnosis of PCa. PCA3 is considered to not be related with prostate size and serum PSA levels [65].

Although the function of PCA3 remains unknown, multiple studies have demonstrated that PCA3 is not expressed outside the prostate. Therefore, PCA3 levels in malignant tissue generally far exceed levels in benign tissue [66]. Multiple studies have evaluated the clinical role of the PCA3 assay in the early detection of PCa [67]. Some researchers suggest that at a cut-off value of 25, PCA3 has a predictive value in the presence of significant cancer and can be a useful tool in selecting better candidates for active surveillance [68].

During PCA3 testing, urine samples are collected after a digital rectal examination. This procedure can enhance the sensitivity of PCA3 detection [69].

Gene Fusion

The fusion of the enzyme transmembrane protease serine 2 (TMPRSS2) gene is a type of commonly detected genetic damage in PCa patients. Gene fusion can be observed in up to 50% of cases. In a previous study, it was reported that gene fusion led to a significant decrease in overall survival (8-year survival; 25% to 90%) in patients with PCa [70]. However, the gene fusion test has not yet been adapted in routine clinical practice for the management of the disease.

SelectMDx

This is a new biomarker approved by the National Comprehensive Cancer Network. It is regarded to be a valuable biomarker in patients with an abnormal PSA and/or digital rectal examination who are being considered for a prostate biopsy. SelectMDx is able to accurately assess the risk of PCa, especially the clinically significant form [71]. Therefore, it can reduce the overdiagnosis and overtreatment of PCa. The measurement is performed by collecting urine after a digital rectal examination.

Annexin A3

This novel urine-based prostate cancer biomarker has an inverse relationship with the presence of PCa. Annexin A3 contains calcium- and phospholipid-binding proteins that have been shown to be altered in cancer [72]. Determining the concentrations of annexin A3 before biopsy can have predictive value [73].

Tissue-Based Biomarkers

α-Methylacyl Coenzyme A Racemase (AMACR)

AMACR functions as an enzyme that cause the beta-oxidation of fatty acids obtained from diets rich in beef and dairy products. Immunohistochemical studies have shown that AMACR expression in biopsy tissue can provide 97% sensitivity and 100% specificity for the detection of PCa [74]. Although such studies offer promising results, the routine clinical use of these tests has not been accepted.

Epigenetic Modifications

Changes in gene expression may occur as a result of damage in DNA sequences. These changes, called epigenetic modifications, contain DNA methylation and histone acetylation status. As a result of epigenetic modification, carcinogenesis-associated cascade occurs [75].

ConfirmMDx is based on the quantification of the methylation of known epigenetic markers of PCa in prostate tissue (epigenetic modification). The PCR test is used for this purpose. A multi-center study found a negative predictive value of 88% in the absence of methylation [76]. This helps avoid unnecessary repeat biopsies in these patients [77].

Genomic Expression Profiles

Recent advances in genetic studies have improved the evaluation of gene expression in cancer tissue. For optimal gene expression profiling, many different methods have been developed. Although these methods are increasingly used in the evaluation of patients with early stage PCa, recent studies suggest that these methods may not be useful to assay tumor aggressiveness. For example, in a previous study, discordant RNA expression profiles were observed across the distinct localization of PCa tissue obtained from the same patient [78].

Inherited Genetic Markers

Familial and hereditary factors are considered to be related to nearly 50% of PCa risk [79]. Single nucleotide polymorphisms are commonly detected genetic variants. that can play a critical role in carcinogenesis and are associated with an increased risk of PCa, especially in many of these familial cases [80].

A number of genes, such as BRCA1, BRCA2, CHEK2, ATM, and PALB2 have been found related to PCa [81]. Among these genes, the close relationship of BRCA with PCa aggressiveness has been shown. A 1.8-fold and 4.5-fold increase in the risk of PCa has been reported in patients with a BRCA1 and BRCA2 mutation, respectively [82].

SUMMARY

Molecular studies in the oncologic area mostly focus on the detection of cancer at an early stage. Early detection before cancer progresses to an advanced stage not only improves cure rates but also decreases mortality and morbidity associated with PCa. Although PSA is a landmark in this area, other biomarkers also have promising results. The patient-specific use of these biomarkers will lead to significant improvements in the efficiency of PCa management.

BIOMARKERS IN RENAL CELL CARCINOMA

Renal cell carcinoma (RCC) accounts for approximately 3% of all cancers. The highest incidence of RCC is observed in Western countries. RCC is the sixth most common cancer in men and the eighth most common cancer type in women, accounting for 5% and 3% of all new cancer diagnoses, respectively. In Europe, the mortality rates of RCC were on the rise until the early 1990s, after which they first stabilized and then started to decline. With the widespread use of cross-sectional imaging methods worldwide, the detection of small-sized renal masses has increased in the last three decades. However, metastatic disease still accounts for 17% of all diagnoses [83].

RCC is categorized into several subtypes according to the 2016 classification of the World Health Organization. There are three common types of RCC: clear cell, papillary, and chromophobe. Other subtypes exist but they are much rarer.

Along with the increase in the incidental diagnoses of small renal masses with the use of advanced imaging methods, the identification of the subtypes and differentiation of benign and malignant masses have become critical for the treatment of RCC cases, for which the first-line treatment is surgery. However, since 17% of patients continue to present with metastatic disease, additional markers are needed to predict prognosis and plan treatment effectively in this group of patients.

Biomarkers are defined as objectively measurable indicators of normal biological states, pathological processes, and pharmacological responses to therapy [84]. As revealed by the definition, biomarkers can be used for many purposes, especially in cancer patients. For instance, while diagnostic biomarkers are important for diagnostic and follow-up purposes, prognostic biomarkers assist in the identification of medium- and high-risk patients to make an early decision on transition to adjuvant therapy. Another group comprises predictive biomarkers that evaluate the potential of a patient for benefiting from a particular treatment [85].

Many tissue-, blood- and urine-based biomarkers have been identified in RCC. However, when it comes to daily use, each biomarker has particular limitations. For example, the heterogeneity of RCC pathologies particularly limits the use of tissue-based biomarkers [86].

Diagnostic Biomarkers

In RCC, these biomarkers are used to guide the differentiation of benign and malignant masses, especially in small kidney tumors in order to avoid unnecessary surgical interventions and follow up the outcomes of curative treatments.

Carbonic Anhydrase IX (CAIX)

This HIF-1α-regulated transmembrane protein has been associated with neoplastic growth, aggressive tumor phenotype, and poor prognosis in a large group of tumor types (lung, breast, esophagus, and brain tumors) [87 - 89] . CAIX is thought to regulate intra and extracellular pH levels in hypoxic states. CAIX has been found to be expressed in more than 80% of RCCs. It has been found out that the expression of CAIX often increases in clear cell RCC. In many studies, high levels of CAIX expression have been shown to indicate a good prognosis in local and metastatic RCC [90 - 93]. In the REDECT study, which was carried out to provide a method to be used for differentiating small renal masses, the postoperative pathology reports were compared with the results obtained from preoperative PET/CT examinations radiologically labeled with girentuximab (a monoclonal antibody against CAIX). A total of 195 patients with clear cell RCC were included in the analysis. The sensitivity and the specificity of the method were reported as 86.2% and 85.9%, respectively [94].

Gene Expression Profile

Gene expression sequences have been assembled to differentiate RCC histological subtypes and determine the aggressiveness and metastatic potential of tumors. Numerous studies have associated the gene expression profiles of different types and grades of RCC with respective categories in the morphological classification [95 - 97]. The analysis of the genes obtained from early-stage clear cell RCC has provided an insight into additional pathways in tumor formation. Such additional pathways also include the down-regulation of transcription factors, such as GATA3, TFCP2L1, TFAP2B, and DMRT2, which are essential for normal kidney development [98].

Urine Biomarkers

Urine is an easily accessible source for biomarker assays. In previous studies, nuclear matrix protein 22 (NMP-22) was investigated as a urine biomarker and defined as a marker of urothelial cancer. It has also recently been found at significantly high levels in the urine samples of RCC patients. However, there are no recent studies or results indicating an association between NMP-22 and RCC [99 - 101]. Other biomarkers, such as kidney injury molecule-1, neutrophil gelatinase-associated lipocalin, and matrix metalloproteinases have failed to differentiate the malignant tumors of the kidney [102, 103].

Tissue Biomarkers

The pathological diagnosis and classification of RCC are generally based on a morphological evaluation. A large panel of antibodies, including CK7, CD10, Pax 2, Pax 8, vimentin, and AMACR is available and currently used for diagnostic purposes [104]. However, due to the particular difficulties in differentiating benign and malignant eosinophilic tumors (chromophobe RCC and oncocytoma), several additional biomarkers, such as cadherin and BCA2 have also been studied [105].

Prognostic Biomarkers

The importance of proper prognostic biomarkers in cancer treatment is indisputable. As in all cancer types, prognostic biomarkers are indispensable in determining the need for adjuvant therapies and developing appropriate follow-up strategies in RCC.

Routine Blood Biomarkers

Routine blood tests can be very useful in providing prognostic information on RCC. Serum lactate dehydrogenase, calcium, and hemoglobin have been widely reported to have independent prognostic significance in metastatic disease and included in the International Metastatic RCC Database Consortium risk model and the Memorial Sloan-Kettering Cancer Center scoring system [106, 107]. Hypercalcemia, anemia, and a high erythrocyte sedimentation rate have also been reported as independent prognostic factors of cancer-specific survival in localized clear cell RCC [108]. Another frequently used routine blood test parameter, CRP, has also been evaluated for its prognostic significance by many researchers. In several studies, CRP has been shown to be a predictor for metastasis and associated with total mortality after nephrectomy in local RCC [109, 110]. Thrombocytosis is considered as an unreliable marker to predict total survival in metastatic RCC patients using drugs targeting vascular endothelial growth factor (VEGF) [111].

Von Hippel-Lindau (VHL) Gene

This is a tumor suppressor gene located on chromosome 3. The absence of the VHL gene is often seen in familial RCC but also detected in 70% of sporadic clear cell RCCs. The mutation or hypermethylation of the VHL gene has been associated with lower mortality and long progression-free survival in stage 1-3

RCCs [112]. However, there are studies reporting no differences in survival rates across patients with and without VHL gene mutations [113].

VEGF

This dimeric glycoprotein belongs to the platelet-derived growth factor family, and it is involved in tumor angiogenesis in normal and pathological conditions and induces tumor angiogenesis. It has been reported in several studies that increased VEGF expression in clear cell RCC correlates with the size, stage and Fuhrman grade of the tumor, tumor necrosis, microvascular invasion, and progression rate [114 - 116].

Survivin

This is a protein belonging to the inhibitor of an apoptosis gene family. It is undetectable in normal adult tissues but high levels of survivin expression are observed in many types of cancer [117, 118]. High levels of survivin expression can be observed in all RCC variants [119] and have been associated with poor differentiation, aggressive tumor behavior, and low survival. It has also been associated with disease progression in local RCC [120].

p53

The p53 protein is a DNA-binding molecule involved in transcription. It induces apoptosis when DNA damage occurs and is significantly involved in cell growth and proliferation [121]. High levels of p53 expression are more common in papillary RCC compared to the other types [122] and have also been associated with metastasis-free survival in localized RCC but there is insufficient evidence to support this finding [122].

Ki-67

Ki-67 is a cell proliferation marker associated with an aggressive phenotype in RCC. A high level of Ki-67 expression has been associated with high recurrence rates and low survival [123]. There are also studies showing that the combined use of Ki-67 and CAIX is superior to nuclear grading for predicting cancer-specific mortality [91].

Phosphatase and Tensin Homolog (PTEN)

This is a protein encoded by a tumor suppressor gene and associated with the mTOR pathway [124]. PTEN mutations are rare. PTEN loss occurs during carcinogenesis and is associated with a poor prognosis [125].

Predictive Biomarkers

Immune Checkpoint Inhibitors

The most studied biomarker in this group is the expression of PD-L1. However, the predictive value of PD-L1, especially in metastatic RCC has not yet been determined. In the literature, PD-L1 expression was associated with a worse prognosis but was not found to be related to the patient response to checkpoint inhibitors [106, 126].

mTOR Inhibitors

A prospective study conducted with 77 patients in 2018 by the Spanish Oncology Genitourinary Group (SOGUG) examined immunohistochemical biomarkers, such as p-S6, p-AKT, p21, and PBRM1 and their effects on the key genes of the mTOR pathway. In that study, the multivariate analysis revealed that only the expression of BAP1 and PBRM1 predicted better response to mTOR inhibitors [127].

Pediatric Renal Tumors:

Renal tumors comprise 7% of all pediatric malignancies and they occur different histologic features [128]. Wilms tumor (WT) is the most commonly diagnosed pediatric renal tumor. Clear cell sarcoma, malignant rhabdoid tumor, RCC, and the other histologic types are rare. Although the clinical characteristics of children with renal tumors are similar; age distribution may be helpful in the differential diagnosis [129]. The diagnosis and classification of relevant subtypes still relies on histopathologic examination.

Widespread use of molecular techniques such as cytogenetic analysis and sequencing has enhanced the understanding of molecular characterization of pediatric renal tumors. Molecular analysis has importance for the current diagnosis and determining the risk of invasive nature of the tumor. Also, molecular investigation directs the treatment protocols [129]. Thanks to this development, progress has been achieved in the diagnosis, classification, and treatment.

Renal cell carcinomas are uncommon in pediatric patients, (5% of renal tumors) [129]. According to the World Health Organization most of the cases are defined in the category of microphthalmia-associated transcription (MiT) factor family of translocation RCCs (tRCCs). Also, a wide spectrum of fusions involving genes in the MiT family has seen in tRCCs [130]. As seen in other fusion-dependent tumor, NGS has expanded the known fusion spectrum in tRCC. In tRCCs, TFE3 fusions with primary partners of ASPSCR1 and PRCC are most commonly seen [131]. Besides, TFEB fusions with MALAT1 less commonly seen in this tumor group [132].

Pediatric clear cell sarcoma of the kidney is one of the aggressive pediatric renal tumors with a rate of approximately 5%. Internal tandem duplication of the BCOR gene which may be detected by PCR-based fragment analysis or NGS technologies has observed in most of the cases [130]. It constitutes the single most effective molecular test for the diagnosis of this malignancy.

The most aggressive pediatric renal tumors are rhabdoid tumors. Somatic or germline SMARCB1 mutations in rhabdoid tumors are well described in the literature, the functions and downstream interactions of the SMARCB1 / INI1 protein in epigenetic regulation of other genes are now directing research into new therapeutic strategies [133].

Cystic nephroma has been used to determine neoplasms such as mixed epithelial stromal tumor (MEST). However pediatric cystic nephroma is a little bit different component that associated with somatic or germline mutations in the DICER1 gene [130]. A direct detailed morphologic, immunohistochemical and genetic determinations of these lesions has not been well documented.

SUMMARY

Our knowledge of the molecular biology of RCC has greatly increased in recent years, leading to the development of targeted therapies. Nevertheless, there is still a need for further advancement to develop novel clinical tools to guide physicians through decision-making processes for the diagnosis, management and treatment of this highly heterogeneous disease both histologically and clinically. The literature contains many studies on the diagnostic, prognostic and predictive biomarkers of RCC but none of these biomarkers has yet been accepted for routine clinical use.

Obstacles to the acceptance of the investigated biomarkers for use in routine clinical practice can be listed as the lack of standardization, lack of statistical strength, and inclusion of a small number of patients in relevant studies. For the discovery of cost-effective, highly sensitive and specific RCC biomarkers, further large-scale, multi-center, and prospective studies are needed.

BIOMARKERS IN TESTICULAR CANCER

Testicular cancer (TC) is the most common malignancy among young adult men (aged 15-34 years), representing 1.5% of male neoplasms and 5% of all urologic tumors. The incidence of testis cancer can be as high as 10 new cases per 100,000 males/per year in western societies [134]. An annual increase up to 6% is reported in Caucasian populations, with the highest incidence seen in the north of Europe [135].

Testicular cancer is divided in two main parts. Approximately 95% testicular germ cell tumors (TGCTs) originate from the testis, and 5% are extragonadal in origin. Non-TGCT of the testis are rare and include sex-cord stromal tumors, lymphoid and hematopoietic tumors, and the tumors of the collecting duct, rete testis, testicular adnexa. TGCTs encompass five histologic entities: seminoma, embryonal carcinoma, yolk sac carcinoma, choriocarcinoma, and teratoma. Histologically, they are grouped into seminoma representing 60% of all tumors, and non-seminomatous germ cell tumors (NSGCT) comprising the remaining 40% [136]. With the development of cisplatin-based chemotherapy and the integration of surgery, TGCTs have become a model of curable neoplasm and serve as a paradigm for the multidisciplinary treatment of cancer.

Testicular cancer is one of the few malignancies associated with serum tumor markers (TMs) that are essential in its management. These serum proteins are alpha-fetoprotein (AFP) produced by yolk sac cells, human chorionic gonadotrophin (hCG) expressed by trophoblasts, and lactate dehydrogenase (LDH). TM levels should be obtained at the time of diagnosis and after orchidectomy to monitor patient response to chemotherapy and determine relapse in patients that are under surveillance or have completed therapy. The tumor, node, metastasis (TNM) classification of the International Union Against Cancer (UICC) recommends assessing the level of serum markers after radical orchidectomy (defined as 'S' class) in addition to the anatomical extent of the disease (Table **2**) [137].

Table 2. S classification in addition to TNM for staging [137].

Serum Tumor Markers (Pre-Chemotherapy)			
SX	Serum marker studies not available or not performed		
S0	Serum marker study levels within normal		
	LDH (U/L)	**hCG** (mIU/mL)	**AFP** (ng/mL)

(Table 2) cont.....

Serum Tumor Markers (Pre-Chemotherapy)			
S1	< 1.5 x N* and	< 5,000 and	< 1,000
S2	1.5-10 x N or	5,000-50,000 or	1,000 – 10,000
S3	>10 x N or	>50,000 or	>10,000

*N indicates the upper limit of the normal range.

For metastatic TC, the International Germ Cell Cancer Collaborative Group (IGCCCG) defined a prognostic factor-based staging system based on the identification of clinically independent adverse factors. This widely-used classification uses histology, location of the primary tumor, location of metastases, and pre-chemotherapy serum TM levels as prognostic factors to categories patients into those with 'good', 'intermediate' or 'poor' prognosis (Table **3**) [138]. The TM assessment should be included in the follow-up schedule regardless of levels at the time of diagnosis.

Patients suspected of having TC should have a blood test to investigate serum AFP, HCG, and LDH before orchiectomy to assist in diagnosis and the interpretation of post-orchiectomy TM levels. For staging purposes, it is relevant to know whether pre-orchiectomy serum marker levels are declining after orchiectomy and if so, how quickly.

Table 3. Prognostic-based staging system for metastatic germ cell cancer (IGCCCG) [138].

Good Prognosis	
Non-seminoma (56% of cases) 5-year PFS* 89% 5-year PFS 89% 5-year survival 92%	*All the following criteria:* •Testis/retro-peritoneal primary • No non-pulmonary visceral metastases • AFP < 1,000 ng/mL • hCG < 5,000 IU/L (1,000 ng/mL) • LDH < 1.5 x ULN**
Seminoma (90% of cases) 5-year PFS 82% 5-year PFS 82% 5-year survival 86%	*All the following criteria:* • Any primary site • No non-pulmonary visceral metastases. • Normal AFP, any hCG, any LDH
Intermediate Prognosis	
Non-seminoma (28% of cases) 5-year PFS 75% 5-year PFS 75% 5-year survival 80%	*Any of the following criteria:* • Testis/retro-peritoneal primary • No non-pulmonary visceral metastases • AFP 1,000 - 10,000 ng/mL or hCG 5,000 - 50,000 IU/L or LDH 1.5 - 10 x ULN

(Table 3) cont.....

Seminoma (10% of cases) 5-year PFS 67% 5-year PFS 67% 5-year survival 72%	*All the following criteria:* • Any primary site • Non-pulmonary visceral metastases • Normal AFP, any hCG, any LDH
	Poor Prognosis
Non-seminoma (16% of cases) 5-year PFS 41% 5-year PFS 41% 5-year survival 48%	*Any of the following criteria:* • Mediastinal primary • Non-pulmonary visceral metastases • AFP > 10,000 ng/mL or hCG > 50,000 IU/L (10,000 ng/mL) or LDH > 10 x ULN
Seminoma	No patients classified as poor prognosis

*PFS = Progression-free survival ***ULN = Upper limit of normal*

Alfa-fetoprotein (AFP)

This is a glycoprotein produced primarily in the fetal yolk sac but also in the liver and gastrointestinal tract, and its secretion from germ cell tumors is restricted to non-seminomatous histology. Consequently, any patient with histological pure seminomas and an increased serum AFP are classified and treated like those with non-seminomatous germ cell tumors. AFP is increased in approximately 60% of patients with metastatic non-seminoma and 20% of patients with clinical stage I non-seminoma [139]. The serum half-life of AFP is approximately 5–7 days. With complete tumor excision, elevated serum levels return to normal within 25–30 days [140]. Pure seminomas and choriocarcinomas never express elevated AFP values. AFP is not specific to TC.

Human Chorionic Gonodotropin (HCG)

HCG is mainly used for the detection and monitoring of pregnancy and pregnancy-related disorders, but it is also an extremely sensitive and specific marker for the trophoblastic tumors of placental and germ cell origin. HCG is a glycoprotein produced by the syncytiotrophoblasts and increased in approximately 15% of pure seminomas and in 40% of advanced non-seminomas [139]. Levels above 5,000 IU/L are usually associated with NSGCT. The serum half-life of HCG is approximately 24–48 hours. With complete tumor excision, even markedly elevated serum levels return to normal within 5–7 days.[7] Similar to AFP, HCG is not specific to TC.

Lactate Dehydrogenase (LDH)

This protein enzyme is known to be a useful marker for cell turnover and is often raised in malignancy. LDH is expressed in smooth, cardiac and skeletal muscles. Tissue levels are 500 times greater than those in serum, and thus even a small mass of damaged tissue causes the leakage of the enzyme and consequently increases its level in serum significantly. The total LDH can be further separated into five components. LDH elevation, especially LDH-1 elevation is noted in patients with the large volume disease of all histologic types of germ cell tumors, even in pure seminoma. LDH-1 levels are correlated with the chromosome arm 12p copy number, which is frequently amplified in TGCTs. Increases in serum LDH correlate with tumor burden, growth rate, and cellular proliferation. Elevated LDH is present in approximately 60% of patients with advanced NSGCT and 80% of those with metastatic seminoma [141]. The half-life of LDH is 24 hours. LDH is a valuable marker in the surveillance of cases with advanced seminomas to indicate tumor bulk. However, LDH alone has limited sensitivity, specificity, and positive predictive value for detecting the relapse of germ cell tumors, and since false-positive increases are common, the results should be interpreted cautiously [142].

There are also numerous conditions that can result in the serum elevation of the above-mentioned markers: hepatocellular carcinoma, viral hepatitis, hereditary tyrosinemia, cirrhosis, gastric and pancreatic cancer, bronchial cancer, antiepileptic, and anesthetics for AFP; hypogonadism, BC, marijuana use, gonadotroph adenomas, and bladder, kidney, liver gastric and lung cancers for hCG; and skeletal muscle disease, pulmonary embolism, leukemia, hemolysis, thalassemia, and myocardial infarction for LDH [143]. Thus, there is a considerable low sensitivity and false positive rate that should be considered when using these markers in the clinical setting.

Potential Biomarkers in Future

In recent years, a significant effort has been made to identify and validate new markers to better prognosticate TGCTs. About 90% of NSGCTs present with a rise in 1 or both of the markers, but only 30% of seminomas present or develop an elevated hCG during the course of the disease [141]. In addition to the relatively low expression of these tumor markers in TGCTs (only 60% of TGCTs demonstrate elevated tumor markers at initial diagnoses), these markers are non-specific and unreliable for the follow-up and monitoring of advanced disease. Therefore, clinicians need better biomarkers to predict prognosis and detect metastasis after first-line treatment.

Placental Alkaline Phosphatase (PLAP)

PLAP is a membrane-bound protein expressed in fetal germ cells. The role of PLAP in TGCT prognosis is not well-defined, but it is known to be specific to testicular tumors when compared with normal testicular tissue. The Current European Association of Urology guidelines consider PLAP as an optional biomarker for patients with seminoma (50% of seminomas have elevated levels) [144]. Despite the low false-positive rates (1.6%), the serum levels are frequently elevated in smokers even in the absence of testicular cancer [145].

Oct-3/4, Sall 4, and CD 117

These are some other immunohistochemical markers clinically used in the presence of suspicious findings. Oct 3/4 is expressed in all testicular germ cell tumors tested, even in carcinoma *in situ* (CIS). Additionally, high levels have been suggested to potentially correlate with an increased malignant potential in CIS, but further studies are required for a definitive conclusion [146].

miRNAs

miRNAs are potential new biomarkers for the TC diagnosis. Preliminary evidence reveals their higher discriminatory accuracy than conventional markers [147]. miRNAs were first described in 1993 by Lee *et al.*, and since then, a growing amount of evidence has been presented in relation to their involvement in regulatory functions related to cell growth, development and differentiation [148]. In 2006, Voorhoeve *et al.* performed the first miRNA expression study of GCT tissue, observing that 2 miRNAs (miR), miR372a-3p and miR373a-3p were particularly abundant in GCT tissue and cell lines [149]. The first prospective study of serum miR371 in patients with GCT was published in 2019. The serum expression of miR371 *(M371 Test)* in patients with TGCT was compared with 258 male controls [150]. According to that study, the sensitivity of miR371 in detecting GCT prior to the orchiectomy was 90.1%, specificity 94.0%, and positive predictive value 97.2%.

Literature data confirmed that miR371 expression levels were associated with clinical stage, primary tumor size, and response to treatment. miR371 also was overexpressed in patients who presented with GCT relapse and normalized to normal levels on remission. Following orchiectomy, preliminary results suggest that miRNAs may indicate metastatic disease [151]. Before miRNAs can be considered for use in routine clinical practice, issues concerning laboratory standardization and availability of the test need to be resolved.

SUMMARY

Tumor markers play an important role in the diagnosis and staging of TC, as well as for the treatment and long-term follow-up of the disease. Although current serum tumor markers have had a significant impact on the diagnosis and treatment of TGCTs, they fundamentally lack sensitivity and specificity. miRNAs are promising biomarkers that can help clinicians more accurately determine the natural history of the disease and its heterogeneity, and offer the best assistance in decision-making for young patients suffering from the disease.

NEXT GENERATION SEQUENCING

The use of next-generation sequencing (NGS) leads to the determination of the genomic and transcriptomic structure of malignancies. The clinical approach of NGS aims to assess the heterogeneity of patient and cancer-associated factors. In this way, the patient-specific optimal treatment and follow-up protocol may be possible.

When bladder cancer is considered with the perspective of NGS, studies have been conducted to evaluate immunotherapy options, neoadjuvant chemotherapy efficiency, and radiotherapy success. Thanks to NGS, researchers have cataloged mutational profiles of most human malignancies. "Mutation burden" profiles of a broad spectrum of common malignancies, including bladder cancer has determined [152]. Mutation burden has evaluated as a biomarker of response to immunotherapy. In phase 2 study which assessed treatment with atezolizumab (IMvigor210), mutational burden was assessed in 150 patients with post-platinum metastatic urothelial carcinoma utilizing the 315-gene DNA-based FoundationOne® pannel. Patients demonstrating responses to the atezolizumab treatment had significantly higher mutational burdens compared to non-responders (12.4 vs. 6.4 mutations/megabase $p < 0.0001$) [153].

As explained in the related section T-cells play a crucial role in the immune system anti-tumor effects. T-cell maturation is characterized by antigen-specific recognition by their cell membrane associated receptors (TCRs). NGS provide analysis and identification of the each individual TCRs. In a study conducted by Choudhury et al, patients with low TCR variety demonstrated significantly longer recurrence-free survival compared to those with high ($p = 0.018$) [154]. Between the 29 patients with metastatic post-platinum urothelial carcinoma in cohort 2 of the IMVigor210 trial of atezolizumab, patients with lower baseline TCR variety identified on the ImmunoSEQ® platform had significantly higher overall survival ($p = 0.0116$) and progression-free survival ($p = 0.0514$) [155].

Alterations in cell signaling such as fibroblast growth factor pathway (FGFR) mutation is found in multiple malignancies [156]. Mutations of FGFR3 have been observed in up to 70% of non-muscle-invasive bladder cancers [156]. Identification of FGFR3 mutations in bladder cancer has generated significant interest in the therapeutic use of these mutations. In a clinical trial of dovitinib, an oral inhibitor of FGFR3 and other proangiogenic receptor tyrosine kinases, failed to reach its primary endpoint of overall response rate in a two-stage designed study [157].

In a study that evaluates somatic mutations that affect the neoadjuvant chemotherapy results, the authors performed NGS on pretreatment tumor and germline DNA from patients with muscle-invasive urothelial carcinoma who received neoadjuvant chemotherapy followed by cystectomy. They found that ERCC2, a nucleotide excision repair gene, was the only significantly somatic gene mutation enriched in the cisplatin responders compared with nonresponders ($q < 0.01$) [158]. In another study NGS identifies that carriage of at least one of six rare variants of the MRE11A gene was associated with the worse radiotherapy outcomes in muscle-invasive bladder cancer (hazard ratio [HR] 4.04, 95% confidence interval [95% CI] 1.42–11.51, P = 0.009) [159].

In prostate cancer, the genomic information can be used for better patient identification, minimal side effects and maximum tumor response. In PCa, the most emphasized molecular pathway in carcinogenesis is Androgen Receptor (AR) and Androgen Receptor Splice Variants (ARV576) signaling [160]. In this context, it is recommended to use new generation chemotherapeutic agents in addition to hormone therapy in metastatic prostate cancer patients with AR or ARV576 signalling alteration.

Other identified NGS in PCa are PI3K/PTEN/AKT and RTK/RAS/MAPK [161]. These alterations have been reported with a higher frequency in patients with castration resistant prostate cancer (CRPC). For example, PTEN signaling alterations have been identified in more than 40% of CRPC patients [162]. New agents, such as everolimus (A mTOR Inhibitör), dovitinib (A Trosine Kinase Inhibitör) may be used in the treatment of these patients [162]. The inherited genetic markers, genomic expression profile, epigenetic modification mentioned in the prostate cancer chapter are among the parameters evaluated in this context. Further randomized controlled clinical trials are needed to assess the potential effects and side effects of new agents in this area.

Renal cell cancers may occur in a variable range in terms of histopathology, tumor behavior and prognosis. Therefore, different therapeutic agents have identified for RCC treatment. With these new agents, the treatment approach has also changed. The Cancer Genome Atlas (TCGA) consortium assessed nearly 400 RCC samples and reported altered genomic structures of renal cell cancer subgroups [163]. Initial next-generation sequencing technologies have led to identifying 3p loss,14q loss and 5q gain in RCC [164]. Now, there are 14 genes in which germ line mutation are related with a predisposition to renal cancer. However, genetic origin of most inherited renal cancer remains unclear [165]. Fourteen known renal cancer predisposition genes are VHL, tuberous sclerosis 1 (TSC1), TSC2, phosphatase and tensin homolog (PTEN), MET, FH, succinate dehydrogenase complex B (SDHB), SDHC, SDHD, folliculin (FLCN), BAP1, melanogenesis-associated transcription factor (MITF), HNF1 homeobox B (HNF1B), and polybromo 1 (PBRM1) [165]. Additionally, germ line mutations in DNA repair genes as BRCA 1-2, and other potential predisposition genes KDM5C, KDM6A, NF2, SETD2, CDKN2A, TCEB1, and TP53) was determined [165].

The most common genetic aberration useful in the diagnosis of WT is WT1, CTNNB1 and WTX, which can determine with NGS. Fish analysis should be performed to determine TFEB and TFE3 genetic aberration in the diagnosis of RCC in pediatric patients [166 - 168]. Although Next generation sequencing (NGS) based multi gene panels provide a possibility for investing germ line mutations on a larger scale. Genetic sequencing is important to define the risk of early onset of the tumoral presentation and invasive nature of the renal tumor [165].

Testicular cancer development is a polygenic, multistep process; initial events occur in the embryo stage of development. Hormonal stimulation involved in spermatogenesis results in further genetic events in the clinical manifestation of TGCT [169]. Thus, molecular genetic studies may help to predict metastases in early-stage and responders to chemotherapy in advanced disease. Adult TGCTs are characterized by frequent chromosomal anomalies and low rates of somatic mutations. The KIT gene, which encodes a proto-oncogene receptor tyrosine kinase protein and plays crucial roles in cell survival and proliferation, is well characterized in TGCT [170]. Imatinib (thyrozine kinase inhibitor) sensitivity were reported in TGCTs [171], but a small phase II study of imatinib in heavily pretreated GCTs demonstrating KIT-positivity on immunhistochemical staining failed to demonstrate any response [172]. Recent next generation sequencing (NGS) efforts identified that most patients with TGCT harbor exon 17 mutations, which are generally imatinib-resistant, emphasizing the need for precise genotyping and potentially explaining the dissapointing results of the phase II study [173]. KRAS and NRAS are proto-oncogene receptor tyrosine kinase proteins that activate multiple downstream pathways including the Raf/MEK/ERK and PI3Kinase pathways [174]. RAS mutations and FGFR3 mutations were identified at a higher frequency in chemoresistant *versus* sensitive tumors [175]. In a study, five patients with cisplatin-resistant TGCT treated with sunitinib (vascular endothelial growth factor receptor inhibitor), NGS identified a rare RET proto-oncogene amplification known to be sensitizing to sunitinib in one responder [176]. NGS in future can provide important information for decision making about treatment protocol selection.

CONCLUSION

The use of cancer biomarkers in clinical practice is an important aspect of the treatment of urological cancers. These biomarkers can help to improve the detection, treatment, and follow-up of urological cancers. Thanks to next-generation sequencing, patient management has also progressed day by day. Today, PSA and testicular biomarkers (AFP, hCG, and LDH) are the few well-established biomarkers in daily practice. Although, developments that occurred in the field of biomarkers led to new opportunities to improve more accurate treatment and follow-up alternatives. Future studies will bring us new biomarkers that we can use widely in clinical practice.

CONSENT FOR PUBLICATION

Not applicable.

CONFLICT OF INTEREST

The authors declare no conflict of interest, financial or otherwise.

ACKNOWLEDGEMENT

Declared none.

REFERENCES

[1] Ferlay J, Soerjomataram I, Ervik M, *et al.* GLOBOCAN 2012 v1.0: Estimated cancer incidence, mortality and prevalence worldwide in 2012. 2013. 2015. Accessed December 2019. 2015.

[2] Alishahi S, Byrne D, Goodman CM, Baxby K. Haematuria investigation based on a standard protocol: emphasis on the diagnosis of urological malignancy. J R Coll Surg Edinb 2002; 47(1): 422-7. [PMID: 11874263]

[3] Rink M, Chun FK, Dahlem R, *et al.* Prognostic role and HER2 expression of circulating tumor cells in peripheral blood of patients prior to radical cystectomy: a prospective study. Eur Urol 2012; 61(4): 810-7. [http://dx.doi.org/10.1016/j.eururo.2012.01.017] [PMID: 22277196]

[4] Lin CT, Tung CL, Tsai YS, *et al.* Prognostic relevance of preoperative circulating CD8-positive lymphocytes in the urinary bladder recurrence of urothelial carcinoma. Urol Oncol 2012; 30(5): 680-7. [http://dx.doi.org/10.1016/j.urolonc.2010.08.009] [PMID: 21420334]

[5] Wu S, Zhao X, Wang Y, *et al.* Pretreatment Neutrophil-Lymphocyte Ratio as a Predictor in Bladder Cancer and Metastatic or Unresectable Urothelial Carcinoma Patients: a Pooled Analysis of Comparative Studies. Cell Physiol Biochem 2018; 46(4): 1352-64. [http://dx.doi.org/10.1159/000489152] [PMID: 29689562]

[6] Ojerholm E, Smith A, Hwang WT, *et al.* Neutrophil-to-lymphocyte ratio as a bladder cancer biomarker: Assessing prognostic and predictive value in SWOG 8710. Cancer. 2017 ;1; 123(5): 794-801.

[7] Svatek RS, Shah JB, Xing J, *et al.* A multiplexed, particle-based flow cytometric assay identified plasma matrix metalloproteinase-7 to be associated with cancer-related death among patients with

bladder cancer. Cancer 2010; 116(19): 4513-9.
[http://dx.doi.org/10.1002/cncr.25401] [PMID: 20572047]

[8] Yates DR, Rehman I, Abbod MF, *et al.* Promoter hypermethylation identifies progression risk in bladder cancer. Clin Cancer Res 2007; 13(7): 2046-53.
[http://dx.doi.org/10.1158/1078-0432.CCR-06-2476] [PMID: 17404085]

[9] Catto JW, Azzouzi AR, Rehman I, *et al.* Promoter hypermethylation is associated with tumor location, stage, and subsequent progression in transitional cell carcinoma. J Clin Oncol 2005; 23(13): 2903-10.
[http://dx.doi.org/10.1200/JCO.2005.03.163] [PMID: 15753461]

[10] Weinstein JN, Verhaak RGV, McConkey D, *et al.* Comprehensive molecular characterization of urothelial bladder carcinoma. Nature 2014; 507(7492): 315-22.
[http://dx.doi.org/10.1038/nature12965] [PMID: 24476821]

[11] Shariat SF, Chade DC, Karakiewicz PI, *et al.* Combination of multiple molecular markers can improve prognostication in patients with locally advanced and lymph node positive bladder cancer. J Urol 2010; 183(1): 68-75.
[http://dx.doi.org/10.1016/j.juro.2009.08.115] [PMID: 19913255]

[12] Lotan Y, Bagrodia A, Passoni N, *et al.* Prospective evaluation of a molecular marker panel for prediction of recurrence and cancer-specific survival after radical cystectomy. Eur Urol 2013; 64(3): 465-71.
[http://dx.doi.org/10.1016/j.eururo.2013.03.043] [PMID: 23571005]

[13] Shariat SF, Passoni N, Bagrodia A, *et al.* Prospective evaluation of a preoperative biomarker panel for prediction of upstaging at radical cystectomy. BJU Int 2014; 113(1): 70-6.
[http://dx.doi.org/10.1111/bju.12343] [PMID: 24053584]

[14] Bellmunt J, Mullane SA, Werner L, *et al.* Association of PD-L1 expression on tumor-infiltrating mononuclear cells and overall survival in patients with urothelial carcinoma. Ann Oncol 2015; 26(4): 812-7.
[http://dx.doi.org/10.1093/annonc/mdv009] [PMID: 25600565]

[15] Schultz L, Albadine R, Hicks J, *et al.* Expression status and prognostic significance of mammalian target of rapamycin pathway members in urothelial carcinoma of urinary bladder after cystectomy. Cancer 2010; 116(23): 5517-26.
[http://dx.doi.org/10.1002/cncr.25502] [PMID: 20939013]

[16] Hoffmann AC, Wild P, Leicht C, *et al.* MDR1 and ERCC1 expression predict outcome of patients with locally advanced bladder cancer receiving adjuvant chemotherapy. Neoplasia 2010; 12(8): 628-36.
[http://dx.doi.org/10.1593/neo.10402] [PMID: 20689757]

[17] Zhao J, Xu W, Zhang Z, *et al.* Prognostic role of HER2 expression in bladder cancer: a systematic review and meta-analysis. Int Urol Nephrol 2015; 47(1): 87-94.
[http://dx.doi.org/10.1007/s11255-014-0866-z] [PMID: 25384433]

[18] Kluck RM, Bossy-Wetzel E, Green DR, Newmeyer DD. The release of cytochrome c from mitochondria: a primary site for Bcl-2 regulation of apoptosis. Science 1997; 275(5303): 1132-6.
[http://dx.doi.org/10.1126/science.275.5303.1132] [PMID: 9027315]

[19] Liang Z, Xin R, Yu Y, Wang R, Wang C, Liu X. Diagnostic value of urinary survivin as a biomarker for bladder cancer: a systematic review and meta-analysis of published studies. World J Urol 2018; 36(9): 1373-81.
[http://dx.doi.org/10.1007/s00345-018-2285-8] [PMID: 29610963]

[20] Karam JA, Lotan Y, Karakiewicz PI, *et al.* Use of combined apoptosis biomarkers for prediction of bladder cancer recurrence and mortality after radical cystectomy. Lancet Oncol 2007; 8(2): 128-36.
[http://dx.doi.org/10.1016/S1470-2045(07)70002-5] [PMID: 17267327]

[21] Klatte T, Seitz C, Rink M, *et al.* ERCC1 as a prognostic and predictive biomarker for urothelial

carcinoma of the bladder following radical cystectomy. J Urol 2015; 194(5): 1456-62.
[http://dx.doi.org/10.1016/j.juro.2015.06.099] [PMID: 26162296]

[22] Choudhury A, Nelson LD, Teo MT, *et al.* MRE11 expression is predictive of cause-specific survival following radical radiotherapy for muscle-invasive bladder cancer. Cancer Res 2010; 70(18): 7017-26.
[http://dx.doi.org/10.1158/0008-5472.CAN-10-1202] [PMID: 20843819]

[23] Font A, Taron M, Gago JL, *et al.* BRCA1 mRNA expression and outcome to neoadjuvant cisplatin-based chemotherapy in bladder cancer. Ann Oncol 2011; 22(1): 139-44.
[http://dx.doi.org/10.1093/annonc/mdq333] [PMID: 20603439]

[24] Fristrup N, Ulhøi BP, Birkenkamp-Demtröder K, *et al.* Cathepsin E, maspin, Plk1, and survivin are promising prognostic protein markers for progression in non-muscle invasive bladder cancer. Am J Pathol 2012; 180(5): 1824-34.
[http://dx.doi.org/10.1016/j.ajpath.2012.01.023] [PMID: 22449953]

[25] Smith SC, Baras AS, Dancik G, *et al.* A 20-gene model for molecular nodal staging of bladder cancer: development and prospective assessment. Lancet Oncol 2011; 12(2): 137-43.
[http://dx.doi.org/10.1016/S1470-2045(10)70296-5] [PMID: 21256081]

[26] Marrone M, Potosky AL, Penson D, Freedman AN. A22 gene-expression assay, decipher(R) (genomeDx biosciences) to predict five-year risk of metastatic prostate cancer in men treated with radical prostatectomy. PLoS Curr 2015; 7: 7.
[PMID: 26664778]

[27] Partin AW, Dmochowski RR, Kavoussi LR, *et al.* Tumors of the Bladder. Campbell-Walsh-Wein Urology Twelfth Edition. Philadelphia, Elsevier, ISBN:978-0-323-54642-32020.

[28] Babjuk M, Burger M, Comperat E, *et al.* EAU Guidelines on non-muscle invasive bladder cancer. EAU Annual Congress Amsterdam. 978-94-92671-07-3.

[29] Chang SS, Boorjian SA, Chou R, *et al.* Diagnosis and treatment of non-muscle invasive bladder cancer: AUA/SUO guideline. J Urol. 2016;196:1021.28. 2016.

[30] Chou R, Gore JL, Buckley D, *et al.* Urinary biomarkers for diagnosis of bladder cancer. A systematic review and meta-analysisurinary biomarkers for diagnosis of bladder cancer. Ann Intern Med 2015; 163(12): 922-31.
[http://dx.doi.org/10.7326/M15-0997] [PMID: 26501851]

[31] Schlomer BJ, Ho R, Sagalowsky A, Ashfaq R, Lotan Y. Prospective validation of the clinical usefulness of reflex fluorescence *in situ* hybridization assay in patients with atypical cytology for the detection of urothelial carcinoma of the bladder. J Urol 2010; 183(1): 62-7.
[http://dx.doi.org/10.1016/j.juro.2009.08.157] [PMID: 19913822]

[32] Kim PH, Sukhu R, Cordon BH, *et al.* Reflex fluorescence *in situ* hybridization assay for suspicious urinary cytology in patients with bladder cancer with negative surveillance cystoscopy. BJU Int 2014; 114(3): 354-9.
[PMID: 24128299]

[33] Gayed BA, Seideman C, Lotan Y. Cost-effectiveness of fluorescence *in situ* hybridization in patients with atypical cytology for the detection of urothelial carcinoma. J Urol 2013; 190(4): 1181-6.
[http://dx.doi.org/10.1016/j.juro.2013.03.117] [PMID: 23583531]

[34] Kinders R, Jones T, Root R, *et al.* Complement factor H or a related protein is a marker for transitional cell cancer of the bladder. Clin Cancer Res 1998; 4(10): 2511-20.
[PMID: 9796985]

[35] Lokeshwar VB, Soloway MS. Current bladder tumor tests: does their projected utility fulfill clinical necessity? J Urol 2001; 165(4): 1067-77.
[http://dx.doi.org/10.1016/S0022-5347(05)66428-2] [PMID: 11257640]

[36] Lüdecke G, Pilatz A, Hauptmann A, Bschleipfer T, Weidner W. Comparative analysis of sensitivity to blood in the urine for urine-based point-of-care assays (UBC rapid, NMP22 BladderChek and BTA-

stat) in primary diagnosis of bladder carcinoma. Interference of blood on the results of urine-based POC tests. Anticancer Res 2012; 32(5): 2015-8.
[PMID: 22593481]

[37] Soloway MS, Briggman V, Carpinito GA, *et al.* Use of a new tumor marker, urinary NMP22, in the detection of occult or rapidly recurring transitional cell carcinoma of the urinary tract following surgical treatment. J Urol 1996; 156(2 Pt 1): 363-7.
[http://dx.doi.org/10.1097/00005392-199608000-00008] [PMID: 8683680]

[38] Toma MI, Friedrich MG, Hautmann SH, *et al.* Comparison of the ImmunoCyt test and urinary cytology with other urine tests in the detection and surveillance of bladder cancer. World J Urol 2004; 22(2): 145-9.
[http://dx.doi.org/10.1007/s00345-003-0390-8] [PMID: 14991320]

[39] Messing EM, Teot L, Korman H, *et al.* Performance of urine test in patients monitored for recurrence of bladder cancer: a multicenter study in the United States. J Urol 2005; 174(4 Pt 1): 1238-41.
[http://dx.doi.org/10.1097/01.ju.0000173918.84006.4d] [PMID: 16145377]

[40] Holyoake A, O'Sullivan P, Pollock R, *et al.* Development of a multiplex RNA urine test for the detection and stratification of transitional cell carcinoma of the bladder. Clin Cancer Res 2008; 14(3): 742-9.
[http://dx.doi.org/10.1158/1078-0432.CCR-07-1672] [PMID: 18245534]

[41] O'Sullivan P, Sharples K, Dalphin M, *et al.* A multigene urine test for the detection and stratification of bladder cancer in patients presenting with hematuria. J Urol 2012; 188(3): 741-7.
[http://dx.doi.org/10.1016/j.juro.2012.05.003] [PMID: 22818138]

[42] Breen V, Kasabov N, Kamat AM, *et al.* A holistic comparative analysis of diagnostic tests for urothelial carcinoma: a study of Cxbladder Detect, UroVysion® FISH, NMP22® and cytology based on imputation of multiple datasets. BMC Med Res Methodol 2015; 15(1): 45.
[http://dx.doi.org/10.1186/s12874-015-0036-8] [PMID: 25962444]

[43] Ferlay J, Soerjomataram I, Dikshit R, *et al.* Cancer incidence and mortality worldwide: sources, methods and major patterns in GLOBOCAN 2012. Int J Cancer 2015; 136(5): E359-86.
[http://dx.doi.org/10.1002/ijc.29210] [PMID: 25220842]

[44] Haas GP, Delongchamps N, Brawley OW, Wang CY, de la Roza G. The worldwide epidemiology of prostate cancer: perspectives from autopsy studies. Can J Urol 2008; 15(1): 3866-71.
[PMID: 18304396]

[45] Bell KJ, Del Mar C, Wright G, Dickinson J, Glasziou P. Prevalence of incidental prostate cancer: A systematic review of autopsy studies. Int J Cancer 2015; 137(7): 1749-57.
[http://dx.doi.org/10.1002/ijc.29538] [PMID: 25821151]

[46] Stamey TA, Yang N, Hay AR, McNeal JE, Freiha FS, Redwine E. Prostate-specific antigen as a serum marker for adenocarcinoma of the prostate. N Engl J Med 1987; 317(15): 909-16.
[http://dx.doi.org/10.1056/NEJM198710083171501] [PMID: 2442609]

[47] De Angelis G, Rittenhouse HG, Mikolajczyk SD, Blair Shamel L, Semjonow A. Twenty Years of PSA: From Prostate Antigen to Tumor Marker. Rev Urol 2007; 9(3): 113-23.
[PMID: 17934568]

[48] Hong SK. Kallikreins as biomarkers for prostate cancer. BioMed Res Int 2014; 2014: 526341.
[http://dx.doi.org/10.1155/2014/526341] [PMID: 24809052]

[49] van der Toom EE, Axelrod HD, de la Rosette JJ, de Reijke TM, Pienta KJ, Valkenburg KC. Prostate-specific markers to identify rare prostate cancer cells in liquid biopsies. Nat Rev Urol 2019; 16(1): 7-22.
[http://dx.doi.org/10.1038/s41585-018-0119-5] [PMID: 30479377]

[50] Loeb S, Roehl KA, Antenor JA, Catalona WJ, Suarez BK, Nadler RB. Baseline prostate-specific antigen compared with median prostate-specific antigen for age group as predictor of prostate cancer

risk in men younger than 60 years old. Urology 2006; 67(2): 316-20.
[http://dx.doi.org/10.1016/j.urology.2005.08.040] [PMID: 16442597]

[51] Mejak SL, Bayliss J, Hanks SD. Long distance bicycle riding causes prostate-specific antigen to increase in men aged 50 years and over. PLoS One 2013; 8(2): e56030.
[http://dx.doi.org/10.1371/journal.pone.0056030] [PMID: 23418500]

[52] Wang MC, Valenzuela LA, Murphy GP, Chu TM. Purification of a human prostate specific antigen. Invest Urol 1979; 17(2): 159-63.
[PMID: 89106]

[53] Salinas CA, Tsodikov A, Ishak-Howard M, Cooney KA. Prostate cancer in young men: an important clinical entity. Nat Rev Urol 2014; 11(6): 317-23.
[http://dx.doi.org/10.1038/nrurol.2014.91] [PMID: 24818853]

[54] Hugosson J, Roobol MJ, Månsson M, *et al.* A 16-yr followup of the european randomized study of screening for prostate cancer. Eur Urol 2019; 76(1): 43-51.
[http://dx.doi.org/10.1016/j.eururo.2019.02.009] [PMID: 30824296]

[55] Woodrum DL, Brawer MK, Partin AW, Catalona WJ, Southwick PC. Interpretation of free prostate specific antigen clinical research studies for the detection of prostate cancer. J Urol 1998; 159(1): 5-12.
[http://dx.doi.org/10.1016/S0022-5347(01)63996-X] [PMID: 9400426]

[56] Hoffman RM, Clanon DL, Littenberg B, Frank JJ, Peirce JC. Using the free-to-total prostate-specific antigen ratio to detect prostate cancer in men with nonspecific elevations of prostate-specific antigen levels. J Gen Intern Med 2000; 15(10): 739-48.
[http://dx.doi.org/10.1046/j.1525-1497.2000.90907.x] [PMID: 11089718]

[57] Stephan C, Vincendeau S, Houlgatte A, Cammann H, Jung K, Semjonow A. Multicenter evaluation of [-2]proprostate-specific antigen and the prostate health index for detecting prostate cancer. Clin Chem 2013; 59(1): 306-14.
[http://dx.doi.org/10.1373/clinchem.2012.195784] [PMID: 23213080]

[58] Catalona WJ, Partin AW, Sanda MG, *et al.* A multicenter study of [-2]pro-prostate specific antigen combined with prostate specific antigen and free prostate specific antigen for prostate cancer detection in the 2.0 to 10.0 ng/ml prostate specific antigen range. J Urol 2011; 185(5): 1650-5.
[http://dx.doi.org/10.1016/j.juro.2010.12.032] [PMID: 21419439]

[59] Punnen S, Pavan N, Parekh DJ. Finding the Wolf in Sheep's Clothing: The 4Kscore Is a Novel Blood Test That Can Accurately Identify the Risk of Aggressive Prostate Cancer. Rev Urol 2015; 17(1): 3-13.
[PMID: 26028995]

[60] Vickers AJ, Cronin AM, Aus G, *et al.* A panel of kallikrein markers can reduce unnecessary biopsy for prostate cancer: data from the European Randomized Study of Prostate Cancer Screening in Göteborg, Sweden. BMC Med 2008; 6(1): 19.
[http://dx.doi.org/10.1186/1741-7015-6-19] [PMID: 18611265]

[61] Danila DC, Fleisher M, Scher HI. Circulating tumor cells as biomarkers in prostate cancer. Clin Cancer Res 2011; 17(12): 3903-12.
[http://dx.doi.org/10.1158/1078-0432.CCR-10-2650] [PMID: 21680546]

[62] Scher HI, Jia X, de Bono JS, *et al.* Circulating tumour cells as prognostic markers in progressive, castration-resistant prostate cancer: a reanalysis of IMMC38 trial data. Lancet Oncol 2009; 10(3): 233-9.
[http://dx.doi.org/10.1016/S1470-2045(08)70340-1] [PMID: 19213602]

[63] Bouchelouche K, Choyke PL. Prostate-specific membrane antigen positron emission tomography in prostate cancer: a step toward personalized medicine. Curr Opin Oncol 2016; 28(3): 216-21.
[http://dx.doi.org/10.1097/CCO.0000000000000277] [PMID: 26967720]

[64] Emmett L, Willowson K, Violet J, Shin J, Blanksby A, Lee J. Lutetium [177] PSMA radionuclide therapy for men with prostate cancer: a review of the current literature and discussion of practical aspects of therapy. J Med Radiat Sci 2017; 64(1): 52-60. [http://dx.doi.org/10.1002/jmrs.227] [PMID: 28303694]

[65] Hessels D, Klein Gunnewiek JM, van Oort I, *et al.* DD3(PCA3)-based molecular urine analysis for the diagnosis of prostate cancer. Eur Urol 2003; 44(1): 8-15. [http://dx.doi.org/10.1016/S0302-2838(03)00201-X] [PMID: 12814669]

[66] Popa I, Fradet Y, Beaudry G, Hovington H, Beaudry G, Têtu B. Identification of PCA3 (DD3) in prostatic carcinoma by *in situ* hybridization. Mod Pathol 2007; 20(11): 1121-7. [http://dx.doi.org/10.1038/modpathol.3800963] [PMID: 17873893]

[67] Wei JT, Feng Z, Partin AW, *et al.* Can urinary PCA3 supplement PSA in the early detection of prostate cancer? J Clin Oncol 2014; 32(36): 4066-72. [http://dx.doi.org/10.1200/JCO.2013.52.8505] [PMID: 25385735]

[68] Ploussard G, Durand X, Xylinas E, *et al.* Prostate cancer antigen 3 score accurately predicts tumour volume and might help in selecting prostate cancer patients for active surveillance. Eur Urol 2011; 59(3): 422-9. [http://dx.doi.org/10.1016/j.eururo.2010.11.044] [PMID: 21156337]

[69] Sokoll LJ, Ellis W, Lange P, *et al.* A multicenter evaluation of the PCA3 molecular urine test: pre-analytical effects, analytical performance, and diagnostic accuracy. Clin Chim Acta 2008; 389(1-2): 1-6. [http://dx.doi.org/10.1016/j.cca.2007.11.003] [PMID: 18061575]

[70] Attard G, Clark J, Ambroisine L, *et al.* Duplication of the fusion of TMPRSS2 to ERG sequences identifies fatal human prostate cancer. Oncogene 2008; 27(3): 253-63. [http://dx.doi.org/10.1038/sj.onc.1210640] [PMID: 17637754]

[71] Dijkstra S, Govers TM, Hendriks RJ, *et al.* Cost-effectiveness of a new urinary biomarker-based risk score compared to standard of care in prostate cancer diagnostics - a decision analytical model. BJU Int 2017; 120(5): 659-65. [http://dx.doi.org/10.1111/bju.13861] [PMID: 28370948]

[72] Köllermann J, Schlomm T, Bang H, *et al.* Expression and prognostic relevance of annexin A3 in prostate cancer. Eur Urol 2008; 54(6): 1314-23. [http://dx.doi.org/10.1016/j.eururo.2008.01.001] [PMID: 18222597]

[73] Schostak M, Schwall GP, Poznanović S, *et al.* Annexin A3 in urine: a highly specific noninvasive marker for prostate cancer early detection. J Urol 2009; 181(1): 343-53. [http://dx.doi.org/10.1016/j.juro.2008.08.119] [PMID: 19012935]

[74] Rubin MA, Zhou M, Dhanasekaran SM, *et al.* alpha-Methylacyl coenzyme A racemase as a tissue biomarker for prostate cancer. JAMA 2002; 287(13): 1662-70. [http://dx.doi.org/10.1001/jama.287.13.1662] [PMID: 11926890]

[75] Jones PA, Baylin SB. The epigenomics of cancer. Cell 2007; 128(4): 683-92. [http://dx.doi.org/10.1016/j.cell.2007.01.029] [PMID: 17320506]

[76] Blute ML Jr, Damaschke NA, Jarrard DF. The epigenetics of prostate cancer diagnosis and prognosis: update on clinical applications. Curr Opin Urol 2015; 25(1): 83-8. [http://dx.doi.org/10.1097/MOU.0000000000000132] [PMID: 25405932]

[77] Wojno KJ, Costa FJ, Cornell RJ, *et al.* Reduced rate of repeated prostate biopsies observed in ConfirmMDx clinical utility field study. Am Health Drug Benefits 2014; 7(3): 129-34. [PMID: 24991397]

[78] Wei L, Wang J, Lampert E, *et al.* Intratumoral and Intertumoral Genomic Heterogeneity of Multifocal Localized Prostate Cancer Impacts Molecular Classifications and Genomic Prognosticators. Eur Urol 2017; 71(2): 183-92.

[http://dx.doi.org/10.1016/j.eururo.2016.07.008] [PMID: 27451135]

[79] Lichtenstein P, Holm NV, Verkasalo PK, *et al.* Environmental and heritable factors in the causation of cancer--analyses of cohorts of twins from Sweden, Denmark, and Finland. N Engl J Med 2000; 343(2): 78-85.
[http://dx.doi.org/10.1056/NEJM200007133430201] [PMID: 10891514]

[80] Severi G, Byrnes GB, Hopper JL. Five genetic variants associated with prostate cancer. N Engl J Med 2008; 358(25): 2739-40.
[PMID: 18572450]

[81] Yoshida K, Miki Y. Role of BRCA1 and BRCA2 as regulators of DNA repair, transcription, and cell cycle in response to DNA damage. Cancer Sci 2004; 95(11): 866-71.
[http://dx.doi.org/10.1111/j.1349-7006.2004.tb02195.x] [PMID: 15546503]

[82] Thompson D, Easton DF. Cancer Incidence in BRCA1 mutation carriers. J Natl Cancer Inst 2002; 94(18): 1358-65.
[http://dx.doi.org/10.1093/jnci/94.18.1358] [PMID: 12237281]

[83] Capitanio U, Bensalah K, Bex A, *et al.* Epidemiology of Renal Cell Carcinoma. Eur Urol 2019; 75(1): 74-84.
[http://dx.doi.org/10.1016/j.eururo.2018.08.036] [PMID: 30243799]

[84] Mallappallil M, Friedman EA, Delano BG, McFarlane SI, Salifu MO. Chronic kidney disease in the elderly: evaluation and management. Clin Pract (Lond) 2014; 11(5): 525-35.
[http://dx.doi.org/10.2217/cpr.14.46] [PMID: 25589951]

[85] Davis JC, Furstenthal L, Desai AA, *et al.* The microeconomics of personalized medicine: today's challenge and tomorrow's promise. Nat Rev Drug Discov 2009; 8(4): 279-86.
[http://dx.doi.org/10.1038/nrd2825] [PMID: 19300459]

[86] Ngo TC, Wood CG, Karam A. Biomarkers of renal cell carcinoma. Urol. Oncol. Semin. Orig. Investig 2013; pp. 1-9.

[87] Driessen A, Landuyt W, Pastorekova S, *et al.* Expression of carbonic anhydrase IX (CA IX), a hypoxia-related protein, rather than vascular-endothelial growth factor (VEGF), a pro-angiogenic factor, correlates with an extremely poor prognosis in esophageal and gastric adenocarcinomas. Ann Surg 2006; 243(3): 334-40.
[http://dx.doi.org/10.1097/01.sla.0000201452.09591.f3] [PMID: 16495697]

[88] Måseide K, Kandel RA, Bell RS, *et al.* Carbonic anhydrase IX as a marker for poor prognosis in soft tissue sarcoma. Clin Cancer Res. 2004;1;10(13):4464-71.
[http://dx.doi.org/10.1158/1078-0432.CCR-03-0541]

[89] Cianchi F, Vinci MC, Supuran CT, *et al.* Selective inhibition of carbonic anhydrase IX decreases cell proliferation and induces ceramide-mediated apoptosis in human cancer cells. J Pharmacol Exp Ther 2010; 334(3): 710-9.
[http://dx.doi.org/10.1124/jpet.110.167270] [PMID: 20519553]

[90] Phuoc NBA, Ehara H, Gotoh T, *et al.* Prognostic value of the co-expression of carbonic anhydrase IX and vascular endothelial growth factor in patients with clear cell renal cell carcinoma. Oncol Rep 2008; 20(3): 525-30.
[PMID: 18695901]

[91] Bui MH, Visapaa H, Seligson D, *et al.* Prognostic value of carbonic anhydrase IX and KI67 as predictors of survival for renal clear cell carcinoma. J Urol 2004; 171(6 Pt 1): 2461-6.
[http://dx.doi.org/10.1097/01.ju.0000116444.08690.e2] [PMID: 15126876]

[92] Patard JJ, Fergelot P, Karakiewicz PI, *et al.* Low CAIX expression and absence of VHL gene mutation are associated with tumor aggressiveness and poor survival of clear cell renal cell carcinoma. Int J Cancer 2008; 123(2): 395-400.
[http://dx.doi.org/10.1002/ijc.23496] [PMID: 18464292]

[93] Atkins M, Regan M, McDermott D, *et al.* Carbonic anhydrase IX expression predicts outcome of interleukin 2 therapy for renal cancer. Clin Cancer Res 2005; 11(10): 3714-21. [http://dx.doi.org/10.1158/1078-0432.CCR-04-2019]

[94] Divgi CR, Uzzo RG, Gatsonis C, *et al.* Positron emission tomography/computed tomography identification of clear cell renal cell carcinoma: results from the REDECT trial. J Clin Oncol 2013; 10; 31(2): 187-94.

[95] Takahashi M, Rhodes DR, Furge KA, *et al.* Gene expression profiling of clear cell renal cell carcinoma: gene identification and prognostic classification. Proc Natl Acad Sci USA 2001; 98(17): 9754-9. [http://dx.doi.org/10.1073/pnas.171209998] [PMID: 11493696]

[96] Kosari F, Parker AS, Kube DM, *et al.* Clear cell renal cell carcinoma: gene expression analyses identify a potential signature for tumor aggressiveness. Clin Cancer Res 2005; 11(14): 5128-39. [http://dx.doi.org/10.1158/1078-0432.CCR-05-0073] [PMID: 16033827]

[97] Hsieh JJ, Purdue MP, Signoretti S, *et al.* Renal cell carcinoma. Nat Rev Dis Primers 2017; 3(1): 17009. [http://dx.doi.org/10.1038/nrdp.2017.9] [PMID: 28276433]

[98] Cooper SJ, Tun HW, Roper SM, *et al.* Current Status of Biomarker Discovery in Human Clear Cell Renal Cell Carcinoma. J Mol Biomark Diagn 2012; 2: 5. [http://dx.doi.org/10.4172/2155-9929.S2-005]

[99] Ozer G, Altinel M, Kocak B, Yazicioglu A, Gonenc F. Value of urinary NMP-22 in patients with renal cell carcinoma. Urology 2002; 60(4): 593-7. [http://dx.doi.org/10.1016/S0090-4295(02)01857-5] [PMID: 12385914]

[100] Pastore AL, Palleschi G, Silvestri L, *et al.* Serum and urine biomarkers for human renal cell carcinoma. Dis Markers 2015; 2015: 251403. [http://dx.doi.org/10.1155/2015/251403] [PMID: 25922552]

[101] Kaya K, Ayan S, Gokce G, Kilicarslan H, Yildiz E, Gultekin EY. Urinary nuclear matrix protein 22 for diagnosis of renal cell carcinoma. Scand J Urol Nephrol 2005; 39(1): 25-9. [http://dx.doi.org/10.1080/00365590410002500] [PMID: 15764267]

[102] Morrissey JJ, London AN, Lambert MC, Kharasch ED. Sensitivity and specificity of urinary neutrophil gelatinase-associated lipocalin and kidney injury molecule-1 for the diagnosis of renal cell carcinoma. Am J Nephrol 2011; 34(5): 391-8. [http://dx.doi.org/10.1159/000330851] [PMID: 21912102]

[103] Di Carlo A. Matrix metalloproteinase-2 and -9 in the sera and in the urine of human oncocytoma and renal cell carcinoma. Oncol Rep 2012; 28(3): 1051-6. [http://dx.doi.org/10.3892/or.2012.1864] [PMID: 22711190]

[104] Tan PH, Cheng L, Rioux-Leclercq N, *et al.* Renal tumors: diagnostic and prognostic biomarkers. Am J Surg Pathol 2013; 37(10): 1518-31. [http://dx.doi.org/10.1097/PAS.0b013e318299f12e] [PMID: 24025522]

[105] Ng KL, Rajandram R, Morais C, *et al.* Differentiation of oncocytoma from chromophobe renal cell carcinoma (RCC): can novel molecular biomarkers help solve an old problem? J Clin Pathol 2014; 67(2): 97-104. [http://dx.doi.org/10.1136/jclinpath-2013-201895] [PMID: 24170213]

[106] Motzer RJ, Escudier B, McDermott DF, *et al.* Nivolumab *versus* Everolimus in Advanced Renal-Cell Carcinoma. N Engl J Med 2015; 373(19): 1803-13. [http://dx.doi.org/10.1056/NEJMoa1510665] [PMID: 26406148]

[107] Heng DY, Xie W, Regan MM, *et al.* External validation and comparison with other models of the International Metastatic Renal-Cell Carcinoma Database Consortium prognostic model: a population-based study. Lancet Oncol 2013; 14(2): 141-8.

[http://dx.doi.org/10.1016/S1470-2045(12)70559-4] [PMID: 23312463]

[108] Grivas N, Kafarakis V, Tsimaris I, Raptis P, Hastazeris K, Stavropoulos NE. Clinico-pathological prognostic factors of renal cell carcinoma: A 15-year review from a single center in Greece. Urol Ann 2014; 6(2): 116-21.
[http://dx.doi.org/10.4103/0974-7796.130552] [PMID: 24833821]

[109] Johnson TV, Abbasi A, Owen-Smith A, *et al.* Absolute preoperative C-reactive protein predicts metastasis and mortality in the first year following potentially curative nephrectomy for clear cell renal cell carcinoma. J Urol 2010; 183(2): 480-5.
[http://dx.doi.org/10.1016/j.juro.2009.10.014] [PMID: 20006861]

[110] Karakiewicz PI, Hutterer GC, Trinh QD, *et al.* C-reactive protein is an informative predictor of renal cell carcinoma-specific mortality: a European study of 313 patients. Cancer 2007; 110(6): 1241-7.
[http://dx.doi.org/10.1002/cncr.22896] [PMID: 17639589]

[111] Heng DYC, Xie W, Regan MM, *et al.* Prognostic factors for overall survival in patients with metastatic renal cell carcinoma treated with vascular endothelial growth factor-targeted agents: results from a large, multicenter study. J Clin Oncol 2009; 27(34): 5794-9.
[http://dx.doi.org/10.1200/JCO.2008.21.4809] [PMID: 19826129]

[112] Yao M, Yoshida M, Kishida T, *et al.* VHL tumor suppressor gene alterations associated with good prognosis in sporadic clear-cell renal carcinoma. J Natl Cancer Inst 2002; 94(20): 1569-75.
[http://dx.doi.org/10.1093/jnci/94.20.1569] [PMID: 12381710]

[113] Schraml P, Struckmann K, Hatz F, *et al.* VHL mutations and their correlation with tumour cell proliferation, microvessel density, and patient prognosis in clear cell renal cell carcinoma. J Pathol 2002; 196(2): 186-93.
[http://dx.doi.org/10.1002/path.1034] [PMID: 11793370]

[114] Paradis V, Lagha NB, Zeimoura L, *et al.* Expression of vascular endothelial growth factor in renal cell carcinomas. Virchows Arch 2000; 436(4): 351-6.
[http://dx.doi.org/10.1007/s004280050458] [PMID: 10834538]

[115] Rioux-Leclercq N, Fergelot P, Zerrouki S, *et al.* Plasma level and tissue expression of vascular endothelial growth factor in renal cell carcinoma: a prospective study of 50 cases. Hum Pathol 2007; 38(10): 1489-95.
[http://dx.doi.org/10.1016/j.humpath.2007.02.014] [PMID: 17597181]

[116] Yildiz E, Gokce G, Kilicarslan H, Ayan S, Goze OF, Gultekin EY. Prognostic value of the expression of Ki-67, CD44 and vascular endothelial growth factor, and microvessel invasion, in renal cell carcinoma. BJU Int 2004; 93(7): 1087-93.
[http://dx.doi.org/10.1111/j.1464-410X.2004.04786.x] [PMID: 15142169]

[117] Altieri DC. Survivin, versatile modulation of cell division and apoptosis in cancer. Oncogene 2003; 22(53): 8581-9.
[http://dx.doi.org/10.1038/sj.onc.1207113] [PMID: 14634620]

[118] Shariat SF, Karakiewicz PI, Godoy G, *et al.* Survivin as a prognostic marker for urothelial carcinoma of the bladder: a multicenter external validation study. Clin Cancer Res 2009; 15(22): 7012-9.
[http://dx.doi.org/10.1158/1078-0432.CCR-08-2554] [PMID: 19903782]

[119] Zamparese R, Pannone G, Santoro A, *et al.* Survivin expression in renal cell carcinoma. Cancer Invest 2008; 26(9): 929-35.
[http://dx.doi.org/10.1080/07357900802017553] [PMID: 19034775]

[120] Parker AS, Kosari F, Lohse CM, *et al.* High expression levels of survivin protein independently predict a poor outcome for patients who undergo surgery for clear cell renal cell carcinoma. Cancer 2006; 107(1): 37-45.
[http://dx.doi.org/10.1002/cncr.21952] [PMID: 16736510]

[121] Lane DP, Benchimol S. p53: oncogene or anti-oncogene? Genes Dev 1990; 4(1): 1-8.

[http://dx.doi.org/10.1101/gad.4.1.1] [PMID: 2137806]

[122] Zigeuner R, Ratschek M, Rehak P, Schips L, Langner C. Value of p53 as a prognostic marker in histologic subtypes of renal cell carcinoma: a systematic analysis of primary and metastatic tumor tissue. Urology 2004; 63(4): 651-5.
[http://dx.doi.org/10.1016/j.urology.2003.11.011] [PMID: 15072872]

[123] Tollefson MK, Thompson RH, Sheinin Y, *et al.* Ki-67 and coagulative tumor necrosis are independent predictors of poor outcome for patients with clear cell renal cell carcinoma and not surrogates for each other. Cancer 2007; 110(4): 783-90.
[http://dx.doi.org/10.1002/cncr.22840] [PMID: 17594714]

[124] Brenner W, Färber G, Herget T, Lehr HA, Hengstler JG, Thüroff JW. Loss of tumor suppressor protein PTEN during renal carcinogenesis. Int J Cancer 2002; 99(1): 53-7.
[http://dx.doi.org/10.1002/ijc.10303] [PMID: 11948491]

[125] Pantuck AJ, Seligson DB, Klatte T, *et al.* Prognostic relevance of the mTOR pathway in renal cell carcinoma: implications for molecular patient selection for targeted therapy. Cancer 2007; 109(11): 2257-67.
[http://dx.doi.org/10.1002/cncr.22677] [PMID: 17440983]

[126] Motzer RJ, Tannir NM, McDermott DF, *et al.* Nivolumab plus Ipilimumab *versus* Sunitinib in Advanced Renal-Cell Carcinoma. N Engl J Med 2018; 378(14): 1277-90.
[http://dx.doi.org/10.1056/NEJMoa1712126] [PMID: 29562145]

[127] García-Donas Js, Roldan JM, Lainez N, *et al.* Comprehensive molecular and immunohistochemical analysis of advanced renal cell carcinoma patients treated with mTOR inhibitors. J Clin Oncol 2018; 36: 4559. 2018.

[128] Perlman EJ. Pediatric renal tumors: practical updates for the pathologist. Pediatr Dev Pathol 2005; 8(3): 320-38.
[http://dx.doi.org/10.1007/s10024-005-1156-7] [PMID: 16010493]

[129] Van den HE, Grundy P, Graf N, *et al.* Characteristics and Survival of 750 Children Diagnosed with a Renal Tumor in the First Seven Months of Life: A Collaborative Study by the SIOP/GPOH/SFOP, NWTSG, and UKCCSG Wilms Tumor Study Groups. Pediatr. Blood Cancer 2008;50, 1130–1134..

[130] Moch H, Humphrey PA, Ulbright TM, Eds. WHO classification of tumours of the urinary system and male genital organs. 4th ed., Lyon, France: International Agency for Research on Cancer 2016.

[131] Xia Q, Shi S, Shen Q, *et al.* [Renal cell carcinoma with t(6;11)(p21.2;q13)/MALAT1-TFEB fusion: a clinical and pathological analysis]. Zhonghua Bing Li Xue Za Zhi 2015; 44(12): 895-9.
[PMID: 26888508]

[132] Cajaiba MM, Dyer LM, Geller JI, *et al.* The classification of pediatric and young adult renal cell carcinomas registered on the children's oncology group (COG) protocol AREN03B2 after focused genetic testing. Cancer 2018; 124(16): 3381-9.
[http://dx.doi.org/10.1002/cncr.31578] [PMID: 29905933]

[133] Roy A, Kumar V, Zorman B, *et al.* Recurrent internal tandem duplications of BCOR in clear cell sarcoma of the kidney. Nat Commun 2015; 6(1): 8891.
[http://dx.doi.org/10.1038/ncomms9891] [PMID: 26573325]

[134] Le Cornet C, Lortet-Tieulent J, Forman D, *et al.* Testicular cancer incidence to rise by 25% by 2025 in Europe? Model-based predictions in 40 countries using population-based registry data. Eur J Cancer 2014; 50(4): 831-9.
[http://dx.doi.org/10.1016/j.ejca.2013.11.035] [PMID: 24369860]

[135] Horwich A, Shipley J, Huddart R. Testicular germ-cell cancer. Lancet 2006; 367(9512): 754-65.
[http://dx.doi.org/10.1016/S0140-6736(06)68305-0] [PMID: 16517276]

[136] Williamson SR, Delahunt B, Magi-Galluzzi C, *et al.* The World Health Organization 2016 classification of testicular germ cell tumours: a review and update from the International Society of

Urological Pathology Testis Consultation Panel. Histopathology 2017; 70(3): 335-46. [http://dx.doi.org/10.1111/his.13102] [PMID: 27747907]

[137] Brierley JE, Gospodarowicz MK, Wittekind C. The TNM Classification of Malignant Tumours. 8th ed., 2016.

[138] Mead GM, Stenning SP. The International Germ Cell Consensus Classification: a new prognostic factor-based staging classification for metastatic germ cell tumours. Clin Oncol (R Coll Radiol) 1997; 9(4): 207-9. [http://dx.doi.org/10.1016/S0936-6555(97)80001-5] [PMID: 9315391]

[139] Carver BS, Sheinfeld J. Germ cell tumors of the testis. Ann Surg Oncol 2005; 12(11): 871-80. [http://dx.doi.org/10.1245/ASO.2005.01.013] [PMID: 16184443]

[140] Fakouri B, Coogan C. Testicular tumors.Surgical oncology: an algorith- mic approach for the general surgeon. New York: Springer 2003. [http://dx.doi.org/10.1007/0-387-21701-0_51]

[141] Gilligan TD, Seidenfeld J, Basch EM, *et al.* American Society of Clinical Oncology Clinical Practice Guideline on uses of serum tumor markers in adult males with germ cell tumors. J Clin Oncol 2010; 28(20): 3388-404. [http://dx.doi.org/10.1200/JCO.2009.26.4481] [PMID: 20530278]

[142] Venkitaraman R, Johnson B, Huddart RA, Parker CC, Horwich A, Dearnaley DP. The utility of lactate dehydrogenase in the follow-up of testicular germ cell tumours. BJU Int 2007; 100(1): 30-2. [http://dx.doi.org/10.1111/j.1464-410X.2007.06905.x] [PMID: 17552950]

[143] Oosterhuis JW, Looijenga LH. Testicular germ-cell tumours in a broader perspective. Nat Rev Cancer 2005; 5(3): 210-22. [http://dx.doi.org/10.1038/nrc1568] [PMID: 15738984]

[144] European Association of Urology Guidelines on Testicular Cancer. Edition 2020.

[145] Neumann A, Keller T, Jocham D, Doehn C. [Human placental alkaline phosphatase (hPLAP) is the most frequently elevated serum marker in testicular cancer]. Aktuelle Urol 2011; 42(5): 311-5. [http://dx.doi.org/10.1055/s-0031-1271545] [PMID: 21809268]

[146] Rajpert-De Meyts E. Developmental model for the pathogenesis of testicular carcinoma *in situ*: genetic and environmental aspects. Hum Reprod Update 2006; 12(3): 303-23. [http://dx.doi.org/10.1093/humupd/dmk006] [PMID: 16540528]

[147] Murray MJ, Huddart RA, Coleman N. The present and future of serum diagnostic tests for testicular germ cell tumours. Nat Rev Urol 2016; 13(12): 715-25. [http://dx.doi.org/10.1038/nrurol.2016.170] [PMID: 27754472]

[148] Lee RC, Feinbaum RL, Ambros V. The C. elegans heterochronic gene lin-4 encodes small RNAs with antisense complementarity to lin-14. Cell 1993; 75(5): 843-54. [http://dx.doi.org/10.1016/0092-8674(93)90529-Y] [PMID: 8252621]

[149] Voorhoeve PM, le Sage C, Schrier M, *et al.* A genetic screen implicates miRNA-372 and miRNA-373 as oncogenes in testicular germ cell tumors. Cell 2006; 124(6): 1169-81. [http://dx.doi.org/10.1016/j.cell.2006.02.037] [PMID: 16564011]

[150] Dieckmann KP, Radtke A, Geczi L, *et al.* Serum levels of MicroRNA-371a-3p (M371 Test) as a new biomarker of testicular germ cell tumors: results of a prospective multicentric study. J Clin Oncol 2019; 37(16): 1412-23. Epub ahead of print [http://dx.doi.org/10.1200/JCO.18.01480] [PMID: 30875280]

[151] Leão R, van Agthoven T, Figueiredo A, *et al.* Serum miRNA Predicts Viable Disease after Chemotherapy in Patients with Testicular Nonseminoma Germ Cell Tumor. J Urol 2018; 200(1): 126-35. [http://dx.doi.org/10.1016/j.juro.2018.02.068] [PMID: 29474847]

[152] Lawrence MS, Stojanov P, Polak P, *et al.* Mutational heterogeneity in cancer and the search for new cancer-associated genes. Nature 2013; 499(7457): 214-8.
[http://dx.doi.org/10.1038/nature12213] [PMID: 23770567]

[153] Rosenberg JE, Hoffman-Censits J, Powles T, *et al.* Atezolizumab in patients with locally advanced and metastatic urothelial carcinoma who have progressed following treatment with platinum-based chemotherapy: a single-arm, multicentre, phase 2 trial. Lancet 2016; 387(10031): 1909-20.
[http://dx.doi.org/10.1016/S0140-6736(16)00561-4] [PMID: 26952546]

[154] Choudhury NJ, Kiyotani K, Yap KL, *et al.* Low T-cell receptor diversity, high somatic mutation burden, and high neoantigen load as predictors of clinical outcome in muscle-invasive bladder cancer. Eur Urol Focus 2016; 2(4): 445-52.
[http://dx.doi.org/10.1016/j.euf.2015.09.007] [PMID: 28723478]

[155] Funt S, Charen AS, Yusko E, *et al.* Correlation of peripheral and intratumoral T-cell receptor (TCR) clonality with clinical outcomes in patients with metastatic urothelial cancer (mUC) treated with atezolizumab. J Clin Oncol 2016; 34(suppl): abstr 3005.

[156] Iyer G, Milowsky MI. Fibroblast growth factor receptor-3 in urothelial tumorigenesis. Urol Oncol 2013; 31(3): 303-11.
[http://dx.doi.org/10.1016/j.urolonc.2011.12.001] [PMID: 22285006]

[157] Milowsky MI, Dittrich C, Duran Martinez I, *et al.* Final results of a multicenter, open-label phase II trial of dovitinib (TKI258) in patients with advanced urothelial carcinoma with either mutated or nonmutated FGFR3. J Clin Oncol 2013; 31(6_suppl): 255.

[158] Van Allen EM, Mouw KW, Kim P, *et al.* Somatic ERCC2 mutations correlate with cisplatin sensitivity in muscle-invasive urothelial carcinoma. Cancer Discov 2014; 4(10): 1140-53.
[http://dx.doi.org/10.1158/2159-8290.CD-14-0623] [PMID: 25096233]

[159] Teo MTW, Dyrskjøt L, Nsengimana J, *et al.* Next-generation sequencing identifies germline MRE11A variants as markers of radiotherapy outcomes in muscle-invasive bladder cancer. Ann Oncol 2014; 25(4): 877-83.
[http://dx.doi.org/10.1093/annonc/mdu014] [PMID: 24623370]

[160] Yadav SS, Li J, Lavery HJ, Yadav KK, Tewari AK. Next-generation sequencing technology in prostate cancer diagnosis, prognosis, and personalized treatment. Urologic Oncology: Seminars and Original Investigations 2015; 33(6): 267.e1-267.e13.
[http://dx.doi.org/10.1016/j.urolonc.2015.02.009] [PMID: 25791755]

[161] Beltran H, Yelensky R, Frampton GM, *et al.* Targeted next-generation sequencing of advanced prostate cancer identifies potential therapeutic targets and disease heterogeneity. Eur Urol 2013; 63(5): 920-6.
[http://dx.doi.org/10.1016/j.eururo.2012.08.053] [PMID: 22981675]

[162] Donovan MJ, Cordon-Cardo C. Predicting high-risk disease using tissue biomarkers. Curr Opin Urol 2013; 23(3): 245-51.
[http://dx.doi.org/10.1097/MOU.0b013e32835f89cc] [PMID: 23478499]

[163] Beroukhim R, Brunet JP, Di Napoli A, *et al.* Patterns of gene expression and copy-number alterations in von-hippel lindau disease-associated and sporadic clear cell carcinoma of the kidney. Cancer Res 2009; 69(11): 4674-81.
[http://dx.doi.org/10.1158/0008-5472.CAN-09-0146] [PMID: 19470766]

[164] Chen M, Ye Y, Yang H, *et al.* Genome-wide profiling of chromosomal alterations in renal cell carcinoma using high-density single nucleotide polymorphism arrays. Int J Cancer 2009; 125(10): 2342-8.
[http://dx.doi.org/10.1002/ijc.24642] [PMID: 19521957]

[165] Wu J, Wang H, Ricketts CJ, *et al.* Germline mutations of renal cancer predisposition genes and clinical relevance in Chinese patients with sporadic, early-onset disease. Cancer 2019; 125(7): 1060-9.

[http://dx.doi.org/10.1002/cncr.31908] [PMID: 30548481]

[166] Glick RD, Hicks MJ, Nuchtern JG, Wesson DE, Olutoye OO, Cass DL. Renal tumors in infants less than 6 months of age. J Pediatr Surg 2004; 39(4): 522-5. [http://dx.doi.org/10.1016/j.jpedsurg.2003.12.007] [PMID: 15065020]

[167] Argani P, Yonescu R, Morsberger L, *et al.* Molecular confirmation of t(6;11)(p21;q12) renal cell carcinoma in archival paraffin-embedded material using a break-apart TFEB FISH assay expands its clinicopathologic spectrum. Am J Surg Pathol 2012; 36(10): 1516-26. [http://dx.doi.org/10.1097/PAS.0b013e3182613d8f] [PMID: 22892601]

[168] Thorner PS, Shago M, Marrano P, Shaikh F, Somers GR. TFE3-positive renal cell carcinomas are not always Xp11 translocation carcinomas: Report of a case with a TPM3-ALK translocation. Pathol Res Pract 2016; 212(10): 937-42. [http://dx.doi.org/10.1016/j.prp.2016.07.004] [PMID: 27450657]

[169] Oliver RT. Clues from natural history and results of treatment supporting the monoclonal origin of germ cell tumours. Cancer Surv 1990; 9(2): 333-68. [PMID: 1964626]

[170] Vliagoftis H, Worobec AS, Metcalfe DD. The protooncogene c-kit and c-kit ligand in human disease. J Allergy Clin Immunol 1997; 100(4): 435-40. [http://dx.doi.org/10.1016/S0091-6749(97)70131-3] [PMID: 9338533]

[171] Pectasides D, Nikolaou M, Pectasides E, Koumarianou A, Valavanis C, Economopoulos T. Complete response after imatinib mesylate administration in a patient with chemoresistant stage IV seminoma. Anticancer Res 2008; 28(4C): 2317-20. [PMID: 18751412]

[172] Einhorn LH, Brames MJ, Heinrich MC, Corless CL, Madani A. Phase II study of imatinib mesylate in chemotherapy refractory germ cell tumors expressing KIT. Am J Clin Oncol 2006; 29(1): 12-3. [http://dx.doi.org/10.1097/01.coc.0000195086.47548.ef] [PMID: 16462496]

[173] Bagrodia A, Kaffenberger SD, Lee B, *et al.* Actionable targets in patients with cisplatin-resistant advanced germ cell tumors. J Clin Oncol 2016; 34(2_suppl): 473. [http://dx.doi.org/10.1200/jco.2016.34.2_suppl.473]

[174] Colicelli J. Human RAS superfamily proteins and related GTPases. Sci STKE 2004; 2004(250): RE13. [http://dx.doi.org/10.1126/stke.2502004re13] [PMID: 15367757]

[175] Feldman DR, Iyer G, Van Alstine L, *et al.* Presence of somatic mutations within PIK3CA, AKT, RAS, and FGFR3 but not BRAF in cisplatin-resistant germ cell tumors. Clin Cancer Res 2014; 20(14): 3712-20. [http://dx.doi.org/10.1158/1078-0432.CCR-13-2868] [PMID: 24812411]

[176] Subbiah V, Meric-Bernstam F, Mills GB, *et al.* Next generation sequencing analysis of platinum refractory advanced germ cell tumor sensitive to Sunitinib (Sutent®) a VEGFR2/PDGFRβ/c-kit/ FLT3/RET/CSF1R inhibitor in a phase II trial. J Hematol Oncol 2014; 7(1): 52. [http://dx.doi.org/10.1186/s13045-014-0052-x] [PMID: 25085632]

CHAPTER 4

Circulating Biomarkers in Thyroid Cancer

Lebriz Uslu-Beşli[1,*]

[1] *Department of Nuclear Medicine, Cerrahpasa Medical Faculty, Istanbul University-Cerrahpasa, Istanbul, Turkey*

Abstract: Thyroid cancer is the most important endocrine cancer with increasing incidence. While thyroid cancers, especially papillary thyroid cancers, are known to exhibit generally a favorable outcome with excellent survival rates, some thyroid cancers are more aggressive with a poor prognosis. Several different biomarkers have been introduced for the diagnosis of disease, identification of tumor load, assessment of therapy response, and the detection of recurrence during follow-up of the thyroid cancer patients. This chapter gives a brief overview of the circulating biomarkers used in thyroid cancer patients.

Keywords: Biomarkers, Calcitonin, Carcinoembryonic antigen, Liquid biopsy, Thyroglobulin, Thyroid cancer.

INTRODUCTION

Thyroid cancer incidence has increased significantly over the last few decades, although the mortality rates have stayed relatively constant [1]. The demographics of thyroid cancer have also changed, with decreasing proportion of anaplastic thyroid cancers (ATC) and an increasing proportion of papillary thyroid cancers (PTC). Relatively stable mortality rates despite an increasing incidence have been attributed to the over-diagnosis of thyroid cancer as a result of better diagnostic tests, as well as to advancements in the treatment strategies of thyroid cancer.

Thyroid cancers are a heterogeneous group of tumors with cell types of different origin, developing from thyroid nodules and exhibiting different genetic alterations, biological behavior, and clinical course. Therefore, different biomarkers exist, each reflecting the unique biology of the selected tumor type. To date, there is not a single circulating biomarker that can be used to diagnose thyroid cancer has been found; rather, the diagnosis relies mainly on ultrasonography and fine-needle aspiration biopsy.

* **Corresponding author Lebriz Uslu-Beşli.** Department of Nuclear Medicine, Cerrahpasa Medical Faculty, Istanbul University-Cerrahpasa, Istanbul, Turkey; Tel: +90 (532) 494 05 96; E-mail: lebriz.uslu@istanbul.edu.tr

Pınar Atukeren and Hafize Uzun (Eds.)

Thyroglobulin (Tg) is the most important biomarker for differentiated thyroid cancer (DTC), which originates from follicular epithelium, and calcitonin is the primary biomarker for medullary thyroid cancer (MTC), which stems from parafollicular cells of the thyroid gland. Basal serum calcitonin is a useful biomarker to determine the extent and presence of lymph node metastases in MTC patients.

Thyroid cancer is a malignant disease with generally a favorable prognosis, and relatively low rates of disease-related mortality compared to other cancers, and several studies have been conducted regarding biomarkers for diagnosis, therapy response assessment, and the follow-up of thyroid cancer patients. This chapter gives a brief overview on the histology of the thyroid gland, histopathology of thyroid cancers, their medical management and reviews the most important serum biomarkers that are used in the clinical management of thyroid cancer patients, as well as the liquid biopsy technique, which is a promising tool for the diagnosis and follow-up of thyroid cancer patients.

THYROID GLAND

The thyroid gland is located anterior to the larynx and consists of two lobes and a connecting isthmus in between. It is mainly composed of follicles, which consist of cuboidal-columnar epithelium filled with Tg-rich colloid. The major function of the thyroid follicles is to synthesize thyroid hormones, mainly thyroxine (T_4) and to a lesser extent triiodothyronine (T_3), both of which function in the regulation of metabolic rate and are required throughout human life. During fetal development and childhood, thyroid hormones are crucial for normal brain development, and they are required as regulators of intermediary metabolism in adults [2]. T_4 and T_3 are stored in great quantity in the extracellular colloid within Tg to avoid deleterious effects of thyroid hormone deficiency [2, 3].

Thyroid hormone synthesis and storage involve a unique process where iodide is necessary. Synthesis of thyroid hormones mainly comprises four stages: iodide trapping, which involves uptake of the circulating iodide by a membrane transport protein called sodium/iodide symporter (NIS) located in the basolateral membrane of follicular cells; synthesis of Tg within follicular cells; oxidation of iodide to form oxidized form of iodine by the enzyme thyroid peroxidase (TPO) located in the apical cell membrane and transport of iodine into the follicle cavity by pendrin; finally organification of Tg, which involves iodination of tyrosine residues of Tg within the colloid. Tyrosine is first iodized to monoiodotyrosine (MIT) and then to diiodotyrosine (DIT) by the enzyme TPO. The coupling of two DITs forms T_4, whereas the coupling of one MIT and one DIT forms T_3.

Thyroid follicle function is regulated by pituitary thyroid-stimulating hormone (TSH), also called thyrotropin, which stimulates both synthesis of thyroid hormones and their release into the systemic circulation. Upon TSH stimulation, the colloid is taken into the follicular cells and digested in lysosomes to liberate T_4 and T_3, and finally, T_4 and T_3 are released into the systemic circulation. MIT and DIT, which are freed from Tg during digestion, are not released into the blood; instead, they are recycled for new hormone synthesis.

Apart from thyroid follicles, the thyroid gland also has another cell type, namely parafollicular cells, or C cells, which are located as part of the follicular epithelium or as isolated clusters between thyroid follicles. Parafollicular cells are neuroendocrine cells that synthesize mainly calcitonin, which is a hormone responsible for calcium metabolism. When the serum calcium level is elevated, calcitonin is secreted, which in turn decreases bone resorption and increases calcium absorption by the skeletal system to normalize serum calcium levels. Carcinoembryonic antigen (CEA), adrenocorticotropic hormone (ACTH), B-melanocyte stimulating hormone, chromogranin, histaminase, neurotensin, and somatostatin are the other hormones or biological amines secreted by parafollicular cells.

CLASSIFICATION OF THYROID CANCER

Major subtypes of thyroid carcinomas are PTC (75-85%), follicular thyroid carcinoma (FTC) (10-20%), MTC (5%), and ATC (<5%) [4].

DTCs, including PTC and FTC, are derived from the follicular epithelium, and they preserve some of the normal thyrocyte functions, such as Tg synthesis and response to TSH stimulation. PTC is the most common thyroid cancer subtype in populations with sufficient iodine intake, whereas the incidence of FTC is higher in areas with iodine deficiency. Most patients with PTCs have favorable outcomes after appropriate disease management, whereas FTCs have a higher risk for recurrence and distant metastasis and lower survival rates compared to PTCs [5, 6]. Mutations in RET, RAS, or BRAF proto-oncogenes, all of which trigger the activation of mitogen-activated protein kinase (MAPK) cascade, are present in nearly 70% of PTC patients, and RAS mutated PTCs are typically follicular variant subtype of PTC [7 - 9]. RAS mutations and PAX8-PPARγ1 fusion are seen frequently in FTC patients [10, 11]. While cervical lymph nodes are the most common site of metastasis for PTC, hematogenous metastases to lungs and bones are more frequent in FTC [5].

MTCs are derived from parafollicular cells, therefore, unlike DTCs, they do not produce Tg; instead, they produce calcitonin, and they do not respond to TSH stimulation. Most MTCs are seen sporadically (70-80%), but familial autosomal dominant inheritance is present in a significant percentage of patients (20-30%) [12]. Hereditary MTCs can also be seen together with other endocrine tumors in syndromes called multiple endocrine neoplasia (MEN). MEN type 2A is characterized by MTC, pheochromocytoma and/or hyperparathyroidism, whereas MEN type 2B has MTC, pheochromocytoma, mucosal neuromas, and intestinal ganglioneuromas [13]. Mutations in RET proto-oncogene, which is located on chromosome 10q11.2, encodes receptor tyrosine kinase protein and plays essential roles in cell migration, differentiation, and proliferation, which are the most common genetic mutations associated with MTC and are seen approximately in 40-60% of sporadic cases and over 95% of hereditary MTCs [12]. In 68% of sporadic MTCs without RET mutations, RAS mutation (HRAS, KRAS, or rarely NRAS) was documented [14]. MTC metastasize frequently to cervical lymph nodes; 70% of MTC patients with palpable thyroid nodules have cervical metastasis at surgery [15]. Common sites of distant metastases are the liver, lungs, bones, with the brain and skin being less common [12].

Anaplastic thyroid carcinoma (ATC) is a rare but the most aggressive type of thyroid cancer with poor overall survival rates [16]. ATCs originate from follicular cells either de novo or from pre-existing DTC or poorly differentiated thyroid carcinomas [17]. While different gene mutations exist in the pathogenesis of ATC; MAP kinase and PI3CA kinase pathways are the most affected pathways [17]. While ATC can exhibit mutations typical for differentiated thyroid cancer, such as BRAF and RAS, TP53 mutation is found exclusively in ATC [18 - 20]. The most common sites of distant metastases are lungs, bones and brain [5].

MEDICAL MANAGEMENT OF THYROID CANCER

Differentiated Thyroid Cancer

Although several risk factors for the development of malignancy in thyroid tissue have been introduced, such as radiation exposure, family history and goiter, many thyroid cancer patients are diagnosed incidentally during medical imaging. Diagnosis of thyroid cancer is based on ultrasonography and fine needle aspiration biopsy. To date, screening people for thyroid cancer is not recommended unless there is a history of increased risk for thyroid cancer, such as radiation exposure.

Although the therapy strategy of DTC varies according to the size and extent of the nodules and associated risk factors, generally, the initial therapy of DTC involves surgical resection of the primary tumor and clinically significant lymph nodes, followed by post-operative radioiodine administration for removal of all thyroid tissues left, where appropriate and TSH suppression during follow-up [21]. Depending on the size and extension of the nodule and associated lymph node involvement, either total thyroidectomy, subtotal thyroidectomy or lobectomy is performed. Lobectomy is generally reserved for low-risk group patients with unifocal small-sized nodules. Depending on the stage and post-operative risk stratification of patients, radioiodine can be used in the post-operative period for patients who had either total thyroidectomy or subtotal thyroidectomy operation for normal remnant thyroid tissue ablation or as adjuvant therapy for residual or metastatic disease. TSH suppression during follow-up is recommended, especially for the high-risk group of patients.

Medullary Thyroid Cancer

MTC is more aggressive than DTC and unfortunately, treatment options are more limited for MTC. As it is not derived from thyroid follicular cells, MTC does not respond to TSH suppression or radioactive iodine therapy. Also, chemotherapy options are limited for MTC. Therefore, surgery is currently the best treatment option for MTC and it is used both for curative treatment or for reduction of tumor burden and palliation [15, 22]. Total thyroidectomy with bilateral lymphadenectomy is the most common initial surgical approach in MTC. External beam radiotherapy is reserved for extensive regional disease or residual tumor mass and systemic chemotherapy preferentially with tyrosine kinase inhibitors, radiotherapy or metastasectomy are used for metastatic disease control [22].

Anaplastic Thyroid Cancer

ATC is an undifferentiated and aggressive form of thyroid malignancy with a poor prognosis and limited therapeutic options. Unlike poorly differentiated thyroid cancer, which is another aggressive type of thyroid cancer, ATC does not retain follicular elements or produce thyroglobulin [23, 24]. Generally, a multimodal therapeutic approach is recommended, including surgical resection, external beam radiotherapy and a combination of chemotherapies [23]. The first-line treatment of ATC generally involves surgical resection, unless immediate life threatening disease elsewhere, such as brain or spine metastases or pulmonary hemorrhage, is present [25]. Most patients have an extensively invasive primary tumor at the time of diagnosis, complicating total resection of the primary tumor, therefore neoadjuvant external beam radiotherapy and/or chemotherapy followed by surgical resection are recommended for those patients who have unresectable primary tumors at the initial assessment [25]. Systemic chemotherapy involves a combination of taxane (paclitaxel or docetaxel) and/or anthracyclines (doxorubicin) and/or platin (cisplatin, carboplatin) either in combination with radiotherapy or as first-line therapy in advanced ATC [25]. Second-line treatments generally include tyrosine kinase inhibitors, anti-angiogenic drugs and multi-kinase inhibitors [23]. Despite early and aggressive treatment, most of the

ATC patients have disease progression within a few months [25].

SEROLOGICAL BIOMARKERS

Thyroglobulin

Tg is a large dimeric glycoprotein with a monomeric mass of approximately 330 kDa and approximately 2750 amino acid residues [2]. Tg is stored in the thyroid colloid and can be secreted into systemic circulation in small amounts as a result of TSH stimulation. Tg levels in serum are proportional to the thyroid size, with concentrations between 0.5 and 1.0 mg/L of Tg per gram of thyroid tissue [26, 27]. Serum Tg population is heterogeneous as a result of mRNA splicing [28]. Half-life of Tg ranges from 3 to 65 hours and it is cleared mainly by the liver [28]. Tg is specific for thyroid cells, as it is synthesized only by the thyroid follicles and is the most abundant protein of the thyroid gland. Therefore, serum Tg measurement can be used as a valuable parameter for therapy plan, therapy response assessment and follow-up of thyroid cancer patients in the post-operative period. However, preoperative elevated serum Tg levels are an insensitive and non-specific test for thyroid cancer screening in patients with thyroid nodules [21]. Although, elevated preoperative serum Tg levels increase the probability of thyroid cancer, Tg measurement has limited sensitivity and specificity for the differentiation of benign or malignant thyroid nodules [29, 30]. Also, elevated serum Tg levels can also be seen in the presence of benign thyroid disorders, such as subacute thyroiditis, Hashimoto's thyroiditis and iodine deficiency, as well as in non-thyroidal etiologies, such as pregnancy and renal failure [28, 31, 32]. Therefore serum Tg is accepted as a poor preoperative diagnostic biomarker of malignancy in DTC and its usage as a cancer screening test or its routine preoperative measurement is not recommended [21].

Tg is a valuable marker in the post-operative period for assessment of residual thyroid tissue mass, risk-stratification and therapy plan of DTC patients [21, 33]. Elevated post-operative serum Tg levels were shown to be associated with the presence of residual or metastatic disease and are prognostic indicators of disease recurrence during follow-up [34, 35]. Tg is also used as a useful biomarker for therapy response assessment and detection of recurrent or metastatic disease. Since Tg is produced upon TSH stimulation, Tg measurement under TSH stimulation increases the sensitivity of serum Tg analysis [36]. Excellent therapy response is generally defined for patients who had total thyroidectomy and radioiodine ablation as TSH stimulated Tg of less than 1 ng/ml in the absence of interfering anti-thyroglobulin (anti-Tg) antibodies, or any structural or functional evidence of residual disease, whereas the incomplete response is defined as TSH stimulated Tg value of higher than 10 ng/ml or non-stimulated Tg values of higher than 1 ng/ml [21, 37 - 39]. In case of recurrent or residual disease, progressive rising Tg levels is generally observed. Conversely, in case of no structural evidence of disease, serum Tg levels were shown to decrease on serial Tg measurements for patients under follow-up after radioactive iodine ablation [33]. Therefore, for the follow-up of thyroid cancer patients, serum Tg measurement is recommended for every 6-24 months interval, depending on the risk assessment of patients [21].

Serum Tg measurement can be performed using immunoassays, including radioimmunoassay (RIA) and immunometric assay (IMA), or using liquid chromatography-tandem mass spectrometry (LC-MS/MS) assay [26]. Each method has different sensitivity and specificity and different levels of interference with anti-Tg antibodies. LC-MS/MS has few interference problems compared to immunoassays, but their sensitivity is also lower.

Anti-thyroglobulin Antibodies

Anti-Tg antibodies, which are mainly IgG and to a lesser extent IgM and IgA antibodies, are produced as a result of the immune response against different epitopes in the Tg molecule [28]. While anti-Tg antibody levels are elevated in DTC and several different autoimmune thyroid diseases, such as Hashimoto's thyroiditis and Graves' disease, elevated titers were also documented in other autoimmune diseases, such as type-I diabetes mellitus, Addison's disease, celiac disease, and vitiligo and also in non-disease states, such as tobacco smoking and elderly women [40]. The prevalence of anti-Tg antibodies in the normal population varies between 4.2%-91% depending on the analysis method [28].

In DTC patients, in addition to serum Tg level measurement, anti-Tg levels are also assessed. Anti-Tg antibodies are known to interfere with serum Tg measurement, and limit Tg utility and cause false-negative results in nearly 30% of anti-Tg antibody positive patients [26, 41]. Therefore serum Tg levels are recommended to be measured together with anti-Tg antibody levels in the same laboratory using the same assay given for the patient [21]. Anti-Tg antibody levels can also be used as a biomarker for both detections of recurrent thyroid cancer and therapy response assessment in the post-operative period. While rising or new-appearance of anti-Tg antibodies reflect the presence of recurrent disease, initially elevated anti-Tg antibody levels are expected to be decreased during follow-up in patients with complete therapy response [42 - 44].

Several recent studies have shown that preoperative anti-Tg antibody level measurement helps in the prediction of malignancy or adverse tumor characteristics and the assessment of prognosis in nodular goiter patients [45]. Serum anti-Tg antibody levels are increased two-fold in DTC patients compared to the general population [43]. The role of anti-Tg antibodies as a prognostic indicator is controversial. Some studies suggest that recurrent cancer is more frequent in patients with anti-Tg antibodies, whereas other studies found no significant association of anti-Tg antibodies with the stage of disease, disease-free survival or overall survival [45 - 47]. Nevertheless, serum anti-Tg antibody level measurement is not recommended as a screening test for the detection of thyroid malignancy due to limited sensitivity and specificity.

Calcitonin and Procalcitonin

Calcitonin is a 3.4 kDa polypeptide hormone with 32 amino acids [48]. It is secreted by thyroid parafollicular cells, and its main action is to reduce serum calcium levels by inhibiting osteoclastic activity and enhancing calcium and phosphate excretion by the kidney [48]. Procalcitonin, which is derived from a 141 amino acid peptide pre-procalcitonin, is the precursor peptide of calcitonin [48].

Calcitonin is the gold-standard biomarker for diagnosis and follow-up of MTC. Preoperative basal serum calcitonin level is useful to determine the disease extent and presence of lymph node metastases [22]. Calcitonin levels are generally correlated with tumor burden. Most patients with calcitonin levels of higher than 100 pg/ml have a diagnosis of MTC (normal level<10 pg/ml), lymph node metastases are seen at calcitonin levels of 10-40 pg/ml, and distant metastases are seen at calcitonin levels higher than 150 pg/ml [49]. Postoperative low serum calcitonin level is a reliable indicator of adequate surgery and patients with post-operative serum calcitonin levels of less than 10 pg/ml have a 10-year survival rate of 97.7% [50]. Conversely, postoperative high serum calcium level is an early indicator of persistent and recurrent disease [51]. However, calcitonin levels can remain detectable in many patients despite an adequate surgical approach, complicating the disease management. To overcome this problem, calcitonin doubling-time measurement was introduced as a reliable prognostic indicator of persistent or recurrent disease [52]. Also, the pentagastrin test is used for stimulation of calcitonin secretion for patients who have familial MTC and to ascertain disease-free status in MTC patients who have normal postoperative basal calcitonin levels [12].

Although calcitonin measurement is a highly sensitive method for MTC, elevated calcitonin levels are not specific for MTC, rather several stimuli such as hypercalcemia, gastrin, glucagon and catecholamines can increase calcitonin production of parafollicular cells and apart from MTC other etiologies such as non-thyroidal malignancies including lung, colon, breast, pancreas and gastric cancer, renal failure and lactation may cause elevated calcitonin levels [12, 53 -

56]. Non-thyroidal malignancies do not respond to pentagastrin or calcium stimulation and they produce less calcitonin per gram of tissue compared to MTC [22]. Procalcitonin is also not specific for MTC, but it can be produced by other neuroendocrine cells in the lung and gastrointestinal system and elevated procalcitonin levels can be seen in the presence of inflammatory stimuli, playing an important role in the diagnosis and prognosis of sepsis [53, 55]. Both somatostatin and vitamin D suppresses calcitonin production [55]. Procalcitonin is more stable and has less circadian variability compared to calcitonin [53].

Different assays are used for serum calcitonin measurement and intra-assay or inter-assay variations may lead to conflicting results [57]. Also cross-reactions with procalcitonin, the 'hook-effect', which is defined as falsely low calcitonin measurement as a result of very high serum calcitonin levels or interference with heterophilic antibodies or anti-calcitonin can lead to false measurements [57]. Immunochemiluminometric assays (ICMAs) are recommended as a highly sensitive and specific method for calcitonin measurement, and cross-reactions are less frequent with ICMA [22].

Carcinoembryonic Antigen

CEA is a highly glycosylated glycoprotein with a molecular weight of 200 kDa [58]. It was first postulated to be produced only in fetal tissue, but later it was found in small amounts in normal adult tissues as well. CEA was initially discovered in human colon cancer tissue extracts, but it is not specific to colon cancer rather it is expressed in several different cancers, including gastric, pancreatic, breast, non-small-cell lung cancer and MTC [52, 58]. Elevated CEA levels were also documented in benign diseases, such as inflammatory bowel diseases, pancreatitis, renal and hepatic insufficiency and chronic obstructive pulmonary disease, but CEA levels are generally lower in patients with benign diseases (<10 ng/ml) compared to those with malignant diseases [59 - 62]. Also, the false elevation of CEA can occur due to smoking or the presence of heterophilic antibodies [22].

Together with calcitonin, CEA is also used as a tumor marker in MTC management. Although calcitonin was shown to be better correlated with tumor burden, high basal CEA levels also reflect larger primary tumor and more lymph node metastases [63]. While a preoperative serum CEA level of higher than 30 ng/mL often suggests an extrathyroidal extension of MTC, levels higher than 100 ng/ml are generally associated with lymph node or distant organ metastases. An increasing CEA level despite stable calcitonin levels suggests dedifferentiation of MTC and an indicator of a worse prognosis [49]. MTC growth rate and disease prognosis can be assessed by measuring CEA doubling time. Therefore, together with calcitonin, CEA level measurement is recommended every 6 months to determine their doubling time [22].

Other Peptides

In MTC, increased expression of several genes encoding several hormones and peptides, including somatostatin, pro-opiomelanocortin, vasoactive intestinal peptide, gastrin-releasing peptide, neurotensin, prostaglandin, kinin, serotonin and histaminase was reported [12]. Also chromogranin-A levels were shown to be increased in MTC patients with large metastases [64]. However, these hormones or peptides are neither sensitive nor specific for the detection of MTC.

LIQUID BIOPSY

Liquid biopsy is the term called for the technique used for detection and isolation of tumor cells or their cancer-derived components circulating in body fluids to yield genomic and proteomic information about the tumor in cancer patients [65 - 67]. This new technique stems from the fact that whole tumor cells or cell-free nucleic acids released from apoptotic or necrotic cancer cells can be found in peripheral blood or other body fluids such as urine, saliva, pleural effusions or cerebrospinal fluid and can be detected using the new technologies called next-generation sequencing (NGS) and digital polymerase chain reaction (PCR) [68, 69]. Either circulating tumor cells (CTC), cell-free nucleic acids, such as cell-free tumor deoxyribonucleic acid (cfDNA), or a fraction of cfDNA called cell-free circulating tumor DNA (ctDNA), cell-free ribonucleic acids (cfRNAs), including messenger RNAs (mRNAs), ribosomal RNAs (rRNAs), transfer RNAs (tRNAs), small nuclear RNAs (snRNAs), microRNAs (miRNAs) and long noncoding RNAs (lncRNAs), also exosomes (EXO), which are nano-sized membrane-encapsulated vesicles containing proteins and nucleic acids released by tumor cells into body fluids, can be isolated using liquid biopsy technique [65, 70].

Liquid biopsy was introduced as a non-invasive alternative to current invasive sampling and diagnostic methods, such as tissue biopsy and surgical procedures, and provide additional information about the tissue heterogeneity of tumor subclones that cannot be acquired using the conventional sampling methods from a limited part of the tumor [66, 68]. To date, liquid biopsy is not used as a routine clinical test but has the potential to serve in detection and characterization of the primary tumor, as well as give prognostic information about therapy response or resistance to therapy and enable disease monitoring during follow-up [65].

The studies regarding the role of liquid biopsy in thyroid cancer detection and disease management are still limited, and study findings are heterogeneous due to different methodologies used for obtaining samples, sample storage, RNA isolation, primer selection and amplification cycles in each institution [71]. Tg mRNA and TSH receptor (TSHR) mRNA are the most common liquid biopsy biomarkers assessed in thyroid cancer. Although there is not enough evidence for the role of Tg mRNA in the preoperative diagnosis of DTC, it may have value in the follow-up of patients who have positive anti-Tg antibodies interfering with serum Tg analysis [71]. Studies regarding the role of TSHR mRNA for preoperative assessment of thyroid nodules revealed encouraging results for the diagnosis of DTC, with a sensitivity and specificity of 61-72% and 82.5-83%, respectively [71 - 74]. In one of these studies, sensitivity and specificity were reported to increase to 90% and 80%, respectively, in the subgroup with indeterminate biopsy results [72]. Also, combined usage of TSHR mRNA, Tg mRNA and fine needle aspiration biopsy results increased the sensitivity and specificity to 95% and 83%, respectively, which may help to refrain from unnecessary surgeries in nodular goiter patients [74]. Therefore, TSHR mRNA may have an additional role in patients with indeterminate biopsy results [71].

TSHR mRNA has a short half-life and preoperative high TSHR mRNA levels normalize early after total thyroidectomy. Patients with preoperative high TSHR mRNA levels were shown to have a dramatic decrease in the early postoperative period (on day 1) and post-operative persistently elevated TSHR mRNA levels predicted residual or metastatic disease in DTC patients [72, 73]. Therefore, TSHR mRNA is also a promising biomarker indicating adequate surgery and may provide additional information to Tg in patients with DTC.

The miRNAs are small non-coding RNAs with 19-25 nucleotides in length and regulate gene expression at the posttranscriptional level in both eukaryotic and prokaryotic cells [75]. Primary miRNAs (pri-miRNAs) are transcribed by the enzymes RNA polymerase II and III, and further, undergo a series of cleavage to form mature miRNA [76]. The miRNA targets mRNA and negatively regulates its expression [76]. Circulating miRNAs are either secreted by passive leakage from dead cells, by active secretion in exosomes or with RNA-binding proteins and they are protected from RNase degradation either through active secretion in vesicles or by binding to RNA-binding proteins [75, 77, 78]. Therefore, compared to mRNAs, miRNAs are more preferable for liquid biopsy as they are more resistant to degradation and more stable in biological fluids [70, 79].

Most of the studies regarding the role of circulating miRNAs in thyroid cancer are limited to DTC. Several different circulating miRNAs, including miR-146b, miR-155, miR-222, miR-151-5p, let-7e and miR-21 were found to be higher in DTC patients compared to healthy controls [70, 80 - 83]. Also, high preoperative levels of miR-146b, miR-151-5p and miR-222 were shown to decrease after thyroidectomy [80, 81]. Serum level measurements of miR-146a-5p and miR-22-

-3p were suggested to be a promising candidate for post-treatment monitoring of PTC patients [84].

CONCLUSION

Several different circulating biomarkers have been introduced for the management of thyroid cancers. While thyroglobulin is the most important biomarker for DTC, namely PTC and FTC, which originates from the follicular epithelium and retain thyroglobulin synthesis function, calcitonin is the most important biomarker for MTC, which originates from parafollicular cells. ATC is an aggressive and undifferentiated form of thyroid cancer with a poor prognosis and does not retain follicular elements or thyroglobulin production. Diagnosis of thyroid cancer relies mainly on ultrasonography and fine-needle aspiration biopsy and surgical resection is the primary treatment modality for each thyroid cancer. Circulating biomarkers have a limited role in the diagnosis of the disease and are mainly used for assessment of adequate surgery, treatment monitoring and detection of recurrent disease during follow-up of thyroid cancer patients.

CONSENT FOR PUBLICATION

Not applicable.

CONFLICT OF INTEREST

The author declares no conflict of interest, financial or otherwise.

ACKNOWLEDGEMENT

Declared none.

REFERENCES

[1] Olson E, Wintheiser G, Wolfe KM, Droessler J, Silberstein PT. Epidemiology of Thyroid Cancer: A Review of the National Cancer Database, 2000-2013. Cureus 2019; 11(2): e4127. [http://dx.doi.org/10.7759/cureus.4127] [PMID: 31049276]

[2] Di Jeso B, Arvan P. Thyroglobulin From Molecular and Cellular Biology to Clinical Endocrinology. Endocr Rev 2016; 37(1): 2-36. [http://dx.doi.org/10.1210/er.2015-1090] [PMID: 26595189]

[3] Junqueira LC, Carneirai J. Adrenals, Islets of Langerhans, Thyroid, Parathyroids and Pineal Gland. Tenth Edit Basic Histol. Tenth Ed., McGraw-Hill: Lange Medical Books 2003.

[4] Maitra A, Kumar V. The Endocrine System. In: Kumar V, Cotran R, Robbins S, Eds. Robbins Basic Pathol. 7th ed. Saunders 2003; pp. 719-54.

[5] Cabanillas ME, McFadden DG, Durante C. Thyroid cancer. Lancet 2016; 388(10061): 2783-95. [http://dx.doi.org/10.1016/S0140-6736(16)30172-6] [PMID: 27240885]

[6] Kitamura Y, Shimizu K, Nagahama M, *et al.* Immediate causes of death in thyroid carcinoma: clinicopathological analysis of 161 fatal cases. J Clin Endocrinol Metab 1999; 84(11): 4043-9.

[http://dx.doi.org/10.1210/jcem.84.11.6115] [PMID: 10566647]

[7] Abdullah MI, Junit SM, Ng KL, Jayapalan JJ, Karikalan B, Hashim OH. Papillary thyroid cancer: genetic alterations and molecular biomarker investigations. Int J Med Sci 2019; 16(3): 450-60. [http://dx.doi.org/10.7150/ijms.29935] [PMID: 30911279]

[8] Nikiforova MN, Tseng GC, Steward D, Diorio D, Nikiforov YE. MicroRNA expression profiling of thyroid tumors: biological significance and diagnostic utility. J Clin Endocrinol Metab 2008; 93(5): 1600-8. [http://dx.doi.org/10.1210/jc.2007-2696] [PMID: 18270258]

[9] Younis E. Oncogenesis of thyroid cancer. Asian Pac J Cancer Prev 2017; 18(5): 1191-9. [PMID: 28610401]

[10] Nikiforova MN, Lynch RA, Biddinger PW, *et al.* RAS point mutations and PAX8-PPAR γ rearrangement in thyroid tumors: evidence for distinct molecular pathways in thyroid follicular carcinoma. J Clin Endocrinol Metab 2003; 88(5): 2318-26. [http://dx.doi.org/10.1210/jc.2002-021907] [PMID: 12727991]

[11] Nikiforov YE. Thyroid carcinoma: molecular pathways and therapeutic targets. Mod Pathol 2008; 21(S2) (Suppl. 2): S37-43. [http://dx.doi.org/10.1038/modpathol.2008.10] [PMID: 18437172]

[12] Leboulleux S, Baudin E, Travagli J-P, Schlumberger M. Medullary thyroid carcinoma. Clin Endocrinol (Oxf) 2004; 61(3): 299-310. [http://dx.doi.org/10.1111/j.1365-2265.2004.02037.x] [PMID: 15355445]

[13] A T, F S, G P, M B. Genetic alterations in medullary thyroid cancer: diagnostic and prognostic markers. Curr Genomics 2011; 12(8): 618-25. [http://dx.doi.org/10.2174/138920211798120835] [PMID: 22654561]

[14] Marques AR, Espadinha C, Catarino AL, *et al.* Expression of PAX8-PPAR γ 1 rearrangements in both follicular thyroid carcinomas and adenomas. J Clin Endocrinol Metab 2002; 87(8): 3947-52. [http://dx.doi.org/10.1210/jc.87.8.3947] [PMID: 12161538]

[15] Moley JF. Medullary thyroid carcinoma: management of lymph node metastases. J Natl Compr Canc Netw 2010; 8(5): 549-56. [http://dx.doi.org/10.6004/jnccn.2010.0042] [PMID: 20495084]

[16] Hundahl SA, Fleming ID, Fremgen AM, Menck HR. A National Cancer Data Base report on 53,856 cases of thyroid carcinoma treated in the U.S., 1985-1995 [see commetns]. Cancer 1998; 83(12): 2638-48. [http://dx.doi.org/10.1002/(SICI)1097-0142(19981215)83:12<2638::AID-CNCR31>3.0.CO;2-1] [PMID: 9874472]

[17] Rashid M, Agarwal A, Pradhan R, *et al.* Genetic alterations in anaplastic thyroid carcinoma. Indian J Endocrinol Metab 2019; 23(4): 480-5. [http://dx.doi.org/10.4103/ijem.IJEM_321_19] [PMID: 31741910]

[18] Charles R-P, Silva J, Iezza G, Phillips WA, McMahon M. Activating BRAF and PIK3CA mutations cooperate to promote anaplastic thyroid carcinogenesis. Mol Cancer Res 2014; 12(7): 979-86. [http://dx.doi.org/10.1158/1541-7786.MCR-14-0158-T] [PMID: 24770869]

[19] Santarpia L, El-Naggar AK, Cote GJ, Myers JN, Sherman SI. Phosphatidylinositol 3-kinase/akt and ras/raf-mitogen-activated protein kinase pathway mutations in anaplastic thyroid cancer. J Clin Endocrinol Metab 2008; 93(1): 278-84. [http://dx.doi.org/10.1210/jc.2007-1076] [PMID: 17989125]

[20] Manzella L, Stella S, Pennisi MS, *et al.* New Insights in Thyroid Cancer and p53 Family Proteins. Int J Mol Sci 2017; 18(6): 1325. [http://dx.doi.org/10.3390/ijms18061325] [PMID: 28635633]

[21] Haugen BR, Alexander EK, Bible KC, *et al.* 2015 American Thyroid Association Management

Guidelines for Adult Patients with Thyroid Nodules and Differentiated Thyroid Cancer: The American Thyroid Association Guidelines Task Force on Thyroid Nodules and Differentiated Thyroid Cancer. Thyroid 2016; 26(1): 1-133.
[http://dx.doi.org/10.1089/thy.2015.0020] [PMID: 26462967]

[22] Wells SA Jr, Asa SL, Dralle H, *et al.* Revised American Thyroid Association guidelines for the management of medullary thyroid carcinoma. Thyroid 2015; 25(6): 567-610.
[http://dx.doi.org/10.1089/thy.2014.0335] [PMID: 25810047]

[23] Saini S, Tulla K, Maker AV, Burman KD, Prabhakar BS. Therapeutic advances in anaplastic thyroid cancer: a current perspective. Mol Cancer 2018; 17(1): 154.
[http://dx.doi.org/10.1186/s12943-018-0903-0] [PMID: 30352606]

[24] Burman KD. Is poorly differentiated thyroid cancer poorly characterized? J Clin Endocrinol Metab 2014; 99(4): 1167-9.
[http://dx.doi.org/10.1210/jc.2014-1549] [PMID: 24702015]

[25] Smallridge RC, Ain KB, Asa SL, *et al.* American Thyroid Association guidelines for management of patients with anaplastic thyroid cancer. Thyroid 2012; 22(11): 1104-39.
[http://dx.doi.org/10.1089/thy.2012.0302] [PMID: 23130564]

[26] Algeciras-Schimnich A. Thyroglobulin measurement in the management of patients with differentiated thyroid cancer. Crit Rev Clin Lab Sci 2018; 55(3): 205-18.
[http://dx.doi.org/10.1080/10408363.2018.1450830] [PMID: 29546779]

[27] Grebe SK. Diagnosis and management of thyroid carcinoma: a focus on serum thyroglobulin. Expert Rev Endocrinol Metab 2009; 4(1): 25-43.
[http://dx.doi.org/10.1586/17446651.4.1.25]

[28] Torréns JI, Burch HB. Serum thyroglobulin measurement. Utility in clinical practice. Endocrinol Metab Clin North Am 2001; 30(2): 429-67.
[http://dx.doi.org/10.1016/S0889-8529(05)70194-8] [PMID: 11444170]

[29] Lee EK, Chung K-W, Min HS, *et al.* Preoperative serum thyroglobulin as a useful predictive marker to differentiate follicular thyroid cancer from benign nodules in indeterminate nodules. J Korean Med Sci 2012; 27(9): 1014-8.
[http://dx.doi.org/10.3346/jkms.2012.27.9.1014] [PMID: 22969246]

[30] Suh I, Vriens MR, Guerrero MA, *et al.* Serum thyroglobulin is a poor diagnostic biomarker of malignancy in follicular and Hurthle-cell neoplasms of the thyroid. Am J Surg 2010; 200(1): 41-6.
[http://dx.doi.org/10.1016/j.amjsurg.2009.08.030] [PMID: 20637335]

[31] Ma ZF, Skeaff SA. Thyroglobulin as a biomarker of iodine deficiency: a review. Thyroid 2014; 24(8): 1195-209.
[http://dx.doi.org/10.1089/thy.2014.0052] [PMID: 24762031]

[32] Glinoer D, Puttemans N, Van Herle AJ, Camus M, Ermans AM. Sequential study of the impairment of thyroid function in the early stage of subacute thyroiditis. Acta Endocrinol (Copenh) 1974; 77(1): 26-34.
[http://dx.doi.org/10.1530/acta.0.0770026] [PMID: 4137186]

[33] Yim JH, Kim EY, Bae Kim W, *et al.* Long-term consequence of elevated thyroglobulin in differentiated thyroid cancer. Thyroid 2013; 23(1): 58-63.
[http://dx.doi.org/10.1089/thy.2011.0487] [PMID: 22973946]

[34] Pelttari H, Välimäki MJ, Löyttyniemi E, Schalin-Jäntti C. Post-ablative serum thyroglobulin is an independent predictor of recurrence in low-risk differentiated thyroid carcinoma: a 16-year follow-up study. Eur J Endocrinol 2010; 163(5): 757-63. https://eje.bioscientifica.com/view/journals/eje/163/5/757.xml
[http://dx.doi.org/10.1530/EJE-10-0553] [PMID: 20813788]

[35] Klubo-Gwiezdzinska J, Burman KD, Van Nostrand D, Wartofsky L. Does an undetectable rhTSH-

stimulated Tg level 12 months after initial treatment of thyroid cancer indicate remission? Clin Endocrinol (Oxf) 2011; 74(1): 111-7. [http://dx.doi.org/10.1111/j.1365-2265.2010.03898.x] [PMID: 21054476]

[36] Pellegriti G, Scollo C, Regalbuto C, *et al.* The diagnostic use of the rhTSH/thyroglobulin test in differentiated thyroid cancer patients with persistent disease and low thyroglobulin levels. Clin Endocrinol (Oxf) 2003; 58(5): 556-61. [http://dx.doi.org/10.1046/j.1365-2265.2003.01744.x] [PMID: 12699436]

[37] Tuttle RM, Tala H, Shah J, *et al.* 2010. https://www.liebertpub.com/doi/10.1089/thy.2010.0178

[38] Vaisman F, Momesso D, Bulzico DA, *et al.* Spontaneous remission in thyroid cancer patients after biochemical incomplete response to initial therapy. Clin Endocrinol (Oxf) 2012; 77(1): 132-8. [http://dx.doi.org/10.1111/j.1365-2265.2012.04342.x] [PMID: 22248037]

[39] Pacini F, Schlumberger M, Dralle H, Elisei R, Smit JWA, Wiersinga W. European consensus for the management of patients with differentiated thyroid carcinoma of the follicular epithelium. Eur J Endocrinol 2006; 154(6): 787-803. https://eje.bioscientifica.com/view/journals/eje/154/6/1540787.xml [http://dx.doi.org/10.1530/eje.1.02158] [PMID: 16728537]

[40] Torréns JI, Burch HB. Serum thyroglobulin measurement. Utility in clinical practice. Endocrinol Metab Clin North Am 2001; 30(2): 429-67. [http://dx.doi.org/10.1016/S0889-8529(05)70194-8] [PMID: 11444170]

[41] Spencer C, Petrovic I, Fatemi S, LoPresti J. Serum thyroglobulin (Tg) monitoring of patients with differentiated thyroid cancer using sensitive (second-generation) immunometric assays can be disrupted by false-negative and false-positive serum thyroglobulin autoantibody misclassifications. J Clin Endocrinol Metab 2014; 99(12): 4589-99. https://academic.oup.com/jcem/article/99/12/4589/2834392 [http://dx.doi.org/10.1210/jc.2014-1203] [PMID: 25226290]

[42] Spencer CA, Bergoglio LM, Kazarosyan M, Fatemi S, LoPresti JS. Clinical impact of thyroglobulin (Tg) and Tg autoantibody method differences on the management of patients with differentiated thyroid carcinomas. J Clin Endocrinol Metab 2005; 90(10): 5566-75. [http://dx.doi.org/10.1210/jc.2005-0671] [PMID: 15985472]

[43] Spencer CA, Takeuchi M, Kazarosyan M, *et al.* Serum thyroglobulin autoantibodies: prevalence, influence on serum thyroglobulin measurement, and prognostic significance in patients with differentiated thyroid carcinoma. J Clin Endocrinol Metab 1998; 83(4): 1121-7. [http://dx.doi.org/10.1210/jc.83.4.1121] [PMID: 9543128]

[44] Chung J-K, Park YJ, Kim TY, *et al.* Clinical significance of elevated level of serum antithyroglobulin antibody in patients with differentiated thyroid cancer after thyroid ablation. Clin Endocrinol (Oxf) 2002; 57(2): 215-21. [http://dx.doi.org/10.1046/j.1365-2265.2002.01592.x] [PMID: 12153600]

[45] Jo K, Lim DJ. Clinical implications of anti-thyroglobulin antibody measurement before surgery in thyroid cancer. Korean J Intern Med (Korean Assoc Intern Med) 2018; 33(6): 1050-7. [http://dx.doi.org/10.3904/kjim.2018.289] [PMID: 30396251]

[46] Durante C, Tognini S, Montesano T, *et al.* Clinical aggressiveness and long-term outcome in patients with papillary thyroid cancer and circulating anti-thyroglobulin autoantibodies. Thyroid 2014; 24(7): 1139-45. [http://dx.doi.org/10.1089/thy.2013.0698] [PMID: 24702238]

[47] McLeod DSA, Cooper DS, Ladenson PW, *et al.* Prognosis of differentiated thyroid cancer in relation to serum thyrotropin and thyroglobulin antibody status at time of diagnosis. Thyroid 2014; 24(1): 35-42. [http://dx.doi.org/10.1089/thy.2013.0062] [PMID: 23731273]

[48] Masi L, Brandi ML. Calcitonin and calcitonin receptors. Clin Cases Miner Bone Metab 2007; 4(2): 117-22.

[PMID: 22461211]

[49] Chen H, Sippel RS, O'Dorisio MS, Vinik AI, Lloyd RV, Pacak K. The North American Neuroendocrine Tumor Society consensus guideline for the diagnosis and management of neuroendocrine tumors: pheochromocytoma, paraganglioma, and medullary thyroid cancer. Pancreas 2010; 39(6): 775-83. [http://dx.doi.org/10.1097/MPA.0b013e3181ebb4f0] [PMID: 20664475]

[50] Modigliani E, Cohen R, Campos J-M, *et al.* Prognostic factors for survival and for biochemical cure in medullary thyroid carcinoma: results in 899 patients. The GETC Study Group. Groupe d'étude des tumeurs à calcitonine. Clin Endocrinol (Oxf) 1998; 48(3): 265-73. [http://dx.doi.org/10.1046/j.1365-2265.1998.00392.x] [PMID: 9578814]

[51] Pellegriti G, Leboulleux S, Baudin E, *et al.* Long-term outcome of medullary thyroid carcinoma in patients with normal postoperative medical imaging. Br J Cancer 2003; 88(10): 1537-42. [http://dx.doi.org/10.1038/sj.bjc.6600930] [PMID: 12771918]

[52] Meijer JAA, le Cessie S, van den Hout WB, *et al.* Calcitonin and carcinoembryonic antigen doubling times as prognostic factors in medullary thyroid carcinoma: a structured meta-analysis. Clin Endocrinol (Oxf) 2010; 72(4): 534-42. [http://dx.doi.org/10.1111/j.1365-2265.2009.03666.x] [PMID: 19563448]

[53] Karagiannis AKA, Girio-Fragkoulakis C, Nakouti T. Procalcitonin: A new biomarker for medullary thyroid cancer? A systematic review. Anticancer Res 2016; 36(8): 3803-10. [PMID: 27466480]

[54] Schwartz KE, Wolfsen AR, Forster B, Odell WD. Calcitonin in nonthyroidal cancer. J Clin Endocrinol Metab 1979; 49(3): 438-44. [http://dx.doi.org/10.1210/jcem-49-3-438] [PMID: 468977]

[55] Maruna P, Nedelníková K, Gürlich R. Physiology and genetics of procalcitonin. Physiol Res 2000; 49 (Suppl. 1): S57-61. [Internet]. [PMID: 10984072]

[56] Toledo SP, Lourenço DM Jr, Santos MA, Tavares MR, Toledo RA, Correia-Deur JE de M. Hypercalcitoninemia is not pathognomonic of medullary thyroid carcinoma. Clinics (São Paulo) 2009; 64(7): 699-706. [http://dx.doi.org/10.1590/S1807-59322009000700015] [PMID: 19606248]

[57] Camacho CP, Lindsey SC, Melo MCC, *et al.* Measurement of calcitonin and calcitonin gene-related peptide mRNA refines the management of patients with medullary thyroid cancer and may replace calcitonin-stimulation tests. Thyroid 2013; 23(3): 308-16. [http://dx.doi.org/10.1089/thy.2012.0361] [PMID: 23259706]

[58] Hammarström S. The carcinoembryonic antigen (CEA) family: structures, suggested functions and expression in normal and malignant tissues. Semin Cancer Biol 1999; 9(2): 67-81. [http://dx.doi.org/10.1006/scbi.1998.0119] [PMID: 10202129]

[59] Loewenstein MS, Zamcheck N. Carcinoembryonic antigen (CEA) levels in benign gastrointestinal disease states. Cancer 1978; 42(3) (Suppl.): 1412-8. [http://dx.doi.org/10.1002/1097-0142(197809)42:3+<1412::AID-CNCR2820420805>3.0.CO;2-8] [PMID: 361200]

[60] Ruibal Morell A. CEA serum levels in non-neoplastic disease. Int J Biol Markers 1992; 7(3): 160-6. [http://dx.doi.org/10.1177/172460089200700307] [PMID: 1431339]

[61] Rule AH, Goleski-Reilly C, Sachar DB, Vandevoorde J, Janowitz HD. Circulating carcinoembryonic antigen (CEA): relationship to clinical status of patients with inflammatory bowel disease. Gut 1973; 14(11): 880-4. [http://dx.doi.org/10.1136/gut.14.11.880] [PMID: 4761608]

[62] Bulut I, Arbak P, Coskun A, *et al.* Comparison of serum CA 19.9, CA 125 and CEA levels with

severity of chronic obstructive pulmonary disease. Med Princ Pract 2009; 18(4): 289-93. [http://dx.doi.org/10.1159/000215726] [PMID: 19494536]

[63] Machens A, Dralle H. Biomarker-based risk stratification for previously untreated medullary thyroid cancer. J Clin Endocrinol Metab 2010; 95(6): 2655-63. [http://dx.doi.org/10.1210/jc.2009-2368] [PMID: 20339026]

[64] Guignat L, Bidart JM, Nocera M, Comoy E, Schlumberger M, Baudin E. Chromogranin A and the α-subunit of glycoprotein hormones in medullary thyroid carcinoma and phaeochromocytoma. Br J Cancer 2001; 84(6): 808-12. [http://dx.doi.org/10.1054/bjoc.2000.1677] [PMID: 11259096]

[65] Palmirotta R, Lovero D, Cafforio P, *et al.* Liquid biopsy of cancer: a multimodal diagnostic tool in clinical oncology. Ther Adv Med Oncol 2018; 10: 1758835918794630. [http://dx.doi.org/10.1177/1758835918794630] [PMID: 30181785]

[66] Perakis S, Speicher MR. Emerging concepts in liquid biopsies. BMC Med 2017; 15(1): 75. [http://dx.doi.org/10.1186/s12916-017-0840-6] [PMID: 28381299]

[67] Alix-Panabières C. The future of liquid biopsy. Nature 2020; 579(7800): S9. [http://dx.doi.org/10.1038/d41586-020-00844-5] [PMID: 32214259]

[68] Siravegna G, Marsoni S, Siena S, Bardelli A. Integrating liquid biopsies into the management of cancer. Nat Rev Clin Oncol 2017; 14(9): 531-48. [http://dx.doi.org/10.1038/nrclinonc.2017.14] [PMID: 28252003]

[69] Cai X, Janku F, Zhan Q, Fan JB. Accessing genetic information with liquid biopsies. Trends Genet 2015; (10): 564-75. [http://dx.doi.org/10.1016/j.tig.2015.06.001]

[70] Durante C, Rosignolo F, Sponziello M, Verrienti A, Filetti S. Circulating molecular biomarkers in thyroid cancer.Atlas Thyroid Neuroendocr Tumor Markers. Cham: Springer International Publishing 2018; pp. 93-104. [http://dx.doi.org/10.1007/978-3-319-62506-5_6]

[71] Nixon AM, Provatopoulou X, Kalogera E, Zografos GN, Gounaris A. Circulating thyroid cancer biomarkers: Current limitations and future prospects. Clin Endocrinol (Oxf) 2017; 87(2): 117-26. [http://dx.doi.org/10.1111/cen.13369] [PMID: 28493290]

[72] Chia S-Y, Milas M, Reddy SK, *et al.* Thyroid-stimulating hormone receptor messenger ribonucleic acid measurement in blood as a marker for circulating thyroid cancer cells and its role in the preoperative diagnosis of thyroid cancer. J Clin Endocrinol Metab 2007; 92(2): 468-75. [http://dx.doi.org/10.1210/jc.2006-2088] [PMID: 17118994]

[73] Milas M, Shin J, Gupta M, *et al.* Circulating thyrotropin receptor mRNA as a novel marker of thyroid cancer: clinical applications learned from 1758 samples. Ann Surg 2010; 252(4): 643-51. [http://dx.doi.org/10.1097/SLA.0b013e3181f5ba51] [PMID: 20881771]

[74] Wagner K, Arciaga R, Siperstein A, *et al.* Thyrotropin receptor/thyroglobulin messenger ribonucleic acid in peripheral blood and fine-needle aspiration cytology: diagnostic synergy for detecting thyroid cancer. J Clin Endocrinol Metab 2005; 90(4): 1921-4. [http://dx.doi.org/10.1210/jc.2004-1793] [PMID: 15687333]

[75] Boufraqech M, Klubo-Gwiezdzinska J, Kebebew E. MicroRNAs in the thyroid. Best Pract Res Clin Endocrinol Metab 2016; 30(5): 603-19. [http://dx.doi.org/10.1016/j.beem.2016.10.001] [PMID: 27923454]

[76] Macfarlane L-A, Murphy PR. MicroRNA: Biogenesis, Function and Role in Cancer. Curr Genomics 2010; 11(7): 537-61. [http://dx.doi.org/10.2174/138920210793175895] [PMID: 21532838]

[77] Valadi H, Ekström K, Bossios A, Sjöstrand M, Lee JJ, Lötvall JO. Exosome-mediated transfer of mRNAs and microRNAs is a novel mechanism of genetic exchange between cells. Nat Cell Biol 2007;

9(6): 654-9.
[http://dx.doi.org/10.1038/ncb1596] [PMID: 17486113]

[78] Kosaka N, Iguchi H, Ochiya T. Circulating microRNA in body fluid: a new potential biomarker for cancer diagnosis and prognosis. Cancer Sci 2010; 101(10): 2087-92.
[http://dx.doi.org/10.1111/j.1349-7006.2010.01650.x] [PMID: 20624164]

[79] El-Hefnawy T, Raja S, Kelly L, *et al.* Characterization of amplifiable, circulating RNA in plasma and its potential as a tool for cancer diagnostics. Clin Chem 2004; 50(3): 564-73.
[http://dx.doi.org/10.1373/clinchem.2003.028506] [PMID: 14718398]

[80] Lee JC, Zhao JT, Clifton-Bligh RJ, *et al.* MicroRNA-222 and microRNA-146b are tissue and circulating biomarkers of recurrent papillary thyroid cancer. Cancer 2013; 119(24): 4358-65.
[http://dx.doi.org/10.1002/cncr.28254] [PMID: 24301304]

[81] Yu S, Liu Y, Wang J, *et al.* Circulating microRNA profiles as potential biomarkers for diagnosis of papillary thyroid carcinoma. J Clin Endocrinol Metab 2012; 97(6): 2084-92.https://academic.oup.com/jcem/article/97/6/2084/2536713
[http://dx.doi.org/10.1210/jc.2011-3059] [PMID: 22472564]

[82] Lee YS, Lim YS, Lee J-C, *et al.* Differential expression levels of plasma-derived miR-146b and miR-155 in papillary thyroid cancer. Oral Oncol 2015; 51(1): 77-83.
[http://dx.doi.org/10.1016/j.oraloncology.2014.10.006] [PMID: 25456009]

[83] Yoruker EE, Terzioglu D, Teksoz S, Uslu FE, Gezer U, Dalay N. MicroRNA Expression Profiles in Papillary Thyroid Carcinoma, Benign Thyroid Nodules and Healthy Controls. J Cancer 2016; 7(7): 803-9.http://www.jcancer.org/v07p0803.htm
[http://dx.doi.org/10.7150/jca.13898] [PMID: 27162538]

[84] Rosignolo F, Sponziello M, Giacomelli L, *et al.* Identification of Thyroid-Associated Serum microRNA Profiles and Their Potential Use in Thyroid Cancer Follow-Up. J Endocr Soc 2017; 1(1): 3-13.
[PMID: 29264441]

CHAPTER 5

DNA Methylation Biomarkers in Cancer: Current Clinical Utility and Future Perspectives

Sinem Durmus[1,*], **Remise Gelisgen**[1] and **Hafize Uzun**[2]

[1] *Department of Biochemistry, Cerrahpasa Faculty of Medicine, Istanbul University-Cerrahpasa, Istanbul, Turkey*

[2] *Department of Biochemistry, Faculty of Medicine, Istanbul Atlas University, Istanbul, Turkey*

Abstract: Epigenetic alterations are related to inherited but reversible changes in modifications that regulate gene activity beyond the DNA sequence. DNA methylation is the best characterized epigenetic modification, controlling DNA stability, DNA structure, transcription, and regulation, contributing to normal development and differentiation. In this section, we first discuss the cellular functions of DNA methylation and focus on how this fundamental biological process is impaired in cancer. Changes in DNA methylation status in cancer have been heralded as promising targets for the development of diagnostic, prognostic, and predictive biomarkers due to their noninvasive accessibility in bodily fluids (such as blood, urine, stool), reversibility, stability, and frequency. The absence of markers for definitive diagnosis of most types of cancer and, in some cases, DNA methylation biomarkers being more specific and sensitive than commonly used protein biomarkers indicate a strong need for continued research to expand DNA methylation markers. Although the information on changes in DNA methylation status in cancer and research on its clinical relevance is rapidly increasing, the number of DNA methylation biomarkers currently available as commercial tests is very small. Here, we focus on the importance of DNA methylation location and target genes likely to be developed in the future for the development of biomarkers in addition to existing commercial tests. Following a detailed study of possible target genes, we summarize the current clinical application status of the most studied and validated DNA methylation biomarkers, including *SEPT9, SDC2, BMP3, NDRG4, SFRP2, TFPI2, VIM* and *MGMT*.

Keywords: Biomarkers, Cancer, Carcinogenesis, CpG islands, DNA Damage, DNA Methylation, Genes, Gene Expression, Gene Silencing, Hypomethylation, Hypermethylation, Methyltransferases, Neoplasms, Oncogenes, Tumor, Tumor Suppressor.

[*] **Corresponding author Sinem Durmus:** Department of Medical Biochemistry, Cerrahpaşa Faculty of Medicine, Istanbul University-Cerrahpasa Istanbul, Turkey; Tel: +90 (538) 729 79 29; E-mail: durmus.sinem@gmail.com

Pınar Atukeren and Hafize Uzun (Eds.)

INTRODUCTION

Although cancer has been defined as a disease caused by mutations in driver genes for a long time, it is now accepted that epigenetic modifications also play a role in both the development and progression of the tumor [1]. Epigenetics is the field of science that studies gene activation or genetic changes in its function through processes, such as histone modification, DNA methylation, and acetylation without changing the DNA sequence. Although all cells in an organism contain the same genetic material, the changes that lead to the emergence of a wide variety of cell types are due to these epigenetic mechanisms [2]. In other words, epigenetic mechanisms are very important regulators for normal development and differentiation [3]. However, as can be predicted, the defects in this regulatory mechanism also lead to the emergence of various diseases such as cancer. If the cancer is to be mentioned, which has a complicated etiology, it should be noted that epigenome dysregulation, especially DNA methylation, is quite common in cancer and is associated with tumorigenesis and cancer phenotype [4]. Identifying epigenetic changes, such as DNA methylation and determining parameters that can be used as biomarkers, are extremely important, especially in determining the focal points of promising personalized treatment approaches. The main point of importance for these therapies is that, unlike genetic changes, such as mutations, epigenetic changes are inherited but reversible. In addition, DNA methylation markers are very important in terms of detecting diseases, such as cancer at an early stage.

DNA METHYLATION AND REGULATION MECHANISM

The best-known type of DNA methylation is defined as the formation of 5-methylcytosine (5mC) by attaching the methyl group (-CH_3) to the 5'-position of the pyrimidine ring of the cytosine (Fig. **1**). This methylation can occur in cytosines in the context of CpG, CHG, and CHH (where H refers to either A, C or T nucleotides), but in humans, it usually occurs in palindromic CG (or CpG) dinucleotides, and more rarely in non-CpG sequences [5, 6]. In recent years, it has been shown that N^6-methyladenine (N^6-mA) exists in the mammalian genome. Although it has been stated that this methylation may play a role in epigenetic regulation, its function has not been clearly clarified yet [7, 8]. 5-methylcytosine tends to spontaneously deaminate and undergo C-T transition. Because of this high mutagenicity, the distribution of CpG dinucleotides in the genome is thought to be unevenly distributed throughout the evolutionary process (also called CpG suppression) and is observed less frequently in the genome than expected. However, there are clusters of CpG sites, and such clusters are referred to as CpG islands (CGIs). CGIs are found in the promoter regions of 50% of the genes in the genome and are not methylated in normal cells, thereby transcription occurs.

Approximately 70-80% of the CpG sites are methylated; however, only 10% of the CpGs are methylated in the CGIs and active regulatory regions. These are CGIs in the promoter region of repressed genes such as X chromosome inactivated genes, imprinted genes and some tissue specific genes [6, 9 - 11]. Methylated cytosines do not occur randomly in the genome, they are largely found in transposable elements and repetitive DNA sequences, including centromeric and pericentric satellite DNA [12].

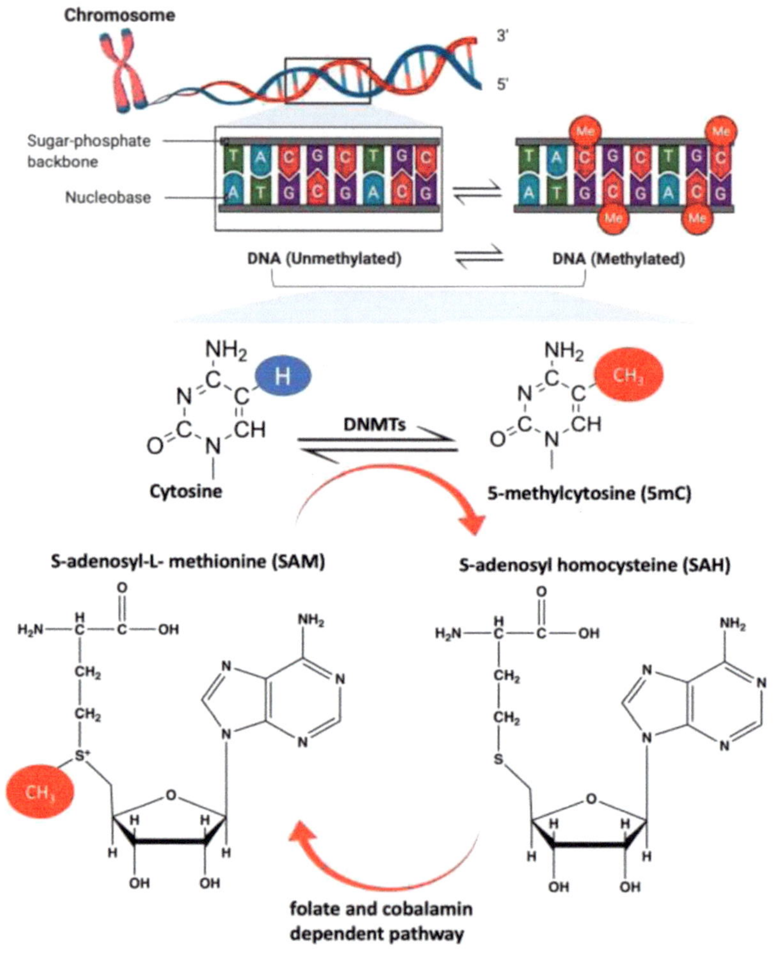

Fig. (1). DNA methylation. Formation of 5-methylcytosine (5mC) by attaching the methyl group from the methyl carrier, S-adenosyl-L- methionine (SAM), to the 5'-position of the pyrimidine ring of the cytosine.

DNA methylation is catalyzed by the DNA methyltransferase (DNMT) family that mediate the covalent addition of a methyl group from the methyl carrier, S-adenosyl-L- methionine (SAM) to the 5th carbon of a cytosine residue to form 5mC. After SAM donates the methyl group, it is converted to S-adenosyl homocysteine (SAH), which inhibits activity of DNMTs [13]. SAH is also recycled to SAM in a folate and cobalamin-dependent pathway (Fig. **1**).

There are 5 known DNMT enzymes in this family and these enzymes have different biochemical and biological functions due to their structural differences: DNMT1, DNMT2, DNMT3A, DNMT3B and DNMT3L. Among these five enzymes, the catalytically active ones are DNMT1, DNMT3A and DNMT3B. Of the others, DNMT2 functions as RNA methyltransferase while DNMT3L lacks the catalytic domain and regulates the activity of DNMT3A as a stimulating factor [4]. DNMT1 methylated hemi methylated DNA during replication, whereas DNMT3A and DNMT3B catalyze de novo DNA methylation. After DNA methylation patterns are determined by de novo type DNA methyltransferases (DNMT3A and DNMT3B) during early embryogenesis, they are transferred by maintenance DNMT (DNMT1) in somatic cells in a cell type specific manner [12, 14] (Fig. **2**).

The existence and working principle of the DNMT1 enzyme also explains the inheritance of methylation patterns; if only one strand remains methylated, the signal is lost in half of the daughter cells after replication. Also, another function of the DNMT1 enzyme is to repair DNA methylation, so it is also known as maintenance DNMT [2, 4]. DNMT3A and DNMT3B enzymes differ in their expression levels. Although DNMT3A is expressed in relatively most adult tissues, DNMT3B expression levels are low in most differentiated tissues except thyroid, testicles and bone marrow [15].

DNA methylation patterns formed by DNA demethylation, sex-specific de novo DNA methylation and cell-type specific de novo DNA methylations during gametogenesis and embryogenesis are maintained by the DNMT1 throughout lifetime [16]. These enzymes have different specialized regions to bind DNA: In DNMT3, the PWWP region binds DNA, while the zinc finger domain (CXXC) of DNMT1 binds unmethylated CpGs, and residues in the catalytic region are in sequence independent interactions with the negatively charged DNA backbone [17]. In addition, one important protein is required for DNMT1 binding to DNA: UHRF1. The SRA domain of UHRF1 recognizes and binds newly replicated hemi-methylated DNA. This interaction promotes the recruitment of DNMT1, and methylation occurs *via* the DNMT1/UHRF1 complex [18].

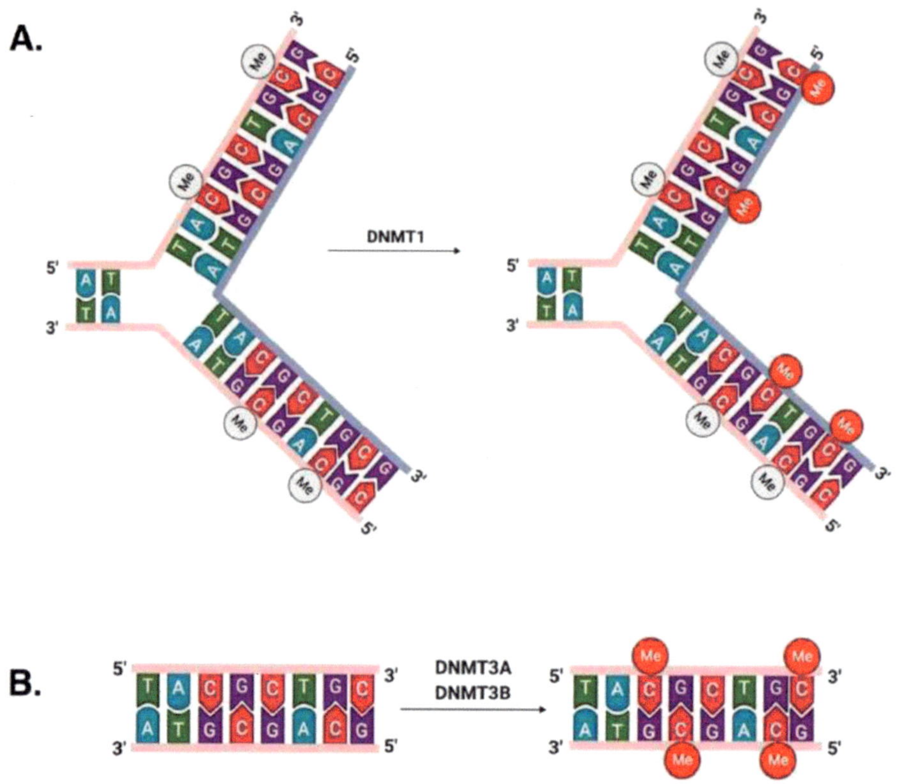

Fig. (2). DNA methylation pathways. A. During replication, DNA undergoes semi-conservative replication, hemi-methylated parental DNA (pink) retains the original DNA methylation pattern (gray methyl groups). DNMT1 (maintenance DNMT) adds methyl groups (red methyl groups) to the newly synthesized daughter strand. B. DNMT3A and DNMT3B catalyze de novo DNA methylation, and transfer methyl groups (red) onto naked DNA.

One of the regulation mechanisms is DNA demethylation. Although its mechanism is still not clearly known, it has been reported that DNA demethylation takes place in two different forms: active and passive demethylation. Passive 'DNA demethylation means that DNA methylation patterns cannot be maintained during several rounds of replication of newly synthesized DNA strands and is believed to result in replication-dependent dilution of 5mC [18]. The passive process takes place due to reduced activation or absence of DNMTs [19]. “Active” DNA demethylation refers to the conversion of 5mC to unmodified cytosine by the sequence of enzymatic reactions. In this process, ten eleven translocation (TET) enzymes, thymine – DNA – glycosylase (TDG) and DNA base excision repair (BER) mechanisms are involved. First, TET enzymes oxidize 5mC to catalyze the formation of 5-hydroxymethylcytosine (5hmC), 5-formylcytosine (5fC) and 5- carboxylcytosine (5caC), respectively.

TDG then excises 5fC and 5caC (not 5mC and 5hmC) from DNA. The resulting abasic region is repaired with BER [18, 20] (Fig. **3**).

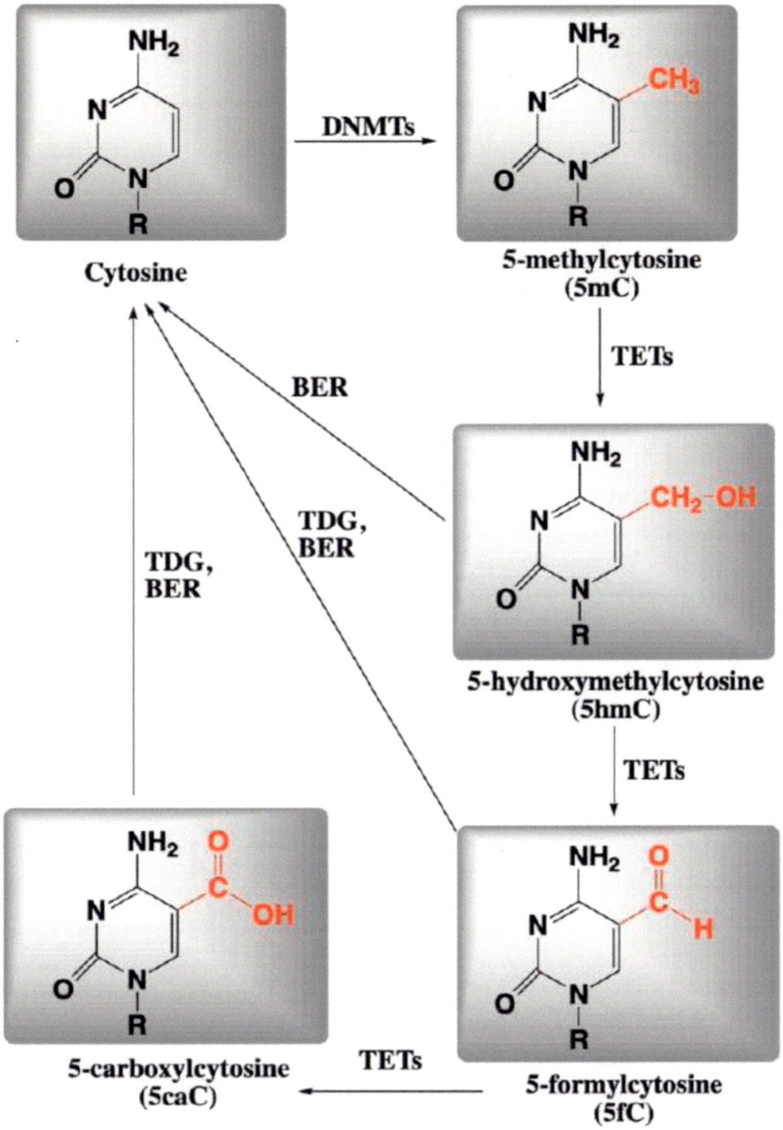

Fig. (3). DNA demethylation.

DNA Methylation in Cancer

DNA methylation regulation is known to be impaired in cancer, and the resulting altered patterns are categorized into two types: regional hypermethylation and global hypomethylation. These altered patterns can occur in CGI promoters, CpG island shore, enhancers, repetitive sequences (retrotransposons, *Arthrobacter luteus* (Alu) and long interspersed nuclear element 1 (LINE1), and centromeric satellite DNA, α-satellite) and gene bodies [21, 22].

Regional Hypermethylation

Aberrant hypermethylation (regional hypermethylation) of CGI promoters is by far the best categorized epigenetic change occurring in cancer cells and is associated with silencing of genes involved in cell proliferation, promoting apoptosis or cell differentiation, DNA repair, invasion and metastasis, genomic stability and other possible mechanisms [11, 23]. Examples of genes identified in various cancer types, frequently hypermethylated by CGI promoters are shown in Table **1**. However, hypermethylation does not always cause gene silencing. While it has been associated with gene silencing when it occurs in promoter regions, hypermethylation in gene bodies has been associated with higher gene expression, which triggers cancer development if it occurs on an oncogenic gene [23].

Table 1. Promoter hypermethylation in different types of cancer.

Cancer	Genes	Refs.
Breast	*RARβ2, DAPK, hMLH1, p14, p15, RASSF1, BRCA1, PITX2, CDH1, RARB, PCDH10, PGR, MSH2, ESR1β, AKR1β1, COL6A2, GPX7, HIST1H3C, HOXβ4, RASGRF2, TM6SF1, ARHGEF7, TMEFF2, RASSF1, BRCA1, SFN, DKK3, ITIH5, SFRP1, SFRP2, SFRP5, WIF1, APC*	[24 - 28]
Colorectal	*SEPT9, hMLH1, CDKN2A/p16, FOXN2, ALX4, TMEFF2, NGFR, SFRP2, NEUROG1, RUNX3, UBE2Q1, MSH2, MGMT, APC*	[26, 28, 29]
Lung	*RARβ2, CHFR, SFN, SHOX2, RASSF1, APC1, MGMT, CDH1, WIF1, CDH13,*	[26, 28]
Gastric	*RUNX3, MGMT, CDH1, DAPK, MLH1,*	[26, 28]
Hepatocellular	*CDH, CDKN2A, RUNX3, SOCS1*	[28]

If several thousand CpG islands appear to be methylated at the same time in the genome, this is called the CpG island methylator phenotype (CIMP). The CIMP property has been described in some types of cancer such as colorectal, gastric, glioblastoma, lung, and renal cancers; and it has been found to be associated with various clinical, histopathological and epidemiological features such as older age, female gender, prognosis, proximal tumor location, poorly differentiated or mucinous histology, drug sensitivity, and high microsatellite instability and some mutations rates. However, there is a complex situation in the relationship of CIMP with prognosis. In fact, CIMP-positive patients with neuroblastoma, lung cancer and renal cell carcinoma have a worse prognosis, while patients with CIMP-positive glioblastoma and breast cancer have a better prognosis [22, 30].

We mentioned that two important enzymes are involved in the regulation of DNA methylation. It is obvious that the activity or expression levels of these enzymes will be associated with aberrant hypermethylation. It has been reported that DNMT enzymes are up-regulated in some cancer types, and gain-of-function mutations are detected in the genes encoding these enzymes in some cancer types. In addition, it has been reported that mutations causing the formation of products that inhibit TET enzymes have been detected in patients with glioblastoma and AML. It is also argued that DNA methylation is in communication with other epigenetic mechanisms, and that there is a directly proportional relationship between histone modification and aberrant DNA methylation. In addition, aberrant DNA methylation is also induced by environmental stimuli such as aging, bacterial infection (Helicobacter pylori), viral infection (Hepatitis B and C virus, Epstein-Barr virus, papillomavirus), smoking and hormone (estrogen) [22].

Global Hypomethylation

Global hypomethylation, or DNA methylation loss, is the first epigenetic alteration found in cancer, but its role in carcinogenesis is still unclear [31]. It usually occurs in highly methylated centromeric repetitions and repetitive sequences in normal cells. There are various opinions and studies on the role of DNA hypomethylation in carcinogenesis. The best known of these is that DNA hypomethylation occurs more frequently in repetitive sequences and centromeric repeats, increasing genomic instability, damaging chromatin dynamics and cause loss of imprinting. Hypomethylation occurring in these regions also disrupts the function of genes and causes the activation of transposable elements that can cause mutagenesis [32]. It causes the rate of mutation to increase, which contributes to the disruption of chromosomal balance [22]. About 18% of the human genome is LINE-1 sequences, and 10% is Alu repetitive elements, and they are usually methylated in normal cells. For this reason, these regions are used especially to measure global DNA methylation [32]. Although hypomethylation is thought to have an effect on oncogenes as well as the direct effect of hypermethylation on the expression of tumor suppressor genes; this was not clarified until recent years. Although it is known that oncogenes are hypomethylated in various cancers, it is not clear whether this is due to an effect such as a direct effect of hypermethylation on the transcription of tumor suppressor genes. However, in recent years, aberrant promoter hypomethylation of the well-known oncogene *CYP1B1* gene has been demonstrated in prostate cancer [33]. In addition, the promoter hypomethylation of the *PLAU* gene, which is known to have a role in tumor invasion and metastasis development, is associated with increased expression levels and causes oncogenic effect; also similarly the promoter hypomethylation of the *HPSE*, which might contribute to altering the tumor-preferred microenvironment, is associated with increased expression levels [34]. It is also known that hypomethylation increases the activation of genes called cancer-germline *(CG)* or cancer-testis *(CT)* genes. Antigens produced by CT genes were first cloned in 1991 and significantly improved tumor biomarker development [35]. The reason these genes are called cancer testicular genes is that they are germline specific genes and are normally found in methylated form in somatic tissues. In other words, while it is expressed at very low levels or not at all in normal somatic tissues; expressed in cancerous cells [32]. More than 100 *CT* gene families have been identified to date [36]. It is possible to examine the functions of these genes under two main headings: the stimulation of oncogenic pathways (cell proliferation, angiogenesis, immortality, metastasis) and the inhibition of tumor suppressor pathways (controlling growth inhibition signals, apoptosis, and genome integrity) [37]. The fact that they are not expressed (or low expressed) in normal somatic tissues shows that these genes have a promise as a biomarker in cancer diagnosis. Some of the *CT* genes and their functions are shown in Table **2**.

Table 2. Some of the CT genes and their functions [36, 37].

Function	Gene	Chromosomal Localization	Expression Level	Cancer
Stimulate Oncogenic Pathways				
Transcriptional regulators	*SSX*	Xp11.23-p11.22	mRNA and protein	A variety of cancers
Nucleic acid binding and ATP-dependent RNA helicase activity	*DDX43*	6q12-q13	mRNA	Hematologic malignancies, Melanocytic lesion, Brain and Lung cancer
Gene silencing during spermatogenesis	*MAEL*	1q24.1	mRNA	A variety of cancers
Stem cell self-renewal	*PIWIL1*	12q24.33	mRNA and protein	A variety of cancers
Cellular growth and proliferation	*DDX53*	Xp22.11	mRNA	Brain, Gastric, Lung and Gynecological cancers
Transcriptional regulator/the regulation of the cell cycle	*CTCFL*	20q13.31	mRNA and protein	Melanocytic lesion, Brain, Breast, Gastric, Lung, Gynecological, Esophageal, Head and Neck cancer
The major inhibitory neurotransmitter	*GABRA3*	Xq28	mRNA	Colon cancer
Inhibit Tumor Suppressor Pathways				
An androgen receptor coregulator	*MAGEA11*	Xq28	mRNA	Hepatobiliary and lung cancers

(Table 2) cont.....

Function	Gene	Chromosomal Localization	Expression Level	Cancer
Activate RING E3 ubiquitin ligases	*MAGEB2*	Xp21.3	mRNA	Liver, lung, melanocytic, and testicular cancer
	MAGEA2	Xq28	mRNA	Melanocytic lesion, Sarcoma Brain, Hepatobiliary, Lung and Ovarian cancer,
	MAGEC2	Xq27	mRNA and protein	Hematologic malignancies, Melanocytic lesion, Brain, Breast, Prostate, Testicular, Lung, Head and Neck, Hepatobiliary cancer
Chromatin binding and cell cycle regulation	*HORMAD1*	1q21.2	mRNA	Breast, esophageal, gastric and lung cancer
RNA export activity/regulation of mitotic fidelity	*NXF2*	Xq22.1	mRNA	Hematologic malignancies, Melanocytic lesion, lung and testicular cancer

DNA Methylation Biomarkers in Cancer

Most of the aberrations in DNA methylation regulation occur early in tumorigenesis and are quite common in various cancers. Also, these epigenetic aberrations are reversible (such as the use of demethylating agents for hypermethylation). This allows DNA methylation markers to have use in a number of areas related to early cancer detection, cancer diagnosis, prognosis, therapy response and targeting epigenetic drugs. In addition, DNA methylation markers are attractive biomarkers because they are cost-effective, have high specificity and sensitivity, and allow examining multiple target regions in a single assay. As a cancer biomarker, DNA methylation can be detected from tumor tissue or body fluids (such as urine, bronchial lavage fluid, sputum, plasma and serum, and peripheral blood) [38].

There are many studies in the literature on DNA methylation markers that can be used in cancer. Some of these biomarkers (Fig. **4**) are approved for clinical use [39].

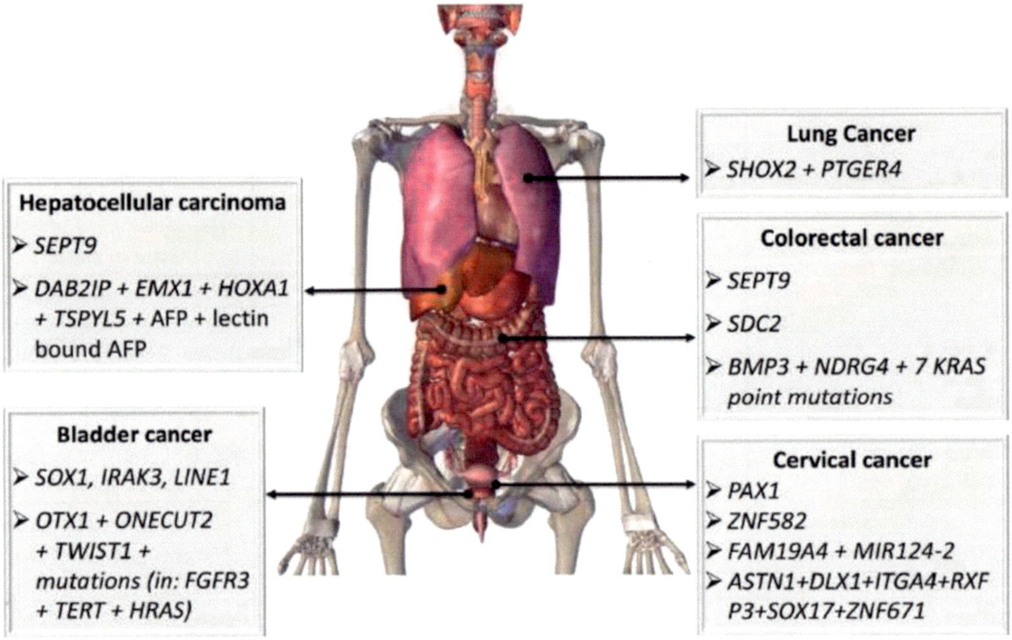

Fig. (4). DNA methylation biomarkers in cancers.

Among these biomarkers, *SEPT9* used in colorectal screening is the first methylation-based biomarker approved by the Food and Drug Administration (FDA) [40]. The *SEPT9* gene encodes septin-9, a GTP-binding protein, one of the septin proteins involved in cytokinesis, cytoskeleton, and cell cycle control. It is stated that this gene, which can act as an oncogenic or tumor suppressor in various cancers, undergoes promoter hypermethylation. In addition to colorectal cancer, it has been reported that this gene is hypermethylated in head and neck squamous cell carcinoma and gastric cancer [41]. Plasma is used as a biospecimen in the SEPT9 methylation. Until now, different assays that determine the *SEPT9* methylation have been put on the market and various studies have been conducted to examine their diagnostic accuracy. Although there are small differences between them, the specificity and sensitivity values of *SEPT9* methylation alone in the diagnosis of colorectal cancer have been reported as 95.9-97.5% and 73.0-76.6%, respectively, with a negative predictive value of over 90%, and a positive predictive value of close to 80%. It is also stated to have higher sensitivity especially in early-stage CRC than routinely used glycoprotein antigens (such as CEA, CA19-9, CA242, CA72-4 and CA125). As a result of the combination with some of these antigens, it is reported that the positive detection rate reaches 97% [42]. *SDC2* is the target gene of another methylation biomarker approved for use in the diagnosis of colorectal cancer. The biospecimen used for this test in colorectal cancer is stool. It has been reported that both the specificity and the sensitivity are 90.2% in the detection of colorectal cancer regardless of the tumor stage [39]. The syndecan-2 protein encoded by *SDC2* gene is a transmembrane heparan sulfate proteoglycan which functions in cell proliferation, cell migration and cell-matrix interactions *via* its receptor for extracellular matrix proteins. In another study, it was reported that *SDC2* hypermethylation was in 84% of CRC patients, and it could detect tumor samples with 87.2% sensitivity and 100% specificity [43]. Methylations of various genes such as *BMP3*, *NDRG4*, *SDC2*, *SFRP2*, *TFPI2* and *VIM* can be detected in the stool, especially early detection of colorectal cancer, there is a view that they may be potential biomarkers. These markers also have sensitivity of 46 - 90% and specificities of 76.8 - 93%. A multi-target panel that consist of a assay for two of these markers (*BMP3 and NDRG4*) and seven point mutations in the *KRAS* gene together has also been approved [44]. This panel can also detect colorectal cancer with 92% sensitivity and 87% specificity [39]. In addition, some another potential DNA methylation-based biomarkers that can be used in the diagnosis of colorectal cancer have been proposed in recent years: *TMEFF2* (transmembrane protein with an EGF-like and two Follistatin-like domains 2; tumor suppressor) hypermethylation [45], *RASSF1A* (Ras association domain family 1 isoform A; a complex tumor suppressor regulating apoptosis and cell cycle control) hypermethylation [46], *LINE* hypomethylation [47]. In addition, it is stated that *LINE-1* hypomethylation and *CDKN2A* (tumor suppressor) hypermethylation may be associated with poor survival outcome; by contrast *MGMT* (methylated-DNA-protein-cysteine methyltransferase) and *FAP2E* hypermethylation may be associated with good prognosis and improved clinical outcomes, and therefore they are thought to be prognostic and predictive biomarkers [48].

The *SEPT9* methylation assay for plasma specimen has also been approved for use in the diagnosis of hepatocellular carcinoma patients with cirrhosis. The other methylation assay approved for hepatocellular carcinoma is a panel test that includes 4 different gene (*DAB2IP, EMX1, HOXA1, and TSPYL5*) methylations and 2 proteins (AFP and lectin bound AFP) in the blood with 80.3% sensitivity and 90.0% specificity [39]. DAB2IP is a Ras GTPase-activating protein that acts as a tumor suppressor; and probably inactivated by promoter hypermethylation in hepatocellular carcinoma [49]. *HOXA1*, homeobox A1, encodes homeodomain proteins that act as DNA binding transcription factors that can regulate cell adhesion, proliferation and differentiation [50]. There are studies that CpG island of the *HOXA1* gene are hypermethylated in hepatocellular cancer and are overexpressed in hepatocellular cancer, and even high protein levels are associated with poor prognosis [51, 52]. In the literature, there are many studies investigating the role of many genes in hepatocellular carcinogenesis by undergoing hypermethylation (*CDKN1C, CRABP1, DLEC1, p16, CCND2,*

CACNA1G, RUNX3, PTGS2, BCL2, GRIN2B, NEUROG1, GSTP1, SYK, SFRP1, CALCA, SOCS3, APC, TERT, TNFRSF10C, HOXA9, ZIC4, OTX1, SIX2, VASH2, IL17D, TLX1, and etc.) or hypomethylation (*SAT2, LINE-1, ALU, HRNBP3,* etc.) [51, 53]. TSPYL5, which encodes testicular Y-like protein 5, has been shown to be tumor suppressor and hypermethylated in hepatocellular cancer [54]. In another study conducted in the past year and registered in a clinical trial, they reported that a panel that examined four DNA methylation markers (*HOXA1*, *EMX1*, *TSPYL5* and *B3GALT6*) and 2 protein markers (AFP and AFP-L3) together could be used in the detection of early-stage HCC (with 90% specificity and 71% sensitivity) [55].

Two different DNA methylation marker panels in the voided urine have been approved for use in the diagnosis of bladder cancer: 1) methylation of *OTX1*, *ONECUT2*, *TWIST1* combined with mutations in: *FGFR3*, *TERT*, *HRAS; 2)* methylation status of *SOX1, IRAK3, LINE1* [39]. Until today, many genes with changing DNA methylation status have been identified in bladder cancer. Some of these genes are *APC, SFRP1, SFRP2, SFRP3, FRP4, TBX2, TBX3, PAX5A, PAX6, RASSF1A, TMEFF2, CACNA1G, KISS1, TIMP3, VIM, LAMC2, RARB, RUNX3, HOXA9, BRCA1, MLH1, MGMT, CDH1, CDH13, CDKN2A, CHFR,* etc [56].

The cancer type that has received the most DNA methylation biomarker approval for detection is cervical cancer. There are two different assays in cervical or oral scrapes indicating the methylation status of *PAX1* and *ZNF582*. In addition, there are two different panels recommended for use in people with HPV positive: 1) *ASTN1, DLX1, ITGA4, RXFP3, SOX17, ZNF671* methylation in cervical smear; 2) *FAM19A4* and *MIR124-2* in cervical or vaginal specimens [39]. In a study examining the performance of the tests, the overall sensitivity of *ASTN1, DLX1, ITGA4, RXFP3, SOX17, ZNF671* panel for detecting severe cervical lesions (CIN3) or cervical cancer was found to be 67.7% (95% CI 57.3-77.1%). In addition, for women with no CIN, the overall false positive rate was 17.4% (95% CI 12.5-23.1), and for HPV-positive women was 24.0% (95% CI 16.0-33.6%) [57].

There are also approved DNA methylation assays used as disease management markers for glioblastoma (*MGMT* methylation assay) and breast cancer (*PITX2* methylation assay). Alkylating agents are used as standard chemotherapy in glioblastoma. The efficacy of this alkylating agent therapy has been shown to depend on the methylation state of the *MGMT* gene. MGMT protein is a key component in the process of repairing DNA damage caused by alkylating agents. Thus, tumors with *MGMT* gene promoter methylation and down-regulation of the MGMT protein respond to therapy. It has been accepted as an assay disease management marker that determines the methylation of the *PITX2* gene, which is stated to be a prognostic gene in breast cancer [39].

CONCLUSION

DNA methylation profiles can be used to obtain information on cancer risk, detection, treatment monitoring and clinical outcomes. Early detection and intervention can greatly improve cancer survival and reduce hospitalization and treatment costs. DNA methylation markers are highly stable and robust, making them ideal candidates for non-invasive diagnosis, prognosis, and treatment markers for malignancies. The study and approval of DNA methylation biomarkers does not take a long history. This shows that the field of study is open and needs improvement. Especially developing non-invasive detection methods is very important for cancer patients and practitioners. In addition, highly effective personalized treatment, which has attracted much attention in recent years but is still considered a utopia, can soon become a reality.

CONSENT FOR PUBLICATION

Not applicable.

CONFLICT OF INTEREST

The authors declare no conflict of interest, financial or otherwise

ACKNOWLEDGEMENT

Declared none.

REFERENCES

[1] Chatterjee A, Rodger EJ, Eccles MR. Epigenetic drivers of tumourigenesis and cancer metastasis. Seminars in Cancer Biology. Academic Press. 2018; 51:149–59. [http://dx.doi.org/10.1016/j.semcancer.2017.08.004]

[2] Moore LD, Le T, Fan G. DNA methylation and its basic function. Neuropsychopharmacology 2013; 38(1): 23-38. [http://dx.doi.org/10.1038/npp.2012.112] [PMID: 22781841]

[3] Handy DE, Castro R, Loscalzo J. Epigenetic modifications: basic mechanisms and role in cardiovascular disease. Circulation 2011; 123(19): 2145-56. [http://dx.doi.org/10.1161/CIRCULATIONAHA.110.956839] [PMID: 21576679]

[4] Cheng Y, He C, Wang M, *et al.* Targeting epigenetic regulators for cancer therapy: mechanisms and advances in clinical trials. Signal Transduct Target Ther 2019; 4(1): 62. [http://dx.doi.org/10.1038/s41392-019-0095-0] [PMID: 31871779]

[5] Michalak EM, Burr ML, Bannister AJ, Dawson MA. The roles of DNA, RNA and histone methylation in ageing and cancer. Nat Rev Mol Cell Biol 2019; 20: 573-89. [http://dx.doi.org/10.1038/s41580-019-0143-1]

[6] Xin Y, O'Donnell AH, Ge Y, *et al.* Role of CpG context and content in evolutionary signatures of brain DNA methylation. Epigenetics 2011; 6(11): 1308-18. [http://dx.doi.org/10.4161/epi.6.11.17876] [PMID: 22048252]

[7] Wu TP, Wang T, Seetin MG, *et al.* DNA methylation on N(6)-adenine in mammalian embryonic stem cells. Nature 2016; 532(7599): 329-33. [http://dx.doi.org/10.1038/nature17640] [PMID: 27027282]

[8] Li Z, Zhao S, Nelakanti RV, *et al.* N^6-methyladenine in DNA antagonizes SATB1 in early development. Nature 2020; 583(7817): 625-30. [http://dx.doi.org/10.1038/s41586-020-2500-9] [PMID: 32669713]

[9] Robertson KD, Wolffe AP. DNA methylation in health and disease. Nat Rev Genet 2000; 1: 11-9. [http://dx.doi.org/10.1038/35049533]

[10] Du J, Johnson LM, Jacobsen SE, Patel DJ. DNA methylation pathways and their crosstalk with histone methylation. Nat Rev Mol Cell Biol 2015; 16(9): 519-32. [http://dx.doi.org/10.1038/nrm4043] [PMID: 26296162]

[11] Jones PA, Baylin SB. The fundamental role of epigenetic events in cancer. Nat Rev Genet 2002; 3: 415-28. [http://dx.doi.org/10.1038/nrg816]

[12] Veland N, Chen T. Mechanisms of DNA methylation and demethylation during mammalian development.Handbook of Epigenetics: The New Molecular and Medical Genetics. Elsevier 2017; pp. 11-24. [http://dx.doi.org/10.1016/B978-0-12-805388-1.00002-X]

[13] Deguchi T, Barchas J. Inhibition of transmethylations of biogenic amines by S-adenosylhomocysteine. Enhancement of transmethylation by adenosylhomocysteinase. J Biol Chem 1971; 246(10): 3175-81. [http://dx.doi.org/10.1016/S0021-9258(18)62211-2] [PMID: 5574392]

[14] Kim JG, Takeshima H, Niwa T, *et al.* Comprehensive DNA methylation and extensive mutation analyses reveal an association between the CpG island methylator phenotype and oncogenic mutations in gastric cancers. Cancer Lett 2013; 330(1): 33-40. [http://dx.doi.org/10.1016/j.canlet.2012.11.022] [PMID: 23196062]

[15] Xie S, Wang Z, Okano M, *et al.* Cloning, expression and chromosome locations of the human DNMT3 gene family. Gene 1999; 236(1): 87-95. [http://dx.doi.org/10.1016/S0378-1119(99)00252-8] [PMID: 10433969]

[16] Unoki M. Recent Insights into the Mechanisms of *De Novo* and Maintenance of DNA Methylation in Mammals. In: Budak M, Yildiz M, Eds. DNA Methylation Mechanism. London: IntechOpen 2019. Available from: www.intechopen.com10.5772/intechopen.89238

[17] Laisné M, Gupta N, Kirsh O, Pradhan S, Defossez P-A. Mechanisms of DNA Methyltransferase Recruitment in Mammals. Genes (Basel) 2018; 9(12): 617. [http://dx.doi.org/10.3390/genes9120617] [PMID: 30544749]

[18] Rasmussen KD, Helin K. Role of TET enzymes in DNA methylation, development, and cancer. Genes Dev 2016; 30(7): 733-50. [http://dx.doi.org/10.1101/gad.276568.115] [PMID: 27036965]

[19] Monk M, Adams RLP, Rinaldi A. Decrease in DNA methylase activity during preimplantation development in the mouse. Development 1991; 112(1): 189-92. [http://dx.doi.org/10.1242/dev.112.1.189] [PMID: 1769327]

[20] Hashimoto H, Zhang X, Cheng X. Selective excision of 5-carboxylcytosine by a thymine DNA glycosylase mutant. J Mol Biol 2013; 425(6): 971-6. [http://dx.doi.org/10.1016/j.jmb.2013.01.013] [PMID: 23337108]

[21] Hashemi R, Arefhosseini SR, Folkerts G, Varasteh S, Morshedi M. Targeted nutrigenomics as a means

of breast cancer management : from DNA methylation to microRNAs. Integr Food Nutr Metab 2020; 7(3).
[http://dx.doi.org/10.15761/IFNM.1000288]

[22] Takeshima H, Ushijima T. DNA methylation changes in cancer: Mechanisms. Encyclopedia of Cancer. Third Edit. Elsevier 2018; 1: pp. 520-9.
[http://dx.doi.org/10.1016/B978-0-12-801238-3.65058-4]

[23] Pfeifer GP. Defining driver DNA methylation changes in human cancer. Int J Mol Sci 2018; 19(4): 1166.
[http://dx.doi.org/10.3390/ijms19041166] [PMID: 29649096]

[24] Qi M, Xiong X. Promoter hypermethylation of RARβ2, DAPK, hMLH1, p14, and p15 is associated with progression of breast cancer: A PRISMA-compliant meta-analysis. Medicine (Baltimore) 2018; 97(51): e13666.
[http://dx.doi.org/10.1097/MD.0000000000013666] [PMID: 30572486]

[25] de Ruijter TC, van der Heide F, Smits KM, Aarts MJ, van Engeland M, Heijnen VCG. Prognostic DNA methylation markers for hormone receptor breast cancer: a systematic review. Breast Cancer Res 2020; 22(1): 13.
[http://dx.doi.org/10.1186/s13058-020-1250-9] [PMID: 32005275]

[26] Moosavi A, Motevalizadeh Ardekani A. Role of epigenetics in biology and human diseases. Iran Biomed J 2016; 20(5): 246-58.
[PMID: 27377127]

[27] Kloten V, Becker B, Winner K, *et al.* Promoter hypermethylation of the tumor-suppressor genes ITIH5, DKK3, and RASSF1A as novel biomarkers for blood-based breast cancer screening. Breast Cancer Res 2013; 15(1): R4.
[http://dx.doi.org/10.1186/bcr3375] [PMID: 23320751]

[28] Bouras E, Karakioulaki M, Bougioukas KI, Aivaliotis M, Tzimagiorgis G, Chourdakis M. Gene promoter methylation and cancer: An umbrella review. Gene 2019; 710: 333-40.
[http://dx.doi.org/10.1016/j.gene.2019.06.023] [PMID: 31202904]

[29] Lønning PE, Eikesdal HP, Løes IM, Knappskog S. Constitutional Mosaic Epimutations - a hidden cause of cancer? Cell Stress 2019; 3(4): 118-35.
[http://dx.doi.org/10.15698/cst2019.04.183] [PMID: 31225507]

[30] Liu J, Tang L, Yi J, *et al.* Unique characteristics of CpG island methylator phenotype (CIMP) in a Chinese population with colorectal cancer. BMC Gastroenterol 2019; 19(1): 173.
[http://dx.doi.org/10.1186/s12876-019-1086-x] [PMID: 31690257]

[31] Feinberg AP, Vogelstein B. Hypomethylation distinguishes genes of some human cancers from their normal counterparts. Nature 1983; 301(5895): 89-92.
[http://dx.doi.org/10.1038/301089a0] [PMID: 6185846]

[32] Toraño EG, Petrus S, Fernandez AF, Fraga MF. Global DNA hypomethylation in cancer: review of validated methods and clinical significance. Clin Chem Lab Med 2012; 50(10): 1733-42.
[http://dx.doi.org/10.1515/cclm-2011-0902] [PMID: 23089701]

[33] Tokizane T, Shiina H, Igawa M, *et al.* Cytochrome P450 1B1 is overexpressed and regulated by hypomethylation in prostate cancer. Clin Cancer Res 2005; 11(16): 5793-801.
[http://dx.doi.org/10.1158/1078-0432.CCR-04-2545] [PMID: 16115918]

[34] Sugiura M, Sato H, Kanesaka M, *et al.* Epigenetic modifications in prostate cancer. Int J Urol 2020; 28(2): 140-9.
[PMID: 33111429]

[35] van der Bruggen P, Traversari C, Chomez P, *et al.* A gene encoding an antigen recognized by cytolytic T lymphocytes on a human melanoma. Science 1991; 254(5038): 1643-7. http://science.sciencemag.org/content/254/5038/1643

[http://dx.doi.org/10.1126/science.1840703]

[36] CTpedia. [cited 2021 Jan 12]. Available from: http://www.cta.lncc.br/modelo.php

[37] Van Tongelen A, Loriot A, De Smet C. Oncogenic roles of DNA hypomethylation through the activation of cancer-germline genes. Cancer Lett 2017; 396: 130-7. [http://dx.doi.org/10.1016/j.canlet.2017.03.029] [PMID: 28342986]

[38] Li L, Choi JY, Lee KM, *et al.* DNA methylation in peripheral blood: a potential biomarker for cancer molecular epidemiology. J Epidemiol 2012; 22(5): 384-94. [http://dx.doi.org/10.2188/jea.JE20120003] [PMID: 22863985]

[39] Taryma-Leśniak O, Sokolowska KE, Wojdacz TK. Current status of development of methylation biomarkers for *in vitro* diagnostic IVD applications. Clin Epigenetics 2020; 12(1): 100. [http://dx.doi.org/10.1186/s13148-020-00886-6] [PMID: 32631437]

[40] Kim H, Wang X, Jin P. Developing DNA methylation-based diagnostic biomarkers. J Genet Genomics 2018; 45(2): 87-97. [http://dx.doi.org/10.1016/j.jgg.2018.02.003] [PMID: 29496486]

[41] Shen N, Wang T, Li D, Zhu Y, Xie H, Lu Y. Hypermethylation of the SEPT9 Gene Suggests Significantly Poor Prognosis in Cancer Patients: A Systematic Review and Meta-Analysis. Front Genet 2019; 10: 887. [http://dx.doi.org/10.3389/fgene.2019.00887] [PMID: 31608117]

[42] Wu D, Zhou G, Jin P, *et al.* Detection of Colorectal Cancer Using a Simplified SEPT9 Gene Methylation Assay Is a Reliable Method for Opportunistic Screening. J Mol Diagn 2016; 18(4): 535-45. [http://dx.doi.org/10.1016/j.jmoldx.2016.02.005] [PMID: 27133379]

[43] Barták BK, Kalmár A, Péterfia B, *et al.* Colorectal adenoma and cancer detection based on altered methylation pattern of SFRP1, SFRP2, SDC2, and PRIMA1 in plasma samples. Epigenetics 2017; 12(9): 751-63. [http://dx.doi.org/10.1080/15592294.2017.1356957] [PMID: 28753106]

[44] Han YD, Oh TJ, Chung T-H, *et al.* Early detection of colorectal cancer based on presence of methylated syndecan-2 (SDC2) in stool DNA. Clin Epigenetics 2019; 11(1): 51. [http://dx.doi.org/10.1186/s13148-019-0642-0] [PMID: 30876480]

[45] Masood M, Grimm S, El-Bahrawy M, Yagüe E. TMEFF2: A Transmembrane Proteoglycan with Multifaceted Actions in Cancer and Disease. Cancers (Basel) 2020; 12(12): 3862. [http://dx.doi.org/10.3390/cancers12123862] [PMID: 33371267]

[46] Hu F, Chen L, Bi MY, *et al.* Potential of RASSF1A promoter methylation as a biomarker for colorectal cancer: Meta-analysis and TCGA analysis. Elsevier GmbH 2020; 216(8): 153009. [http://dx.doi.org/10.1016/j.prp.2020.153009] [PMID: 32703486]

[47] Kerachian MA, Kerachian M. Long interspersed nucleotide element-1 (LINE-1) methylation in colorectal cancer. Elsevier BV 2019; 488: 209-14. [http://dx.doi.org/10.1016/j.cca.2018.11.018] [PMID: 30445031]

[48] Jung G, Hernández-Illán E, Moreira L, Balaguer F, Goel A. Epigenetics of colorectal cancer: biomarker and therapeutic potential. Nat Rev Gastroenterol Hepatol 2020; 17(2): 111-30. [http://dx.doi.org/10.1038/s41575-019-0230-y] [PMID: 31900466]

[49] Qiu GH, Xie H, Wheelhouse N, *et al.* Differential expression of hDAB2IPA and hDAB2IPB in normal tissues and promoter methylation of hDAB2IPA in hepatocellular carcinoma. J Hepatol 2007; 46(4): 655-63. [http://dx.doi.org/10.1016/j.jhep.2006.11.012] [PMID: 17258345]

[50] Rodrigues MFSD, Esteves CM, Xavier FCA, Nunes FD. Methylation status of homeobox genes in common human cancers. Genomics 2016; 108(5-6): 185-93. [http://dx.doi.org/10.1016/j.ygeno.2016.11.001] [PMID: 27826049]

[51] Lee HS, Kim BH, Cho NY, *et al.* Prognostic implications of and relationship between CpG island hypermethylation and repetitive DNA hypomethylation in hepatocellular carcinoma. Clin Cancer Res 2009; 15(3): 812-20.
[http://dx.doi.org/10.1158/1078-0432.CCR-08-0266] [PMID: 19188151]

[52] Zha TZ, Hu BS, Yu HF, Tan YF, Zhang Y, Zhang K. Overexpression of HOXA1 correlates with poor prognosis in patients with hepatocellular carcinoma. Tumour Biol 2012; 33(6): 2125-34.
[http://dx.doi.org/10.1007/s13277-012-0472-6] [PMID: 22864671]

[53] Kuo CC, Lin CY, Shih YL, *et al.* Frequent methylation of HOXA9 gene in tumor tissues and plasma samples from human hepatocellular carcinomas. Clin Chem Lab Med 2014; 52(8): 1235-45.
[http://dx.doi.org/10.1515/cclm-2013-0780] [PMID: 24681432]

[54] Qiu X, Hu B, Huang Y, Deng Y, Wang X, Zheng F. Hypermethylation of ACP1, BMP4, and TSPYL5 in Hepatocellular Carcinoma and Their Potential Clinical Significance. Dig Dis Sci 2016; 61(1): 149-57.
[http://dx.doi.org/10.1007/s10620-015-3878-3] [PMID: 26386860]

[55] Chalasani NP, Ramasubramanian TS, Bhattacharya A, *et al.* A Novel Blood-Based Panel of Methylated DNA and Protein Markers for Detection of Early-Stage Hepatocellular Carcinoma. Clin Gastroenterol Hepatol 2021; 19(12): 2597-2605.e4.
[http://dx.doi.org/10.1016/j.cgh.2020.08.065] [PMID: 32889146]

[56] Sharma A, Reutter H, Ellinger J. DNA Methylation and Bladder Cancer: Where Genotype does not Predict Phenotype. Curr Genomics 2020; 21(1): 34-6.
[http://dx.doi.org/10.2174/1389202921666200102163422] [PMID: 32655296]

[57] Schmitz M, Wunsch K, Hoyer H, *et al.* Performance of a methylation specific real-time PCR assay as a triage test for HPV-positive women. Clin Epigenetics 2017; 9(1): 118.
[http://dx.doi.org/10.1186/s13148-017-0419-2] [PMID: 29090037]

CHAPTER 6

Circulating Biomarkers in Predicting Pathological Response to Neoadjuvant Therapy for Colorectal Cancer

Ravindri Jayasinghe[1], Umesh Jayarajah[1,*] and Sanjeewa Seneviratne[1]

[1] *Department of Surgery, Faculty of Medicine, University of Colombo, Colombo, Sri Lanka*

Abstract: Circulating biomarkers show promise in the management of many cancers. They have become the novel non-invasive approach to complement the current strategies in colorectal cancer (CRC) management. Their ability in guiding diagnosis, evaluating response to treatment, screening and prognosis is phenomenal, especially when it comes to their minimally invasive nature. These "liquid biopsies," which show potential for replacing invasive surgical biopsies, provide useful information on the primary and metastatic disease by providing an insight into cancer biology. Analysis of blood and body fluids for circulating tumour DNA (ctDNA), carcinoembryonic antigen (CEA), circulating tumour cells (CTC), or circulating micro RNA (miRNA) shows potential for improving CRC management. Recognizing a predictive model to assess response to neoadjuvant chemotherapy would help in better patient selection. This review was conducted with the aim of outlining the use of circulatory biomarkers in current practice and their effectiveness in the management of patients having CRC with a focus on response to neoadjuvant therapy.

Keywords: Biomarkers, Colon cancer, Neoadjuvant therapy.

INTRODUCTION

Colorectal cancer (CRC) is the second commonest cancer diagnosed in females (approximately 820,000 cases in 2018) and the third commonest cancer in males (approximately 1,020,000 cases in 2018) [1]. It is the second most commonest cause of cancer-related death globally, with approximately 900,000 deaths occurring annually [1]. There is a rising incidence of CRC in both developing and developed countries [2, 3]. The pathogenesis includes a series of genetic and epigenetic mutations causing alterations in cell growth, differentiation, and apoptosis, transforming the normal colonic epithelium into an adenocarcinoma. Loss of oncosuppressor genes, such as *APC*, *TP53,* and a series of molecular alte-

[*] **Corresponding author Umesh Jayarajah:** Department of Surgery, Faculty of Medicine, University of Colombo, Colombo, Sri Lanka; Tel: +94 76 680 86 25; E-mail: umeshe.jaya@gmail.com

Pınar Atukeren and Hafize Uzun (Eds.)

rations in proto-oncogenes, such as *KRAS,* lead to progression of an adenomatous polyp to invasive carcinoma in a stepwise fashion, which is known as the adenoma-carcinoma sequence [4].

The prognosis of CRC largely depends on the stage of the disease at diagnosis, and therefore, early detection is known to improve morbidity and mortality. Conventional means of disease assessment such as radiological imaging and histopathological analysis have minimum sensitivity in terms of early detection of systemic spread. Therefore, in recent years, the concept of liquid biopsy, which includes the analysis of circulatory biomarkers in blood and body fluids, has emerged as a promising diagnostic tool [5]. The diagnosis and management of CRC have been generally based on biopsy-based techniques for decades. However, determining the individualised response to treatment and the prognosis is often difficult based solely on the histological assessment [6]. This has led to the discovery of biomarkers for early detection of relapse and treatment resistance leading to the improved overall quality of life and patient survival [7]. There is ample evidence to suggest that biomarkers such as CTC in peripheral blood are associated with a negative impact on patient survival [8]. In contrast to the conventional tissue biopsies, liquid biopsies are associated with considerable benefits such as minimal invasiveness, less pain and lower procedural risks. The term "liquid biopsy" reflects a much broader perspective which refers to the utility of all circulating tumour traits, including CTC, ctDNA, circulating miRNA, exosomes, proteins, and mRNAs in the peripheral blood and body fluids [9]. An ideal biomarker should have properties including the ability to quantify, high specificity, reliability, measurability, predictability, and sensitivity [10].

Neoadjuvant chemotherapy revolutionised the management of CRC by substantially reducing the risk of local recurrence and improving outcomes, especially in locally advanced cancers. However, the response to neoadjuvant chemotherapy varies ranging from pathological complete response (pCR) to no response. Pre-operative chemotherapy has its own associated disadvantages, including significant adverse effects which may interfere with timely surgery. The use of circulatory biomarkers in predicting response to neoadjuvant therapy would help select patients who are more likely to respond to such therapy.

CRC restaging after neoadjuvant chemotherapy is equally important, as it will determine the presence of nodal metastasis, involvement of the circumferential resection margin, and the depth of invasion in rectal carcinoma [11]. Conventional imaging modalities, namely computed tomography (CT), magnetic resonance imaging (MRI), endoscopic ultrasonography (EUS), and positron emission tomography (PET) are used in the process of restaging. However, the overall accuracy of these investigations is reduced in restaging after neoadjuvant therapy in comparison to the initial staging. This is mainly due to failure of differentiating tumour infiltration and residual tumour from radiation-induced fibrosis. Radiation related complications, including proctitis, may also be misinterpreted as local invasion [11, 12].

Among patients receiving neoadjuvant CRT for rectal carcinoma, a significant proportion (15-40%) has been shown to achieve a pCR. Such patients tend to have an excellent prognosis with an almost non-existent risk of local recurrence and more than 95% five-year overall survival [13, 14]. There is an ongoing debate regarding the possibility of managing these patients without surgery, sparing the postoperative and long-term morbidity associated with radical surgical resections [15]. However, classification as pCR mandates a histological examination of the entire specimen which is not possible without surgical resection. Therefore, a complete clinical response (cCR) which is the absence of clinically detectable tumour after neoadjuvant CRT is used as a surrogate for pCR. Assessing for cCR would depend on clinical examination findings correlated with endoscopy radiological investigations and may also include histological examination of biopsies from the previous tumour site [16]. Surgery is usually timed within 10 to 12 weeks following completion of CRT, which allows adequate time for tumour regression. Management of patients who have achieved cCR to neoadjuvant therapy still remains a controversial topic, where non-operative management is becoming a popular alternative in this group, especially among patients who are at high risk for surgery. Although controversial, a combination of frequent clinical, endoscopic and radiographic evaluation have been shown to be helpful in ensuring local tumour control and detecting early recurrences [17]. Patients after completion of 10-12 weeks of neoadjuvant CRT with evidence of residual tumour, should undergo surgery. The rest with a cCR may be given the option for surgical resection or watch and wait strategy. Those who opt for non-operative management should be monitored closely for 1-year after neoadjuvant CRT followed by a long-term follow up protocol [13, 18]. As cCR is only a surrogate of pCR, the utility of circulatory biomarkers in this scenario is a matter of great interest.

The objective of this comprehensive review is to outline the use of circulatory biomarkers in current practice and their effectiveness in the management of patients with colorectal cancer. This review summarises tissue-based biomarkers in blood, blood-based biomarkers, circulating peptides, and other novel biomarkers in CRC with a special emphasis on predicting response to neoadjuvant therapy in CRC.

ESTABLISHED BIOMARKERS

Tissue Based Biomarkers

A biopsy from the tumour during colonoscopy or rectal examination is routinely performed before commencing therapy for a suspected CRC. However, disparities exist in sample collection, storage, and processing. Recent literature had highlighted the impact of tumour microenvironment in modulating response to chemoradiotherapy. Molecular factors expressed in the neighbouring cells and tumour microenvironments such as chemotactic molecules, growth factors, and factors associated with cell death are known contributors to tumour resistance to therapy and regulating cell recruitment in the immune system [19, 20].

DNA Alterations: Polymorphisms

DNA alterations are known to determine tumour sensitivity to chemotherapy. Antineoplastic agents including oxaliplatin, inhibit cell replication by promoting DNA damage. Both advanced CRCs and metastatic CRCs show striking responses to oxaliplatin therapy. However, treatment failures are common, which is mainly due to genetic variations, and hence, evaluation of pharmacogenetic markers may help determine individualized patient prognosis [21]. Polymorphisms such as *GSTP1* (Ile105Val) and *XRCC1* (Arg399Gln) have been recognised as prognostic factors determining response to oxaliplatin based chemotherapy [22]. The *XRCC1* polymorphism has been shown to intensify the sensitivity to chemotherapy. However, it also reduces DNA repair, which leads to increased accumulation of DNA damage and induction of mutations [23]. Increasing evidence shows that the tumour response rate to oxaliplatin-based chemotherapy is significantly lower in patients who carry polymorphisms in *XRCC1* [24]. Furthermore, genetic variations play a role in the tolerance of drug dependant DNA adducts, drug metabolism, and DNA repair protein functions which have negative impacts on the patients' prognosis [23].

Gene Expression

Gene expression is a known predictor of tumour response after preoperative chemoradiotherapy. Recent literature elaborates its ability to differentiate patients undergoing neoadjuvant or adjuvant chemotherapy as responders and non-responders [25]. Gene expression profiles have been studied in biopsies from rectal cancer patients after neoadjuvant therapy and they have shown promising results. However, the differences in tissue types, treatment protocols, type of platform used, profiling technologies, tissue preparations, and the gene lists make it difficult to isolate a specific predictor gene from these studies [26 - 28].

MicroRNA (miRNA)

miRNAs are non-coding RNAs that control diverse metabolic and cellular pathways and regulate gene expression by post-transcriptional gene regulation and RNA silencing via RNA interference. They show altered expression based on the tumour type. They have shown promise in cancer diagnosis, prognostication, and monitoring of treatment efficacy [29]. Their impact on protein expression is by translational inhibition through a sequence dependant process and by *RISC* (RNA-induced silencing complex), which cleaves mRNA [30]. Some circulating miRNAs have been identified in biological fluids, although miRNAs are mostly found in the cellular microenvironment.

The miRNAs occur as complexes in combination with lipoproteins to maintain their stability in the extracellular environment by degradation from RNases. Their stability in extracellular environments makes them show promise as non-invasive biomarkers. A liquid biopsy biomarker for response to chemoradiotherapy would help classify patients as treatment non-responders and responders, which potentially would help avoid exposure to inefficient toxic treatment. Circulating miRNAs are difficult to study due to their low concentrations in plasma. However, recent technological advancements have resulted in test kits that are capable of detecting these at low concentrations, which have helped with accurate detection [11].

Several types of miRNA have been identified as predictive markers to check response to CRT. Furthermore, miRNA-125b's increased expression in tumour also enables detectability in serum in higher concentrations. Certain miRNA like miR-345 are shown to be correlated with unfavourable response to CRT. Furthermore, lower levels of the same correlate positively with local recurrence-free survival [31, 32]. Dysregulation of expression of miRNA is shown to correlate with the tumour response to CRT. Studies have also shown an association with longer disease-free survival and higher expression of miR-29. Furthermore, an upregulation of miR-29 was shown in tumour tissue from patients with liver metastases in comparison to patients having non-metastatic CRC. Additionally, it has been shown to be helpful in predicting early recurrence of CRC [33, 34]. Studies have shown that miR-143 plays a role in tumours resistant to CRT. A study by Simmer *et al* showed that a low expression of miR-143 in patients with metastatic colon cancer after treatment with fluoropyrimidine–based chemotherapy caused an increased median progression free survival compared to those with high miR-143 expression [35]. Furthermore, miR-143 was shown to be a protective factor for cells from DNA damage induced killing [36]. Therefore, miRNA can be considered a new, non-invasive predictive biomarker for assessing the treatment response allowing development of individualised treatment strategies.

Blood Based Biomarkers

Patients with cancer release DNA, mRNA, and miRNA into their bloodstream and their levels are known to be associated with tumour burden and disease progression [37]. There is increasing evidence highlighting the use of circulatory tumour markers in screening, monitoring efficacy of therapy and prognostication.

cfDNA and ctDNA

Healthy individuals have circulating cfDNA. However, it is a known marker of malignancy useful in diagnosis and prognosis [37]. A limitation of cfDNA is the reporting of contradictory findings, in relation to proportion of tumour derived and non-tumour derived cfDNA. However, the tumour derived cfDNA which is hypothesised to be released into circulation due to tumour necrosis is known to vary in length compared to those released from non-tumour related apoptosis in healthy individuals. Therefore, an integrity index which is a ratio of longer and shorter DNA fragments which reflects the presence of malignancy is calculated [38, 39]. Evidence has shown that for a tumour weighing 100g, nearly 3.3% of the DNA from the tumour would pass in to the bloodstream daily [40]. A study by Umetani *et al* demonstrated real-time polymerase chain reaction (qPCR) as a quantitative technique to measure the integrity of cfDNA in the serum of CRC patients [41].

In comparison with non-responders, patients showing tumour regression after neoadjuvant therapy has shown significantly lower levels of circulating cfDNA [42]. Furthermore, the therapeutic response is negatively associated with post-treatment cfDNA integrity index. The variations in the long fragments and the integrity index of cfDNA is clinically significant than single pre or post-op CRT assessment [43]. However, cfDNA is shown to quantitatively vary from patient to patient and therefore, further studies are needed to identify its use as a clinical marker. This variability is known to occur due to factors affecting its release such as tumour necrosis and apoptosis, mitochondrial and T-cell origin, low DNase activity, spontaneous and active release of DNA by multiplying tumour cells and activated lymphocytes [44, 45]. Although basal levels are not used in measuring tumour response, responders and non-responders are identified during treatment in correlation with other conventional investigations such as CT, transrectal ultrasound, MRI *etc.* These help to treat patients in real time where non-responders can undergo surgery by stopping CRT which are likely to be of minimal benefit [46].

cfmRNA and hTERT

Cell free mRNA (cfmRNA) has been found in the plasma of patients with malignancies. It is also found in non-malignant conditions in normal subjects such as in maternal circulation derived from placenta [47]. Human telomerase reverse transcriptase (h*TERT*) is activated inappropriately in most tumours and therefore, it can be considered as a marker specific for malignancy [46]. Studies have shown significant variations in both these markers correlating to a response to CRT [48]. A response to therapy is indicated by a decrease in plasma levels while increasing or unchanged levels are seen in non-responders. Higher levels of cfmRNA are hypothesised to be associated with active tumour necrosis or a major tumour extension in non-responders. This is due to release of these biomarkers following necrosis of advanced, large tumours or cell apoptosis [49, 50]. Surprisingly, h*TERT* which is absent in plasma of healthy subjects has been detected in patients showing a pathological complete response which could be due to microscopic disease or reduced clearance of h*TERT* from plasma [48, 51].

cfmiRNA

Tumour specific mRNA have been detected in blood in stable forms even under adverse conditions such as temperature and pH variations [52]. Circulating miR-21 was found to be a marker of resistance to chemotherapy in prostate cancer where it is found to be over-expressed in patients showing resistance to docetaxel [53]. It is also shown to have a predictive role in NSCLC (non-small cell lung carcinoma) and breast cancer. However, studies on CRC are limited and further research is warranted in determining its role in personalised therapy in patients with CRC.

CIRCULATING PROTEINS/PEPTIDES

Circulating proteins and plasma peptides (the peptidome) are promising biomarkers in the light of new evidence [54]. These proteins are released as fragments from tumour tissue. These are known to originate from the tumour microenvironment due to events such as apoptosis and necrosis. Studies have indicated that these peptide signatures are shown to pin point tumour pathology and tumour site [55 - 57]. The study of peptidome would provide information on various peptides differentiating them from common peptides available in plasma such as albumin and immunoglobulins.

Although these peptides have shown promise in diagnosis, evidence is limited on their application in assessing response to CRT in CRC. The available evidence includes specific patterns of diagnosis in different cancer pathologies [58, 59].

Studies had been done to identify the clinical predictive value of circulating peptides in CRC patients treated with neoadjuvant CRT. A study performed by Crotti *et al* explored the peptidome of patients with locally advanced rectal cancer who underwent CRT. Non-porous silica chips (NSCs) technology was used coupled with MALDI-TOF analysis to detect these peptides in plasma of locally advanced rectal cancer patients undergoing CRT. They had identified certain peptides like complement C3 protein and its fragments to be involved with resistance to preoperative CRT [60]. A similar study by Smith *et al* on serum profiling in patients with rectal cancer CRT before and after each radiation therapy has revealed that the proteomic profile shows specific features helping to discriminate patients depending on their histological response to therapy. Furthermore, the study revealed 14 differentially abundant proteins that are capable of predicting the ultimate pathological response with a specificity of 80% and a sensitivity of 87.5% [61]. Helgason *et al* found two proteins which are tentatively identified as haemoglobulin alpha-chain fragment (MW 2kDa) and the acute phase Apo A-I protein (MW 28kDa) to correlate with response to therapy in patients with CRC treated with oxaliplatin and capecitabine [62]. Although the evidence is promising, further studies are needed to establish their use in clinical practice.

OTHER PREDICTIVE BIOMARKERS

Markers such as epidermal growth factor receptor (EGFR), p53, ki-67, thymidylate synthase (TS), cyclooxygenase 2, p21 and bax/bcl-2, circulating tumour DNA (ctDNA), and carcinoembryonic antigen (CEA) are frequently studied and investigated for a potential role in prediction in response to CRT. However, for some of these biomarkers evidence seems to be inadequate to determine their use as a predictor of response. For example, p53 is one such marker that is widely studied for its expression in patients with rectal cancer receiving CRT. However, to date, evidence to consider p53 as a predictive marker for tumour response to CRT is limited. Therefore, further research is needed to determine their place in clinical practice [63].

Thymidylate Synthase (TS)

Thymidylate synthase (TS) is the main intracellular target for drugs such as 5-flurouracil (5-FU), a widely used chemotherapy agent in CRC. TS involved in DNA synthesis and it plays a vital part by phosphorylating deoxyuridine monophosphate to deoxyuridine triphosphate. Overexpression of TS in CRC is known to be associated with a resistance to 5-FU [64]. Studies have been done in rectal cancer patients undergoing CRT to evaluate the expression of TS by evaluating the protein levels using immunohistochemistry (IHC) and polymerase chain reaction (PCR). Genetic polymorphisms are also studied in relation with tumour response and expression of TS. A comprehensive review done by Spolverato *et al* have identified that responders to CRT had lower TS levels whereas a majority (75%) of non-responders to CRT had high levels of TS. Furthermore, they managed to determine that ki-67 plays a role in prediction of response in patients with rectal cancer who are treated with 5-FU based therapy [63]. Although TS was associated with tumour response after CRT, there is considerable variability in determining the exact response which highlights the necessity for further research.

Epidermal Growth Factor Receptor (EGFR)

EGFR is a GTPase protein playing a role in transduction of signals in PI3K and MAP kinase cascades driving cell proliferation, apoptosis, and differentiation [63]. EGFR gene polymorphisms have been investigated to determine their role in regulating gene expression as a predictive factor in response to CRT. A study by Spindler *et al* has shown increased gene expression in locally advanced rectal cancer patients [65]. Similar study by Bengala *et al* has shown that the wild type KRAS and the number of high EGFR gene copies are associated with response to treatment [66]. The level of expression of EGFR can be a useful predictive marker of pathological response of CRC to CRT. However, more research needs to be performed to confirm these findings and to ascertain their utility in clinical practice.

Cyclooxygenase 2 (COX-2)

Cyclooxygenase 2 is an enzyme which is responsible for the conversion of arachidonic acid to prostaglandins (PG). Tumour cells produce PG using COX-2, protecting themselves against cell death induced by radiation thus, indicating tumour invasiveness. Expression of COX-2 was studied in patients undergoing CRT for rectal cancer, and the non-responders showed an overexpression of COX-2 in comparison to the normal patients [67]. A study by Min *et al* further clarified this fact by showing that overexpression of COX-2 was indicative of a poor response to CRT and showed its role in predicting a poor response to therapy [68]. However, further research is needed in determining the role of COX-2 in evaluating the tumour response rate.

Bax/bcl-2

Bax and bcl-2 are proteins in the apoptotic pathway. Studies have shown bcl-2 to be an inhibitory protein which in response to apoptotic signals inhibits cell death [69]. Bax in CRC is associated with chemotherapy resistance, where its loss would suggest resistance and its induction would suggest increased chemosensitivity [70, 71]. Studies have shown that a majority of treatment responders were bcl-2 positive and the Bax expression was found to be greater in responders in comparison with non-responders. Furthermore, Bax response was found to be increased in complete responders than in partial responders [63]. These findings show the role of tumour apoptosis in response to CRT and the expression of Bax is a potential biomarker for chemo-radio sensitivity. A study done by Huerta *et al* revealed that increased Bax and a lower bcl-2 level would indicate good response whereas, a similar study by Kikuchi *et al* revealed the specificity and sensitivity of Bax to be 83.9% and 82.8%, respectively [72, 73].

P21

P21 is a cyclin dependant kinase which is known to arrest the cell cycle at G1 to S phase transition in mammals. P53 is known to mediate its tumour suppressor activity by inducing P21 [74, 75]. However, it is shown to respond to cellular DNA damage independent of p53 [76]. A study by Sturm *et al* found p21 to be positively correlated with tumour response to treatment. Furthermore, tumours showing pre-treatment p21 expression indicated a better local tumour response. High expression of p21 in tumour cells after completion of neoadjuvant therapy for a prolong duration of four to six weeks has been observed to be associated with worse prognosis and survival. Resistant tumour cells to neoadjuvant therapy have shown persistence of p21 expression and the reason for the observation is unclear [77].

Ki-67

It is a marker of cellular proliferation in the cell cycle. Although most studies failed to show an association between response to CRT and Ki-67 expression in patients with rectal cancer, some studies have shown evidence to the contrary. A study by Jakob *et al* revealed that rectal cancers having low Ki-67 expression treated with 5-FU, had an improved response rate than those who had a high Ki-67 expression [78]. Confirming this further, Kikuchi *et al* recognised Ki-67 as an independent factor with a sensitivity and specificity of 82.8% and 83.9%, respectively in predicting response to treatment. Furthermore, a positive correlation was found with a high Ki-67 and the effects of treatment [73]. However, further research is necessary in order to confirm the utility in clinical practice.

CEA

CEA is expressed in normal mucosal cells and overexpressed by adenocarcinomas of the colon, rectum, breast, pancreas, and lungs. It is a serum-based glycosylated tumour marker, and an oncofoetal glycoprotein [79, 80]. The domain structure of

CEA proteins closely resembles the gamma heavy chain of the immunoglobulin IgG [81]. Although the exact functions of CEA have not been identified yet, recent studies portray an involvement in cancer growth, invasion and metastasis by a playing a role in cell migration [82]. Park *et al* investigated the role of CEA in predicting response to CRT in patients with rectal cancer where they found that levels greater than 5ng/ml to be significantly associated with poor response to CRT and nodal infiltration. Furthermore, elevated pre-CRT serum CEA levels were found to be the only predictor of poor response to CRT [83]. A baseline CEA level of less than 2.5ng/ml was found to be predictive of pCR where it is associated with a higher pCR rate than those with higher CEA levels [84]. A study by Moureau-Zabotto *et al* had shown that a pre-treatment CEA level of less than 5ng/ml to be specifically associated with tumour downstaging [85]. Furthermore, evidence suggests a post-CRT level of less than 5 ng/ml to be associated with a more favourable prognosis in rectal cancer such as early stage and complete tumour regression. Overall, patients with lower CEA levels were shown to respond to therapy better [86, 87].

ctDNA

Tumour derived ctDNA or circulating tumour DNA are a component of cfDNA detected in the bloodstream. They are known to carry similar genomic alterations to that of the primary tumour [88]. Hypoxia caused by the growth of tumour exceeding limits of vasculature is known to cause cell death and inefficient phagocytosis would allow ctDNA to escape into blood [89]. The levels of ctDNA in blood appear to rise with advancing disease stage. In patients treated with CRT followed by surgery for rectal cancer, a pCR was associated with overall increase in survival. However, imaging is incapable of evaluating the response accurately as the post-CRT fibrotic changes may be confused with residual tumour. Therefore, assessment of biomarkers could potentially be used to complement imaging and provide a better predictive model for complete response to CRT. This would enable better selection of patients for watchful waiting. Furthermore, ctDNA can be used to detect recurrences in patients undergoing watch and wait approach with regular invasive investigations such as endoscopy. A growing body of evidence shows postoperative detection of ctDNA to predict recurrences irrespective of adjuvant chemotherapy [88]. A recent study by Tie *et al* had shown the potential for patients to be stratified depending on the risk of recurrence as high risk and low risk, after analysing ctDNA for locally advanced rectal cancer following curative intent surgery. They identified ctDNA status to be a strong prognostic marker independent of other pathological risk factors. Furthermore, they concluded that the detection of ctDNA at diagnosis before initiating treatment was not a predictive factor for disease recurrence as the baseline levels of ctDNA is from the primary tumour and not from the residual micro-metastatic disease. They concluded that ctDNA is a promising guide in selecting patients for adjuvant chemotherapy [90].

Circulating Tumour Cells (CTC)

Most studies since the discovery of CTCs were hampered by the technological constraints in detection of these rare cells in blood. However, with the advancements in technology, CTC detection was initiated in clinical practice for breast cancer in 2004, followed by CRC in 2007 [91]. Detection of CTC in blood of patients with advanced metastatic CRC was found to be related to their overall survival. A meta-analysis provided strong evidence on the prognostic utility of CTC in metastatic disease which showed a worse overall survival and progression free survival [92]. Conforming these findings, a study by Cohen *et al* on patients receiving first, second and third lines of chemotherapy for metastatic CRC showed that CTC isolated in blood was found to be prognostic and predictive [93]. Furthermore, the detection of ≥3 CTC / 7.5ml after commencing chemotherapy was shown to have a shorter progression free survival and overall survival [94]. These findings depict its potential role in monitoring patients and aiding selection of patients in order to sustain longer treatment free duration, minimising treatment toxicity without compromising patient outcomes.

Response to chemotherapy and its association with CTC in patients having metastatic CRC was studied by Kawahara *et al* where they identified three groups of patients with different biological states. Patients in group 1, had localised metastasis and no CTCs prior to treatment. Patients in group 2, had baseline positive CTCs which later became untraceable after chemotherapy. Group 3, included patients with persisting CTCs at baseline and after chemotherapy. This classification was an aid in obtaining informed decisions in patient management where patients in group 2, would likely be benefited by further chemotherapy as they showed good response. For non-responders in group 3, intensification or change of treatment should be considered. Furthermore, patients in group 1 would benefit from radical curative surgery [95].

Further studies have shown that the tumour stage is the single clinicopathological factor significantly associated with CTCs which showed that 61% of patients with metastatic CRC had positive CTCs compared to the minority of 15% of non-metastatic patients [96]. CTCs have been found to provide better accuracy when used in combination with standard assessment tools. Studies which evaluated the suitability of using CTCs as a predictive tool to complement imaging modalities like CT demonstrated a higher accuracy for predictions on progression of disease where CTCs can be used in defining patients as having stable disease or progressive disease. This would further stratify treatment response and guide therapy. This correlation with imaging is helpful because targeted therapy induces necrosis and cavitation of the tumour internally without an actual shrinkage of the size [91].

CONCLUSION

Prediction of response to treatment in patients undergoing neoadjuvant chemoradiotherapy for CRC is vital in identifying and stratifying the patients who would benefit from therapy and to identify the non-responders who will experience drug toxicity and other side effects of treatment which could otherwise be prevented. Studies focusing on identifying gene expression signatures and phenotypes of CRC would help in individualising therapy with the use of biomarkers to assess the response to therapy. Although promising, the clinical use of circulatory biomarkers requires further studies. Bridging the knowledge gap on application of peptidome profiling, quantification, and gene expression of these circulatory biomarkers would help in formulating a clinical predictive model for therapeutic response.

CONSENT FOR PUBLICATION

Not applicable.

CONFLICT OF INTEREST

The authors declare no conflict of interest, financial or otherwise.

ACKNOWLEDGEMENT

Declared none.

REFERANCES

[1] World Health Organization. Cancer Today - Globocan. 2018. https://gco.iarc.fr/today/home

[2] Jayarajah U, Udayanga V, Fernando A, Samarasekera DN, Seneviratne S. The incidence and patterns of colorectal cancers in Sri Lanka from 2001 to 2010: Analysis of national cancer registry data. Eur J Cancer Care (Engl) 2020; 29(4): e13247. [http://dx.doi.org/10.1111/ecc.13247] [PMID: 32432404]

[3] Waidyasekera RH, Jayarajah U, Samarasekera DN. Quality and scientific accuracy of patient-oriented information on the internet on minimally invasive surgery for colorectal cancer. Health Policy Technol 2020; 9(1): 86-93. [http://dx.doi.org/10.1016/j.hlpt.2019.11.004]

[4] Winder T, Lenz H-J. Molecular predictive and prognostic markers in colon cancer. Cancer Treat Rev 2010; 36(7): 550-6. [http://dx.doi.org/10.1016/j.ctrv.2010.03.005] [PMID: 20363564]

[5] Norcic G. Liquid biopsy in colorectal cancer-current status and potential clinical applications. Micromachines (Basel) 2018; 9(6): 300.

[http://dx.doi.org/10.3390/mi9060300] [PMID: 30424233]

[6] Amin MB, Greene FL, Edge SB, *et al.* The eighth edition AJCC cancer staging manual: continuing to build a bridge from a population-based to a more "personalized" approach to cancer staging. CA: a cancer journal for clinicians. 2017; 67: p. (2)93.

[7] Marcuello M, Vymetalkova V, Neves RPL, *et al.* Circulating biomarkers for early detection and clinical management of colorectal cancer. Mol Aspects Med 2019; 69: 107-22. [http://dx.doi.org/10.1016/j.mam.2019.06.002] [PMID: 31189073]

[8] Jia S, Zhang R, Li Z, Li J. Clinical and biological significance of circulating tumor cells, circulating tumor DNA, and exosomes as biomarkers in colorectal cancer. Oncotarget 2017; 8(33): 55632-45. [http://dx.doi.org/10.18632/oncotarget.17184] [PMID: 28903450]

[9] Marrugo-Ramírez J, Mir M, Samitier J. Blood-based cancer biomarkers in liquid biopsy: a promising non-invasive alternative to tissue biopsy. Int J Mol Sci 2018; 19(10): 2877. [http://dx.doi.org/10.3390/ijms19102877] [PMID: 30248975]

[10] Link A, Balaguer F, Shen Y, *et al.* Fecal MicroRNAs as novel biomarkers for colon cancer screening. Cancer Epidemiol Biomarkers Prev 2010; 19(7): 1766-74. [http://dx.doi.org/10.1158/1055-9965.EPI-10-0027] [PMID: 20551304]

[11] Machackova T, Prochazka V, Kala Z, Slaby O. Translational Potential of MicroRNAs for Preoperative Staging and Prediction of Chemoradiotherapy Response in Rectal Cancer. Cancers (Basel) 2019; 11(10): 1545. [http://dx.doi.org/10.3390/cancers11101545] [PMID: 31614848]

[12] Jhaveri KS, Hosseini-Nik H. MRI of rectal cancer: an overview and update on recent advances. AJR Am J Roentgenol 2015; 205(1): W42-55. [http://dx.doi.org/10.2214/AJR.14.14201] [PMID: 26102418]

[13] Maas M, Nelemans PJ, Valentini V, *et al.* Long-term outcome in patients with a pathological complete response after chemoradiation for rectal cancer: a pooled analysis of individual patient data. Lancet Oncol 2010; 11(9): 835-44. [http://dx.doi.org/10.1016/S1470-2045(10)70172-8] [PMID: 20692872]

[14] Martin S, Heneghan H, Winter D. Systematic review and meta-analysis of outcomes following pathological complete response to neoadjuvant chemoradiotherapy for rectal cancer. Database of Abstracts of Reviews of Effects (DARE): Quality-assessed Reviews. UK: Centre for Reviews and Dissemination 2012.

[15] Hendren SK, O'Connor BI, Liu M, *et al.* Prevalence of male and female sexual dysfunction is high following surgery for rectal cancer. Ann Surg 2005; 242(2): 212-23. [http://dx.doi.org/10.1097/01.sla.0000171299.43954.ce] [PMID: 16041212]

[16] Habr-Gama A, Perez RO, Wynn G, Marks J, Kessler H, Gama-Rodrigues J. Complete clinical response after neoadjuvant chemoradiation therapy for distal rectal cancer: characterization of clinical and endoscopic findings for standardization. Dis Colon Rectum 2010; 53(12): 1692-8. [http://dx.doi.org/10.1007/DCR.0b013e3181f42b89] [PMID: 21178866]

[17] Willett CG, Ryan DP. Neoadjuvant chemoradiotherapy, radiotherapy, and chemotherapy for rectal adenocarcinoma 2020.

[18] Habr-Gama A, Perez RO, Proscurshim I, *et al.* Patterns of failure and survival for nonoperative treatment of stage c0 distal rectal cancer following neoadjuvant chemoradiation therapy. J Gastrointest Surg 2006; 10(10): 1319-28. [http://dx.doi.org/10.1016/j.gassur.2006.09.005] [PMID: 17175450]

[19] Mothersill C, Moriarty M, Seymour C. Radiotherapy and the potential exploitation of bystander effects. International Journal of Radiation Oncology* Biology* Physics 2004; 58(2): 575-9. [http://dx.doi.org/10.1016/j.ijrobp.2003.09.038]

[20] Mothersill C, Seymour RJ, Seymour CB. Bystander effects in repair-deficient cell lines. Radiat Res

2004; 161(3): 256-63.
[http://dx.doi.org/10.1667/RR3136] [PMID: 14982489]

[21] Paré L, Marcuello E, Altés A, *et al.* Pharmacogenetic prediction of clinical outcome in advanced colorectal cancer patients receiving oxaliplatin/5-fluorouracil as first-line chemotherapy. Br J Cancer 2008; 99(7): 1050-5.
[http://dx.doi.org/10.1038/sj.bjc.6604671] [PMID: 18797464]

[22] Liang J, Jiang T, Yao RY, Liu ZM, Lv HY, Qi WW. The combination of ERCC1 and XRCC1 gene polymorphisms better predicts clinical outcome to oxaliplatin-based chemotherapy in metastatic colorectal cancer. Cancer Chemother Pharmacol 2010; 66(3): 493-500.
[http://dx.doi.org/10.1007/s00280-009-1186-3] [PMID: 19960344]

[23] Lv H, Li Q, Qiu W, *et al.* Genetic polymorphism of XRCC1 correlated with response to oxaliplatin-based chemotherapy in advanced colorectal cancer. Pathol Oncol Res 2012; 18(4): 1009-14.
[http://dx.doi.org/10.1007/s12253-012-9536-6] [PMID: 22549274]

[24] Lunn RM, Helzlsouer KJ, Parshad R, *et al.* XPD polymorphisms: effects on DNA repair proficiency. carcinogenesis 2000; 21(4): 551-5.

[25] Schauer M, Janssen K-P, Rimkus C, *et al.* Microarray-based response prediction in esophageal adenocarcinoma. Clin Cancer Res 2010; 16(1): 330-7.
[http://dx.doi.org/10.1158/1078-0432.CCR-09-1673] [PMID: 20028767]

[26] Rimkus C, Friederichs J, Boulesteix AL, *et al.* Microarray-based prediction of tumor response to neoadjuvant radiochemotherapy of patients with locally advanced rectal cancer. Clin Gastroenterol Hepatol 2008; 6(1): 53-61.
[http://dx.doi.org/10.1016/j.cgh.2007.10.022] [PMID: 18166477]

[27] Kim I-J, Lim S-B, Kang HC, *et al.* Microarray gene expression profiling for predicting complete response to preoperative chemoradiotherapy in patients with advanced rectal cancer. Dis Colon Rectum 2007; 50(9): 1342-53.
[http://dx.doi.org/10.1007/s10350-007-277-7] [PMID: 17665260]

[28] Larsson O, Wennmalm K, Sandberg R. Comparative microarray analysis. OMICS 2006; 10(3): 381-97.
[http://dx.doi.org/10.1089/omi.2006.10.381] [PMID: 17069515]

[29] Mei M, Ren Y, Zhou X, *et al.* Downregulation of miR-21 enhances chemotherapeutic effect of taxol in breast carcinoma cells. Technol Cancer Res Treat 2010; 9(1): 77-86.
[http://dx.doi.org/10.1177/153303461000900109] [PMID: 20082533]

[30] Yang L, Belaguli N, Berger DH. MicroRNA and colorectal cancer. World J Surg 2009; 33(4): 638-46.
[http://dx.doi.org/10.1007/s00268-008-9865-5] [PMID: 19123024]

[31] D'Angelo E, Fassan M, Maretto I, *et al.* Serum miR-125b is a non-invasive predictive biomarker of the pre-operative chemoradiotherapy responsiveness in patients with rectal adenocarcinoma. Oncotarget 2016; 7(19): 28647-57.
[http://dx.doi.org/10.18632/oncotarget.8725] [PMID: 27081702]

[32] Yu J, Li N, Wang X, *et al.* Circulating serum microRNA-345 correlates with unfavorable pathological response to preoperative chemoradiotherapy in locally advanced rectal cancer. Oncotarget 2016; 7(39): 64233-43.
[http://dx.doi.org/10.18632/oncotarget.11649] [PMID: 27572313]

[33] Wang LG, Gu J. Serum microRNA-29a is a promising novel marker for early detection of colorectal liver metastasis. Cancer Epidemiol 2012; 36(1): e61-7.
[http://dx.doi.org/10.1016/j.canep.2011.05.002] [PMID: 22018950]

[34] Brunet Vega A, Pericay C, Moya I, *et al.* microRNA expression profile in stage III colorectal cancer: circulating miR-18a and miR-29a as promising biomarkers. Oncol Rep 2013; 30(1): 320-6.
[http://dx.doi.org/10.3892/or.2013.2475] [PMID: 23673725]

[35] Simmer F, Venderbosch S, Dijkstra JR, *et al.* MicroRNA-143 is a putative predictive factor for the response to fluoropyrimidine-based chemotherapy in patients with metastatic colorectal cancer. Oncotarget 2015; 6(26): 22996-3007.
[http://dx.doi.org/10.18632/oncotarget.4035] [PMID: 26392389]

[36] Lin YX, Yu F, Gao N, Sheng JP, Qiu JZ, Hu BC. microRNA-143 protects cells from DNA damage-induced killing by downregulating FHIT expression. Cancer Biother Radiopharm 2011; 26(3): 365-72.
[http://dx.doi.org/10.1089/cbr.2010.0914] [PMID: 21711110]

[37] Taback B, Hoon DS. Circulating nucleic acids in plasma and serum: past, present and future. Curr Opin Mol Ther 2004; 6(3): 273-8.
[PMID: 15264429]

[38] Gormally E, Caboux E, Vineis P, Hainaut P. Circulating free DNA in plasma or serum as biomarker of carcinogenesis: practical aspects and biological significance. Mutat Res 2007; 635(2-3): 105-17.
[http://dx.doi.org/10.1016/j.mrrev.2006.11.002] [PMID: 17257890]

[39] Wang BG, Huang H-Y, Chen Y-C, *et al.* Increased plasma DNA integrity in cancer patients. Cancer Res 2003; 63(14): 3966-8.
[PMID: 12873992]

[40] Diehl F, Li M, Dressman D, *et al.* Detection and quantification of mutations in the plasma of patients with colorectal tumors. Proc Natl Acad Sci USA 2005; 102(45): 16368-73.
[http://dx.doi.org/10.1073/pnas.0507904102] [PMID: 16258065]

[41] Umetani N, Kim J, Hiramatsu S, *et al.* Increased integrity of free circulating DNA in sera of patients with colorectal or periampullary cancer: direct quantitative PCR for ALU repeats. Clin Chem 2006; 52(6): 1062-9.
[http://dx.doi.org/10.1373/clinchem.2006.068577] [PMID: 16723681]

[42] Agostini M, Pucciarelli S, Enzo MV, *et al.* Circulating cell-free DNA: a promising marker of pathologic tumor response in rectal cancer patients receiving preoperative chemoradiotherapy. Ann Surg Oncol 2011; 18(9): 2461-8.
[http://dx.doi.org/10.1245/s10434-011-1638-y] [PMID: 21416156]

[43] Zitt M, Müller HM, Rochel M, *et al.* Circulating cell-free DNA in plasma of locally advanced rectal cancer patients undergoing preoperative chemoradiation: a potential diagnostic tool for therapy monitoring. Dis Markers 2008; 25(3): 159-65.
[http://dx.doi.org/10.1155/2008/598071] [PMID: 19096128]

[44] Kuhl M, Schalhorn Lsmba, Seidel D. Apoptosis in serum of patients with solid tumours. Anticancer Res 1999; 9: 2721-4.

[45] Leon SA, Shapiro B, Sklaroff DM, Yaros MJ. Free DNA in the serum of cancer patients and the effect of therapy. Cancer Res 1977; 37(3): 646-50.
[PMID: 837366]

[46] Agostini M, Crotti S, Bedin C, *et al.* Predictive response biomarkers in rectal cancer neoadjuvant treatment. Front Biosci (Schol Ed) 2014; 6(1): 110-9.
[http://dx.doi.org/10.2741/S418] [PMID: 24389265]

[47] Tsui NB, Chim SS, Chiu RW, *et al.* Systematic micro-array based identification of placental mRNA in maternal plasma: towards non-invasive prenatal gene expression profiling. J Med Genet 2004; 41(6): 461-7.
[http://dx.doi.org/10.1136/jmg.2003.016881] [PMID: 15173234]

[48] Pucciarelli S, Rampazzo E, Briarava M, *et al.* Telomere-specific reverse transcriptase (hTERT) and cell-free RNA in plasma as predictors of pathologic tumor response in rectal cancer patients receiving neoadjuvant chemoradiotherapy. Ann Surg Oncol 2012; 19(9): 3089-96.
[http://dx.doi.org/10.1245/s10434-012-2272-z] [PMID: 22395986]

[49] Tzimagiorgis G, Michailidou EZ, Kritis A, Markopoulos AK, Kouidou S. Recovering circulating

extracellular or cell-free RNA from bodily fluids. Cancer Epidemiol 2011; 35(6): 580-9. [http://dx.doi.org/10.1016/j.canep.2011.02.016] [PMID: 21514265]

[50] El-Hefnawy T, Raja S, Kelly L, *et al.* Characterization of amplifiable, circulating RNA in plasma and its potential as a tool for cancer diagnostics. Clin Chem 2004; 50(3): 564-73. [http://dx.doi.org/10.1373/clinchem.2003.028506] [PMID: 14718398]

[51] Terrin L, Rampazzo E, Pucciarelli S, *et al.* Relationship between tumor and plasma levels of hTERT mRNA in patients with colorectal cancer: implications for monitoring of neoplastic disease. Clin Cancer Res 2008; 14(22): 7444-51. [http://dx.doi.org/10.1158/1078-0432.CCR-08-0478] [PMID: 19010861]

[52] Lawrie CH, Gal S, Dunlop HM, *et al.* Detection of elevated levels of tumour-associated microRNAs in serum of patients with diffuse large B-cell lymphoma. Br J Haematol 2008; 141(5): 672-5. [http://dx.doi.org/10.1111/j.1365-2141.2008.07077.x] [PMID: 18318758]

[53] Zhang HL, Yang LF, Zhu Y, *et al.* Serum miRNA-21: elevated levels in patients with metastatic hormone-refractory prostate cancer and potential predictive factor for the efficacy of docetaxel-based chemotherapy. Prostate 2011; 71(3): 326-31. [http://dx.doi.org/10.1002/pros.21246] [PMID: 20842666]

[54] Karpova MA, Moshkovskii SA, Toropygin IY, Archakov AI. Cancer-specific MALDI-TOF profiles of blood serum and plasma: biological meaning and perspectives. J Proteomics 2010; 73(3): 537-51. [http://dx.doi.org/10.1016/j.jprot.2009.09.011] [PMID: 19782778]

[55] Pietrowska M, Polańska J, Suwiński R, *et al.* Comparison of peptide cancer signatures identified by mass spectrometry in serum of patients with head and neck, lung and colorectal cancers: association with tumor progression. Int J Oncol 2012; 40(1): 148-56. [PMID: 21894432]

[56] He K, Wen X-Y, Li A-L, *et al.* Serum peptidome variations in a healthy population: reference to identify cancer-specific peptides. PLoS One 2013; 8(5): e63724. [http://dx.doi.org/10.1371/journal.pone.0063724] [PMID: 23667664]

[57] Seraglia R, Ragazzi E, Vogliardi S, *et al.* Search of plasma markers for colorectal cancer by matrix-assisted laser desorption/ionization mass spectrometry. J Mass Spectrom 2005; 40(1): 123-6. [http://dx.doi.org/10.1002/jms.769] [PMID: 15584010]

[58] Villanueva J, Shaffer DR, Philip J, *et al.* Differential exoprotease activities confer tumor-specific serum peptidome patterns. J Clin Invest 2006; 116(1): 271-84. [http://dx.doi.org/10.1172/JCI26022] [PMID: 16395409]

[59] Shen Y, Tolić N, Liu T, *et al.* Blood peptidome-degradome profile of breast cancer. PLoS One 2010; 5(10): e13133. [http://dx.doi.org/10.1371/journal.pone.0013133] [PMID: 20976186]

[60] Crotti S, Enzo MV, Bedin C, *et al.* Clinical predictive circulating peptides in rectal cancer patients treated with neoadjuvant chemoradiotherapy. J Cell Physiol 2015; 230(8): 1822-8. [http://dx.doi.org/10.1002/jcp.24894] [PMID: 25522009]

[61] Smith FM, Gallagher WM, Fox E, *et al.* Combination of SELDI-TOF-MS and data mining provides early-stage response prediction for rectal tumors undergoing multimodal neoadjuvant therapy. Ann Surg 2007; 245(2): 259-66. [http://dx.doi.org/10.1097/01.sla.0000245577.68151.bd] [PMID: 17245180]

[62] Helgason HH, Engwegen JY, Zapatka M, *et al.* Identification of serum proteins as prognostic and predictive markers of colorectal cancer using surface enhanced laser desorption ionization-time of flight mass spectrometry. Oncol Rep 2010; 24(1): 57-64. [http://dx.doi.org/10.3892/or_00000828] [PMID: 20514444]

[63] Spolverato G, Pucciarelli S, Bertorelle R, De Rossi A, Nitti D. Predictive factors of the response of rectal cancer to neoadjuvant radiochemotherapy. Cancers (Basel) 2011; 3(2): 2176-94.

[http://dx.doi.org/10.3390/cancers3022176] [PMID: 24212803]

[64] Kuremsky JG, Tepper JE, McLeod HL. Biomarkers for response to neoadjuvant chemoradiation for rectal cancer. International Journal of Radiation Oncology* Biology* Physics 2009; 74(3): 673-88. [http://dx.doi.org/10.1016/j.ijrobp.2009.03.003]

[65] Spindler K-LG, Nielsen JN, Lindebjerg J, Brandslund I, Jakobsen A. Prediction of response to chemoradiation in rectal cancer by a gene polymorphism in the epidermal growth factor receptor promoter region. International Journal of Radiation Oncology* Biology* Physics 2006; 66(2): 500-4. [http://dx.doi.org/10.1016/j.ijrobp.2006.03.030]

[66] Bengala C, Bettelli S, Bertolini F, *et al.* Epidermal growth factor receptor gene copy number, K-ras mutation and pathological response to preoperative cetuximab, 5-FU and radiation therapy in locally advanced rectal cancer. Ann Oncol 2009; 20(3): 469-74. [http://dx.doi.org/10.1093/annonc/mdn647] [PMID: 19095777]

[67] Smith FM, Reynolds JV, Kay EW, Crotty P, Murphy JO, Hollywood D, *et al.* COX-2 overexpression in pretreatment biopsies predicts response of rectal cancers to neoadjuvant radiochemotherapy. International Journal of Radiation Oncology* Biology* Physics 2006; 64(2): 466-72.

[68] Min BS, Choi YJ, Pyo HR, *et al.* Cyclooxygenase-2 expression in pretreatment biopsy as a predictor of tumor responses after preoperative chemoradiation in rectal cancer. Arch Surg 2008; 143(11): 1091-7. [http://dx.doi.org/10.1001/archsurg.143.11.1091] [PMID: 19015468]

[69] Pasche B, Mulcahy M, Benson AB III. Molecular markers in prognosis of colorectal cancer and prediction of response to treatment. Best Pract Res Clin Gastroenterol 2002; 16(2): 331-45. [http://dx.doi.org/10.1053/bega.2002.0289] [PMID: 11969242]

[70] Wagener C, Bargou RC, Daniel PT, *et al.* Induction of the death-promoting gene bax-α sensitizes cultured breast-cancer cells to drug-induced apoptosis. Int J Cancer 1996; 67(1): 138-41. [http://dx.doi.org/10.1002/(SICI)1097-0215(19960703)67:1<138::AID-IJC22>3.0.CO;2-9] [PMID: 8690514]

[71] Yamaguchi H, Bhalla K, Wang H-G. Bax plays a pivotal role in thapsigargin-induced apoptosis of human colon cancer HCT116 cells by controlling Smac/Diablo and Omi/HtrA2 release from mitochondria. Cancer Res 2003; 63(7): 1483-9. [PMID: 12670894]

[72] Huerta S, Hrom J, Gao X, *et al.* Tissue microarray constructs to predict a response to chemoradiation in rectal cancer. Dig Liver Dis 2010; 42(10): 679-84. [http://dx.doi.org/10.1016/j.dld.2010.02.003] [PMID: 20227932]

[73] Kikuchi M, Mikami T, Sato T, *et al.* High Ki67, Bax, and thymidylate synthase expression well correlates with response to chemoradiation therapy in locally advanced rectal cancers: proposal of a logistic model for prediction. Br J Cancer 2009; 101(1): 116-23. [http://dx.doi.org/10.1038/sj.bjc.6605105] [PMID: 19491899]

[74] Harper JW, Adami GR, Wei N, Keyomarsi K, Elledge SJ. The p21 Cdk-interacting protein Cip1 is a potent inhibitor of G1 cyclin-dependent kinases. Cell 1993; 75(4): 805-16. [http://dx.doi.org/10.1016/0092-8674(93)90499-G] [PMID: 8242751]

[75] el-Deiry WS, Harper JW, O'Connor PM, *et al.* WAF1/CIP1 is induced in p53-mediated G1 arrest and apoptosis. Cancer Res 1994; 54(5): 1169-74. [PMID: 8118801]

[76] Roninson IB. Oncogenic functions of tumour suppressor p21(Waf1/Cip1/Sdi1): association with cell senescence and tumour-promoting activities of stromal fibroblasts. Cancer Lett 2002; 179(1): 1-14. [http://dx.doi.org/10.1016/S0304-3835(01)00847-3] [PMID: 11880176]

[77] Sturm I, Rau B, Schlag PM, *et al.* Genetic dissection of apoptosis and cell cycle control in response of colorectal cancer treated with preoperative radiochemotherapy. BMC Cancer 2006; 6(1): 124.

[http://dx.doi.org/10.1186/1471-2407-6-124] [PMID: 16686938]

[78] Jakob C, Liersch T, Meyer W, Becker H, Baretton GB, Aust DE. Predictive value of Ki67 and p53 in locally advanced rectal cancer: correlation with thymidylate synthase and histopathological tumor regression after neoadjuvant 5-FU-based chemoradiotherapy. World J Gastroenterol 2008; 14(7): 1060-6.
[http://dx.doi.org/10.3748/wjg.14.1060] [PMID: 18286688]

[79] Harris L, Fritsche H, Mennel R, *et al.* American Society of Clinical Oncology 2007 update of recommendations for the use of tumor markers in breast cancer. J Clin Oncol 2007; 25(33): 5287-312.
[http://dx.doi.org/10.1200/JCO.2007.14.2364] [PMID: 17954709]

[80] Berling B, Kolbinger F, Grunert F, *et al.* Cloning of a carcinoembryonic antigen gene family member expressed in leukocytes of chronic myeloid leukemia patients and bone marrow. Cancer Res 1990; 50(20): 6534-9.
[PMID: 2208113]

[81] Marić P, Ozretić P, Levanat S, Oresković S, Antunac K, Beketić-Oresković L. Tumor markers in breast cancer--evaluation of their clinical usefulness. Coll Antropol 2011; 35(1): 241-7.
[PMID: 21661378]

[82] Blumenthal RD, Hansen HJ, Goldenberg DM. Inhibition of adhesion, invasion, and metastasis by antibodies targeting CEACAM6 (NCA-90) and CEACAM5 (Carcinoembryonic Antigen). Cancer Res 2005; 65(19): 8809-17.
[http://dx.doi.org/10.1158/0008-5472.CAN-05-0420] [PMID: 16204051]

[83] Park YA, Sohn SK, Seong J, *et al.* Serum CEA as a predictor for the response to preoperative chemoradiation in rectal cancer. J Surg Oncol 2006; 93(2): 145-50.
[http://dx.doi.org/10.1002/jso.20320] [PMID: 16425302]

[84] Das P, Skibber JM, Rodriguez-Bigas MA, *et al.* Predictors of tumor response and downstaging in patients who receive preoperative chemoradiation for rectal cancer. Cancer 2007; 109(9): 1750-5.
[http://dx.doi.org/10.1002/cncr.22625] [PMID: 17387743]

[85] Moureau-Zabotto L, Farnault B, de Chaisemartin C, *et al.* Predictive factors of tumor response after neoadjuvant chemoradiation for locally advanced rectal cancer. International Journal of Radiation Oncology* Biology* Physics 2011; 80(2): 483-91.
[http://dx.doi.org/10.1016/j.ijrobp.2010.02.025]

[86] Perez RO, São Julião GP, Habr-Gama A, *et al.* The role of carcinoembriogenic antigen in predicting response and survival to neoadjuvant chemoradiotherapy for distal rectal cancer. Dis Colon Rectum 2009; 52(6): 1137-43.
[http://dx.doi.org/10.1007/DCR.0b013e31819ef76b] [PMID: 19581858]

[87] Moral M, Fdez-Aceñero MJ, Cuberes R, Suárez A. Factors influencing prognosis after neo-adjuvant chemoradiation therapy for rectal carcinoma. Acta Chir Belg 2009; 109(3): 345-51.
[http://dx.doi.org/10.1080/00015458.2009.11680437] [PMID: 19943591]

[88] Khakoo S, Georgiou A, Gerlinger M, Cunningham D, Starling N. Circulating tumour DNA, a promising biomarker for the management of colorectal cancer. Crit Rev Oncol Hematol 2018; 122: 72-82.
[http://dx.doi.org/10.1016/j.critrevonc.2017.12.002] [PMID: 29458792]

[89] Jahr S, Hentze H, Englisch S, *et al.* DNA fragments in the blood plasma of cancer patients: quantitations and evidence for their origin from apoptotic and necrotic cells. Cancer Res 2001; 61(4): 1659-65.
[PMID: 11245480]

[90] Tie J, Cohen JD, Wang Y, *et al.* Serial circulating tumour DNA analysis during multimodality treatment of locally advanced rectal cancer: a prospective biomarker study. Gut 2019; 68(4): 663-71.
[http://dx.doi.org/10.1136/gutjnl-2017-315852] [PMID: 29420226]

[91] Lim SH, Becker TM, Chua W, *et al.* Circulating tumour cells and circulating free nucleic acid as prognostic and predictive biomarkers in colorectal cancer. Cancer Lett 2014; 346(1): 24-33. [http://dx.doi.org/10.1016/j.canlet.2013.12.019] [PMID: 24368189]

[92] Groot Koerkamp B, Rahbari NN, Büchler MW, Koch M, Weitz J. Circulating tumor cells and prognosis of patients with resectable colorectal liver metastases or widespread metastatic colorectal cancer: a meta-analysis. Ann Surg Oncol 2013; 20(7): 2156-65. [http://dx.doi.org/10.1245/s10434-013-2907-8] [PMID: 23456317]

[93] Cohen SJ, Punt CJ, Iannotti N, *et al.* Relationship of circulating tumor cells to tumor response, progression-free survival, and overall survival in patients with metastatic colorectal cancer. J Clin Oncol 2008; 26(19): 3213-21. [http://dx.doi.org/10.1200/JCO.2007.15.8923] [PMID: 18591556]

[94] Matsusaka S, Suenaga M, Mishima Y, *et al.* Circulating tumor cells as a surrogate marker for determining response to chemotherapy in Japanese patients with metastatic colorectal cancer. Cancer Sci 2011; 102(6): 1188-92. [http://dx.doi.org/10.1111/j.1349-7006.2011.01926.x] [PMID: 21401804]

[95] Kawahara H, Watanabe K, Toyama Y, Yanagisawa S, Kobayashi S, Yanaga K. Determination of circulating tumor cells for prediction of recurrent colorectal cancer progression. Hepatogastroenterology 2012; 59(119): 2115-8. [http://dx.doi.org/10.5754/hge11970] [PMID: 22328299]

[96] Maestro LM, Sastre J, Rafael SB, *et al.* Circulating tumor cells in solid tumor in metastatic and localized stages. Anticancer Res 2009; 29(11): 4839-43. [PMID: 20032444]

CHAPTER 7

Circulating Biomarkers in the Management of Breast Cancer

Ravindri Jayasinghe[1], **Umesh Jayarajah**[1,*] and **Sanjeewa Seneviratne**[1]

[1] *Department of Surgery, Faculty of Medicine, University of Colombo, Colombo, Sri Lanka*

Abstract: Circulating biomarkers have become a promising modality in the management of many cancers. Similarly, in breast cancer, circulatory biomarkers are useful, non-invasive methods in the diagnosis, prognostication, and evaluation of response to treatment. Invasive surgical biopsies can be potentially replaced by "liquid biopsy," which involves analysing circulatory biomarkers that may reveal features of primary and metastatic disease. Therefore, providing an insight into the cancer biology can be utilised to monitor treatment response, treatment-induced adaptation and tumour and disease progression through non-invasive means. The objective of this review is to provide an overview of the current status of the circulating biomarkers highlighting their promising impact on the management of patients with breast cancer.

Keywords: Biomarkers, Breast cancer, Circulatory biomarkers, Diagnosis, Prognostication, Liquid biopsy.

INTRODUCTION

Breast cancer is the most frequently diagnosed cancer worldwide, being a leading cause of cancer death among women in both developing and developed countries [1, 2]. However, the survival rates are improving due to the advances in screening, diagnosis, understanding of tumour biology and targeted systemic therapy [3]. Furthermore, the role of surgery and invasive approaches in the management of breast cancer has evolved substantially over the past century, where radical approaches have gradually been replaced by more conservative approaches [4]. The triple assessment is used as the 'gold standard' in the assessment of symptomatic breast lumps, which comprises a clinical examination, radiological investigations and histopathological analyses [5]. Circulating biomarkers are promising non-invasive surrogates for tissue diagnosis. They are defined as soluble molecules released to the bloodstream by tumour cells or cells in the

* **Corresponding author Umesh Jayarajah:** Department of Surgery, Faculty of Medicine, University of Colombo, Colombo, Sri Lanka; E-mail: umeshe.jaya@gmail.com

Pınar Atukeren and Hafize Uzun (Eds.)

tumour microenvironment or response to the tumour [6]. Analysis of these biomarkers would provide useful insight into the molecular makeup of cancer.

The study of circulatory biomarkers in the form of circulating tumour cells (CTC) dates back to the late 1800s although, they were studied extensively after the mid-20th century. Their presence was generally considered to be associated with a poorer prognosis [7]. Similar to other solid malignancies, breast cancer is associated with changes in immune [8], glycoproteomic [9], proteomics [10], and nucleic acid [11] biomarkers in the blood, which arise directly from the tumour cells. A measure of these indicators can be acquired by simply drawing peripheral blood, which is less invasive and more feasible compared with tissue biopsy. With the advancements of molecular diagnostics, this method of "liquid biopsy" for biomarkers helps immensely in clinical management [12].

Although many potential biomarkers have been identified, only a few are used in clinical practice which include carcinoembryonic antigen (CEA), MUC-1 protein, human epidermal growth factor receptor-2 (HER-2), and cytokeratins [7]. Recently the interest has shifted towards other markers, including cell-free plasma DNA (cfDNA), circulating tumour cells (CTCs) and microRNAs (miRNAs) [13]. However, there is still an ongoing debate concerning their usefulness in patient monitoring, follow-up and assessment of response to therapy. These biomarkers lack reliability when considered singularly due to their intrinsic lack of specificity and sensitivity. However, certain biomarkers such as circulating tumour cells likely will have a predictive value in specific situations including diagnosis of metastatic carcinoma [14]. With the advancing interest in these markers, it is evident that they provide valuable information on the status of malignancy both qualitatively and quantitatively [14]. These biomarkers are useful in analysing the molecular characteristics of cancer cells in instances where tissue samples cannot be obtained or are not available and as a valuable tool in monitoring the therapeutic response of the tumour and to assess residual disease after treatment [13]. The purpose of this review is to objectively summarize the current recommendations on different types of circulating biomarkers in breast cancer, their usefulness in the diagnosis, prognosis, and monitoring of response to treatment.

ESTABLISHED BIOMARKERS

Carcinoembryonic Antigen (CEA)

CEA is a serum-based glycosylated oncofoetal glycoprotein expressed in normal mucosal cells and overexpressed by adenocarcinomas of the colon, rectum, breast, pancreas, and lungs [15, 16]. CEA is useful in the monitoring metastatic breast cancer during treatment [15, 16]. Immunobiological studies have revealed that the domain structure of CEA proteins belonging to a large family of glycoproteins closely resembles the gamma heavy chain of the immunoglobulin IgG [6]. Being a member of the immunoglobulin superfamily, CEA attaches on to cell membranes by a glycosyl phosphatidylinositol (GPI) anchor [17]. The exact mechanism on how it is released to the extracellular matrix still remains unclear. However, experiments have revealed that similar to other GPI anchored proteins, CEA could be released through a GPI anchor cleavage catalysis, mediated by an endogenous enzyme called glycosylphosphatidylinositol-specific phospholipase D [18].

Although the exact function of CEA has not been identified yet, recent studies portray an involvement in cancer growth, invasion and metastasis by playing a role in cell migration [19]. Elevated CEA levels are observed in approximately 50-60% of patients with metastatic breast cancer [20]. However, this marker can only be accurately interpreted in combination with normal levels of CA15-3 and CA 27.29 [20]. Increased levels of CEA and CA 15-3 have been shown to be independent prognostic factors of metastatic breast cancer. CA 15-3 elevation is more marked in younger patients while CEA is a good prognostic marker in the context of older patients with estrogen receptor (ER) negative disease. The two biomarkers were also shown to provide information on recurrence and overall survival in metastatic breast cancer [21].

Elevated levels of CEA in the preoperative context generally devise a poorer prognosis when interpreted in combination with CA 15-3 [22]. Ebeling at al studied a sample of 1046 women on the association of CEA and CA 15-3 with disease free survival and death from breast cancer [23]. In that study, the postoperative decrease of CEA was an independent prognostic factor for both these study outcomes. Furthermore, multivariate comparison of preoperative and post-operative CEA levels showed a decrease in postoperative levels by more than 33% to be associated with a higher risk of relapse and death [23]. This finding was further confirmed by a similar study which also showed that raised preoperative values of CEA was associated with overall unfavourable cancer status with regards to tumour burden, lymph-node metastasis and tumour stage leading to poorer clinical outcomes [21]. Furthermore, this study also showed that reduced CEA and CA 15-3 levels predicted an overall better outcome [21]. It is quite noteworthy that the reliability and the prognostic values improved when the two biomarkers are used in combination rather than in singular where both the sensitivity and the specificity dropped substantially [21]. Although these classic biomarkers were found to be of considerable clinical importance, they may not be objectively representative of a reliable panel in the context of a liquid biopsy.

MUC-1

Cancer antigen 15-3 (CA15-3), a soluble form of MUC-1 protein, is a type I transmembrane glycoprotein expressed normally on the external layer of epithelial cells of ducts and lumens [24]. It physiologically lines ducts and lumens in different parts of the body and is found to be related to protection, lubrication, and hydration of external surfaces of epithelial tissue layers [24]. MUC-1 is found strongly expressed in epithelial surfaces of ducts and glands and also in columnar and goblet cells of the epithelial tissues where it blocks the degenerative enzymatic activity of microbes preventing them from invading the cell wall [25]. MUC-1 levels are upregulated in epithelial cancer cells and continue to rise with cancer progression and metastasis [26]. Furthermore, it plays an important role in cancer immune surveillance by blocking the immune cells from detecting the tumour cells, protecting the cancer cells from immune mediated clearance [27]. MUC-1 is overexpressed in breast cancer cells with aberrant glycosylated patterns helping in identification. CA 15-3 is thus considered the soluble biomarker of MUC-1 antigen [22]. Therefore, the expression of MUC-1 is considered to be associated with the aggressiveness of breast cancer and its clinical use depends on the serum levels of its soluble form CA 15-3 [7].

CA 15-3 lacks both specificity and sensitivity when considered in isolation. CA 15-3 is detected in increased levels in the presence of other neoplasms including lung, gastric, pancreatic, and ovarian cancers along with other conditions including chronic hepatitis, hypothyroidism and cirrhosis [22, 28]. Nevertheless, this marker had been shown to be of moderate prognostic value in metastatic breast cancer. Increased preoperative levels are predictive of unfavourable outcomes in relation to disease free survival and overall survival independent of lymph node involvement and tumour size. Therefore, CA 15-3 has a possible use in the early detection of micrometastases [29, 30]. Another potential and more reliable use of this biomarker is in the monitoring of response to treatment following systemic therapies. It can predict whether a patient would benefit and response to systemic therapies when the disease status cannot be evaluated by conventional methods. However, expert panels have recommended serial measurements only in correlation with conventional diagnostic and prognostic procedures due to lack of clinical evidence. Furthermore, at present, as some panels recommend its use in follow-up of advanced breast cancer by measuring preoperative levels and 3-monthly thereafter in patients receiving endocrine therapy [31]. A reduction by 50% of serum CA15-3 levels imply treatment efficacy [28].

HER-2

Human epidermal growth factor receptor-2 (HER2) is expressed in 20-30% of all breast cancers [32]. HER2 gene may be overexpressed and amplified in certain types of breast cancers which result in 40-100 fold increase in the expression of HER2 protein on the tumour cell surface. This abnormal activation results in a cascade of signalling events with the activation of a series of transcription factors which regulate genes involving cell proliferation, survival, differentiation and invasion [32]. It is known to predict the aggressiveness of the tumour, and also acts as a marker of therapeutic efficacy of targeted therapies including monoclonal antibodies such as trastuzumab and pertuzumab demonstrating clinical efficacy in HER2 positive breast cancer treatment [33].

In the light of new research, it was shown that the extracellular domain (ECD) of HER2 can shed and therefore, be detected in the serum of breast cancer patients which may provide useful information on tumour status. Quantitative determination of HER2 is performed by determination of its serum ECD. This is useful in explaining the mechanisms of HER2 negative breast cancer patients responding to trastuzumab-based therapies [34]. Therefore, the quantitative assessment of serum HER2 in primary breast cancer will increase the sensitivity of tissue determination tests using immunohistochemical (IHC) and fluorescence *in-situ* hybridization (FISH) methods [35]. Recent studies have also shown that the ECD levels correlate with overexpression of tissue HER2 and the postmenopausal status. Furthermore, increased levels of serum ECD and CA15-3 are considered as independent prognostic factors than determination of HER2 in tissue samples [36]. The ECD measurements could potentially be of benefit in the context of advanced breast cancer. It is shown that reduction of serum HER2-ECD levels following trastuzumab-based therapy can become a real-time indicator predicting the success of therapy. In addition to HER2 tissue status, serum ECD levels are also shown to be clearly associated with the presence of metastasis, the CA15-3 and CEA levels, and the number of metastases [37]. Even though serial measurements are not currently recommended, a study done by Lee *et al* had showed augmented ECD serum levels in tissues of 15% of HER2 positive patients with a linear correlation with the increased aggressiveness of tumours [38] Furthermore, the increase in serum ECD levels were shown to be an independent prognostic factor in disease free survival and cancer specific survival specifically in HER2 positive cases [38]. Therefore, ECD that is shed from HER2 can be considered valuable in the context of HER2 positive breast cancer patients and hence, both the tissue and serum HER2 levels are of importance.

Cytokeratins (CK)

They are intermediate filaments playing a role in the organization of the cytoskeleton in epithelial cells for maintenance of cellular morphology, fixation of the nucleus and protection of the cell from both mechanical and non-mechanical stressors [39]. The expression of CKs are differentiation dependant. They comprise 20 related polypeptides grouped as Type I and Type II. Keratins are used as immunohistochemical markers in diagnosing tumour pathology as the pattern of keratin expression of normal epithelia is maintained to a larger extent even by their neoplastic counterparts [39, 40]. CKs comprise a group of widely used tumour markers such as tissue peptide antigen (TPA), tissue polypeptide specific antigen (TPS), and Cyfra 21.1 [41]. Usually, combinations of CK8/CK18 and CK9/CK19 are commonly expressed in epithelial cells from many tissues and are also found in the majority of breast carcinomas [42]. A growing body of evidence over the last few years has disclosed the role of CKs in the pathophysiology of cancer particularly in the case of hormonally responsive breast cancer. It is shown that CK18 plays a regulatory role in estrogen receptor-alpha mediated signalling and estrogen-stimulated cell cycle progression in breast cancer cells [43].

Specific Classes of Circulatory Biomarkers

Proteins

Circulating biomarkers in breast cancer have been identified using various techniques such as targeted sandwich ELISA micro assays, and plasma proteomic profiling. Biomarkers that are differentially detected in plasma have been identified using plasma proteomic profiling [12]. Raised levels of epidermal growth factor receptor (EGFR) have been detected in post-menopausal hormone therapy users up to 17 months before the diagnosis of oestrogen receptor positive breast cancer [44]. Serum levels of peptides have been shown to change a year after completing chemotherapy after surgical resection of breast cancer. The growing body of research has shown detection of elevated levels of proteins such as afamin, apolipoprotein E, and the isoform-1 of inter-alpha trypsin inhibitor heavy chain H4 (ITIH4) as well as reduced levels of ceruloplasmin and alpha 2-macroglobulin prior to breast cancer diagnosis [45]. Exploration of the tumour interstitial fluid (TIF) which is a rich source of tumour secretome is becoming increasingly popular where it facilitates detection biomarkers specific to breast cancer locally, in increased concentration before they get diluted in blood [46, 47]. Othcr potential biomarkers include cytokines, angiogenic factors, metalloproteases, dermcidin and circulating c-reactive protein [12]. Post translational modifications (PTMs) are also potential circulating biomarkers which include oxidised ceruloplasmin, alterations in mucin glycoproteins, and platelet derived growth factor (PDGF) which are differentially detected in plasma from breast cancer patients [12]. However, the majority of these circulating protein biomarkers have to be validated using clinical trials to provide better evidence to be used in clinical practice.

Autoantibodies

Autoantibodies to tumour associated antigens in patients with breast cancer are induced by an alteration in structure or expression of proteins or glycans. They belong to a class of circulating biomarkers that rapidly adapts to clinical diagnostic platforms reflecting the immune surveillance of tumours due to their increased specificity, stability, and ease of detection in patients' sera. Similar to most other biomarkers, when taken in isolation they have low sensitivity and specificity. Therefore, autoantibody combinations are analysed in order to improve the sensitivity of detection. These include plasminogen, centrosomal antigens, nuclear antigens, and alpha-1 antitrypsin [12]. Although these potential biomarkers show promise, these have to be validated using clinical trials prior to routine use in clinical practice.

Circulating Tumour Cells (CTC)

CTC are rare cancer cells that are found in peripheral blood that has a role in tumour dissemination and progression [48]. Diverse methods are used in detecting these CTCs in the bloodstream and disseminated tumour cells (DTC) in bone marrow samples in patients with breast cancer. Their extremely low concentrations in blood or bone marrow makes it challenging to detect [49, 50]. The methods used in detection are, immunohistochemical analysis, immunofluorescence, fluorescence *in situ* hybridization (FISH), flow cytometry, Southern blot, Northern blot, polymerase chain reaction (PCR), reverse transcription PCR (RT-PCR), cell culture, and proteomic techniques [51]. Despite the advances in these fields, the important challenges in using this in clinical practice include the complexity and lack of standardization in assessment of CTCs [13].

CTC enumeration is a known prognostic marker in breast cancer backed by level I evidence [52]. CTCs were shown to correlate closely with the imaging results compared before and after treatment thus show potential benefit in predicting the effectiveness of treatment before detectable changes appear in imaging [52]. CTC are used for cancer characterization and diagnostic purposes. They can also be used in identification of drug targets in selection of suitable therapy, monitoring the therapeutic response, and to acquire important information on the intra-tumour and inter-tumour heterogeneity in the mechanisms of metastasis [53]. Increasing evidence supports the relevance in prognosis to qualitative and quantitative assessment of CTCs in the context on non-metastatic breast cancer. CTCs were detected in adjuvant and neoadjuvant settings in approximately 20-25% of patients with breast cancer [13].

SUCCESS trial comparing the disease free survival (DFS) in patients treated with anthracycline-based and taxane-based chemotherapy in early breast cancer, demonstrated CTCs to persist even after neoadjuvant chemotherapy which correlated significantly with shorter DFS [54]. CTCs are associated with an increased risk of breast cancer related deaths and they continue to be a surrogate biomarker for the presence of residual disease and early tumour dissemination [55, 56]. For patients with metastatic breast cancer, the change in the number of CTCs in therapy is shown to be associated with patient outcomes including disease-free and overall survival. The prediction of clinical outcome in these patients depends on the methodology used for harvesting CTCs. For example the cell search system is known to be more accurate in predicting clinical outcomes in patients with metastatic breast cancer [13]. Furthermore, detection of CTCs are shown to be a reliable prognostic marker in both early and late stage breast cancer [57]. Currently efforts are being made in identifying the molecular characteristics and the enumeration of CTCs which in future may provide promising opportunities in risk-stratification of patients with breast cancer. Furthermore, it may set the trend in using prognostic, predictive and pharmacodynamics biomarkers in personalized therapy [58].

Exosomes

These are nano-sized vesicles in almost all body fluids released from all types of cells, playing a key role in intercellular communication [59]. They comprise of proteins, DNA, mRNA, IncRNA and miRNA which also provide an insight to their cells of origin [60]. Exosomes derived from tumours will enhance cancer progression and metastasis [61] and are shown to be present in increased concentrations in sera from patients with cancer when compared to that of healthy individuals. Breast cancer related exosomes are found to contain unique tRNAs and are found to convert pre-miRNA into miRNA [62, 63].

MicroRNA (miRNA)

These are small RNAs acting in gene regulatory networks that are highly conserved and show altered expression depending on tumour type. They have become promising robust biomarkers in cancer diagnosis, prognostication and monitoring treatment efficacy. MicroRNAs are deregulated in breast cancer with regards to their expression, biological role, and are associated with cancer outcomes and response to treatment [64]. Studies have demonstrated their usefulness in treatment by recognising specific miRNAs responsible for resistance to drugs such as taxol, epirubicin, tyrosine kinase inhibitors, and in HER2 positive breast cancer showing resistance to trastuzumab [65]. In recent studies, miRNAs have been found to be quantifiable, detectable, and stable in cell free components of blood and body fluids making them ideal cancer biomarkers. Furthermore, in a study done by Heneghan *et al*, it was found that breast cancer specific miRNAs were detected in elevated levels in breast cancer patients which was shown to decrease postoperatively to levels similar to that of control subjects indicating curative tumour resection. These specific miRNAs are shown to correlate with estrogen receptor status and nodal status [66].

A growing body of evidence shows miRNAs as promising markers in early detection of breast cancer which involves a single blood test as a pre-screening tool in younger women which would facilitate future decisions on invasive diagnostic tests [67]. In a comprehensive review by van Schooneveld *et al* on the potential role of miRNAs as prognostic and predictive biomarkers in patient management, it was found that miRNAs played a role in predicting individualized therapy and diagnosis correlating with response to treatment. These markers are also shown to be involved in resistance to endocrine treatment. Therefore, finding a correlation with miRNA expression patterns and histological subtypes of breast cancer can lead to future developments in the non-invasive detection of biological characteristics of breast cancer [68].

Circulating Cell-Free DNA (cfDNA)

Small amounts of cell free nucleic acids are present in serum and plasma which likely originates from cancer cells and can be considered as potential genetic material helping in tumour identification associated with molecular alterations [69]. Genetic and epigenetic alterations found in tumour tissue in breast cancer can be identified using cfDNA as a surrogate marker [70]. Furthermore, cfDNA in the plasma of breast cancer patients may have the potential to reveal the HER2 status by detection of HER2 amplification of cfDNA [71]. Studies have also shown that analysis of cfDNA provide information regarding tumour dormancy which can be used in monitoring breast cancer patients before the onset of clinically detectable disease [72]. The use of cfDNA has shown promise in developing personalized treatment at the genomic level for advanced breast cancer [73]. Furthermore, circulating tumour DNA (ctDNA) is identified as a consistent method in monitoring the tumour burden in patients undergoing systemic therapy for metastatic breast cancer [73]. Fluctuating amounts of ctDNA copies were found to be associated with response to therapy, and studies have shown a significant relationship between patient outcomes and ctDNA, where rising levels of ctDNA were shown to be associated with poor overall survival [73].

The use of ctDNA as a biomarker is already established for assessing response to therapy. However, expanding research on tumour related somatic arrangements in blood derived specimens may provide a perspective on the utility as real time liquid biopsy for management of patients with breast cancer in the future. However, the main limitation in the use of cfDNA is in cancers with increased levels of intra-tumour genetic heterogenicity. In such situations with alterations in the genome, using cfDNA will be restricted only to a subpopulation of cells which may hinder the effectiveness of this approach altogether in assessing residual disease [13]. Further studies are needed in assessing the sensitivity, specificity and predictive values with regard to using these biomarkers.

BIOMARKERS FOR DIAGNOSIS AND PROGNOSIS OF BREAST CANCER

In the early stage of the disease, the sensitivity of the circulatory biomarkers is likely to be low. Therefore, it is unlikely that they may become superior to imaging modalities including mammography as a screening modality for breast cancer in the near future [74]. Studies have demonstrated a sensitivity of about 10-15% with CA15-3 in stage I disease. However, the sensitivity increased to 20-25% in stage II disease and 20-45% in stage III disease [75]. However, in combination, CA15-2 and CEA will allow early diagnosis of metastasis in breast cancer in 60-80% of patients [76]. By combining this with cytokeratins, the sensitivity can be increased to at least 90% in metastatic disease [76].

Besides the use of traditional prognostic factors such as the number of positive regional lymph nodes, tumour size and grade, circulatory biomarkers when used in combination with traditional prognostic factors will increase the reliability and may help with advancement towards the era of personalised treatment [77]. A growing interest has been shown in recent years in the validation and development of molecular biomarkers which would provide prognostic information and predict the response to therapy [78]. These prognostic tests developed for breast cancer are referred to as multi-gene, multi-analyte or multi-parameter tests [79]. The validated multigene tests currently used in clinical practice include Oncotype DX®, MammaPrint® and uPA/PAI-1. Oncotype DX is widely used in predicting the outcome of breast cancer which uses RT-PCR to measure the expression of 21 genes at the mRNA level. MammaPrint is validated for giving the prognosis of recurrence which helps guide treatment decisions in breast cancer patients which use expression of 70 hallmark genes [80]. Future studies are needed to evaluate the clinical utility of gene analysis in combination with circulatory biomarkers.

In addition to the above molecular tests, commonly used biomarkers including CEA, CA15-3, HER2 levels can be used in the diagnosis and prognostication of breast cancer. CEA sensitivity increases in symptomatic breast cancer patients. As a tool of prognosis, positive pre-therapeutic levels of CEA will predict worse prognosis and recurrence after primary therapy [7, 21]. Increasing levels of CA15-3 are used to predict advanced disease. Although HER2 level is a poor test for diagnosis, it can be used in monitoring response following trastuzumab treatment in HER2 positive breast cancer patients [81].

MONITORING RESPONSE TO THERAPY AND SURVEILLANCE

The advantage of using biomarkers is that they can be repeatedly assessed *via* non-invasive approaches providing real time information on treatment response and disease progression. CTC can be used in predicting overall disease free survival and early assessment of response to therapy in advanced breast cancer [82]. Treatment resistance is demonstrated to be associated with changes in CTC epithelial and mesenchymal composition [83]. Circulating miRNAs are shown to predict and monitor treatment response. However further research is needed in determining their actual validity [84]. Circulating DNA has shown promise in the management of cancer in allowing the development of minimally invasive and sensitive biomarkers in measuring treatment response and monitoring disease progression [13]. Furthermore, ctDNA has been shown to mirror the clinical course of disease in patients with metastatic breast cancer showing a correlation with tumour burden and is also associated with acquired drug resistance [65]. The main limitations of circulatory biomarkers are the cost and lack of availability in both developed and developing countries. The routine use of such biomarkers with repeated assays during surveillance would substantially increase the cost of cancer care which may not be bearable to many of the health systems, especially in the developing world. Further studies should focus on the cost effectiveness in the utility of these biomarkers in clinical practice.

CONCLUSION

Assessment of circulating biomarkers involves detection of tumour related markers in blood which is a real time non-invasive surrogate for tissue diagnosis. Due to the considerably low sensitivity and specificity, established biomarkers have limited usefulness in the diagnosis and screening of breast cancer. However, they show proven usefulness in relation to monitoring patients during follow-up, assessing response to treatment and detecting recurrence.

CONSENT FOR PUBLICATION

Not applicable.

CONFLICT OF INTEREST

The authors declare no conflict of interest, financial or otherwise.

ACKNOWLEDGEMENTS

Declared none.

REFERENCES

[1] Taghian A, El-Ghamry MN, Merajver S. Overview of the treatment of newly diagnosed, non-metastatic breast cancer. UpToDate. New York: Wolters Kluwer 2016.

[2] Fernando A, Jayarajah U, Prabashani S, Fernando EA, Seneviratne SA. Incidence trends and patterns of breast cancer in Sri Lanka: an analysis of the national cancer database. BMC Cancer 2018; 18(1): 482.
[http://dx.doi.org/10.1186/s12885-018-4408-4] [PMID: 29703165]

[3] Nounou MI, ElAmrawy F, Ahmed N, Abdelraouf K, Goda S, Syed-Sha-Qhattal H. Breast Cancer: Conventional Diagnosis and Treatment Modalities and Recent Patents and Technologies. Breast Cancer: Basic and Clinical Research 2015; 9s2: S29420.
[http://dx.doi.org/10.4137/BCBCR.S29420]

[4] Basnayake O, Jayarajah U, Seneviratne S. Management of axilla in breast cancer: the past, present and the future. Sri Lanka Journal of Surgery 2018; 36(4): 25.
[http://dx.doi.org/10.4038/sljs.v36i4.8555]

[5] Clarke D, Sudhakaran N, Gateley CA. Replace fine needle aspiration cytology with automated core biopsy in the triple assessment of breast cancer. Ann R Coll Surg Engl 2001; 83(2): 110-2.
[PMID: 11320918]

[6] Marić P, Ozretić P, Levanat S, Oresković S, Antunac K, Beketić-Oresković L. Tumor markers in breast cancer--evaluation of their clinical usefulness. Coll Antropol 2011; 35(1): 241-7.
[PMID: 21661378]

[7] Mirabelli P, Incoronato M. Usefulness of traditional serum biomarkers for management of breast cancer patients. Biomed Res Int 2013; 2013: 685641.
[http://dx.doi.org/10.1155/2013/685641]

[8] Kroemer G, Senovilla L, Galluzzi L, André F, Zitvogel L. Natural and therapy-induced immunosurveillance in breast cancer. Nat Med 2015; 21(10): 1128-38.
[http://dx.doi.org/10.1038/nm.3944] [PMID: 26444637]

[9] Boersema PJ, Geiger T, Wiśniewski JR, Mann M. Quantification of the N-glycosylated secretome by super-SILAC during breast cancer progression and in human blood samples. Mol Cell Proteomics 2013; 12(1): 158-71.
[http://dx.doi.org/10.1074/mcp.M112.023614] [PMID: 23090970]

[10] Pernikářová V, Bouchal P. Targeted proteomics of solid cancers: from quantification of known biomarkers towards reading the digital proteome maps. Expert Rev Proteomics 2015; 12(6): 651-67.
[http://dx.doi.org/10.1586/14789450.2015.1094381] [PMID: 26456120]

[11] Schwarzenbach H. Circulating nucleic acids as biomarkers in breast cancer. Breast Cancer Res 2013; 15(5): 211.
[http://dx.doi.org/10.1186/bcr3446] [PMID: 24090167]

[12] Ewaisha R, Gawryletz CD, Anderson KS. Crucial considerations for pipelines to validate circulating biomarkers for breast cancer. Expert Rev Proteomics 2016; 13(2): 201-11.
[http://dx.doi.org/10.1586/14789450.2016.1132170] [PMID: 26653344]

[13] De Mattos-Arruda L, Cortes J, Santarpia L, *et al.* Circulating tumour cells and cell-free DNA as tools for managing breast cancer. Nat Rev Clin Oncol 2013; 10(7): 377-89. [http://dx.doi.org/10.1038/nrclinonc.2013.80] [PMID: 23712187]

[14] Ravelli A, Reuben JM, Lanza F, *et al.* Breast cancer circulating biomarkers: advantages, drawbacks, and new insights. Tumour Biol 2015; 36(9): 6653-65. [http://dx.doi.org/10.1007/s13277-015-3944-7] [PMID: 26307395]

[15] Harris L, Fritsche H, Mennel R, *et al.* American Society of Clinical Oncology 2007 update of recommendations for the use of tumor markers in breast cancer. J Clin Oncol 2007; 25(33): 5287-312. [http://dx.doi.org/10.1200/JCO.2007.14.2364] [PMID: 17954709]

[16] Berling B, Kolbinger F, Grunert F, *et al.* Cloning of a carcinoembryonic antigen gene family member expressed in leukocytes of chronic myeloid leukemia patients and bone marrow. Cancer Res 1990; 50(20): 6534-9. [PMID: 2208113]

[17] Hammarström S, Ed. The carcinoembryonic antigen (CEA) family: structures, suggested functions and expression in normal and malignant tissues. Seminars in cancer biology. Elsevier 1999.

[18] Pakdel A, Naghibalhossaini F, Mokarram P, Jaberipour M, Hosseini A. Regulation of carcinoembryonic antigen release from colorectal cancer cells. Mol Biol Rep 2012; 39(4): 3695-704. [http://dx.doi.org/10.1007/s11033-011-1144-0] [PMID: 21725636]

[19] Blumenthal RD, Hansen HJ, Goldenberg DM. Inhibition of adhesion, invasion, and metastasis by antibodies targeting CEACAM6 (NCA-90) and CEACAM5 (Carcinoembryonic Antigen). Cancer Res 2005; 65(19): 8809-17. [http://dx.doi.org/10.1158/0008-5472.CAN-05-0420] [PMID: 16204051]

[20] Hayes DF, Zurawski VR Jr, Kufe DW. Comparison of circulating CA15-3 and carcinoembryonic antigen levels in patients with breast cancer. J Clin Oncol 1986; 4(10): 1542-50. [http://dx.doi.org/10.1200/JCO.1986.4.10.1542] [PMID: 2428949]

[21] Lee JS, Park S, Park JM, Cho JH, Kim SI, Park B-W. Elevated levels of serum tumor markers CA 15-3 and CEA are prognostic factors for diagnosis of metastatic breast cancers. Breast Cancer Res Treat 2013; 141(3): 477-84. [http://dx.doi.org/10.1007/s10549-013-2695-7] [PMID: 24072270]

[22] Duffy MJ. Serum tumor markers in breast cancer: are they of clinical value? Clin Chem 2006; 52(3): 345-51. [http://dx.doi.org/10.1373/clinchem.2005.059832] [PMID: 16410341]

[23] Ebeling FG, Stieber P, Untch M, *et al.* Serum CEA and CA 15-3 as prognostic factors in primary breast cancer. Br J Cancer 2002; 86(8): 1217-22. [http://dx.doi.org/10.1038/sj.bjc.6600248] [PMID: 11953875]

[24] Brayman M, Thathiah A, Carson DD. MUC1: a multifunctional cell surface component of reproductive tissue epithelia. Reprod Biol Endocrinol 2004; 2(1): 4. [http://dx.doi.org/10.1186/1477-7827-2-4] [PMID: 14711375]

[25] Kufe DW. Mucins in cancer: function, prognosis and therapy. Nat Rev Cancer 2009; 9(12): 874-85. [http://dx.doi.org/10.1038/nrc2761] [PMID: 19935676]

[26] Kufe DW. MUC1-C oncoprotein as a target in breast cancer: activation of signaling pathways and therapeutic approaches. Oncogene 2013; 32(9): 1073-81. [http://dx.doi.org/10.1038/onc.2012.158] [PMID: 22580612]

[27] Suzuki Y, Sutoh M, Hatakeyama S, *et al.* MUC1 carrying core 2 O-glycans functions as a molecular shield against NK cell attack, promoting bladder tumor metastasis. Int J Oncol 2012; 40(6): 1831-8. [PMID: 22446589]

[28] Duffy MJ, Evoy D, McDermott EW. CA 15-3: uses and limitation as a biomarker for breast cancer.

Clin Chim Acta 2010; 411(23-24): 1869-74.
[http://dx.doi.org/10.1016/j.cca.2010.08.039] [PMID: 20816948]

[29] Duffy MJ, Duggan C, Keane R, *et al.* High preoperative CA 15-3 concentrations predict adverse outcome in node-negative and node-positive breast cancer: study of 600 patients with histologically confirmed breast cancer. Clin Chem 2004; 50(3): 559-63.
[http://dx.doi.org/10.1373/clinchem.2003.025288] [PMID: 14726467]

[30] Gion M, Boracchi P, Dittadi R, *et al.* Prognostic role of serum CA15.3 in 362 node-negative breast cancers. An old player for a new game. Eur J Cancer 2002; 38(9): 1181-8.
[http://dx.doi.org/10.1016/S0959-8049(01)00426-9] [PMID: 12044503]

[31] Trillet-Lenoir V, Freyer G. [Advantage of using tumor markers in colorectal and breast cancers]. Bull Cancer 1997; 84(7): 767-8.
[PMID: 9339206]

[32] Tafe LJ, Tsongalis GJ. The human epidermal growth factor receptor 2 (HER2). Clin Chem Lab Med 2011; 50(1): 23-30. [CCLM].
[PMID: 21919545]

[33] Maly JJ, Macrae ER. Pertuzumab in combination with trastuzumab and chemotherapy in the treatment of HER2-positive metastatic breast cancer: safety, efficacy, and progression free survival. Breast Cancer: Basic and Clinical Research 2014; 8
[http://dx.doi.org/10.4137/BCBCR.S9032]

[34] Paik S, Kim C, Wolmark N. HER2 status and benefit from adjuvant trastuzumab in breast cancer. N Engl J Med 2008; 358(13): 1409-11.
[http://dx.doi.org/10.1056/NEJMc0801440] [PMID: 18367751]

[35] Wolff AC, Hammond MEH, Allison KH, *et al.* Human epidermal growth factor receptor 2 testing in breast cancer: American Society of Clinical Oncology/College of American Pathologists clinical practice guideline focused update. Arch Pathol Lab Med 2018; 142(11): 1364-82.
[http://dx.doi.org/10.5858/arpa.2018-0902-SA] [PMID: 29846104]

[36] Di Gioia D, Dresse M, Mayr D, Nagel D, Heinemann V, Stieber P. Serum HER2 in combination with CA 15-3 as a parameter for prognosis in patients with early breast cancer. Clin Chim Acta 2015; 440: 16-22.
[http://dx.doi.org/10.1016/j.cca.2014.11.001] [PMID: 25444743]

[37] Zhou J, Liu Y, Wang T, *et al.* [Serum HER2 ECD level and its clinical significance in advanced breast cancer patients with different molecular subtypes]. Zhonghua Yi Xue Za Zhi 2014; 94(18): 1384-7.
[PMID: 25142988]

[38] Lee SB, Lee JW, Yu JH, *et al.* Preoperative serum HER2 extracellular domain levels in primary invasive breast cancer. BMC Cancer 2014; 14(1): 929.
[http://dx.doi.org/10.1186/1471-2407-14-929] [PMID: 25491647]

[39] Karantza V. Keratins in health and cancer: more than mere epithelial cell markers. Oncogene 2011; 30(2): 127-38.
[http://dx.doi.org/10.1038/onc.2010.456] [PMID: 20890307]

[40] Lehr H-A, Folpe A, Yaziji H, Kommoss F, Gown AM. Cytokeratin 8 immunostaining pattern and E-cadherin expression distinguish lobular from ductal breast carcinoma. Am J Clin Pathol 2000; 114(2): 190-6.
[http://dx.doi.org/10.1309/CPUX-KWEH-7B26-YE19] [PMID: 10941333]

[41] Nagle RB. Intermediate filaments: a review of the basic biology. Am J Surg Pathol 1988; 12 (Suppl. 1): 4-16.
[PMID: 2451443]

[42] Moll R. Cytokeratins in the histological diagnosis of malignant tumors. Int J Biol Markers 1994; 9(2): 63-9.

[http://dx.doi.org/10.1177/172460089400900201] [PMID: 7523543]

[43] Meng Y, Wu Z, Yin X, *et al.* Keratin 18 attenuates estrogen receptor α-mediated signaling by sequestering LRP16 in cytoplasm. BMC Cell Biol 2009; 10(1): 96. [http://dx.doi.org/10.1186/1471-2121-10-96] [PMID: 20035625]

[44] Pitteri SJ, Amon LM, Busald Buson T, *et al.* Detection of elevated plasma levels of epidermal growth factor receptor before breast cancer diagnosis among hormone therapy users. Cancer Res 2010; 70(21): 8598-606. [http://dx.doi.org/10.1158/0008-5472.CAN-10-1676] [PMID: 20959476]

[45] Opstal-van Winden AW, Krop EJ, Kåredal MH, *et al.* Searching for early breast cancer biomarkers by serum protein profiling of pre-diagnostic serum; a nested case-control study. BMC Cancer 2011; 11(1): 381. [http://dx.doi.org/10.1186/1471-2407-11-381] [PMID: 21871081]

[46] Gromov P, Gromova I, Bunkenborg J, *et al.* Up-regulated proteins in the fluid bathing the tumour cell microenvironment as potential serological markers for early detection of cancer of the breast. Mol Oncol 2010; 4(1): 65-89. [http://dx.doi.org/10.1016/j.molonc.2009.11.003] [PMID: 20005186]

[47] Haslene-Hox H, Tenstad O, Wiig H. Interstitial fluid—a reflection of the tumor cell microenvironment and secretome. Biochimica et Biophysica Acta (BBA)-. Proteins and Proteomics 2013; 1834(11): 2336-46. [http://dx.doi.org/10.1016/j.bbapap.2013.01.028] [PMID: 23376185]

[48] Mego M, Mani SA, Cristofanilli M. Molecular mechanisms of metastasis in breast cancer--clinical applications. Nat Rev Clin Oncol 2010; 7(12): 693-701. [http://dx.doi.org/10.1038/nrclinonc.2010.171] [PMID: 20956980]

[49] Zippelius A, Pantel K. RT-PCR-based detection of occult disseminated tumor cells in peripheral blood and bone marrow of patients with solid tumors. An overview. Ann N Y Acad Sci 2000; 906(1): 110-23. [http://dx.doi.org/10.1111/j.1749-6632.2000.tb06600.x] [PMID: 10818606]

[50] Cristofanilli M, Budd GT, Ellis MJ, *et al.* Circulating tumor cells, disease progression, and survival in metastatic breast cancer. N Engl J Med 2004; 351(8): 781-91. [http://dx.doi.org/10.1056/NEJMoa040766] [PMID: 15317891]

[51] Ross JS, Slodkowska EA. Circulating and disseminated tumor cells in the management of breast cancer. Am J Clin Pathol 2009; 132(2): 237-45. [http://dx.doi.org/10.1309/AJCPJI7DEOLKCS6F] [PMID: 19605818]

[52] Nakamura S, Yagata H, Ohno S, *et al.* Multi-center study evaluating circulating tumor cells as a surrogate for response to treatment and overall survival in metastatic breast cancer. Breast Cancer 2010; 17(3): 199-204. [http://dx.doi.org/10.1007/s12282-009-0139-3] [PMID: 19649686]

[53] Parkinson DR, Dracopoli N, Petty BG, *et al.* Considerations in the development of circulating tumor cell technology for clinical use. J Transl Med 2012; 10(1): 138. [http://dx.doi.org/10.1186/1479-5876-10-138] [PMID: 22747748]

[54] Rack B, Schindlbeck C, Andergassen U, Lorenz R, Zwingers T, Schneeweiss A, *et al.* Prognostic relevance of circulating tumor cells in the peripheral blood of primary breast cancer patients. Cancer Res 2010; 70(24) (Suppl.): 93s.

[55] Franken B, de Groot MR, Mastboom WJ, *et al.* Circulating tumor cells, disease recurrence and survival in newly diagnosed breast cancer. Breast Cancer Res 2012; 14(5): R133. [http://dx.doi.org/10.1186/bcr3333] [PMID: 23088337]

[56] Pierga J-Y, Bidard F-C, Mathiot C, *et al.* Circulating tumor cell detection predicts early metastatic relapse after neoadjuvant chemotherapy in large operable and locally advanced breast cancer in a

phase II randomized trial. Clin Cancer Res 2008; 14(21): 7004-10.
[http://dx.doi.org/10.1158/1078-0432.CCR-08-0030] [PMID: 18980996]

[57] Zhang L, Riethdorf S, Wu G, *et al.* Meta-analysis of the prognostic value of circulating tumor cells in breast cancer. Clin Cancer Res 2012; 18(20): 5701-10.
[http://dx.doi.org/10.1158/1078-0432.CCR-12-1587] [PMID: 22908097]

[58] Devriese LA, Voest EE, Beijnen JH, Schellens JH. Circulating tumor cells as pharmacodynamic biomarker in early clinical oncological trials. Cancer Treat Rev 2011; 37(8): 579-89.
[http://dx.doi.org/10.1016/j.ctrv.2011.04.006] [PMID: 21592671]

[59] O'Driscoll L. Expanding on exosomes and ectosomes in cancer. N Engl J Med 2015; 372(24): 2359-62.
[http://dx.doi.org/10.1056/NEJMcibr1503100] [PMID: 26061842]

[60] Yu S, Cao H, Shen B, Feng J. Tumor-derived exosomes in cancer progression and treatment failure. Oncotarget 2015; 6(35): 37151-68.
[http://dx.doi.org/10.18632/oncotarget.6022] [PMID: 26452221]

[61] O'Brien K, Rani S, Corcoran C, *et al.* Exosomes from triple-negative breast cancer cells can transfer phenotypic traits representing their cells of origin to secondary cells. Eur J Cancer 2013; 49(8): 1845-59.
[http://dx.doi.org/10.1016/j.ejca.2013.01.017] [PMID: 23453937]

[62] Melo SA, Sugimoto H, O'Connell JT, *et al.* Cancer exosomes perform cell-independent microRNA biogenesis and promote tumorigenesis. Cancer Cell 2014; 26(5): 707-21.
[http://dx.doi.org/10.1016/j.ccell.2014.09.005] [PMID: 25446899]

[63] Guzman N, Agarwal K, Asthagiri D, *et al.* Breast cancer–specific miR signature unique to extracellular vesicles includes "microRNA-like" tRNA fragments. Mol Cancer Res 2015; 13(5): 891-901.
[http://dx.doi.org/10.1158/1541-7786.MCR-14-0533] [PMID: 25722304]

[64] Mei M, Ren Y, Zhou X, *et al.* Downregulation of miR-21 enhances chemotherapeutic effect of taxol in breast carcinoma cells. Technol Cancer Res Treat 2010; 9(1): 77-86.
[http://dx.doi.org/10.1177/153303461000900109] [PMID: 20082533]

[65] Cappelletti V, Appierto V, Tiberio P, Fina E, Callari M, Daidone MG. Circulating biomarkers for prediction of treatment response. J Natl Cancer Inst Monogr 2015; 2015(51): 60-3.
[http://dx.doi.org/10.1093/jncimonographs/lgv006] [PMID: 26063889]

[66] Heneghan HM, Miller N, Lowery AJ, Sweeney KJ, Newell J, Kerin MJ. Circulating microRNAs as novel minimally invasive biomarkers for breast cancer. Ann Surg 2010; 251(3): 499-505.
[http://dx.doi.org/10.1097/SLA.0b013e3181cc939f] [PMID: 20134314]

[67] Cuk K, Zucknick M, Heil J, *et al.* Circulating microRNAs in plasma as early detection markers for breast cancer. Int J Cancer 2013; 132(7): 1602-12.
[http://dx.doi.org/10.1002/ijc.27799] [PMID: 22927033]

[68] van Schooneveld E, Wildiers H, Vergote I, Vermeulen PB, Dirix LY, Van Laere SJ. Dysregulation of microRNAs in breast cancer and their potential role as prognostic and predictive biomarkers in patient management. Breast Cancer Res 2015; 17(1): 21.
[http://dx.doi.org/10.1186/s13058-015-0526-y] [PMID: 25849621]

[69] Schwarzenbach H, Hoon DS, Pantel K. Cell-free nucleic acids as biomarkers in cancer patients. Nat Rev Cancer 2011; 11(6): 426-37.
[http://dx.doi.org/10.1038/nrc3066] [PMID: 21562580]

[70] Schwarzenbach H, Eichelser C, Kropidlowski J, Janni W, Rack B, Pantel K. Loss of heterozygosity at tumor suppressor genes detectable on fractionated circulating cell-free tumor DNA as indicator of breast cancer progression. Clin Cancer Res 2012; 18(20): 5719-30.
[http://dx.doi.org/10.1158/1078-0432.CCR-12-0142] [PMID: 23014523]

[71] Page K, Hava N, Ward B, *et al.* Detection of HER2 amplification in circulating free DNA in patients with breast cancer. Br J Cancer 2011; 104(8): 1342-8. [http://dx.doi.org/10.1038/bjc.2011.89] [PMID: 21427727]

[72] Shaw JA, Page K, Blighe K, *et al.* Genomic analysis of circulating cell-free DNA infers breast cancer dormancy. Genome Res 2012; 22(2): 220-31. [http://dx.doi.org/10.1101/gr.123497.111] [PMID: 21990379]

[73] Dawson S-J, Tsui DW, Murtaza M, *et al.* Analysis of circulating tumor DNA to monitor metastatic breast cancer. N Engl J Med 2013; 368(13): 1199-209. [http://dx.doi.org/10.1056/NEJMoa1213261] [PMID: 23484797]

[74] Lumachi F, Basso SM. Serum tumor markers in breast cancer Progress in Tumor Marker Research New York. NY: Nova Biomedical Books 2007; pp. 83-100.

[75] Seregni E, Coli A, Mazzucca N. Circulating tumour markers in breast cancer. Eur J Nucl Med Mol Imaging 2004; 31(1) (Suppl. 1): S15-22. [http://dx.doi.org/10.1007/s00259-004-1523-z] [PMID: 15127239]

[76] Molina R, Barak V, van Dalen A, *et al.* Tumor markers in breast cancer- European Group on Tumor Markers recommendations. Tumour Biol 2005; 26(6): 281-93. [http://dx.doi.org/10.1159/000089260] [PMID: 16254457]

[77] Nicolini A, Ferrari P, Duffy MJ, Eds. Prognostic and predictive biomarkers in breast cancer: past, present and future. Seminars in cancer biology. Elsevier 2018.

[78] Duffy MJ, O'Donovan N, McDermott E, Crown J. Validated biomarkers: The key to precision treatment in patients with breast cancer. Breast 2016; 29: 192-201. [http://dx.doi.org/10.1016/j.breast.2016.07.009] [PMID: 27521224]

[79] Harris LN, Ismaila N, McShane LM, *et al.* Use of biomarkers to guide decisions on adjuvant systemic therapy for women with early-stage invasive breast cancer: American Society of Clinical Oncology Clinical Practice Guideline. J Clin Oncol 2016; 34(10): 1134-50. [http://dx.doi.org/10.1200/JCO.2015.65.2289] [PMID: 26858339]

[80] van de Vijver MJ, He YD, van't Veer LJ, *et al.* A gene-expression signature as a predictor of survival in breast cancer. N Engl J Med 2002; 347(25): 1999-2009. [http://dx.doi.org/10.1056/NEJMoa021967] [PMID: 12490681]

[81] Carney WP, Neumann R, Lipton A, Leitzel K, Ali S, Price CP. Monitoring the circulating levels of the HER2/neu oncoprotein in breast cancer. Clin Breast Cancer 2004; 5(2): 105-16. [http://dx.doi.org/10.3816/CBC.2004.n.014] [PMID: 15245613]

[82] Pantel K, Riethdorf S. Pathology: are circulating tumor cells predictive of overall survival? Nat Rev Clin Oncol 2009; 6(4): 190-1. [http://dx.doi.org/10.1038/nrclinonc.2009.23] [PMID: 19333222]

[83] Yu M, Bardia A, Wittner BS, *et al.* Circulating breast tumor cells exhibit dynamic changes in epithelial and mesenchymal composition. science 2013; 339(6119): 580-4.

[84] Leidner RS, Li L, Thompson CL. Dampening enthusiasm for circulating microRNA in breast cancer. PLoS One 2013; 8(3): e57841. [http://dx.doi.org/10.1371/journal.pone.0057841] [PMID: 23472110]

CHAPTER 8

Clinical Application of Biomarkers for Hematologic Malignancies

Nurgül Özgür Yurttaş[1] and **Ahmet Emre Eşkazan**[1,*]

[1] *Division of Hematology, Department of Internal Medicine, Cerrahpasa Faculty of Medicine, Istanbul University-Cerrahpasa, Istanbul, Turkey*

Abstract: Over the last decade, significant advancements have been made in the molecular mechanisms, diagnostic methods, prognostication, and treatment options in hematologic malignancies. As the treatment landscape continues to expand, personalized treatment is much more important.

With the development of new technologies, more sensitive evaluation of residual disease using flow cytometry and next generation sequencing is possible nowadays. Although some conventional biomarkers preserve their significance, novel potential biomarkers accurately detect the mutational landscape of different cancers, and also, serve as prognostic and predictive biomarkers, which can be used in evaluating therapy responses and relapses. It is likely that we will be able to offer a more targeted and risk-adapted therapeutic approach to patients with hematologic malignancies guided by these potential biomarkers. This chapter summarizes the biomarkers used (or proposed to be used) in the diagnosis and/or monitoring of hematologic neoplasms.

Keywords: Chromosomal, Classification, Cytogenetic, Epigenetic, Genomic biomarkers, Hematologic neoplasm, Immunohistochemistry, lncRNA, Microenvironment, MicroRNA, Minimal residual disease, Molecular, Mutations, Scoring system.

INTRODUCTION

There are many markers (both traditional and novel) described for hematologic malignancies that might play a role in the pathogenesis, diagnosis, and prognosis. These biomarkers may also act as targets that might have the potential for providing new therapeutic approaches.

* **Corresponding author Ahmet Emre Eskazan**: Division of Hematology, Department of Internal Medicine, Cerrahpasa Faculty of Medicine, Istanbul University-Cerrahpasa, Istanbul, Turkey; Tel: 905337227376; Fax: 902125897934; E-mail: emre.eskazan@iuc.edu.tr

Pınar Atukeren and Hafize Uzun (Eds.)

Hematologic malignancies can be categorized as lymphoid and myeloid. The most recent World Health Organization (WHO) classification of these malignancies includes many subgroups [1, 2]. In this chapter, we try to summarize the biomarkers used (or proposed to be used) in the diagnosis and/or monitoring of these neoplasms.

BIOMARKERS IN MYELOID NEOPLASMS

This part of the chapter mainly focuses on the biomarkers for the myeloid neoplasms such as acute myeloid leukemia (AML), myelodysplastic syndrome (MDS), and myeloproliferative neoplasms (MPNs), including chronic myeloid leukemia (CML).

Acute Myeloid Leukemia

AML arises from the clonal expansion of malignant hematopoietic precursor cells in the bone marrow. AML is the second most common leukemia in adults and the most common type of acute leukemia. The median age at diagnosis is approximately 65 years, and the incidence increases with age. There is a slight predominance in men and a higher incidence in non-Hispanic whites than other racial and ethnic groups.

AML has been associated with environmental factors like exposure to chemicals, radiation, tobacco, chemotherapy, and retroviruses. In some cases, AML precedes other clonal hematopoietic disorders such as MDS, MPNs, paroxysmal nocturnal hemoglobinuria, and aplastic anemia.

Initial genetic analyses are not practical in AML because of the genetic heterogeneity of the disease. This makes the diagnosis difficult and as a result, influences therapeutic strategy negatively [3]. AML is lethal if untreated, and with intensive treatment, 60-80% of younger adults achieve a complete remission (CR), and almost one-third are finally cured. However, a 5-year relative survival rate of AML is approximately 25%. Outcomes are heterogenous and overall survival (OS) rates range between 5-70% [4, 5]. Therefore, prognostic markers to guide therapy and predict outcome are still needed. Prognostic markers can be clinical, disease-related, and molecular. However, the strongest prognostic factor for predicting therapeutic response and survival is cytogenetics.

Multiparameter flow cytometry (MFC) immunophenotyping gives information for AML diagnosis, classification, and monitoring. MFC allows identification, quantification, and lineage assessment of blasts and disease classification according to the maturation stage. AML confers very heterogeneous immunophenotypic features, mainly due to the genetic diversity of the disease [3].

Performing precise risk stratification due to immunophenotypic profiles can be difficult. There are not any exact recommendations and/or any reliable approaches for the use of new biomarkers in the immunophenotypic panels for AML as well as for the markers that might have an effect on prognosis and survival in clinical practice [3].

CYTOGENETIC CLASSIFICATION OF AML

The WHO criteria for the diagnosis and classification of AML are based on morphologic, immunophenotypic, cytogenetic, and molecular features [2, 6]. This categorization provides information for prognosis and may be useful for the selection of the best therapy [7]. The WHO classification has replaced the former French-American-British (FAB) classification system, which mainly depended on morphology and cytogenetic features [2, 8] (Tables **1** and **2**).

Table 1. WHO classification of AML and related malignancies (adapted from reference [2]).

AML with Recurrent Genetic Abnormalities	AML, NOS
AML with t(8;21) (q22;q22.1); *RUNX1-RUNX1T1*	AML with minimal differantiation
AML with inv(16) (p13.1q22) or t(16;16) (p13.1;q22); *CBFB-MYH11*	AML without maturation
APL with *PML-RARA*	AML with maturation
AML with t(9;11) (p21.3;q23.3); *MLLT3-KMT2A*	Acute myelomonocytic leukemia
AML with t(6;9) (p23;q34.1); *DEK-NUP214*	Acute monoblastic/monocytic leukemia
AML with inv(3) (q21.3q26.2) or t(3;3) (q21.3;q26.2); *GATA2, MECOM*	Pure erythroid leukemia
AML (megakaryoblastic) with t(1;22) (p13.3;q13.3); *RBM15-MKL1*	Acute megakaryoblastic leukemia
AML with *BCR-ABL1*	Acute basophilic leukemia
AML with mutated *NPM1*	Acute panmyelosis with myelofibrosis
AML with biallelic mutations of *CEBPA*	Myeloid sarcoma
AML with mutated *RUNX1*	Myeloid proliferations related to Down syndrome
AML with myelodysplasia-related changes	Transient abnormal myelopoiesis (TAM)
Therapy-related myeloid neoplasms	Myeloid leukemia associated with Down syndrome

Table 2. French-American-British (FAB) classification of AML (adapted from reference [8]).

M0 AML with no Romanowsky or Cytochemical Evidence of Differentiation
M1 Myeloblastic leukemia with little maturation
M2 Myeloblastic leukemia with maturation
M3 Acute promyelocytic leukemia (APL)
M3h APL, hypergranular variant
M3v APL, microgranular variant
M4 Acute myelomonocytic leukemia (AMML)
M4eo AMML with dysplastic marrow eosinophils
M5 Acute monoblastic leukemia (AMoL)
M5a AMoL, poorly differentiated
M5b AMoL, differentiated
M6 Erythroleukemia
M6a AML with erythroid dysplasia
M6b Erythroleukemia
M7 Acute megakaryoblastic leukemia (AMkL)

Myeloid blasts should be more than 20% of the whole bone marrow biopsy cellularity for the diagnosis of AML. However, the presence of myeloid sarcoma or the following genetic abnormalities directly refer to AML diagnosis independent from the blast count [6]:

- t(8;21)(q22;q22.1); *RUNX1-RUNX1T1* (previously called *AML1-ETO*)
- inv(16)(p13.1q22) or t(16;16)(p13.1;q22); *CBFB-MYH11*
- t(15;17)(q24.1;q21.2); *PML-RARA*

The most common karyotypes in AML are [9]:

- Normal (41%)
- t(15;17)(q24.1;q21.2) (13%)
- Trisomy 8 (10%)
- t(8;21)(q22;q22.1) (7%)
- 11q23.3 rearrangements (6%)
- inv(16)(p13.1q22)/t(16;16)(p13.1;q22) (5%)

AML with Normal Karyotype: AML with a normal karyotype contains almost half of de novo and 10% of therapy-related AML cases. Many genetic mutations are associated with normal karyotypes, including *FLT3-ITD/TKD*, *KMT2A-PTD*,

NPM1, *NRAS*, *CEBPA*, *BAALC*, and *WT1* [10]. Detection of mutations in AML patients with normal cytogenetics is essential for prognosis.

AML With Recurrent Genetic Abnormalities: The prognostically significant categories of cytogenetic abnormalities in AML are:

- t(9;11)(p21.3;q23.3); *KMT2A-MLLT3*
- t(6;9)(p23;q34.1); *DEK-NUP214*
- inv(3)(q21.3q26.2) or t(3;3)(q21.3;q26.2); *MECOM* (*EVI1*)
- t(1;22)(p13.3;q13.1); *RBM15-MKL1*

Also, the following rare groups were marked out as provisional categories in 2016 WHO classification [2]:

- *BCR::ABL1* fusion/t(9;22). Patients with this entity may benefit from treatment with tyrosine kinase inhibitors (TKIs).
- Mutated *RUNX1*. Patients with *RUNX1* mutations have an inherited predisposition to leukemia and it is associated with poor prognosis [11].

t(8;21) in AML: t(8;21)(q22;q22.1), *RUNX1-RUNX1T1* (previously called *AML1-ETO*), is observed in almost 7% of adults with newly diagnosed AML [9], and it is the most frequent abnormality in children with AML. When this translocation is detected, there is no need for the bone marrow blast count for the diagnosis of AML [6]. AML with the t(8;21) has a good prognosis in adults, but children have a poor outcome, especially those with additional cytogenetic abnormalities. The median age in adults is approximately 25-30 years, significantly younger than the total group of adults with AML. The CR rates are high and, with intensive consolidation chemotherapy, the expected disease-free survival is more than two years, after which relapses are uncommon [9].

Cases that have remained in continuous CR for at least eight years might still have *RUNX1-RUNX1T1* mRNA detectable by polymerase chain reaction (PCR) techniques in leukocytes within the circulation [12]. The reason for this long-term outcome in such patients is not known, but the persistence of this rearrangement may not always be related to a progressive course. Continuously increasing values for measurable/minimal residual disease (MRD) over time may be associated with a higher risk of relapse, so repetitive measurements may be helpful in these patients [13]. Nonetheless, therapeutic decisions cannot be only based on a positive signal with reverse transcriptase-PCR (RT-PCR). However, patients who are RT-PCR negative are most likely cured [14].

t(15;17) in APL: Acute promyelocytic leukemia (APL; AML-M3 in the FAB classification) is mainly characterized by balanced translocation t(15;17)(q24.1;q21.2) [7]. This rearrangement is seen in 13% of newly diagnosed AML and is very specific for APL [9].

APL is characterized by the infiltration of bone marrow by promyelocytes. Clinical or laboratory evidence of disseminated intravascular coagulation (DIC) and fibrinolysis, which may get worse during the initial response to chemotherapy, may be characteristic in this disease. A typical folded, kidney-shaped or bilobed nucleus is always found in some of the promyelocytes; coarse azurophilic granules and Auer rods are also very common [6].

APL is a medical emergency with a high rate of mortality, mainly because of bleeding diathesis due to DIC. It is crucial to start treatment with all trans-retinoic acid (ATRA) without any delay as soon as the diagnosis is suspected according to the cytologic features and before certain cytogenetic diagnoses. Patients have an exceptional prognosis when ATRA and chemotherapy together are initiated immediately. *PML-RARA* gene can be routinely detected by PCR or FISH methods. Thus, the diagnosis of APL is confirmed.

Some variant chromosome translocations that include *RARA*, but not *PML*, were detected in APL, like t(5;17), and t(11;17) [6]. When the *PML-RARA* fusion is not identified with PCR in a patient with suspected APL, FISH analysis with a *RARA* break-apart probe may provide information regarding these uncommon translocations. Cases of APL that do not carry t(15;17) generally do not respond to ATRA; the *PLZF-RARA* fusion protein that is produced in the t(11;17) rearrangement in rare cases shows low sensitivity to retinoic acid. So, standard doses of ATRA alone may not work in this very rare entity [15].

Inv(16) and t(16;16): Abnormalities of chromosome 16 are observed in almost 7% of adults with newly diagnosed AML. These abnormalities can be mainly divided into two groups:

- AML with inv(16)(p13.1q22) or t(16;16)(p13.1;q22); *CBFB-MYH11* (also known as acute myelomonocytic leukemia, AMML, FAB M4Eo)
- AML with other abnormalities of chromosome 16.

Patients with inv(16)(p13.1q22) or t(16;16)(p13.1;q22) have a good prognosis with standard chemotherapy, while those with other abnormalities of chromosome 16 do not [9]. RT-PCR can be used to monitor patients for MRD.

Rearrangements of 11q23.3: Rearrangements of 11q are observed in almost 6% of young adults with newly diagnosed AML and 12% of children with AML [6, 9]. Translocations involving 11q23.3 have an uncommon age distribution. This abnormality consists of about three-quarters of all chromosomal abnormalities of infants under one year of age with the diagnosis of acute leukemia. These leukemias are either myelomonocytic or monocytic (AML-M4 and AML-M5) or acute lymphoblastic leukemia (ALL) [16].

t(3;3) and inv(3) in AML with thrombocytosis: The t(3;3) and inv(3) account for almost 1% of AML patients [9]. This abnormality is observed in de novo AML and in therapy-related MDS/AML. Cytogenetic abnormalities of 3q are related to thrombocytosis and increased atypical megakaryocytes in the bone marrow of patients with AML [6].

t(6;9): AML with t(6;9) is detected in almost 1% of newly diagnosed AML cases [9]. This subtype is typically associated with basophilia, pancytopenia, and dysplasia. Patients with the t(6;9) may show single or multilineage dysplasia, variable FAB morphology, and a high incidence (71%) of *FLT3* internal tandem duplications. Therefore, they usually have a poor outcome with standard chemotherapy [6].

Chromosomal gain and loss, complex karyotype, and clonal heterogeneity: In addition to chromosomal translocations, there is a repetitive pattern of specific chromosomal loss or gain in patients with AML. The most common abnormalities are trisomy 8 (13%), monosomy 7 (9%), and monosomy 5 (6%). Abnormalities of chromosomes 5 and 7 are especially related to therapy-related AML, mainly caused by alkylating agents and/or radiation therapy [6, 17].

Complex karyotypes with three or more numerical and/or structural chromosomal abnormalities are frequently detected in AML secondary to MDS or MPN, especially in older patients. Using FISH, genomic microarray assays, and mutation analysis, *TP53* deletions and/or mutations are detected in almost 70% of patients with AML with a complex karyotype [18]. This subgroup is frequently associated with del(5q), or monosomy 7, 16, 18, and trisomy 1 and 11. *TP53* alterations are associated with a poor prognosis in patients with AML with a complex karyotype. Additionally, 17p and/or *TP53* abnormalities are predictors of a high treatment failure risk in AML [7].

MRD IN AML

In most patients with AML, leukemic blasts become undetectable after chemotherapy. Even so, leukemia may relapse afterward because of the chemoresistant blasts indistinctive from normal hematopoietic progenitor cells by conventional morphologic tests, *i.e.*, MRD. MRD is a powerful and independent prognostic factor in AML [19].

MRD was measured by either PCR amplification or MFC detection of leukemia-related cell markers. In approximately 20% of adult and 35% of pediatric AML cases, cells carry gene fusions, like *RUNX1-RUNX1T1*, *CBFB-MYH11*, or *MLL* [20, 21]; *NPM1* mutations occur in almost 30% of adults and <10% of pediatric patients [22]. Detection of these abnormalities during treatment is associated with disease relapse. MFC monitoring of MRD is also prognostically illuminating. Unlike PCR, it is not limited to patients with specific genetic abnormalities. The sensitivity of MRD by MFC monitoring is generally less than 0.1% (1 leukemic cell in 1000 normal bone marrow cells) [19].

Some antigens can be used as markers for MRD, and as potential therapeutic targets. ILT3 is an antigen expressed in AML subtypes with monocytic differentiation. It supports differentiation, and in addition, it is helpful for the diagnosis of AML. It may also be used as a marker for MRD detection in AML cases because of its high sensitivity, specificity, and stable expression. This antigen may also be a target for therapy in AML with monocytic differentiation resulting from the inhibition of ILT3 signaling with specific antibodies or antagonists [23]. CD93 is a marker in AML with *MLL* rearrangements. And it may also be a useful therapeutic target for stem cell therapy and a prognostic marker for quantitation of MRD [24]. CD82 might be an interesting target for leukemia stem cell elimination because of its considerable role in the regulation of the AML cell survival and their adhesion to the bone marrow microenvironment [25].

Myelodysplastic Syndrome

MDS is a heterogeneous clonal pre-malignant hematopoietic entity that is mainly associated with peripheral blood cytopenia, ineffective hematopoiesis, hypercellular bone marrow with dysplasia of one or more lineages and increased risk of progression to AML. del(3q/5q/7q/11q/12p/20q), monosomy 5/7, trisomy 8/19, i(17q), -Y are detected in MDS patients with variable frequencies [26]. Some of the chromosomal abnormalities, such as del(5q), -7/del(7q), del(9q), -13/del(13q), del(11q)/t(11q), del(12p)/t(12p), 17/del(17p)/i(17q), +19/t(19) and idic(Xq13) designate a presumptive diagnosis of MDS in the absence of morphological dysplasia. However, +8, del(20q) and -Y are not assessed as presumptive diagnostic markers of MDS, because they can also be detected as positive in other myeloid neoplasms. Therapeutic response and karyotypic evolution of chromosomal rearrangements are important for the classification and risk assessment of MDS [27]. However, normal chromosomal patterns in more than 30% de novo and 50% therapy-related MDS resulted in silent progression of MDS to AML [26].

In addition to chromosomal alterations, whole genome/exome sequencing and targeted deep sequencing detected an aspect of mutated genes that encode signal transduction proteins (*NRAS* at 1p13, *FLT3-ITD* at 13q12, *CBL* at 11q23, *JAK2* at 9p24, *KIT* at 4q12), transcription factors (*RUNX1* at 21q22, *TP53* at 17p13, *ETV6* at 12p13), tumor suppressors (*TP53*, *WT1*), epigenetic modifiers (*TET2* at 4q24, *ASXL1* at 20q11, *IDH1* at 2q34, *IDH2* at 15q26, *EZH2* at 7q36, *DNMT3A* at 2p23), RNA splicing machinery (*SF3B1* at 2q33, *U2AF1* at 21q22, *SRSF2* at 17q25, *ZRSR2* at Xp22) and components of the cohesin complex (*STAG2*, *RAD21*, *SMC3*, *SMC1A*). However, no specific mutation was identified in almost 20% of MDS patients [26, 28 - 31].

Regardless of low mutation frequency and interpatient changes, some mutations may show independent expression and different prognostic values. For example, *SF3B1* mutation with ring sideroblasts and decreased *TET2* activity in stem and progenitor cells can result in an increase in DNA methylation, clonal dominance, and transformation to AML. *DNMT3A*+ MDS patients may show lower OS and faster progression to AML. Various molecular tests including array-CGH (aCGH), array-SNP (SNP-array), multiple ligation dependent probe amplification (MLPA), NGS facilitated the molecular understanding of MDS [32, 33].

CHROMOSOMAL ABNORMALITIES IN MDS

Patients with MDS may show single or multiple chromosomal abnormalities at the time of diagnosis, or abnormal clones may occur during the course of the disease. Simple chromosome changes may include numerical changes (*i.e.*, monosomy or trisomy), structural abnormalities consisting of one chromosome (*e.g.*, inversion and interstitial deletion), or, less commonly, a balanced translocation including two chromosomes. 10-15% of the patients show complex karyotypes with multiple abnormalities. Additional chromosomal changes may develop during the course of MDS, or an abnormal clone may occur in a patient with a normal karyotype; these changes seem to be the indicators of progression to acute leukemia [34 - 36].

Clonal chromosomal abnormalities can be identified in bone marrow cells in almost 50% of patients with primary MDS [34]. The possibility of chromosomal abnormalities is increased in advanced MDS. The most common chromosomal abnormalities in MDS are del(5q), -7 or del(7q), trisomy 8, del(20q), and loss of Y chromosome. Loss of Y chromosome is common in males, especially in the elderly without any hematologic disease, and it is not thought to play a major role in the pathogenesis of MDS [34, 35] (Table **3**).

Table 3. WHO classification of myelodysplastic syndromes (MDS) (adapted from reference [2]).

MDS with Single Lineage Dysplasia	MDS with Excess Blasts In Transformation
MDS with ring siderablasts	Chronic myelomonocytic leukemia (CMML)-1
MDS with multilineage dysplasia	CMML-2
MDS with excess blasts-1	Atypical chronic myeloid leukemia, BCR-ABL1 negative
MDS with excess blasts-2	Chronic neutrophilic leukemia
MDS, unclassifiable	Juvenile myelomonocytic leukemia
MDS with isolated del(5q)	MDS/MPN unclassifiable
Refractory cytopenias of childhood	MDS/MPN with ring sideroblasts and thrombocytosis

There are some characteristics that differentiate cytogenetic abnormalities in primary MDS from those observed in de novo AML:

- Although +8, del(5q), -7/del(7q), and del(20q) are common in both MDS and AML, balanced translocations that are closely related to de novo AML, are not commonly detected in MDS. However, t(3;21), inv(3)/t(3;3), and t(11;16) are the exceptions. They are mainly observed in primary MDS, AML with myelodysplastic changes, therapy-related myeloid neoplasms (t-MN), and in rare cases of accelerated or blastic phase CML [34, 37].
- Deletions of chromosomes 5 and -7/del(7q) are especially characteristic (up to 70%) for the therapy-related MDS/AML triggered by alkylating agents and/or radiation therapy.
- With rare exceptions (like the 5q- syndrome/MDS with isolated del(5q)), chromosomal abnormalities in MDS are not related to specific clinical or morphological subsets using the WHO classification system.

Deletions of Chromosome 5

Deletion of the long arm of chromosome 5 (5q) is the most common chromosomal abnormality in MDS (almost 15%) [35]. The detection of an isolated del(5q), or del(5q) plus one other abnormality, except for -7/del(7q), defines the 5q- syndrome in WHO classification system of MDS (Table **3**).

There are two small commonly deleted regions detected [38, 39]:

- 5q32-33.1 – Deletion of this locus is most related to 5q- syndrome and is associated with a good prognosis. All patients with the 5q- syndrome carry this deletion.

- 5q31.2 – Deletion of this locus is more commonly observed in high-risk or therapy-related MDS and is mainly related to a complex karyotype, *TP53* loss or mutation, and a more aggressive disease.

MDS patients with single del(5q) usually have a relatively good prognosis and better treatment response to lenalidomide. However, monosomy 5 or del(5q) with other chromosomal abnormalities are associated with advanced MDS and a poorer outcome [38].

Monosomy 7 and Deletions of Chromosome 7

Almost 10% of patients with de novo MDS and nearly 50% of patients with therapy-related MDS have -7/del(7q) [40]. Approximately 90% of cases have a loss of a whole chromosome 7 (-7), and 10% are del(7q). Chromosome arm 7q contains three genes involved in MDS and myeloid leukemogenesis [41 - 44]:

- 7q22, including cut-like homeobox 1 *(CUX1)* is a tumor suppressor gene that is mutated recurrently in hematopoietic malignancies and solid tumors. *CUX1* inactivating mutations are independently related to a worse outcome.
- 7q31.31: It is detected in patients with MDS, AML, and other myeloid neoplasia with del(7q) as a single abnormality.
- 7q34
- 7q35-q36

Trisomy 8: Trisomy 8 is seen in <10% of patients with MDS and is thought to be in the intermediate-risk group of the International Prognostic Scoring System (IPSS). It is related to increased expression of anti-apoptotic genes and may result in a discriminating advantage over normal hematopoietic precursor cells. Furthermore, *MYC* oncogene overexpression is involved in the pathogenesis of myeloid disorders with trisomy 8 [27].

Deletions of chromosome 20: Deletions of the long arm of chromosome 20, del(20q), occur in <5% of MDS patients and are also observed in patients with AML and MPNs. When detected as the only chromosomal abnormality, it is associated with a favourable prognosis [27].

Diagnosis and Classification

The diagnosis of MDS is made based on an evaluation of the bone marrow and peripheral blood smear. Identification of certain chromosomal abnormalities, by conventional cytogenetic analysis or FISH, takes role in the classification of MDS and determination of prognostic risk groups (Table **3**).

Moreover, the presence of one of the following chromosomal changes suggest a presumptive diagnosis of MDS in cases with unexplained refractory cytopenia and no morphologic evidence of dysplasia: -7/del(7q); del(5q)/t(5q); del(13q); del(11q); del(12p) or t(12p); del(9q); idic(X)(q13); del(17p)/t(17p) (unbalanced translocations) or i(17q) (ie, loss of 17p); t(11;16)(q23.3;p13.3); t(3;21)(q26.2;q22.1); t(1;3)(p36.3;q21.3); t(2;11)(p21;q23.3); inv(3)(q21.3q26.2); t(6;9)(p23.3;q34.1) [2].

5q-syndrome/MDS with an Isolated del(5q)

5q- syndrome is a subtype of primary MDS that is identified with an isolated del(5q), or del(5q) plus one other abnormality (apart from -7/del(7q)) in WHO classification system (Table **3**). It generally occurs in older women (opposite to a male predominance in other forms of MDS), and the median age at diagnosis is 65 to 70 years. Patients typically present with refractory macrocytic anemia normal or elevated platelet counts with the absence of significant neutropenia [1, 38]. Due to the typical lack of thrombocytopenia and significant neutropenia, there is a low incidence of bleeding and infection. Red blood cell transfusions are frequently needed. This specific syndrome might show relatively benign progress that extends over several years. The possibility of progression to AML is very low and it usually responds to oral lenalidomide therapy quickly [45]. With this treatment, most of patients with 5q- syndrome accomplish red blood cell transfusion independence and complete cytogenetic remissions.

Role of Cytogenetics in the Prognosis of MDS

Certain cytogenetic abnormalities are helpful for estimating survival in MDS and progression to AML (Table **4**) [27]. The presence or absence of such chromosomal abnormalities are integrated within the commonly used prognostic scoring systems for MDS [46].

Table 4. Common cytogenetics and their prognostications in MDS [27].

Prognosis	Cytogenetic abnormality		
-	Single	Double	Complex
Very good	Del(11q), -Y	-	-
Good	Normal, del(5q), del(12p), del(20q)	Any including del(5q)	-
Intermediate	Del(7q), +8, i(17q), +19, any other independent clones	Any other	-
Poor	Inv(3)/t(3q)/del(3q)	Including -7/del(7q)	3
Very poor	-	-	>3

GENETIC MUTATIONS IN MDS

Gene mutations may be observed in patients with MDS and AML, with or without associated chromosomal changes. Mutations are found in genes that encode proteins that affect epigenetic regulation (*e.g.,* DNA methylation, histone modification), RNA-splicing machinery, transcription factors, and cytokine signaling pathways. Most of these are driver mutations in MDS, and have independent prognostic importance (*TP53*, *EZH2*, *ETV6*, *RUNX1*, and *ASXL1*) and are related to progression to AML [47]. The most frequently mutated genes are *SF3B1*, *TET2*, *SRSF2*, and *ASXL1* (each of them mutated in >10% of cases), followed by *DNMT3A* and *RUNX1* (each of them mutated in 5-10% of cases). The number and pattern of genes mutated varies between MDS subtypes. Most of the common mutations are detected more frequently in high-risk MDS subtypes. Particularly, the hierarchy of these gene mutations in MDS may change during treatment with lenalidomide and become more complex, which is mostly followed by disease progression and treatment failure [48]. Repetitive gene mutations related to isolated -7 or del(7q) -7/del(7q) may have prognostic significance in MDS.

TET2: DNA methylation is a prognostic marker and predictor of response to therapy in patients with MDS and seems to have a role in progression to AML. *TET* (ten-eleven translocation) genes encode proteins taking part in the control of DNA expression through demethylation. Somatic mutations in *TET2* occur in almost 15% of myeloid cancers and up to 30% in MDS. Loss of function mutations of *TET2* causes increased methylation and a decrease in the genes that are normally expressed. *TET2* mutations are associated with a better prognosis when detected in MDS [28, 49, 50].

DNMT3A: The *DNMT3A* gene is one of three genes that encode DNA methyltransferase enzymes catalyzing the addition of a methyl group to the cytosine residue of CpG dinucleotides, which regulate gene expression. Mutations of *DNMT3A* are observed in 8-13% of primary MDS patients, and in 22% of de novo AML, especially in AML with normal karyotype. Mutations of *DNMT3A* are related to decreased OS in MDS, and increased risk of progression to AML [51, 52].

IDH: Isocitrate dehydrogenase oncogene mutations (*i.e.*, *IDH1* and *IDH2*) are detected in some cases of MDS that cause DNA hypermethylation and changes in gene expression and are suggested to have a poor prognosis [53].

ASXL1: The additional sex-comb like-1 (*ASXL1*) gene encodes a protein that takes part in the epigenetic regulation of gene expression and is mutated in almost

10-20% of MDS patients. *ASXL1* mutations usually result in decreased OS and less time to progress to AML [28, 47, 54].

SF3B1: The *SF3B1* gene encodes for a ribonucleoprotein that is responsible for splicing messenger RNA. Repetitive somatic point mutations in this gene correspond to the most common mutations in MDS. Mutations of *SF3B1* are more common in MDS with ring sideroblasts (65%), but less common in MDS with refractory anemia (10%), refractory cytopenia and multilineage dysplasia (6%), and refractory anemia with excess blasts (5%) [55].

SRSF2: The *SRSF2* gene encodes a splicing factor 2 that is important in the processing of precursor mRNA to mature mRNA. Mutations of *SRSF2* are detected in 12 to 15% of MDS patients, with a higher frequency in elderly males, and are related to an inferior prognosis [56].

RUNX1: Mutations of *RUNX1* gene are detected in 7-15% of patients with de novo MDS, are more common in therapy-related MDS, and show a worse outcome [28, 34].

TP53: *TP53* tumor suppressor gene is located on 17p; *TP53* protein mediates cell cycle arrest. In MDS, about 5-15% of patients have *TP53* mutations at the time of diagnosis, frequently associated with del(5q) [28, 34]. *TP53* abnormalities are more common in patients with secondary MDS related to exposure to alkylating agents and/or radiation (*i.e.*, therapy-related MDS). Loss of *wild-type TP53* is related to treatment resistance and is a poor prognostic marker independent of the IPSS risk score. *TP53* mutations are associated with adverse outcomes in patients with MDS with complex karyotype [35, 47].

RAS: The *RAS* (*HRAS*, *KRAS*, *NRAS*) proto-oncogenes play a significant role in signal transduction. Mutations of *RAS* are detected in 10-35% of MDS patients and in a subgroup of patients with AML [28, 34, 47]. Most of the mutations in MDS are in the *NRAS* gene, whereas, *KRAS* mutations are observed less frequently. The importance of *RAS* mutations in MDS is still unclear; nevertheless, *RAS* mutations are related to MDS subgroup carrying -7/del(7q) mutation [40].

FLT3: Even though *FLT3* mutations are rare in MDS, they are related to a poor outcome [57].

Myeloproliferative Neoplasms

Among the four classic MPNs (CML, polycythemia vera (PV), primary myelofibrosis (PMF), essential thrombocythemia (ET)), only CML is characterized by a reciprocal translocation between chromosomes 9 and 22. This translocation is related to a shorter chromosome 22 (Philadelphia chromosome) in 95% of the cases [58]. In the other ones, t(9;22) can be demonstrated by using FISH or RT-PCR for the detection of *BCR::ABL1*. As the cytogenetic abnormalities are common in PMF (almost 30% at the time of diagnosis), unlike in ET (almost 5% at the time of diagnosis), no specific cytogenetic abnormality has been associated with MPNs other than CML [59].

Since *JAK2* mutations are found in almost all patients with PV and nearly 50% of those with ET and PMF, the diagnostic criteria of the classic MPNs are revised [2, 60, 61]:

- PV is considered to be present when there is no reason for an unexplained increase in hematocrit levels and/or red blood cell mass together with the presence of a *JAK2* mutation.
- PMF (agnogenic myeloid metaplasia, chronic idiopathic myelofibrosis) is characterized by bone marrow fibrosis that cannot be explained by another myeloid disorder like CML, PV, ET, or MDS.
- ET is a diagnosis of exclusion, representing clonal thrombocytosis when there are no other diagnostic criteria for the other MPNs.

Mutations in PV, ET, and PMF

The most commonly mutated genes in PV, ET, and PMF are *JAK2*, *CALR*, and *MPL*; estimates of their frequency follow [62 - 65]:

- PV – ≈ 100% *JAK2* mutations (exon 14 or 12)
- ET – 60-65% *JAK2* mutation, 20-25% *CALR* mutation, 3% *MPL* mutation
- PMF – 60-65% *JAK2* mutation, 20-25% *CALR* mutation, 7% *MPL* mutation

Approximately 8-10% of PMF cases and 10-15% of ET cases lack conventional mutations in *JAK2*, *CALR*, and *MPL* (*i.e.*, "triple-negative" MPNs).

***JAK2* Mutations:** Mutations of the tyrosine kinase Janus kinase 2 (*JAK2*), a gene on the short arm of chromosome 9, appear to account for hypersensitivity of hematopoietic progenitors in MPNs to growth factors and other cytokines [66 - 68]. *JAK2* mutations are most abundant in erythroid and myeloid cells. A single gain-of-function point mutation in *JAK2* in either exon 14 or 12 has been identified in almost all patients with PV, and in 60-65% of those with ET and PMF [67, 69]. The presence of *JAK2* mutations in EPO-independent erythroid colonies connects this mutation to growth factor hypersensitivity [70]. Dysregulation in the tyrosine kinase activity also proposes a possible therapeutic approach with the use of *JAK* inhibitors [71 - 73], like those used in CML and others. Documenting the presence of a *JAK2* mutation may help to distinguish PV from secondary polycythemia and ET from reactive thrombocytosis [74]. Mutation testing can be performed with equal sensitivity and specificity using either peripheral blood granulocytes or bone marrow [75]. Such testing may be used as a marker to measure MRD following treatment with interferon or hematopoietic cell transplantation and stratify the risk of thrombosis or evolution to post-PV or post-ET myelofibrosis [76 - 79].

Calreticulin (*CALR*) Mutations: Calreticulin (also known as calregulin) is a Ca^{++} binding protein localized primarily in the endoplasmic reticulum. The encoding gene (*CALR*) is located on chromosome 19p13.2. *CALR* mutations have been observed in almost 70% of patients with ET or PMF who do not carry a mutation in either *JAK2* or *MPL* and have only rarely been observed in patients with PV, all known *CALR* mutations involve mutations within exon 9 [63, 64]. Normal functions of calreticulin include proper folding of newly synthesized glycoproteins and modulating calcium homeostasis, mutations associated with ET and PMF disrupt the calcium-binding and endoplasmic reticulum retention domains. These mutations appear to alter the functions of the protein, resulting in the growth and activation of JAK/STAT signaling that is sensitive to *JAK2* inhibition [63].

***MPL* Mutations:** Activating mutations of *MPL*, which encodes the thrombopoietin receptor (TPO-R), are observed in patients with familial ET. Nevertheless, these kinds of mutations are not commonly found in patients with MPNs, and their clinical significance is unclear [80].

Other Mutations: An increased number of other somatic mutations are found in patients with MPNs, but their role in the pathogenesis of MPNs is not certain. The genes that are most commonly mutated in PV, ET, and PMF are [62, 81, 82]:

- *TET2*
- *ASXL1*
- *DNMT3A*
- *CBL* (Casitas B-lineage lymphoma)
- *TP53*
- *PPMD1* (p53–inducible protein phosphatase 1)
- *SRSF2* (Splicing factor, arginine/serine-rich 2)
- *U2AF1* (Splicing factor U2AF1)
- *KMT2C/MLL3* (Lysine N-methyltransferase 2C/mixed-lineage leukemia protein 3)
- *SF3B1*

- *NFE2* (Nuclear factor E2)
- *EZH2* (histone H3 lysine 27 methyltransferase)
- *LNK* (Src homology 2 B3, SH2B3)
- *IDH 1/2*
- *IKZF1* (IKAROS family zinc finger 1)

The significance of *JAK2*, *MPL*, *CALR*, and other mutations in these disorders is unclear. Nevertheless, it is suggested that patients with PV, ET, or PMF, despite the diagnosis or *JAK2* status, are characterized by a precise gene expression with upregulation of JAK-STAT target genes, showing the significance of this pathway [83]. Mutations of *TET2*, *DNMT3A*, *SF3B1*, *ASXL1*, and other genes are often seen in older patients with MPNs, but their importance is uncertain. However, neoplasms that carry acquired *JAK2* mutation prior to *TET2* mutation is more common at younger ages, have clinical features of PV, and these cases may develop thrombosis. In contrast, the presence of a *TET2* mutation prior to the acquisition of a *JAK2* mutation may alter the impact of the *JAK2* mutation, perhaps through methylation-based silencing of genes that are normally expressed [83, 84].

Homozygosity or heterozygosity for a *JAK2* mutation, the presence of other mutations (*MPL*, *CALR*, or *TET2* mutations), the order of the mutations, the allele burden, MPL receptor expression, loss of heterozygosity in *TP53*, and/or other host genetic variations may determine predisposition to the development of one of these mutations, the clinical phenotype (*e.g.,* degree of thrombocytosis, leukocytosis, and erythrocytosis, splenic size, pruritus, transformation to myelofibrosis or AML) as well as the response to treatment [85 - 88].

As a result, the diagnostic approach to the MPNs must now include evaluation of the mutational status of *JAK2*, *MPL*, and *CALR*. Such information will also be worthy for determining prognosis as well as individual risk assessment. Quantitative assays for the *JAK2V617F* mutation are preferred due to the better evaluation of allele burden (*i.e.*, lower in ET, higher in PV, and highest in post-PV myelofibrosis) in addition to the molecular monitoring of treatment response [89].

BIOMARKERS IN CHRONIC MYELOID LEUKEMIA

CML is a clonal hematopoietic disorder that is originated from CML stem cells that contain a *BCR::ABL1* fusion gene. During the chronic phase (CP) of the disease, these cells can differentiate into almost all blood cell types. In most CML patients, the *BCR::ABL1* fusion gene demonstrates a simple reciprocal translocation between the long arms of chromosomes 9 and 22 (t(9;22)) that forms the Philadelphia (Ph) chromosome. The *BCR::ABL1* fusion gene encodes a chimeric oncoprotein exhibiting an exaggerated tyrosine kinase activity. These characteristics dysregulate cellular proliferation and apoptosis control [90].

An understanding of the role of *BCR::ABL1* oncogene has permitted the development of famous "targeted" therapies that selectively block the tyrosine kinase activity of *BCR::ABL1*-encoded fusion proteins. The first TKI was imatinib. Clinical trials exhibited complete cytogenetic response rates of 87% in CML-CP patients treated with imatinib for 60 months [91]. However, although these striking initial responses are promising with imatinib therapy, imatinib-resistant clones continue to become a significant problem [92]. In fact, almost 10–15% of patients with early CML-CP show imatinib resistance ultimately [90]. Some of these cases may respond to higher doses of imatinib or to other TKIs like dasatinib, nilotinib, bosutinib, and ponatinib. Various patient- and drug-related factors, as well as the financial issues, take a role in the treatment selection. Certain clinical characteristics at diagnosis, the depth of response, the presence of certain kinase domain mutations, and additional molecular abnormalities may affect the prognosis of CML [93].

Ph chromosome in CML was discovered in 1960. t(9;22) juxtaposes most of ABL1 to 5' regions of BCR that is concluded by an increased tyrosine kinase activity and malignant transformation. The breakpoints in BCR are more varying and typically occur in neither major nor minor breakpoint regions (M- or m-bpr). M-bpr fusions constitute a p210 fusion protein, which is the major type found in Ph+ CML. A quite large number of patients with Ph+ B-ALL (40% of adults and 10% of children) also carry the p210 product. However, m-bpr translocations form a p190 fusion protein observed in most Ph+ B-ALL cases but rarely in Ph+ CML. Very rarely, BCR breakpoints fall in the microregion (μ-bcr), causing the p230 fusion product that is related to chronic neutrophilic leukemia [94]. t(9;22) detection is most commonly performed by conventional cytogenetics, fluorescence *in situ* hybridization (FISH), or RT-PCR. RT-PCR is used for MRD testing. Conventional cytogenetic tests can identify t(9;22) in 95% of CML patients at the time of the diagnosis; however, an additional 2.5% of cases with other rare translocations can not be detected with the standard techniques. If both cytogenetic and FISH results are negative, then an alternative diagnosis should be considered. Repetitive t(9;22) assays can be used to monitor the response of TKIs and to detect replases [4, 94].

Prognostic Factors at the Time of Diagnosis

Prognostic scoring systems: CML prognostic scoring systems differentiate the patients into risk groups according to the patient- and disease-related properties at the time of diagnosis. Up to a short while ago, there were two widely used scoring systems, Sokal and Hasford. Sokal score separates the patients treated with standard chemotherapy (busulfan or hydroxyurea) into three risk categories and each of them shows significantly different survival [95]. With the use of interferon-α for early-stage CML, the Hasford score was developed [96]. In 2011, the European Treatment and Outcome Study (EUTOS) score was created to reveal the effects of first-line imatinib [97]. In a comparing assessment, the EUTOS score was found to be better than Sokal or Hasford scores to reflect the most exact complete cytogenetic response at 18 months (positive predictive value, 34%; sensitivity, 23%; specificity, 92%) and PFS at 5 years (sensitivity, 16%; specificity, 91%) [97]. Currently, ELTS (EUTOS long-term survival score) was derived from survival data that reflect the treatment of CML with TKIs and provides the best discrimination for the probability of CML-specific death compared to other scoring systems [98].

Use of the Mutational Analysis in CML

Point mutations in the *BCR::ABL1* kinase are the major mechanism of resistance [99]. Most kinase domain mutations obtained during TKI treatment result in resistance and are associated with a poor prognosis. The presence of kinase domain mutations usually concluded with disease progression, especially when there is no response to second-line TKI therapy. Identification of the kinase domain mutations at the time of diagnosis to find out the presence of primary resistance is almost impossible due to the low levels of these TKI-resistant subclones to be detected with the conventional measurement methods [100]. A more sensitive method for the detection of TKI-resistant subclones was developed and called a 'multiplexed mass spectrometry assay'. This method was used to identify nilotinib and/or dasatinib resistant mutations in patients with imatinib resistance, including clinically related mutations that were not detectable by conventional methods [101]. Upgrading this technique for the detection of low-level mutations at the time of diagnosis would help to guide the most appropriate TKI and, identify a multidrug resistant mutation and in order to determine the most appropriate therapy [93].

Identification of Molecular Markers of Progression

Identification of potential disease markers with advanced techniques like gene microarray profiling and proteomic analysis is a popular area of research. Microarray profiling was previously used to determine if the specific gene expression at the time of diagnosis could show a response to TKI treatment [102]. Gene expression profiling may be useful to evaluate the patients with a worse outcome and clarify the biological basis of CML progression [103]. Proteomic analysis detected a group of proteins expressed at low levels in CML-CP, but at high levels in CML-BC such as HSP90, RB, AIF, PP2A, BCL2, XIAP, SMAD1, SSBP2α, PARP, GAB2, and TRIM24. On the other hand, PKCp664, AKTpT308, actin, p70S6Kp, Rac1.2.3, PDK1p, MEK, and CDK4 proteins were found to be high in CML-CP, but low in BC [104].

BIOMARKERS IN LYMPHOID NEOPLASMS

In this part, we tried to focus on the biomarkers for the lymphoid neoplasms, including ALL, and other lymphoproliferative neoplasms.

BIOMARKERS IN ACUTE LYMPHOBLASTIC LEUKEMIA/ LYMPHOMA

ALL is a heterogeneous neoplasm of lymphoid progenitors. Almost 85% of cases are of B-cell lineage, whereas, 15% are of T-cell lineage. Although childhood ALL patients are mostly cured (approximately 90%), adult patients with ALL generally exhibit a worse prognosis with a long-term survival rate of less than 35-40%, even with allogeneic hematopoietic stem cell transplantation. The information related to the biologic markers and the reasons for this poor prognosis in especially adult ALL patients is insufficient. Thus, treatment of refractory or relapsed ALL remains a major challenge [105]. ALL is more frequent in males. T-ALL subtypes are much more common than the other subtypes, especially in adolescents and young adults. The classification of ALL is based upon different genetic abnormalities. ALL is divided into two major groups as B- or T-cell subtypes, which are defined by immunophenotyping. B-ALL is characterized by the expression of B-cell–specific antigens, including PAX (B cell–specific activator protein), CD19, CD20, CD22 (on the cell surface and/or in the cytoplasm), CD24, and CD79a (in cytoplasm). CD20 is one of the most known mature B-cell antigens, but the leukemic lymphoblasts may show weak or no CD20 positivity. CD34 can be detected in leukemic cells, but it is typically expressed in early stages of B-cell precursors during the lymphopoiesis [106]. B-ALL can be stratified according to the immunoglobulin (Ig) pattern. Pre-B ALL is the most common subtype in adults, whereas mature B-ALL exhibits similar features with Burkitt lymphoma (BL). T-ALL blasts express T-lineage antigens such as CD2, CD3, CD4, CD5, CD7, and CD8, but also CD1a, CD10, CD34, CD99, and HLA-DR. A higher incidence of B-ALL than T-ALL is also detected with immunophenotyping [107] (Tables **5** and **6**).

Table 5. WHO classification of acute lymphoblastic leukemia/lymphoma (adapted from reference [2]).

Blastic plasmacytoid dendritic cell neoplasm
Acute leukemias of ambiguous lineage

(Table 5) cont.....

Blastic plasmacytoid dendritic cell neoplasm
Acute undifferentiated leukemia
Mixed phenotype acute leukemia (MPAL) with t(9;22) (q34.1;q11.2); *BCR::ABL1*
MPAL with t(v;11q23.3); *KMT2A* rearranged
MPAL, B/myeloid, NOS
MPAL, T/myeloid, NOS
B-lymphoblastic leukemia/lymphoma
B-lymphoblastic leukemia/lymphoma, NOS
B-lymphoblastic leukemia/lymphoma with recurrent genetic abnormalitites
B-lymphoblastic leukemia/lymphoma with t(9;22) (q34.1;q11.2); *BCR::ABL1*
B-lymphoblastic leukemia/lymphoma with t(v;11q23.3); *KMT2A* rearranged
B-lymphoblastic leukemia/lymphoma with t(12;21) (p13.2;q22.1); *ETV6-RUNX1*
B-lymphoblastic leukemia/lymphoma with hyperdiploidy
B-lymphoblastic leukemia/lymphoma with hypodiploidy
B-lymphoblastic leukemia/lymphoma with t(5;14) (q31.1;q32.3) *IL3-IGH*
B-lymphoblastic leukemia/lymphoma with t(1;19) (q23;p13.3); *TCF3-PBX1*
B-lymphoblastic leukemia/lymphoma, *BCR::ABL1-like*
B-lymphoblastic leukemia/lymphoma with *iAMP21*
T- lymphoblastic leukemia/lymphoma
Early T-cell precursor lymphoblastic leukemia
Natural killer cell lymphoblastic leukemia/lymphoma

Table 6. FAB and EGIL classification of ALL (adapted from references [108, 109]).

FAB classification of ALL (morphological)		**EGIL classification of ALL (immunological)**
L1	Small uniform blasts with scanty cytoplasm	**B-lineage ALL (CD19+ and/or CD79a+ and/or CD22+)**
L2	Large heterogeneous blasts with nucleoli and low nuclear-cytoplasmic ratio	**Pro-B-ALL (B-1):** No expression of other differentiation markers of B-cell antigens
		Common B-ALL (B-2): CD10+
		Pre-B-ALL (B-3): Cytoplasmic IgM+
		Mature B-ALL (B-4): Cytoplasmic or surface kappa or lambda +
L3	Basophilic vacuolated blasts	**T-lineage ALL (cytoplasmic/membrane CD3+**

(Table 6) cont.....

FAB classification of ALL (morphological)	EGIL classification of ALL (immunological)
	Pro-T-ALL (T-1): CD7+
	Pre-T-ALL (T-2): CD2 and/or CD5 and/or CD8+
	Cortical T-ALL (T-3): CD1a+
	Mature T-ALL (T-4): Membrane CD3+, CD1a-

The genetic profile and/or the choice of treatments may determine the different prognoses in different age groups. Younger patients are usually more prone to be resistant to high-dose chemotherapy, such as high-dose methotrexate, asparaginase, or reinduction therapy, and demonstrate higher compliance. But even in this favorable group, the results may be disappointing. Pro-B ALL phenotype and poor prognostic molecular markers are the most important factors associated with treatment failure [110]. t(12;21) and hyperdiploidy are usually associated with low-risk disease, whereas *KMT2A* (*MLL*) translocations, haploidy or low hypodiploidy are the examples of high-risk biomarkers. t(9;22) is also a high-risk factor; however, it enables the use of TKIs, against *BCR::ABL1* fusion protein [111, 112]. Different analytic techniques are used to identify the features of the disease, such as FISH, gene expression microarrays, gene set enrichment analysis, NGS or single nucleotide polymorphism (SNP) microarrays [113]. The application of high-resolution genomic profiling used in leukemogenesis is a popular issue nowadays for the evaluation of the treatment effect as well as the determination of the treatment strategy.

CYTOGENETIC AND MOLECULAR ABNORMALITIES IN ACUTE LYMPHOBLASTIC LEUKEMIA/LYMPHOMA

Most cases of ALL/LBL have non-random cytogenetic and/or molecular abnormalities that contribute to the phenotype, natural history, and/or prognosis of the disease. B cell ALL/LBL (B-ALL/LBL) and T cell ALL/LBL (T-ALL/LBL) differ regarding their characteristic, recurring cytogenetic, and molecular features. Conventional karyotyping and FISH reveal non-random cytogenetic and/or molecular abnormalities in approximately 80% of cases of B-ALL/LBL and 50-70% of T-ALL/LBL [2, 6]. Individual cases may manifest numerical abnormalities (*e.g.,* hyperdiploidy or hypodiploidy), structural changes (*e.g.,* translocations, inversions, or deletions), or both. Most chromosomal findings in ALL/LBL are associated with specific molecular/genetic abnormalities [114].

Genes that are important in the regulation of lymphocyte growth and development are frequently mutated in ALL/LBL. However, except for those associated with specific chromosomal findings in B-ALL/LBL, specific molecular abnormalities are not used in categorizing ALL/LBL. Gene expression profiles, microRNA expression, and epigenetic changes appear to have prognostic value in patients with ALL/LBL. However, except for the distinctive gene expression profile of *BCR::ABL1*-like B cell ALL, these findings are not presently used in the categorization of ALL/LBL, and their clinical value is not well-defined [115].

The 2016 WHO classification of ALL/LBL is based on immunophenotyping, cytogenetic abnormalities, and molecular findings (Table **5**). ALL/LBL is broadly divided into tumors of B-cell, T-cell, and natural killer (NK)-cell lineage, based on immunophenotyping. In this system, cytogenetic and molecular features are relied on to subclassify B-cell ALL/LBL (B-ALL/LBL), but not T cell ALL/LBL (T-ALL/LBL), early T-cell precursor ALL/LBL, or NK-cell ALL/LBL [2, 6].

B-ALL/LBL with Recurrent Genetic Abnormalities

t(9;22); *BCR::ABL1*: t(9;22) generates the *BCR::ABL1* fusion gene. The resultant *BCR::ABL1* protein is a constitutively active tyrosine kinase, and this category of ALL/LBL requires distinctive treatment that includes a TKI. The t(9;22) is present in one-third of all adult ALL/LBL cases and half of the adults >50 years but in approximately 5% of children [114]. The clinical and morphologic features of t(9;22) are similar to other types of B-ALL/LBL. Although there may be organ involvement, purely lymphomatous presentations of B-ALL/LBL with t(9;22) are rare. The blasts in this subtype frequently express myeloid-associated markers, such as CD13 and CD33, and expression of CD25 is common in adults [116]. In both adults and children, the presence of *BCR::ABL1* is associated with adverse outcomes, but combination chemotherapies that include a TKI have improved the outcomes. The higher frequency of t(9;22) in adults may account for the overall adverse prognosis of B-ALL/LBL in this age group. Approximately half of tumors in adults with B-ALL/LBL with t(9;22) produce a 210 kilodalton (kD) fusion protein (p210) and the remainder produce a p190 fusion protein; most tumors in children produce the p190 isoform [6]. The clinical behaviors of tumors expressing p210 or p190 do not appear to differ. B-ALL/LBL with t(9;22) may be associated with other cytogenetic or molecular abnormalities. Approximately two-thirds had abnormalities in addition to the Ph chromosome; gain of a Ph chromosome, monosomy 7, +8, +X, and del(9p) are among the commonly associated secondary abnormalities [117].

t(11q23): The category of B-ALL/LBL with t(11q23) is the most common category of B-ALL/LBL in infants. It is defined by the presence of a translocation between *KMT2A* (also called *MLL*) at 11q23.3 [6]. *KMT2A* regulates gene transcription; target genes include *HOX* genes and *IKZF1*, both of which have critical roles in normal lymphocyte development [117]. Translocations involving

11q23.3 are observed in 5-7% of all patients with ALL/LBL. B-ALL/LBL with t(v;11q23.3) is found in 60-80% of infants <1 year; it is less common in older children, but it again becomes slightly more common in adults [6]. Clinically, it typically presents with very high white blood cell (WBC) counts (*e.g.*, >100 x 10^9/L) and central nervous system (CNS) involvement is frequent [117, 118].

t(12;21); *ETV6-RUNX1*: The t(12;21) is the most common genetic lesion in childhood ALL/LBL. It occurs in 15-25% of childhood precursor B-ALL/LBL, but it is less common in adults. There are no distinctive clinical, morphologic, or cytochemical features that distinguish t(12;21) from other types of ALL/LBL. FISH is generally required to detect this chromosomal rearrangement, and its presence can also be inferred by using RT-PCR to detect the *ETV6-RUNX1* fusion transcript. Most patients with B-ALL/LBL with t(12;21) are 1 to 10 years old and have a favorable prognosis, in which >90% of the cases can be cured [6].

Hyperdiploidy: B-ALL/LBL with hyperdiploidy is characterized by lymphoblasts that have >50 chromosomes and lack translocations or other structural alterations. The clinical, morphologic, and cytochemical features of B-ALL/LBL with hyperdiploidy are similar to those of other categories [6]. Hyperdiploid B-ALL/LBL can be demonstrated by conventional karyotype, FISH, or flow cytometric DNA index. This group accounts for up to one-quarter of B-ALL/LBL in children; it is not detected in infants, the frequency decreases in older children, and it accounts for <10% of B-ALL/LBL in adults [119]. Hyperdiploid ALL/LBL has a very favorable outcome in children, with cure in >90% of children. Children with hyperdiploid ALL typically have other favorable clinical features, such as aged one to nine years, low WBC counts, and favorable immunophenotypic features (early pre-B or pre-B) [118].

Hypodiploidy: Hypodiploid B-ALL/LBL has lymphoblasts with <46 chromosomes and is divided into subtypes based upon the number of chromosomes, whereas, there may also be associated structural karyotypic abnormalities. Three distinct subgroups of hypodiploid ALL/LBL have been identified, which have different genetic findings and prognoses [120]:

- Near-haploid (23 to 29 chromosomes)
- Low hypodiploid (33 to 39 chromosomes)
- High hypodiploid (40 to 45 chromosomes)

There are no distinctive clinical, morphologic, or cytochemical characteristics of this category. Hypodiploid B-ALL/LBL is uncommon, accounting for <5% of cases. Hypodiploid B-ALL/LBL generally carries a poor prognosis, especially in patients with near-haploid and low-hypodiploid clones [121].

t(5;14); *IGH/IL3*: The category of B-ALL/LBL with t(5;14); *IGH/IL3* has blasts that harbor a translocation between *IL3* and an *IGH* (immunoglobulin heavy chain) gene and is associated with variable degrees of eosinophilia. The clinical features can resemble other types of B-ALL/LBL, but some cases present with asymptomatic eosinophilia and few blasts in peripheral blood [6]. The blasts have typical ALL/LBL morphology, but there is an increase of eosinophils, which are not part of the leukemic clone (*i.e.*, they constitute a reactive population of eosinophils).

t(1;19): The category of B-ALL/LBL with t(1;19) has blasts that harbor a translocation between *TCF3* (*E2A*) on chromosome 19 and *PBX1* on chromosome 1. It is relatively common in children (6%) and less common in adults [6]. The morphologic and cytochemical features are like other ALL/LBL categories. The blasts have a pre-B phenotype, and this rearrangement accounts for almost one-third of cases of childhood pre-B cell ALL/LBL. Clinically, there is an increased risk for CNS relapse [122]. Although the t(1;19) rearrangement was formerly considered to carry a poor prognosis, outcomes using contemporary intensive therapy are now comparable or more favorable than other categories of ALL/LBL [118].

***iAMP21*:** B-ALL/LBL with *iAMP21* is a provisional category in the WHO classification in which blasts have amplification of a portion of chromosome 21. It constitutes 2% of cases of B-ALL/LBL and is usually found in older children who present with low WBC counts [6]. There are no distinctive morphologic and cytochemical features. It is usually detected by FISH. ALL/LBL with *iAMP21* has been considered to have a poor prognosis. In one-fifth of cases, *iAMP21* is the only cytogenetic abnormality, but in other cases, there are other karyotypic or molecular findings, such as gains of X chromosome, chromosome 7 abnormalities, del(9p), del(12p), del(13q)/*RB1*, deletion of *IKZF1, P2YB8-CRLF2* fusion, rearrangements of *CRLF2,* and deletions of *RB1* and *ETV6*. However, it remains uncertain if these additional changes contribute to the pathogenesis of this leukemia or not [123].

***BCR::ABL1*-like B-cell ALL:** *BCR::ABL1*-like B-cell ALL is a provisional diagnosis in the 2016 WHO classification that contains a similar gene expression profile with B-ALL/LBL with t(9;22), but lacks the *BCR::ABL1* fusion or t(9;22) by cytogenetic, FISH, or molecular tests [2, 6]. This specific subtype is observed in 10-13% of pediatric patients, 21% of adolescents, and up to 33% of adults with B-ALL/LBL; it is resistant to standard chemotherapy and related to a poor prognosis [124, 125]. Detection of *BCR::ABL1*-like B-cell ALL is important for prognosis and treatment stratification. Compared with other cases of precursor B-cell ALL, this subgroup has a higher WBC count at presentation, higher levels of

MRD at the end of induction therapy, and inferior five-year EFS and OS. Genes that are mutated in *BCR::ABL1*-like B-ALL generally encode proteins involved in B-cell development, proliferation and differentiation, cell cycle regulation, and cell signaling. These mutations cause kinase activation and signaling *via* activation of the ABL1 and JAK-STAT pathways, which are sensitive to TKIs or JAK kinase inhibitors [124, 126].

Chromosome 9p abnormalities: Abnormalities of the short arm of chromosome 9 are found in almost 5-10% of childhood and adult ALL patients [6]. Deletion of 9p is an unfavorable prognostic factor related to an increased risk of relapse in precursor B cell ALL in children [127].

T-ALL/LBL

T cell ALL/LBL (T-ALL/LBL) is a hematologic malignancy of precursor lymphoid cells that are differentiated to T-cell lineage. T-ALL/LBL constitutes almost 15% of childhood ALL/LBL. It is more common in adolescents than in younger children and more common in males. It accounts for one-quarter of adult cases of ALL/LBL. T-ALL/LBL accounts for 85-90% of LBL presentations [6].

Cytogenetic Abnormalities in T-ALL/LBL

An abnormal karyotype is detected in 50-70% of T-ALL/LBL patients. But, in many cases, rearrangements are not identified with karyotyping but only with molecular genetic studies. In contrast to the important prognostic impact of recurring cytogenetic abnormalities in B-cell ALL/LBL, the prognostic implication of chromosomal translocations, gene mutations, and fusions is less well-defined in T-ALL/LBL [128].

Molecular Abnormalities in T-ALL/LBL

TCR rearrangements: The most common molecular/genetic abnormalities in T-ALL/LBL are the T-cell receptor (TCR) alpha and delta [129].

Non-TCR loci: Other karyotypic and/or mutational abnormalities in T-ALL/LBL that do not involve the TCR loci are as follows [129]:

- t(10;11); 10% of cases
- *KMT2A* translocations; 8% of cases
- del(9p)
- *NOTCH1* activating mutations
- *JAK1* or *JAK3* mutations
- *NUP214/ABL1* cryptic fusion; 6% of T-ALL/LBL (and rare cases of B-ALL/LBL)

EARLY T-CELL PRECURSOR ALL/LBL

Early T-cell precursor-ALL/LBL is a hematologic malignancy of T-cell lineage but with a unique immunophenotypic feature that shows only early T-cell differentiation [6]. These cells do not express CD1a and CD8 and weakly express CD5. These cases constitute up to 15% of T-ALL/LBL in children and 5-10% in adults. They are generally related to an increased rate of treatment failure [130].

MRD IN ALL

Relapse of leukemia is the major cause of treatment failure in ALL. Relapses originate from leukemic cells that are resistant to chemotherapy but usually become undetectable after initial treatment. Sensitive methods can identify MRD [131, 132]. MRD is recently the most important prognostic factor in both childhood and adult ALL. Therefore, MRD monitoring is commonly used nowadays in treatment protocols to evaluate the risk of the disease and the choice of treatment [133 - 135]. MRD measurements are also used in routine clinical practice to guide the clinician for the timing of hematopoetic stem cell transplantation in relapsed ALL patients that had a second remission as well as the timing of donor lymphocyte infusion after stem cell transplantation [136, 137]. Among methods for detecting MRD in ALL, PCR amplification of antigen receptor genes is proven to be more valuable and reliable among the other techniques. However, PCR amplification of fusion transcripts may also give useful clinical information, but its use in ALL is restricted because molecular targets recently adaptable to MRD studies are present in only a minority of patients [138]. MFC detection of leukemia-specific markers was also shown to predict outcome [134]. MRD studies by MFC withstand antibodies to identify immunophenotypic patterns of leukemic blasts, which must separate blasts from the normal CD19+ CD10+ lymphoid precursors of the bone marrow known as 'hematogones' [132]. Thus, acknowledgement of new markers that are easily detectable and are stably highly expressed in ALL patients should facilitate the use of MRD studies, widen the benefit comprising to all patients and possibly increase the sensitivity of MRD tests [131].

BIOMARKERS IN LYMPHOMAS

Lymphomas are a group of hematological malignancies derived from lymphocytes and occur mainly in lymph nodes and/or other lymphoid tissues. Lymphomas are affected by abnormal genetic changes, epigenetic dysregulation, aberrant pathway activation, and tumor–microenvironment interactions. So, the diagnosis and classification of lymphomas can be difficult. Besides, the molecular heterogeneity that determines the aggressiveness and progression of lymphomas causes different outcomes with similar treatment strategies. Even though biomarkers identified primarily by immunohistochemistry and/or MFC, are used widely and subscribe to diagnosis, classification, and prognosis of lymphomas, new clinically applicable, reliable, and reproducible biomarkers are still needed [139].

Biomarkers of Genetic Changes

Genomic aberrations and appropriate dysregulated oncogenic regulatory pathways are responsible for many lymphoma subtypes [140]. Microarray-based methods such as gene expression profiling and massively parallel sequencing methods like NGS, facilitated the exploration of new biomarkers and detection of underlying molecular mechanisms of lymphomas. Furthermore, these results support better diagnosis and classification of patients who may benefit from potential targeted therapeutic options [141].

B-Cell Lymphoma 6 (*BCL6*)

BCL6 encodes a transcriptional factor that takes part in germinal center B-cell differentiation and pathogenesis of germinal center-related lymphomas. *BCL6* activity is mostly dysregulated with translocations and mutations. Both rearrangement and somatic mutation of *BCL6* can be detected in diffuse large B-cell lymphoma (DLBCL) at the same time. However, somatic mutations occur unaccompanied with *BCL6* rearrangement. Increased number of expressions of *BCL6* mRNA and protein are observed to be related to a better outcome in patients with DLBCL, compatible with immunohistochemical staining [142 - 144].

Tumor Protein 53 (*TP53*)

TP53 mutation is an independent prognostic biomarker in patients with DLBCL, follicular lymphoma (FL), and mantle cell lymphoma (MCL). In DLBCL patients, rather than *TP53* deletion and loss of heterozygosity, *TP53* mutation is found to be associated with a worse outcome in patients with either germinal center B-cel--like DLBCL or activated B-cell like DLBCL [145].

V-Myc Avian Myelocytomatosis Viral Oncogene Homolog (*MYC*)

MYC is included in nearly all stages of cell biology. In BL, *MYC* that is located on chromosome 8q24, is aligned with the *IGH* on chromosome 14 or with the 'IG light chains' on chromosome 8. t(8;14), t(2;8), or t(8;22) that cause *MYC* overexpression. *MYC* translocations may also be detected in DLBCL and in other aggressive B-cell lymphomas, which are mainly related to an inferior prognosis [146, 147].

B-Cell Lymphoma 2 (*BCL2*)

t(14;18) involving *IGH* and *BCL2* is an important molecular hallmark of FL, and it also can occur in germinal center B-cell-like DLBCL. Conversely, *BCL2* is seldom translocated but usually amplified in activate B-cell-like DLBCL. FISH analysis confirms a strong relationship between *BCL2* rearrangement and germinal center B-cell-like DLBCL. The prognostic effect of *BCL2* changes in DLBCL is still uncertain, and it may depend on subtype, *BCL2* expression, *BCL2* gene polymorphism, and the treatment type [148 - 150]. A small number of DLBCL subtypes related to both *MYC* and t(14;18)/*IGH/BCL2* and/or *BCL6* translocation are usually defined as 'double-hit or triple-hit' lymphoma, and these patients generally have an aggressive clinical course and worse outcome [149].

Myeloid Differentiation Primary Response 88 (*MYD88*)

MYD88 has a role in catalyzing Toll-like and interleukin 1 receptor signaling. Somatic mutation of *MYD88* is the most frequent mutation that is observed in lymphoplasmacytic lymphoma/Waldenström macroglobulinemia and activated B-cell-like DLBCL [151]. They also occur in a small subtype of marginal zone lymphoma (MZL), MCL, and chronic lymphocytic leukemia/small lymphocytic lymphoma (CLL/SLL). This mutation is not shown to have a prognostic significance on patients with DLBCL, but *MYD88* protein expression is revealed to be significantly related to recurrence and shorter disease-free survival [152].

Other Genetic Biomarkers in B-Cell Lymphomas

Quantification of gene expression for activated B-cell-like/germinal center B-cel--like subgroups using formalin-fixed and paraffin-embedded tissues are clinically applicable [153]. For example, it is discovered that *ID3* is mutated in BL, but not in most of the cases with DLBCL, implying *ID3* as a new tumor suppressor gene and a novel potential therapeutic target in BL. Besides, *ID3, MYC, TPST2,* and *RET* mutations are also detected mostly in BL. However, *PIM1, CECR1* and *MYD88* mutations are mainly observed primarily in DLBCL. *MLL3, TP53,* and *LAMA3* exhibited super-imposed patterns of mutation in BL and DLBCL [154, 155]. The chromosomal translocation t(11:14) results in overexpression of cyclin D1 and it is an important molecular hallmark of MCL [156].

The diagnosis of primary mediastinal B-cell lymphoma may be difficult due to the morphological and immunophenotypic overlapping with classical Hodgkin lymphoma (cHL) [157]. Special features of primary mediastinal B-cell lymphoma involve *STAT6* integral activation and *JAK2* amplification. On the other hand,

NF-κB1A mutation is specific for classical HL, proposing that *NF-κB* activation in classical HL and primary mediastinal B-cell lymphoma may be related to different gene mutations [158].

Other Genetic Biomarkers in Peripheral T-Cell Lymphoma (PTCL)

Almost 30–50% cases of PTCL are classified as PTCL, not otherwise specified (PTCL-NOS), due to the absence of specific morphological features and biomarkers. However, studies using gene expression profiling identified that a subgroup of patients with PTCL-NOS has features more appropriate with angioimmunoblastic T-cell lymphoma, adult T-cell leukemia/lymphoma, anaplastic large T-cell lymphoma (ALCL) and/or extra-nodal NK/T-cell lymphoma [159, 160]. Furthermore, cases of PTCL-NOS with no specific feature can be stratified into two subgroups based upon different oncogenic pathways and related to varying outcomes. One subtype, defined with increased expression of *GATA3* and its target genes such as *CCR4, IL18RA, CXCR7,* and *IK* are associated with worse survival. Another subtype, defined with increased expression of cytotoxic gene features like *TBX21, EOMES,* and their target genes (*CXCR3, IL2RB, CCL3,* and *IFNγ*) are related to a better prognosis [159]. Chromosomal translocations may also be detected in some the T-cell lymphoma subtypes. In anaplastic lymphoma kinase (*ALK*)-positive ALCL, a characteristic chromosomal translocation, t(2;5) involving *NPM/ALK*, or other *ALK* translocations are identified. As a result, *ALK* is aberrantly expressed by lymphoma cells [161]. In *ALK*- ALCL, almost 20% of cases carry t(6;7) involving *DUSSP22. DUSP22* takes part in the inhibition of T-cell receptor signaling. Translocations including *DUSP22* cause a reduction in *DUSP22* expression. Thus, translocation disturbs the tumor suppressor gene functions of *DUSP22*. The second subgroup of nearly 10% of cases of *ALK*- ALCL contains *TP63* rearrangements. FISH may be used to detect these translocations. The remaining cases of *ALK*- ALCL may either have *DUSP22* or *TP63* translocations. *ALK*+ and *ALK*- ALCL have a different prognoss, and in general, *ALK*+ ALCL patients have a better prognosis than negative ones. However, patients with *DUSSP22/IRF4* abnormalities also have a reasonably favorable outcome [162]. For this reason, these translocations can be used as biomarkers in the diagnosis as well as treatment targets. *ALK*+ ALCL may benefit from *ALK*-targeting agents [163]. Evaluation of the T-cell receptor rearrangement most commonly with PCR tests, is needed in defining the diagnosis in cases with T-cell lymphoma. On the other hand, NK-cell lymphoma is an exception because NK-cells do not rearrange their T-cell receptor genes. Killer cell immunoglobulin-like receptor (*KIR*), which is one of NK-cell receptors, can be used as a representative biomarker of clonal cellular expansion to differentiate reactive NK-cells from malignant cells. Limited *KIR* expression can be immunohistochemically identified in extranodal NK/T-cell lymphoma

[164]. In general, genetic changes specific for a subgroup of lymphoma may be a characteristic feature or they might have prognostic effects. So, they are very important in the evaluation of lymphomas [139].

Biomarkers Related to Epigenetic Changes

Epigenetic changes are generally described as inherited changes independent from the DNA sequence alterations. Epigenetic changes are observed to alter genes included in signal transduction, DNA repair, cell cycle regulation, differentiation, invasion, and apoptosis. Epigenetic changes, including DNA methylation, histone modification, and noncoding RNA expression, are enticing potential biomarkers for diagnosis, evaluation of prognosis, and prediction of treatment responses [165, 166].

***P16/INK4*:** One of the most frequent molecular changes of *P16/INK4A* is homozygous deletion. Nevertheless, methylation also has an important role in the inactivation of *P16/INK4A* and it is related to an inferior outcome in lymphomas [167].

DNA Repair Enzyme O6-Methylguanine DNA Methyltransferase (*MGMT*): *MGMT*, even though seldom deleted or mutated, often loses its function due to epigenetic changes. *MGMT* promoter hypermethylation is related to a good prognosis in DLBCL patients treated with multidrug chemotherapy regimens [168].

Kruppel-Like Factor 4 (*KLF4*): *KLF4* plays role as a tumor suppressor gene in T- and B-cell lymphomas and it is aberrantly hypermethylated in most of the lymphoma subtypes involving FL, DLBCL, BL, and cHL. Epigenetic silencing of *KLF4* might have a benefit in patients with B-cell lymphomas and especially in cHL [167, 169].

Myeloid/Lymphoid or Mixed-Lineage Leukemia 2 (*MLL2*): *MLL2* (also known as *KMT2D*) is inactivated by mutations in almost 90% of FL cases and in 32% of patients with DLBCL [167, 170].

Myocyte Enhancer Factor 2 (*MEF2*): *MEF2*, a calcium-regulating gene, encodes transcription factors and enrolls histone-modifying enzymes. *MEF2B* mutations are connected to lymphoma generation. These mutations are somatic, and they are identified frequently in germinal center B-cell-like DLBCL and FL, proposing that *MEF2B* mutations take part at improving the malignant transformation of germinal center B cells to lymphoma cells [167, 170].

Other Epigenetic Gene Features

Other aberrant epigenetic gene features are detected in some studies. For instance, five hypermethylated genes are known as *SOX9, HOXA9, AHR, NR2F2*, and *ROBO1* are found to be related to increased proliferation and chromosomal abnormalities, and adverse outcomes in MCL patients [171]. Also, promoter methylation of *CNRIP1* is associated with inferior survival in patients with DLBCL and is a potential prognostic factor. Hypermethylation of *SMAD* family member 1 (*SMAD1*) was reported to predict treatment resistance in patients with DLBCL [172]. It is shown that frequently mutated genes involved in epigenetic changes in PTCL were ten-eleven translocation 2 (*TET2*), DNA methyltransferase 3A (*DNMT3A*), isocitrate dehydrogenase 2 (*IDH2*), and *RHOA* detected by using whole-exome sequencing, RNA sequencing analysis, and targeted deep sequencing tests [173]. *RHOA* is mutated in nearly 67% of angioimmunoblastic T-cell lymphoma and 18% of patients with PTCL-NOS. *TET2* mutations are identified in both angioimmunoblastic T-cell lymphoma and PTCL, and they are associated with advanced-stage disease, thrombocytopenia, high International Prognostic Index (IPI) scores, and a decreased PFS. *IDH2 R172* mutation is detected especially in angioimmunoblastic T-cell lymphoma, and it is a specific biomarker of the disease [173]. *IDH2* mutations in angioimmunoblastic T-cell lymphoma do not correlate with OS [174]. Biomarkers associated with epigenetic regulation have great potential in making the diagnosis and in the prediction of prognosis in lymphoma patients.

MicroRNA and LncRNA in Diagnosis and Prognosis of Lymphomas

MicroRNAs, a subtype of small noncoding RNA, regulate many biological events at the posttranscriptional stage by directing to messenger RNAs and affecting cellular transcripts and pathways [175]. Long noncoding RNAs (lncRNAs) have certain expression patterns. So, they display an important role in gene regulation at different levels and in malignant transformation, especially including the regulation of epigenetic events. It is suggested that expression defects in microRNAs and lncRNAs might play a role in the development and progression of lymphomas. Therefore, they may be promising biomarkers in the diagnosis and prognosis of cases with lymphoma [175, 176].

miR-155: miR-155 is one of the most investigated miRNAs in lymphomas, especially DLBCL. miR-155 takes part in the pathogenesis and clinical aggressiveness of DLBCL. Activated B-cell-like DLBCL expresses much higher levels of miR-155 than that observed in germinal center B-cell-like DLBCL. So, miR-155 can be used in differentiating the activated B-cell-like subtype from germinal center B-cell-like DLBCL at the time of diagnosis. Also, increased levels of miR-155 may explain the inferior outcome observed in this subtype. Moreover, a relation between miR-155 expression and resistance to treatment in DLBCL was previously detected [176, 177].

miR-17-92 Cluster: The miR-17-92 cluster, consisting of six microRNAs known as miR-17, miR-18a, miR-19a, miR-20a, miR-19b1, and miR-92-1, is overexpressed in some forms of lymphomas such as germinal center B-cell-like DLBCL, FL, MCL, ALCL, and BL, but not in activated B-cell like DLBCL [178 - 180]. In addition, miR-17-92 is expressed higher in *ALK*+ ALCL than in the negative cases. Overexpression of miR-17-92 was found to be associated with inferior survival rates in patients with MCL [178, 179].

miR-15a/16-1: Dysregulation of miR-15a/16-1 has a role in the pathogenesis of CLL/SLL. Low expression of miR-15a/16-1 is associated with a good prognosis in CLL/SLL, while other chromosomal changes like 17p deletion, 12q trisomy, and 11q deletion are associated with worse outcomes [181].

Other MicroRNAs and MicroRNA-Based Classifiers

Increased expression of miR-18b and dysregulation of miR-29 are independent predictors of poor prognosis in MCL. Particularly, miR-127-3p and Ki67 combination may propose a new prognostic model for MCL. In the same way, a combination of miR-615-3p and MCL IPI has a prognostic value [182]. Iqbal *et al.* [176] created a 27-miRNA classifier that could differentiate BL from DLBCL, and an 8-miRNA classifier that could distinguish DLBCL from other subgroups. Six miRNAs – miR-28-3p, miR-28-5p, miR-129-3p, miR-589, miR-331-5p, and miR-597 – are highly expressed in germinal center B-cell-like DLBCL. This miRNA categorization in DLBCL subtypes is found to be 90% concordant with gene expression profiling-based categorization [176]. A group of miRNAs such as miR-222, miR-181a, miR-129-5p, and miR-18a has prognostic importance in DLBCL [183], while another miRNA group is characteristic in hepatitis C virus (HCV)-related DLBCL [184]. In this subtype, miR-138-5p was associated with better survival, whereas the patient with miR-147a, miR-147b, and miR-511-5p had inferior survival rates. miR-222 expression was found to be uniformly related to poor survival in lymphoma [176]. In addition, a new *TP53/MIR34A* 'double hit' DLBCL subtype was defined, which had concurrent *MIR34A* methylation and *TP53* mutation. This subtype was associated with inferior outcomes. Therefore, *MIR34A* methylation together with *TP53* mutation can distinguish an aggressive subtype of DLBCL, which may respond to epigenetic therapy [185]. Other than tumor biopsy samples, microRNA can be evaluated in serum, plasma, and bone marrow smears. Tissue miRNAs seem to be more trustworthy in the identification and staging of malignancy, while circulating miRNAs are more stable and can be measured noninvasively [186]. In PTCLs, some remarkable diagnostic miRNA classifiers were identified. One 7-miRNA signature – miR-512-3p, miR-886-5p, miR-886-3p, miR-708, miR-135b, miR-146a, and miR-155 – was primarily detected in *ALK*+ ALCL. miR-155 is much more common in *ALK*- than *ALK*+ ALCL. Another 11-miRNA signature –miR-210, miR-197, miR-191, miR-512-3p, miR-451, miR-146a, miR-22, miR-455-3p, miR-455-5p, miR-143, and miR-494 – may differentiate *ALK*- ALCL from other PTCLs [187].

To conclude, some miRNA-based diagnostic and prognostic classifiers were identified in many types of lymphomas and they are expected to help in improving clinical decisions.

Immunophenotypic Biomarkers

The primary diagnosis, differential diagnosis, and stratifications of lymphomas are mainly based upon histopathological characteristics and results of immunohistochemistry or MFC immunophenotypic tests. Some immunophenotypic biomarkers that are routinely used are classified as T-cell markers (CD2, CD3, CD4, CD5, CD7, CD8, and T-cell receptors), B-cell markers (CD19, CD20, CD22, CD79, and Pax-5), cytotoxic markers (like TIA-1, granzyme B, and perforin), germinal center markers (CD10, *Bcl-6*, and *LMO2*), and plasma cell markers (*i.e.*, CD38, CD138). Also, some biomarkers are extremely related to specific entities like CD15, CD30 in cHL; *ALK-1*, CD30 in ALCL; cyclin D1 and SOX-11 in MCL; and follicular T helper markers consisting of CXCL13 and PD-1 in angioimmunoblastic T-cell lymphoma. A large number of protein biomarkers were also revealed to predict prognosis or were found to be useful in targeting specialized treatment strategies [139].

CD30

CD30 is a cell surface marker expressed by activated B- or T-cells normally and it is expressed in large amounts by lymphoma cells in cHL and ALCL, and a subgroup of DLBCL and Epstein-Barr virus (EBV)-related lymphoproliferative disorders. CD30 is a worthy biomarker for diagnosing cHL and ALCL, and the presence of soluble CD30 in serum and/or in body fluids can independently predict disease progression and prognosis in CD30+ lymphomas. Due to the expression of CD30 in lymphoma subgroups, brentuximab vedotin, an antibody drug conjugate made up of anti-CD30 is revealed to be effective in cases with CD30+ lymphomas, particularly in patients with relapsed/refractory cHL, systemic ALCL, and also in a subtype of patients with relapsed/refractory CD30+ DLBCL [188].

Bcl-6, Bcl-2, and Myc

Bcl-6 has a critical role in the regulation of germinal center B cells, and it is thought to be a remarkable biomarker for B-cells at the germinal center stage. Anti-*bcl-6* antibody is usually used to stain the germinal center B-cells at lymphoma diagnosis, but this is not specific for germinal center B-cell-like DLBCL, because it is also expressed by post-germinal center B cells [189]. *Bcl-6* is also expressed by T-cells of angioimmunoblastic T-cell lymphoma, which are derived from normal follicular T helper cells, which can be used as a biomarker in this lymphoma subtype [143]. *Bcl-6* expression can also be utilized as a good prognostic marker in cases with IG-MYC positive high-grade B-cell lymphomas [190]. *Myc* and *Bcl-2* overexpression can be ascribed to chromosomal translocations like t(8;14) and t(14;18), and they might have specific roles in the diagnosis of BL and FL, respectively [148]. However, overexpression of *Myc* and/or *Bcl-2* are usually detected in lymphomas, in which these translocations do not exist. Immunohistochemically, coexpression of *Myc/Bcl-2* in DLBCL is related to a more aggressive clinical form with worse outcomes, independent from the patients who have cytogenetically proven *Myc* and *Bcl-2* double-hit DLBCL. *Myc/Bcl-2* coexpression is more common in activated B-cell-like DLBCL, and this may account for the worse prognosis of activated B-cell like DLBCL, while *Myc/Bcl-6* coexpression is detected in both germinal center B-cell-like DLBCL and activated B-cell-like DLBCL, whereas this is not associated with outcome [191]. Identically, *Myc* expression in MCL is usually associated with more aggressive subtypes and an inferior outcome [192].

Cyclin D1 and SOX-11

t(11;14) results in the overexpression of cyclin D1 protein, and it is used as a diagnostic hallmark of MCL [193]. Infrequently, some MCL subtypes are reported as negative for cyclin D1, but they have a gene expression profiling consistent with the diagnosis of MCL, in which overexpression of cyclin D2 or D3 can be detected. Also, SOX-11 is a marker of MCL independent of cyclin D1. Thus, the combination of cyclin D1 and SOX-11 is useful for the diagnosis of MCL, while SOX-11 is especially helpful for the cyclin D1-negative cases [194]. SOX-11 expression in MCL cases treated with intensive chemotherapy is associated with better survival rates. Cyclin D1 is also overexpressed in almost 3% of DLBCL cases, which are generally the post-germinal center or activated B-cell subtypes, negative for CD10 and positive for *Bcl-6* and multiple myeloma (MM) oncogene 1(*MUM-1/IRF4*). t(11;14) is absent in DLBCL, so cyclin D1 expression in DLBCL is not related to t(11;14) [193].

TP53 and MDM2

The impact of *TP53* expression on the survival rates of lymphoma patients, which is generally evaluated with immunohistochemistry, remains uncertain. *TP53* overexpression is associated with inferior survival in DLBCL, and this can be used to classify cases into different prognostic subgroups [195]. In DLBCL, *MDM2* expression was found not to be an independent predictor of outcome, but *MDM2* expression together with mutated-type *TP53* expression can predict significantly inferior survival rates [196].

Immunophenotypic Biomarkers

Expression of some parts of oncogenic signaling pathways also have prognostic importance. Expression of cyclin D2, cyclin D3, protein kinase c-B, caspase-9, and survivin are associated with poor prognosis in DLBCL [197]. Activation of the *NF-κB* pathway participates to lymphoma pathogenesis in some subtypes including DLBCL, cHL, primary mediastinal B-cell lymphoma, and ALCL [198]. Expression of parts of *NF-κB* pathway, like p50 *(NFκB1)* and p65 *(RelA)*, are related to activated B-cell-like DLBCL and they might have an inferior outcome, while p52/RelB may distinguish a subtype of germinal center B-cell-like DLBCL with a better prognosis [199]. Aberrant activation of the B-cell receptor signaling pathway that leads to the activation of *NF-kB, MAPK,* and *NFAT* pathways, has a significant role in lymphoma pathogenesis in most of the lymphoma subtypes, especially in CLL/SLL and DLBCL. Bruton tyrosine kinase (BTK), one of the important constituents of the B-cell receptor pathway, became a major therapeutic target. Ibrutinib is a BTK inhibitor, which is effective in cases with CLL/SLL, MCL, FL, activated B-cell-like DLBCL, lymphoplasmacytic lymphoma/Waldenström macroglobulinemia, primary CNS DLBCL, and MZL which also have aberrant BTK activity [200].

Immunohistochemical Biomarkers of Peripheral T-Cell Lymphoma

Biomarkers evaluated by immunohistochemistry are used in differentiating two entities sharing morphological and histopathological features. As an example, cHL and angioimmunoblastic T-cell lymphoma may overlap, and angioimmunoblastic T-cell lymphoma may include EBV-positive cells resembling Reed-Sternberg cells both morphologically and phenotypically. Thus, immunophenotypic expression of CD10, PD-1, and CXCL13 in angioimmunoblastic T-cell lymphoma may be helpful in the diagnosis [201]. Also, CXCL13 and PD-1 are utilized in the differential diagnosis of angioimmunoblastic T-cell lymphoma *versus* EBV+ DLBCL, angioimmunoblastic T-cell lymphoma *versus* EBV+ cHL, and angioimmunoblastic T-cell lymphoma *versus* a follicular variant of PTCL-NOS.

Patients with a diagnosis of follicular variant of PTCL-NOS do not have an expansion of follicular dendritic cell groups outside the follicles and major inflammation, but Hodgkin-like cells can also be detected [201]. SAP, an additional immunohistochemical marker of germinal center T cells, is critical for the diagnosis of angioimmunoblastic T-cell lymphoma, primarily together with PD-1 and CXCL13 [202]. In practice, these immunohistochemical biomarkers are very significant for the diagnosis of angioimmunoblastic T-cell lymphoma and in the differential diagnosis with other lymphoma subtypes. T-cell-restricted intracellular antigen-1 (TIA-1), granzyme B, and perforin are cytotoxic biomarkers that are mainly used to diagnose NK/T-cell lymphomas. ALCL is also another type of lymphoma that expresses cytotoxic biomarkers; clusterin is a potentially helpful diagnostic biomarker in the diagnosis of ALCL. Currently, granzyme H is found to be a biomarker of NK/T-cell lymphomas, and it may be used in the differential diagnosis of NK/T-cell lymphomas with other PTCLs [203].

In summary, the immunophenotypic biomarkers, mostly evaluated by immunohistochemistry, take part in the diagnosis and prognosis of lymphomas in addition to becoming targets for specific treatment strategies.

Microenvironment-related Biomarkers for Prognosis and Immunotherapy

The tumor microenvironment is the extracellular area, in which the tumor cells live together with benign cells and the extracellular matrix. It is detected that benign and malignant cells in the microenvironment interact through cytokines and chemokines secreted or expressed by malignant or non-malignant cells. Thus, biomarkers reflecting the microenvironment of lymphomas are explored and they can be used to predict prognosis. Quantitative immunohistochemistry, FISH, and gene expression profiling are strong assessment tools used while evaluating microenvironmental biomarkers [139].

CD21+ Follicular Dendritic Cells: cHL is usually characterized by scarce malignant cells (usually <1%) with numerous reactive cells consisting of follicular dendritic cells that constitute the microenvironment. It was shown that the absence of CD21+ follicular dendritic cells in cHL was associated with an unfavorable prognosis [204].

CD117+ Mast Cells: CD117, expressed vigorously by mast cells, can be useful in the assessment of mast cells infiltration in the microenvironment. It is revealed that the number of mast cells was related to an inferior outcome in cHL cases [205]. On the other hand, another study observed that mast cell infiltration in DLBCL was associated with favorable prognosis [206].

CD68+ or CD163+ Macrophages: CD68 and CD163 are expressed primarily by macrophages and identified by immunohistochemistry. Several studies revealed that a high number of CD68+ or CD163+ macrophages was associated with an unfavorable outcome in patients with cHL or FL [207]. Nevertheless, CD68+ macrophage count together with high CD68 mRNA levels prior to treatment can predict a better outcome in DLBCL patients treated with chemoimmunotherapy [208]. CD163+ macrophages were also shown to be helpful in the prognosis of pediatric cHL patients [209].

***FOXP3+* Regulatory T Cells:** A high number of *FOXP3+* regulatory T-cells was found to be associated with a favorable prognosis in FL, HL and NK/T-cell lymphomas [210 - 212].

PD-1 and PD-L1: PD-1 (CD279) is expressed by T-cells in several subtypes of lymphoma, while PD-L1 and PD-L2 (CD274 and CD273), the ligands of PD-1, are expressed by neoplastic and antigen presenting cells. Increased numbers of PD-1+ infiltrating T-cells are associated with inferior survival in patients with cHL [213]. Besides, the presence of rosettes with PD-1+ reactive T-cells is a property of nodular lymphocyte predominant HL [214]. Importantly, immunotherapeutic agents inhibiting PD-1 are successfully used in lymphomas as well as in solid tumors. Agents that block PD-L1 are in comprehensive development [215 - 218].

CD58 and Tumor Necrosis Factor Receptor Superfamily Member 14 (*TNFRSF14*): CD58 is the receptor of CD2, which is expressed primarily by T- and NK-cells. Low expression of CD58 can be detected in two-thirds of activated B-cell-like DLBCL, and one third of germinal center B-cell-like DLBCL. Inactive CD58 may help in immune escape, and as a result, it facilitates other genetic defects [219]. Another genetic change having a role in immune escape is *TNFRSF14*, which functions as a tumor suppressor and regulates immune responses. *TNFRSF14* inactivation can be associated with worse outcomes in FL cases [220].

EBV and Other Viruses: Although the pathological role of EBV infection in some lymphoma subtypes is commonly accepted, the role of EBV infection in prognosis is uncertain. EBV positivity is usually associated with an inferior outcome in elderly patients with cHL, while the outcome is better among young cases [221]. EBV positivity is observed in almost 5% of DLBCL cases, which has an unfavorable outcome. Interestingly, this finding was revealed in Asian, East European, and African American patients but not in North Americans [222]. In addition to EBV, other viruses are also related to lymphoma pathogenesis. As an example, patients with hepatitis B virus-related DLBCL have specific properties and inferior outcomes. DLBCL patients with HCV infection also have good responses to antiviral agents. Human herpesvirus 8 (HHV8) infection is a characteristic of primary effusion lymphoma, which is especially detected in immunosuppressed patients. HHV8 infection is also common in multicentric Castleman disease [223 - 225].

BIOMARKERS IN MULTIPLE MYELOMA

Multiple myeloma (MM) is a clonal plasma cell malignant disease that constitutes 10% of all hematologic neoplasms. Biomarkers have an important role in the diagnosis and management of MM. The first biomarker in MM was the Bence Jones protein, which was followed by others, like M protein, plasmacytosis of the bone marrow, and β2 microglobulin. The 5-year survival rate of patients with MM is 48.5% and, through the use of immunomodulatory drugs (IMIDs) and proteasome inhibitors (PIs), most patients with high-risk properties still have low PFS and OS rates [226]. MM generally starts with asymptomatic precursor stages, known as monoclonal gammopathy of undetermined significance (MGUS) and smoldering multiple myeloma (SMM). MGUS can proceed to MM at a rate of 1% per year, while SMM has a progression rate of approximately 10% per year. Rapid progression from MGUS/SMM to MM may occur in some patients, while the others continue to be indolent and progress minimally lifelong. Prognostic categorizations standardly used in newly diagnosed MM, like the Durie-Salmon (DS), International Staging System (ISS) and revised ISS classifications, are implemented with the aid of risk stratification and act in the assessment of OS. But they do not fully explain the genomic complexity of the disease [226].

STAGING SYSTEMS

Durie-Salmon Staging System: DS staging system is one of the first discovered standard staging systems for MM. Patients are categorized as stages of I (low), II (intermediate), and III (high), together with subtypes of A and B based upon renal function. This system evaluates hemoglobin, M protein, serum calcium, and creatinine levels and examines bone lesions with x-ray imaging [227]. Nevertheless, the DS staging system has some restrictions in predicting prognosis and survival, since bone lesion interpretation changes accoring to the clinician.

ISS and Revised ISS: In contrast to DS staging, ISS is more precise in the evaluation of prognosis in MM on given grounds of serum β2 microglobulin and albumin levels used in the prognosis. ISS categorizes newly diagnosed patients into three stages based on the prognosis [228]. Even though ISS seems to be more practical than the DS staging, it relies on albumin level that is not disease-specific, and both albumin and β2-microglobulin can be affected by factors like renal failure and other comorbidities. Nevertheless, with the development of further biomarkers in MM, a revised version of the ISS was replaced consisting of lactate dehydrogenase levels and cytogenetics [229]. The revised ISS is nowadays the gold-standard staging system in newly diagnosed MM cases, but these classifications are not used in SMM. Prognostic systems distinguishing the risk of progression in SMM are suggested to classify patients at higher risk of developing the active disease [230, 231].

Plasma Cell Percentage: The normal plasma cell percentage is <3%, whereas it increases from 10% to 100% in MM patients depending on the disease status. Determination of plasma cell percentage with bone marrow biopsy is performed to screen the patients. Although it is invasive and painful, it is used as a confirmatory test for the MM diagnosis [232].

Serum M-protein and Light Chains: Monoclonal antibody (M protein) is the primary feature of plasma cell diseases. Even though M protein alone does not involve in the diagnostic criteria of MM, it is mostly increased except in oligo and non-secretory myeloma patients. Previously, repeated measurements of M protein were performed during and after treatment to monitor response and to detect disease relapse. Nevertheless, its role as a prognostic or predictive biomarker is uncertain. However, serum free light chain (sFLC) ratio has a role in the diagnosis, monitoring of disease response, and prognosis [233]. sFLC ratio is one of the MM describing phenomena, and it is a precise predictor of progression from SMM to MM [234]. Abnormal sFLC ratio before autologous stem cell transplantation is related to early disease progression [235]. The prognostic effect of abnormal sFC on PFS and OS is examined to be independent of the t(4;14) and t(14;16) related to high-risk disease. Patients who relapse with sFLC ± M protein have a poorer outcome than those who relapse with M protein alone [236].

Bence-Jones Protein: The urinary analysis of MM patients proposed the presence of specific proteins called as 'Bence Jones protein'. The urinary Bence Jones protein is one of the pertinent parameters for the diagnosis and prognosis of MM [232].

Serum Protein Electrophoresis (SPE): MM exhibits the proliferation of a single clone of malignant plasma cells, which produces a single type of immunoglobulins, heavy and/or light chains. These antibodies (M proteins or paraproteins) are identified as a spike (called M spike) containing α, β or γ globulins in the serum protein electrophoresis of the patient. M−spike in SPE is an indicator for MM and is used to monitor disease progression, remission, or recurrence [232].

GENOMIC BIOMARKERS

Previously, genetic defects identified by FISH, were used widely as diagnostic and prognostic markers in MM. Hyperdiploidy and focal or chromosome arm gains or losses, together with translocations including immunoglobulin heavy chain locus on chromosome 14, are hallmarks of MM [237]. Hyperdiploidy and chromosomal translocations are the most common genetic defects and are considered the main events in MM. Hyperdiploidy alone has a good prognosis in contrast to nonhyperdiploidy and hypodiploidy [238]. Chromosomal translocations achieve a fusion product, in which the partner genes become under the control of the immunoglobulin heavy chain enhancer. This entity causes the overexpression of oncogenes like *MMSET/FGFR3, CCND3, CCND1, MAF,* and *MAFB* in translocations t(4;14), t(6;14), t(11;14), t(14;16), and t(14;20), respectively [239 - 241]. Neverthclcss, they have different impacts on the prognosis and progression of SMM to MM. For instance, t(4:14) was found to be related to a poor prognosis in previous studies; however, survival outcomes in patients horboring this translocation enhanced significantly with the use of PIs. On the other hand, patients who present with high-risk cytogenetics such as 17p loss affecting *TP53*, t(14;20), and t(14;16), resume having inferior outcomes in spite of therapeutic advances in MM [239]. Furthermore, *MAF* protein which is the product of t(14;16) and its paralogue, *MAFB*, in t(14;20), are considered to have an inherent resistance to PIs and indirectly enhance the stability of *MAF* protein. Interestingly, though they are associated with poor prognosis in MM, these two defects are related to a standard risk of progression in MGUS and SMM [242]. Translocations activating the *MYC* oncogene are associated with disease progression. These translocations are found in 15% of MM patients and may result in decreased PFS and OS [243]. The presence of *MYC* translocations alone or together with 1q amplification, defines a poor prognostic subgroup in MM with hyperdiploidism [244]. *BET*-bromodomain inhibitors acting by interfering with the super-enhancers demonstrate that patients with *MYC* overexpression can benefit from this targeted therapy [245]. Cytogenetics are critical prognostic biomarkers that are used in the risk stratification of patients. They can also participate in the selection of treatment modalities.

Gene Expression Profiling

Gene expression profiling (GEP) evaluates the molecular heterogeneity of MM and explores the expression level of a group of genes and correlates them with the outcome. The University of Arkansas for Medical Sciences model was introduced in a study mainly focused on one of the most common genetic defects in MM known as chromosome 1 mutations [246]. An expression profile of 70 genes, 30% of which are on chromosome 1 (1p and 1q), in CD138+ plasma cells isolated from newly diagnosed cases, discloses fifty-one upregulated and 19 downregulated genes. This model score can predict disease-free survival in most patients and is associated with MM staging, cytogenetic defects, and serum β2 microglobulin levels [246]. Furthermore, another fifteen-gene model was developed that predicted survival in newly diagnosed MM and the relapsed setting [247]. Most of the GEP studies with diverse MM cohorts concluded that GEP might be a potential method for the risk assessment in MM [248, 249].

NOVEL BIOMARKERS

Minimal Residual Disease

MRD is an important marker in MM. Response to treatment was previously based upon paraprotein measurements to evaluate residual disease. Although there are many important advances in the treatment of MM, there are still many patients who demonstrate an initially favorable response, but relapse later. This proposes that there may be remaining disease that can not be detected by standard biomarkers of MRD. It is shown that MRD is better than the prognostic value of CR accomplishment of PFS and OS [250]. Novel techniques in the measurement of MRD, like MFC and NGS, were also developed. MRD monitoring may serve as a tool for risk-adapted monitoring [251]. MFC, allele-specific oligonucleotide qPCR, NGS are the major methods to monitor MRD [250].

Next-Generation Biomarkers in MM

Blood Biopsy: Blood biopsy includes detecting circulating tumor cells (CTCs) or circulating tumor DNA (ctDNA) in the peripheral blood. It is a noninvasive encouraging tool. Bone marrow biopsy is the gold standard for diagnosis in MM; but it is an invasive and painful procedure. It is also limited to the first diagnosis and is less likely to confirm the progression of the disease. Moreover, as MM is a multifocal disease with patchy bone marrow infiltration and remarkable clonal heterogeneity, bone marrow biopsy is unlikely to represent a mutational view of the disease. A blood biopsy can be an attainable noninvasive alternative method that depends on blood samples [226]. The results evaluated that ctDNA and CTCs can exactly identify the mutational view of different malignancies compared with the standard methods. Furthermore, they can act as prognostic and predictive markers for defining both therapy response and disease relapse [252].

CTCs: CTCs are released from the primary tumor or metastatic sites into the blood. MM CTCs are detected by MFC and may act as a predictor for survival in newly diagnosed cases and in relapsed disease [253, 254]. In patients with the newly diagnosed disease, the presence of more than four hundred CTCs was found to be related to increased proliferation, unfavorable cytogenetics, the shorter time between treatments, and inferior OS [255]. Another study tried to define the prognostic value of CTCs in the relapsed setting, and the presence of more than one hundred CTCs predicted poor survival in patients experiencing active disease relapse [253]. Furthermore, high levels of CTCs in SMM were associated with a higher risk of progression to MM in the first 2-3 years [256]. However, the sensitivity of standard FC is not enough to identify MM CTCs in almost 25% of patients, even in cases with a high tumor burden. A study revealed the suitability of CTCs detected by MFC, and the findings proved the viability of CTCs as an agent to assess the clonal form and evolution of MM [257].

ctDNA: ctDNA can be observed in more than one form; either in tumor cells or present as cell-free DNA (cfDNA). Nevertheless, cfDNA is a general term that includes both ctDNA and nontumor DNA [258]. ctDNA is released by tumor cells into the blood and therefore harbors the mutations of the tumor. In MM, two studies analyzed the importance of ctDNA as a biomarker. The sensitivity rates of ctDNA in harboring clonal and subclonal mutations in the bone marrow are 100% and 96%, respectively. Moreover, the use of ctDNA in monitoring disease progression and response to treatment was evaluated by measuring the allele fraction in patient samples. According to the results of these studies, ctDNA can be a more encouraging biomarker, given the technical difficulties included in isolating and characterizing CTCs by conventional MFC methods [259, 260]. It is stated that NGS of clonal immunoglobulin VDJ gene rearrangements can be performed on cfDNA for MRD detection [261]. Besides, sequencing methods, such as cancer personalized profiling by deep sequencing, are very sensitive and specific in quantifying ctDNA [262]. Unlike other protein markers, the half-life of ctDNA is less than two hours, meaning that it can reflect a real-time tumor burden in treated patients and identify tumor dynamics changes before conventional protein biomarkers and imaging methods [263].

Immunotherapy-related Biomarkers

The striking results of immune checkpoint inhibitors in melanoma, non–small cell lung cancer and bladder cancers give rise to the examination of these medications in MM. Preclinical data showed that PD-L1 is expressed on plasma cells at different stages of MM and is extremely expressed in relapsed disease [264]. Furthermore, PD-1 expression is upregulated in the natural killer or T cells in MM patients [265]. Nevertheless, the elementary results of a study that used anti–PD-1 as a single agent in relapsed myeloma were disappointing and showed the need for combination therapies to have any response [266, 267]. Studies that involve immunotherapy together with other new agents revealed better preliminary results [268]. The expression of these biomarkers in MM, even though the mutational load is not high as in other neoplasms, should be investigated for any possible predictive or prognostic value [265].

CONCLUSION

The application of new detection platforms provides an opportunity for the assessment of genomic and transcriptomic biomarkers in hematologic malignancies. Exploration of molecular biomarkers increases our understanding of pathogenetic mechanisms and gives opportunities to refine classification, improve the precision of diagnosis, and categorize the patient outcomes, especially in combination with classical immunophenotypic biomarkers. However, the discovery and exploration of additional novel biomarkers lead to challenges as we need to select, validate, and translate these markers into clinical practice to serve as targets in order to provide new therapeutic approaches. The integration of diagnosis, prognostication, and the evaluation of therapeutic targets will become the norm in the assessment of hematologic neoplasms, and these tasks will depend on the effective application of diagnostically reliable and therapeutically meaningful biomarkers.

CONSENT FOR PUBLICATION

Not applicable.

CONFLICT OF INTEREST

The authors declare no conflict of interest, financial or otherwise.

ACKNOWLEDGEMENT

Declared none.

REFERENCES

[1] Swerdlow SH, Campo E, Pileri SA, *et al.* The 2016 revision of the World Health Organization classification of lymphoid neoplasms. Blood 2016; 127(20): 2375-90.
[http://dx.doi.org/10.1182/blood-2016-01-643569] [PMID: 26980727]

[2] Arber DA, Orazi A, Hasserjian R, *et al.* The 2016 revision to the World Health Organization classification of myeloid neoplasms and acute leukemia. Blood 2016; 127(20): 2391-405.
[http://dx.doi.org/10.1182/blood-2016-03-643544] [PMID: 27069254]

[3] Costa AFO, Menezes DL, Pinheiro LHS, *et al.* Role of new immunophenotypic markers on prognostic and overall survival of acute myeloid leukemia: a systematic review and meta-analysis. Sci Rep 2017; 7(1): 4138.
[http://dx.doi.org/10.1038/s41598-017-00816-2] [PMID: 28646224]

[4] Hussaini M. Biomarkers in Hematological Malignancies: A Review of Molecular Testing in Hematopathology. Cancer Contr 2015; 22(2): 158-66.
[http://dx.doi.org/10.1177/107327481502200206] [PMID: 26068760]

[5] Grimwade D, Walker H, Oliver F, *et al.* The importance of diagnostic cytogenetics on outcome in AML: analysis of 1,612 patients entered into the MRC AML 10 trial. Blood 1998; 92(7): 2322-33. [http://dx.doi.org/10.1182/blood.V92.7.2322] [PMID: 9746770]

[6] Swerdlow SH, Campo E, Harris NL, Eds. World health organization classification of tumours of haematopoietic and lymphoid tissues. revised 4th edition., Lyon: IARC 2017.

[7] Döhner H, Estey E, Grimwade D, *et al.* Diagnosis and management of AML in adults: 2017 ELN recommendations from an international expert panel. Blood 2017; 129(4): 424-47. [http://dx.doi.org/10.1182/blood-2016-08-733196] [PMID: 27895058]

[8] Bennett JM, Catovsky D, Daniel MT, *et al.* Proposed revised criteria for the classification of acute myeloid leukemia. A report of the French-American-British Cooperative Group. Ann Intern Med 1985; 103(4): 620-5. [http://dx.doi.org/10.7326/0003-4819-103-4-620] [PMID: 3862359]

[9] Grimwade D, Hills RK, Moorman AV, *et al.* Refinement of cytogenetic classification in acute myeloid leukemia: determination of prognostic significance of rare recurring chromosomal abnormalities among 5876 younger adult patients treated in the United Kingdom Medical Research Council trials. Blood 2010; 116(3): 354-65. [http://dx.doi.org/10.1182/blood-2009-11-254441] [PMID: 20385793]

[10] Lindsley RC, Mar BG, Mazzola E, *et al.* Acute myeloid leukemia ontogeny is defined by distinct somatic mutations. Blood 2015; 125(9): 1367-76. [http://dx.doi.org/10.1182/blood-2014-11-610543] [PMID: 25550361]

[11] Churpek JE, Pyrtel K, Kanchi KL, *et al.* Genomic analysis of germ line and somatic variants in familial myelodysplasia/acute myeloid leukemia. Blood 2015; 126(22): 2484-90. [http://dx.doi.org/10.1182/blood-2015-04-641100] [PMID: 26492932]

[12] Nucifora G, Larson RA, Rowley JD. Persistence of the 8;21 translocation in patients with acute myeloid leukemia type M2 in long-term remission. Blood 1993; 82(3): 712-5. [http://dx.doi.org/10.1182/blood.V82.3.712.712] [PMID: 8338940]

[13] Tobal K, Yin JA. Monitoring of minimal residual disease by quantitative reverse transcriptase-polymerase chain reaction for AML1-MTG8 transcripts in AML-M2 with t(8; 21). Blood 1996; 88(10): 3704-9. [http://dx.doi.org/10.1182/blood.V88.10.3704.bloodjournal88103704] [PMID: 8916934]

[14] Satake N, Maseki N, Kozu T, *et al.* Disappearance of AML1-MTG8(ETO) fusion transcript in acute myeloid leukaemia patients with t(8;21) in long-term remission. Br J Haematol 1995; 91(4): 892-8. [http://dx.doi.org/10.1111/j.1365-2141.1995.tb05406.x] [PMID: 8547135]

[15] de Thé H, Pandolfi PP, Chen Z. Acute Promyelocytic Leukemia: A Paradigm for Oncoprotein-Targeted Cure. Cancer Cell 2017; 32(5): 552-60. [http://dx.doi.org/10.1016/j.ccell.2017.10.002] [PMID: 29136503]

[16] Winters AC, Bernt KM. MLL-Rearranged Leukemias-An Update on Science and Clinical Approaches. Front Pediatr 2017; 5: 4. [http://dx.doi.org/10.3389/fped.2017.00004] [PMID: 28232907]

[17] Breems DA, Van Putten WL, De Greef GE, *et al.* Monosomal karyotype in acute myeloid leukemia: a better indicator of poor prognosis than a complex karyotype. J Clin Oncol 2008; 26(29): 4791-7. [http://dx.doi.org/10.1200/JCO.2008.16.0259] [PMID: 18695255]

[18] Rücker FG, Schlenk RF, Bullinger L, *et al.* TP53 alterations in acute myeloid leukemia with complex karyotype correlate with specific copy number alterations, monosomal karyotype, and dismal outcome. Blood 2012; 119(9): 2114-21. [http://dx.doi.org/10.1182/blood-2011-08-375758] [PMID: 22186996]

[19] Coustan-Smith E, Song G, Shurtleff S, *et al.* Universal monitoring of minimal residual disease in acute myeloid leukemia. JCI Insight 2018; 3(9): e98561.

[http://dx.doi.org/10.1172/jci.insight.98561] [PMID: 29720577]

[20] Grimwade D, Vyas P, Freeman S. Assessment of minimal residual disease in acute myeloid leukemia. Curr Opin Oncol 2010; 22(6): 656-63. [http://dx.doi.org/10.1097/CCO.0b013e32833ed831] [PMID: 20805746]

[21] Harrison CJ, Hills RK, Moorman AV, *et al.* Cytogenetics of childhood acute myeloid leukemia: United Kingdom Medical Research Council Treatment trials AML 10 and 12. J Clin Oncol 2010; 28(16): 2674-81. [http://dx.doi.org/10.1200/JCO.2009.24.8997] [PMID: 20439644]

[22] Falini B, Martelli MP, Bolli N, *et al.* Acute myeloid leukemia with mutated nucleophosmin (NPM1): is it a distinct entity? Blood 2011; 117(4): 1109-20. [http://dx.doi.org/10.1182/blood-2010-08-299990] [PMID: 21030560]

[23] Dobrowolska H, Gill KZ, Serban G, *et al.* Expression of immune inhibitory receptor ILT3 in acute myeloid leukemia with monocytic differentiation. Cytometry B Clin Cytom 2013; 84(1): 21-9. [http://dx.doi.org/10.1002/cyto.b.21050] [PMID: 23027709]

[24] Iwasaki M, Liedtke M, Gentles AJ, Cleary ML. CD93 marks a non-quiescent human leukemia stem cell population and is required for development of MLL-rearranged acute myeloid leukemia. Cell Stem Cell 2015; 17(4): 412-21. [http://dx.doi.org/10.1016/j.stem.2015.08.008] [PMID: 26387756]

[25] Nishioka C, Ikezoe T, Furihata M, *et al.* $CD34^{+}/CD38^{-}$ acute myelogenous leukemia cells aberrantly express CD82 which regulates adhesion and survival of leukemia stem cells. Int J Cancer 2013; 132(9): 2006-19. [http://dx.doi.org/10.1002/ijc.27904] [PMID: 23055153]

[26] Ganguly BB, Kadam NN. Mutations of myelodysplastic syndromes (MDS): An update. Mutat Res Rev Mutat Res 2016; 769: 47-62. [http://dx.doi.org/10.1016/j.mrrev.2016.04.009] [PMID: 27543316]

[27] Greenberg PL, Tuechler H, Schanz J, *et al.* Revised international prognostic scoring system for myelodysplastic syndromes. Blood 2012; 120(12): 2454-65. [http://dx.doi.org/10.1182/blood-2012-03-420489] [PMID: 22740453]

[28] Haferlach T, Nagata Y, Grossmann V, *et al.* Landscape of genetic lesions in 944 patients with myelodysplastic syndromes. Leukemia 2014; 28(2): 241-7. [http://dx.doi.org/10.1038/leu.2013.336] [PMID: 24220272]

[29] Lukackova R, Gerykova Bujalkova M, Majerova L, Mladosievicova B. Molecular genetic methods in the diagnosis of myelodysplastic syndromes. A review. Biomed Pap Med Fac Univ Palacky Olomouc Czech Repub 2014; 158(3): 339-45. [http://dx.doi.org/10.5507/bp.2013.084] [PMID: 24263214]

[30] Kulasekararaj AG, Mohamedali AM, Mufti GJ. Recent advances in understanding the molecular pathogenesis of myelodysplastic syndromes. Br J Haematol 2013; 162(5): 587-605. [http://dx.doi.org/10.1111/bjh.12435] [PMID: 23869491]

[31] Nybakken GE, Bagg A. The genetic basis and expanding role of molecular analysis in the diagnosis, prognosis, and therapeutic design for myelodysplastic syndromes. J Mol Diagn 2014; 16(2): 145-58. [http://dx.doi.org/10.1016/j.jmoldx.2013.11.005] [PMID: 24457119]

[32] Bejar R, Stevenson KE, Caughey B, *et al.* Somatic mutations predict poor outcome in patients with myelodysplastic syndrome after hematopoietic stem-cell transplantation. J Clin Oncol 2014; 32(25): 2691-8. [http://dx.doi.org/10.1200/JCO.2013.52.3381] [PMID: 25092778]

[33] Valencia A, Masala E, Rossi A, *et al.* Expression of nucleoside-metabolizing enzymes in myelodysplastic syndromes and modulation of response to azacitidine. Leukemia 2014; 28(3): 621-8. [http://dx.doi.org/10.1038/leu.2013.330] [PMID: 24192812]

[34] Olney HJ, Le Beau MM. Evaluation of recurring cytogenetic abnormalities in the treatment of myelodysplastic syndromes. Leuk Res 2007; 31(4): 427-34. [http://dx.doi.org/10.1016/j.leukres.2006.10.023] [PMID: 17161457]

[35] Schanz J, Tüchler H, Solé F, *et al.* New comprehensive cytogenetic scoring system for primary myelodysplastic syndromes (MDS) and oligoblastic acute myeloid leukemia after MDS derived from an international database merge. J Clin Oncol 2012; 30(8): 820-9. [http://dx.doi.org/10.1200/JCO.2011.35.6394] [PMID: 22331955]

[36] Weinberg OK, Hasserjian RP. The current approach to the diagnosis of myelodysplastic syndromes. Semin Hematol 2019; 56(1): 15-21. [http://dx.doi.org/10.1053/j.seminhematol.2018.05.015] [PMID: 30573039]

[37] Bacher U, Haferlach T, Kern W, Weiss T, Schnittger S, Haferlach C. The impact of cytomorphology, cytogenetics, molecular genetics, and immunophenotyping in a comprehensive diagnostic workup of myelodysplastic syndromes. Cancer 2009; 115(19): 4524-32. [http://dx.doi.org/10.1002/cncr.24501] [PMID: 19569249]

[38] List A, Ebert BL, Fenaux P. A decade of progress in myelodysplastic syndrome with chromosome 5q deletion. Leukemia 2018; 32(7): 1493-9. [http://dx.doi.org/10.1038/s41375-018-0029-9] [PMID: 29445113]

[39] McNerney ME, Godley LA, Le Beau MM. Therapy-related myeloid neoplasms: when genetics and environment collide. Nat Rev Cancer 2017; 17(9): 513-27. [http://dx.doi.org/10.1038/nrc.2017.60] [PMID: 28835720]

[40] Bejar R, Levine R, Ebert BL. Unraveling the molecular pathophysiology of myelodysplastic syndromes. J Clin Oncol 2011; 29(5): 504-15. [http://dx.doi.org/10.1200/JCO.2010.31.1175] [PMID: 21220588]

[41] McNerney ME, Brown CD, Wang X, *et al.* CUX1 is a haploinsufficient tumor suppressor gene on chromosome 7 frequently inactivated in acute myeloid leukemia. Blood 2013; 121(6): 975-83. [http://dx.doi.org/10.1182/blood-2012-04-426965] [PMID: 23212519]

[42] Wong CC, Martincorena I, Rust AG, *et al.* Inactivating CUX1 mutations promote tumorigenesis. Nat Genet 2014; 46(1): 33-8. [http://dx.doi.org/10.1038/ng.2846] [PMID: 24316979]

[43] Hartmann L, Haferlach C, Meggendorfer M, Kern W, Haferlach T, Stengel A. Myeloid malignancies with isolated 7q deletion can be further characterized by their accompanying molecular mutations. Genes Chromosomes Cancer 2019; 58(10): 698-704. [http://dx.doi.org/10.1002/gcc.22761] [PMID: 30994218]

[44] Hosono N, Makishima H, Jerez A, *et al.* Recurrent genetic defects on chromosome 7q in myeloid neoplasms. Leukemia 2014; 28(6): 1348-51. [http://dx.doi.org/10.1038/leu.2014.25] [PMID: 24429498]

[45] List A, Dewald G, Bennett J, *et al.* Lenalidomide in the myelodysplastic syndrome with chromosome 5q deletion. N Engl J Med 2006; 355(14): 1456-65. [http://dx.doi.org/10.1056/NEJMoa061292] [PMID: 17021321]

[46] Pfeilstöcker M, Tuechler H, Sanz G, *et al.* Time-dependent changes in mortality and transformation risk in MDS. Blood 2016; 128(7): 902-10. [http://dx.doi.org/10.1182/blood-2016-02-700054] [PMID: 27335276]

[47] Bejar R, Stevenson K, Abdel-Wahab O, *et al.* Clinical effect of point mutations in myelodysplastic syndromes. N Engl J Med 2011; 364(26): 2496-506. [http://dx.doi.org/10.1056/NEJMoa1013343] [PMID: 21714648]

[48] Mossner M, Jann JC, Wittig J, *et al.* Mutational hierarchies in myelodysplastic syndromes dynamically adapt and evolve upon therapy response and failure. Blood 2016; 128(9): 1246-59. [http://dx.doi.org/10.1182/blood-2015-11-679167] [PMID: 27268087]

[49] Delhommeau F, Dupont S, Della Valle V, *et al.* Mutation in TET2 in myeloid cancers. N Engl J Med 2009; 360(22): 2289-301.
[http://dx.doi.org/10.1056/NEJMoa0810069] [PMID: 19474426]

[50] Kosmider O, Gelsi-Boyer V, Cheok M, *et al.* TET2 mutation is an independent favorable prognostic factor in myelodysplastic syndromes (MDSs). Blood 2009; 114(15): 3285-91.
[http://dx.doi.org/10.1182/blood-2009-04-215814] [PMID: 19666869]

[51] Walter MJ, Ding L, Shen D, *et al.* Recurrent DNMT3A mutations in patients with myelodysplastic syndromes. Leukemia 2011; 25(7): 1153-8.
[http://dx.doi.org/10.1038/leu.2011.44] [PMID: 21415852]

[52] Roller A, Grossmann V, Bacher U, *et al.* Landmark analysis of DNMT3A mutations in hematological malignancies. Leukemia 2013; 27(7): 1573-8.
[http://dx.doi.org/10.1038/leu.2013.65] [PMID: 23519389]

[53] Kosmider O, Gelsi-Boyer V, Slama L, *et al.* Mutations of IDH1 and IDH2 genes in early and accelerated phases of myelodysplastic syndromes and MDS/myeloproliferative neoplasms. Leukemia 2010; 24(5): 1094-6.
[http://dx.doi.org/10.1038/leu.2010.52] [PMID: 20376084]

[54] Boultwood J, Perry J, Pellagatti A, *et al.* Frequent mutation of the polycomb-associated gene ASXL1 in the myelodysplastic syndromes and in acute myeloid leukemia. Leukemia 2010; 24(5): 1062-5.
[http://dx.doi.org/10.1038/leu.2010.20] [PMID: 20182461]

[55] Papaemmanuil E, Cazzola M, Boultwood J, *et al.* Somatic SF3B1 mutation in myelodysplasia with ring sideroblasts. N Engl J Med 2011; 365(15): 1384-95.
[http://dx.doi.org/10.1056/NEJMoa1103283] [PMID: 21995386]

[56] Wu SJ, Kuo YY, Hou HA, *et al.* The clinical implication of SRSF2 mutation in patients with myelodysplastic syndrome and its stability during disease evolution. Blood 2012; 120(15): 3106-11.
[http://dx.doi.org/10.1182/blood-2012-02-412296] [PMID: 22932795]

[57] Kennedy JA, Ebert BL. Clinical Implications of Genetic Mutations in Myelodysplastic Syndrome. J Clin Oncol 2017; 35(9): 968-74.
[http://dx.doi.org/10.1200/JCO.2016.71.0806] [PMID: 28297619]

[58] Nowell PC, Hungerford DA. Chromosome studies on normal and leukemic human leukocytes. J Natl Cancer Inst 1960; 25: 85-109.
[PMID: 14427847]

[59] Reilly JT. Pathogenetic insight and prognostic information from standard and molecular cytogenetic studies in the BCR-ABL-negative myeloproliferative neoplasms (MPNs). Leukemia 2008; 22(10): 1818-27.
[http://dx.doi.org/10.1038/leu.2008.218] [PMID: 18754027]

[60] Vannucchi AM, Guglielmelli P, Tefferi A. Advances in understanding and management of myeloproliferative neoplasms. CA Cancer J Clin 2009; 59(3): 171-91.
[http://dx.doi.org/10.3322/caac.20009] [PMID: 19369682]

[61] Bench AJ, White HE, Foroni L, *et al.* Molecular diagnosis of the myeloproliferative neoplasms: UK guidelines for the detection of JAK2 V617F and other relevant mutations. Br J Haematol 2013; 160(1): 25-34.
[http://dx.doi.org/10.1111/bjh.12075] [PMID: 23057517]

[62] Grinfeld J, Nangalia J, Baxter EJ, *et al.* Classification and Personalized Prognosis in Myeloproliferative Neoplasms. N Engl J Med 2018; 379(15): 1416-30.
[http://dx.doi.org/10.1056/NEJMoa1716614] [PMID: 30304655]

[63] Klampfl T, Gisslinger H, Harutyunyan AS, *et al.* Somatic mutations of calreticulin in myeloproliferative neoplasms. N Engl J Med 2013; 369(25): 2379-90.
[http://dx.doi.org/10.1056/NEJMoa1311347] [PMID: 24325356]

[64] Tefferi A, Lasho TL, Finke CM, *et al.* CALR *vs* JAK2 *vs* MPL-mutated or triple-negative myelofibrosis: clinical, cytogenetic and molecular comparisons. Leukemia 2014; 28(7): 1472-7. [http://dx.doi.org/10.1038/leu.2014.3] [PMID: 24402162]

[65] Cazzola M, Kralovics R. From Janus kinase 2 to calreticulin: the clinically relevant genomic landscape of myeloproliferative neoplasms. Blood 2014; 123(24): 3714-9. [http://dx.doi.org/10.1182/blood-2014-03-530865] [PMID: 24786775]

[66] James C, Ugo V, Le Couédic JP, *et al.* A unique clonal JAK2 mutation leading to constitutive signalling causes polycythaemia vera. Nature 2005; 434(7037): 1144-8. [http://dx.doi.org/10.1038/nature03546] [PMID: 15793561]

[67] Baxter EJ, Scott LM, Campbell PJ, *et al.* Acquired mutation of the tyrosine kinase JAK2 in human myeloproliferative disorders. Lancet 2005; 365(9464): 1054-61. [http://dx.doi.org/10.1016/S0140-6736(05)71142-9] [PMID: 15781101]

[68] Levine RL, Pardanani A, Tefferi A, Gilliland DG. Role of JAK2 in the pathogenesis and therapy of myeloproliferative disorders. Nat Rev Cancer 2007; 7(9): 673-83. [http://dx.doi.org/10.1038/nrc2210] [PMID: 17721432]

[69] Kralovics R, Passamonti F, Buser AS, *et al.* A gain-of-function mutation of JAK2 in myeloproliferative disorders. N Engl J Med 2005; 352(17): 1779-90. [http://dx.doi.org/10.1056/NEJMoa051113] [PMID: 15858187]

[70] Goerttler PS, Steimle C, März E, *et al.* The Jak2V617F mutation, PRV-1 overexpression, and EEC formation define a similar cohort of MPD patients. Blood 2005; 106(8): 2862-4. [http://dx.doi.org/10.1182/blood-2005-04-1515] [PMID: 15985544]

[71] Pardanani A. JAK2 inhibitor therapy in myeloproliferative disorders: rationale, preclinical studies and ongoing clinical trials. Leukemia 2008; 22(1): 23-30. [http://dx.doi.org/10.1038/sj.leu.2404948] [PMID: 17882282]

[72] Chen AT, Prchal JT. JAK2 kinase inhibitors and myeloproliferative disorders. Curr Opin Hematol 2010; 17(2): 110-6. [http://dx.doi.org/10.1097/MOH.0b013e3283366b91] [PMID: 20087176]

[73] Stein BL, Crispino JD, Moliterno AR. Janus kinase inhibitors: an update on the progress and promise of targeted therapy in the myeloproliferative neoplasms. Curr Opin Oncol 2011; 23(6): 609-16. [http://dx.doi.org/10.1097/CCO.0b013e32834d1b22] [PMID: 21993415]

[74] Spivak JL. Narrative review: Thrombocytosis, polycythemia vera, and JAK2 mutations: The phenotypic mimicry of chronic myeloproliferation. Ann Intern Med 2010; 152(5): 300-6. [http://dx.doi.org/10.7326/0003-4819-152-5-201003020-00008] [PMID: 20194236]

[75] Takahashi K, Patel KP, Kantarjian H, *et al.* JAK2 p.V617F detection and allele burden measurement in peripheral blood and bone marrow aspirates in patients with myeloproliferative neoplasms. Blood 2013; 122(23): 3784-6. [http://dx.doi.org/10.1182/blood-2013-07-515676] [PMID: 24068492]

[76] Kiladjian JJ, Cassinat B, Turlure P, *et al.* High molecular response rate of polycythemia vera patients treated with pegylated interferon alpha-2a. Blood 2006; 108(6): 2037-40. [http://dx.doi.org/10.1182/blood-2006-03-009860] [PMID: 16709929]

[77] Lange T, Edelmann A, Siebolts U, *et al.* JAK2 p.V617F allele burden in myeloproliferative neoplasms one month after allogeneic stem cell transplantation significantly predicts outcome and risk of relapse. Haematologica 2013; 98(5): 722-8. [http://dx.doi.org/10.3324/haematol.2012.076901] [PMID: 23300178]

[78] Austin SK, Lambert JR. The JAK2 V617F mutation and thrombosis. Br J Haematol 2008; 143(3): 307-20. [http://dx.doi.org/10.1111/j.1365-2141.2008.07258.x] [PMID: 19004076]

[79] Alvarez-Larrán A, Bellosillo B, Pereira A, *et al.* JAK2V617F monitoring in polycythemia vera and essential thrombocythemia: clinical usefulness for predicting myelofibrotic transformation and thrombotic events. Am J Hematol 2014; 89(5): 517-23.
[http://dx.doi.org/10.1002/ajh.23676] [PMID: 24458835]

[80] Rumi E, Pietra D, Guglielmelli P, *et al.* Acquired copy-neutral loss of heterozygosity of chromosome 1p as a molecular event associated with marrow fibrosis in MPL-mutated myeloproliferative neoplasms. Blood 2013; 121(21): 4388-95.
[http://dx.doi.org/10.1182/blood-2013-02-486050] [PMID: 23575445]

[81] Tefferi A. Mutations galore in myeloproliferative neoplasms: would the real Spartacus please stand up? Leukemia 2011; 25(7): 1059-63.
[http://dx.doi.org/10.1038/leu.2011.92] [PMID: 21750560]

[82] Vainchenker W, Delhommeau F, Constantinescu SN, Bernard OA. New mutations and pathogenesis of myeloproliferative neoplasms. Blood 2011; 118(7): 1723-35.
[http://dx.doi.org/10.1182/blood-2011-02-292102] [PMID: 21653328]

[83] Rampal R, Al-Shahrour F, Abdel-Wahab O, *et al.* Integrated genomic analysis illustrates the central role of JAK-STAT pathway activation in myeloproliferative neoplasm pathogenesis. Blood 2014; 123(22): e123-33.
[http://dx.doi.org/10.1182/blood-2014-02-554634] [PMID: 24740812]

[84] Tefferi A, Vainchenker W. Myeloproliferative neoplasms: molecular pathophysiology, essential clinical understanding, and treatment strategies. J Clin Oncol 2011; 29(5): 573-82.
[http://dx.doi.org/10.1200/JCO.2010.29.8711] [PMID: 21220604]

[85] Lundberg P, Karow A, Nienhold R, *et al.* Clonal evolution and clinical correlates of somatic mutations in myeloproliferative neoplasms. Blood 2014; 123(14): 2220-8.
[http://dx.doi.org/10.1182/blood-2013-11-537167] [PMID: 24478400]

[86] Vannucchi AM, Antonioli E, Guglielmelli P, *et al.* Clinical profile of homozygous JAK2 617V>F mutation in patients with polycythemia vera or essential thrombocythemia. Blood 2007; 110(3): 840-6.
[http://dx.doi.org/10.1182/blood-2006-12-064287] [PMID: 17379742]

[87] Williams DM, Kim AH, Rogers O, Spivak JL, Moliterno AR. Phenotypic variations and new mutations in JAK2 V617F-negative polycythemia vera, erythrocytosis, and idiopathic myelofibrosis. Exp Hematol 2007; 35(11): 1641-6.
[http://dx.doi.org/10.1016/j.exphem.2007.08.010] [PMID: 17920755]

[88] Sirhan S, Lasho TL, Hanson CA, Mesa RA, Pardanani A, Tefferi A. The presence of JAK2V617F in primary myelofibrosis or its allele burden in polycythemia vera predicts chemosensitivity to hydroxyurea. Am J Hematol 2008; 83(5): 363-5.
[http://dx.doi.org/10.1002/ajh.21149] [PMID: 18266209]

[89] Vannucchi AM, Antonioli E, Guglielmelli P, Pardanani A, Tefferi A. Clinical correlates of JAK2V617F presence or allele burden in myeloproliferative neoplasms: a critical reappraisal. Leukemia 2008; 22(7): 1299-307.
[http://dx.doi.org/10.1038/leu.2008.113] [PMID: 18496562]

[90] Jiang X, Zhao Y, Forrest D, Smith C, Eaves A, Eaves C. Stem cell biomarkers in chronic myeloid leukemia. Dis Markers 2008; 24(4-5): 201-16.
[http://dx.doi.org/10.1155/2008/146871] [PMID: 18525114]

[91] Druker BJ, Guilhot F, O'Brien SG, *et al.* Five-year follow-up of patients receiving imatinib for chronic myeloid leukemia. N Engl J Med 2006; 355(23): 2408-17.
[http://dx.doi.org/10.1056/NEJMoa062867] [PMID: 17151364]

[92] Lahaye T, Riehm B, Berger U, *et al.* Response and resistance in 300 patients with BCR-ABL-positive leukemias treated with imatinib in a single center: a 4.5-year follow-up. Cancer 2005; 103(8): 1659-69.

[http://dx.doi.org/10.1002/cncr.20922] [PMID: 15747376]

[93] Sweet K, Zhang L, Pinilla-Ibarz J. Biomarkers for determining the prognosis in chronic myelogenous leukemia. J Hematol Oncol 2013; 6(1): 54. [http://dx.doi.org/10.1186/1756-8722-6-54] [PMID: 23870290]

[94] Leonard DG, Ed. Molecular Pathology in Clinical Practice: Genetics. New York: Springer 2008.

[95] Sokal JE, Cox EB, Baccarani M, *et al.* Prognostic discrimination in "good-risk" chronic granulocytic leukemia. Blood 1984; 63(4): 789-99. [http://dx.doi.org/10.1182/blood.V63.4.789.789] [PMID: 6584184]

[96] Hasford J, Pfirrmann M, Hehlmann R, *et al.* A new prognostic score for survival of patients with chronic myeloid leukemia treated with interferon alfa. J Natl Cancer Inst 1998; 90(11): 850-8. [http://dx.doi.org/10.1093/jnci/90.11.850] [PMID: 9625174]

[97] Hasford J, Baccarani M, Hoffmann V, *et al.* Predicting complete cytogenetic response and subsequent progression-free survival in 2060 patients with CML on imatinib treatment: the EUTOS score. Blood 2011; 118(3): 686-92. [http://dx.doi.org/10.1182/blood-2010-12-319038] [PMID: 21536864]

[98] Pfirrmann M, Clark RE, Prejzner W, *et al.* The EUTOS long-term survival (ELTS) score is superior to the Sokal score for predicting survival in chronic myeloid leukemia. Leukemia 2020; 34(8): 2138-49. [http://dx.doi.org/10.1038/s41375-020-0931-9] [PMID: 32601376]

[99] Soverini S, Hochhaus A, Nicolini FE, *et al.* BCR-ABL kinase domain mutation analysis in chronic myeloid leukemia patients treated with tyrosine kinase inhibitors: recommendations from an expert panel on behalf of European LeukemiaNet. Blood 2011; 118(5): 1208-15. [http://dx.doi.org/10.1182/blood-2010-12-326405] [PMID: 21562040]

[100] Soverini S, Martinelli G, Rosti G, *et al.* ABL mutations in late chronic phase chronic myeloid leukemia patients with up-front cytogenetic resistance to imatinib are associated with a greater likelihood of progression to blast crisis and shorter survival: a study by the GIMEMA Working Party on Chronic Myeloid Leukemia. J Clin Oncol 2005; 23(18): 4100-9. [http://dx.doi.org/10.1200/JCO.2005.05.531] [PMID: 15867198]

[101] Parker WT, Lawrence RM, Ho M, *et al.* Sensitive detection of BCR-ABL1 mutations in patients with chronic myeloid leukemia after imatinib resistance is predictive of outcome during subsequent therapy. J Clin Oncol 2011; 29(32): 4250-9. [http://dx.doi.org/10.1200/JCO.2011.35.0934] [PMID: 21990409]

[102] Yong AS, Melo JV. The impact of gene profiling in chronic myeloid leukaemia. Best Pract Res Clin Haematol 2009; 22(2): 181-90. [http://dx.doi.org/10.1016/j.beha.2009.04.002] [PMID: 19698927]

[103] McWeeney SK, Pemberton LC, Loriaux MM, *et al.* A gene expression signature of CD34+ cells to predict major cytogenetic response in chronic-phase chronic myeloid leukemia patients treated with imatinib. Blood 2010; 115(2): 315-25. [http://dx.doi.org/10.1182/blood-2009-03-210732] [PMID: 19837975]

[104] Quintas-Cardama A, Qiu YH, Post S, *et al.* Proteomic profiling of 112 proteins in chronic myeloid leukemia patient samples using reverse phase proteins arrays (RPPA) reveals distinct protein expression signatures associated with advanced phase disease and the CD34+ compartment. Blood 2011; 118(21): 2503. [abstract]. [ASH Annual Meeting Abstracts]. [http://dx.doi.org/10.1182/blood.V118.21.2503.2503]

[105] Zhao Y, Huang H, Wei G. Novel agents and biomarkers for acute lymphoid leukemia. J Hematol Oncol 2013; 6(1): 40. [http://dx.doi.org/10.1186/1756-8722-6-40] [PMID: 23773228]

[106] Simioni C, Bergamini F, Ferioli M, Rimondi E, Caruso L, Neri LM. New biomarkers and therapeutic strategies in acute lymphoblastic leukemias: Recent advances. Hematol Oncol 2020; 38(1): 22-33.

[http://dx.doi.org/10.1002/hon.2678] [PMID: 31487068]

[107] Chiaretti S, Vitale A, Cazzaniga G, *et al.* Clinico-biological features of 5202 patients with acute lymphoblastic leukemia enrolled in the Italian AIEOP and GIMEMA protocols and stratified in age cohorts. Haematologica 2013; 98(11): 1702-10. [http://dx.doi.org/10.3324/haematol.2012.080432] [PMID: 23716539]

[108] Bennett JM, Catovsky D, Daniel MT, *et al.* Proposals for the classification of the acute leukaemias. French-American-British (FAB) co-operative group. Br J Haematol 1976; 33(4): 451-8. [http://dx.doi.org/10.1111/j.1365-2141.1976.tb03563.x] [PMID: 188440]

[109] Bene MC, Castoldi G, Knapp W, *et al.* Proposals for the immunological classification of acute leukemias. Leukemia 1995; 9(10): 1783-6. [PMID: 7564526]

[110] Huguet F, Leguay T, Raffoux E, *et al.* Pediatric-inspired therapy in adults with Philadelphia chromosome-negative acute lymphoblastic leukemia: the GRAALL-2003 study. J Clin Oncol 2009; 27(6): 911-8. [http://dx.doi.org/10.1200/JCO.2008.18.6916] [PMID: 19124805]

[111] Jacobson S, Tedder M, Eggert J. Adult acute lymphoblastic leukemia: a genetic overview and application to clinical practice. Clin J Oncol Nurs 2016; 20(6): E147-54. [http://dx.doi.org/10.1188/16.CJON.E147-E154] [PMID: 27857258]

[112] Moorman AV. New and emerging prognostic and predictive genetic biomarkers in B-cell precursor acute lymphoblastic leukemia. Haematologica 2016; 101(4): 407-16. [http://dx.doi.org/10.3324/haematol.2015.141101] [PMID: 27033238]

[113] Wei MC, Cleary ML. Novel methods and approaches to acute lymphoblastic leukemia drug discovery. Expert Opin Drug Discov 2014; 9(12): 1435-46. [http://dx.doi.org/10.1517/17460441.2014.956720] [PMID: 25212992]

[114] Moorman AV, Chilton L, Wilkinson J, Ensor HM, Bown N, Proctor SJ. A population-based cytogenetic study of adults with acute lymphoblastic leukemia. Blood 2010; 115(2): 206-14. [http://dx.doi.org/10.1182/blood-2009-07-232124] [PMID: 19897583]

[115] Mullighan CG, Goorha S, Radtke I, *et al.* Genome-wide analysis of genetic alterations in acute lymphoblastic leukaemia. Nature 2007; 446(7137): 758-64. [http://dx.doi.org/10.1038/nature05690] [PMID: 17344859]

[116] Paietta E, Racevskis J, Neuberg D, Rowe JM, Goldstone AH, Wiernik PH. Expression of CD25 (interleukin-2 receptor alpha chain) in adult acute lymphoblastic leukemia predicts for the presence of BCR/ABL fusion transcripts: results of a preliminary laboratory analysis of ECOG/MRC Intergroup Study E2993. Leukemia 1997; 11(11): 1887-90. [http://dx.doi.org/10.1038/sj.leu.2400836] [PMID: 9369422]

[117] Wetzler M, Dodge RK, Mrózek K, *et al.* Additional cytogenetic abnormalities in adults with Philadelphia chromosome-positive acute lymphoblastic leukaemia: a study of the Cancer and Leukaemia Group B. Br J Haematol 2004; 124(3): 275-88. [http://dx.doi.org/10.1046/j.1365-2141.2003.04736.x] [PMID: 14717774]

[118] Pui CH, Robison LL, Look AT. Acute lymphoblastic leukaemia. Lancet 2008; 371(9617): 1030-43. [http://dx.doi.org/10.1016/S0140-6736(08)60457-2] [PMID: 18358930]

[119] Mrózek K, Harper DP, Aplan PD. Cytogenetics and molecular genetics of acute lymphoblastic leukemia. Hematol Oncol Clin North Am 2009; 23(5): 991-1010, v. [http://dx.doi.org/10.1016/j.hoc.2009.07.001] [PMID: 19825449]

[120] Harrison CJ, Moorman AV, Broadfield ZJ, *et al.* Three distinct subgroups of hypodiploidy in acute lymphoblastic leukaemia. Br J Haematol 2004; 125(5): 552-9. [http://dx.doi.org/10.1111/j.1365-2141.2004.04948.x] [PMID: 15147369]

[121] Safavi S, Paulsson K. Near-haploid and low-hypodiploid acute lymphoblastic leukemia: two distinct

subtypes with consistently poor prognosis. Blood 2017; 129(4): 420-3.
[http://dx.doi.org/10.1182/blood-2016-10-743765] [PMID: 27903530]

[122] Jeha S, Pei D, Raimondi SC, *et al.* Increased risk for CNS relapse in pre-B cell leukemia with the t(1;19)/TCF3-PBX1. Leukemia 2009; 23(8): 1406-9.
[http://dx.doi.org/10.1038/leu.2009.42] [PMID: 19282835]

[123] Harrison CJ, Moorman AV, Schwab C, *et al.* An international study of intrachromosomal amplification of chromosome 21 (iAMP21): cytogenetic characterization and outcome. Leukemia 2014; 28(5): 1015-21.
[http://dx.doi.org/10.1038/leu.2013.317] [PMID: 24166298]

[124] Roberts KG, Li Y, Payne-Turner D, *et al.* Targetable kinase-activating lesions in Ph-like acute lymphoblastic leukemia. N Engl J Med 2014; 371(11): 1005-15.
[http://dx.doi.org/10.1056/NEJMoa1403088] [PMID: 25207766]

[125] Jain N, Roberts KG, Jabbour E, *et al.* Ph-like acute lymphoblastic leukemia: a high-risk subtype in adults. Blood 2017; 129(5): 572-81.
[http://dx.doi.org/10.1182/blood-2016-07-726588] [PMID: 27919910]

[126] Reshmi SC, Harvey RC, Roberts KG, *et al.* Targetable kinase gene fusions in high-risk B-ALL: a study from the Children's Oncology Group. Blood 2017; 129(25): 3352-61.
[http://dx.doi.org/10.1182/blood-2016-12-758979] [PMID: 28408464]

[127] Kuchinskaya E, Heyman M, Nordgren A, *et al.* Interphase fluorescent *in situ* hybridization deletion analysis of the 9p21 region and prognosis in childhood acute lymphoblastic leukaemia (ALL): results from a prospective analysis of 519 Nordic patients treated according to the NOPHO-ALL 2000 protocol. Br J Haematol 2011; 152(5): 615-22.
[http://dx.doi.org/10.1111/j.1365-2141.2010.08532.x] [PMID: 21241277]

[128] Graux C, Cools J, Michaux L, Vandenberghe P, Hagemeijer A. Cytogenetics and molecular genetics of T-cell acute lymphoblastic leukemia: from thymocyte to lymphoblast. Leukemia 2006; 20(9): 1496-510.
[http://dx.doi.org/10.1038/sj.leu.2404302] [PMID: 16826225]

[129] Belver L, Ferrando A. The genetics and mechanisms of T cell acute lymphoblastic leukaemia. Nat Rev Cancer 2016; 16(8): 494-507.
[http://dx.doi.org/10.1038/nrc.2016.63] [PMID: 27451956]

[130] Coustan-Smith E, Mullighan CG, Onciu M, *et al.* Early T-cell precursor leukaemia: a subtype of very high-risk acute lymphoblastic leukaemia. Lancet Oncol 2009; 10(2): 147-56.
[http://dx.doi.org/10.1016/S1470-2045(08)70314-0] [PMID: 19147408]

[131] Coustan-Smith E, Song G, Clark C, *et al.* New markers for minimal residual disease detection in acute lymphoblastic leukemia. Blood 2011; 117(23): 6267-76.
[http://dx.doi.org/10.1182/blood-2010-12-324004] [PMID: 21487112]

[132] Campana D. Role of minimal residual disease monitoring in adult and pediatric acute lymphoblastic leukemia. Hematol Oncol Clin North Am 2009; 23(5): 1083-1098, vii.
[http://dx.doi.org/10.1016/j.hoc.2009.07.010] [PMID: 19825454]

[133] Cavé H, van der Werff ten Bosch J, Suciu S, *et al.* Clinical significance of minimal residual disease in childhood acute lymphoblastic leukemia. European Organization for Research and Treatment of Cancer--Childhood Leukemia Cooperative Group. N Engl J Med 1998; 339(9): 591-8.
[http://dx.doi.org/10.1056/NEJM199808273390904] [PMID: 9718378]

[134] Holowiecki J, Krawczyk-Kulis M, Giebel S, *et al.* Status of minimal residual disease after induction predicts outcome in both standard and high-risk Ph-negative adult acute lymphoblastic leukaemia. The Polish Adult Leukemia Group ALL 4-2002 MRD Study. Br J Haematol 2008; 142(2): 227-37.
[http://dx.doi.org/10.1111/j.1365-2141.2008.07185.x] [PMID: 18492099]

[135] Bassan R, Spinelli O, Oldani E, *et al.* Improved risk classification for risk-specific therapy based on

the molecular study of MRD in adult ALL. Blood 2009; 113(18): 4153-62.
[http://dx.doi.org/10.1182/blood-2008-11-185132] [PMID: 19141862]

[136] Paganin M, Zecca M, Fabbri G, *et al.* Minimal residual disease is an important predictive factor of outcome in children with relapsed 'high-risk' acute lymphoblastic leukemia. Leukemia 2008; 22(12): 2193-200.
[http://dx.doi.org/10.1038/leu.2008.227] [PMID: 18754029]

[137] Bader P, Kreyenberg H, Henze GH, *et al.* Prognostic value of minimal residual disease quantification before allogeneic stem-cell transplantation in relapsed childhood acute lymphoblastic leukemia: the ALL-REZ BFM Study Group. J Clin Oncol 2009; 27(3): 377-84.
[http://dx.doi.org/10.1200/JCO.2008.17.6065] [PMID: 19064980]

[138] Brüggemann M, Schrauder A, Raff T, *et al.* Standardized MRD quantification in European ALL trials: proceedings of the Second International Symposium on MRD assessment in Kiel, Germany, 18-20 September 2008. Leukemia 2010; 24(3): 521-35.
[http://dx.doi.org/10.1038/leu.2009.268] [PMID: 20033054]

[139] Sun R, Medeiros LJ, Young KH. Diagnostic and predictive biomarkers for lymphoma diagnosis and treatment in the era of precision medicine. Mod Pathol 2016; 29(10): 1118-42.
[http://dx.doi.org/10.1038/modpathol.2016.92] [PMID: 27363492]

[140] Soldini D, Campo E. New insights into the diagnosis of lymphomas. Ann Oncol 2012; 23 (Suppl. 10): x83-8.
[http://dx.doi.org/10.1093/annonc/mds304] [PMID: 22987998]

[141] Campo E. Whole genome profiling and other high throughput technologies in lymphoid neoplasms--current contributions and future hopes. Mod Pathol 2013; 26(S1) (Suppl. 1): S97-S110.
[http://dx.doi.org/10.1038/modpathol.2012.179] [PMID: 23281439]

[142] Basso K, Dalla-Favera R. BCL6: master regulator of the germinal center reaction and key oncogene in B cell lymphomagenesis. Adv Immunol 2010; 105: 193-210.
[http://dx.doi.org/10.1016/S0065-2776(10)05007-8] [PMID: 20510734]

[143] Wagner SD, Ahearne M, Ko Ferrigno P. The role of BCL6 in lymphomas and routes to therapy. Br J Haematol 2011; 152(1): 3-12.
[http://dx.doi.org/10.1111/j.1365-2141.2010.08420.x] [PMID: 21083654]

[144] Iqbal J, Greiner TC, Patel K, *et al.* Distinctive patterns of BCL6 molecular alterations and their functional consequences in different subgroups of diffuse large B-cell lymphoma. Leukemia 2007; 21(11): 2332-43.
[http://dx.doi.org/10.1038/sj.leu.2404856] [PMID: 17625604]

[145] Xu-Monette ZY, Medeiros LJ, Li Y, *et al.* Dysfunction of the TP53 tumor suppressor gene in lymphoid malignancies. Blood 2012; 119(16): 3668-83.
[http://dx.doi.org/10.1182/blood-2011-11-366062] [PMID: 22275381]

[146] Ott G, Rosenwald A, Campo E. Understanding MYC-driven aggressive B-cell lymphomas: pathogenesis and classification. Blood 2013; 122(24): 3884-91.
[http://dx.doi.org/10.1182/blood-2013-05-498329] [PMID: 24009228]

[147] Barrans S, Crouch S, Smith A, *et al.* Rearrangement of MYC is associated with poor prognosis in patients with diffuse large B-cell lymphoma treated in the era of rituximab. J Clin Oncol 2010; 28(20): 3360-5.
[http://dx.doi.org/10.1200/JCO.2009.26.3947] [PMID: 20498406]

[148] Freedman A. Follicular lymphoma: 2014 update on diagnosis and management. Am J Hematol 2014; 89(4): 429-36.
[http://dx.doi.org/10.1002/ajh.23674] [PMID: 24687887]

[149] Kendrick SL, Redd L, Muranyi A, *et al.* BCL2 antibodies targeted at different epitopes detect varying levels of protein expression and correlate with frequent gene amplification in diffuse large B-cell

lymphoma. Hum Pathol 2014; 45(10): 2144-53.
[http://dx.doi.org/10.1016/j.humpath.2014.06.005] [PMID: 25090918]

[150] Visco C, Tzankov A, Xu-Monette ZY, *et al.* Patients with diffuse large B-cell lymphoma of germinal center origin with BCL2 translocations have poor outcome, irrespective of MYC status: a report from an International DLBCL rituximab-CHOP Consortium Program Study. Haematologica 2013; 98(2): 255-63.
[http://dx.doi.org/10.3324/haematol.2012.066209] [PMID: 22929980]

[151] Ngo VN, Young RM, Schmitz R, *et al.* Oncogenically active MYD88 mutations in human lymphoma. Nature 2011; 470(7332): 115-9.
[http://dx.doi.org/10.1038/nature09671] [PMID: 21179087]

[152] Kim Y, Ju H, Kim DH, *et al.* CD79B and MYD88 mutations in diffuse large B-cell lymphoma. Hum Pathol 2014; 45(3): 556-64.
[http://dx.doi.org/10.1016/j.humpath.2013.10.023] [PMID: 24444466]

[153] Scott DW, Wright GW, Williams PM, *et al.* Determining cell-of-origin subtypes of diffuse large B-cell lymphoma using gene expression in formalin-fixed paraffin-embedded tissue. Blood 2014; 123(8): 1214-7.
[http://dx.doi.org/10.1182/blood-2013-11-536433] [PMID: 24398326]

[154] Alizadeh AA, Gentles AJ, Alencar AJ, *et al.* Prediction of survival in diffuse large B-cell lymphoma based on the expression of 2 genes reflecting tumor and microenvironment. Blood 2011; 118(5): 1350-8.
[http://dx.doi.org/10.1182/blood-2011-03-345272] [PMID: 21670469]

[155] Love C, Sun Z, Jima D, *et al.* The genetic landscape of mutations in Burkitt lymphoma. Nat Genet 2012; 44(12): 1321-5.
[http://dx.doi.org/10.1038/ng.2468] [PMID: 23143597]

[156] Fernàndez V, Salamero O, Espinet B, *et al.* Genomic and gene expression profiling defines indolent forms of mantle cell lymphoma. Cancer Res 2010; 70(4): 1408-18.
[http://dx.doi.org/10.1158/0008-5472.CAN-09-3419] [PMID: 20124476]

[157] Eberle FC, Rodriguez-Canales J, Wei L, *et al.* Methylation profiling of mediastinal gray zone lymphoma reveals a distinctive signature with elements shared by classical Hodgkin's lymphoma and primary mediastinal large B-cell lymphoma. Haematologica 2011; 96(4): 558-66.
[http://dx.doi.org/10.3324/haematol.2010.033167] [PMID: 21454882]

[158] Steidl C, Gascoyne RD. The molecular pathogenesis of primary mediastinal large B-cell lymphoma. Blood 2011; 118(10): 2659-69.
[http://dx.doi.org/10.1182/blood-2011-05-326538] [PMID: 21700770]

[159] Iqbal J, Wright G, Wang C, *et al.* Gene expression signatures delineate biological and prognostic subgroups in peripheral T-cell lymphoma. Blood 2014; 123(19): 2915-23.
[http://dx.doi.org/10.1182/blood-2013-11-536359] [PMID: 24632715]

[160] Iqbal J, Weisenburger DD, Greiner TC, *et al.* Molecular signatures to improve diagnosis in peripheral T-cell lymphoma and prognostication in angioimmunoblastic T-cell lymphoma. Blood 2010; 115(5): 1026-36.
[http://dx.doi.org/10.1182/blood-2009-06-227579] [PMID: 19965671]

[161] Amin HM, Lai R. Pathobiology of ALK+ anaplastic large-cell lymphoma. Blood 2007; 110(7): 2259-67.
[http://dx.doi.org/10.1182/blood-2007-04-060715] [PMID: 17519389]

[162] Parrilla Castellar ER, Jaffe ES, Said JW, *et al.* ALK-negative anaplastic large cell lymphoma is a genetically heterogeneous disease with widely disparate clinical outcomes. Blood 2014; 124(9): 1473-80.
[http://dx.doi.org/10.1182/blood-2014-04-571091] [PMID: 24894770]

[163] Vose J, Armitage J, Weisenburger D. International peripheral T-cell and natural killer/T-cell lymphoma study: pathology findings and clinical outcomes. J Clin Oncol 2008; 26(25): 4124-30. [http://dx.doi.org/10.1200/JCO.2008.16.4558] [PMID: 18626005]

[164] Morice WG. The immunophenotypic attributes of NK cells and NK-cell lineage lymphoproliferative disorders. Am J Clin Pathol 2007; 127(6): 881-6. [http://dx.doi.org/10.1309/Q49CRJ030L22MHLF] [PMID: 17509985]

[165] Shi H, Guo J, Duff DJ, *et al.* Discovery of novel epigenetic markers in non-Hodgkin's lymphoma. Carcinogenesis 2007; 28(1): 60-70. [http://dx.doi.org/10.1093/carcin/bgl092] [PMID: 16774933]

[166] Esteller M. Profiling aberrant DNA methylation in hematologic neoplasms: a view from the tip of the iceberg. Clin Immunol 2003; 109(1): 80-8. [http://dx.doi.org/10.1016/S1521-6616(03)00208-0] [PMID: 14585279]

[167] Jiang Y, Hatzi K, Shaknovich R. Mechanisms of epigenetic deregulation in lymphoid neoplasms. Blood 2013; 121(21): 4271-9. [http://dx.doi.org/10.1182/blood-2012-12-451799] [PMID: 23704048]

[168] Esteller M, Gaidano G, Goodman SN, *et al.* Hypermethylation of the DNA repair gene O(6)-methylguanine DNA methyltransferase and survival of patients with diffuse large B-cell lymphoma. J Natl Cancer Inst 2002; 94(1): 26-32. [http://dx.doi.org/10.1093/jnci/94.1.26] [PMID: 11773279]

[169] Guan H, Xie L, Leithäuser F, *et al.* KLF4 is a tumor suppressor in B-cell non-Hodgkin lymphoma and in classic Hodgkin lymphoma. Blood 2010; 116(9): 1469-78. [http://dx.doi.org/10.1182/blood-2009-12-256446] [PMID: 20519630]

[170] Morin RD, Mendez-Lago M, Mungall AJ, *et al.* Frequent mutation of histone-modifying genes in non-Hodgkin lymphoma. Nature 2011; 476(7360): 298-303. [http://dx.doi.org/10.1038/nature10351] [PMID: 21796119]

[171] Enjuanes A, Fernàndez V, Hernández L, *et al.* Identification of methylated genes associated with aggressive clinicopathological features in mantle cell lymphoma. PLoS One 2011; 6(5): e19736. [http://dx.doi.org/10.1371/journal.pone.0019736] [PMID: 21603610]

[172] Clozel T, Yang S, Elstrom RL, *et al.* Mechanism-based epigenetic chemosensitization therapy of diffuse large B-cell lymphoma. Cancer Discov 2013; 3(9): 1002-19. [http://dx.doi.org/10.1158/2159-8290.CD-13-0117] [PMID: 23955273]

[173] Palomero T, Couronné L, Khiabanian H, *et al.* Recurrent mutations in epigenetic regulators, RHOA and FYN kinase in peripheral T cell lymphomas. Nat Genet 2014; 46(2): 166-70. [http://dx.doi.org/10.1038/ng.2873] [PMID: 24413734]

[174] Wang C, McKeithan TW, Gong Q, *et al.* IDH2R172 mutations define a unique subgroup of patients with angioimmunoblastic T-cell lymphoma. Blood 2015; 126(15): 1741-52. [http://dx.doi.org/10.1182/blood-2015-05-644591] [PMID: 26268241]

[175] Marques SC, Laursen MB, Bødker JS, *et al.* MicroRNAs in B-cells: from normal differentiation to treatment of malignancies. Oncotarget 2015; 6(1): 7-25. [http://dx.doi.org/10.18632/oncotarget.3057] [PMID: 25622103]

[176] Iqbal J, Shen Y, Huang X, *et al.* Global microRNA expression profiling uncovers molecular markers for classification and prognosis in aggressive B-cell lymphoma. Blood 2015; 125(7): 1137-45. [http://dx.doi.org/10.1182/blood-2014-04-566778] [PMID: 25498913]

[177] Jardin F, Figeac M. MicroRNAs in lymphoma, from diagnosis to targeted therapy. Curr Opin Oncol 2013; 25(5): 480-6. [http://dx.doi.org/10.1097/CCO.0b013e328363def2] [PMID: 23852382]

[178] Rao E, Jiang C, Ji M, *et al.* The miRNA-17 to 92 cluster mediates chemoresistance and enhances

tumor growth in mantle cell lymphoma *via* PI3K/AKT pathway activation. Leukemia 2012; 26(5): 1064-72.
[http://dx.doi.org/10.1038/leu.2011.305] [PMID: 22116552]

[179] Merkel O, Hamacher F, Laimer D, *et al.* Identification of differential and functionally active miRNAs in both anaplastic lymphoma kinase (ALK)+ and ALK- anaplastic large-cell lymphoma. Proc Natl Acad Sci USA 2010; 107(37): 16228-33.
[http://dx.doi.org/10.1073/pnas.1009719107] [PMID: 20805506]

[180] Olive V, Bennett MJ, Walker JC, *et al.* miR-19 is a key oncogenic component of mir-17-92. Genes Dev 2009; 23(24): 2839-49.
[http://dx.doi.org/10.1101/gad.1861409] [PMID: 20008935]

[181] Döhner H, Stilgenbauer S, Benner A, *et al.* Genomic aberrations and survival in chronic lymphocytic leukemia. N Engl J Med 2000; 343(26): 1910-6.
[http://dx.doi.org/10.1056/NEJM200012283432602] [PMID: 11136261]

[182] Goswami RS, Atenafu EG, Xuan Y, *et al.* MicroRNA signature obtained from the comparison of aggressive with indolent non-Hodgkin lymphomas: potential prognostic value in mantle-cell lymphoma. J Clin Oncol 2013; 31(23): 2903-11.
[http://dx.doi.org/10.1200/JCO.2012.45.3050] [PMID: 23835716]

[183] Mazan-Mamczarz K, Gartenhaus RB. Role of microRNA deregulation in the pathogenesis of diffuse large B-cell lymphoma (DLBCL). Leuk Res 2013; 37(11): 1420-8.
[http://dx.doi.org/10.1016/j.leukres.2013.08.020] [PMID: 24054860]

[184] Augello C, Gianelli U, Savi F, *et al.* MicroRNA as potential biomarker in HCV-associated diffuse large B-cell lymphoma. J Clin Pathol 2014; 67(8): 697-701.
[http://dx.doi.org/10.1136/jclinpath-2014-202352] [PMID: 24914240]

[185] Asmar F, Hother C, Kulosman G, *et al.* Diffuse large B-cell lymphoma with combined TP53 mutation and MIR34A methylation: Another "double hit" lymphoma with very poor outcome? Oncotarget 2014; 5(7): 1912-25.
[http://dx.doi.org/10.18632/oncotarget.1877] [PMID: 24722400]

[186] Cheng G. Circulating miRNAs: roles in cancer diagnosis, prognosis and therapy. Adv Drug Deliv Rev 2015; 81: 75-93.
[http://dx.doi.org/10.1016/j.addr.2014.09.001] [PMID: 25220354]

[187] Liu C, Iqbal J, Teruya-Feldstein J, *et al.* MicroRNA expression profiling identifies molecular signatures associated with anaplastic large cell lymphoma. Blood 2013; 122(12): 2083-92.
[http://dx.doi.org/10.1182/blood-2012-08-447375] [PMID: 23801630]

[188] Sotomayor EM, Young KH, Younes A. Clinical roundtable monograph: CD30 in lymphoma: its role in biology, diagnostic testing, and targeted therapy. Clin Adv Hematol Oncol 2014; 12(4) (Suppl. 10): 1-22.
[PMID: 24870054]

[189] Capello D, Vitolo U, Pasqualucci L, *et al.* Distribution and pattern of BCL-6 mutations throughout the spectrum of B-cell neoplasia. Blood 2000; 95(2): 651-9.
[PMID: 10627476]

[190] Seegmiller AC, Garcia R, Huang R, Maleki A, Karandikar NJ, Chen W. Simple karyotype and bcl-6 expression predict a diagnosis of Burkitt lymphoma and better survival in IG-MYC rearranged high-grade B-cell lymphomas. Mod Pathol 2010; 23(7): 909-20.
[http://dx.doi.org/10.1038/modpathol.2010.76] [PMID: 20348878]

[191] Hu S, Xu-Monette ZY, Tzankov A, *et al.* MYC/BCL2 protein coexpression contributes to the inferior survival of activated B-cell subtype of diffuse large B-cell lymphoma and demonstrates high-risk gene expression signatures: a report from The International DLBCL Rituximab-CHOP Consortium Program. Blood 2013; 121(20): 4021-31.
[http://dx.doi.org/10.1182/blood-2012-10-460063] [PMID: 23449635]

[192] Aukema SM, Siebert R, Schuuring E, *et al.* Double-hit B-cell lymphomas. Blood 2011; 117(8): 2319-31.
[http://dx.doi.org/10.1182/blood-2010-09-297879] [PMID: 21119107]

[193] Vela-Chávez T, Adam P, Kremer M, *et al.* Cyclin D1 positive diffuse large B-cell lymphoma is a post-germinal center-type lymphoma without alterations in the CCND1 gene locus. Leuk Lymphoma 2011; 52(3): 458-66.
[http://dx.doi.org/10.3109/10428194.2010.540361] [PMID: 21281227]

[194] Mozos A, Royo C, Hartmann E, *et al.* SOX11 expression is highly specific for mantle cell lymphoma and identifies the cyclin D1-negative subtype. Haematologica 2009; 94(11): 1555-62.
[http://dx.doi.org/10.3324/haematol.2009.010264] [PMID: 19880778]

[195] Xu-Monette ZY, Wu L, Visco C, *et al.* Mutational profile and prognostic significance of TP53 in diffuse large B-cell lymphoma patients treated with R-CHOP: report from an International DLBCL Rituximab-CHOP Consortium Program Study. Blood 2012; 120(19): 3986-96.
[http://dx.doi.org/10.1182/blood-2012-05-433334] [PMID: 22955915]

[196] Xu-Monette ZY, Møller MB, Tzankov A, *et al.* MDM2 phenotypic and genotypic profiling, respective to TP53 genetic status, in diffuse large B-cell lymphoma patients treated with rituximab-CHOP immunochemotherapy: a report from the International DLBCL Rituximab-CHOP Consortium Program. Blood 2013; 122(15): 2630-40.
[http://dx.doi.org/10.1182/blood-2012-12-473702] [PMID: 23982177]

[197] Muris JJF, Cillessen SAGM, Vos W, *et al.* Immunohistochemical profiling of caspase signaling pathways predicts clinical response to chemotherapy in primary nodal diffuse large B-cell lymphomas. Blood 2005; 105(7): 2916-23.
[http://dx.doi.org/10.1182/blood-2004-07-2716] [PMID: 15576477]

[198] Jost PJ, Ruland J. Aberrant NF-kappaB signaling in lymphoma: mechanisms, consequences, and therapeutic implications. Blood 2007; 109(7): 2700-7.
[http://dx.doi.org/10.1182/blood-2006-07-025809] [PMID: 17119127]

[199] Bavi P, Uddin S, Bu R, *et al.* The biological and clinical impact of inhibition of NF-κB-initiated apoptosis in diffuse large B cell lymphoma (DLBCL). J Pathol 2011; 224(3): 355-66.
[http://dx.doi.org/10.1002/path.2864] [PMID: 21506127]

[200] Wiestner A. Targeting B-Cell receptor signaling for anticancer therapy: the Bruton's tyrosine kinase inhibitor ibrutinib induces impressive responses in B-cell malignancies. J Clin Oncol 2013; 31(1): 128-30.
[http://dx.doi.org/10.1200/JCO.2012.44.4281] [PMID: 23045586]

[201] Nicolae A, Pittaluga S, Venkataraman G, *et al.* Peripheral T-cell lymphomas of follicular T-helper cell derivation with Hodgkin/Reed-Sternberg cells of B-cell lineage: both EBV-positive and EBV-negative variants exist. Am J Surg Pathol 2013; 37(6): 816-26.
[http://dx.doi.org/10.1097/PAS.0b013e3182785610] [PMID: 23598959]

[202] Roncador G, García Verdes-Montenegro JF, Tedoldi S, *et al.* Expression of two markers of germinal center T cells (SAP and PD-1) in angioimmunoblastic T-cell lymphoma. Haematologica 2007; 92(8): 1059-66.
[http://dx.doi.org/10.3324/haematol.10864] [PMID: 17640856]

[203] Bisig B, Gaulard P, de Leval L. New biomarkers in T-cell lymphomas. Best Pract Res Clin Haematol 2012; 25(1): 13-28.
[http://dx.doi.org/10.1016/j.beha.2012.01.004] [PMID: 22409820]

[204] Alavaikko MJ, Blanco G, Aine R, *et al.* Follicular dendritic cells have prognostic relevance in Hodgkin's disease. Am J Clin Pathol 1994; 101(6): 761-7.
[http://dx.doi.org/10.1093/ajcp/101.6.761] [PMID: 8209866]

[205] Molin D, Edström A, Glimelius I, *et al.* Mast cell infiltration correlates with poor prognosis in

Hodgkin's lymphoma. Br J Haematol 2002; 119(1): 122-4.
[http://dx.doi.org/10.1046/j.1365-2141.2002.03768.x] [PMID: 12358914]

[206] Hedström G, Berglund M, Molin D, *et al.* Mast cell infiltration is a favourable prognostic factor in diffuse large B-cell lymphoma. Br J Haematol 2007; 138(1): 68-71.
[http://dx.doi.org/10.1111/j.1365-2141.2007.06612.x] [PMID: 17555448]

[207] Tan KL, Scott DW, Hong F, *et al.* Tumor-associated macrophages predict inferior outcomes in classic Hodgkin lymphoma: a correlative study from the E2496 Intergroup trial. Blood 2012; 120(16): 3280-7.
[http://dx.doi.org/10.1182/blood-2012-04-421057] [PMID: 22948049]

[208] Riihijärvi S, Fiskvik I, Taskinen M, *et al.* Prognostic influence of macrophages in patients with diffuse large B-cell lymphoma: a correlative study from a Nordic phase II trial. Haematologica 2015; 100(2): 238-45.
[http://dx.doi.org/10.3324/haematol.2014.113472] [PMID: 25381134]

[209] Barros MH, Hassan R, Niedobitek G. Tumor-associated macrophages in pediatric classical Hodgkin lymphoma: association with Epstein-Barr virus, lymphocyte subsets, and prognostic impact. Clin Cancer Res 2012; 18(14): 3762-71.
[http://dx.doi.org/10.1158/1078-0432.CCR-12-0129] [PMID: 22645050]

[210] Carreras J, Lopez-Guillermo A, Fox BC, *et al.* High numbers of tumor-infiltrating FOXP3-positive regulatory T cells are associated with improved overall survival in follicular lymphoma. Blood 2006; 108(9): 2957-64.
[http://dx.doi.org/10.1182/blood-2006-04-018218] [PMID: 16825494]

[211] Alvaro T, Lejeune M, Salvadó MT, *et al.* Outcome in Hodgkin's lymphoma can be predicted from the presence of accompanying cytotoxic and regulatory T cells. Clin Cancer Res 2005; 11(4): 1467-73.
[http://dx.doi.org/10.1158/1078-0432.CCR-04-1869] [PMID: 15746048]

[212] Kim WY, Jeon YK, Kim TM, *et al.* Increased quantity of tumor-infiltrating FOXP3-positive regulatory T cells is an independent predictor for improved clinical outcome in extranodal NK/T-cell lymphoma. Ann Oncol 2009; 20(10): 1688-96.
[http://dx.doi.org/10.1093/annonc/mdp056] [PMID: 19542249]

[213] Muenst S, Hoeller S, Dirnhofer S, Tzankov A. Increased programmed death-1+ tumor-infiltrating lymphocytes in classical Hodgkin lymphoma substantiate reduced overall survival. Hum Pathol 2009; 40(12): 1715-22.
[http://dx.doi.org/10.1016/j.humpath.2009.03.025] [PMID: 19695683]

[214] Nam-Cha SH, Roncador G, Sanchez-Verde L, *et al.* PD-1, a follicular T-cell marker useful for recognizing nodular lymphocyte-predominant Hodgkin lymphoma. Am J Surg Pathol 2008; 32(8): 1252-7.
[http://dx.doi.org/10.1097/PAS.0b013e318165b0d6] [PMID: 18594468]

[215] Ansell SM, Lesokhin AM, Borrello I, *et al.* PD-1 blockade with nivolumab in relapsed or refractory Hodgkin's lymphoma. N Engl J Med 2015; 372(4): 311-9.
[http://dx.doi.org/10.1056/NEJMoa1411087] [PMID: 25482239]

[216] Wu X, Gu Z, Chen Y, *et al.* Application of PD-1 Blockade in Cancer Immunotherapy. Comput Struct Biotechnol J 2019; 17: 661-74.
[http://dx.doi.org/10.1016/j.csbj.2019.03.006] [PMID: 31205619]

[217] Xu-Monette ZY, Zhou J, Young KH. PD-1 expression and clinical PD-1 blockade in B-cell lymphomas. Blood 2018; 131(1): 68-83.
[http://dx.doi.org/10.1182/blood-2017-07-740993] [PMID: 29118007]

[218] Jelinek T, Mihalyova J, Kascak M, Duras J, Hajek R. PD-1/PD-L1 inhibitors in haematological malignancies: update 2017. Immunology 2017; 152(3): 357-71.
[http://dx.doi.org/10.1111/imm.12788] [PMID: 28685821]

[219] Pasqualucci L, Trifonov V, Fabbri G, *et al.* Analysis of the coding genome of diffuse large B-cell lymphoma. Nat Genet 2011; 43(9): 830-7.
[http://dx.doi.org/10.1038/ng.892] [PMID: 21804550]

[220] Cheung KJ, Johnson NA, Affleck JG, *et al.* Acquired TNFRSF14 mutations in follicular lymphoma are associated with worse prognosis. Cancer Res 2010; 70(22): 9166-74.
[http://dx.doi.org/10.1158/0008-5472.CAN-10-2460] [PMID: 20884631]

[221] Jarrett RF, Stark GL, White J, *et al.* Impact of tumor Epstein-Barr virus status on presenting features and outcome in age-defined subgroups of patients with classic Hodgkin lymphoma: a population-based study. Blood 2005; 106(7): 2444-51.
[http://dx.doi.org/10.1182/blood-2004-09-3759] [PMID: 15941916]

[222] Nicolae A, Pittaluga S, Abdullah S, *et al.* EBV-positive large B-cell lymphomas in young patients: a nodal lymphoma with evidence for a tolerogenic immune environment. Blood 2015; 126(7): 863-72.
[http://dx.doi.org/10.1182/blood-2015-02-630632] [PMID: 25999451]

[223] Deng L, Song Y, Young KH, *et al.* Hepatitis B virus-associated diffuse large B-cell lymphoma: unique clinical features, poor outcome, and hepatitis B surface antigen-driven origin. Oncotarget 2015; 6(28): 25061-73.
[http://dx.doi.org/10.18632/oncotarget.4677] [PMID: 26314957]

[224] Peveling-Oberhag J, Arcaini L, Hansmann ML, Zeuzem S. Hepatitis C-associated B-cell non-Hodgkin lymphomas. Epidemiology, molecular signature and clinical management. J Hepatol 2013; 59(1): 169-77.
[http://dx.doi.org/10.1016/j.jhep.2013.03.018] [PMID: 23542089]

[225] Christenson ES, Teply B, Agrawal V, Illei P, Gurakar A, Kanakry JA. Human herpesvirus 8-related primary effusion lymphoma after liver transplantation. Am J Transplant 2015; 15(10): 2762-6.
[http://dx.doi.org/10.1111/ajt.13321] [PMID: 25988353]

[226] Bustoros M, Mouhieddine TH, Detappe A, Ghobrial IM. Established and Novel Prognostic Biomarkers in Multiple Myeloma. Am Soc Clin Oncol Educ Book 2017; 37(37): 548-60.
[http://dx.doi.org/10.1200/EDBK_175175] [PMID: 28561668]

[227] Durie BG, Salmon SE. A clinical staging system for multiple myeloma. Correlation of measured myeloma cell mass with presenting clinical features, response to treatment, and survival. Cancer 1975; 36(3): 842-54.
[http://dx.doi.org/10.1002/1097-0142(197509)36:3<842::AID-CNCR2820360303>3.0.CO;2-U] [PMID: 1182674]

[228] Greipp PR, San Miguel J, Durie BG, *et al.* International staging system for multiple myeloma. J Clin Oncol 2005; 23(15): 3412-20.
[http://dx.doi.org/10.1200/JCO.2005.04.242] [PMID: 15809451]

[229] Palumbo A, Avet-Loiseau H, Oliva S, *et al.* Revised International Staging System for Multiple Myeloma: A Report From International Myeloma Working Group. J Clin Oncol 2015; 33(26): 2863-9.
[http://dx.doi.org/10.1200/JCO.2015.61.2267] [PMID: 26240224]

[230] Kastritis E, Terpos E, Roussou M, *et al.* Evaluation of the Revised International Staging System in an independent cohort of unselected patients with multiple myeloma. Haematologica 2017; 102(3): 593-9.
[http://dx.doi.org/10.3324/haematol.2016.145078] [PMID: 27789676]

[231] Rajkumar SV, Landgren O, Mateos MV. Smoldering multiple myeloma. Blood 2015; 125(20): 3069-75.
[http://dx.doi.org/10.1182/blood-2014-09-568899] [PMID: 25838344]

[232] Gupta N, Sharma A, Sharma A. Emerging biomarkers in Multiple Myeloma: A review. Clin Chim Acta 2020; 503: 45-53.
[http://dx.doi.org/10.1016/j.cca.2019.12.026] [PMID: 31901479]

[233] Lê GN, Bones J, Coyne M, *et al.* Current and future biomarkers for risk-stratification and treatment personalisation in multiple myeloma. Mol Omics 2019; 15(1): 7-20. [http://dx.doi.org/10.1039/C8MO00193F] [PMID: 30652172]

[234] Rajkumar SV. Myeloma today: Disease definitions and treatment advances. Am J Hematol 2016; 91(1): 90-100. [http://dx.doi.org/10.1002/ajh.24236] [PMID: 26565896]

[235] Özkurt ZN, Sucak GT, Akı ŞZ, Yağcı M, Haznedar R. Early Prognostic Value of Monitoring Serum Free Light Chain in Patients with Multiple Myeloma Undergoing Autologous Stem Cell Transplantation. Cancer Invest 2017; 35(3): 195-201. [http://dx.doi.org/10.1080/07357907.2016.1270958] [PMID: 28112977]

[236] Brioli A, Giles H, Pawlyn C, *et al.* Serum free immunoglobulin light chain evaluation as a marker of impact from intraclonal heterogeneity on myeloma outcome. Blood 2014; 123(22): 3414-9. [http://dx.doi.org/10.1182/blood-2013-12-542662] [PMID: 24733348]

[237] Fonseca R, Barlogie B, Bataille R, *et al.* Genetics and cytogenetics of multiple myeloma: a workshop report. Cancer Res 2004; 64(4): 1546-58. [http://dx.doi.org/10.1158/0008-5472.CAN-03-2876] [PMID: 14989251]

[238] Sawyer JR, Tian E, Shaughnessy JD Jr, *et al.* Hyperhaploidy is a novel high-risk cytogenetic subgroup in multiple myeloma. Leukemia 2017; 31(3): 637-44. [http://dx.doi.org/10.1038/leu.2016.253] [PMID: 27694925]

[239] Avet-Loiseau H, Leleu X, Roussel M, *et al.* Bortezomib plus dexamethasone induction improves outcome of patients with t(4;14) myeloma but not outcome of patients with del(17p). J Clin Oncol 2010; 28(30): 4630-4. [http://dx.doi.org/10.1200/JCO.2010.28.3945] [PMID: 20644101]

[240] Fonseca R, Blood E, Rue M, *et al.* Clinical and biologic implications of recurrent genomic aberrations in myeloma. Blood 2003; 101(11): 4569-75. [http://dx.doi.org/10.1182/blood-2002-10-3017] [PMID: 12576322]

[241] Boyd KD, Ross FM, Chiecchio L, *et al.* A novel prognostic model in myeloma based on co-segregating adverse FISH lesions and the ISS: analysis of patients treated in the MRC Myeloma IX trial. Leukemia 2012; 26(2): 349-55. [http://dx.doi.org/10.1038/leu.2011.204] [PMID: 21836613]

[242] Qiang YW, Ye S, Chen Y, *et al.* MAF protein mediates innate resistance to proteasome inhibition therapy in multiple myeloma. Blood 2016; 128(25): 2919-30. [http://dx.doi.org/10.1182/blood-2016-03-706077] [PMID: 27793878]

[243] Walker BA, Leone PE, Chiecchio L, *et al.* A compendium of myeloma-associated chromosomal copy number abnormalities and their prognostic value. Blood 2010; 116(15): e56-65. [http://dx.doi.org/10.1182/blood-2010-04-279596] [PMID: 20616218]

[244] Weinhold N, Kirn D, Seckinger A, *et al.* Concomitant gain of 1q21 and MYC translocation define a poor prognostic subgroup of hyperdiploid multiple myeloma. Haematologica 2016; 101(3): e116-9. [http://dx.doi.org/10.3324/haematol.2015.136929] [PMID: 26611471]

[245] Lovén J, Hoke HA, Lin CY, *et al.* Selective inhibition of tumor oncogenes by disruption of super-enhancers. Cell 2013; 153(2): 320-34. [http://dx.doi.org/10.1016/j.cell.2013.03.036] [PMID: 23582323]

[246] Shaughnessy JD Jr, Zhan F, Burington BE, *et al.* A validated gene expression model of high-risk multiple myeloma is defined by deregulated expression of genes mapping to chromosome 1. Blood 2007; 109(6): 2276-84. [http://dx.doi.org/10.1182/blood-2006-07-038430] [PMID: 17105813]

[247] Decaux O, Lodé L, Magrangeas F, *et al.* Prediction of survival in multiple myeloma based on gene expression profiles reveals cell cycle and chromosomal instability signatures in high-risk patients and

hyperdiploid signatures in low-risk patients: a study of the Intergroupe Francophone du Myélome. J Clin Oncol 2008; 26(29): 4798-805.
[http://dx.doi.org/10.1200/JCO.2007.13.8545] [PMID: 18591550]

[248] Hermansen NE, Borup R, Andersen MK, *et al.* Gene expression risk signatures maintain prognostic power in multiple myeloma despite microarray probe set translation. Int J Lab Hematol 2016; 38(3): 298-307.
[http://dx.doi.org/10.1111/ijlh.12486] [PMID: 27027250]

[249] Chng WJ, Chung TH, Kumar S, *et al.* Gene signature combinations improve prognostic stratification of multiple myeloma patients. Leukemia 2016; 30(5): 1071-8.
[http://dx.doi.org/10.1038/leu.2015.341] [PMID: 26669975]

[250] Levin A, Hari P, Dhakal B. Novel biomarkers in multiple myeloma. Transl Res 2018; 201: 49-59.
[http://dx.doi.org/10.1016/j.trsl.2018.05.003] [PMID: 30301522]

[251] Kumar S, Paiva B, Anderson KC, *et al.* International Myeloma Working Group consensus criteria for response and minimal residual disease assessment in multiple myeloma. Lancet Oncol 2016; 17(8): e328-46.
[http://dx.doi.org/10.1016/S1470-2045(16)30206-6] [PMID: 27511158]

[252] Bettegowda C, Sausen M, Leary RJ, *et al.* Detection of circulating tumor DNA in early- and late-stage human malignancies. Sci Transl Med 2014; 6(224): 224ra24.
[http://dx.doi.org/10.1126/scitranslmed.3007094] [PMID: 24553385]

[253] Gonsalves WI, Morice WG, Rajkumar V, *et al.* Quantification of clonal circulating plasma cells in relapsed multiple myeloma. Br J Haematol 2014; 167(4): 500-5.
[http://dx.doi.org/10.1111/bjh.13067] [PMID: 25113422]

[254] Paiva B, Paino T, Sayagues JM, *et al.* Detailed characterization of multiple myeloma circulating tumor cells shows unique phenotypic, cytogenetic, functional, and circadian distribution profile. Blood 2013; 122(22): 3591-8.
[http://dx.doi.org/10.1182/blood-2013-06-510453] [PMID: 24072855]

[255] Gonsalves WI, Rajkumar SV, Gupta V, *et al.* Quantification of clonal circulating plasma cells in newly diagnosed multiple myeloma: implications for redefining high-risk myeloma. Leukemia 2014; 28(10): 2060-5.
[http://dx.doi.org/10.1038/leu.2014.98] [PMID: 24618735]

[256] Bianchi G, Kyle RA, Larson DR, *et al.* High levels of peripheral blood circulating plasma cells as a specific risk factor for progression of smoldering multiple myeloma. Leukemia 2013; 27(3): 680-5.
[http://dx.doi.org/10.1038/leu.2012.237] [PMID: 22902364]

[257] Mishima Y, Paiva B, Shi J, *et al.* The mutational landscape of circulating tumor cells in multiple myeloma. Cell Rep 2017; 19(1): 218-24.
[http://dx.doi.org/10.1016/j.celrep.2017.03.025] [PMID: 28380360]

[258] Batth IS, Mitra A, Manier S, *et al.* Circulating tumor markers: harmonizing the yin and yang of CTCs and ctDNA for precision medicine. Ann Oncol 2017; 28(3): 468-77.
[http://dx.doi.org/10.1093/annonc/mdw619] [PMID: 27998963]

[259] Manier S, Park S, Freeman S, *et al.* Whole-exome sequencing and targeted deep sequencing of cfDNA enables a comprehensive mutational profiling of multiple myeloma. Blood 2016; 128(22): 197.
[http://dx.doi.org/10.1182/blood.V128.22.197.197]

[260] Mithraprabhu S, Khong T, Ramachandran M, *et al.* Circulating tumour DNA analysis demonstrates spatial mutational heterogeneity that coincides with disease relapse in myeloma. Leukemia 2017; 31(8): 1695-705.
[http://dx.doi.org/10.1038/leu.2016.366] [PMID: 27899805]

[261] Logan AC, Zhang B, Narasimhan B, *et al.* Minimal residual disease quantification using consensus primers and high-throughput IGH sequencing predicts post-transplant relapse in chronic lymphocytic

leukemia. Leukemia 2013; 27(8): 1659-65.
[http://dx.doi.org/10.1038/leu.2013.52] [PMID: 23419792]

[262] Newman AM, Bratman SV, To J, *et al.* An ultrasensitive method for quantitating circulating tumor DNA with broad patient coverage. Nat Med 2014; 20(5): 548-54.
[http://dx.doi.org/10.1038/nm.3519] [PMID: 24705333]

[263] Alix-Panabières C, Pantel K. Clinical applications of circulating tumor cells and circulating tumor DNA as liquid biopsy. Cancer Discov 2016; 6(5): 479-91.
[http://dx.doi.org/10.1158/2159-8290.CD-15-1483] [PMID: 26969689]

[264] Paiva B, Azpilikueta A, Puig N, *et al.* PD-L1/PD-1 presence in the tumor microenvironment and activity of PD-1 blockade in multiple myeloma. Leukemia 2015; 29(10): 2110-3.
[http://dx.doi.org/10.1038/leu.2015.79] [PMID: 25778100]

[265] Ray A, Das DS, Song Y, *et al.* Targeting PD1-PDL1 immune checkpoint in plasmacytoid dendritic cell interactions with T cells, natural killer cells and multiple myeloma cells. Leukemia 2015; 29(6): 1441-4.
[http://dx.doi.org/10.1038/leu.2015.11] [PMID: 25634684]

[266] Minn AJ, Wherry EJ. Combination Cancer Therapies with Immune Checkpoint Blockade: Convergence on Interferon Signaling. Cell 2016; 165(2): 272-5.
[http://dx.doi.org/10.1016/j.cell.2016.03.031] [PMID: 27058661]

[267] Badros A, Hyjek E, Ma N, *et al.* Pembrolizumab, pomalidomide, and low-dose dexamethasone for relapsed/refractory multiple myeloma. Blood 2017; 130(10): 1189-97.
[http://dx.doi.org/10.1182/blood-2017-03-775122] [PMID: 28461396]

[268] Ziogas DC, Dimopoulos MA, Kastritis E. Prognostic factors for multiple myeloma in the era of novel therapies. Expert Rev Hematol 2018; 11(11): 863-79.
[http://dx.doi.org/10.1080/17474086.2018.1537776] [PMID: 30334460]

CHAPTER 9

Applications of Biomarkers in Cancer Surgery

Berrin Papila Kundaktepe[1,*]

[1] *Department of General Surgery, Cerrahpasa Medical Faculty, Istanbul University-Cerrahpasa, İstanbul, Turkey*

Abstract: The most problematic and difficult cases of general surgery applications are cancers. Cancer can be diagnosed in many ways, including the presence of some symptoms and signs, screening tests, and imaging methods. The diagnosis of cancer is confirmed anatomically by microscopic examination of tissue samples. Biomarker research has become increasingly interesting for many medical fields today. Biomarkers are indicators of the biological process that can show changes in disease and health status and reveal pathological conditions. The sensitivity of markers used for early and painless diagnosis is weak in all types of cancer, including thyroid, breast, stomach, pancreatic, and colorectal cancers. Biochemical and molecular markers have been found to be of great importance in cancer research in recent years in the formation mechanisms and treatment steps of cancer. Several biomarkers have been studied in the circulating body fluids and tissue of preoperative, intraoperative, and postoperative patients. The main use of tumor markers is the follow-up of the response to treatment and early detection of recurrences in patients diagnosed with cancer. Tumor markers correctly selected will guide the patient's monitoring and treatment planning. But, still, the field of cancer surgery needs new biomarkers.

Keywords: Biomarkers, Breast cancers, Colorectal cancers, Pancreatic cancers, Stomach cancers, Surgery, Thyroid cancers.

INTRODUCTION

Biomarkers are molecules that can evaluate biological and pathological processes and are present in measurable amounts in blood, body fluids, or tissues. Biomarkers used in surgery may play a role in early diagnosis, risk stratification, evaluation, and metastasis and prognosis prediction [1]. An ideal biomarker should be determined easily in body fluids as plasma and urine samples, predict relevant clinical outcomes and reflect prognosis and treatment efficacy. These evaluations are essential (essential/indispensable/ crucial) to establish an appropriate treatment and recruit the patient. The properties sought in an ideal biomarker are as follows: it is biochemically stable, it can be studied with low

* **Corresponding author Berrin Papila Kundaktepe:** Department of General Surgery, Cerrahpasa Medical Faculty, Istanbul University-Cerrahpasa, İstanbul, Turkey; Tel: +90 (532) 493 34 74; E-mail: papilaberrin@yahoo.com

Pınar Atukeren and Hafize Uzun (Eds.)

blood amounts or body fluids, its sampling time is wide, it can be detected with a fast, cheap, simple, and automated system, its sensitivity is 100%, and its specificity is> 85%, it can distinguish bacterial and viral pathogens, and it is associated with inflammatory response [2].

Clinical use of circulating tumor markers include neoplasia screening, risk identification, cancer diagnosis, clinical classification of cancer, treatment selection, monitoring response to treatment, determination of the prognosis, recurrence detection, and metastasis tracking. Tumor markers are rather used after the initial treatment is completed to define the disease state and to evaluate the treatment options. Detection of cancer recurrence allows re-initiation of treatment or changes in treatment [3].

Although high levels of tumor markers are seen in malignancies, many hormonal and metabolic changes can be encountered in benign diseases. Sometimes a tumor may not produce the substance it is associated with until the advanced stage. A substance may not be specific to a single tumor type. In other words, these items are not sensitive and specific enough. However, they are useful in monitoring a malignancy with a tissue diagnosis, treatment selection, and survival prediction [4 - 6]. Tumor markers are often used in combination with immunohistochemical tests to diagnose cancer.

The most known and used tumor markers are shown in Tables **1**, **2** and **3**.

Table 1. The most known and used enzyme tumor markers in routine clinical use.

ENZYME TUMOR MARKERS			
-	**Malignancy**	**Non-neoplastic Causes**	**Clinical Utility**
Prostate specific antigen (PSA)	Prostate cancer	Prostatitis, prostate hypertrophy	Screening, therapy monitoring, and recurrence
Lactate dehydrogenase (LDH)	Hematologic malignancies, lymphoma, Ewing sarcoma, Germ cell	Hepatitis, hemolytic anemia	Prognostic indicator and elevated nonspecifically in numerous cancers.
Alkaline phosphatase (ALP)	Metastatic carcinoma of bone, hepatocellular carcinoma, osteosarcoma, lymphoma, leukemia.	Growth-development, pregnancy, bone diseases, liver diseases	Determination of liver and bone involvement; nonspecific elevation in many bone-related and liver cancers.
Neuron-specific enolase (NSE)	Neuroendocrine tumors, Small cell lung cancer, neuroblastoma	Septic shock, pneumonia, and nervous system trauma	Prognostic indicator and monitoring disease progression for neuroendocrine tumors.

Table 2. The most known and used endocrine tumor markers in routine clinical use.

h-CG	**Gestational trophoblastic diseases, germ cell tumors**	**Pregnancy, renal failure, heterophile antibody**	**Diagnosis of choriocarcinoma, testicular cancer**
Adrenocorticotropic hormone (ACTH)	Pituitary adenoma, ectopic ACTH-producing tumor	Cushing syndrome	Diagnosis of ectopic ACTH-producing tumor
Calcitonin	MTC and neuroendocrine tumors	Chronic renal failure, Zollinger-Ellison syndrome, pernicious anemia	Screening, response to therapy, and monitoring recurrence of MTC
Growth hormone (GH)	Pituitary adenoma, ectopic GH-secreting tumors	Acromegaly, dwarfism, gigantism	Diagnosis and post monitoring of acromegaly
Prolactin	Pituitary adenoma	Irregular or no periods, infertility, breast milk discharge in non-pregnant, tenderness in the breast	Diagnosis and postsurgical monitoring of prolactinoma
Antidiuretic hormone (ADH)	Posterior pituitary tumors	Guillain-Barré syndrome, multiple sclerosis, epilepsy, and acute intermittent porphyria	Diagnosis of SIADH
Cortisol	Adrenal tumors	Cushing syndrome, excess cortisol, adrenal insufficiency or Addison disease, deficient cortisol.	Diagnosis of Cushing's syndrome, adrenal adenoma

Table 3. The most known and used tumor markers as oncofetal proteins in routine clinical. use.

α- fetoprotein (AFP)	**Hepatocellular carcinoma, Hepatoblastoma, Non-semitomatous germ cell tumours**	**Hepatitis, cirrhosis, biliary obstruction,**	**Hepatocellular carcinoma (screening, diagnosis, treatment monitoring), hepatoblastoma (diagnosis), non-semitomatous germ cell tumours (prognosis, treatment monitoring)**
Carcinoembryonic antigen (CEA)	Colorectal carcinoma (main), Other gastrointestinal malignancies, breast and ovarian carcinoma.	Cigarette smoking, bronchitis, hepatitis, alcoholic liver disease, ulcerative colitis, Crohn's disease and pancreatitis.	Prognosis, treatment monitoring and recurrence of colorectal carcinoma.

(Table 3) cont.....

α- fetoprotein (AFP)	**Hepatocellular carcinoma, Hepatoblastoma, Non-semitomatous germ cell tumours**	**Hepatitis, cirrhosis, biliary obstruction,**	**Hepatocellular carcinoma (screening, diagnosis, treatment monitoring), hepatoblastoma (diagnosis), non-semitomatous germ cell tumours (prognosis, treatment monitoring)**
CA 125	Ovarian cancer	Endometriosis, peritonitis, ascites, pelvic inflammatory disease	Monitoring therapy
CA 15.3	Metastatic breast cancer	Vitamin B12 deficiency, kidney failure	Response to therapy and detecting recurrence
CA 19.9	Pancreatic cancer (main), Gastric, colorectal and cholangiocarcinoma	Pancreatitis, cholangitis, hepatitis, cirrhosis.	Monitoring pancreatic cancer
CA 27-29	Metastatic breast carcinoma	Chronic hepatitis, liver cirrhosis, tuberculosis, sarcoidosis, benign breast disease	Response to therapy and detecting recurrence

CIRCULATING TUMOR MARKERS: CLINICAL APPLICATIONS

Applications Of Tumor Markers In Colorectal Cancer

Carcinoembryonic antigen (CEA) is an oncofetal glycoprotein and is released from normal mucosa cells. It is released in excessive amounts, especially in colon adenocarcinomas. It may increase in other malignancies. In addition, high values can be found in smoking, peptic ulcer, inflammatory bowel disease, pancreatitis, hypothyroidism, biliary obstruction and common diseases. However, over 10 ng/mL should suggest malignancy. The probability of being detected increases with the spread of the tumor. While CEA is high in 25% of the cases with local tumors, half of those with lymph node metastases; Levels are high in 75% of cases with distant metastases. However, it is less expressed in poorly differentiated tumors. It returns to normal levels 4-6 weeks after a successful surgical procedure [5]. CEA is not recommended for colorectal cancer screening [7 - 9]. Preoperative CEA values correlate positively with disease stage and negatively correlated with disease-free survival, thus having prognostic value in colorectal cancer patients [10]. CEA is the most commonest used for postoperative follow-up in the preoperative period. Serial estimation of CEA can be used to monitor disease recurrence [11, 12].

Applications of Tumor Markers in Hepatocellular Carcinoma (HCC)

Alpha feto protein (AFP) is the main protein in fetal serum and decreases to immeasurable levels after birth. Malignancies with high AFP are HCC and germ cell tumors. Although increased levels are also seen in other gastrointestinal system cancers, it does not exceed 1000 ng/mL. In cases of cirrhosis and hepatitis, it may increase to not exceed 500 ng/mL. AFP is abnormal in 80% of patients with HCC and at levels exceeding 1000 units in 40%. In nonalcoholic liver cirrhosis cases, the presence of a liver mass and AFP > 500 ng/mL are considered equivalent to biopsy for the diagnosis of HCC [5]. In patients with hepatocellular carcinoma, rates of increase in AFP levels are proportional to tumor burden. However, there is no similar relationship between life span and AFP levels. AFP is useful in monitoring the response and clinical relapse of patients with HCC. After complete surgical resection, it immediately decreases to normal. AFP that does not decrease to normal levels is considered to indicate that the tumor cannot be removed completely, and its gradual increases are considered to indicate relapse. Although not as much as surgery in evaluating the effectiveness of chemotherapy, a 50% reduction compared to the initial values is considered a success. AFP is a very useful tumor markers in primary liver cancers, beta HCG and AFP support the diagnosis of germ cell tumors, evaluate response to treatment and follow-up [13].

Applications of Tumor Markers in Pancreatic Adenocarcinoma

CA 19-9 is an intracellular adhesion molecule that is increased in pancreatic and biliary tract cancers. It may also increase non-specific in other malignancies. The specificity and sensitivity for pancreatic cancer are 90% and 80%, respectively; for biliary tract, it shows 60-70% sensitivity. It may also increase in benign conditions such as cirrhosis, cholestasis, cholangitis and pancreatitis. However, under these conditions, the value does not exceed 1000 units/mL. Lewis O group patients do not carry this antigen. Serum CA19-9 levels at diagnosis is an indicator for tumour resectability, the effectiveness of chemotherapy and patient survival. Since its positive predictive value is less than 1%, it cannot be used for screening. However, at values above 1000 units, its positive predictivity increases up to 97% and is usually associated with metastases. Response of CA 19-9 to treatment can also be useful for stratification [5, 12 - 15].

Applications of Tumor Markers In Breast Cancer

Estrogen and progesterone receptors, CA15-3, BR27.29 (CA27.29), CEA and HER-2/neu (c-erb B2) oncoprotein, are used as tumor markers in breast cancer

[16]. CA 15-3 is a high molecular weight mucin glycoprotein (MUC-1). CA 15-3 and associated BR 27.29 are widely used in the follow-up of patients with breast cancer. Mucin-type breast cancer markers are CA15.3, BR27.29, CA549, MCA, CA-M26 and CA-M29. However, in terms of specificity and sensitivity, only CA 5.3 and CA27.29 have been approved by the FDA for the clinical follow-up of patients with metastatic breast cancer. In a study in which 352 patients were followed, it was shown that CA 15-3 levels increased in 67% of patients with cancer recurrence. Again, the levels were found to be normal in 92% of 1320 people who were found to have no recurrence of the disease. Although only small groups of cases have been shown to be asymptomatic and have a better prognosis due to elevations in CA15-3 levels, these data did not provide sufficient evidence for treating asymptomatic individuals. While CEA was used in the follow-up of metastatic breast cancers, it was later shown that CA15.3 was superior in many studies. CA15.3 levels are increased in 60-80% of patients with metastatic breast cancer, and these levels are parallel to changes in clinical status and tumor treatment response. It also seems to be useful in evaluating the clinical course of patients in the treatment of metastatic disease. Many studies have shown that this tumor marker is useful as an indicator of progression and regression in the follow-up of treatment response in breast cancer. Serial marker measurements are particularly sensitive in the early detection of liver and bone metastases [3, 17 - 19].

BR27.29 (CA27.29) is a monoclonal antibody that develops against a glycoprotein (MUCI) on the surface of normal epithelial cells. Although it is detected especially in breast cancers, higher values may be found in other malignancies. It may increase benign conditions such as breast, liver, and kidney cysts. However, values above 100 U/mL should suggest malignancy. This marker, which replaced CA15-3 due to its better sensitivity and specificity, increased in one-third of early breast cancer patients and two-thirds of advanced cases. Therefore, it cannot be used for scanning purposes. Its specificity and sensitivity is high in terms of detecting preclinical metastases [5].

Estrogen and progesterone receptors are members of a nuclear hormone receptor family that includes receptors that bind androgen, retinoid, and thyroid hormones. These receptors are known as ligand-dependent transcription factors [20]. In breast cancer, hormone receptors such as tumor size and lymph node involvement are also effective markers on prognosis. Especially with anti-estrogens such as tamoxifen, toremifene and drolifixene and progesterone such as megestrol acetate to medroxyprogesterone acetate It is used with similar drugs as an indicator of response to endocrine therapy. In general, estrogen receptor positive tumors have a better prognosis than estrogen receptor negative tumors. However, the level of estrogen and progesterone receptors should only be measured just before menopause and in primary and metastatic breast cancers in menopausal patients. If viable tissue could not be obtained during breast biopsy or surgery, or if the amount of tissue biopsy prevents receptor count, it is possible to determine these receptors immunohistochemically. Although extended endocrine therapy should be considered for high-risk estrogen receptor -positive patients, whether other subgroups could benefit in view of toxicity risks, such as endometrial cancer and embolism, is unclear [16, 21 - 23]. However, both estrogen and progesterone receptors status were significantly prognostic, as their expression correlated with disease-free survival [24].

HER-2 (human epidermal growth factor) (c-erbB-2) is an overactive protooncogene in 15-30% of invasive breast cancers. HER-2 t is useful in determining prognosis and predicting response to treatment. The levels of HER-2 are also determined in the tissue. However, the extracellular domain (ECD) can be mixed into the blood from the cells spilled from the tumor tissue, and the blood level of HER-2 ECD may be found to be high in some patients with metastatic breast cancer. Those with high blood levels have a worse clinical picture and less response to treatment, and may be useful in detecting disease recurrence in the preclinical period. It is currently not recommended to routinely measure circulating HER-2/neu (c-erb B2) levels [25 - 27].

Various markers such as HER-2/neu (c-erb B2), p53, Ki67, ER and PR were found to be guiding in predicting the prognosis of breast cancer and determining the treatment modality [28 - 33].

HER-2/neu (c-erb B2) oncoprotein, the receptor and inhibitor of cathepsin-D and urokinase-type plasminogen activator, are new tumor markers currently under investigation in breast cancer, and their clinical use is not recommended outside of research [18].

Circulating Tumor Cells

The gold standard for cancer diagnosis is tumor tissue biopsy sampling and pathological examination. Difficulties in obtaining sufficient tumor cells for analysis, the need for invasive serial sampling to monitor response, the potential for selection bias due to intratumoral and inter-tumoral heterogeneity [34 - 36].

These difficulties create the need for more effective biopsy methods.

"Liquid biopsy" analysis using circulating tumor cells (CTCs) or plasma-derived circulating tumor DNA (ctDNA) offers a non-invasive alternative to tissue biopsies. CTCs were defined by ThomasAshworth in 1869. CTCs are found in very low amounts in the blood (1/106-107 core blood cells) [37]. CTC is mainly studied on common epithelial malignant neoplasms such as colorectal, lung, breast, pancreatic and prostate cancer, and melanoma. There are also studies on gynecological malignancies, head and neck cancers and even Hodgkin lymphoma. The results obtained are promising for the clinical use of this technology [38, 39]. Although CTC has not yet entered practice as a routine clinical practice, areas that have the potential to find a place in the future can be examined under two main headings:

1- Determination of disease severity and prognosis, response to treatment follow-up and personalized treatment modalities.
2- Targeted therapy applications.

ENDOCRINE TUMOR MARKERS

The best known endocrine tumors are thyroid, adrenal, pituitary, pancreatic and gonadal carcinoma. Endocrine surgery generally includes thyroid, parathyroid, adrenal and neuroendocrine disease or pancreatic or GI tract disease.

Surgical treatment has an important place in thyroid gland diseases, especially in cancers. Thyroid cancer is one of the most common endocrine malignancies, accounting for 0.5-1.5% of all cancers [40]. Differentiated thyroid cancers (DTC) originating from follicular epithelial cells constitute more than 95% of all thyroid cancers. Papillary thyroid carcinomas (PTC) are the most common type, accounting for more than 85% of all thyroid cancers. Although it has a good prognosis, it requires lifelong follow-up as the long-term mortality rate can be as high as 10% [41, 42]. Pre-operative diagnostic tests such as U/S, CT scan, and fine needle aspirate (FNA) biopsy often do not provide a definitive diagnosis. In these indeterminate cases, diagnostic hemithyroidectomy surgery is often performed and up to 75% of these patients have the benign disease [43].

In some patients, papillary cancer may transform into poorly differentiated type or anaplastic cancer. FNA biopsy is the gold standard in diagnosis, and its sensitivity is 98% and its positive predictive value is 99% in experienced hands [44, 45]. In differential thyroid cancers (DTCs), there is no optimum cytological test that can definitively show malignancy or be ruled out. However, Galectin-3 (Gal-3), which is examined immunohistochemically according to the results of many studies, is positive in differentiated thyroid cancers, especially in papillary thyroid cancers (PTC) [46]. Although its role in cancer development is not known exactly, it is reported that the increased expression of Gal-3 increases the adhesion and motility of tumor cells and increases the metastasis potential [47]. Studies are being conducted on its place in diagnostic evaluation to rule out malignancy, especially in suspicious thyroid nodules [48]. However, although the sensitivity of Gal-3 in thyroid malignancies is high, its specificity is low [49]. It is known that Gal-3, which is examined immunohistochemically, can help in the diagnosis of classical PTC. However, there are rare studies to show the diagnosis and prognosis of DTC in serum Gal-3 levels. Inohara *et al.* [50] found that serum Gal-3 was higher in PTCs compared to patients with other thyroid nodules, and there was no difference between benign thyroid nodules and healthy individuals. Xue *et al.* [51] reported that a positive Gal-3 serum could be used in the diagnosis of thyroid malignancies. However, some researchers found that there was no significant difference in serum Gal-3 levels between patients with benign thyroid nodules and patients with PTC [52, 53].

According to the results of Išić *et al.* [53], there was no statistically significant difference in serum YKL-40, galactin-3 (Gal-3), CK19, TIMP-1 and Ang-1 levels between PTC patients and benign subjects. Although many of these gene products may be upregulated in cancer tissue, overexpression of these proteins in serum has not been demonstrated. With the increasing incidence of thyroid cancer, the need for reliable testing for an accurate diagnosis is increasing. Development of such a test can significantly reduce patient morbidity and unnecessary surgery in benign thyroid disease. Detailed studies are needed to verify these results.

miRNA profiles have been identified as miRNA-221, −222 and -181b in thyroid cancer [54, 55]. This miRNA requires various studies to be used in diagnosis, prognosis and monitoring response to treatment in thyroid carcinoma. Graham *et al.* [56], when comparing miRNAs in patients with benign thyroid tumors and controls, serum miR-150-5p and miR-342-3p were determined statistically significantly down-regulated and serum miR-let7b-5p, and miR-191-5p was up-regulated. They also demonstrated deregulation of these serum miRNA by down-regulation of miR-146a-5p and up-regulation of miR-93-5p in patients with PTC. There was statistically significant down-regulation of miR-146a-5p and miR-199b-3p and up-regulation of let7b-5p and miR-10a-5p when comparing PTC serum to that of patients with benign tumors using limma and t-tests. Igci *et al.* [57] reported that serum miR-30a-5p might be considered as a new prognostic marker in the diagnosis of PTC. Therefore, revealing the role of miR-30a-5p will provide the therapeutic potential of this miRNA in PTC, and studies using miRNA mimics and anti-miRs of miR-30a-5p will be valuable in the future.

Thyroglobulin (Tg) is among the most important glycoproteins produced by thyrocytes in the thyroid. Tg gene expression is regulated by thyroid-stimulating hormone (also known as thyrotropin, thyrotropic hormone, TSH). Serum Tg concentrations are increased in many thyroid diseases such as thyroiditis, nodular goiter, Graves' disease [60]. Tg measurement has two important indications [58, 59]:

1. Detection of recurrent or residual thyroid epithelial cancers (papillary, follicular, Hürthle cell carcinomas) after thyroidectomy.
2. Differentiation of exogenous thyroid hormone intake from true hyperthyroidism (there is Tg suppression in exogenous thyroid hormone intake, hyperthyroidism is normal or high).

European and American guidelines recommend using Tg levels to follow-up patients with differentiated thyroid carcinomas. Preoperative serum Tg value is a nonspecific diagnostic tool for thyroid cancers, according to the European and American Thyroid Associations. The serum Tg level may be higher in patients with malignant thyroid lesions, as Tg is released during the conversion to malignancy in the follicular cells of thyroid gland [60]. Preoperative serum Tg level is an independent risk factor for malignancy [61 - 65]. Although the preoperative serum Tg level has not yet been proven to be a specific diagnostic marker for distinguishing malignant nodules from benign nodules, some clinicians use it as a diagnostic tool [66].

Calcitonin, a hypocalcemic and hypophosphatemic hormone, is synthesized from parafollicular cells. Medullary thyroid carcinoma (MTC) originates from calcitonin-producing parafollicular cells and constitutes approximately 5-7% of thyroid cancers [67, 68]. While high calcitonin levels are considered a sensitive marker for MTC, they are not specific. Because, in non-thyroid pathologies (such as hypercalcemia, hyperparathyroidism, renal failure, neuroendocrine tumors) and the use of some drugs (such as omeprazole, glucocorticoids, beta-blockers, glucagon), calcitonin levels may increase [69]. Although serum calcitonin measurement is not generally recommended in patients with all thyroid nodules, it is also offered as a recommendation by some groups [70]. Thyroid malignancies present as thyroid nodules. Although the evaluation of suspicious thyroid nodules shows similarity (ultrasonography, cytology) in both patients with differentiated thyroid cancer and MTC patients, the diagnosis of MTC requires some specific and different methods (calcitonin, carcinoembrogenic antigen, the REarranged during Transfection (RET) germline mutation) [71, 72].

In a recent study, serum calcitonin levels were evaluated in 10,864 patients with nodular thyroid disease, and MTC was diagnosed in 0.40% of patients [73]. In these and other studies, it has been shown that serum calcitonin measurement has higher diagnostic sensitivity and specificity when compared to FNA findings [73 - 75]. In the largest study in which patients diagnosed with calcitonin screening were evaluated, MTC patients were diagnosed at an earlier stage in the calcitonin screening group compared to the control group without screening, and at the end of the follow-up period, complete remission was detected as 59% in the screening group and 2.7% in the control group ($p = 0.0001$) [73]. These findings are related to the low cure rate when the disease spreads outside the thyroid gland and support the role of serum calcitonin screening in the early diagnosis of MTC in patients with thyroid nodules. In another study evaluating patients with thyroid nodules, basal serum calcitonin level was found below 10 pg/mL in 5535 (95.15%) of 5817 patients [75]. When thyroidectomy was performed in 9 out of 66 patients with basal serum calcitonin level > 20 pg/mL when calcitonin was > 100 pg/mL, the diagnosis of MTH was histologically confirmed in each. When thyroidectomy was performed in 9 of 66 patients with a basal serum calcitonin level > 20 pg/mL when calcitonin was > 100 pg/mL, the diagnosis of MTC was histologically confirmed in each [75]. In the light of this information, it may be useful to measure serum calcitonin in patients with thyroid nodules in the decision of surgical indication or safe follow-up.

Preoperative and postoperative parathyroid hormone and blood calcium values can be compared in order to be sure of the removal of adenoma in surgeries performed for adenomas secreting parathormone, or to predict permanent hypocalcemia complication that may develop as a result of unintentional paratidectomies during thyroidectomies. A decrease of more than 10% calcium in the blood or more than 60% in parathyroid hormone (PTH) level in the postoperative 5-6. hours compared to the initial value was found to be significant for persistent hypocalcemia [76]. If hormone monitoring will be performed after surgery, it should not be forgotten to take blood samples to determine the basic values before starting the surgery.

Neuroendocrine cells are found in many organs. Tumors that develop from these cells are different from each other in terms of clinical course and prognosis. Neuroendocrine tumors (NETs) are had been a little-known and heterogeneous group of tumors known for many years. Carcinoid, non-carcinoid gastroenteropancreatic tumors, paraganglioma, pheochromocytoma, thyroid medullary carcinoma, small cell lung carcinoma, Merkel cell carcinoma are examples of this group. The most important part of the diffuse neuroendocrine system is the gastrointestinal neuroendocrine system [77].

Endocrine cells instead neuroendocrine called the reason is that there are proteins such as synaptophysin, neuron-specific enolase (NSE) and chromogranin (Cg), which are properties belonging to cells of neural origin (particularly embryogenic as ectodermal). The gastroenteropancreatic neuroendocrine system consists of cells with specific neuronal and endocrine features in mucosa and pancreatic islets in the gastrointestinal system. Gastroenteropancreatic neuroendocrine tumor is the tumor arising from possible immature stem cells of this system. NETs of the gastrointestinal tract and pancreas can together be called as GEP-NETs. The functional distinction of endocrine carcinomas is made in a simple way. If the tumor secretes a peptide hormone that produces symptoms, a functional endocrine tumor is called a non-functional endocrine tumor if the patient is asymptomatic [78].

Pancreatic neuroendocrine tumors (PNETs) are rare neuroendocrine neoplasms. A multidisciplinary approach is best suited for the diagnosis and treatment of PNET. When a neuroendocrine tumor is suspected, a complete biochemical evaluation should be performed and hormones secreted from the pancreas should be included in this examination. In this way, it will be revealed whether the tumor is functional or not. Apart from biochemical analysis, tumor markers have an important role in diagnosis. These markers are determinative in terms of both diagnosis and prognosis, especially in non-functional PNETs [79 - 81]. Chromogranin A (Cg A) and pancreatic polypeptide have been proposed as circulating tumor markers in non-functional-PNETs. However, the percentage of patients with increased pancreatic polypeptide (PP) levels is lower than those with high Cg A levels. Measuring Cg A may be useful for non-functional-PNET patients to reach a diagnosis. If it is high, it is potentially useful for assessing treatment response and detecting progression and recurrence at an early stage. CgA level is the most commonly used. It is detected in 88-100% of the cases [79, 80].

In 5-10% of patients with sporadic gastric carcinoid, characterized by a patchy 'geographic' flush, cutaneous oedema, bronchospasm, salivary gland swelling and lacrimation may develop "atypical carcinoid syndrome" due to histamine release. Patients may have a slight increase in urinary 5-hydroxy-indoleacetic acid (5-HIAA) values and urinary predictions of the histamine metabolite methylimidazole acetic acid (MelmAA) serve as a tumor marker [82, 83].

Immunohistochemical and Circulating Tumor Markers of NETs

Endocrine carcinomas have markers secreted into the blood. There are numerous kinds of NETs markers are used for their diagnosis and follow-up. The most important of these are chromogranins. Biochemical markers can be used to evaluate the functionality of NETs (Table **4**).

Chromogranins are peptides found in vesicles in endocrine cells and are precursors of hormones called vasostatin. Chromogranin level may be more than three times the normal value in the blood in endocrine tumors. It has importance at the stage of diagnosis. The high level of chromogranin also suggests that the potential for metastasis may be high. Cg A can be used to predict response to medical treatment in patients with GEP-NET. In the test with octreotide, serum levels of Cg A are evaluated before and after injection of the drug. If there is a decrease of more than 30% in Cg A levels compared to baseline, these patients will have a high probability of response to treatment [84].

Table 4. Circulating tumor markers of most common neuroendocrine tumors.

General Markers	Tumor Specific Markers
Chromogranins • Kromogranin A, B (Cg A, B)	Carcinoid tumors • 5-HIAA in 24-hour urine • 5-Hydroxy tryptophan in 24-hour urine • Serotonin
Pancreatic polypeptide (PP)	Nonfunctioning pancreatic NET • CgA, • Neuron Specific Enolase (NSE)
Secratogranins	Insulinoma • Fasting insulin, proinsulin, C peptide
Neuron Specific Enolase (NSE)	Glucagonoma • Fasting glucagon
Pancreatic polypeptide (PP)	VIP-oma • Fasting VIP
Chorionic Gonadotropin Hormone (CGH)	Somatostatinoma • Fasting somatostatin Gastrinoma • Fasting and induced gastrin
CgA	Nonfunctioning carcinoid tumor • NSE
Adrenokortikotropik hormone (ACTH)	ACTH oma [Cushing syndrome (%100)]
Calcitonin	Medullary thyroid carcinoma • Carcinoembryonic antigen (CEA) • CgA
Parathormone(PTH)	Parathyroid tumour
Urinary/plasma metanephrine & normetanephrine	Pheochromocytoma, paraganglioma • CgA, • NSE

Neuron specific enolase (NSE) are neuron specific isomers of the glycolytic enzyme 2-phospho-D-glycerate hydrolase or enolase. NSE level is often high in small cell lung cancers and 30 to 50% of NETs (even if serum Cg A level is normal) despite poor differentiation [85]. Its sensitivity increases especially when combined with Cg A [86]. It cannot be used as a prognostic marker since it is not associated with tumor burden [87].

Pancreatic Polypeptide (PP) is a 36 amino acid oligopeptide secreted by PP cells in the intestinal mucosa and posterior lobe of the pancreatic head. It is released especially after meals, exercise, and acute hypoglycemia. Its release is reduced with somatostatin and intravenous glucose [88]. It is used as a neuroendocrine differentiation marker in the diagnosis of non-functional NETs in endocrinological oncology. It generally increases with sensitivity of 50-80% in metastases [89].

Serotonin and its metabolites (5-HT and 5-HIAA) are the most widely used markers in the diagnosis and follow-up of NETs. Serotonin [5-Hydroxytryptamine (5-HT)] is synthesized from tryptophan in enterochromaffin cells. Serotonin (50-80%) is metabolized in the liver and is excreted in the urine as 5-hydroxyindolacetic acid (5-HIAA) [90, 91]. Carcinoid syndrome characterized by flushing, diarrhea, abdominal pain, right heart failure and skin findings may develop in patients with liver metastasis. The most important test in the diagnosis of this syndrome is the measurement of the serotonin metabolite 5-HIAA in 24-hour urine. For example, measurement of the serotonin metabolite 5-HIAA in 24-hour urine is useful in confirming the diagnosis and monitoring of serotonin secreting tumors [92, 93]. Indeed, carcinoid syndrome has the features related to excess secretion of serotonin, kallikrein, histamine, tachykinins (neurokinin A and substance P), and others [90]. Though the prominent product seems to be serotonin, some patients with carcinoid syndrome may have low or normal levels of 5-HIAA [94].

CgA levels are also elevated in most of carcinoid tumors [92, 95, 96]. Whereas CgA has 98.4% and 62.9% to detect the carcinoid tumors which are functioning, 5-HIAA has a sensitivity and specificity of 100% and 73%, respectively [95]. But, CgA is certainly more useful than 5-HIAA for carcinoid tumors which are non-functioning [92].

About 10% of PNETs are functional with symptoms related to the type of hormone secretion; In this group, insulinomas are the most common (30-40%), followed by gastrinomas (16-30%), glucagonomas (<10%), VIPomas (<10%) and somatostatinomas (<5%) [97]. Insulinomas originate from the insulin-secreting beta cells of the pancreas. Insulinomas are the most common functional-PNET and occur at a rate of 0.1-0.3 per million per year. Among the PNETs, functional-PNETs are the most benign. It has been reported that less than 10% are malignant in diagnosis and more than 95% are suitable for surgical treatment [98, 99]. Sometimes, beta cell hyperplasia can clinically mimic insulinoma [100].

It manifests itself with the symptoms of hypoglycemia due to excessive insulin release. Findings called the Whipple triad are diagnostic. These findings are symptoms of hypoglycemia, low blood glucose level, and improvement of symptoms with glucose administration. The diagnosis of insulinoma is made using 6 criteria: (1) blood glucose levels <2.2 mmol/L (40 mg/dL) (2) concurrent insulin levels > 36 pmol/L; (3) C-peptide levels > 200 pmol/L (4) proinsulin levels > 5 pmol/L, (5) β-hydroxybutyrate levels <2.7 mmol/L (6) absence of sulfonylurea (metabolites) in plasma or urine [99].

Glucagonomas as pancreatic tumor and accompanying skin disease have been defined. Its classic clinical picture consists of diabetes, dermatitis, depression and deep vein thrombosis. Its annual incidence is 0.18 to 0.12 per million. It represents less than 10% of PNETs. 20% of glucagonomas are associated with multiple endocrine neoplasia syndrome type 1 (MEN1) syndrome. Diagnosis requires serum glucagon levels higher than 500 pg/mL. However, this is not specific to glucagonoma. Because most glucagonomas are large masses, their localization is easy to determine by CT/MRI, octreotide scanning, and endoscopic ultrasound [100 - 104].

Vipoma are PNETs that secrete vasoactive intestinal peptides (VIP). VIP stimulates fluid and electrolyte secretion into the intestinal lumen. Excessive secretion of VIP leads to watery diarrhea and loss of water, NaCl and potassium from the body. They cause a clinical syndrome known as Verner-Morrison syndrome (watery diarrhea, hypokalemia, and achlorhydria) or pancreatic cholera syndrome. Diagnosis of vipoma is made by large amounts of watery diarrhea (> 700 mL/day) and plasma VIP level higher than (500 pg/mL). Most are easily localized by contrast-enhanced CT or somatostatin receptor scintigraphy [89].

Somatostatinoma (SSoma) are rare PNETs with an incidence of 1 in 40 million. Somatostatin is produced from pancreatic delta cells. In the case of hypersecretion, diabetes leads to malabsorption steatorrhea and gallstones as the contractility of the gallbladder decreases. Approximately half occurs in the pancreas, the other half in the duodenum and jejunum. 60 - 70% of them are malignant. These tumors are rarely associated with von Hippel-Lindau disease (VHL) or MEN1. The diagnosis of SSoma is rarely made before surgery or biopsy. The diagnosis is most strongly supported by a pancreatic or duodenal mass and an elevated serum somatostatin level. The diagnosis is confirmed when the somatostatin level is detected above 14 mol/L [101, 105].

Growth hormone releasing factor neuroendocrine tumours (GRFoma) is a rare pancraetic endocrine tumor, the incidence of which is unknown. 30-50% of the cases are malignant. GRFoma occurs with acromegaly that is clinically indistinguishable from those seen in pituitary adenomas. GRFoma should be suspected in acromegaly patients with abdominal mass, without pituitary adenoma. The diagnosis is confirmed by plasma growth hormone (GH) levels and elevated serum growth hormone releasing hormone (GHRH) [101, 106].

ACTHoma causes Cushing's syndrome by releasing adrenocorticotropic hormone (ACTH). ACTHoma is very rare. 95% of the cases are malignant. ACTHoma can occur simultaneously in patients with Zollinger-Ellisonsyndrome; gastrinoma (ZES). It has been reported that 8% of the patients with ZES previously accompanied ACTHoma. Cushings disease in MEN1/ZES patients is usually treated with pituitary surgery, whereas the ACTHoma in sporadic cases is treated medically initially and then frequently with adrenalectomy, as covered in the later section on treatment of ACTHomas [107 - 109].

PTHrpomas are rare functional-PNETs that secrete peptides associated with parathyroid hormone (PTH). PTHrpomas are found in the pancreas. More than 85% is malignant. PTHrpoma should be considered in the differential diagnosis of patients with hypercalcemia with pancreatic lesions. As the main secretory products of NETs are peptides or amines, almost all cases of NET-related hypercalcaemia are associated with hypophosphataemia suggesting a parathyroid hormone (PTH)-like effect, implicating as mediators either PTH or PTH-related protein (PTHrP) [110 - 112].

Rectal carcinoids have a rising incidence, comprising w11% of all carcinoids and 1.5% of rectal tumours. The carcinoid syndrome does not occur with rectal carcinoids. Octreoscan is rarely positive in rectal carcinoids due to lack of somatostatin receptors, and chromogranin A is often not secreted and is not valuable as a tumour marker. Carcinoembryonic antigen (CEA) and prostate-specific antigen (PSA) levels are raised by 25% and 80%, respectively [83, 113, 114].

The prognosis of poorly differentiated neuroendocrine tumors of unknown primary may be different. In the histopathological diagnosis of these tumors, the positivity of neuroendocrine markers such as neuron-specific enolase, synaptophysin, Cg A and epithelial markers such as pancytokeratin should be determined immunohistochemically. CT/MR can be used to detect the primary [115].

CONCLUSION

In breast, colon and pancreatic cancers, there are tumor markers that are useful in demonstrating response to treatment and monitoring the disease. As a result, today there is no ideal tumor marker. New biomarkers are needed in the field of surgery, especially in cancer surgery. Further studies are needed to assess the role of serum tumor marker testing in the detection of cancer and to optimize surveillance strategies. Biomarkers have not been identified that reliably reflect elevated tumor tissue levels. A tumor marker panel should not be desired in patients with non-specific symptoms. In patients with specific symptoms, a marker for cancer, which is considered in the differential diagnosis, may be requested, but it contributes little to the diagnosis. The main use of tumor markers is the follow-up of the response to treatment and early detection of recurrences in patients diagnosed with cancer. There is unnecessary and overuse of tumor markers in all countries. There are well-defined restrictions on the use of tumor markers. For example, many of the tumor markers increase in benign and malignant diseases and cannot allow these conditions to be distinguished from each other. Similarly, many of these markers are not reliable in diagnosing early disease and may not even detect recurrence or advanced disease. The use of non-targeted tumor markers causes anxiety and unhelpful examinations in healthy individuals and patients. It should be kept in mind that tumor markers can give false positive and negative results.

CONSENT FOR PUBLICATION

Not applicable.

CONFLICT OF INTEREST

The author declares no conflict of interest, financial or otherwise

ACKNOWLEDGEMENT

Declared none.

REFERENCES

[1] Biomarkers Definitions Working Group.. Biomarkers and surrogate endpoints: preferred definitions and conceptual framework. Clin Pharmacol Ther 2001; 69(3): 89-95. [http://dx.doi.org/10.1067/mcp.2001.113989] [PMID: 11240971]

[2] Aronson JK. Biomarkers and surrogate endpoints. Br J Clin Pharmacol 2005; 59(5): 491-4. [http://dx.doi.org/10.1111/j.1365-2125.2005.02435.x] [PMID: 15842546]

[3] Duffy MJ. Evidence for the clinical use of tumour markers. Ann Clin Biochem 2004; 41(Pt 5): 370-7. [http://dx.doi.org/10.1258/0004563041731529] [PMID: 15333188]

[4] McGinley PJ, Kilpatrick ES. Tumour markers: their use and misuse by clinicians. Ann Clin Biochem 2003; 40(Pt 6): 643-7. [http://dx.doi.org/10.1258/000456303770367234] [PMID: 14629802]

[5] Perkins GL, Slater ED, Sanders GK, Prichard JG. Serum tumor markers. Am Fam Physician 2003; 68(6): 1075-82. [PMID: 14524394]

[6] Vittoraki A, Alexiou G, Karakatsanis A, *et al.* Tumour marker prescriptions for cancer screening in the Hellenic primary care. Eur J Cancer Care (Engl) 2007; 16(1): 86-9. [http://dx.doi.org/10.1111/j.1365-2354.2006.00710.x] [PMID: 17227357]

[7] McGinley PJ, Kilpatrick ES. Tumour markers: their use and misuse by clinicians. Ann Clin Biochem 2003; 40(Pt 6): 643-7.
[http://dx.doi.org/10.1258/000456303770367234] [PMID: 14629802]

[8] Kilpatrick ES, Lind MJ. Appropriate requesting of serum tumour markers. BMJ 2009; 339(sep22 1): b3111.
[http://dx.doi.org/10.1136/bmj.b3111] [PMID: 19773324]

[9] Duffy MJ. CEA as a marker for colorectal cancer: is it clinically useful? Clin Chem 2001; 47: 624-30.
[http://dx.doi.org/10.1093/clinchem/47.4.624] [PMID: 11274010]

[10] Cunningham D, Atkin W, Lenz HJ, *et al.* Colorectal cancer. Lancet 2010; 375(9719): 1030-47.
[http://dx.doi.org/10.1016/S0140-6736(10)60353-4] [PMID: 20304247]

[11] Bast RC Jr, Ravdin P, Hayes DF, *et al.* 2000 update of recommendations for the use of tumor markers in breast and colorectal cancer: clinical practice guidelines of the American Society of Clinical Oncology. J Clin Oncol 2001; 19(6): 1865-78.
[http://dx.doi.org/10.1200/JCO.2001.19.6.1865] [PMID: 11251019]

[12] Klapdor R, Aronsson AC, Duffy MJ, *et al.* Tumor markers in gastrointestinal cancers: EGTM recommendations. Anticancer Res 1999; 119: 2811-5.

[13] Johnson PJ. The role of serum alpha-fetoprotein estimation in the diagnosis and management of hepatocellular carcinoma. Clin Liver Dis 2001; 5(1): 145-59.
[http://dx.doi.org/10.1016/S1089-3261(05)70158-6] [PMID: 11218912]

[14] Schlieman MG, Ho HS, Bold RJ. Utility of tumor markers in determining resectability of pancreatic cancer. Arch Surg 2003; 138(9): 951-5.
[http://dx.doi.org/10.1001/archsurg.138.9.951] [PMID: 12963650]

[15] Montgomery RC, Hoffman JP, Riley LB, Rogatko A, Ridge JA, Eisenberg BL. Prediction of recurrence and survival by post-resection CA 19-9 values in patients with adenocarcinoma of the pancreas. Ann Surg Oncol 1997; 4(7): 551-6.
[http://dx.doi.org/10.1007/BF02305535] [PMID: 9367020]

[16] Duffy MJ. Biochemical markers in breast cancer: which ones are clinically useful? Clin Biochem 2001; 34(5): 347-52.
[http://dx.doi.org/10.1016/S0009-9120(00)00201-0] [PMID: 11522269]

[17] Cheung KL, Graves CR, Robertson JF. Tumour marker measurements in the diagnosis and monitoring of breast cancer. Cancer Treat Rev 2000; 26(2): 91-102.
[http://dx.doi.org/10.1053/ctrv.1999.0151] [PMID: 10772967]

[18] Duffy MJ, van Dalen A, Haglund C, *et al.* Clinical utility of biochemical markers in colorectal cancer: European Group on Tumour Markers (EGTM) guidelines. Eur J Cancer 2003; 39(6): 718-27.
[http://dx.doi.org/10.1016/S0959-8049(02)00811-0] [PMID: 12651195]

[19] Gion M, Mione R, Leon AE, Dittadi R. Comparison of the diagnostic accuracy of CA27.29 and CA15.3 in primary breast cancer. Clin Chem 1999; 45(5): 630-7.
[PMID: 10222349]

[20] Mangelsdorf DJ, Thummel C, Beato M, *et al.* The nuclear receptor superfamily: the second decade. Cell 1995; 83(6): 835-9.
[http://dx.doi.org/10.1016/0092-8674(95)90199-X] [PMID: 8521507]

[21] Fisher B, Redmond C, Brown A, *et al.* Treatment of primary breast cancer with chemotherapy and tamoxifen. N Engl J Med 1981; 305(1): 1-6.
[http://dx.doi.org/10.1056/NEJM198107023050101] [PMID: 7015139]

[22] Järvinen TA, Pelto-Huikko M, Holli K, Isola J. Estrogen receptor beta is coexpressed with ERalpha and PR and associated with nodal status, grade, and proliferation rate in breast cancer. Am J Pathol 2000; 156(1): 29-35.

[http://dx.doi.org/10.1016/S0002-9440(10)64702-5] [PMID: 10623650]

[23] Matsumoto A, Jinno H, Ando T, *et al.* Biological markers of invasive breast cancer. Jpn J Clin Oncol 2016; 46(2): 99-105.
[PMID: 26486826]

[24] Bartlett JM, Brookes CL, Robson T, *et al.* Estrogen receptor and progesterone receptor as predictive biomarkers of response to endocrine therapy: a prospectively powered pathology study in the tamoxifen and exemestane adjuvant multinational trial. J Clin Oncol 2011; 29(12): 1531-8.
[http://dx.doi.org/10.1200/JCO.2010.30.3677] [PMID: 21422407]

[25] Ross JS, Slodkowska EA, Symmans WF, Pusztai L, Ravdin PM, Hortobagyi GN. The HER-2 receptor and breast cancer: ten years of targeted anti-HER-2 therapy and personalized medicine. Oncologist 2009; 14: 320-68.19.

[26] Owens MA, Horten BC, Da Silva MM. HER2 amplification ratios by fluorescence in situ hybridization and correlation with immunohistochemistry in a cohort of 6556 breast cancer tissues. Clin Breast Cancer 2004; 5(1): 63-9.
[http://dx.doi.org/10.3816/CBC.2004.n.011]

[27] Ridolfi RL, Jamehdor MR, Arber JM. HER-2/neu testing in breast carcinoma: a combined immunohistochemical and fluorescence in situ hybridization approach. Mod Pathol 2000; 13(8): 866-73.
[http://dx.doi.org/10.1038/modpathol.3880154] [PMID: 10955453]

[28] Gutierrez C, Schiff R. HER2: biology, detection, and clinical implications. Arch Pathol Lab Med 2011; 135(1): 55-62.
[http://dx.doi.org/10.5858/2010-0454-RAR.1] [PMID: 21204711]

[29] Tsutsui S, Ohno S, Murakami S, Kataoka A, Kinoshita J, Hachitanda Y. Prognostic significance of the coexpression of p53 protein and c-erbB2 in breast cancer. Am J Surg 2003; 185(2): 165-7.
[http://dx.doi.org/10.1016/S0002-9610(02)01203-5] [PMID: 12559449]

[30] Anderson WF, Rosenberg PS, Prat A, Perou CM, Sherman ME. How many etiological subtypes of breast cancer: two, three, four, or more? J Natl Cancer Inst 2014; 106(8): dju165.
[http://dx.doi.org/10.1093/jnci/dju165] [PMID: 25118203]

[31] Wesolowski R, Ramaswamy B. Gene expression profiling: changing face of breast cancer classification and management. Gene Expr 2011; 15(3): 105-15.
[http://dx.doi.org/10.3727/105221611X13176664479241] [PMID: 22268293]

[32] Putti TC, El-Rehim DMA, Rakha EA, *et al.* Estrogen receptor-negative breast carcinomas: a review of morphology and immunophenotypical analysis. Mod Pathol 2005; 18(1): 26-35.
[http://dx.doi.org/10.1038/modpathol.3800255] [PMID: 15332092]

[33] Kesse-Adu R, Shousha S. Myoepithelial markers are expressed in at least 29% of oestrogen receptor negative invasive breast carcinoma. Mod Pathol 2004; 17(6): 646-52.
[http://dx.doi.org/10.1038/modpathol.3800103] [PMID: 15073596]

[34] Bedard PL, Hansen AR, Ratain MJ, Siu LL. Tumour heterogeneity in the clinic. Nature 2013; 501(7467): 355-64.
[http://dx.doi.org/10.1038/nature12627] [PMID: 24048068]

[35] Burrell RA, McGranahan N, Bartek J, Swanton C. The causes and consequences of genetic heterogeneity in cancer evolution. Nature 2013; 501(7467): 338-45.
[http://dx.doi.org/10.1038/nature12625] [PMID: 24048066]

[36] Meacham CE, Morrison SJ. Tumour heterogeneity and cancer cell plasticity. Nature 2013; 501(7467): 328-37.
[http://dx.doi.org/10.1038/nature12624] [PMID: 24048065]

[37] Yang D, Wang L, Tian X. Application of circulating tumor cells scope technique on circulating tumor cell research. Mol Cell Ther 2014; 2(1): 8.

[http://dx.doi.org/10.1186/2052-8426-2-8] [PMID: 26056577]

[38] Gharbaran R, Park J, Kim C, Goy A, Suh KS. Circulating tumor cells in Hodgkin's lymphoma - a review of the spread of HL tumor cells or their putative precursors by lymphatic and hematogenous means, and their prognostic significance. Crit Rev Oncol Hematol 2014; 89(3): 404-17. [http://dx.doi.org/10.1016/j.critrevonc.2013.09.004] [PMID: 24176672]

[39] Mocellin S, Keilholz U, Rossi CR, Nitti D. Circulating tumor cells: the 'leukemic phase' of solid cancers. Trends Mol Med 2006; 12(3): 130-9. [http://dx.doi.org/10.1016/j.molmed.2006.01.006] [PMID: 16488189]

[40] Schlumberger MJ. Papillary and follicular thyroid carcinoma. N Engl J Med 1998; 338(5): 297-306. [http://dx.doi.org/10.1056/NEJM199801293380506] [PMID: 9445411]

[41] Sherman SI. Thyroid carcinoma. Lancet 2003; 361(9356): 501-11. [http://dx.doi.org/10.1016/S0140-6736(03)12488-9] [PMID: 12583960]

[42] Kuo SF, Lin SF, Chao TC, Hsueh C, Lin KJ, Lin JD. Prognosis of multifocal papillary thyroid carcinoma. Int J Endocrinol 2013; 2013 [http://dx.doi.org/10.1155/2013/809382]

[43] Wiseman SM, Baliski C, Irvine R, *et al.* Hemithyroidectomy: the optimal initial surgical approach for individuals undergoing surgery for a cytological diagnosis of follicular neoplasm. Ann Surg Oncol 2006; 13(3): 425-32.. [http://dx.doi.org/10.1245/ASO.2006.03.089] [PMID: 16485160]

[44] Malloy KM, Cunnane MF. Pathology and cytologic features of thyroid neoplasms. Surg Oncol Clin N Am 2008; 17(1): 57-70, viii. [http://dx.doi.org/10.1016/j.soc.2007.10.012] [PMID: 18177800]

[45] Suliburk J, Delbridge L. Surgical management of well-differentiated thyroid cancer: state of the art. Surg Clin North Am 2009; 89(5): 1171-91. [http://dx.doi.org/10.1016/j.suc.2009.06.013] [PMID: 19836491]

[46] Fernández PL, Merino MJ, Gómez M, *et al.* Galectin-3 and laminin expression in neoplastic and non-neoplastic thyroid tissue. J Pathol 1997; 181(1): 80-6. [http://dx.doi.org/10.1002/(SICI)1096-9896(199701)181:1<80::AID-PATH699>3.0.CO;2-E] [PMID: 9072007]

[47] Liu Z, Li X, Shi L, *et al.* Cytokeratin 19, thyroperoxidase, HBME-1 and galectin-3 in evaluation of aggressive behavior of papillary thyroid carcinoma. Int J Clin Exp Med 2014; 7(8): 2304-8. [PMID: 25232425]

[48] Maruta J, Hashimoto H, Yamashita H, Yamashita H, Noguchi S. Immunostaining of galectin-3 and CD44v6 using fine-needle aspiration for distinguishing follicular carcinoma from adenoma. Diagn Cytopathol 2004; 31(6): 392-6. [http://dx.doi.org/10.1002/dc.20156] [PMID: 15540177]

[49] Haugen BR, Alexander EK, Bible KC, *et al.* 2015 American Thyroid Association Management Guidelines for Adult Patients with Thyroid Nodules and Differentiated Thyroid Cancer: The American Thyroid Association Guidelines Task Force on Thyroid Nodules and Differentiated Thyroid Cancer. Thyroid 2016; 26(1): 1-133. [http://dx.doi.org/10.1089/thy.2015.0020] [PMID: 26462967]

[50] Inohara H, Segawa T, Miyauchi A, *et al.* Cytoplasmic and serum galectin-3 in diagnosis of thyroid malignancies. Biochem Biophys Res Commun 2008; 376(3): 605-10. [http://dx.doi.org/10.1016/j.bbrc.2008.09.041] [PMID: 18809382]

[51] Xue G, Liu J, Huang J, *et al.* [Detection of galectin-3 in both serum and tissue for early diagnosis of thyroid carcinoma]. Nan Fang Yi Ke Da Xue Xue Bao 2013; 33(7): 1027-30. [PMID: 23895846]

[52] Makki FM, Taylor SM, Shahnavaz A, *et al.* Serum biomarkers of papillary thyroid cancer. J

Otolaryngol Head Neck Surg 2013; 42(1): 16.
[http://dx.doi.org/10.1186/1916-0216-42-16] [PMID: 23663694]

[53] Išić T, Savin S, Cvejić D, *et al.* Serum Cyfra 21.1 and galectin-3 protein levels in relation to immunohistochemical cytokeratin 19 and galectin-3 expression in patients with thyroid tumors. J Cancer Res Clin Oncol 2010; 136(12): 1805-12.
[http://dx.doi.org/10.1007/s00432-010-0838-3] [PMID: 20204403]

[54] Li X, Abdel-Mageed AB, Mondal D, Kandil E. MicroRNA expression profiles in differentiated thyroid cancer, a review. Int J Clin Exp Med 2013; 6(1): 74-80.
[PMID: 23236561]

[55] de la Chapelle A, Jazdzewski K. MicroRNAs in thyroid cancer. J Clin Endocrinol Metab 2011; 96(11): 3326-36.
[http://dx.doi.org/10.1210/jc.2011-1004] [PMID: 21865360]

[56] Graham ME, Hart RD, Douglas S, *et al.* Serum microRNA profiling to distinguish papillary thyroid cancer from benign thyroid masses. J Otolaryngol Head Neck Surg 2015; 44(1): 33.
[http://dx.doi.org/10.1186/s40463-015-0083-5] [PMID: 26341226]

[57] Igci YZ, Ozkaya M, Korkmaz H, *et al.* Expression Levels of miR-30a-5p in Papillary Thyroid Carcinoma: A Comparison Between Serum and Fine Needle Aspiration Biopsy Samples. Genet Test Mol Biomarkers 2015; 19(8): 418-23.
[http://dx.doi.org/10.1089/gtmb.2015.0062] [PMID: 26047355]

[58] Cooper SD, Greenspan FS, Ladenson PW. The thyroid gland. In: Gardner DG, Shoback D, Eds. Greenspan's Basic and Clinical Endocrinology. New York: Lange Mc Graw Hill 2007; pp. 209-80.

[59] Spencer CA, Wang CC. Thyroglobulin measurement. Techniques, clinical benefits, and pitfalls. Endocrinol Metab Clin North Am 1995; 24(4): 841-63.
[http://dx.doi.org/10.1016/S0889-8529(18)30023-9] [PMID: 8608781]

[60] Suh I, Vriens MR, Guerrero MA, *et al.* Serum thyroglobulin is a poor diagnostic biomarker of malignancy in follicular and Hurthle-cell neoplasms of the thyroid. Am J Surg 2010; 200(1): 41-6.
[http://dx.doi.org/10.1016/j.amjsurg.2009.08.030] [PMID: 20637335]

[61] Hocevar M, Auersperg M. Role of serum thyroglobulin in the pre-operative evaluation of follicular thyroid tumours. Eur J Surg Oncol 1998; 24(6): 553-7.
[http://dx.doi.org/10.1016/S0748-7983(98)93644-7] [PMID: 9870733]

[62] Sugino K, Ito K, Mimura T, Kameyama K, Iwasaki H, Ito K. Hürthle cell tumor of the thyroid: analysis of 188 cases. World J Surg 2001; 25(9): 1160-3.
[http://dx.doi.org/10.1007/BF03215865] [PMID: 11571953]

[63] Lee EK, Chung KW, Min HS, *et al.* Preoperative serum thyroglobulin as a useful predictive marker to differentiate follicular thyroid cancer from benign nodules in indeterminate nodules. J Korean Med Sci 2012; 27(9): 1014-8.
[http://dx.doi.org/10.3346/jkms.2012.27.9.1014] [PMID: 22969246]

[64] Petric R, Besic H, Besic N. Preoperative serum thyroglobulin concentration as a predictive factor of malignancy in small follicular and Hürthle cell neoplasms of the thyroid gland. World J Surg Oncol 2014; 12(1): 282.
[http://dx.doi.org/10.1186/1477-7819-12-282] [PMID: 25213012]

[65] Kars A, Aktan B, Kilic K, *et al.* Preoperative Serum Thyroglobulin Level as a Useful Predictive Marker to Differentiate Thyroid Cancer. ORL J Otorhinolaryngol Relat Spec 2018; 80(5-6): 290-5.
[http://dx.doi.org/10.1159/000491932] [PMID: 30253396]

[66] Suh I, Vriens MR, Guerrero MA, *et al.* Serum thyroglobulin is a poor diagnostic biomarker of malignancy in follicular and Hurthle-cell neoplasms of the thyroid. Am J Surg 2010; 200(1): 41-6.
[http://dx.doi.org/10.1016/j.amjsurg.2009.08.030] [PMID: 20637335]

[67] Sarlis NJ, Gourgiotis L. Unresolved issues, dilemmas and points of interest in thyroid cancer: a current

perspective. Hormones (Athens) 2004; 3(3): 149-70.
[http://dx.doi.org/10.14310/horm.2002.11123] [PMID: 16982589]

[68] Sand M, Gelos M, Sand D, *et al.* Serum calcitonin negative medullary thyroid carcinoma. World J Surg Oncol 2006; 4(1): 97-101.
[http://dx.doi.org/10.1186/1477-7819-4-97] [PMID: 17184544]

[69] Toledo SP, Lourenço DM Jr, Santos MA, Tavares MR, Toledo RA, Correia-Deur JE. Hypercalcitoninemia is not pathognomonic of medullary thyroid carcinoma. Clinics (São Paulo) 2009; 64(7): 699-706.
[http://dx.doi.org/10.1590/S1807-59322009000700015] [PMID: 19606248]

[70] Tuttle RM, Ball DW, Byrd D, *et al.* Medullary carcinoma. J Natl Compr Canc Netw 2010; 8(5): 512-30.
[http://dx.doi.org/10.6004/jnccn.2010.0040] [PMID: 20495082]

[71] Wells SA Jr, Asa SL, Dralle H, *et al.* Revised American Thyroid Association guidelines for the management of medullary thyroid carcinoma. Thyroid 2015; 25(6): 567-610.
[http://dx.doi.org/10.1089/thy.2014.0335] [PMID: 25810047]

[72] Haugen BR, Alexander EK, Bible KC, *et al.* 2015 American thyroid association management guidelines for adult patients with thyroid nodules and differentiated thyroid cancer. Thyroid 2016; 26(1): 1-133.
[http://dx.doi.org/10.1089/thy.2015.0020] [PMID: 26462967]

[73] Elisei R, Bottici V, Luchetti F, *et al.* Impact of routine measurement of serum calcitonin on the diagnosis and outcome of medullary thyroid cancer: experience in 10,864 patients with nodular thyroid disorders. J Clin Endocrinol Metab 2004; 89(1): 163-8.
[http://dx.doi.org/10.1210/jc.2003-030550] [PMID: 14715844]

[74] Vierhapper H, Raber W, Bieglmayer C, Kaserer K, Weinhäusl A, Niederle B. Routine measurement of plasma calcitonin in nodular thyroid diseases. J Clin Endocrinol Metab 1997; 82(5): 1589-93.
[http://dx.doi.org/10.1210/jcem.82.5.3949] [PMID: 9141554]

[75] Costante G, Meringolo D, Durante C, *et al.* Predictive value of serum calcitonin levels for preoperative diagnosis of medullary thyroid carcinoma in a cohort of 5817 consecutive patients with thyroid nodules. J Clin Endocrinol Metab 2007; 92(2): 450-5.
[http://dx.doi.org/10.1210/jc.2006-1590] [PMID: 17119000]

[76] Jumaily JS, Noordzij JP, Dukas AG, *et al.* Prediction of hypocalcemia after using 1- to 6-hour postoperative parathyroid hormone and calcium levels: an analysis of pooled individual patient data from 3 observational studies. Head Neck 2010; 32(4): 427-34.
[PMID: 19780054]

[77] Rehfeld JF. A centenary of gastrointestinal endocrinology. Horm Metab Res 2004; 36(11-12): 735-41.
[http://dx.doi.org/10.1055/s-2004-826154] [PMID: 15655701]

[78] Modlin IM, Moss SF, Gustafsson BI, Lawrence B, Schimmack S, Kidd M. The archaic distinction between functioning and nonfunctioning neuroendocrine neoplasms is no longer clinically relevant. Langenbecks Arch Surg 2011; 396(8): 1145-56.
[http://dx.doi.org/10.1007/s00423-011-0794-7] [PMID: 21523447]

[79] Massironi S, Sciola V, Peracchi M, Ciafardini C, Spampatti MP, Conte D. Neuroendocrine tumors of the gastro-entero-pancreatic system. World J Gastroenterol 2008; 14(35): 5377-84.
[http://dx.doi.org/10.3748/wjg.14.5377] [PMID: 18803349]

[80] Falconi M, Bartsch DK, Eriksson B, *et al.* ENETS Consensus Guidelines for the management of patients with digestive neuroendocrine neoplasms of the digestive system: well-differentiated pancreatic non-functioning tumors. Neuroendocrinology 2012; 95(2): 120-34.
[http://dx.doi.org/10.1159/000335587] [PMID: 22261872]

[81] Falconi M, Eriksson B, Kaltsas G, *et al.* ENETS Consensus Guidelines Update for the Management of

Patients with Functional Pancreatic Neuroendocrine Tumors and Non-Functional Pancreatic Neuroendocrine Tumors. Neuroendocrinology 2016; 103(2): 153-71. [http://dx.doi.org/10.1159/000443171] [PMID: 26742109]

[82] Akerström G, Hellman P, Hessman O. Gastrointestinal carcinoids. In: Lennard TWJ, Ed. Endocrine Surgery. Elsevier 2006; pp. 163-98.

[83] Akerström G, Hellman P. Surgery on neuroendocrine tumours. Best Pract Res Clin Endocrinol Metab 2007; 21(1): 87-109. [http://dx.doi.org/10.1016/j.beem.2006.12.004] [PMID: 17382267]

[84] Massironi S, Conte D, Sciola V, *et al.* Plasma chromogranin A response to octreotide test: prognostic value for clinical outcome in endocrine digestive tumors. Am J Gastroenterol 2010; 105(9): 2072-8. [http://dx.doi.org/10.1038/ajg.2010.154] [PMID: 20372113]

[85] Nobels FR, Kwekkeboom DJ, Coopmans W, *et al.* Chromogranin A as serum marker for neuroendocrine neoplasia: comparison with neuron-specific enolase and the alpha-subunit of glycoprotein hormones. J Clin Endocrinol Metab 1997; 82(8): 2622-8. [PMID: 9253344]

[86] Korse CM, Taal BG, Vincent A, *et al.* Choice of tumour markers in patients with neuroendocrine tumours is dependent on the histological grade. A marker study of Chromogranin A, Neuron specific enolase, Progastrin-releasing peptide and cytokeratin fragments. Eur J Cancer 2012; 48(5): 662-71. [http://dx.doi.org/10.1016/j.ejca.2011.08.012] [PMID: 21945100]

[87] Baudin E, Gigliotti A, Ducreux M, *et al.* Neuron-specific enolase and chromogranin A as markers of neuroendocrine tumours. Br J Cancer 1998; 78(8): 1102-7. [http://dx.doi.org/10.1038/bjc.1998.635] [PMID: 9792158]

[88] Lonovics J, Devitt P, Watson LC, Rayford PL, Thompson JC. Pancreatic polypeptide. A review. Arch Surg 1981; 116(10): 1256-64. [http://dx.doi.org/10.1001/archsurg.1981.01380220010002] [PMID: 7025798]

[89] Metz DC, Jensen RT. Gastrointestinal neuroendocrine tumors: pancreatic endocrine tumors. Gastroenterology 2008; 135(5): 1469-92. [http://dx.doi.org/10.1053/j.gastro.2008.05.047] [PMID: 18703061]

[90] van der Lely AJ, de Herder WW. Carcinoid syndrome: diagnosis and medical management. Arq Bras Endocrinol Metabol 2005; 49(5): 850-60. [http://dx.doi.org/10.1590/S0004-27302005000500028] [PMID: 16444370]

[91] Salazar R, Wiedenmann B, Rindi G, Ruszniewski P. ENETS 2011 Consensus Guidelines for the Management of Patients with Digestive Neuroendocrine Tumors: an update. Neuroendocrinology 2012; 95(2): 71-3. [http://dx.doi.org/10.1159/000335600] [PMID: 22262042]

[92] Oberg K, Knigge U, Kwekkeboom D, Perren A. Neuroendocrine gastroentero-pancreatic tumors: ESMO Clinical Practice Guidelines for diagnosis, treatment and follow-up. Ann Oncol 2012; 23: 124-30. [http://dx.doi.org/10.1093/annonc/mds295]

[93] Ferolla P, Faggiano A, Mansueto G, *et al.* The biological characterization of neuroendocrine tumors: the role of neuroendocrine markers. J Endocrinol Invest 2008; 31(3): 277-86. [http://dx.doi.org/10.1007/BF03345602] [PMID: 18401212]

[94] Eriksson B, Oberg K, Stridsberg M. Tumor markers in neuroendocrine tumors. Digestion 2000; 62 (Suppl. 1): 33-8. [http://dx.doi.org/10.1159/000051853] [PMID: 10940685]

[95] Maroun J, Kocha W, Kvols L, *et al.* Guidelines for the diagnosis and management of carcinoid tumours. Part 1: the gastrointestinal tract. A statement from a Canadian National Carcinoid Expert Group. Curr Oncol 2006; 13(2): 67-76.

[http://dx.doi.org/10.3390/curroncol13020006] [PMID: 17576444]

[96] Ramage JK, Ahmed A, Ardill J, *et al.* Guidelines for the management of gastroenteropancreatic neuroendocrine (including carcinoid) tumours (NETs). Gut 2012; 61(1): 6-32. [http://dx.doi.org/10.1136/gutjnl-2011-300831] [PMID: 22052063]

[97] Orditura M, Petrillo A, Ventriglia J, *et al.* Pancreatic neuroendocrine tumors: Nosography, management and treatment. Int J Surg 2016; 28 (Suppl. 1): S156-62. [http://dx.doi.org/10.1016/j.ijsu.2015.12.052] [PMID: 26708853]

[98] de Herder WW, Niederle B, Scoazec JY, *et al.* Well-differentiated pancreatic tumor/carcinoma: insulinoma. Neuroendocrinology 2006; 84(3): 183-8. [http://dx.doi.org/10.1159/000098010] [PMID: 17312378]

[99] Jensen RT, Cadiot G, Brandi ML, *et al.* ENETS Consensus Guidelines for the management of patients with digestive neuroendocrine neoplasms: functional pancreatic endocrine tumor syndromes. Neuroendocrinology 2012; 95(2): 98-119. [http://dx.doi.org/10.1159/000335591] [PMID: 22261919]

[100] Service FJ, Natt N, Thompson GB, *et al.* Noninsulin oma pancreatogenous hypoglycemia: a novel syndrome of hyperinsulinemic hypoglycemia in adults independent of mutations in Kir6. 2 a nd SUR1 genes. J Clin Endocri nol Metab 1999; 8(4): 1582-9.

[101] Anderson CW, Bennett JJ. Clinical presentation and diagnosis of pancreatic neuroendocrine tumors. Surg Oncol Clin N Am 2016; 25(2): 363-74. [http://dx.doi.org/10.1016/j.soc.2015.12.003] [PMID: 27013370]

[102] Wermers RA, Fatourechi V, Kvols LK. Clinical spectrum of hyperglucagonemia associated with malignant neuroendocrine tumors. Mayo Clin Proc 1996; 71(11): 1030-8. [http://dx.doi.org/10.4065/71.11.1030] [PMID: 8917287]

[103] Jensen RT, Norton JA. Endocrine tumors of the pancreas and gastrointestinaltract. In: Feldman M, Friedman LS, Brandt LJ, Eds. Sleisenger and Fordtran'sgastrointestinal and liver disease. Philadelphia: Saunders 2010; pp. 491-522. [http://dx.doi.org/10.1016/B978-1-4160-6189-2.00032-9]

[104] Vinik A, Feliberti E, Perry RR. Glucagonoma syndrome. In: De Groot LJ, Beck-Peccoz P, Chrousos G, Eds. Endotext. South Dartmouth(MA): MDText.com, Inc 2014; p. 2000.

[105] Nesi G, Marcucci T, Rubio CA, Brandi ML, Tonelli F. Somatostatinoma: clinico-pathological features of three cases and literature reviewed. J Gastroenterol Hepatol 2008; 23(4): 521-6. [http://dx.doi.org/10.1111/j.1440-1746.2007.05053.x] [PMID: 17645474]

[106] Garby L, Caron P, Claustrat F, *et al.* Clinical characteristics and outcome of acromegaly induced by ectopic secretion of growth hormone-releasing hormone (GHRH): a French nationwide series of 21 cases. J Clin Endocrinol Metab 2012; 97(6): 2093-104. [http://dx.doi.org/10.1210/jc.2011-2930] [PMID: 22442262]

[107] Jensen RT, Berna MJ, Bingham DB, Norton JA. Inherited pancreatic endocrine tumor syndromes: advances in molecular pathogenesis, diagnosis, management, and controversies. Cancer 2008; 113(7) (Suppl.): 1807-43. [http://dx.doi.org/10.1002/cncr.23648] [PMID: 18798544]

[108] Gibril F, Schumann M, Pace A, Jensen RT. Multiple endocrine neoplasia type 1 and Zollinger-Ellison syndrome: a prospective study of 107 cases and comparison with 1009 cases from the literature. Medicine (Baltimore) 2004; 83(1): 43-83. [http://dx.doi.org/10.1097/01.md.0000112297.72510.32] [PMID: 14747767]

[109] Ito T, Igarashi H, Uehara H, Jensen RT. Pharmacotherapy of Zollinger-Ellison syndrome. Expert Opin Pharmacother 2013; 14(3): 307-21. [http://dx.doi.org/10.1517/14656566.2013.767332] [PMID: 23363383]

[110] Kaltsas G, Androulakis II, de Herder WW, Grossman AB. Paraneoplastic syndromes secondary to

neuroendocrine tumours. Endocr Relat Cancer 2010; 17(3): R173-93.
[http://dx.doi.org/10.1677/ERC-10-0024] [PMID: 20530594]

[111] Kanakis G, Kaltsas G, Granberg D, *et al.* Unusual complication of a pancreatic neuroendocrine tumor presenting with malignant hypercalcemia. J Clin Endocrinol Metab 2012; 97(4): E627-31.
[http://dx.doi.org/10.1210/jc.2011-2592] [PMID: 22319031]

[112] DeLellis RA, Xia L. Paraneoplastic endocrine syndromes: a review. Endocr Pathol 2003; 14(4): 303-17.
[http://dx.doi.org/10.1385/EP:14:4:303] [PMID: 14739488]

[113] Modlin IM, Lye KD, Kidd M. A 5-decade analysis of 13,715 carcinoid tumors. Cancer 2003; 97(4): 934-59.
[http://dx.doi.org/10.1002/cncr.11105] [PMID: 12569593]

[114] Stinner B, Rothmund M. Neuroendocrine tumours (carcinoids) of the appendix. Best Pract Res Clin Gastroenterol 2005; 19(5): 729-38.
[http://dx.doi.org/10.1016/j.bpg.2005.06.003] [PMID: 16253897]

[115] Solcia E, Rindi G, Paolotti D, La Rosa S, Capella C, Fiocca R. Clinicopathological profile as a basis for classification of the endocrine tumours of the gastroenteropancreatic tract. Ann Oncol 1999; 10 (Suppl. 2): S9-S15.
[http://dx.doi.org/10.1093/annonc/10.suppl_2.S9] [PMID: 10399027]

CHAPTER 10

A Pathophysiological Approach To Current Biomarkers

Aykut Oruc[1,*] and **Gonul Simsek**[1]

[1] *Department of Physiology,Cerrahpasa Faculty of Medicine, Istanbul University-Cerrahpaşa, Istanbul, Turkey*

Abstract: Biomarkers are necessary for screening and diagnosing numerous diseases, predicting the prognosis of patients, and following-up treatment and the course of the patient. Everyday new biomarkers are being used in clinics for these purposes. This section will discuss the physiological roles of the various current biomarkers in a healthy person and the pathophysiological mechanisms underlying the release of these biomarkers. This chapter aims to gain a new perspective for evaluating and interpreting the most current biomarkers.

Keywords: AFP, Alpha-fetoprotein, CA15-3, CA27.29, Carcinoembryonic antigen, CEA, GFAP, GPBB, GSTP-1, H-FABP, MUC1, NSE, NT-proBNP, Pathophysiological mechanisms, Physiology, S100B, Troponin.

CARDIAC BIOMARKERS

Cardiac biomarkers are widely used to detect functional heart problems and predict possible heart disease that may clinically show up later. The most common use of cardiac biomarkers is for acute myocardial infarction and congestive heart diseases, but they can also be used for a wide variety of diseases [1].

Cardiac biomarkers are, in general, intracellular components of myocytes or enzymes, formally. This title will discuss the pathophysiological mechanisms of typical and recent cardiac biomarkers, and the intracellular pathways of the release of these biomarkers will be explained in detail.

Cardiac Troponins

Troponin functions as a Ca^{+2} receptor and regulates Ca^{+2} levels during myosin and actin's interaction for contraction in skeletal and cardiac muscles. It is shown that

* **Corresponding author Aykut ORUC:** Department of Physiology,Cerrahpasa Faculty of Medicine, Istanbul University-Cerrahpaşa, Istanbul, Turkey; Tel: +90 212 414 30 56; Fax: +90 212 633 29 87; E-mail: aykutoruc86@istanbul.edu.tr

Pınar Atukeren and Hafize Uzun (Eds.)

troponin spreads through the thin filaments of a muscle with a length of 38 nm. Troponin that runs along the thin filament and attaches to active sites of tropomyosin settles on the grooves of actin. When intracellular Ca^{+2} levels decrease, troponin suppresses the contractile interaction between actin and myosin with the help of tropomyosin. On the contrary, this suppression is removed when intracellular Ca^{+2} levels increase as Ca^{+2} ions bind to troponin [2].

Troponin complex consists of three different subunits: troponin I, C, and T. They all have different functions such as removing of interaction between actin and myosin (troponin I), binding to Ca^{+2} (troponin C), and binding to tropomyosin (troponin T). Also, studies have shown that several mutations in genes that encode cardiac troponin subtypes are associated with inherited cardiomyopathies [2].

Cardiac troponin subtypes are widely used for myocardial ischemia. Before the discovery of troponin, different biomarkers were used to identify myocardial ischemia, such as aspartate transaminase (AST), lactate dehydrogenase (LDH), and creatinine kinase (CK). These biomarkers are still in use, but none show specificity for cardiac muscle. The later found biomarkers showed more specificity to cardiac muscle, such as CK-MB and LDH, but they still had a high false positivity rate. Troponins were first found in 1965, but the routine laboratory usage as a marker started in the 1990s. Troponin measurements have up to 100% sensitivity when observed 6 – 12 hours after the start of chest pain, and they also have high specificity for damaged cardiac muscle. Because of these benefits, high troponin levels were added as a criterion of myocardial infarction in "Third Universal Definition of Myocardial Infarction", which is the current definition used by the American College of Cardiology [3 - 5].

Troponin Subunits

Troponin consists of three different subunits: troponin I, C, and T. The primary function of the troponin is inhibiting the interaction between actin and myosin, which is directly related to intracellular Ca^{+2} levels.

Troponin I: Troponin I directly inhibits the contractile interaction between myosin and actin. This inhibition occurs with the absence of Ca+2 and is also very strong in the presence of tropomyosin. A After discovering troponin I, studies with synthetic troponins showed a region composed of residues 105–114 (minimum inhibitory peptide), called the 'inhibitory region.' It is relieved that this region is essential for showing the inhibitory activity with the help of a secondary region of residues 140–148 (second actin-binding region). The presence of this secondary region is also necessary for total inhibition. The N-terminus of troponin I, which is close to the inhibitory site, attaches to troponin C and together with troponin T, they all build a helix-coil structure.

On the other hand, the C-terminus attaches to the tropomyosin and actin complex. The N-terminal region that binds to troponin C- Ca^{+2} dependent troponin – is called the 'switch region.' This switch region is constantly phosphorylated and dephosphorylated *via* protein kinase A (PKA). When PKA phosphorylates the switch region, the affinity of troponin C with Ca^{+2} decreases, which inhibits the contraction of muscle as well. This mechanism is essential for the sympathetic nervous system to decrease diastole duration with increasing intracellular cyclic adenosine monophosphate (cAMP) levels and cAMP-related PKA activity *via* β_2 receptors [6].

Troponin C: Troponin C is the main subunit that reacts with Ca and forms a complex structure with troponin I and troponin T. The primary function of troponin C is suppressing the inhibition of the interaction between actin and myosin formed mainly by tropomyosin. Troponin C can bind calcium *via* four active sites located in the N terminus. Sites 1 and 2 are formally low-affinity Ca+2 receptors, while sites 3 and 4 are both high-affinity Ca^{+2} and Mg^{+2} receptors. This explains the necessity of magnesium in striated muscle contraction as well as in cardiac contraction. The binding of Ca^{+2} to these sites induces a conformational change resulting from a hydrophobic site that connects to troponin I to open (open conformation). After binding to troponin I, they form a helical structure and inhibit the suppression of troponin I-troponin T-tropomyosin complex on actin [6].

Troponin T: Troponin T is the subunit that binds to tropomyosin and also attracts other troponins. Troponin T is a tropomyosin-binding subunit of troponin. Troponin T is the largest of the three troponins and interacts with both troponins I and troponin C. Troponin T consists of two subunits: TnT1 and TnT2, which bind to N-terminus and C-terminus of tropomyosin, respectively. The role of troponin T is not directly regulating contractile interaction, but it works as an indirect factor. Troponin T adjusts the sensitivity of troponin C to Ca^{+2} with its subunit TnT2. The N-terminal part of cardiac troponin T (cTnT) contains 30 more residues than troponin T in striated muscle. These residues contain mainly exons, and by alternative splicing of these exons, four different isoforms can be produced in the human heart, and all these variants show different sensitivity to Ca^{+2}. Also, aberrant splicing of these exons due to genetic and epigenetic changes may alter the ventricular pumping function [7].

Regulatory Function of Troponin

The primary function of the troponin complex is to regulate Ca^{+2} with two related mechanisms: inhibiting actin-tropomyosin complex *via* troponin T and interacting between troponin C and troponin I, depending on intracellular Ca^{+2} levels. The contractile interaction between actin and myosin is suppressed *via* troponin I in the absence of Ca^{+2}. On the other hand, if $Ca^{+}2$ is present, this inhibition is removed *via* interactions with troponin C and troponin I (Fig. **1**). To summarize, the force of contractile elements in striated or cardiac muscle is suppressed in the absence of Ca^{+2} by Troponin I directly as well as Troponin C and Troponin T indirectly. Also, when the intracellular Ca^{+2} levels increase, and the Troponin C-Ca^{+2} attraction occurs, this results in increased myosin adenosine triphosphate (ATP)ase activity [6].

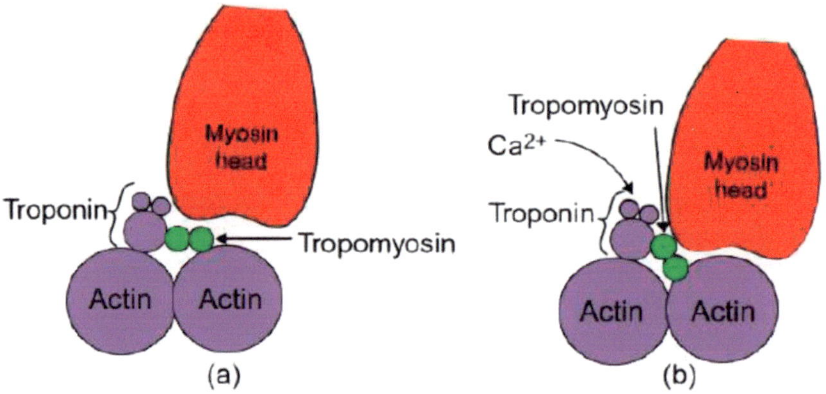

Fig. (1). Inhibition of contractile interaction between actin and myosin *via* troponin-tropomyosin complex (a); disinhibition of contractile interaction between actin and myosin in the presence of calcium ions (b).

Cardiac Troponins as a Biomarker for Acute Myocardial Infarction: The Pathophysiology

cTnI and cTnT are the most accurate biomarkers for detecting acute myocardial infarction, and they are widely used even they have a short half-life (90 min approximately) [8]. Blood cTnC levels are not used to detect acute myocardial infarction as the cTnC isoform is very similar to those of slow-twitch striated muscles, so cTnC does not have specificity for cardiomyocytes [9].

Acute myocardial infarction is a condition with a lack of blood supply for cardiomyocytes - decreased coronary blood flow results with hypoxia and a mismatch between the oxygen demand and the supply. Increased demand and decreased supply will disrupt the mechanisms dependent on ATP- primarily Na-K-ATPase pump. The Na-K-ATPase pump is responsible for many vital functions of a cell as well as a cardiomyocyte. These functions include maintaining membrane potentials and intracellular-extracellular electrolyte equilibrium. The failure of the Na-K-ATPase pump due to lack of ATP will cause an electrolyte disequilibrium and an increase in intracellular Na+2 concentrations, which will lead to swelling of the cell because of high hydration force and solvation of sodium ions. After this process, membranes of swelled cells will rupture, and intracellular contents will spill into the extracellular space and the blood, respectively. These contents include troponins as well, and the cardiac troponin levels get high to detect from blood [10, 11].

The regular turnover of cardiomyocytes will cause a basal cardiac troponin level in healthy individuals. There must be a 99% increase in cardiac troponin levels to indicate pathological cardiomyocyte damage, which means three standard derivations from the mean values. Also, there must be a specific elevation and fall of troponin levels according to the Third Universal Definition of Myocardial Infarction. The elevation starts 2-3 hours after the chest pain and will reach peak levels in 12-48 hours. In 4-10 days, the levels should decrease gradually and be normal. This elevation and fall pattern is helpful to distinguish acute myocardial infarction from other pathologies that will increase the blood troponin levels [11].

Cardiac Troponins as a Biomarker in Congestive Heart Failure

Missov *et al.* reported that cTnI might be a plasma biomarker in congestive heart failure (CHF) even without ischemia [12]. Several possibilities may cause these elevated plasma troponin levels in a non-ischemic CHF patient. Apoptosis and necrosis are usually seen in a heart with failure. In CHF, cardiac mural stress increases due to concentric or eccentric hypertrophy, resulting in diminished oxygenation. Also, increased renal reabsorption and altered renal clearance may disrupt the excretion of troponin and may contribute to the increased plasma troponin levels. Additionally, it has been reported that troponin plasma concentrations inversely correlate with left ventricular ejection fraction. A. Savic-Radojevic *et al.* reported that in patients with acute HF, plasma levels of cTnI and cTnT are higher than normal individuals (51% *vs.* 30%, respectively) elevations are correlated with increased mortality [36]. While cTn level evaluation has a high level of evidence in CHF diagnosis, it is also helpful for determining prognosis [13 - 17]. In CHF patients, cTn levels are significantly lower compared to patients with acute myocardial infarction. The absence of infarction is the main reason for these low levels in CHF patients. Because of this, several years ago, high-sensitive troponin (hs-cTn) assays were developed [18]. Hs-cTn assay is able to detect shallow levels of troponin and intends to measure cytoplasmic troponin levels [19]. After developing hs-cTn assay, troponin level determinations in CHF patients increased dramatically [20].

Heart-Type Fatty Acid Binding Protein

Heart-type fatty acid binding protein (H-FABP) is a protein encoded in humans by the FABP3 gene located on the chromosome [21]. It can be used as a biomarker during myocardial ischemia as it is released from myocardial cells after an ischemic attack [22]. It is responsible for transporting poly-unsaturated fatty acids (PUFA) from membrane to mitochondria before oxidation and metabolism. H-FABP plays an essential role in fatty acid metabolism [23]. It is also called a mammary-derived growth inhibitor because they also modulate cell proliferation and arrest the replication [21, 23]. H-FABP synthesis was detected in ventricles and atria, as well as in renal epithelial cells, Leydig cells of the testis, acinar and ductal cells of the breast, striated muscles, parietal cells, corpus luteum, adipocytes, and vascular endothelial cells [23, 24]

H-FABP is believed to protect cardiomyocytes from the oxidative stress derived from increased and unbalanced saturated fatty acid (SFA) oxidation. It is found that exposure to saturated fatty acids (C16:0 and C18:0) but not mono-unsaturated fatty acids (C16:1 or C18:1) or poly-unsaturated fatty acids (C18:2, C18:3, and C20:4) resulted in cell death in primary cultures of rat neonatal myocytes [25, 26]. As H-FBAP increases the transport of PUFA to cells, it balances SFA levels with increased PUFA levels and reverses the effect of SFA, and preserves the vital functions of cardiomyocytes [25 - 27]. In addition, PUFAs are the primary source of energy for cardiomyocytes. It has been shown that mice lacking H-FABP showed impaired uptake of PUFA from plasma as the main energy source [27]. This situation also caused a cardiac intolerance to exercise and cardiac hypertrophy [27].

Heart-Type Fatty Acid Binding Protein in Acute Myocardial Ischemia

As told above, H-FABP plays a significant role in metabolism and survival, so another situation that causes increased levels of H-FABP is ischemia [28]. Two research groups showed that evaluating plasma H-FABP levels may be beneficial in diagnosing acute myocardial infarction [22, 29]. During ischemia, H-FABP secretion is increased to aggravate the intake of PUFAs for survival and maintain vital functions of cardiomyocytes [26]. Also, cardiomyocytes with unbalanced levels of SFA are more likely to get more damage from acute stress such as ischemia. SFAs are toxic to cardiomyocytes, and SFA accumulation may cause cell membrane oxidation due to free radical production during reperfusion following ischemia. Increased PUFAs may alter the toxic effects of SFAs thanks to H-FABP.

Glycogen Phosphorylase Isoenzyme BB

Glycogen phosphorylase (GP) regulates carbohydrate metabolism by maintaining the mobilization of glycogen. It is an essential intracellular enzyme for energy production required to maintain vital functions and contraction in myocytes as well [30, 31]. The first enzymatic step in glycogenolysis is catalyzed by GP that converts glycogen into glucose-I-phosphate. This activity of GP is regulated by phosphorylation *via* phosphorylase kinase. Phosphorylase kinase activates GP to achieve its function by converting it into an active form: GP-alpha. The necessary phosphorylase kinase is usually stored in the sarcoplasmic reticulum in cardiomyocytes with glycogen deposits in a form called sarcoplasmic reticulum glycogenolysis complex [31, 32]. The formation of this complex directly depends on the metabolic demands of the myocardium, and if metabolic demands increase or hypoxia occurs, glycogen will be released from this complex for compensation of energy demand, and GP will be converted into a soluble form and released to the cytoplasm [31 - 33].

GP consists of 2 subunits, which are made of identical isoforms. Three isoforms are synthesized in humans by three distinct genes and named by the location where they are mostly found: GP-LL, GP-MM, and GP-BB. GP-LL is mainly found in the liver, GP-MM in muscle, and GP-BB in the brain [30, 34]. On the other hand, both BB and MM isotypes are found in human cardiomyocytes, while BB is the dominant isoform [35].

Myocardial Oxygen Deficiency and GPBB Release: The Pathophysiology

GP-BB is used as an early and highly responsive cardiac biomarker. As told above, GP plays an important metabolic role during anaerobic conditions as during the hypoxia, glycogen, and GP releases from the endoplasmic reticulum to provide energy to cells by anaerobic pathways. S Similarly, GP-BB is used to degrade glycogen in cardiomyocytes in hypoxic/ischemic conditions. Ischemia decreases ATP synthesis and increases AMP concentrations in the cell. Increased AMP levels trigger the phosphorylase kinase, the enzyme that converts GP-BB into active form "alpha" (a) for a short time [36 - 39]. Within minutes, intracytoplasmic GB-BB- a turns into "beta" (b) -the water-soluble- form by dephosphorylation. With increased membrane permeability due to ischemia, this form can freely diffuse from the cell to the bloodstream as well as other intracellular components of the myocardium. As a result, the plasma levels of GP-BB increase, indicating hypoxia/ischemia, and these explain why GP-BB is a novel biomarker for acute coronary syndromes [40].

Glutathione S-Transferase P1

Glutathione S-transferase P1 (GSTP1) is the most common isoform of all the glutathione S-transferase enzymes. It mainly maintains responses to inflammation by inhibiting apoptosis, eliminating reactive oxygen species (ROS), and remaining vital functions in the cell [41 - 43]. Its anti-apoptotic functions are related to inhibiting tumor necrosis factor α and receptor-associated factor 2(TRAF2), which are responsible for activating mitogen-activated protein kinase (MAPK) signaling, a major pro-apoptotic pathway [44, 45].

CHF is a complex disease that has multiple pathophysiological mechanisms. Chronic inflammation and excessive apoptosis are usually observed in the myocytes of a failing heart. As triggered apoptosis is usually a result of ROS accumulation and inflammation, increased cellular responses (such as increased GSTP1 levels) to ROS and inflammation are expected consequently [46]. Because of this, GSTP1 has gained much importance as a potential biomarker in CHF patients. In fact, Andrukhova *et al.* found that high GSTP1 levels may be related to the poor prognosis of CHF. Also, they found that GSTP1 may be more valuable and sensitive than N-Terminal pro-brain natriuretic peptide (BNP), a marker commonly used for monitoring CHF patients [47]. As GSTP1 is a very recent biomarker, further studies are needed for routine clinical use in patients with CHF.

N-Terminal pro-Brain Natriuretic Peptide

The natriuretic peptides, especially BNP, are still the gold standard biomarkers in monitoring and predicting prognosis for CHF patients. Three main types of natriuretic peptides are produced in humans: atrial natriuretic peptide (ANP), BNP, and C-type natriuretic peptide (CNP) but as CNP functions as a local factor, especially in renal capillary endothelium and is not released to bloodstream, only ANP and BNP can be used as a biomarker [8]. The myocytes of the atrium produce ANP, and the brain and ventricular myocytes produce BNP. These peptides are secreted in response to an increased intravascular volume. Increased intravascular volume induces the stretch receptors located in the atria and the ventricles, causing ANP and BNP secretion. These natriuretic peptides try to compensate for high volume load by inducing diuresis and natriuresis and decreasing peripheral vascular resistance *via* vasodilation. This compensation mechanism is also constantly active in CHF patients due to volume overload [48, 49]. Despite many diuretic peptides, the recommendation of current HF guidelines for monitoring and diagnosis CHF patients is determining only the BNP levels and its precursor N-terminal-proBNP (NT- proBNP) [50]. This may be explained by the short half-life of ANP (2-5min) compared to BNP (20min) and especially NT-proBNP (60-120min), which makes ANP levels hard to detect in the blood. Also, it has been shown that BNP and NT-proBNP concentrations are more correlated with the prognosis of CHF patients than ANP levels, especially in the terminal period [48]. Besides, BNP levels may vary due to systemic factors and diseases such as age, sex, obesity, renal functions, pulmonary diseases. This makes NT-proBNP hard to use as a single biomarker in monitoring CHF patients (level of evidence: B) [8].

Pathophysiological Mechanisms in Synthesis and Secretion of BNP and NT-proBNP

Human BNP consists of 32 amino acids, including a ring structure with 17 amino acids with a disulfide bond connecting two cysteine residues [50, 51]. Secretion of BNP is directly related to left ventricular (LV) volume. Increased intraventricular volume or pressure stimulates the stretch receptors in LV, causing the release of BNP from cardiomyocytes [52 - 56]. BNP differs from other peptide hormones as it is usually not stored in high amounts. In physiological conditions, when stretch receptors of ventricles are stimulated, the transcription of mRNA, synthesis, post-transcriptional modification, and secretion of BNP occurs rapidly, and it is released to surrounding tissues as well as blood [57, 58]. Nevertheless, on pathological conditions, such as CHF, mRNA of BNP gets unstable and rapidly transcript a precursor form with 134 aminoacids: pre-pro-BNP. This precursor form is rapidly converted to proBNP by removing 26 aminoacids from N-terminus and to active and inactive forms (BNP and NT-proBNP, respectively) by convertases called proNP convertases [54, 59]. Both BNP and NT-proBNP are secreted to blood after this process and can be evaluated from plasma [60, 61].

The Receptors and The Mechanisms of Action

In humans, natriuretic peptides have three types of receptors: natriuretic peptide receptor (NPR)-A, NPR-B, and NPR-C. All three natriuretic peptides (ANP, BNP, and CNP) can bind to all these three receptors but with varied affinities. While ANP and BNP primarily bind to NPR-A and show their effect, CNP predominantly binds to NRP-B. On the contrary, all three NPs can bind to NRP-C, but the main function of this receptor is suppressing the effects of all NPs by removing them from circulation *via* uptake and degradation [59, 62]. The location of these receptors also varies; NPR-A is found chiefly in endothelium, brain, and kidney, NPR-B is primarily found chondrocytes, glial cells, and kidney, NPR-C is found on most of the cells [52].

NPR-A and NPR-B are guanylate cyclase coupled receptors, so the binding of ANP and BNP to these receptors increases cyclic guanylate monophosphate (cGMP) in target cells. Increased cGMP levels trigger vascular smooth muscle relaxation by inhibiting myosin light chain kinase, an essential enzyme for smooth muscle contraction. This process leads to vasodilation [50, 63]. Additionally, binding of BNP to NPR-A inhibits central sympathetic output from rostral ventrolateral medulla (RVLM) and suppresses renin secretion from juxtaglomerular cells, causing the extinction of the renin-angiotensin-aldosterone system. As a result, the glomerular filtration rate and filtration fraction decrease; diuresis, natriuresis, and systemic vasodilation occur [63, 64].

CANCER BIOMARKERS

Cancer biomarkers are used for detecting many types of tumors and can be found in plasma, urine, tissues, or body secretions. Elevated levels of these markers can mostly show the presence of tumors, but sometimes false-positive results can be obtained. In response to cancer, tumor cells and normal cells can produce these biomarkers in the human body. In this topic, the most recent and common cancer biomarkers and their pathophysiological mechanisms related to their secretions will be discussed.

Alpha-Fetoprotein

Alpha-fetoprotein (AFP) is a glycoprotein that consists of 591 amino acids and is encoded by the same-named gene (AFP) on 4q25. It is a member of an albuminoid protein family, and its functions are very similar to albumin, the major plasma protein. It is one of the five members of the albuminoid protein family. In healthy individuals, the normal range of AFP is 5-8 ng/ml [65]. Gitlin *et al.* first discovered AFP as a fetal protein synthesized from fetal hepatocytes and yolk sac [66]. While it is mainly used as a fetal biomarker where its increased levels indicate diminished neural tube and fetal liver development, it is also used for various malignancies such as hepatocarcinoma, yolk sac tumors, intestinal and rectal cancers, and breast cancer [65, 66]. It also functions as a transporter of most organic drugs to cells, and because of this, AFP also is a common biomarker in research that explores the effects of drugs on oncogenesis [67].

Physiological Functions and Oncogenic Roles of Alpha-Fetoprotein

AFP functions are similar to albumin and work as a transporter of various plasma ligands such as bilirubin, heavy metals, steroids, PUFAs, and organic drugs such as phenytoin, warfarin, streptomycin [68 - 71]. Also, it has been shown that AFP also binds to proteins from the steroid/thyroid nuclear receptor superfamily [72 - 74].

Despite the most explained function of AFP -transporting diverse molecules, it has been reported that AFP works as an immune response and cell proliferation modulator too, and these functions revealed the oncogenic roles of AFP [68, 69]. More studies have supported this by showing AFP stimulating cell growth *via* cAMP-PKA pathways *in-vitro* [75, 76]. When cell growth promotion properties of AFP are discovered, the high levels of AFP observed in the fetus made sense. On the other hand, proper fetal development means balancing between triggered and sustained growth during pregnancy. While sustained growth is required for a fetus, bringing regular homeostasis back after experiencing stress such as transient hypoxia, osmolarity, pH changes, hypertension, or physiological development processes such as differentiation transformation need the growing activity to stop. The balance between cessation and continuity of fetal growing is regulated by varying AFP levels [72, 77].

The oncogenic properties of AFP are also studied due to its growth modulation functions. It has been shown that AFP has the potential to modulate the proliferation of hepatocytes, lymphocytes, and gonads [72, 78 - 81]. It has also been reported that AFP is responsible for promotion and initiation during oncogenesis, and at physiological dose, AFP can exacerbate tumor growth both by stimulating the expression of oncogenes (Jun, Ras, Fos) and by triggering neovascularisation [68, 82].

The immune modulation functions of AFP are more likely to suppress immune responses by inhibiting T and B lymphocyte proliferation at their blast forms [83]. Mechanisms underlying this function are not precise yet, but it has been reported that AFP has four human leukocyte antigen (HLA)-A epitopic sites and several minor immunodominant epitopes that can be recognized by lymphocytes [84, 85]. Impairing immune system functions can also suppress apoptotic processes and cause uncontrolled cell proliferation. It also has been shown that AFP may suppress immune responses by diminishing dendritic cell functions, including antigen presentation to T-lymphocytes [86]. In the same study, it has also been reported that this situation may help hepatoblastoma cells to escape immunologic response by over-expressing AFP.

The uptake of AFP by endocytosis is also increased in tumor cells. Studies show both increased intracellular AFP levels conjugated to other proteins and blood secretion. The quantity of uptaken and secreted AFP is reversely correlated with the differentiation. It has been reported that muscle tissue undifferentiated tumor cells show more endocytotic activity than differentiated cells [87]. Specific receptors are responsible for this uptake of AFP on undifferentiated cells and are first discovered on human monocytes, but later they are observed in tumor cells of breast cancer, melanoma, hepatocellular cancer, pulmonary cancers. In fact, these specific-called AFP receptors are several types of scavenger receptors such as LOX-1, CXCL16, SCRA1, and they all show different affinities. Endocytosis of AFP provides delivering the necessary ligands to the tumor for its growth and development, which also leads to more AFP secretion from tumor cells. It has been shown that AFP levels reach a peak when the tumoral expansion gets maximum [88, 89].

MUC1 protein and Related Biomarkers: Cancer Antigens 27.29 and 15-3

In humans, mucins work as a natural protective barrier for pathogens to prevent access to their receptors on host cells. Mucin networks create a thick mucus layer on surfaces of lumens to protect the underlying tissue. Mucins exist in two forms: cell-bound form and secreted form. While goblet cells are responsible for secreting mucins to create a mucus layer, cell-bound mucins are localized on the apical side of the epithelium. If a pathogen succeeds in passing the mucus layer, these cells bound mucins protect the epithelial barrier by covering the receptors recognized by pathogens [90]. MUC1 is the first discovered type of cell-bound mucins, and it has been shown that MUC1 works as an essential barrier for pathogens as well as modulating inflammation caused by pathogens [91 - 93]. It has also been shown that MUC1 expression increases on several types of tumor cells first in 1999 [94].

It has been shown that MUC1 secretion increases in many types of cancer, especially in cancer of organs where mucin is secreted, such as breast, liver, colon, pancreas, and ovary [95]. Because of this, MUC1 is a an excellent biomarker for the diagnosis and follow-up of these types of cancers. In routine, the levels of CA 27.29 and CA 15-3, the different epitopes of the MUC1 protein, are generally used.

Physiological Properties of MUC1 and Its Role in Cancer Development

MUC1 proteins are expressed especially in ductal and glandular epithelia and typically have a 300-600 kDa molecular weight. MUC1 has a very long structure (approx. 500nm) compared to other surface-bound molecules and covers a wide cell membrane area [96, 97]. Mature MUC1 protein has quite many O-linked glycans in its structure. MUC1 maturation is a dynamic process that consists of endocytosis and the following exocytosis processes. During this recycling, intracellular MUC1gains more sialic acid residues get glycosylated and bound to the surface again after exocytosis [98]. Due to this heavy glycosylation, MUC1 gains its roles: lubrication and protection (from pathogens and microenvironment changes such as pH osmolarity).

For this reason, MUC1 is found in breast ductal epithelium, bladder, sweat glands, stomach, and more intensely [99]. The highly negatively charged sialic acid residues bound to MUC1 during maturation also gain MUC1, a very negative charge, and a steric barrier. This negative charge physiologically inhibits pathogen adhesion by electrostatically repulsive effect but also interrupts integrin-mediated cell to cell and cell to extracellular matrix (ECM) adhesion and makes cells more mobile. Decreased bound to ECM makes it tumor cells easier to invade underlying stroma, blood, and lymph vessels [97, 100, 101]. The intracellular component of MUC1 interacts with several molecules involved in cell growth signaling (especially the Wnt signaling pathway), including beta-catenin, growth factor receptor-bound protein 2 (Grb2), glycogen synthase kinase-3-beta (GSK-3B), erbB receptor proteins.

Several studies have shown that increased MUC1 expression on cancer cells increases beta-catenin expression and promotes tumor invasion by increased expression of vimentin and N-cadherin [102]. Increased N-cadherin expression causes the epithelial tumor cells to transform into mesenchymal tumor cells, which are more aggressive and have a capacity to escape from the primary cancer site resulting in metastasis [103].

On the other hand, activation of MUC1 bound erbB receptors *via* growth factors activate tyrosine-kinases located on the erbB receptor, causing phosphorylation of the tyrosine residues of the erbB receptor MUC1. This activates the ERK1/2 pathway, responsible for cell growth and differentiation [104]. Also, phosphorylated tyrosine residues of MUC1 activate another molecule, Grb2, responsible for activating the Wnt signaling pathway [105]. It has also have been shown that MUC1 has a role in modulating immunity by decreasing immune responses during cancer development [106]. In physiological conditions, an enzyme called Core 1 Gal-transferase (C1GT) is responsible for glycosylation of MUC1 with O-linked glycans; this is also a process for maturation of MUC1 [107]. The amount and the glycosylation pattern depend on the cell's physiological state [108, 109]. It has been suggested that oncogenetic changes in cells alter C1GT production, and due to lack of C1GT, glycosylation pattern changes and becomes aberrant [110]. This aberrant glycosylation makes tumor cells escape from CD8 T-lymphocyte-mediated killing and makes them more invasive [109].

Studies have shown that the intracytoplasmic part of MUC1 also interacts with p53 and its response element, p21. P21 and MUC1 interaction get vigorous, especially during molecular stresses, resulting in the arrest of cell cycle instead of p53 mediated apoptosis [101]. It has been suggested that tumoral MUC1 shows this anti-apoptotic function by phosphorylating Akt resulting in Bcl-2 and Bcl-XL phosphorylation. Bcl-2 and Bcl-XL are two major anti-apoptotic molecules of cells and work by inhibiting mitochondrial cytochrome-2 release [97, 101].

MUC1 as a Cancer Biomarker

Especially in breast cancer, plasma levels of CA 27.29 and CA 15-3 -different protein products of distinct epitopes on the MUC1 gene increase [111]. Both CA 15-3 and CA 27.29 may also be elevated in patients with benign ovarian cysts, benign breast disease, and benign liver disease. Elevations may also be seen in cirrhosis, sarcoidosis, and lupus [112].

Carcinoembryonic Antigen

Human carcinoembryonic antigen (CEA) belongs to a protein family consisting of two groups: carcinoembryonic antigen cellular adhesion molecules (CEACAMs) and pregnancy-specific glycoproteins (PSGs) [113]. Cell adhesion molecules (CAMs) are also classified into four subgroups; integrins, the immunoglobulin (Ig) superfamily, cadherins, and selectins, all transmembrane glycoproteins and responsible for cell-to-cell and cell-to-ECM bindings. They all play essential roles in regulating immune responses like controlling leukocyte diapedesis, embryogenesis such as fetal nervous system development, control of cell growth such as inhibiting or promoting cell cycle [114]. CEACAM and CEA belong to the Ig superfamily, the group which also includes major histocompatibility (MHC) antigens, intracellular adhesion molecule 1(ICAM-1), and lymphocyte function-associated antigen-1 [115]. CEA was first discovered in 1972, and many functions such as promoting angiogenesis, activation of the insulin receptor, regulating T lymphocyte proliferation have been shown [116 - 118]. CEA can be found as a membrane surface protein or a secreted protein to the bloodstream. The surface form can be bound to cell membrane either glycophosphatidylinositol (GPI) anchors or a transmembrane domain [119].

The physiological function of carcinoembryonic antigen is not well understood, but as far as known, CEA has various receptors for pathogens and can bind them to protect the underlying stroma from invasion. A healthy person can synthesize about 70mg CEA/day, and it is metabolized in the liver by Kupfer cells. The prominent source is the gut epithelium and goblet cells. In goblet cells, CEA interacts with secreted mucin and helps form the mucus coat of the lumen [120].

It also functions as a regulator of growth signal transduction *via* its transmembrane domain and helps tumor differentiation if overexpressed [121]. In fact, elevated CEA levels have been associated with many types of cancers for a long time, such as colorectal cancers, mucinous ovarian adenocarcinomas, breast cancers, and cervical squamous cell carcinomas [122].

Relation of CEA overexpression with malignity was first shown on colonic polyps and adenomas [123]. CEACAM6, a subtype of CEACAMs, is found to be more responsible for tumoral development. It has been suggested that CEACAM6 overexpression leads tumoral cells to be more invasive and easier to metastasize [124]. On the other hand, it has been shown that monoclonal antibodies that are targetting CEACAM6 suppress the tumoral cell invasion and metastasis [124]. In addition, Hostetter *et al.* showed that liver metastasis ratios increased from %2 to %48 on mice with colorectal cancer when treated with CEA [125].

The Role of CEA in Tumoral Invasion and Metastasis

CEA functions expressed from tumor cells differ in many ways from the one expressed from a healthy cell. While in physiological conditions, CEA has a polarity and is expressed mainly on the apical side of the cell, tumoral CEA loses this polarity and is expressed on the whole surface of the cell. It has been suggested that tumoral CEA starts to work as a homotypic surface adhesion molecule covering tumor cells all-around, making tumor cells easy to aggregate and invade [120].

Elevated serum CEA levels are closely related to liver metastasis [126]. But why especially liver? Moreover, what is the role of CEA in this process?

The microenvironment is essential for a tumor cell to live, and tumor cells are likely to metastasize to organs where the micro-environment is suitable for them [127]. The ideal micro-environment for cancer is where the growth factors are secreted to protect tumor cells from several stresses induced cytotoxicity such as hypoxia; where the immune system is suppressed, and neo-angiogenesis occurs quickly [128]. The crucial important factor providing this micro-environment is local tissue macrophages or the recruited ones, especially tumor-associated macrophages (TAMs) [119]. TAMs can secrete several cytokines and growth factors that induce neo-angiogenesis, such as IL8, vascular endothelial growth factor (VEGF), and tumor necrosis factor (TNF)-alpha. In addition, TAMs also help tumor cells invade the underlying tissue and expand *via* secreting some modulating enzymes like metalloproteinases (MMPs) and collagenases [129, 130].

It is shown that CEA is capable of binding some specific types of macrophages, especially Kupfer cells and TAMs, because of carrying specific receptors for CEA such as heterogeneous nuclear RNA-binding protein M (hnRNP-M) [131]. The binding of CEA to hnRNP-M results in the internalization of the CEA, promoting them to secrete cytokines that makes the microenvironment suitable for tumor cells. Gango- padhyay *et al.* have suggested that CEA-Kupfer cell attraction results with increased secretion of IL-1α and b, IL-6, IL-10, and TNF-α [132, 133]. From these cytokines, the pro-inflammatory ones (IL-1α, IL-6, and TNF-α) induce the upregulation of cell adhesion molecules such as ICAM-1 and E-selectin and MMPs on endothelium, causing enhanced aggregation and invasion of tumoral cells [134, 135]. In addition, anti-inflammatory cytokines such as IL-10 secreted from TAM-like Kupfer cells inhibit inducible nitric oxide synthetase (iNOS). This inhibition protects the tumor cells from NO and ROS damage during hypoxia [136 - 138] (Fig. **2**). In addition, it has also been shown that hypoxia increases CEA expression in colorectal and breast cancers as a defense mechanism of tumor cells [138].

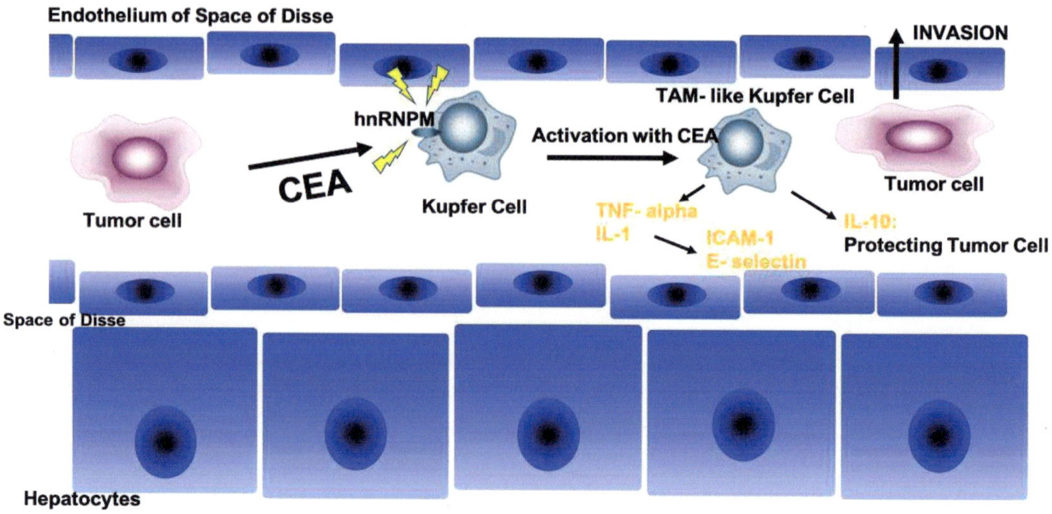

Fig. (2). Tumor cell secretes CEA after reaching the hepatic sinusoids to activate and transform Kupfer cells. Activated Kupfer cells start to act like TAMs and secretes pro-inflammatory cytokines. These pro-inflammatory cytokines upregulate adhesion molecules such as ICAM-1 and E-selectin to ease tumor invasion. Also, the antiinflammatory cytokines released from activated Kupfer cells protect tumor cell from various stress.CEA: carcinoembyronic antigen, hnRNPM: heterogeneous nuclear RNA-binding protein M, TNF: tumor necrosis factor, IL: interleukin, TAM: tumor associated macrophage, ICAM: intracellular adhesion molecule.

NEURONAL BIOMARKERS

Neuronal biomarkers are becoming important determinants in diagnosis and follow-up in several central and peripheric neurological diseases, including neurodegenerative diseases such as Alzheimer's disease, Parkinson's disease; primary brain tumors such as glioblastoma, and oligodendroglioma; cerebrovascular diseases; and demyelinating diseases, especially multiple sclerosis. They also can be critical prognostic factors at the time of diagnosis.

Neuronal biomarkers can be found in the blood as well as cerebrospinal fluid. Significant neuronal biomarkers and their pathological and physiological roles in humans will be discussed in this title.

Glial Fibrillary Acidic Protein

Glial fibrillary acidic protein (GFAP) is a cytoskeletal protein secreted primarily by astrocytes and belongs to the class III intermediate filament (IF) protein family like other cytoskeletal proteins such as desmin, peripherin, and vimentin [139,

140]. GFAP has been discovered as a stabilizer and mobility modulator of astrocytes [141, 142]. However, subsequent studies have shown that GFAP is more than this. It has been found that GFAP also modulates synaptic transmissions, neuronal communications, blood-brain barrier (BBB) formation, and myelination of both peripheric nervous system (PNS) and central nervous system (CNS) neurons [139, 140, 143].

It has been shown that GFAP production is highly increased in a process called astrocytosis (gliosis), which is observed in injuries, degenerations, toxications, and genetic disorders of the CNS [144]. Astrocytosis is a response of CNS to stress and is formed by reactive astrocyte accumulation. As a result of stress and injury, astrocytes transform into their reactive astrocytes and start to synthesize numerous GFAP containing filaments to form a protective structure called glial cicatrix. It has been shown that GFAP secretion is highly correlated with astrocyte activity, which made GFAP a great biomarker for CNS traumas, neurodegenerative diseases, and determining the origin of CNS cancer, whether it is glial or neuronal [145, 146].

The Physiological Properties and Functions of GFAP

GFAP can be secreted in both CNS and PNS with a predominancy of CNS. In a healthy person, GFAP is observed in protoplasmic astrocytes of gray matter – creating glia limitans, in fibrous astrocytes of white matter, and as well as in astrocytes of the cerebellum (Bergmann's glia). The amount of GFAP secreted also differs in astrocytes depending on the location: the fibrous astrocytes in white matter secrete more than the others [139, 141]. On the other hand, neurons of the autonomic nervous system and the Schwann cells of the somatic nerves are the two main sites of PNS where GFAP is secreted [147, 148].

As a member of the IF protein family, the primary functions of GFAP are maintaining the integrity and migration of astrocytes. Additionally, it has been suggested that GFAP is also integrated into protein kinases and may have a role in signal transmission *via* protein kinase activation or inactivation [149]. This modulating function of intracellular signaling is suggested to be important in fetal/neonatal CNS development and blood-brain barrier (BBB) integrity in a study that shows lack of GFAP causes white matter development disorder and excessive water uptake to CNS due to impaired BBB [150]. According to another study, it was suggested that lacking GFAP caused demyelination in PNS, supporting that GFAP has a role in the myelination process of Schwann cells [151]. Impaired metabolic functions of astrocytes also have been observed in GFAP deficiency, supporting that GFAP also plays a role in regulating metabolites in astrocytes [152]. Another vital role of GFAP is obtaining glia-neuron and neuron-neuron communications. It has been shown that GFAP deficiency causes impaired long-term potentiation in neurons of the cerebellum and hippocampus [153].

Reactive Astrocyte Formation and GFAP

Experiencing numerous stresses (inflammation, aging, ionizing radiation, high-frequency magnetic waves, physical or thermogenic traumas) and many pathological states (viral infections, demyelinating diseases such as multiple sclerosis, cryogenic damages, encephalopathies, diabetes mellitus, hyperthyroidism) results in several astrocytic changes in CNS called as reactivation of astrocytes [154 - 156]. Reactive astrocytes aggregate and form a protective tissue, glial cicatrix. This process is also known as gliosis or astrocytosis, and its severity depends on the level of injury [157]. On the other hand, increased Ca+2 uptake to reactive astrocytes activates calcium-dependent proteinases resulting in re-expression and rearrangement of IFs such as vimentin and protecting the integrity of cells [158]. This explains the importance of reactive astrocytes in responses to CNS damage.

While gliosis is a neuroprotective process that leads to the secretion of several neurotrophic factors and cytokines that will help neurons suffer from pathological conditions and maintain the integrity of BBB; excessive gliosis may reverse the benefits and result in deleterious effects because of over-expressed cytokines, loss of neuronal support and communications and causing necrosis [159, 160]. The first and the essential step of the forming reactive astrocytes is increasing GFAP secretion (Fig. **3**). The importance of GFAP in forming glial cicatrix is shown by Ghirnikar *et al.* in their study conducted with GFAP deficit mice. Their study observed insufficient glial cicatrix formation and BBB impairment after CNS damage in mice with GFAP deficiency [161]. It also has been suggested that GFAP has anti-apoptotic functions on neurons by triggering autophagic pathways (such as GSK3B) *via* their integration with protein-kinases, and they enforce neurons to survive [162].

Relation of neurogenerative states with GFAP has also been shown in some neurodegenerative diseases such as amyotrophic lateral sclerosis, Parkinson's disease, Huntington's disease, Pick's disease, and Alzheimer's disease; however, the GFAP levels vary in these diseases. For example, while excessive gliosis formation and elevated GFAP levels are typical signs of Alzheimer's disease due to beta-amyloid accumulation in the brain; decreased GFAP levels are observed in AIDS-dementia complex, as glycoprotein 120 (gp120), a surface protein of HIV, directly inhibits GFAP synthesis [154 - 156, 163, 164]. In addition, numerous viruses other than HIV (varicella-zoster and herpes virus) also inhibits GFAP from turning into an active form by inhibiting phosphorylation of GFAP [165, 166]. In summary, it can be said that decreased GFAP levels are related to impaired astrocyte functions.

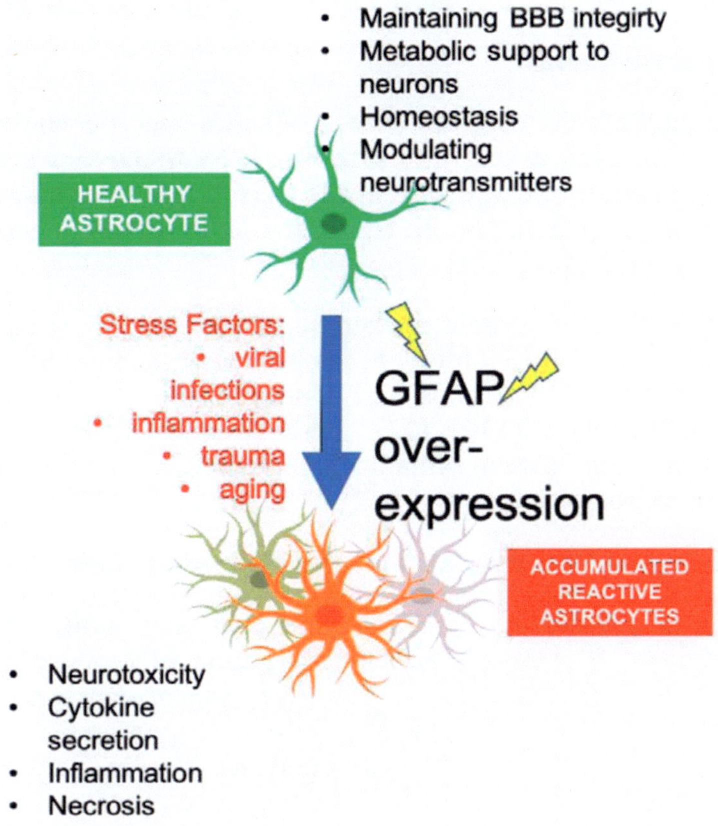

Fig. (3). Excessive reactivation of astrocytes *via* numerous stress factors results in diminished astrocyte function, diminished BBB integrity, and inflammation. Reactive astrocytes synthesize GFAP in large amounts. BBB: blood brain barrier, GFAP: glial fibrillary acidic protein.

Due to the restriction of BBB, increased GFAP levels are observed mainly in cerebrospinal fluid (CSF) and have a high diagnostic value in patients with multiple sclerosis, Alzheimer's disease, stroke, intoxications, subarachnoid hemorrhages [146]. However, GFAP levels also can be measured in the plasma due to impaired BBB functions, especially during CNS traumas. Nevertheless, physical traumas are not the only reason for impaired BBB integrity. There is a dynamic between the pressure of subarachnoid space and hydrostatic capillary pressure in favor of the latter in physiological conditions. CSF production is a passive process that lets the fluid in capillary flows to the subarachnoid space, where the CSF exists. Increased intracranial pressure reverses this process and causes the CSF to flow into the capillary. Intracranial hemorrhages, especially subarachnoid hemorrhage, are the most frequent reasons for increased intracranial pressure, and GFAP can be observed in blood in these pathologies [167].

Neuron Specific Enolase

McGregor and Moore first discovered Neuron-specific enolase in 1965, and Neuron-specific enolase was first discovered by McGregor and Moore in 1965 and named as neuron-specific protein early [168, 169]. The current name, neuron-specific enolase (NSE), is given after demonstrating enolase activity in subsequent studies [170, 171].

NSE is an acidic and soluble protein secreted by neurons as well as by cells undergoing neuronal differentiation, such as neuroendocrine cells. Enolase is a common enzyme of glycolysis and gluconeogenesis, and NSE is a specific isoenzyme of enolase [170, 171]. NSE synthesis starts in late periods of maturation, making NSE a valuable biomarker for showing the maturation of neurons and neuroendocrine cells.

The Physiological Properties of NSE and Pathophysiology

Enolase is a reversible metalloenzyme that acts both on glycolysis and gluconeogenesis. It catalyzes either dehydration of 2-phosphoglycerate (PGA) into phosphoenolpyruvate (PEP) during glycolysis or hydration of PEP into PGA during gluconeogenesis and needs Mg+2 to work appropriately [172, 173]. Mg+2 can bind to two different sites (the site I and site II) of NSE depending on enzyme action. The binding of Mg+2 to site I, known as the conformational site, causes a conformational change in NSE and makes it capable of binding its substrate, PEP, and working as an anabolic enzyme. On the other hand, the binding of Mg+2 to site II, known as catalytic site, makes NSE a catalytic enzyme and processes glycolysis [174, 175].

Enolase consists of a dimer of three subunits (α, β, γ) encoded by different genes in the human body and these three subunits can form five isotypes of enolase (ββ, αα, αβ, αγ and γγ), but only three of them is dominantly produced: αα, ββ and γγ [176]. From these three isotypes, αα (α in short) is ubiquitous, ββ (β in short) is specific to muscle cells, and γγ (γ in short) is specific to neurons, which is also called as NSE [177].

Of all these three isoforms, NSE shows a unique feature showing insensitivity to Cl- ions. In a typical neuron, Cl- ions are responsible for controlling excitability by moving the resting membrane potential to a more negative state and, more importantly, accumulates in neurons during consecutive depolarisation phases [178]. From a different perspective, increased Cl- concentrations of neuron shows their intensive activity and increased metabolic demands. An increased Cl- concentration activates NSE in response to high metabolic activity for meeting the energy requirements [179]. In summary, it can be clearly said that NSE synthesis and activation increase in parallel to metabolic activity. Although NSE is also responsible for catalyzing gluconeogenesis like all enolases, it has been shown that gluconeogenesis only exists on astrocytes in the brain *via* other enzymes than NSE [180]. According to this, NSE prominently works as a catabolic enzyme rather than anabolic [181]. All these make NSE an excellent biomarker for pathological states where metabolic demands of neurons are increased, such as neoplasms or CNS damages. In fact, in several studies, it has been demonstrated that elevation of NSE levels is related to primary tumors of CNS such as neuroblastoma as well as metastatic tumors. Also, it has been suggested that NSE elevation is reversely correlated with tumor differentiation [182 - 184]. Zeltzer *et al.* found that high NSE plasma levels indicate poor prognosis in children with metastatic neuroblastoma [185]. High NSE levels are also found to be correlated with poor functional outcomes and damage severity in stroke patients [186].

S100B Protein

S100 calcium-binding protein, or S100 in short, was first observed in a bovine brain as a protein with advanced ability to bind calcium nearly 50 years ago [187]. Then many subtypes of this protein were discovered and named as S100 protein family. S100 protein family consists of small (approx. 9-14 kDa) and intracellular proteins that bind calcium and act as a signal transmitter in CNS cells such as astrocytes, neurons, oligodendrocytes, and progenitor cells of neurons but as well as in cells of other tissues like chondrocytes, adipocytes, macrophages, myoepithelial cells [188, 189]. Up to now, 21 members of the S100 protein family are specified, and they are all encoded by different genes, which also gives the protein its name [190]. Of all these members, S100B is the most prominent form in CNS and consists of two dimers. In CNS, it is found chiefly as a monomer (S100BB) and less as a heteromer (S100BA), and together these are known as Total S100B, or S100B in practice [191, 192].

Physiological Properties and Pathophysiology of S100B

S100B is primarily an intracellular Ca^{+2} binding protein, and its primary function is maintaining Ca^{+2} homeostasis and transmitting Ca^{+2} dependent intracellular signals, including the signals of cell growth and differentiation [193, 194]. In this context, it regulates; neurogenesis during fetal development gliosis after an injury

[195, 196]. Although S100B is an intracellular protein, it can work extracellularly after secreting, mostly from astrocytes. It has been shown that S100B plays an essential role in processes related to learning and memory and modulating synaptic plasticity [193, 197, 198].

S100B shows these advantageous effects when it is found in physiological levels. On the other hand, overexpression of S100B starts to exert harmful effects rather than beneficial [199, 200]. Higher levels of S100B have been associated with reactivating astrocytes and recruiting active microglia that results in a pro-inflammatory microenvironment for neurons where excessive NO, pro-inflammatory cytokines, and Ca^{+2} levels are found [201, 202]. Studies have shown that CNS infections, traumas, toxications, and metabolic distress may cause this kind of S100B secretion from astrocytes [203]. S100B is stored in astrocytes in normal conditions and can be released quickly after exposure to stress. However, it also has been demonstrated that mRNA levels of S100B also increase in astrocytes in response to stressful factors, so the increased S100B levels depend on both stored and newly synthesized [204]. Increased S100B induces surrounding microglia to secrete pro-inflammatory substances such as IL-1β, TNF-α, causing neuronal damage, subsequently activating more astrocytes, recruiting more microglia, and more neuronal damage [205] (Fig. **4**).

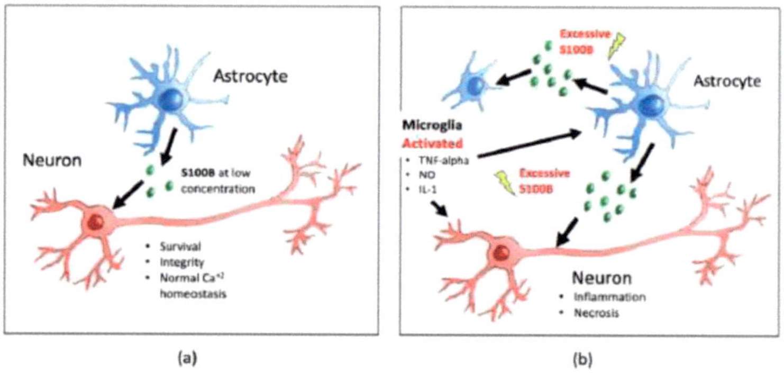

Fig. (4). S100B has many beneficial functions, such as maintaining neuronal Ca^{+2} homeostasis, integrity and development when secreted at physiological levels (a). However, it activates and recruits microglia that secretes numerous pro-inflammatory cytokines and products when it reaches excessive levels. Both pro-inflammatory substances and S100B itself cause neurodegeneration and necrosis (b).

These processes make S100B a suitable biomarker showing neuronal injury. Recent studies have shown that S100B can be used as a reliable biomarker for detecting numerous pathologies such as subdural and epidural hematomas, contusions, cerebrovascular diseases, chronic inflammatory neurodegenerative diseases, and subarachnoid hemorrhages [206 - 209]. S100B levels can be measured directly from CSF as well as from plasma if BBB integrity is disrupted [210]. In this context, it has been suggested that the ratio of S100B levels between serum and CSF can be used for evaluating BBB integrity [211, 212]. It has also been suggested that S100B levels are correlated with the severity of neuronal damage; as the injury gets widespread, S100B levels get higher. Because of this, S100B level monitorization of patients may be important to detect the progression of the disease or the response to treatment [207, 212].

CONCLUSION

The importance of biomarkers used in various diseases for their predictive, diagnostic, and prognostic effects is increasing daily. A biomarker can be any kind of molecule related to cell structure, growth, metabolism, or protection from various dangers. In order to understand the clinical importance of a biomarker, establish a clear relationship with diseases, and fully predict the changes in that biomarker, it is necessary to learn the pathophysiological mechanisms associated with the release of that biomarker. Mastering the pathophysiological mechanisms is also of great importance to make routine the use of current biomarkers and the discovery of new biomarkers.

CONSENT FOR PUBLICATION

Not applicable.

CONFLICT OF INTEREST

The authors declare no conflict of interest, financial or otherwise.

ACKNOWLEDGEMENTS

Declared none.

REFERENCES

[1] Halushka PV, Goodwin AJ, Halushka MK, Lakier JB. Opportunities for microRNAs in the Crowded Field of Cardiovascular Biomarkers. Annu Rev Pathol 2019; 14(1): 211-38. [http://dx.doi.org/10.1146/annurev-pathmechdis-012418-012827] [PMID: 30332561]

[2] Ohtsuki I, Morimoto S. Troponin. In: Lennarz WJ, Lane MD, Eds. Encyclopedia of Biological Chemistry. 2nd ed. Massachusetts: Academic Press 2013; pp. 446-9. [http://dx.doi.org/10.1016/B978-0-12-378630-2.00195-X]

[3] Er TK, Su YF, Chan TH. Analytical Performance of the VIDAS® High-Sensitivity Troponin I Assay and the Beckman Coulter Unicel® DXI AccuTnI+3 Assay in a Stat Laboratory. Clin Lab 2019; 65: 1. [http://dx.doi.org/10.7754/Clin.Lab.2018.180626] [PMID: 30775875]

[4] Black JA, Cheng K, Flood JA, *et al.* Evaluating the benefits of a rapid access chest pain clinic in Australia. Med J Aust 2019; 210(7): 321-5. [http://dx.doi.org/10.5694/mja2.50021] [PMID: 30773636]

[5] Tahir K, Pauley E, Dai X, Smith SC Jr, Sweeney C, Stouffer GA. Mechanisms of ST Elevation Myocardial Infarction in Patients Hospitalized for Noncardiac Conditions. Am J Cardiol 2019; 123(9): 1393-8.
[http://dx.doi.org/10.1016/j.amjcard.2019.01.039] [PMID: 30773247]

[6] Ohtsuki I, Morimoto S. Troponin. In: Lennarz WJ, Lane MD, Eds. Encyclopedia of Biological Chemistry. 2nd ed. Massachusetts: Academic Press 2013; pp. 445-7.
[http://dx.doi.org/10.1016/B978-0-12-378630-2.00195-X]

[7] Feng HZ, Jin JP. Coexistence of cardiac troponin T variants reduces heart efficiency. Am J Physiol Heart Circ Physiol 2010; 299(1): H97-H105.
[http://dx.doi.org/10.1152/ajpheart.01105.2009] [PMID: 20418479]

[8] Savic-Radojevic A, Pljesa-Ercegovac M, Matic M, Simic D, Radovanovic S, Simic T. Novel Biomarkers of Heart Failure. In: Makowski GS, Ed. Advances in Clinical Chemistry. 1st ed. Massachusetts: Academic Press 2017; Vol. 79: pp. 99-100.

[9] Schreier T, Kedes L, Gahlmann R. Cloning, structural analysis, and expression of the human slow twitch skeletal muscle/cardiac troponin C gene. J Biol Chem 1990; 265(34): 21247-53.
[http://dx.doi.org/10.1016/S0021-9258(17)45353-1] [PMID: 2250022]

[10] Mirkin J, Radecki R, Spiegel R. Deriving Peace of Mind: In Search of a Fifth-Generation Troponin Testing Threshold to Safely Rule Out Acute Myocardial Infarction: March 2019 Annals of Emergency Medicine Journal Club. Ann Emerg Med 2019; 73(3): 317-9.
[http://dx.doi.org/10.1016/j.annemergmed.2019.01.029] [PMID: 30797301]

[11] Ramezani F, Ahmadi S, Faridaalee G, Baratloo A, Yousefifard M. Value of Manchester Acute Coronary Syndromes Decision Rule in the Detection of Acute Coronary Syndrome; a Systematic Review and Meta-Analysis. Emergency (Tehran) 2018; 6(1): e61.
[PMID: 30788388]

[12] Missov E, Calzolari C, Pau B. Circulating cardiac troponin I in severe congestive heart failure. Circulation 1997; 96(9): 2953-8.
[http://dx.doi.org/10.1161/01.CIR.96.9.2953] [PMID: 9386162]

[13] Ponikowski P, Voors AA, Anker SD, *et al.* J.R. Gonza´lez-Juanatey, V.-P. Harjola, E.A. Jankowska, M. Jessup, C. Linde, P. Nihoyannopoulos, J.T. Parissis, B. Pieske, J.P. Riley, G.M.C. Rosano, L.M. Ruilope, F. Ruschitzka, F.H. Rutten, P. van der Meer, Authors/Task Force Members, 2016 ESC Guidelines for the diagnosis and treatment of acute and chronic heart failure: the Task Force for the diagnosis and treatment of acute and chronic heart failure of the European Society of Cardiology (ESC). Developed with the special con- tribution of the Heart Failure Association (HFA) of the ESC. Eur Heart J 2016; 37(27): 2129-200.
[http://dx.doi.org/10.1093/eurheartj/ehw128] [PMID: 27206819]

[14] La Vecchia L, Mezzena G, Zanolla L, *et al.* Cardiac troponin I as diagnostic and prognostic marker in severe heart failure. J Heart Lung Transplant 2000; 19(7): 644-52.
[http://dx.doi.org/10.1016/S1053-2498(00)00120-0] [PMID: 10930813]

[15] Horwich TB, Patel J, MacLellan WR, Fonarow GC. Cardiac troponin I is associated with impaired hemodynamics, progressive left ventricular dysfunction, and increased mortality rates in advanced heart failure. Circulation 2003; 108(7): 833-8.
[http://dx.doi.org/10.1161/01.CIR.0000084543.79097.34] [PMID: 12912820]

[16] Latini R, Masson S, Anand IS, *et al.* Prognostic value of very low plasma concentrations of troponin T in patients with stable chronic heart failure. Circulation 2007; 116(11): 1242-9.
[http://dx.doi.org/10.1161/CIRCULATIONAHA.106.655076] [PMID: 17698733]

[17] Peacock WF IV, De Marco T, Fonarow GC, *et al.* Cardiac troponin and outcome in acute heart failure. N Engl J Med 2008; 358(20): 2117-26.
[http://dx.doi.org/10.1056/NEJMoa0706824] [PMID: 18480204]

[18] Giannitsis E, Kurz K, Hallermayer K, Jarausch J, Jaffe AS, Katus HA. Analytical validation of a high-sensitivity cardiac troponin T assay. Clin Chem 2010; 56(2): 254-61. [http://dx.doi.org/10.1373/clinchem.2009.132654] [PMID: 19959623]

[19] Thygesen K, Alpert JS, White HD. Joint ESC/ACCF/AHA/WHF Task Force for the Redefinition of Myocardial Infarction, Universal definition of myocardial infarc- tion. Eur Heart J 2007; 28: 2525-38. [http://dx.doi.org/10.1093/eurheartj/ehm355] [PMID: 17951287]

[20] Chenevier-Gobeaux C, Bonnefoy-Cudraz É, Charpentier S, *et al.* High-sensitivity cardiac troponin assays: answers to frequently asked questions. Arch Cardiovasc Dis 2015; 108(2): 132-49. [http://dx.doi.org/10.1016/j.acvd.2014.11.001] [PMID: 25669958]

[21] Vupputuri A, Sekhar S, Krishnan S, Venugopal K, Natarajan KU. Heart-type fatty acid-binding protein (H-FABP) as an early diagnostic biomarker in patients with acute chest pain. Indian Heart J 2015; 67(6): 538-42. [http://dx.doi.org/10.1016/j.ihj.2015.06.035] [PMID: 26702681]

[22] Kleine AH, Glatz JF, Van Nieuwenhoven FA, Van der Vusse GJ. Release of heart fatty acid-binding protein into plasma after acute myocardial infarction in man. Mol Cell Biochem 1992; 116(1-2): 155-62. [http://dx.doi.org/10.1007/BF01270583] [PMID: 1480144]

[23] Watanabe K, Wakabayashi H, Veerkamp JH, Ono T, Suzuki T. Immunohistochemical distribution of heart-type fatty acid-binding protein immunoreactivity in normal human tissues and in acute myocardial infarct. J Pathol 1993; 170(1): 59-65. [http://dx.doi.org/10.1002/path.1711700110] [PMID: 8326460]

[24] Zschiesche W, Kleine AH, Spitzer E, Veerkamp JH, Glatz JF. Histochemical localization of heart-type fatty-acid binding protein in human and murine tissues. Histochem Cell Biol 1995; 103(2): 147-56. [http://dx.doi.org/10.1007/BF01454012] [PMID: 7634154]

[25] de Vries JE, Vork MM, Roemen TH, *et al.* Saturated but not mono-unsaturated fatty acids induce apoptotic cell death in neonatal rat ventricular myocytes. J Lipid Res 1997; 38(7): 1384-94. [http://dx.doi.org/10.1016/S0022-2275(20)37421-6] [PMID: 9254064]

[26] Van Bilsen M, de Vries JE, Van der Vusse GJ. Long-term effects of fatty acids on cell viability and gene expression of neonatal cardiac myocytes. Prostaglandins Leukot Essent Fatty Acids 1997; 57(1): 39-45. [http://dx.doi.org/10.1016/S0952-3278(97)90491-9] [PMID: 9250607]

[27] Binas B, Danneberg H, McWhir J, Mullins L, Clark AJ. Requirement for the heart-type fatty acid binding protein in cardiac fatty acid utilization. FASEB J 1999; 13(8): 805-12. [http://dx.doi.org/10.1096/fasebj.13.8.805] [PMID: 10224224]

[28] Das UN. Heart-type fatty acid-binding protein (H-FABP) and coronary heart disease. Indian Heart J 2016; 68(1): 16-8. [http://dx.doi.org/10.1016/j.ihj.2015.07.030] [PMID: 26896261]

[29] Tanaka T, Hirota Y, Sohmiya K, Nishimura S, Kawamura K. Serum and urinary human heart fatty acid-binding protein in acute myocardial infarction. Clin Biochem 1991; 24(2): 195-201. [http://dx.doi.org/10.1016/0009-9120(91)90571-U] [PMID: 2040092]

[30] Newgard CB, Hwang PK, Fletterick RJ. The family of glycogen phosphorylases: structure and function. Crit Rev Biochem Mol Biol 1989; 24(1): 69-99. [http://dx.doi.org/10.3109/10409238909082552] [PMID: 2667896]

[31] Meyer F, Heilmeyer LMG Jr, Haschke RH, Fischer EH. Control of phosphorylase activity in a muscle glycogen particle. I. Isolation and characterization of the protein-glycogen complex. J Biol Chem 1970; 245(24): 6642-8. [http://dx.doi.org/10.1016/S0021-9258(18)62582-7] [PMID: 4320610]

[32] Entam ML, Kanike K, Goldstein MA, *et al.* Association of gylcogenolysis with cardiac sarcoplasmic

reticulum. J Biol Chem 1976; 251(10): 3140-6.
[http://dx.doi.org/10.1016/S0021-9258(17)33510-X] [PMID: 5455]

[33] Entman ML, Bornet EP, Van Winkle WB, Goldstein MA, Schwartz A. Association of glycogenolysis with cardiac sarcoplasmic reticulum: II. Effect of glycogen depletion, deoxycholate solubilization and cardiac ischemia: evidence for a phorphorylase kinase membrane complex. J Mol Cell Cardiol 1977; 9(7): 515-28.
[http://dx.doi.org/10.1016/S0022-2828(77)80367-2] [PMID: 408501]

[34] Newgard CB, Littman DR, van Genderen C, Smith M, Fletterick RJ. Human brain glycogen phosphorylase. Cloning, sequence analysis, chromosomal mapping, tissue expression, and comparison with the human liver and muscle isozymes. J Biol Chem 1988; 263(8): 3850-7.
[http://dx.doi.org/10.1016/S0021-9258(18)69003-9] [PMID: 3346228]

[35] Kato K, Shimizu A, Kurobe N, Takashi M, Koshikawa T. Human brain-type glycogen phosphorylase: quantitative localization in human tissues determined with an immunoassay system. J Neurochem 1989; 52(5): 1425-32.
[http://dx.doi.org/10.1111/j.1471-4159.1989.tb09189.x] [PMID: 2651563]

[36] Wollenberger A, Krause E-G, Heier G. Stimulation of 3′,5′-cyclic AMP formation in dog myocardium following arrest of blood flow. Biochem Biophys Res Commun 1969; 36(4): 664-70.
[http://dx.doi.org/10.1016/0006-291X(69)90357-X] [PMID: 4390191]

[37] Dobson JG Jr, Mayer SE. Mechanisms of activation of cardiac glycogen phosphorylase in ischemia and anoxia. Circ Res 1973; 33(4): 412-20.
[http://dx.doi.org/10.1161/01.RES.33.4.412] [PMID: 4355036]

[38] Krause E-G, Wollenberger A. Cyclic nucleotides in heart in acute myocardial ischemia and hypoxia. Adv Cyclic Nucleotide Res 1980; 12: 49-61.
[PMID: 6250374]

[39] Reimer KA, Jennings RB. Myocardial ischemia, hypoxia and infarction. In: Fozzard HA, Haber E, Jennings RB, Katz AM, Morgan HE, Eds. The Heart and Cardiovascular System. 2nd ed. New York: Raven Press 1991; pp. 1875-973.

[40] Krause EG, Rabitzsch G, Noll F. Early Detection of Ischemic Myocardial Damage by Glycogen Phosphorylase Isoenzyme BB: A Biomarker for Evaluation of Chest Pain and Evolving Infarction in Patients. In: Dhalla NS, Takeda N, Singh M, Lukas A, Eds. Myocardial Ischemia and Preconditioning Progress in Experimental Cardiology. Boston, MA: Springer 2003; Vol. 6.
[http://dx.doi.org/10.1007/978-1-4615-0355-2_35]

[41] Moscow JA, Fairchild CR, Madden MJ, *et al.* Expression of anionic glutathione-S-transferase and P-glycoprotein genes in human tissues and tumors. Cancer Res 1989; 49(6): 1422-8.
[PMID: 2466554]

[42] Cumming RC, Lightfoot J, Beard K, Youssoufian H, O'Brien PJ, Buchwald M. Fanconi anemia group C protein prevents apoptosis in hematopoietic cells through redox regulation of GSTP1. Nat Med 2001; 7(7): 814-20.
[http://dx.doi.org/10.1038/89937] [PMID: 11433346]

[43] Shea TC, Kelley SL, Henner WD. Identification of an anionic form of glutathione transferase present in many human tumors and human tumor cell lines. Cancer Res 1988; 48(3): 527-33.
[PMID: 3275499]

[44] Baietto L, D'Avolio A, Marra C, *et al.* Development and validation of a new method to simultaneously quantify triazoles in plasma spotted on dry sample spot devices and analysed by HPLC-MS. J Antimicrob Chemother 2012; 67(11): 2645-9.
[http://dx.doi.org/10.1093/jac/dks285] [PMID: 22872447]

[45] Mak J, Sujishi KK, French D. Development and validation of a liquid chromatography-tandem mass spectrometry (LC-MS/MS) assay to quantify serum voriconazole. J Chromatogr B Analyt Technol Biomed Life Sci 2015; 986-987: 94-9.

[http://dx.doi.org/10.1016/j.jchromb.2015.02.011] [PMID: 25725319]

[46] Egle H, Trittler R, König A, Kümmerer K. Fast, fully automated analysis of voriconazole from serum by LC-LC-ESI-MS-MS with parallel column-switching technique. J Chromatogr B Analyt Technol Biomed Life Sci 2005; 814(2): 361-7. [http://dx.doi.org/10.1016/j.jchromb.2004.10.069] [PMID: 15639460]

[47] Jourdil JF, Tonini J, Stanke-Labesque F. Simultaneous quantitation of azole antifungals, antibiotics, imatinib, and raltegravir in human plasma by two-dimensional high-performance liquid chromatography-tandem mass spectrometry. J Chromatogr B Analyt Technol Biomed Life Sci 2013; 919-920: 1-9. [http://dx.doi.org/10.1016/j.jchromb.2012.12.028] [PMID: 23384531]

[48] Blanchard J. Evaluation of the relative efficacy of various techniques for deproteinizing plasma samples prior to high-performance liquid chromatographic analysis. J Chromatogr, Biomed Appl 1981; 226(2): 455-60. [http://dx.doi.org/10.1016/S0378-4347(00)86080-6] [PMID: 7320173]

[49] Polson C, Sarkar P, Incledon B, Raguvaran V, Grant R. Optimization of protein precipitation based upon effectiveness of protein removal and ionization effect in liquid chromatography-tandem mass spectrometry. J Chromatogr B Analyt Technol Biomed Life Sci 2003; 785(2): 263-75. [http://dx.doi.org/10.1016/S1570-0232(02)00914-5] [PMID: 12554139]

[50] Cheung BM, Kumana CR. Natriuretic peptides--relevance in cardiovascular disease. JAMA 1998; 280(23): 1983-4. [http://dx.doi.org/10.1001/jama.280.23.1983] [PMID: 9863839]

[51] Daniels LB, Maisel AS. Natriuretic peptides. J Am Coll Cardiol 2007; 50(25): 2357-68. [http://dx.doi.org/10.1016/j.jacc.2007.09.021] [PMID: 18154959]

[52] Maalouf R, Bailey S. A review on B-type natriuretic peptide monitoring: assays and biosensors. Heart Fail Rev 2016; 21(5): 567-78. [http://dx.doi.org/10.1007/s10741-016-9544-9] [PMID: 26979601]

[53] Clerico A, Recchia FA, Passino C, Emdin M. Cardiac endocrine function is an essential component of the homeostatic regulation network: physiological and clinical implications. Am J Physiol Heart Circ Physiol 2006; 290(1): H17-29. [http://dx.doi.org/10.1152/ajpheart.00684.2005] [PMID: 16373590]

[54] de Lemos JA, McGuire DK, Drazner MH. B-type natriuretic peptide in cardiovascular disease. Lancet 2003; 362(9380): 316-22. [http://dx.doi.org/10.1016/S0140-6736(03)13976-1] [PMID: 12892964]

[55] Rodeheffer RJ. Measuring plasma B-type natriuretic peptide in heart failure: good to go in 2004? J Am Coll Cardiol 2004; 44(4): 740-9. [PMID: 15312852]

[56] Levin ER, Gardner DG, Samson WK. Natriuretic peptides. N Engl J Med 1998; 339(5): 321-8. [http://dx.doi.org/10.1056/NEJM199807303390507] [PMID: 9682046]

[57] Nakagawa O, Ogawa Y, Itoh H, *et al.* Rapid transcriptional activation and early mRNA turnover of brain natriuretic peptide in cardiocyte hypertrophy. Evidence for brain natriuretic peptide as an "emergency" cardiac hormone against ventricular overload. J Clin Invest 1995; 96(3): 1280-7. [http://dx.doi.org/10.1172/JCI118162] [PMID: 7657802]

[58] Hama N, Itoh H, Shirakami G, *et al.* Rapid ventricular induction of brain natriuretic peptide gene expression in experimental acute myocardial infarction. Circulation 1995; 92(6): 1558-64. [http://dx.doi.org/10.1161/01.CIR.92.6.1558] [PMID: 7664440]

[59] Kerkelä R, Ulvila J, Magga J. Natriuretic Peptides in the Regulation of Cardiovascular Physiology and Metabolic Events. J Am Heart Assoc 2015; 4(10): e002423. [http://dx.doi.org/10.1161/JAHA.115.002423] [PMID: 26508744]

[60] Vanderheyden M, Bartunek J, Goethals M. Brain and other natriuretic peptides: molecular aspects. Eur J Heart Fail 2004; 6(3): 261-8.
[http://dx.doi.org/10.1016/j.ejheart.2004.01.004] [PMID: 14987574]

[61] Yamanouchi S, Kudo D, Endo T, Kitano Y, Shinozawa Y. Blood N-terminal proBNP as a potential indicator of cardiac preload in patients with high volume load. Tohoku J Exp Med 2010; 221(3): 175-80.
[http://dx.doi.org/10.1620/tjem.221.175] [PMID: 20505308]

[62] Volpe M, Rubattu S, Burnett J Jr. Natriuretic peptides in cardiovascular diseases: current use and perspectives. Eur Heart J 2014; 35(7): 419-25.
[http://dx.doi.org/10.1093/eurheartj/eht466] [PMID: 24227810]

[63] Cataliotti A, Boerrigter G, Costello-Boerrigter LC, *et al.* Brain natriuretic peptide enhances renal actions of furosemide and suppresses furosemide-induced aldosterone activation in experimental heart failure. Circulation 2004; 109(13): 1680-5.
[http://dx.doi.org/10.1161/01.CIR.0000124064.00494.21] [PMID: 15023890]

[64] Díez J. Chronic heart failure as a state of reduced effectiveness of the natriuretic peptide system: implications for therapy. Eur J Heart Fail 2017; 19(2): 167-76.
[http://dx.doi.org/10.1002/ejhf.656] [PMID: 27766748]

[65] EASL-EORTC clinical practice guidelines: management of hepatocellular carcinoma. J Hepatol 2012; 56(4): 908-43.
[http://dx.doi.org/10.1016/j.jhep.2011.12.001] [PMID: 22424438]

[66] Abelev GI. Alpha-fetoprotein in ontogenesis and its association with malignant tumors. Adv Cancer Res 1971; 14: 295-358.
[http://dx.doi.org/10.1016/S0065-230X(08)60523-0] [PMID: 4107670]

[67] Anzai H, Kazama S, Kiyomatsu T, *et al.* Alpha-fetoprotein-producing early rectal carcinoma: a rare case report and review. World J Surg Oncol 2015; 13(1): 180.
[http://dx.doi.org/10.1186/s12957-015-0590-x] [PMID: 25962419]

[68] Mizejewski GJ. alpha-fetoprotein as a biologic response modifier: relevance to domain and subdomain structure. Proc Soc Exp Biol Med 1997; 215(4): 333-62.
[http://dx.doi.org/10.3181/00379727-215-44143] [PMID: 9270717]

[69] Mizejewski GJ. Alpha-fetoprotein structure and function: relevance to isoforms, epitopes, and conformational variants. Exp Biol Med (Maywood) 2001; 226(5): 377-408.
[http://dx.doi.org/10.1177/153537020122600503] [PMID: 11393167]

[70] Arsenov DV, Golubeva MB, Kisel' MA, *et al.* Modification of humoral immune response in C57Bl/6 mice with a complex of alpha-fetoprotein and retinoid acid derivatives. Bull Exp Biol Med 2001; 132(4): 963-5.
[http://dx.doi.org/10.1023/A:1013615211043] [PMID: 11782793]

[71] Milligan SR, Khan O, Nash M. Competitive binding of xenobiotic oestrogens to rat alpha-fetoprotein and to sex steroid binding proteins in human and rainbow trout (Oncorhynchus mykiss) plasma. Gen Comp Endocrinol 1998; 112(1): 89-95.
[http://dx.doi.org/10.1006/gcen.1998.7146] [PMID: 9748407]

[72] Mizejewski GJ. An apparent dimerization motif in the third domain of alpha-fetoprotein: molecular mimicry of the steroid/thyroid nuclear receptor superfamily. BioEssays 1993; 15(6): 427-32.
[http://dx.doi.org/10.1002/bies.950150610] [PMID: 7689318]

[73] Dauphinée MJ, Mizejewski GJ. Human alpha-fetoprotein contains potential heterodimerization motifs capable of interaction with nuclear receptors and transcription/growth factors. Med Hypotheses 2002; 58(6): 453-61.
[http://dx.doi.org/10.1054/mehy.2001.1445] [PMID: 12323110]

[74] Bois-Joyeux B, Chauvet C, Nacer-Chérif H, *et al.* Modulation of the far-upstream enhancer of the rat

alpha-fetoprotein gene by members of the ROR alpha, Rev-erb alpha, and Rev-erb beta groups of monomeric orphan nuclear receptors. DNA Cell Biol 2000; 19(10): 589-99. [http://dx.doi.org/10.1089/104454900750019344] [PMID: 11058961]

[75] Wang XW, Xu B. Stimulation of tumor-cell growth by alpha-fetoprotein. Int J Cancer 1998; 75(4): 596-9. [http://dx.doi.org/10.1002/(SICI)1097-0215(19980209)75:4<596::AID-IJC17>3.0.CO;2-7] [PMID: 9466662]

[76] Li MS, Li PF, Yang FY, He SP, Du GG, Li G. The intracellular mechanism of alpha-fetoprotein promoting the proliferation of NIH 3T3 cells. Cell Res 2002; 12(2): 151-6. [http://dx.doi.org/10.1038/sj.cr.7290121] [PMID: 12118941]

[77] Butterstein G, Morrison J, Mizejewski GJ. Effect of alpha-fetoprotein and derived peptides on insulin- and estrogen-induced fetotoxicity. Fetal Diagn Ther 2003; 18(5): 360-9. [http://dx.doi.org/10.1159/000071981] [PMID: 12913349]

[78] Oertel M, Menthena A, Chen YQ, Shafritz DA. Properties of cryopreserved fetal liver stem/progenitor cells that exhibit long-term repopulation of the normal rat liver. Stem Cells 2006; 24(10): 2244-51. [http://dx.doi.org/10.1634/stemcells.2006-0141] [PMID: 16778153]

[79] Schnater JM, Bruder E, Bertschin S, *et al.* Subcutaneous and intrahepatic growth of human hepatoblastoma in immunodeficient mice. J Hepatol 2006; 45(3): 377-86. [http://dx.doi.org/10.1016/j.jhep.2006.03.018] [PMID: 16780998]

[80] Butterstein GM, Mizejewski GJ. Alpha-fetoprotein inhibits frog metamorphosis: implications for protein motif conservation. Comp Biochem Physiol A Mol Integr Physiol 1999; 124(1): 39-45. [http://dx.doi.org/10.1016/S1095-6433(99)00088-4] [PMID: 10605066]

[81] Mizejewski GJ. Biological roles of alpha-fetoprotein during pregnancy and perinatal development. Exp Biol Med (Maywood) 2004; 229(6): 439-63. [http://dx.doi.org/10.1177/153537020422900602] [PMID: 15169963]

[82] Mizejewski GJ. Levels of alpha-fetoprotein during pregnancy and early infancy in normal and disease states. Obstet Gynecol Surv 2003; 58(12): 804-26. [http://dx.doi.org/10.1097/01.OGX.0000099770.97668.18] [PMID: 14668662]

[83] Mizejewski GJ, Muehlemann M, Dauphinee M. Update of alpha fetoprotein growth-inhibitory peptides as biotherapeutic agents for tumor growth and metastasis. Chemotherapy 2006; 52(2): 83-90. [http://dx.doi.org/10.1159/000091728] [PMID: 16498241]

[84] Bui LA, Butterfield LH, Kim JY, *et al. In vivo* therapy of hepatocellular carcinoma with a tumor-specific adenoviral vector expressing interleukin-2. Hum Gene Ther 1997; 8(18): 2173-82. [http://dx.doi.org/10.1089/hum.1997.8.18-2173] [PMID: 9449371]

[85] Hanke P, Rabe C, Serwe M, *et al.* Cirrhotic patients with or without hepatocellular carcinoma harbour AFP-specific T-lymphocytes that can be activated *in vitro* by human alpha-fetoprotein. Scand J Gastroenterol 2002; 37(8): 949-55. [http://dx.doi.org/10.1080/003655202760230928] [PMID: 12229971]

[86] Um SH, Mulhall C, Alisa A, *et al.* Alpha-fetoprotein impairs APC function and induces their apoptosis. J Immunol 2004; 173(3): 1772-8. [http://dx.doi.org/10.4049/jimmunol.173.3.1772] [PMID: 15265907]

[87] Bourreille J, Metayer P, Sauger F, Matray F, Fondimare A. [Existence of alpha feto protein during gastric-origin secondary cancer of the liver]. Presse Med 1970; 78(28): 1277-8. [PMID: 5426134]

[88] Laderoute MP, Pilarski LM. The inhibition of apoptosis by alpha-fetoprotein (AFP) and the role of AFP receptors in anti-cellular senescence. Anticancer Res 1994; 14(6B): 2429-38. [PMID: 7532927]

[89] Li XD, Wu CP, Ji M, *et al.* Characteristic analysis of α-fetoprotein-producing gastric carcinoma in

China. World J Surg Oncol 2013; 11(1): 246.
[http://dx.doi.org/10.1186/1477-7819-11-246] [PMID: 24083471]

[90] Hattrup CL, Gendler SJ. Structure and function of the cell surface (tethered) mucins. Annu Rev Physiol 2008; 70(1): 431-57.
[http://dx.doi.org/10.1146/annurev.physiol.70.113006.100659] [PMID: 17850209]

[91] McAuley JL, Linden SK, Png CW, *et al.* MUC1 cell surface mucin is a critical element of the mucosal barrier to infection. J Clin Invest 2007; 117(8): 2313-24.
[http://dx.doi.org/10.1172/JCI26705] [PMID: 17641781]

[92] Guang W, Ding H, Czinn SJ, Kim KC, Blanchard TG, Lillehoj EP. Muc1 cell surface mucin attenuates epithelial inflammation in response to a common mucosal pathogen. J Biol Chem 2010; 285(27): 20547-57.
[http://dx.doi.org/10.1074/jbc.M110.121319] [PMID: 20430889]

[93] Ng GZ, Menheniott TR, Every AL, *et al.* The MUC1 mucin protects against Helicobacter pylori pathogenesis in mice by regulation of the NLRP3 inflammasome. Gut 2016; 65(7): 1087-99.
[http://dx.doi.org/10.1136/gutjnl-2014-307175] [PMID: 26079943]

[94] Taylor-Papadimitriou J, Burchell J, Miles DW, Dalziel M. MUC1 and cancer. Biochim Biophys Acta 1999; 1455(2-3): 301-13.
[http://dx.doi.org/10.1016/S0925-4439(99)00055-1] [PMID: 10571020]

[95] Lin DC, Genzen JR. Concordance analysis of paired cancer antigen (CA) 15-3 and 27.29 testing. Breast Cancer Res Treat 2018; 167(1): 269-76.
[http://dx.doi.org/10.1007/s10549-017-4513-0] [PMID: 28929449]

[96] Bramwell ME, Wiseman G, Shotton DM. Electron-microscopic studies of the CA antigen, epitectin. J Cell Sci 1986; 86: 249-61.
[http://dx.doi.org/10.1242/jcs.86.1.249] [PMID: 2443520]

[97] Hilkens J, Ligtenberg MJL, VosH L, Litvinov SV. The structure of cell-associated mucin-like molecules and their adhesion modulating property. Trends Biochem Sci 1992.

[98] Cloosen S, Gratama J, van Leeuwen EB, *et al.* von Mensdorff- Pouilly S, Tarp MA, Mandel U, Clausen H, Germeraad WT, Bos GM. Cancer specific Mucin-1 glycoforms are expressed on multiple myeloma. Br J Haematol 2006; 135(4): 513-6.
[http://dx.doi.org/10.1111/j.1365-2141.2006.06331.x]

[99] Baldus SE, Engelmann K, Hanisch FG. MUC1 and the MUCs: a family of human mucins with impact in cancer biology. Crit Rev Clin Lab Sci 2004; 41(2): 189-231.
[http://dx.doi.org/10.1080/10408360490452040] [PMID: 15270554]

[100] Ligtenberg MJ, Kruijshaar L, Buijs F, van Meijer M, Litvinov SV, Hilkens J. KruijshaarL,BuijsF,vanMeijerM,LitvinovSV,HilkensJ.Cell-associated episialin is a complex containing two proteins derived from a common precursor. J Biol Chem 1992; 267(9): 6171-7.
[http://dx.doi.org/10.1016/S0021-9258(18)42677-4]

[101] Wesseling J. vanderValkSW,VosHL,SonnenbergA,HilkensJ.Episialin(MUC1)overex- pression inhibits integrin-mediated cell adhesion to extracellular matrix components. J Cell Biol 1995.
[http://dx.doi.org/10.1083/jcb.129.1.255]

[102] Nan JN, Kim OR, Lee MA. β-Catenin expression is associated with cell invasiveness in pancreatic cancer. Korean J Intern Med (Korean Assoc Intern Med) 2019; 34(3): 618-25.
[http://dx.doi.org/10.3904/kjim.2017.155] [PMID: 30360026]

[103] Mrozik KM, Blaschuk OW, Cheong CM, Zannettino ACW, Vandyke K. N-cadherin in cancer metastasis, its emerging role in haematological malignancies and potential as a therapeutic target in cancer. BMC Cancer 2018; 18(1): 939.

[104] Schroeder JA, Thompson MC, Gardner MM, Gendler SJ. Transgenic MUC1 interacts with epidermal growth factor receptor and correlates with mitogen-activated protein kinase activation in the mouse

mammary gland. J Biol Chem 2001; 276(16): 13057-64.
[http://dx.doi.org/10.1074/jbc.M011248200] [PMID: 11278868]

[105] Chardin P, Camonis JH, Gale NW, *et al.* Human Sos1: a guanine nucleotide exchange factor for Ras that binds to GRB2. Science 1993; 260(5112): 1338-43.
[http://dx.doi.org/10.1126/science.8493579] [PMID: 8493579]

[106] Girling A, Bartkova J, Burchell J, Gendler S, Gillett C, Taylor-Papadimitriou J. A core protein epitope of the polymorphic epithelial mucin detected by the monoclonal antibody SM-3 is selectively exposed in a range of primary carcinomas. Int J Cancer 1989; 43(6): 1072-6.
[http://dx.doi.org/10.1002/ijc.2910430620] [PMID: 2471698]

[107] Piyush T, Rhodes JM, Yu LG. MUC1 *O*-glycosylation contributes to anoikis resistance in epithelial cancer cells. Cell Death Discov 2017; 3(1): 17044.
[http://dx.doi.org/10.1038/cddiscovery.2017.44] [PMID: 28725490]

[108] Tarp MA, Clausen H. Mucin-type O-glycosylation and its potential use in drug and vaccine development. Biochim Biophys Acta 2008; 1780(3): 546-63.
[http://dx.doi.org/10.1016/j.bbagen.2007.09.010] [PMID: 17988798]

[109] Hanisch FG, Ninkovic T. Immunology of O-glycosylated proteins: approaches to the design of a MUC1 glycopeptide-based tumor vaccine. Curr Protein Pept Sci 2006; 7(4): 307-15.
[http://dx.doi.org/10.2174/138920306778018034] [PMID: 16918445]

[110] Ju T, Cummings RD. A unique molecular chaperone Cosmc required for activity of the mammalian core 1 beta 3-galactosyltransferase. Proc Natl Acad Sci USA 2002; 99(26): 16613-8.
[http://dx.doi.org/10.1073/pnas.262438199] [PMID: 12464682]

[111] Gendler SJ. MUC1, the renaissance molecule. J Mammary Gland Biol Neoplasia 2001; 6(3): 339-53.
[http://dx.doi.org/10.1023/A:1011379725811] [PMID: 11547902]

[112] Bearz A, Talamini R, Vaccher E, *et al.* MUC-1 (CA 15-3 antigen) as a highly reliable predictor of response to EGFR inhibitors in patients with bronchioloalveolar carcinoma: an experience on 26 patients. Int J Biol Markers 2007; 22(4): 307-11.
[PMID: 18161663]

[113] Siegel R, Ward E, Brawley O, Jemal A. Cancer statistics, 2011: the impact of eliminating socioeconomic and racial disparities on premature cancer deaths. CA Cancer J Clin 2011; 61(4): 212-36.
[http://dx.doi.org/10.3322/caac.20121] [PMID: 21685461]

[114] Zak Y, Moore HG, Ghosh BC. Patterns of recurrence of colorectal cancer. J Surg Oncol 2008; 97(1): 1-2.
[http://dx.doi.org/10.1002/jso.20882] [PMID: 17680636]

[115] Thompson JA, Grunert F, Zimmermann W. Carcinoembryonic antigen gene family: molecular biology and clinical perspectives. J Clin Lab Anal 1991; 5(5): 344-66.
[http://dx.doi.org/10.1002/jcla.1860050510] [PMID: 1941355]

[116] Bird NC, Mangnall D, Majeed AW. Biology of colorectal liver metastases: A review. J Surg Oncol 2006; 94(1): 68-80.
[http://dx.doi.org/10.1002/jso.20558] [PMID: 16788948]

[117] Fusai G, Davidson BR. Management of colorectal liver metastases. Colorectal Dis 2003; 5(1): 2-23.
[http://dx.doi.org/10.1046/j.1463-1318.2003.00410.x] [PMID: 12780921]

[118] Topal B, Aerts JL, Roskams T, *et al.* Cancer cell dissemination during curative surgery for colorectal liver metastases. Eur J Surg Oncol 2005; 31(5): 506-11.
[http://dx.doi.org/10.1016/j.ejso.2005.01.007] [PMID: 15922887]

[119] Lewis CE, Pollard JW. Distinct role of macrophages in different tumor microenvironments. Cancer Res 2006; 66(2): 605-12.
[http://dx.doi.org/10.1158/0008-5472.CAN-05-4005] [PMID: 16423985]

[120] Thomas P. Carcinoembryonic Antigen. In: Shwab M, Ed. Encyclopedia of Cancer. 4th ed. Germany: Springer 2016; pp. 788-90.

[121] Thompson EW, Newgreen DF, Tarin D. Carcinoma invasion and metastasis: a role for epithelial-mesenchymal transition? Cancer Res 2005; 65(14): 5991-5. [http://dx.doi.org/10.1158/0008-5472.CAN-05-0616] [PMID: 16024595]

[122] van Nagell JR Jr, Donaldson ES, Hanson MB, Gay EC, Pavlik EJ. Biochemical markers in the plasma and tumors of patients with gynecologic malignancies. Cancer 1981; 48(2) (Suppl.): 495-503. [http://dx.doi.org/10.1002/1097-0142(19810715)48:1+<495::AID-CNCR2820481310>3.0.CO;2-F] [PMID: 6168368]

[123] Schölzel S, Zimmermann W, Schwarzkopf G, Grunert F, Rogaczewski B, Thompson J. Carcinoembryonic antigen family members CEACAM6 and CEACAM7 are differentially expressed in normal tissues and oppositely deregulated in hyperplastic colorectal polyps and early adenomas. Am J Pathol 2000; 156(2): 595-605. [http://dx.doi.org/10.1016/S0002-9440(10)64764-5] [PMID: 10666389]

[124] Blumenthal RD, Hansen HJ, Goldenberg DM. Inhibition of adhesion, invasion, and metastasis by antibodies targeting CEACAM6 (NCA-90) and CEACAM5 (Carcinoembryonic Antigen). Cancer Res 2005; 65(19): 8809-17. [http://dx.doi.org/10.1158/0008-5472.CAN-05-0420] [PMID: 16204051]

[125] Hostetter RB, Augustus LB, Mankarious R, *et al.* Carcinoembryonic antigen as a selective enhancer of colorectal cancer metastasis. J Natl Cancer Inst 1990; 82(5): 380-5. [http://dx.doi.org/10.1093/jnci/82.5.380] [PMID: 2304087]

[126] Jessup JM, Thomas P. CEA and metastasis: a facilitator of site specific metastasis. In: Stanners C, Ed. Cell adhesion and communication mediated by the CEA family: basic and clinical perspectives. Amsterdam, Netherlands: Harwood Academic Publishers 1998; pp. 195-222.

[127] Solakoglu O, Maierhofer C, Lahr G, *et al.* Heterogeneous proliferative potential of occult metastatic cells in bone marrow of patients with solid epithelial tumors. Proc Natl Acad Sci USA 2002; 99(4): 2246-51. [http://dx.doi.org/10.1073/pnas.042372199] [PMID: 11854519]

[128] West NP, Dattani M, McShane P, *et al.* The proportion of tumour cells is an independent predictor for survival in colorectal cancer patients. Br J Cancer 2010; 102(10): 1519-23. [http://dx.doi.org/10.1038/sj.bjc.6605674] [PMID: 20407439]

[129] Sunderkötter C, Goebeler M, Schulze-Osthoff K, Bhardwaj R, Sorg C. Macrophage-derived angiogenesis factors. Pharmacol Ther 1991; 51(2): 195-216. [http://dx.doi.org/10.1016/0163-7258(91)90077-Y] [PMID: 1784630]

[130] Lewis CE, Leek R, Harris A, McGee JO. Cytokine regulation of angiogenesis in breast cancer: the role of tumor-associated macrophages. J Leukoc Biol 1995; 57(5): 747-51. [http://dx.doi.org/10.1002/jlb.57.5.747] [PMID: 7539028]

[131] Bajenova OV, Zimmer R, Stolper E, *et al.* HnRNP M4 is a receptor for carcinoembryonic antigen in Kupffer cells. J Biol Chem 2001; 276: 31067-73. [http://dx.doi.org/10.1074/jbc.M104093200] [PMID: 11406629]

[132] Gangopadhyay A, Bajenova O, Kelly TM, Thomas P. Carcinoembryonic antigen induces cytokine expression in Kuppfer cells: implications for hepatic metastasis from colorectal cancer. Cancer Res 1996; 56(20): 4805-10. [PMID: 8841002]

[133] Edmiston KH, Gangopadhyay A, Shoji Y, Nachman AP, Thomas P, Jessup JM. *In vivo* induction of murine cytokine production by carcinoembryonic antigen. Cancer Res 1997; 57(19): 4432-6. [PMID: 9331108]

[134] Minami S, Furui J, Kanematsu T. Role of carcinoembryonic antigen in the progression of colon cancer

cells that express carbohydrate antigen. Cancer Res 2001; 61(6): 2732-5. [PMID: 11289155]

[135] Gangopadhyay A, Lazure DA, Thomas P. Adhesion of colorectal carcinoma cells to the endothelium is mediated by cytokines from CEA stimulated Kupffer cells. Clin Exp Metastasis 1998; 16(8): 703-12. [http://dx.doi.org/10.1023/A:1006576627429] [PMID: 10211983]

[136] Jessup JM, Samara R, Battle P, Laguinge LM. Carcinoembryonic antigen promotes tumor cell survival in liver through an IL-10-dependent pathway. Clin Exp Metastasis 2004; 21(8): 709-17. [http://dx.doi.org/10.1007/s10585-004-7705-z] [PMID: 16035616]

[137] Jessup JM, Laguinge L, Lin S, *et al.* Carcinoembryonic antigen induction of IL-10 and IL-6 inhibits hepatic ischemic/reperfusion injury to colorectal carcinoma cells. Int J Cancer 2004; 111(3): 332-7. [http://dx.doi.org/10.1002/ijc.20264] [PMID: 15221959]

[138] Kokkonen N, Ulibarri IF, Kauppila A, *et al.* Hypoxia upregulates carcinoembryonic antigen expression in cancer cells. Int J Cancer 2007; 121(11): 2443-50. [http://dx.doi.org/10.1002/ijc.22965] [PMID: 17657737]

[139] Eng LF, Ghirnikar RS, Lee YL. Glial fibrillary acidic protein: GFAP-thirty-one years (1969-2000). Neurochem Res 2000; 25(9-10): 1439-51. [http://dx.doi.org/10.1023/A:1007677003387] [PMID: 11059815]

[140] Yang Z, Wang KK. Glial fibrillary acidic protein: from intermediate filament assembly and gliosis to neurobiomarker. Trends Neurosci 2015; 38(6): 364-74. [http://dx.doi.org/10.1016/j.tins.2015.04.003] [PMID: 25975510]

[141] Uyeda CT, Eng LF, Bignami A. Immunological study of the glial fibrillary acidic protein. Brain Res 1972; 37(1): 81-9. [http://dx.doi.org/10.1016/0006-8993(72)90347-2] [PMID: 4621897]

[142] Bignami A, Eng LF, Dahl D, Uyeda CT. Localization of the glial fibrillary acidic protein in astrocytes by immunofluorescence. Brain Res 1972; 43(2): 429-35. [http://dx.doi.org/10.1016/0006-8993(72)90398-8] [PMID: 4559710]

[143] Middeldorp J, Hol EM. GFAP in health and disease. Prog Neurobiol 2011; 93(3): 421-43. [http://dx.doi.org/10.1016/j.pneurobio.2011.01.005] [PMID: 21219963]

[144] Sofroniew MV, Vinters HV. Astrocytes: biology and pathology. Acta Neuropathol 2010; 119(1): 7-35. [http://dx.doi.org/10.1007/s00401-009-0619-8] [PMID: 20012068]

[145] Hol EM, Pekny M. Glial fibrillary acidic protein (GFAP) and the astrocyte intermediate filament system in diseases of the central nervous system. Curr Opin Cell Biol 2015; 32: 121-30. [http://dx.doi.org/10.1016/j.ceb.2015.02.004] [PMID: 25726916]

[146] Petzold A. Glial fibrillary acidic protein is a body fluid biomarker for glial pathology in human disease. Brain Res 2015; 1600: 17-31. [http://dx.doi.org/10.1016/j.brainres.2014.12.027] [PMID: 25543069]

[147] Dahl D, Chi NH, Miles LE, Nguyen BT, Bignami A. Glial fibrillary acidic (GFA) protein in Schwann cells: fact or artifact? J Histochem Cytochem 1982; 30(9): 912-8. [http://dx.doi.org/10.1177/30.9.6182187] [PMID: 6182187]

[148] Suarez-Mier GB, Buckwalter MS. Glial fibrillary acidic protein-expressing glia in the mouse lung. ASN Neuro 2015; 7(5): 1759091415601636. [http://dx.doi.org/10.1177/1759091415601636] [PMID: 26442852]

[149] Chen MH, Hagemann TL, Quinlan RA, Messing A, Perng MD. Caspase cleavage of GFAP produces an assembly-compromised proteolytic fragment that promotes filament aggregation. ASN Neuro 2013; 5(5): e00125. [http://dx.doi.org/10.1042/AN20130032] [PMID: 24102621]

[150] Pekny M, Levéen P, Pekna M, *et al.* Mice lacking glial fibrillary acidic protein display astrocytes devoid of intermediate filaments but develop and reproduce normally. EMBO J 1995; 14(8): 1590-8. [http://dx.doi.org/10.1002/j.1460-2075.1995.tb07147.x] [PMID: 7737111]

[151] Liedtke W, Edelmann W, Bieri PL, *et al.* GFAP is necessary for the integrity of CNS white matter architecture and long-term maintenance of myelination. Neuron 1996; 17(4): 607-15. [http://dx.doi.org/10.1016/S0896-6273(00)80194-4] [PMID: 8893019]

[152] Liedtke W, Edelmann W, Chiu FC, Kucherlapati R, Raine CS. Experimental autoimmune encephalomyelitis in mice lacking glial fibrillary acidic protein is characterized by a more severe clinical course and an infiltrative central nervous system lesion. Am J Pathol 1998; 152(1): 251-9. [PMID: 9422542]

[153] Tardy M, Fages C, Le Prince G, Rolland B, Nunez J. Regulation of the glial fibrillary acidic protein (GFAP) and of its encoding mRNA in the developing brain and in cultured astrocytes. Adv Exp Med Biol 1990; 265: 41-52. [http://dx.doi.org/10.1007/978-1-4757-5876-4_4] [PMID: 2165732]

[154] Hirsch EC, Breidert T, Rousselet E, Hunot S, Hartmann A, Michel PP. The role of glial reaction and inflammation in Parkinson's disease. Ann N Y Acad Sci 2003; 991(1): 214-28. [http://dx.doi.org/10.1111/j.1749-6632.2003.tb07478.x] [PMID: 12846989]

[155] Goodison KL, Parhad IM, White CL III, Sima AA, Clark AW. Neuronal and glial gene expression in neocortex of Down's syndrome and Alzheimer's disease. J Neuropathol Exp Neurol 1993; 52(3): 192-8. [http://dx.doi.org/10.1097/00005072-199305000-00002] [PMID: 8492138]

[156] Ross GW, O'Callaghan JP, Sharp DS, *et al.* Quantification of regional glial fibrillary acidic protein levels in Alzheimer's disease. Acta Neurol Scand 2003; 107(5): 318-23. [http://dx.doi.org/10.1034/j.1600-0404.2003.02098.x] [PMID: 12713522]

[157] Pekny M, Eliasson C, Chien CL, *et al.* GFAP-deficient astrocytes are capable of stellation *in vitro* when cocultured with neurons and exhibit a reduced amount of intermediate filaments and an increased cell saturation density. Exp Cell Res 1998; 239(2): 332-43. [http://dx.doi.org/10.1006/excr.1997.3922] [PMID: 9521851]

[158] Sugaya-Fukasawa M, Watanabe T, Tamura M, Egashira S, Hisatomi H. Glial fibrillary acidic protein is one of the key factors underlying neuron-like elongation in PC12 cells. Exp Ther Med 2011; 2(1): 85-7. [http://dx.doi.org/10.3892/etm.2010.162] [PMID: 22977474]

[159] Eng LF, Ghirnikar RS. GFAP and astrogliosis. Brain Pathol 1994; 4(3): 229-37. [http://dx.doi.org/10.1111/j.1750-3639.1994.tb00838.x] [PMID: 7952264]

[160] Sofroniew MV. Astrogliosis. Cold Spring Harb Perspect Biol 2014; 7(2): a020420. [http://dx.doi.org/10.1101/cshperspect.a020420] [PMID: 25380660]

[161] Pekny M, Wilhelmsson U, Pekna M. The dual role of astrocyte activation and reactive gliosis. Neurosci Lett 2014; 565: 30-8. [http://dx.doi.org/10.1016/j.neulet.2013.12.071] [PMID: 24406153]

[162] Jurado-Arjona J, Rodríguez-Matellán A, Ávila J, Hernández F. GSK3β overexpression driven by GFAP promoter improves rotarod performance. Brain Res 2019; 1712: 47-54. [http://dx.doi.org/10.1016/j.brainres.2019.01.040] [PMID: 30711399]

[163] Panter SS, McSwigan JD, Sheppard JR, Emory CR, Frey WH II. Glial fibrillary acidic protein and Alzheimer's disease. Neurochem Res 1985; 10(12): 1567-76. [http://dx.doi.org/10.1007/BF00988599] [PMID: 4088432]

[164] Levi G, Patrizio M, Bernardo A, Petrucci TC, Agresti C. Human immunodeficiency virus coat protein gp120 inhibits the beta-adrenergic regulation of astroglial and microglial functions. Proc Natl Acad Sci USA 1993; 90(4): 1541-5.

[http://dx.doi.org/10.1073/pnas.90.4.1541] [PMID: 8381971]

[165] Kennedy PG, Major EO, Williams RK, Straus SE. Down-regulation of glial fibrillary acidic protein expression during acute lytic varicella-zoster virus infection of cultured human astrocytes. Virology 1994; 205(2): 558-62.
[http://dx.doi.org/10.1006/viro.1994.1679] [PMID: 7975257]

[166] Danzer E, Zhang L, Radu A, *et al.* Amniotic fluid levels of glial fibrillary acidic protein in fetal rats with retinoic acid induced myelomeningocele: a potential marker for spinal cord injury Am J Obstet Gynecol 2011; 204(2 178): e1-11.
[http://dx.doi.org/10.1016/j.ajog.2010.09.032]

[167] Moore BW, McGregor D. Chromatographic and electrophoretic fractionation of soluble proteins of brain and liver. J Biol Chem 1965; 240(4): 1647-53.
[http://dx.doi.org/10.1016/S0021-9258(18)97483-1] [PMID: 14285503]

[168] Pickel VM, Reis DJ, Marangos PJ, Zomzely-Neurath C. Immunocytochemical localization of nervous system specific protein (NSP-R) in rat brain. Brain Res 1976; 105(1): 184-7.
[http://dx.doi.org/10.1016/0006-8993(76)90936-7] [PMID: 1252955]

[169] Bock E, Dissing J. Demonstration of enolase activity connected to the brainspecific protein 14-3- 2. Scand J Immunol 1975; 4 (Suppl. 2): 31-6.
[http://dx.doi.org/10.1111/j.1365-3083.1975.tb03806.x]

[170] Marangos PJ, Zomzely-Neurath C, York C. Determination and characterization of neuron specific protein (NSP) associated enolase activity. Biochem Biophys Res Commun 1976; 68(4): 1309-16.
[http://dx.doi.org/10.1016/0006-291X(76)90339-9] [PMID: 5081]

[171] Wold F, Ballou CE. Studies on the enzyme enolase. II. Kinetic studies. J Biol Chem 1957; 227(1): 313-28.
[http://dx.doi.org/10.1016/S0021-9258(18)70817-X] [PMID: 13449075]

[172] Brewer JM. Specificity and mechanism of action of metal ions in yeast enolase. FEBS Lett 1985; 182(1): 8-14.
[http://dx.doi.org/10.1016/0014-5793(85)81143-1]

[173] Hanlon DP, Westhead EW. Kinetic studies on the activation of yeast enolase by divalent cations. Biochemistry 1969; 8(11): 4255-60.
[http://dx.doi.org/10.1021/bi00839a005] [PMID: 5353098]

[174] Brewer JM. The increase of yeast enolase fluorescence produced by substrates and competitive inhibitors in the presence of excess Mg 2+. Biochim Biophys Acta 1971; 250(1): 251-7.
[http://dx.doi.org/10.1016/0005-2744(71)90140-9] [PMID: 5141676]

[175] Pancholi V. Multifunctional alpha-enolase: its role in diseases. Cell Mol Life Sci 2001; 58(7): 902-20.
[http://dx.doi.org/10.1007/PL00000910] [PMID: 11497239]

[176] Marangos PJ, Zis AP, Clark RL, Goodwin FK. Neuronal, non-neuronal and hybrid forms of enolase in brain: structural, immunological and functional comparisons. Brain Res 1978; 150(1): 117-33.
[http://dx.doi.org/10.1016/0006-8993(78)90657-1] [PMID: 78744]

[177] Sørensen AT, Ledri M, Melis M, Nikitidou Ledri L, Andersson M, Kokaia M. Published 2017; 2018(Jan): 23.

[178] Marangos PJ, Schmechel DE. Neuron specific enolase, a clinically useful marker for neurons and neuroendocrine cells. Annu Rev Neurosci 1987; 10(1): 269-95.
[http://dx.doi.org/10.1146/annurev.ne.10.030187.001413] [PMID: 3551759]

[179] Yip J, Geng X, Shen J, Ding Y. Cerebral Gluconeogenesis and Diseases. Front Pharmacol 2017; 7: 521.
[http://dx.doi.org/10.3389/fphar.2016.00521]

[180] Chai G, Brewer JM, Lovelace LL, Aoki T, Minor W, Lebioda L. Expression, purification and the 1.8

angstroms resolution crystal structure of human neuron specific enolase. J Mol Biol 2004; 341(4): 1015-21.
[http://dx.doi.org/10.1016/j.jmb.2004.05.068] [PMID: 15289101]

[181] Nobels FR, Kwekkeboom DJ, Coopmans W, *et al.* Chromogranin A as serum marker for neuroendocrine neoplasia: comparison with neuron-specific enolase and the alpha-subunit of glycoprotein hormones. J Clin Endocrinol Metab 1997; 82(8): 2622-8.
[PMID: 9253344]

[182] Eriksson B, Oberg K, Stridsberg M. Tumor markers in neuroendocrine tumors. Digestion 2000; 62 (Suppl. 1): 33-8.
[http://dx.doi.org/10.1159/000051853] [PMID: 10940685]

[183] Baudin E, Gigliotti A, Ducreux M, *et al.* Neuron-specific enolase and chromogranin A as markers of neuroendocrine tumours. Br J Cancer 1998; 78(8): 1102-7.
[http://dx.doi.org/10.1038/bjc.1998.635] [PMID: 9792158]

[184] Zeltzer PM, Marangos PJ, Parma AM, *et al.* Raised neuron-specific enolase in serum of children with metastatic neuroblastoma. A report from the Children's Cancer Study Group. Lancet 1983; 2(8346): 361-3.
[http://dx.doi.org/10.1016/S0140-6736(83)90342-2] [PMID: 6135871]

[185] Zaheer S, Beg M, Rizvi I, Islam N, Ullah E, Akhtar N. Correlation between serum neuron specific enolase and functional neurological outcome in patients of acute ischemic stroke. Ann Indian Acad Neurol 2013; 16(4): 504-8.
[http://dx.doi.org/10.4103/0972-2327.120442] [PMID: 24339568]

[186] Moore BW. A soluble protein characteristic of the nervous system. Biochem Biophys Res Commun 1965; 19(6): 739-44.
[http://dx.doi.org/10.1016/0006-291X(65)90320-7] [PMID: 4953930]

[187] Donato R, Sorci G, Riuzzi F, *et al.* S100B's double life: intracellular regulator and extracellular signal. Biochim Biophys Acta 2009; 1793(6): 1008-22.
[http://dx.doi.org/10.1016/j.bbamcr.2008.11.009] [PMID: 19110011]

[188] Steiner J, Bernstein HG, Bielau H, *et al.* Evidence for a wide extra-astrocytic distribution of S100B in human brain. BMC Neurosci 2007; 8(1): 2.
[http://dx.doi.org/10.1186/1471-2202-8-2] [PMID: 17199889]

[189] Marenholz I, Heizmann CW, Fritz G. S100 proteins in mouse and man: from evolution to function and pathology (including an update of the nomenclature). Biochem Biophys Res Commun 2004; 322(4): 1111-22.
[http://dx.doi.org/10.1016/j.bbrc.2004.07.096] [PMID: 15336958]

[190] Haimoto H, Hosoda S, Kato K. Differential distribution of immunoreactive S100-alpha and S100-beta proteins in normal nonnervous human tissues. Lab Invest 1987; 57(5): 489-98.
[PMID: 3316838]

[191] Isobe T, Ishioka N, Masuda T, Takahashi Y, Ganno S, Okuyama T. A rapid separation of S100 subunits by high performance liquid chromatography: the subunit compositions of S100 proteins. Biochem Int 1983; 6(3): 419-26.
[PMID: 6679332]

[192] Ghanem G, Loir B, Morandini R, *et al.* On the release and half-life of S100B protein in the peripheral blood of melanoma patients. Int J Cancer 2001; 94(4): 586-90.
[http://dx.doi.org/10.1002/ijc.1504] [PMID: 11745448]

[193] Schiavi P, Laccarino C, Servadei F. The value of the calcium binding protein S100 in the management of patients with traumatic brain injury. Acta Biomed 2012; 83(1): 5-20.
[PMID: 22978053]

[194] Schäfer BW, Heizmann CW. The S100 family of EF-hand calcium-binding proteins: functions and

pathology. Trends Biochem Sci 1996; 21(4): 134-40.
[http://dx.doi.org/10.1016/S0968-0004(96)80167-8] [PMID: 8701470]

[195] Brewton LS, Haddad L, Azmitia EC. Colchicine-induced cytoskeletal collapse and apoptosis in N-18 neuroblastoma cultures is rapidly reversed by applied S-100beta. Brain Res 2001; 912(1): 9-16.
[http://dx.doi.org/10.1016/S0006-8993(01)02519-7] [PMID: 11520488]

[196] Haglid KG, Yang Q, Hamberger A, Bergman S, Widerberg A, Danielsen N. S-100beta stimulates neurite outgrowth in the rat sciatic nerve grafted with acellular muscle transplants. Brain Res 1997; 753(2): 196-201.
[http://dx.doi.org/10.1016/S0006-8993(96)01463-1] [PMID: 9125403]

[197] Kleindienst A, Grünbeck F, Buslei R, Emtmann I, Buchfelder M. Intraperitoneal treatment with S100B enhances hippocampal neurogenesis in juvenile mice and after experimental brain injury. Acta Neurochir (Wien) 2013; 155(7): 1351-60.
[http://dx.doi.org/10.1007/s00701-013-1720-2] [PMID: 23649988]

[198] Rothermundt M, Peters M, Prehn JH, Arolt V. S100B in brain damage and neurodegeneration. Microsc Res Tech 2003; 60(6): 614-32.
[http://dx.doi.org/10.1002/jemt.10303] [PMID: 12645009]

[199] Van Eldik LJ, Wainwright MS. The Janus face of glial-derived S100B: beneficial and detrimental functions in the brain. Restor Neurol Neurosci 2003; 21(3-4): 97-108.
[PMID: 14530573]

[200] Hu J, Ferreira A, Van Eldik LJ. S100beta induces neuronal cell death through nitric oxide release from astrocytes. J Neurochem 1997; 69(6): 2294-301.
[http://dx.doi.org/10.1046/j.1471-4159.1997.69062294.x] [PMID: 9375660]

[201] Koppal T, Lam AG, Guo L, Van Eldik LJ. S100B proteins that lack one or both cysteine residues can induce inflammatory responses in astrocytes and microglia. Neurochem Int 2001; 39(5-6): 401-7.
[http://dx.doi.org/10.1016/S0197-0186(01)00047-X] [PMID: 11578775]

[202] Gerlach R, Demel G, König HG, *et al.* Active secretion of S100B from astrocytes during metabolic stress. Neuroscience 2006; 141(4): 1697-701.
[http://dx.doi.org/10.1016/j.neuroscience.2006.05.008] [PMID: 16782283]

[203] Hinkle DA, Baldwin SA, Scheff SW, Wise PM. GFAP and S100beta expression in the cortex and hippocampus in response to mild cortical contusion. J Neurotrauma 1997; 14(10): 729-38.
[http://dx.doi.org/10.1089/neu.1997.14.729] [PMID: 9383091]

[204] Sorci G, Bianchi R, Riuzzi F, *et al.* S100B Protein, A Damage-Associated Molecular Pattern Protein in the Brain and Heart, and Beyond. Cardiovasc Psychiatry Neurol 2010; 2010: 656481.
[http://dx.doi.org/10.1155/2010/656481] [PMID: 20827421]

[205] Ingebrigtsen T, Romner B, Kongstad P, Langbakk B. Increased serum concentrations of protein S-100 after minor head injury: a biochemical serum marker with prognostic value? J Neurol Neurosurg Psychiatry 1995; 59(1): 103-4.

[206] Raabe A, Grolms C, Keller M, Döhnert J, Sorge O, Seifert V. Correlation of computed tomography findings and serum brain damage markers following severe head injury. Acta Neurochir (Wien) 1998; 140(8): 787-91.
[http://dx.doi.org/10.1007/s007010050180] [PMID: 9810445]

[207] Romner B, Ingebrigtsen T, Kongstad P, Børgesen SE. Traumatic brain damage: serum S-100 protein measurements related to neuroradiological findings. J Neurotrauma 2000; 17(8): 641-7.
[http://dx.doi.org/10.1089/089771500415391] [PMID: 10972241]

[208] Undén J, Bellner J, Astrand R, Romner B. Serum S100B levels in patients with epidural haematomas. Br J Neurosurg 2005; 19(1): 43-5.
[http://dx.doi.org/10.1080/02688690500089381] [PMID: 16147582]

[209] Marmarou A. Pathophysiology of traumatic brain edema: current concepts. Acta Neurochir Suppl

(Wien) 2003; 86: 7-10.
[http://dx.doi.org/10.1007/978-3-7091-0651-8_2] [PMID: 14753394]

[210] Blyth BJ, Farahvar A, He H, *et al.* Elevated serum ubiquitin carboxy-terminal hydrolase L1 is associated with abnormal blood-brain barrier function after traumatic brain injury. J Neurotrauma 2011; 28(12): 2453-62.
[http://dx.doi.org/10.1089/neu.2010.1653] [PMID: 21428722]

[211] Lopez NE, Krzyzaniak MJ, Blow C, *et al.* Ghrelin prevents disruption of the blood-brain barrier after traumatic brain injury. J Neurotrauma 2012; 29(2): 385-93.
[http://dx.doi.org/10.1089/neu.2011.2053] [PMID: 21939391]

[212] Pleines UE, Morganti-Kossmann MC, Rancan M, Joller H, Trentz O, Kossmann T. S-100 beta reflects the extent of injury and outcome, whereas neuronal specific enolase is a better indicator of neuroinflammation in patients with severe traumatic brain injury. J Neurotrauma 2001; 18(5): 491-8.
[http://dx.doi.org/10.1089/089771501300227297] [PMID: 11393252]

Biomarkers in Medicine, 2022, 276-308

CHAPTER 11

Biomarkers in Otorhinolaryngology

Ayse Pelin Yigider[1,*] and **Ozgur Yigit**[1]

[1] *Istanbul Research and Training Hospital Otorhinolaryngology, Istanbul, Turkey*

Abstract: Biomarkers of otorhinolaryngologic diseases with higher insult over a person's him/herself and overall health services are summarized in brief. In order to define, diagnose, treat and monitor any disease markers are needed. Otorhinolaryngology (ORL) is interested in special disease entities of the region besides otorhinolaryngologic involvements of the systemic diseases and unique forms of pathologies such as cholesteatoma, Meniere's disease and otosclerosis. Neoplasia is another heading to deal with. In the following chapter, one will find an overview of molecules that have been used as a biomarker as well as the end points of the present research on the issue relevant with ORL. Day by day, new molecules are being named however, the pathways of action are rather the same. Readers will find the headings related to the most common diseases of the field, informing them about where to look for defining new strategies of understanding of each disease.

Keywords: Biomarkers, laryngology, Biomarkers, otology, Biomarkers, rhinology, Biomarkers, thyroid.

INTRODUCTION

Biomarkers are used to diagnose, define the pathophysiology, determine therapeutic alternatives and monitor the outcome by means of both curative attempts and natural history of the disease, which may overall be named as prognosis. In the present chapter, biomarkers of otorhinolaryngologic diseases that have more insult over a person's him/herself and overall health services are summarized in brief. Protein can be monitored *via* their code/decode patterns in the course of a specific disease, namely fragmented RNA particles, mRNA, or DNA's. Owing to the higher cost of the genetic laboratory, it seems more feasible to define biochemical reflections of protein synthesis. It is not rational to list any diseases, but rather frequent diseases are tried to be given with their future-lightening markers by means of diagnosis, disease monitoring and prognosis.

* **Corresponding author Ayşe Pelin Yigider:** Istanbul Research and Training Hospital, Department of Otorhinolaryngology Head and Neck Surgery, İstanbul. Turkey; Tel: +90 (532) 410 90 25; E-mail: pelin.gor@gmail.com

Pınar Atukeren and Hafize Uzun (Eds.)

Rhinology

In rhinology practice, infections, namely sinusitis acute of chronic, space occupying benign masses of sino-nasal tract polys or papilloma, are frequent. Main underlying pathology emerges as allergy. Immune reactions of any type can be monitored *via* re-popularized hematological parameters such as PMN/Lymphocyte ratio, Platelet number or platelet volume, however most of those are far from being specific for an illness. They can be monitored as an adjunctive measure but cannot discriminate a confounding simultaneous inflammatory process from the primary disease of concern.

Sinusitis is basically divided into two as acute or chronic. While acute sinusitis is an infection caused by mainly bacteria, chronic sinusitis is a complex disease entity related to various inflammatory pathways. CRS is classified into CRS with nasal polyps (CRSwNP) and CRS without nasal polyps (CRSsNP). CRS, especially CRSwNP, is further subdivided into eosinophilic CRS (ECRS) and non-eosinophilic CRS. According to the diagnostic criteria established by a multicenter study called the Japan Epidemiological Survey of Refractory Eosinophilic Chronic Rhinosinusitis Study (JESREC Study) in 2015, JESREC score and mucosal eosinophil count have been used to diagnose eosinophilic chronic rhinosinusitis (ECRS) [1].

TH2 inflammation was formerly reported to predominate which is associated with elevated levels of eosinophils and cytokines, including Interleukin (IL)-4, IL-5, and IL-13 [2].

The testing mediums to obtain CRS biomarkers are peripheral blood, nasal secretions and nasal lavage, tissue biopsies from the sinuses, and nasally exhaled breath. CRSwNP is an eosinophilic process and CRSsNP is noneosinophilic. In different parts of the world, eosinophilia may be absent in the pathogenesis of CRSwNP, yet if present is generally associated with higher disease severity. Additionally, eosinophilic polyps are generally more glucocorticoid responsive in comparison to non-eosinophilic counterparts.

IgE, as an inducer of eosinophilia, is utilized as a biomarker for CRS phenotyping. Patients with CRS frequently express high levels of IgE in both serum and local secretions. Allergic sensitization to bacterial antigens is believed to be an important part of the process. Staphylococcus aureus enterotoxins can serve to activate basophil degranulation in those with CRSwNP. Increased local IgE is predictive of recurrence requiring repetitive surgical intervention. The omalizumab, an anti-IgE therapeutic, was also tested and provided benefit up to some extent in patients with CRSwNP. Cytokine profile of CRS w/s NP differs dramatically. In CRSsNP, neutrophilia dominates with the expression of

Transforming Growth Factor β (TGF-β), type I interferons, and IL-6, IL-8, or IL-17. CRSwNP has a TH2 leading microenvironment with marked expression of thymic stromal lymphoprotein (TSLP) and type 2 inflammatory cells. Type 2 Innate Lymphoid (ILC-2) cells, which produce IL-4, IL-5, IL-13, IL-25, and/or IL-33, activate T and B cells that overall result in increased leaking of the epithelial barrier. Evidence nonetheless points to elevated IL-4 playing a role in the overexpression of cysteinyl leukotriene [13]. CD45+ CD4- IL-4 producing cells were found to be elevated in patients with AERD (asprin exacerbating resp disease) [3].

There can be an overlap of cytokine expressions of both TH 1 and 2 pathways. The presence of IL-5 may be defined for a specific endotype of CRSwNP individuals with more severe disease. IL-25 expression is upregulated in CRSwNP tissue is associated with increased severity in CT scores. Anti IL-4,5,13, 25 are used in several trials and provided unneglectable outcomes. Eotaxin-3 as a plasma biomarker for mucosal eosinophil infiltration is closely related to IL-33 and TSLP levels which indicate respiratory diseases [4]. Eotaxin-3 and total IgE in the nasal secretions of patients receiving dupilumab, a monoclonal IgG4 antibody against the alpha subunit of the IL-4 receptor, decreased in comparison to the placebo group, while eosinophilic cathionic protein (ECP) did not decrease significantly. Tissue analysis of nasal polyp revealed that while total IgE, ECP, eotaxin-2, eotaxin-3, IL-13 and PARC are significantly lowered in the dupilumab group; IL-6, IL-1β, IL-4, eotaxin-1, IL-5, IL-10, IL-17, IL-33, TNF-α or TARC are not significantly different.

Longitudinal tracking of biomarkers is important to objectively measure therapeutic efficacy. Cytokines have local and indirect systemic effects of long term. IL 4 and IL-13 has a direct effect on Periostin, that is an extracellular protein taking place in subepithelial fibrosis in airway tract through interacting with integrins. and working with VEGF, resulting in angiogenesis. It was shown that Periostin levels are treatment responsive as the protein works with tissue remodeling factors. IL-16 is an eosinophilic chemotaxis that promotes persistent eosinophil activation with increased levels of serum IL-16 is significantly elevated in eosinophilic CRS [5].

An ATP-dependent transmembrane efflux pump, P-glycoprotein is upregulated in TH2 disease that is thought to take place in CRS-related inflammation through promotion of cytokine secretion. Verapamil is an antagonist of p- glycoprotein, and preliminary trials show that low dose verapamil therapy is safe and effective for CRSwNP treatment [6]. Autoimmunity seems to play a greater role. CXCL-12 and CXCL-13 molecules are present with increased levels, which enhance B-cell chemotaxis. Anti-dsDNA IgG antibody is correlated with a worse clinic. The presence of ANCA in CRS patients may be associated with a poor prognosis [7]. Disturbed nasal gas exchange results in changed concentrations of exhaled air for certain elements. Patients with CRSsNP also have decreased nNO than normal, but significantly higher than those observed in CRSwNP. Exhaled oral nitric oxide is currently used as a marker of asthma control – with lower levels indicating better disease control [8, 9].

Taste receptors are of great importance in inflammatory cascades. When bitter taste receptors on ciliated cells are stimulated, a downstream of antimicrobial nitric oxide response emerges providing direct bacterial killing and improvement in ciliary beat frequency. Patients with a polymorphism in a specific taste receptor, T2R38, have poorer outcomes following surgical intervention. Sweet taste receptors, however, inhibit the antimicrobial cascade in solitary chemosensory cells. Because the bacteria consume sugar rapidly, depletion of airway surface liquid glucose is a sign of bacterial infection. As in diabetic patients, patients with CRS are known to have elevated airway glucose levels. This is because CRS patients with increased airway glucose level carry more gram-negative bacteria, including P. aeruginosa [10].

Microbiome of airway is important and in certain states a dysbiosis takes place damaging the microbial equilibrium. As the number of types of bacteria decreases while the number of overall bacteria increases. CRSwNP are known to be colonized with Staphylococcus aureus, and IgE antibodies to S. aureus enterotoxins are frequently found in diseased tissue specimens. Antimicrobial compounds including lysozyme, S100 proteins, and β-defensins all are decreased in CRSwNP patients compared to matched controls that underlies the dysbiosis.

Genetics keep an important place in the pathogenesis of CRS. It seems differences in gene expression patterns are superior to genetic polymorphism. Transcription factors like Foxp3, SOCS3 that are involved in T cell regulation were shown to be altered. Mucin 5B, lipocalin-1, mucin 5AC, and arfinase-1are the proteins that elevates specifically in patients with nasal polyps. Inducible regulatory T cells (iTreg), namely IL-10^+Foxp3^+CD4^+CD25^+ T cells, are upregulated during the subsequent pollen season after immunotherapy and the extent of upregulation is correlated with the clinical improvements.

Post-immunotherapy serum activities of certain factors, including IgE-blocking factor (IgE-BF) and IgE- facilitated allergen binding (IgE-FAB) imply potential to predict clinical response. It was reported that sIgE/tIgE ratio before immunotherapy has potential to predict the final efficacy of therapy. IgE antibody-specific activity may have direct influence on the extent of degranulation of effector cells such as basophils and mast cells in response to allergen exposure. As the ratio of sIgE/tIgE increases, production of allergen-specific IgG4 (sIgG4) is induced and then result in clinical improvements [11].

Inflammatory processes end up with tissue remodeling and one of the main elements for this is a group of proteases matrix metalloproteinases (MMP). Doxycycline through its anti-MMP effects provides decrease in polyp size in the CRSwNP. Macrolides in general are known with their anti-inflammatory properties which mainly cover neutrophilic processes. An IL-6 related molecule, Oncostatin M, is found to be elevated in CRSwNP and causes increased tissue permeability *in vitro* [12].

FcγIIB, expressed on the surface of human basophils, is a kind of inhibitory low-affinity IgG receptor. FcγIIB can prevent IgG immune complexes from activating basophils and inhibit IgE-induced responses by co- engagement with FcεRI. Therefore IgG-depleted serum lost the ability to decrease basophil sensitivity. Dendritic cells have antigen presenting properties and DC-associated markers reflecting downregulation of pro-allergic response (CD141, GATA3 and receptor-interacting serine/ threonine-protein kinase 4 (RIPK4)) and upregulation of regulatory responses are (complement C1q subcomponent subunit A (C1QA) and FcγRIIIA) can differentiate clinical responders and non-responders for immunotherapy [13].

Sinonasal Inverted Papilloma and Sinonasal Malignancies

Sinonasal inverted papillomas (SNIP) are locally aggressive epithelial lesions with high recurrence rates that may transform into malignancies. Human papilloma virus infection plays an important role in inverted papilloma pathogenesis.

The oncogenicity of high-risk HPV in head and neck cancer have been studied and well described. HPV types 6/11 are more frequent in sinonasal inverted papillomas that are associated with low risk. High risk types are HPV 16/18 are also detected in SNIP. HPV positive tumor cells are dependent on E6 and E7 expression for their malignant potential. Among proteins induced by HPV mediated transcription, E6 and E7 are of particular interest because the ability of E6 to ubiquitination of p53. Owing to the E6 mediated- mechanism, levels of p53 and p21 are decreased in IPs with high risk HPV compared to the IPs without HPV infection. Since decreased p53 levels appear to be a hallmark of high-risk HPV induced tumors [14]. HPV increases expression of EGFR *via* either induced transcription by E6 or altered recycling by E5. Increased expression of EGFR correlates with high grade dysplasia and malignancy. E7 bind Rb a tumor suppressor protein and inhibits its activity resulting in progression of cell cycle from G1 to S phase.

The normal feedback inhibitory protein of this pathway $p16^{INK4a}$ is typically overexpressed in the presence of HPV transforming proteins. Although overexpression of p16 has been shown to be a specific and sensitive marker for HPV involvement in squamous oropharyngeal cancer, p16 could not be proven as a marker of HPV infection and low expression of p16 has also been correlated with malignancy in IPs. HPV-mediated activation on the p53 and pRb pathways, upregulation of EGFR take place which is the rationale for testing EGFR inhibitors as therapy (cetuximab). EGFR mean percentage area staining was the biomarker that was significantly increased in low risk HPV positive set compared to the HPV negative cases.

TrkB can be used as a potential prognostic indicator for the clinical prognosis because high expression of TrkB plays a major role in SNSCC, (14). Another report identified that p53 overexpression might be related to the emergence of SNSCC. It was suggested that the upregulation of miR-21 was involved in the occurrence and prognosis of sinonasal carcinoma. Long non-coding RNAs (lncRNAs) play important roles in various biological progresses of carcinogenesis [15].

The levels of OPA3 protein were significantly downregulated in SCCs compared with SNIPs, suggesting that OPA3 may be involved in the malignant transformation of SNIP into SCC. Plectin is a giant multifunctional cytokine protein that lead the intermediate filament network in cells. PLEC protein levels were significantly upregulated in SCCs compared with SNIPs, suggesting that PLEC may be a marker for the malignant transformation of SNIP. DNA methylation is important in SNIP to SCC progression. The levels of miR-661 mRNA and PLEC protein were significantly higher in SCC arising in SNIP compared with SNIPs; and the levels of OPA3 protein were significantly downregulated in SCCs compared with SNIPs. Gene body methylation (GbM) is a novel biomarker for methylation of certain genes indicating that GbM might play a role in the malignant transformation of SNIP.

Podoplanin (PDPN), a 38-kDa type I transmembrane glycoprotein that consists of 3 structural domains: an extracellular domain that is rich in serine and threonine residues, which are potential O glycosylation sites; a highly conserved hydrophobic transmembrane domain; and a short cytoplasmic tail with putative phosphorylation sites for protein kinase C and cAMP. It was demonstrated that a statistically significant correlation between high PDPN expression in cancer cells and advanced AJCC stage as well as poor histological grade in head and neck SCC (HNSCC). Additionally, high levels of PDPN expression were associated with decreased overall survival and disease-specific survival. Overexpression of

PDPN in primary oral SCC is associated with advanced T stage, lymphatic spread to the cervical region, and a poor clinical outcome [16].

Head and Neck Cancers

Head and neck cancer (HNC) which is related with nicotine and alcohol abuse is reported as the sixth most common malignancy worldwide with a male preponderance [17]. Because the diagnosis is generally at advanced stage, the low 5-year overall survival rate has not changed despite of surgical and oncological innovations [18]. Proteins identified with altered expression in head and neck cancer are listed in brief with the names of the genes that code them [19].

The protein coded by TNC gene is known to activate EGFR due to EGF-like repeats. In case of oral SCC, over expressed in the tumoral tissue in comparison to adjacent normal tissue. The protein takes role in ECM adhesion, angiogenesis, overexpression of the protein is accepted as an adverse prognostic factor in oral SCC. The functions of TMED2 coded protein are trafficking between endoplasmic reticulum and Golgi apparatus, cytoskeletal re-arrangement, cell migration up- and downregulation in oral and hypopharyngeal SCC. The overexpression of TMED2 is seen in HNSCC and mediated by NFE2L2 which is a known transcription factor that may promote carcinogenesis. DYNLL1 gene overexpression is seen HNSCC and the synthesized protein take role in intracellular microtubular vesicle transport, maintenance of cytoskeleton up-regulation under hypoxic conditions on FaDuDD HNSCC cell line (LC-MS/MS).

COTL1 coded protein is an F-actin binding protein that works in cellular motility which is overexpressed in HNSCC. SPRR2D coded protein functions in skin barrier, wound healing, quenching of ROS, terminal differentiation marker of stratified squamous epithelium. The protein is up- regulated in oral SCC and Nasopharyngeal carcinoma. NCL that recruits EGFR mediated signaling pathways, facilitating EGFR cytoplasmic tail dimerization, direct binding with SNRNP200 playing role in RNA metabolism, works as co-factor in transcription regulation and RNA transport and overexpressed in laryngeal SCC. SNRNP200 codes a DeXH box protein takes role in pre-mRNA splicing. The protein directly binds with NCL acting as nuclear interacting partner. Overexpression of SNRNP200 is seen in HNSCC. ALDH3A2 gene codes the protein that plays role in detoxification of aldehydes originating from lipid peroxidation processes and is overexpressed in HNSCC. The protein is known to its direct binding with EGF. DSG1 (codes for the protein that forms desmosome) and KRT9 (codes for the protein that is intermediate filament of intracytoplasmic cytoskeleton) are downregulated in HNSCC in comparison to adjacent normal tissue.

As listed above, anti EGFR regimens provides anticancer effects and the effectivity of the treatment can be pre-evaluated by the detection of expression patterns of certain genes in the tumoral tissue. Inhibition of EGFR causes downregulation of TNC that is an adverse prognostic marker. NCL downregulation is an indicator for efficacy of anti-EGFR therapy. Through TMED2 inhibition or NFE2L2, invasion ability of HNSCC can be reduced. It was formerly reported that IL1β, IL-6, IL-8 and TNF-, which are famous inflammatory cytokines, are increased in saliva of patients suffering oral squamous cell cancer (OSCC) while the plasma levels remained similar with the healthy patients.

Metalloproteinases (MMPs), a family of multifunctional proteins, are up-regulated in various types of cancer including oral cavity cancers. Among MMPs MMP1 is the highest one to increase in the OSCC group. Expression of RNAs and levels of both MMP1 and MMP3 are significantly elevated in tumor tissues and saliva in OSCC patients indicating higher disease stage [20]. MMP1, PADI1, TNC, CSTA and MMP3 are the molecules other than cytokines which selectively increased in saliva of patients with OSCC. Chi *et al* revealed significant increase of three proteins (MMP1, MMP3 and TNC) and decrease of two proteins (CSTA and PADI1) in saliva samples from OSCC patients [21].

According to a study using IL-8p as an anchor marker, double marker analysis was performed with certain mRNA's and proteins. SAT1 and DUSP1 mRNAs had also been worked formerly in different studies [22]. Increased in tenascin-C (TNC) in oral cancer is positively associated with metastasis, indicating that TNC plays role in modulation of the extracellular matrix at tumor front to provide the invasiveness [23].

Overexpression of Cysteine protease inhibitor cystatin A (CSTA) which acts as a tumor suppressor in head and neck cancer, was found in tumor tissues in comparison to the non-tumor part and the risk of tumor recurrence found high in patients with low CSTA level [24, 25]. PADI1, encoding peptidyl arginine deiminase type-1, was shown to be down-regulated in oral cancer tissue specimens. PADI1 plays role in catalyzation of the post-translational deamination of proteins (*i.e.* filaggrin and keratins) through converting arginine residues into citrulline in the last steps of epidermal differentiation, supporting growth and movement of tumor cells [26, 27].

MiRNAs as mentioned before, are used in evaluation of various diseases. Circulating miRNAs (Ci-miRNA) are stable forms of miRNAs. They are present in human plasma and serum and are being utilized as minimally invasive cancer biomarkers. Ci-miRNAs have been described in different anatomical sub-sites in head and neck region including oral cavity, nasopharynx, larynx, salivary glands and cutaneous malignancies [28].

There are several reagents kits to study quantitative reverse transcription polymerase chain reaction (qRT-PCR). Technologies that are utilised include TaqMan, 3D-Gene, SurePrint, miScript, miRCURY, Agilent and Illumina TruSeq. Normalization for data analysis was provided through endogenous (most frequently) snRNA U6 and/or exogenous (most frequently spiked in cel-miR-39) controls [29 - 31].

Patient preparation are sample collection are as important as the technique utilized and the method of normalization. Clotting, extra-hemolysis and denaturation should be avoided in order not to change fractionated cellular material which inevitably alter the mi-RNA level. Type of blood collection tubes (PAXgene RNA tube, EDTA tube) may affect the ci-miRNA. Fasting blood is generally used however in case of ci-mi-RNA evaluation, length of fasting, content of last meal and diabetes status of the patient contribute to ci-mi-RNA which can be considered as technical bias [32]. In anesthetized patients, type of anesthetic should be questioned because some agents may alter ci-miRNA levels. In order to discuss the findings, standardized results are mandatory. Incubation periods for blood samples prior to freezing or centrifugation are additional factors that may affect the reliable outcome. LSCC serum exosomal miRNA profiles are different from those of HC samples [33].

Laryngeal cancer is one of the most common types of neoplasms in the head and neck region accounting for approximately 1–2% of all malignancies. Over 90% of Laryngeal cancer is histologically classified as squamous cell carcinoma. Anatomically glottic lesions take the first raw of order (60–65%), with secondly supraglottis lesions (30–35%), and thirdly subglottic carcinoma (1–2%). Cigarette smoking is the major risk factor that is potentiated with heavy alcohol consumption. Presence of nodal metastases is the most significant prognostic factor that primarily affects the overall survival rate in LCa. Among miRNAs MiR-9 aberrantly upregulated in LCa by influencing malignant development at the larynx through direct regulation of PTEN [34].

While miR-10b overexpression contributes to initiation of cancer invasion and migration in the laryngeal carcinoma cell line Hep-2, E-cadherin (E-cad) targeted miR-10b accelerates the epithelial-mesenchymal transition (EMT) in Hep-2 cells. miR-10b-transfected cells possess reduced epithelial (E-cad and ZO-1) and increased mesenchymal markers (Vim, N-cad, and FN) [35]. miR-21 is upregulated in laryngeal tumor tissues and suppression of BTG2 which is a cell cycle protein targeted by miR-21, causes growth inhibition in HEP2 cell lines. PTEN, TPM1 and PDCD4 are also regulated by miR-21 [36].

It was shown that miR-205 regulated the cyclin-dependent kinase 2-associated protein 1 (CDK2AP1), which is a repressor of cell cycle progression and G1/S transition inhibitor. CDK2AP1 was found to be a negative regulator of LCa and therefore miR-205 acts as an oncogene. miR-205 induced suppression of CDK2AP1 affects cell proliferation and motility through promoting MMP2 and MMP9 and over-expression of both c-Myc and CyclinD1 in LCa cells [37].

While miR-93 accelerates cancer progression through cyclin G2 (CCNG2) inhibition, miR-126 in regulates calmodulin-regulated spectrin-associated protein 1 (Camsap1) that is associated with clinical differentiation of Lca [38]. Human forkhead box O (FOXO) proteins are effectors of PI3K/Akt signaling. Overexpression of FOXO1 targeted miR -132 in LCa, reveals oncogenic role of miR-132 in LCa by mediating PI3K/AKT/FOXO1 pathway [39]. MiR-221 stimulates apoptosis resistance by affecting apoptotic protease activating factor-1 (Apaf-1) hence it works as an oncogene and suppression of it was related with increased survival in the literature [40]. Overexpression of miR-302-3p was found in LCa was shown to be involved in EMT processes. It was identified that Smad4, a potent tumor repressor, is a direct downstream target of miR-302-3p [41]. Up-regulation of miR-423-3p in primary LCa cell lines acts as an oncogene and regulates adiponectin receptor 2 (AdipoR2) [42]. MiR-1 exerts an anti-metastatic effect *via* directly regulating fibronectin 1 (FN1), that has metastasis related function [43]. MiR-24 that binds to the 3'-UTR of XIAP mRNA, inhibits cell growth and colony formation and promotes apoptosis [44]. MiR-34a that regulates CCND1 gene is inversely correlated to nodal metastases, and clinical stage in LCa. It was also found that miR-34a [45]. MiR-34c is also downregulated in LCa and functions as a tumor suppressor. Reintroducing miR-34c in LCa cells results in both inhibition of growth and invasion *via* targeting c-Met [46, 47]. Cancer spreading and metastasis is under control of miR-34a/c directly regulates UDP-N -acetyl-d-galactosamine: polypeptide-N- acetyl galactosaminyl transferase 7 (GALNT7) a member of glycosyl transferases [48]. Downregulated miR-144 has an anti-oncogenic function through the suppression of insulin receptor substrate 1 (IRS1) in Lca. IRS1 induces the phosphatidylinositol 3-kinase (PI3K)/Akt pathway that is then inhibited by miR-144 [49]. MiR-340 acts as an anti-oncogene that limits cell proliferation, metastatic abilities, and apoptosis in LCa through facilitating p27 expression and blocking PI3K/Akt signaling [50]. High-mobility group protein A2 (HMGA2) can change the DNA structure, thus affecting DNA transcription. HMGA2 mRNA overexpression indicates tumor aggressiveness. The expression levels of Let-7a and HMGA2 are associated with differentiation grade nodal metastases in LCa [51].

MiR-21 upregulation is observed in LCa samples. Combination of exosomal serum miR-21 and long non-coding RNA HOTAIR provides high sensitivity (94.2%) and specificity (73.5%), in different neoplastic processes from benign laryngeal lesion to malignancies. Downregulation of miR-375 is observed in LCa samples and ratio of miR-21/miR-375 expression has a sensitivity (94%) and specificity (94%) for the prediction of LCa [52]. Early staged laryngeal cancer can be diagnosed through hsa-miR-657 and hsa-miR-1287 with good specificity and sensitivity [53]. MiR-375 expression differs between advanced LCa (stage III/IV) and early stage (stage I/II) in favor of advanced disease [54, 55]. Five circulating miRNAs (miR-212-3p, miR-331-3p, miR-603, miR-660-5p, and miR-1303) are detectable in patients' plasma [56].

Up to this point diagnostic importance of miRNA and the relevant spectrum of miRNA were tried to be given in brief. This spectrum of markers has potential to be utilized as prognostic parameters. It was shown that aggressiveness, metastatic potential, invasion potential, recurrence and overall survival are correlated with up/down regulation of certain miRNAs. Serum miR-378 expression is highly expressed in larynx ca and correlated with clinical stage [57]. miR-31, miR-33, and LET-7a possess high diagnostic ability for the disease and they can contribute to aggressiveness of the cancer [58, 59]. Higher miR-21 or lower miR-375 in Lca is related with poorer prognosis in comparison to low miR-21 or high miR-375 [53].

MiR-23a overexpression and/ or miR-101 downregulation is found to be related with nodal metastasis and latter one also reveals the advanced stage. Therefore, lower miR-101 expression shows level and poor outcome [60, 61]. MiR-9 overexpression indicates poorer prognosis. Up-regulation of miR-23a enhances cancer migration and invasion [62]. MiR-296-5p was shown to be related with tumor recurrence at early stage. Significant downregulation of miR-200a and miR-200c levels was found in recurred patients [63, 64]. MiR-941 is a human-specific miRNA that has crucial role in human evolution and gene regulation related with neurotransmitter signaling, thus influencing human-specific cognitive functions [65]. MiR-941 was isolated and shown to be a potentially useful LSCC diagnostic marker. miR-941 overexpression promotes LSCC cell proliferation and invasion. miR-941 may play an oncogenic role by promoting LSCC cell proliferation and invasion [33].

Radiosensitivity

Various miRNAs were shown to be related with treatment response in the course of laryngeal cancer. Sensitivity to chemotherapy and radiotherapy can be predicted before the initiation of treatment. Besides duration of post-therapy disease free survival can be monitored *via* those markers. miR-296-5p, miR-452, miR-183, miR-196a and miR-200c were found to be altered in radioresistant patients. For the prediction of radiation sensitivity in HNSCC downregulated let-7e and upregulated miR-16, miR-29b, miR-150, and miR-1254 are useful parameters. Ataxia-telangiectasia mutated expression (ATM) in HNSCC patients is also indicative of radiation sensitivity [66, 67]. While reduced expression of miR-203 was shown to be correlated with resistance to irradiation and local recurrence after radiotherapy, miR-138-2-3p and miR-218 in HNSCC correlated with radio-sensitivity. Level of post radiation miR-15b-5p in HNSCC patients is a prediction marker for RT response [68 - 70].

Chemosensitivity

miR-181a inhibits Twist1 mediating EMT that enhances metastatic potential and induces cisplatin chemo-resistance in TSCC cells [71]. Through directly blocking p53-upregulated modulator of apoptosis (PUMA) expression, miR-222 downregulation by antisense transfection is known to enhance the sensitivity of OSCC to cisplatin [72].

miR-101-3p inhibits cell proliferation and metastatic functions and induced apoptosis in a salivary gland adenoid cystic carcinoma cell (ACC) line. promoted the sensitivity to cisplatin in ACC cell lines [73]. Exosomal miR-196b originated from cancer-associated fibroblasts conferred cisplatin resistance in head and neck cancer (HNC) cells through targeting CDKN1B and ING5. High expression of exosomal miR-196b in plasma is known to correlate with poor overall survival and chemoresistance [74].

HPV Driven HNSCC

In considering HPV driven HNSCC, oropharynx and pharynx are the leading parts. It is seen that most of the infections are of single type and most abundantly HPV 16- driven in all anatomic sub compartments. Multiple types are observed in HPV related oral cavity and laryngeal cancers. Detection of E6*I mRNA, overexpression of $p16^{INK}4a$, and under-expression of pRb, p53, or Cyclin D1 are other markers that were found in formalin-fixed, paraffin-embedded cancer tissues to correlate with the HPV-16 DNA. Because of multiple types of HPV, and confounding factors like tobacco and alcohol consumption in oral cavity and larynx cancers, HPV can be bystander infection. A very recent study focusing recurrence of HPV driven HNSCC pre-print reported that oxidative phosphorylation (OXPHOS) pathway is enriched in recurrent HPV-associated HNSC. Nrf2 regulates the antioxidant response element (ARE) genes in response to oxidative stress yet also plays a critical role in OXPHOS. Activation of Nrf2 increases epithelial-mesenchymal transition (EMT) through mediating a decrease in E-cadherin and stimulating TGF-β1-induced SMAD2/3 activity eventually related with worse survival among HPV-driven HNSCs [75].

HPV related head and neck cancer which has better prognosis and p16 staining is a marker of HPV histological marker used in HNC staging [76]. In HPV driven HNSCC miR-15a/miR-16/miR-195/miR-497 family, miR-143/miR-145, and miR-106-363 cluster may play crucial roles in the pathogenesis of HPV [77]. While overexpression of miR-155 in oral cavity cancers (tonsillar and base of tongue cancer) is known to be associated with HPV positivity and improved survival, low miR-185 expression is associated with HPV negativity and decreased survival for the disease [78].

HPV- driven cancer progression was found to be related with reduced miR-13--3p increased both EGFR and HuR mRNA expression the downregulation of miR-133a-3p is observed in serum and metastatic lymph nodes [79].

Thyroid Pathologies

Head and neck region cover a very fundamental endocrine organ which has lots of functional, inflammatory and neoplastic pathologies. In the following section neoplastic lesions of the thyroid tissue will be given in considering molecules to be used as biomarker.

It is well known that NOTCH signaling pathway participated in the development of a variety of cancers by regulating the tumor microenvironment, tumorigenesis, progression, angiogenesis, invasion and metastasis [80]. Hsa_circ_0058124 suppresses the NOTCH3 pathway *via* miR-218-5p/NUMB. In pathogenesis of papillary thyroid cancer, circRNA hsa_circ_ 0058124 was shown to be important especially in providing resistance to RNase R and it is frequently upregulated in PTC tissues which is associated with poor prognosis. The GATA family is also known to interact with the NOTCH signaling pathway and Hsa_circ_0058124 promotes PTC progression through the NOTCH3/GATAD2A axis [81].

An oncogene, BRAF, which encodes a serine threonine protein kinase, is the key regulator of the mitogen-activated protein kinase (MAPK) pathway. Some mutations and gene fusions activate the receptor tyrosine kinase (RTK)/mitogen-activated protein kinase (MAPK) signaling pathway which regulates cellular proliferation, differentiation, cell-cell adhesion, migration and survival in thyroid tissue. P.V600E-like and RAS-like groups are the major expressions of BRAF in thyroid cancers mostly in papillary thyroid cancer.In order to improve diagnostic accuracy of fine-needle aspiration biopsy of indeterminate thyroid nodules, the presence of p.V600E in the specimen is almost definitive for PTC. RAS genes (NRAS, HRAS, and KRAS), which are the most common oncogenes in human cancers, are codes for small GTPase. Small GTPase group of proteins regulate intracellular signal transduction which activates the MAPK.

The most common gene arrangements in PTC involve RET oncogene with RET/PTC1 (CCDC6-RET is the leading that account for 60% of thyroid cancer with RET rearrangements. RET/PTC3 (NCOA4-RET, 30%), and RET/PTC2 (PRKAR1A-RET, 5%) are the subsequent ones.

Radiation induced papillary thyroid cancer are associated with ETV6/NTRK3 rearrangements which comes as second after the RET/PTCs. NTRK1, 2, and 3 encode the neurotrophin receptor TRKA, B, and C. Other rearrangements found in PTC are NTRK1 with TPM3, TPR, and TFG, first two rearrangement have shown to be related with activation of downstream MAPK, PI3K, and PKC signaling pathways

Although prevalence of ALK rearrangements in PTC is low, STRN-ALK rearrangement results in increased TSH-independent cell proliferation and fibrosarcoma-like transformation. While PTCs with ALK rearrangements have rather a follicular growth pattern and areas of papillae formation, ALK-fusion positive PTCs are not expected to behave aggressively [82, 83].

NF-kb is a ubiquitous transcription factor that regulates cell survival, proliferation, and differentiation as well as cellular and immune responses to cytokines and oxidative stress. The role of NF-B signaling in thyroid cancer cell growth, migration, invasion, and angiogenesis has been shown. Overexpression of NF-B transcription factors is common in thyroid cancer, is associated with larger tumor size, nodal metastases, extrathyroidal extension [84 - 86].

Certain chemokines are abundant in tumor microenvironment and plays crucial role in aggressiveness ot the tumor. Pro-tumorigenic secreted factors include TGF, CCL15, CXCL12, CXCL16, and CXCL8. Hypoxia, that is another key regulator of the tumor microenvironment, commonly seen in central part of solid cancers. Hypoxic micro- environment promotes the secretion of TGF-, bFGF, and PDGF. These hypoxia-driven secretions stimulate the transformation of fibroblasts into cancer-associated fibroblasts (CAFs) which are involved in initiation and progression of cancer by modulating the extracellular matrix. Additionally, BRAF p.V600E mutated PTC exhibits a higher expression of CAFs-associated proteins, suggesting that the aggressiveness of some BRAF mutated tumors might be associated with CAFs. Owing to multicentricity and recurrence tendency of PTC together with high prognosis with long-term survival rates, lots of work has been exhausted toward providing disease-free survival.

Considering miRNA's, miR-95 and miR-190 were found to be accurate and sensitive diagnostic markers for PTC in Caucasians [87].

Upregulation of miR-146b in PTC was demonstrated to be an independent risk factor for worse prognosis in PTC patients [88]. Upregulation of miRNA 221 and miRNA 222 in PTC in comparison to normal thyroid tissue designates increased risk of recurrence [89]. miR-204 acts as a tumor suppressor miRNA and inhibits cell proliferation, arrests the cell cycle and induces apoptosis in PTC cell lines [90]. List is so long for each thyroid pathology, Santiago *et al* reported a recent list for miRNA's [91].

In case of follicular carcinoma of thyroid RAS point mutations is the leading genetic alteration which may account for up to %50 of cases. AX8-PPARG rearrangement is the second most common genetic alteration in FTC and patients of this type are typically younger and vascular invasion is frequently observed [92, 93].

Phosphate and tensin homolog deleted on chromosome ten (PTEN) tumor suppressor gene mutations are detected in sporadic FTCs. TERT promoter mutations and TP53 mutations can be found in a subset of FTCs, thait is found to be associated with aggressive disease and poor outcome. Hurtle cell ca is a rather rare form of thyroid pathology. Unlike other forms, HTC has lower rates of RAS mutation and PAX8–PPARG rearrangement [94].

Salivary Glands, Autoimmune Salivary Adenitis

Salivary gland tumors comprise a relatively small fraction of head and neck malignancies While pleomorphic adenoma is the commonest benign salivary gland lesion, mucoepidermoid ca is the most common malignant salivary gland lesion. Among HNSCCs, miRNAs were mostly studied in adenoid cystic carcinoma.

Down regulation of the miR125b in tissue is related with carcinogenesis. Alterations in miR-125b and miR100 are related with mucoepidermoid ca. MiR-21 which was shown to be expressed in mesenchymal tissue of mammal submandibular gland, is known to be crucial miRNA for embryological development of submandibular gland and its morphogenesis. miR21 expression is up- regulated by epidermal growth factor that is critical in branching process submandibular gland ductus. MiR-21 is the constant miRNA showing up regulation in HNSCC that is related with metastases, invasion and poor prognosis in the end. It was shown that benign tumors are related with down regulation of miR-21, 23a, 27a,223,125b,126, 146a and miR30e in serum. Malignant tumors are associated with up-regulation of miR-21,23,199a,345 in tumor tissue in comparison to benign lesions. MiR-345, which was shown to have a prognostic value in prostate and colorectal cancer, was shown to be increased 18-fold in salivary gland malignant tumors in comparison to benign lesions. MiR-23a may act as a tumor suppressor gene in both benign and malignant salivary gland tumors. Expression of miR-30e is almost 15 fold increased in malignant salivary gland tumors in comparison to benign ones. Genes RPS7 and LIMCH1 are the target genes of miR21 and miR30e. RPS7 gene codes a ribosomal protein belonging to S7E family and LIMCH1 gene codes zinc ion-binding proteins. Zinc ion-binding proteins and ribo-somal proteins belonging to S7E family are potential concerns or future studies for differing biological behavior of the salivary gland tumors [95].

Otology

Otologic diseases are very different in nature in comparison to other subunits of otorhinolaryngology. Middle ear and mastoid cells, which are the transmission routes for data of two major sensations namely hearing and balance toward inner ear are bone covered closed chambers situated nearby intracranial neural parenchyme. Disease spectrum is very broad however in this part basic pathologies are tried to be classified as ear specific infections, function related conditions such as hearing loss and balance problems, inner ear injuries like ototoxicity, ischemia or acoustic trauma, Meniere's disease which is a very unique inner ear pathology in itself. Additionally, temporal bone malignancies will be briefly given in the aspect of biomarkers to detect and follow -up the disease.

Retraction Pockets

Retraction pockets, which are also well studied in the process of cholesteatoma emergence, are much more than pressure defeated places with incapable of self-cleaning. Initial sequestration of keratin debris enlarges and blocks certain air travelling routes of MACS. Bone enclosed chambers with airflow blockage results in variations in gas exchange and acidity, also the microbiome of environment. Lots of molecule were studied and they will be given in brief. If we go back to the pathogenesis of middle ear and mastoid air cell chronic infections namely chronic otitis media, this entity is globally classified as chronic otitis media with or without cholesteatoma.

CD45 LCA (leukocyte common antigen) is a protein in the tyrosine phosphatase category, specific for white blood cells of all categories and presence of the elements was evaluated in both epithelium and stroma of tympanic membrane. Increased incidence of CD45 LCA, CD31, D2-40, MMP9, and Ki67 positivity was shown in retraction pockets.

Greater expression of CD45 has been shown in cholesteatomas in comparison with other forms of chronic otitis without cholesteatoma and count of activated T lymphocytes and macrophage cells are high in the structure of a cholesteatoma, while the number of B lymphocytes is low.

MMP9 (matrix metalloproteinase 9) is a metalloenzyme also known as collagen type IV gelatinase B which is an enzyme participating in degradation of extracellular matrix within fibrous tissue including epithelial basal membrane. MMP9 is important in inflammation driven cell-migration. Its presence was shown in some invasive carcinomas. MMP9's occurrence was observed in moderate to advanced RP stages with high frequency, and its presence is pointing to cholesteatoma formation. Greater MMP9 activity was found in pediatric cholesteatomas which may explain the aggressiveness of pediatric cases.

CD31 (also known as platelet endothelial cell adhesion molecule [PECAM1]) is used as a marker of endothelium with strong sensitivity and specificity. Microvascular density of cholesteatomas was shown to be significantly higher in pediatric cases. CD31 positivity is significantly greater in advanced stage III RPs than in mild to moderate stages of RPs. D2-40 (podoplanin) is a membrane mucoprotein used as a marker of lymphatic vessel endothelium and is used as a marker of tumorous invasions of lymphatic vessels. Upon tympanic membrane perforation, a prompt repair occurs in few days through proliferation of lymphatic vessels into pars tensa, originating from the manubrium area which indicates that lymphatic vessels thus probably play an important role in regenerating capacity in TM. D2-40 occurrence is greater in advanced stage RPs [96]. Ki67 is a nuclear marker of cell proliferation that is expressed in the cell nucleus only in the late G1, S, and G2 phases. Ki67 was not correlated with the aggressiveness of pediatric cholesteatoma, however, increase activity was related with advanced staging of RP and final transformation to cholesteatoma [97].

Cholesteatoma & Chronic Otitis Media

When ear specific infectious and inflammatory conditions are spelled, cholesteatoma is the primadonna of the title. This epidermis- like tissue is hidden and sheltered in the cranny parts of the middle ear and mastoid air cells. (MACS) Sometimes it is operated and totally resected, however most of the infiltrating cases, it tends to recur and invade meninx and neural parenchyme. This invading and recurring feature resembles to tumoral lesions of low-grade. If we go back to the pathogenesis of middle ear and mastoid air cell chronic infections namely chronic otitis media, this entity is globally classified as chronic otitis media with or without cholesteatoma.

Cholesteatoma has been the interest of a lot of scientific study. Recent results are listed below. Theories tried to explain the migration and invasion patterns pertaining to the disease. According to recent findings in biomolecular research, cholesteatoma could be considered as a disordered in cell growth. The growth of cholesteatoma cultures can be explained by a higher growth factor activity, such as keratinocyte growth factor (KGF) and its receptor (KGFR) [5]. These are known to be secreted cytokines by the cholesteatoma perimatrix fibroblasts and initiation of the production is controlled by keratinocyte secreted inflammatory cytokines. Gerdes *et al.* expression of the human Ki-67 protein is strictly associated with cell proliferation. Abnormal behavior of the cholesteatoma seems to be related with paracrine interaction between KGF and KGFR. It was shown that cholesteatoma expressing both KGF and KGFR had high Ki-67, which correlated with its aggressiveness [98]. A long list of growth factor and their receptors were studied in order to predict the prognosis of pathology.

Expression of the human epidermal growth factor (HEGF), all four receptors of HEGF (HER1, HER2, HER3, and HER4), and the ligands EGF, transforming growth factor (TGF)-α, amphiregulin (AR), heparin-binding EGF-like growth factor (HB-EGF), and epiregulin (EPI) are parts of the EGF system. Up-regulation of mRNA encoding EPI and AR and down- regulation of HER1 and EGF were shown indication that increased expression of ligand EPI, suggest that EPI-mediated activation of HER4 might play a role in cholesteatoma growth [99].

The heparin-binding epidermal growth factor-like growth factor (HB-EGF) which plays a crucial role in the invasiveness of cholesteatoma, was shown to be increased dramatically in comparison to normal skin. In the same study correlation was found between bone resorption and HB-EGF level [100].

Inflammatory cytokines like IL-1α, TNF- α, Macrophage -colony stimulating factor, prostaglandin E-2 were all detected in cholesteatoma tissue and related with the bone resorption capability of the pathology. One of bone resorption theories postulate that resorption is performed by osteoclasts themselves are under control of various cytokines. Additional theories came up with acid lysis theory that is cholesteatoma has a weak acidity and acid leakage through cholesteatoma epithelium results in decalcification of neighboring bone.

Expression of transglutaminases (TGase 1-5), that enzymatically form the isopeptide bond in the dermis, differs in different layers of epidermis and cholesteatoma tissue. It was shown that cholesteatoma tissue expresses higher amounts of TGase 2 mRNA and lower amounts of TGase 5 mRNA in comparison to skin. Studies focusing difference between keratinocytes of normal skin and cholesteatoma matrix revealed specific bond formations between pathologic cells.

Corneodesmosins (CDSN) which bind corneocytes together, are important in epidermal barrier functions and expression of CDSN was found to be lower in cholesteatoma tissue in comparison to normal skin. In normal skin degradation of CDSN is catalyzed by Kallikrein (KLK) in the outermost layer of dermis. KLK5, KLK7 and KLK14 were shown to be related with epidermal desquamation. KLK activites are inhibited by serine protease inhibitor Kazal-type 5 (SPINK5) in dermis. While expression of CDSN was lower demonstrated to be lower in cholesteatoma in comparison to normal skin, expression of CDSN-degrading enzymes (KLK) and inhibitors of them (SPINK5) are similar indicating that CDSN level is under control at production level in cholesteatoma tissue.

It was shown that decreased expression of molecules (tight junction proteins namely claudin-1, claudin-3 and tricellulin) that are responsible for epithelial barrier function are all decreased in cholesteatoma tissue. Changes in barrier function can be monitored by the variations in electrical impedance of the cholesteatoma tissue that also reflects the permeability. Electrical impedance is much lower in cholesteatoma tissue [101]. MicroRNA (miR)-203a that has been shown to be essential for the growth and differentiation of keratinocytes, regulate cholesteatoma growth and proliferation by targeting Bmi1. p-Akt, that was shown to be positively correlated with Bmi1, significantly increased in cholesteatoma tissues [102]. MiR-21 functions in the regulation of proliferation and apoptosis. Let- 7 a down- regulates miR-21 resulting in suppression of proliferation and induction of apoptosis [103]. CircRNAs may play crucial roles in cholesteatoma and exonic circRNAs accounted for the largest type among expressed circRNAs. CircRNA-102747, circRNA-101458 are expressed significantly higher in cholesteatoma tissue and associated with metabolic processes like protein kinase activity, phosphatidylinositol 3-kinase (PI3K) binding. PI3Ks are enzymes that catalyze the phosphorylation of phosphatidylinositol (PtdIns). The PI3K pathway is crucial for the regulation of cell survival and proliferation [104]. Activation of the PI3K/Akt signaling pathway protects keratinocytes of cholesteatoma against programmed cell death [105].

Increased PI3K/Akt signaling pathway activation has been proven to be related to cholesteatoma epithelial hyper-proliferation. Competing endogenous RNA (ceRNA) analysis showed that among of the ceRNA network, the most enriched pathway was found to be the MAPK (mitogen-activated protein kinase) signaling pathway. Activation of MAPK (mitogen-activated protein kinase) signaling pathway was proven to play an important role in the terminal differentiation in cholesteatoma epithelium [106, 107].

While the circRNA-102747/ lncRNA-uc001kfc.1/miR-21-3p/targeted mRNAs may act as the regulator of malignant characteristics; circRNA-101458/miR let-

7a-3p/targeted mRNAs plays a role as a benign regulator which overall suggests that circRNA-102747 and circRNA-101458 have ceRNA potential in the pathogenesis of cholesteatoma [108].

We have seen how cholesteatoma is a "wolf in sheep's clothing", resistant infections are of great concerns mainly because of sustained infection origin and close localization of central neural tissue and cranial nerves.

Another ear specific disease affects otic capsule, goes with dysregulated bone formation and resorption, otosclerosis (Os. The disease is still of great concern because a definitive treatment has not been defined yet. We are trying to reverse the clinical insult by surgical intervention; however, we do not know how to stop the progression. Genetic factors that regulate bone-remodeling in otic capsule include receptor activator of nuclear factor k B (RANK), receptor activator of nuclear factor k B ligand (RANKL), transforming growth factor beta 1 (TGF-b1), osteoprotegerin (OPG), collagen type I alpha 1 chain (COL1A1), and bone morphogenetic proteins (BMPs). Increased expression of BMPs, TGF-b, tumor necrosis factor (TNF)-a, and apoptosis inhibitors (hCIAP1/2) supports hypothesis of "embryonic chondrocyte reactivation" in otosclerosis.

Genetic polymorphism studies aimed to define the tendency of certain samples to A recent study revealed the protective role of the GT genotype at the COL1A1 rs1800012 locus, and the disease-promoting role of the BMP4 TC genotype at rs17563 [109]. BMPs and TGF-b are key regulators of new bone formation and increased expression of BMP2, BMP4, BMP5, and BMP7, four members of the TGF-b superfamily plays a role in the disordered bone remodeling in Os [110]. Ear specific conditions goes further from middle ear even with more disabling conditions related with inner ear causing hearing and balance problems. Inner ear damage may be resulted from toxic, ischemic, traumatic, genetic or immune mediated processes. Any parameter related with abovementioned situations can be traced as biomarker for diagnosis and/or follow-up. As the profile extends, new therapeutic strategies are being evolved. Because etiology spectrum is so broad, in this part inner ear specific conditions will be summarized.

Biomarkers for inner ear pathologies are high in count but low in specificity. However, proteome studies enabled us to define the content of inner ear fluids especially perilymph *via* so-called liquid biopsy for specific inner ear pathologies. First of all in certain pathologies like vestibular dysfunction, normal ratio of perilymph elements namely enzymes, immunity, apolipoproteins, protease inhibitors and structural elements change in a way that while in normal perilymph immunity elements are leading with 31 of all proteins, in case of vestibular dysfunction leading proportion becomes enzymes with up to 50% of all proteins.

Analysis of perilymph of normal hearing patients revealed three canonical pathways as follows; firstly proteins belonging to cytokine signaling taking role in acute phase signaling, secondly nuclear receptor signaling proteins that play role in Liver X receptor (LXR) / retinoid X receptors and thirdly another group of nuclear receptor signaling proteins that play role in Farnesoid X receptor (FXR) / retinoid X receptors (RXR) Activation.

Regulation of inner ear inflammation is critical function of perilymph proteins, as increased inflammatory cytokine production is associated with SNHL in meningitis, otitis media, bone diseases that are characterized by pathologic osseous remodeling, such as otosclerosis and Paget's diseas [111, 112]. Nuclear transcription factor κB (NFκB) is important in the degeneration of SGNs. Animals lacking the p50 subunit of NFκB demonstrated accelerated loss of SGNs and age-related hearing loss [113]. One can find the list of genes and coding proteins with spectral counts in perilymphatic fluid in considering normal and diseased ears [114]. According to tissue immunohistochemistry, it was shown that among 228 proteins that constitute the core perilymph proteome, four major ones are of special interest and are candidates for further validation: alpha-1 antichymotrypsin (AACT), EGF-containing fibulin-like extracellular matrix protein 1 (EFEMP1), hepatocyte growth factor activator (HGFAC), transforming growth factor beta induced (TGFBI). Immunoreactivity of AACT is present in inner hair cells (IHCs), outer hair cells (OHCs), the supporting cells and the tectorial membrane. AACT appear to regulate protease-mediated tissue damage and homeostasis [115].

TMPRSS3 may be important for cochlear hair cell survival at the onset of hearing in mice. TMPRSS3, serine protease in which mutations are associated with non-syndromic autosomal recessive deafness (DFNB8/10). Non-coding mutations of Hepatocyte growth factor (HGF) are associated with non-syndromic hearing loss, autosomal recessive deafness-39 (DFNB39) [20], while a mutation in MET is associated with human DFNB97 hearing loss.

Efemp1 RNA expression in sensory and supporting cells of the organ of Corti, the spiral ganglion and spiral ligament overlaps with the expression pattern of EGF receptor. Literature suggests that EFEMP1 may antagonize protease activity to prevent cellular injury in the inner ear [116]. Transforming growth factor beta induced (TGFBI) is an extracellular matrix protein induced by TGF-β and TGFBI is structural homolog of periostin. Localization of TGFBI expression in certain cell types of the adult cochlea validates TGFBI detection in adult human perilymph. Aquaporins 2, 4 and 5 are shown to be important in the pathogenesis of Meniere's diease, and aquaporin 5 is regulated by cyclic adenosine monophosphate (cAMP); other derivatives of adenosine, including adenosine triphosphate (ATP) play important roles in energy transfer [117].

There is increasing evidence of the critical role that miRNAs play in inner ear development and pathologies like sudden and noise-induced hearing loss. MiRNA 4531 and 4688 seem to be linked to regulation of aquaporin expression. KCN, a potassium channel, was found to be related with the MD. Glutathione peroxidase has been shown to be related with the pathophysiology of Meniere's disease, miRNA 4688 and miRNA 3199 can be directly linked to glutathione peroxidase regulation. A recent agent (SPI-1005) is being investigated as a potential treatment for the disease. MiR-1471, miR 3199, miR 3687, miR 4531, miR 4688, miR 4757 and miR 5572 were found to be related with with MD. Each abovementioned miRNA' s was shown to target some following proteins GXP2: glutathione peroxidase, GXP6: glutathione peroxidase, KCNJ13 a potassium channel, AQP9: aquaporin, KCNJ6: potassium channel, AQP3: aquaporin, GPX1: glutathione peroxidase, 6 KCN proteins: potassium channels, KCNC3: potassium channel [118].

Stress factor is important in MD and sympathetic system plays crucial role in MD. Regulation of osmotic balance are of great importance in MD. Osmoregulation was shown to be related with prolactin level as the hormone is a stress hormone. Salivary α-amylase (SAA) is a member of glucosyl hydrolases, and its major production takes place in the parotid glands. SAA takes role in degradation of carbohydrates. SAA was proposed to be a marker of the adrenergic system. Secretion of SAA is mainly related with beta adrenergic and partly to alpha adrenergic stimuli [119]. Antidiuretic hormone (ADH) (Vasopressin) stimulates adenylate cyclase activity both in the stria vascularis and semicircular canal epithelium. ADH was reported to be elevated under psychological stress conditions. Experimental administration of ADH to guinea pigs over 1 week, increases the activity of aquaporin 2 and causes hydrops [120]. When we take a brief look at acoustic trauma, which is a major concern due to both blast type injury, and chronic exposure. CXCL1 mainly expressed in pillar cells which are main supporting cells in inner ear close to the sensory epithelia namely outer and inner hair cells. A recent study revealed that, among the analyzed proinflammatory cytokines in perilymph, CXCL1 experienced the largest noise-induced elevation.

CXCL1, potentially together with IL-6 and TNF-a, is a candidate molecular biomarker of acute permanent threshold shift (PTS) which is permanent elevation of auditory thresholds in audiogram due to acoustic trauma.

Elevation in TNF-a and IL-6 mRNA expression was reported in the stria vascularis, spiral ganglion neurons, and spiral ligament in the early phase after PTS-inducing acoustic trauma (124 dB SPL). Noise Trauma Reduces CXCL1 Immunoreactivity in Pillar Cells.

Tnip1, a gene encoding the "TNF alpha-induced protein 3" was shown to be expressed at higher levels in experimental animals resistant to noise exposure. This protein was formerly shown to inhibit TNF-mediated apoptosis and NF-kappa B activation [121]. A widespread clinical entity called tinnitus is defined as "the perception of sound(s) in the absence of an external sound source" [122].

Tinnitus interferes with stress-associated cytokines and proteins as a result of activation of the hypothalamic-pituitary-adrenal axis and/or the sympathetic nervous system. Dysregulation of the production of cytokines are induced by stressors, such as tumor necrosis factor (TNF)-α, interferon (IFN)-γ and interleukin (IL)-6 that finally modulate N-methyl D-aspartate (NMDA) receptors (NMDARs) *via* glutamate. Salicylate exposure changes the expression levels of IFN-γ, TNF-α, IL-6 and N-methyl D-aspartate receptor subunit 2A (NR2A) genes in the auditory cortex (AC). While expression of TNF-α and NR2A genes increased in the AC following long-term administration of salicylate, the expression of IFN-γ genes decreased. Upon cessation of salicylate administration, however returned to normal levels [123].

NMDAR activation is accompanied by ascending Ca^{2+} influx. Stimulation of Ca2+-dependent enzymes progresses with the ATP depletion and an increase in lipid peroxidation products, contributes to the excitotoxic hazard in the auditory system. Long-term salicylate administration induces TNF-α gene expression reversibly in the AC of rats with tinnitus. TNF-α may serve as a stress marker in tinnitus patients that is TNF-α shifts neurons towards increased excitation and decreased inhibition [124, 125].

Psychological stressors decrease IFN-γ levels during exam stress (41). The decreased IFN-γ has also been attributed to the pharmacological actions of salicylate (36), and an increase in NMDARs (42). It is revealed in these data that chronic salicylate administration induces dysregulation of cytokines and NR2A, which may be involved in tinnitus [126].

Oxidative stress in the inner ear can make individuals more prone to age related hearing loss that is highly related with polymorphisms related to a battery of antioxidant systems [127, 128].

Gene screening analysis for GRM7 which encodes a metabotropic glutamate receptor subtype 7 (mGluR7) and NAT2 which is an isoenzyme of N-acetyltransferase (NAT) are enzymes responsible for the detoxification of exogenic substrates *via* N-acetylation or O-acetylation. In aged patients in order to find factors that might contribute to the diagnosis of presbycusis and tinnitus revealed that T allele in GRM7 gene is the most common allele in the sample of older adults with presbycusis and tinnitus. It was shown that individuals carrying the allele A/T of GRM7 and slow acetylator phenotype of NAT2 (weak statistic relevance) have tendency to develop a more severe form of tinnitus [129].

CONCLUSION

Otorhinolaryngologic diseases contain both surgically curable pathologies as well as idiopathic ones which can be barely treated medically or surgically. Therefore, in understanding the etiology, pathology and the sensible curative paths, biomarkers are being studied. As the technology of detection and tracing of a molecule evolves; most challenging disorders like otosclerosis, cholesteatoma, HPV-driven pathologies *etc.* are getting enlightened. Because this is a continuously evolving process, and new molecules and techniques are being developed every decade, most diseases that are un-stoppable by the date will transform into a curable or repressible state. This chapter shared the recent works in brief to support future works related with the field of otorhinolaryngology-head and neck surgery.

CONSENT FOR PUBLICATION

Not applicable.

CONFLICT OF INTEREST

The authors declare no conflict of interest, financial or otherwise.

ACKNOWLEDGEMENT

Declared none.

REFERENCES

[1] Nakayama T, Sugimoto N, Okada N, *et al.* JESREC score and mucosal eosinophilia can predict endotypes of chronic rhinosinusitis with nasal polyps. Auris Nasus Larynx 2019; 46(3): 374-83. [http://dx.doi.org/10.1016/j.anl.2018.09.004] [PMID: 30243753]

[2] Workman AD, Kohanski MA, Cohen NA. Biomarkers in chronic rhinosinusitis with nasal polyps. Immunology and Allergy Clinics 2018; 38(4): 679-92. [http://dx.doi.org/10.1016/j.iac.2018.06.006] [PMID: 30342588]

[3] Boyle JV, Lam K, Han JK. Dupilumab in the treatment of chronic rhinosinusitis with nasal polyposis. Immunotherapy 2020; 12(2): 111-21. [http://dx.doi.org/10.2217/imt-2019-0191] [PMID: 32075470]

[4] Yamada T, Miyabe Y, Ueki S, *et al.* Eotaxin-3 as a Plasma Biomarker for Mucosal Eosinophil Infiltration in Chronic Rhinosinusitis. Front Immunol 2019; 10: 74.
[http://dx.doi.org/10.3389/fimmu.2019.00074]

[5] Ho J, Earls P, Harvey RJ. Systemic biomarkers of eosinophilic chronic rhinosinusitis. Curr Opin Allergy Clin Immunol 2020; 20(1): 23-9.
[http://dx.doi.org/10.1097/ACI.0000000000000602] [PMID: 31688152]

[6] Nocera AL, Meurer AT, Miyake MM, Sadow PM, Han X, Bleier BS. Secreted P-glycoprotein is a noninvasive biomarker of chronic rhinosinusitis. Laryngoscope 2017; 127(1): E1-4.
[http://dx.doi.org/10.1002/lary.26249] [PMID: 27577924]

[7] Gonçalves C, Pinaffi JV, Carvalho JF, *et al.* Antineutrophil cytoplasmic antibodies in chronic rhinosinusitis may be a marker of undisclosed vasculitis. Am J Rhinol 2007; 21(6): 691-4. [PubMed: 18201448].
[http://dx.doi.org/10.2500/ajr.2007.21.3092] [PMID: 18201448]

[8] Liu C, Zheng M, He F, Wang X, Zhang L. Role of exhaled nasal nitric oxide in distinguishing between chronic rhinosinusitis with and without nasal polyps. Am J Rhinol Allergy 2017; 31(6): 389-94. [PubMed: 28927489].
[http://dx.doi.org/10.2500/ajra.2017.31.4480] [PMID: 28927489]

[9] Jeong JH, Yoo HS, Lee SH, Kim KR, Yoon HJ, Kim SH. Nasal and exhaled nitric oxide in chronic rhinosinusitis with polyps. Am J Rhinol Allergy 2014; 28(1): e11-6. [PubMed: 24717870].
[http://dx.doi.org/10.2500/ajra.2014.28.3984] [PMID: 24717870]

[10] Lee RJ, Kofonow JM, Rosen PL, *et al.* Bitter and sweet taste receptors regulate human upper respiratory innate immunity. J Clin Invest 2014; 124(3): 1393-405. [PubMed: 24531552].
[http://dx.doi.org/10.1172/JCI72094] [PMID: 24531552]

[11] Fokkens WJ, Lund V, Bachert C, *et al.* EUFOREA consensus on biologics for CRSwNP with or without asthma. Allergy 2019; 74(12): 2312-9.
[http://dx.doi.org/10.1111/all.13875] [PMID: 31090937]

[12] Pothoven KL, Norton JE, Hulse KE, *et al.* Oncostatin M promotes mucosal epithelial barrier dysfunction, and its expression is increased in patients with eosinophilic mucosal disease. J Allergy Clin Immunol 2015; 136(3): 737-746.e4. [PubMed: 25840724].
[http://dx.doi.org/10.1016/j.jaci.2015.01.043] [PMID: 25840724]

[13] Wang W, Yin J. Is it worthy to take full-course immunotherapy for allergic rhinitis? Scand J Immunol 2020; 91(1): e12817.
[http://dx.doi.org/10.1111/sji.12817] [PMID: 31650620]

[14] Scheel A, Lin GC, McHugh J. Human papillomavirus infection and biomarkers in sinonasal inverted papillomas: clinical significance and molecular mechanisms. In International forum of allergy & rhinology. 2015; 5: pp. (8)701-7.

[15] Yu B, Qu L, Wu T, *et al.* A novel LncRNA, AC091729. 7 promotes sinonasal squamous cell carcinomas proliferation and invasion through binding SRSF2. Front Oncol 2020; 9: 1575.
[http://dx.doi.org/10.3389/fonc.2019.01575] [PMID: 32039035]

[16] Wang H, Hu C, Song X, *et al.* Expression of podoplanin in sinonasal squamous cell carcinoma and its clinical significance. Am J Rhinol Allergy 2020; 34(6): 800-9. Epub ahead of print
[http://dx.doi.org/10.1177/1945892420930976] [PMID: 32551858]

[17] Pulte D, Brenner H. Changes in survival in head and neck cancers in the late 20th and early 21st century: a period analysis. Oncologist 2010; 15(9): 994-1001.
[http://dx.doi.org/10.1634/theoncologist.2009-0289] [PMID: 20798198]

[18] Nygård M, Aagnes B, Bray F, Møller B, Mork J. Population-based evidence of increased survival in human papillomavirus-related head and neck cancer. Eur J Cancer 2012; 48(9): 1341-6.
[http://dx.doi.org/10.1016/j.ejca.2012.03.014] [PMID: 22516210]

[19] Burian A, Lujber L, Gerlinger I, *et al.* Label-Free Semiquantitative Liquid Chromatography-Tandem Mass Spectrometry Proteomics Analysis of Laryngeal/Hypopharyngeal Squamous Cell Carcinoma on Formalin-Fixed, Paraffin-Embedded Tissue Samples - a Pilot Study. Pathol Oncol Res 2020; 26(4): 2801-7.
[http://dx.doi.org/10.1007/s12253-020-00849-5] [PMID: 32564264]

[20] Stott-Miller M, Houck JR, Lohavanichbutr P, *et al.* Tumor and salivary matrix metalloproteinase levels are strong diagnostic markers of oral squamous cell carcinoma. Cancer Epidemiol Biomarkers Prev 2011; 20(12): 2628-36.
[http://dx.doi.org/10.1158/1055-9965.EPI-11-0503] [PMID: 21960692]

[21] Chi LM, Hsiao YC, Chien KY, *et al.* Assessment of candidate biomarkers in paired saliva and plasma samples from oral cancer patients by targeted mass spectrometry. J Proteomics 2020; 211: 103571.
[http://dx.doi.org/10.1016/j.jprot.2019.103571] [PMID: 31689561]

[22] Gleber-Netto FO, Yakob M, Li F, *et al.* Salivary biomarkers for detection of oral squamous cell carcinoma in a Taiwanese population. Clin Cancer Res 2016; 22(13): 3340-7.
[http://dx.doi.org/10.1158/1078-0432.CCR-15-1761] [PMID: 26847061]

[23] Berndt A, Richter P, Kosmehl H, Franz M. Tenascin-C and carcinoma cell invasion in oral and urinary bladder cancer. Cell Adhes Migr 2015; 9(1-2): 105-11.
[http://dx.doi.org/10.1080/19336918.2015.1005463] [PMID: 25793577]

[24] Strojan P, Budihna M, Smid L, *et al.* Prognostic significance of cysteine proteinases cathepsins B and L and their endogenous inhibitors stefins A and B in patients with squamous cell carcinoma of the head and neck. Clin Cancer Res 2000; 6(3): 1052-62.
[PMID: 10741734]

[25] Ma Y, Chen Y, Li Y, *et al.* Cystatin A suppresses tumor cell growth through inhibiting epithelial to mesenchymal transition in human lung cancer. Oncotarget 2017; 9(18): 14084-98.
[http://dx.doi.org/10.18632/oncotarget.23505] [PMID: 29581829]

[26] Chen C, Méndez E, Houck J, *et al.* Gene expression profiling identifies genes predictive of oral squamous cell carcinoma. Cancer Epidemiol Biomarkers Prev 2008; 17(8): 2152-62.
[http://dx.doi.org/10.1158/1055-9965.EPI-07-2893] [PMID: 18669583]

[27] Méchin MC, Enji M, Nachat R, *et al.* The peptidylarginine deiminases expressed in human epidermis differ in their substrate specificities and subcellular locations. Cell Mol Life Sci 2005; 62(17): 1984-95.
[http://dx.doi.org/10.1007/s00018-005-5196-y] [PMID: 16091842]

[28] Dharmawardana N, Ooi EH, Woods C, Hussey D. Circulating microRNAs in head and neck cancer: a scoping review of methods. Clin Exp Metastasis 2019; 36(3): 291-302.
[http://dx.doi.org/10.1007/s10585-019-09961-6] [PMID: 30877500]

[29] Yan Y, Wang X, Venø MT, *et al.* Circulating miRNAs as biomarkers for oral squamous cell carcinoma recurrence in operated patients. Oncotarget 2017; 8(5): 8206-14.
[http://dx.doi.org/10.18632/oncotarget.14143] [PMID: 28030794]

[30] Summerer I, Unger K, Braselmann H, *et al.* Circulating microRNAs as prognostic therapy biomarkers in head and neck cancer patients. Br J Cancer 2015; 113(1): 76-82.
[http://dx.doi.org/10.1038/bjc.2015.111] [PMID: 26057452]

[31] Victoria Martinez B, Dhahbi JM, Nunez Lopez YO, *et al.* Circulating small non-coding RNA signature in head and neck squamous cell carcinoma. Oncotarget 2015; 6(22): 19246-63.
[http://dx.doi.org/10.18632/oncotarget.4266] [PMID: 26057471]

[32] Lopez S, Bermudez B, Montserrat-de la Paz S, Abia R, Muriana FJG. A microRNA expression signature of the postprandial state in response to a high-saturated-fat challenge. J Nutr Biochem 2018; 57: 45-55.
[http://dx.doi.org/10.1016/j.jnutbio.2018.03.010] [PMID: 29674246]

[33] Zhao Q, Zheng X, Guo H, *et al.* Serum Exosomal miR-941 as a promising Oncogenic Biomarker for Laryngeal Squamous Cell Carcinoma. J Cancer 2020; 11(18): 5329-44. http://www.jcancer.org/v11p5329.htm [http://dx.doi.org/10.7150/jca.45394] [PMID: 32742479]

[34] Lu E, Su J, Zeng W, Zhang C. Enhanced miR-9 promotes laryngocarcinoma cell survival *via* down-regulating PTEN. Biomed Pharmacother 2016; 84: 608-13. [CrossRef]. [PubMed]. [http://dx.doi.org/10.1016/j.biopha.2016.09.047] [PMID: 27694005]

[35] Zhang L, Sun J, Wang B, Ren J-C, Su W, Zhang T. MicroRNA-10b Triggers the Epithelial-Mesenchymal Transition (EMT) of Laryngeal Carcinoma Hep-2 Cells by Directly Targeting the E-cadherin. Appl Biochem Biotechnol 2015; 176(1): 33-44. [http://dx.doi.org/10.1007/s12010-015-1505-6] [PMID: 25875782]

[36] Liu M, Wu H, Liu T, *et al.* Regulation of the cell cycle gene, BTG2, by miR-21 in human laryngeal carcinoma. Cell Res 2009; 19(7): 828-37. [http://dx.doi.org/10.1038/cr.2009.72] [PMID: 19546886]

[37] Zhong G, Xiong X. miR-205 promotes proliferation and invasion of laryngeal squamous cell carcinoma by suppressing CDK2AP1 expression. Boil Res 2015; 48: 60. [http://dx.doi.org/10.1186/s40659-015-0052-5] [PMID: 26515287]

[38] Sun X, Wang ZM, Song Y, Tai XH, Ji WY, Gu H. MicroRNA-126 modulates the tumor microenvironment by targeting calmodulin-regulated spectrin-associated protein 1 (Camsap1). Int J Oncol 2014; 44(5): 1678-84. [http://dx.doi.org/10.3892/ijo.2014.2321] [PMID: 24603804]

[39] Lian R, Lu B, Jiao L, *et al.* MiR-132 plays an oncogenic role in laryngeal squamous cell carcinoma by targeting FOXO1 and activating the PI3K/AKT pathway. Eur J Pharmacol 2016; 792: 1-6. [http://dx.doi.org/10.1016/j.ejphar.2016.10.015] [PMID: 27751825]

[40] Sun X, Liu B, Zhao X-D, Wang L-Y, Ji WY. MicroRNA-221 accelerates the proliferation of laryngeal cancer cell line Hep-2 by suppressing Apaf-1. Oncol Rep 2015; 33(3): 1221-6. [http://dx.doi.org/10.3892/or.2015.3714] [PMID: 25586265]

[41] Lu Y, Gao W, Zhang C, *et al.* Hsa-miR-301a-3p Acts as an Oncogene in Laryngeal Squamous Cell Carcinoma *via* Target Regulation of Smad4. J Cancer 2015; 6(12): 1260-75. [http://dx.doi.org/10.7150/jca.12659] [PMID: 26640587]

[42] Guan G, Zhang D, Zheng Y, *et al.* microRNA-423-3p promotes tumor progression *via* modulation of AdipoR2 in laryngeal carcinoma. Int J Clin Exp Pathol 2014; 7(9): 5683-91. [PMID: 25337209]

[43] Wang F, Song G, Liu M, Li X, Tang H. miRNA-1 targets fibronectin1 and suppresses the migration and invasion of the HEp2 laryngeal squamous carcinoma cell line. FEBS Lett 2011; 585(20): 3263-9. [http://dx.doi.org/10.1016/j.febslet.2011.08.052] [PMID: 21924268]

[44] Xu L, Chen Z, Xue F, *et al.* MicroRNA-24 inhibits growth, induces apoptosis, and reverses radioresistance in laryngeal squamous cell carcinoma by targeting X-linked inhibitor of apoptosis protein. Cancer Cell Int 2015; 15: 61. [http://dx.doi.org/10.1186/s12935-015-0217-x] [PMID: 26106283]

[45] Ye J, Li L, Feng P, Wan J, Li J. Downregulation of miR-34a contributes to the proliferation and migration of laryngeal carcinoma cells by targeting cyclin D1. Oncol Rep 2016; 36(1): 390-8. [http://dx.doi.org/10.3892/or.2016.4823] [PMID: 27220728]

[46] Maroof H, Salajegheh A, Smith RA, Lam AK-Y. MicroRNA-34 family, mechanisms of action in cancer: a review. Curr Cancer Drug Targets 2014; 14(8): 737-51. [http://dx.doi.org/10.2174/1568009614666141020100337] [PMID: 25329673]

[47] Cai K-M, Bao X-L, Kong X-H, *et al.* Hsa-miR-34c suppresses growth and invasion of human laryngeal carcinoma cells *via* targeting c-Met. Int J Mol Med 2010; 25(4): 565-71.

[http://dx.doi.org/10.3892/ijmm_00000378] [PMID: 20198305]

[48] Li W, Ma H, Sun J. MicroRNA□34a/c function as tumor suppressors in Hep□2 laryngeal carcinoma cells and may reduce GALNT7 expression. Mol Med Rep 2014; 9(4): 1293-8. [http://dx.doi.org/10.3892/mmr.2014.1929] [PMID: 24482044]

[49] Wu X, Cui C-L, Chen W-L, Fu ZY, Cui XY, Gong X. miR-144 suppresses the growth and metastasis of laryngeal squamous cell carcinoma by targeting IRS1. Am J Transl Res 2016; 8(1): 1-11. [PMID: 27069535]

[50] Yu W, Zhang G, Lu B, *et al.* MiR-340 impedes the progression of laryngeal squamous cell carcinoma by targeting EZH2. Gene 2016; 577(2): 193-201. [http://dx.doi.org/10.1016/j.gene.2015.11.045] [PMID: 26656176]

[51] Song F-C, Yang Y, Liu J-X. Expression and significances of MiRNA Let-7 and HMGA2 in laryngeal carcinoma. Eur Rev Med Pharmacol Sci 2016; 20(21): 4452-8. [PubMed]. [PMID: 27874955]

[52] Wang J, Zhou Y, Lu J, *et al.* Combined detection of serum exosomal miR-21 and HOTAIR as diagnostic and prognostic biomarkers for laryngeal squamous cell carcinoma. Med Oncol 2014; 31(9): 148. [http://dx.doi.org/10.1007/s12032-014-0148-8] [PMID: 25099764]

[53] Hu A, Huang J-J, Xu W-H, *et al.* miR-21 and miR-375 microRNAs as candidate diagnostic biomarkers in squamous cell carcinoma of the larynx: association with patient survival. Am J Transl Res 2014; 6(5): 604-13. [PMID: 25360224]

[54] Wang Y, Chen M, Tao Z, Hua Q, Chen S, Xiao B. Identification of predictive biomarkers for early diagnosis of larynx carcinoma based on microRNA expression data. Cancer Genet 2013; 206(9-10): 340-6. [http://dx.doi.org/10.1016/j.cancergen.2013.09.005] [PMID: 24238754]

[55] Wu Y, Yu J, Ma Y, Wang F, Liu H. miR-148a and miR-375 may serve as predictive biomarkers for early diagnosis of laryngeal carcinoma. Oncol Lett 2016; 12(2): 871-8. [http://dx.doi.org/10.3892/ol.2016.4707] [PMID: 27446362]

[56] Ayaz L, Görür A, Yaroğlu HY, Ozcan C, Tamer L. Differential expression of microRNAs in plasma of patients with laryngeal squamous cell carcinoma: potential early-detection markers for laryngeal squamous cell carcinoma. J Cancer Res Clin Oncol 2013; 139(9): 1499-506. [http://dx.doi.org/10.1007/s00432-013-1469-2] [PMID: 23817697]

[57] Xu Y, Lin YP, Yang D, Zhang G, Zhou HF. Expression of serum microRNA-378 and its clinical significance in laryngeal squamous cell carcinoma. Eur Rev Med Pharmacol Sci 2016; 20(24): 5137-42. [PMID: 28051256]

[58] Lucas Grzelczyk W, Szemraj J, Kwiatkowska S, Józefowicz-Korczyńska M. Serum expression of selected miRNAs in patients with laryngeal squamous cell carcinoma (LSCC). Diagn Pathol 2019; 14(1): 49. [http://dx.doi.org/10.1186/s13000-019-0823-3] [PMID: 31138255]

[59] Childs G, Fazzari M, Kung G, *et al.* Low-level expression of microRNAs let-7d and miR-205 are prognostic markers of head and neck squamous cell carcinoma. Am J Pathol 2009; 174(3): 736-45. [http://dx.doi.org/10.2353/ajpath.2009.080731] [PMID: 19179615]

[60] Zhang X-W, Liu N, Chen S, *et al.* High microRNA-23a expression in laryngeal squamous cell carcinoma is associated with poor patient prognosis. Diagn Pathol 2015; 10: 22. [http://dx.doi.org/10.1186/s13000-015-0256-6] [PMID: 25879432]

[61] Li M, Tian L, Ren H, *et al.* MicroRNA-101 is a potential prognostic indicator of laryngeal squamous cell carcinoma and modulates CDK8. J Transl Med 2015; 13: 271.

[http://dx.doi.org/10.1186/s12967-015-0626-6] [PMID: 26286725]

[62] Wu S, Jia S, Xu P. MicroRNA-9 as a novel prognostic biomarker in human laryngeal squamous cell carcinoma. Int J Clin Exp Med 2014; 7(12): 5523-8. [PMID: 25664065]

[63] Maia D, de Carvalho AC, Horst MA, Carvalho AL, Scapulatempo-Neto C, Vettore AL. Expression of miR-296-5p as predictive marker for radiotherapy resistance in early-stage laryngeal carcinoma. J Transl Med 2015; 13: 262. [http://dx.doi.org/10.1186/s12967-015-0621-y] [PMID: 26264462]

[64] Cappellesso R, Marioni G, Crescenzi M, *et al.* The prognostic role of the epithelial-mesenchymal transition markers E-cadherin and Slug in laryngeal squamous cell carcinoma. Histopathology 2015; 67(4): 491-500. [http://dx.doi.org/10.1111/his.12668] [PMID: 25684546]

[65] Hu HY, He L, Fominykh K, *et al.* Evolution of the human-specific microRNA miR-941. Nat Commun 2012; 3: 1145. [http://dx.doi.org/10.1038/ncomms2146] [PMID: 23093182]

[66] Liu N, Boohaker RJ, Jiang C, Boohaker JR, Xu B. A radiosensitivity MiRNA signature validated by the TCGA database for head and neck squamous cell carcinomas. Oncotarget 2015; 6(33): 34649-57. [http://dx.doi.org/10.18632/oncotarget.5299] [PMID: 26452218]

[67] de Jong MC, Ten Hoeve JJ, Grénman R, *et al.* Pretreatment microRNA expression impacting on epithelial to mesenchymal transition predicts intrinsic radiosensitivity in head and neck cancer celllines and patients. Clin Cancer Res 2015; 21(24): 5630-8. [http://dx.doi.org/10.1158/1078-0432.CCR-15-0454] [PMID: 26265694]

[68] Zhu Y, Shi LY, Lei Y-M, *et al.* Radiosensitization effect of hsa-miR-138-2-3p on human laryngeal cancer stem cells. PeerJ 2017; 5: e3233. [http://dx.doi.org/10.7717/peerj.3233] [PMID: 28533948]

[69] Yang P-Y, Hsieh P-L, Wang TH, *et al.* Andrographolide impedes cancer stemness and enhances radio-sensitivity in oral carcinomas *via* miR-218 activation. Oncotarget 2017; 8(3): 4196-207. [CrossRef]. [PubMed]. [http://dx.doi.org/10.18632/oncotarget.13755] [PMID: 27926533]

[70] Ahmad P, Sana J, Slavik M, *et al.* MicroRNA-15b-5p Predicts Locoregional Relapse in Head and Neck Carcinoma Patients Treated With Intensity-modulated Radiotherapy. Cancer Genomics Proteomics 2019; 16(2): 139-46. [http://dx.doi.org/10.21873/cgp.20119] [PMID: 30850365]

[71] Liu M, Wang J, Huang H, Hou J, Zhang B, Wang A. miR-181a-Twist1 pathway in the chemoresistance of tongue squamous cell carcinoma. Biochem Biophys Res Commun 2013; 441(2): 364-70. [http://dx.doi.org/10.1016/j.bbrc.2013.10.051] [PMID: 24148247]

[72] Jiang F, Zhao W, Zhou L, Liu Z, Li W, Yu D. MiR-222 targeted PUMA to improve sensitization of UM1 cells to cisplatin. Int J Mol Sci 2014; 15(12): 22128-41. [http://dx.doi.org/10.3390/ijms151222128] [PMID: 25474084]

[73] Liu X-Y, Liu Z-J, He H, Zhang C, Wang Y-L. MicroRNA-101-3p suppresses cell proliferation, invasion and enhances chemotherapeutic sensitivity in salivary gland adenoid cystic carcinoma by targeting Pim-1. Am J Cancer Res 2015; 5(10): 3015-29. [PMID: 26693056]

[74] Qin X, Guo H, Wang X, *et al.* Exosomal miR-196a derived from cancer-associated fibroblasts confers cisplatin resistance in head and neck cancer through targeting CDKN1B and ING5. Genome Biol 2019; 20(1): 12. [http://dx.doi.org/10.1186/s13059-018-1604-0] [PMID: 30642385]

[75] Vyas A, Harbison RA, Faden D. Recurrent human papillomavirus- related head and neck cancer undergoes metabolic re-programming and is driven by oxidative phosphorylation. bioRxiv 2020. Available under a CC-BY-NC-ND 4.0 International license.
[http://dx.doi.org/10.1101/2020.08.21.261206]

[76] Castellsagué X, Alemany L, Quer M, *et al.* HPV Involvement in Head and Neck Cancers: Comprehensive Assessment of Biomarkers in 3680 Patients. J Natl Cancer Inst 2016; 108(6).
[http://dx.doi.org/10.1093/jnci/djv403]

[77] Lajer CB, Garnæs E, Friis-Hansen L, *et al.* The role of miRNAs in human papilloma virus (HPV)-associated cancers: bridging between HPV-related head and neck cancer and cervical cancer. Br J Cancer 2012; 106(9): 1526-34.
[http://dx.doi.org/10.1038/bjc.2012.109] [PMID: 22472886]

[78] Bersani C, Mints M, Tertipis N, *et al.* MicroRNA-155, -185 and -193b as biomarkers in human papillomavirus positive and negative tonsillar and base of tongue squamous cell carcinoma. Oral Oncol 2018; 82: 8-16.
[http://dx.doi.org/10.1016/j.oraloncology.2018.04.021] [PMID: 29909906]

[79] House R, Majumder M, Janakiraman H, *et al.* Smoking-induced control of miR-133a-3p alters the expression of EGFR and HuR in HPV-infected oropharyngeal cancer. PLoS One 2018; 13(10): e0205077.
[http://dx.doi.org/10.1371/journal.pone.0205077] [PMID: 30289952]

[80] Yao Y, Chen X, Yang H, *et al.* Hsa_circ_0058124 promotes papillary thyroid cancer tumorigenesis and invasiveness through the NOTCH3/GATAD2A axis. J Exp Clin Cancer Res 2019; 38(1): 318.
[http://dx.doi.org/10.1186/s13046-019-1321-x] [PMID: 31324198]

[81] Shi E, Ye J, Zhang R, *et al.* A Combination of circRNAs as a Diagnostic Tool for Discrimination of Papillary Thyroid Cancer. Onco Targets Ther 2020; 13: 4365-72.
[http://dx.doi.org/10.2147/OTT.S247796]

[82] Kelly LM, Barila G, Liu P, *et al.* Identification of the transforming STRN-ALK fusion as a potential therapeutic target in the aggressive forms of thyroid cancer. Proc Natl Acad Sci USA 2014; 111(11): 4233-8.
[http://dx.doi.org/10.1073/pnas.1321937111] [PMID: 24613930]

[83] Chou A, Fraser S, Toon CW, *et al.* A detailed clinicopathologic study of ALK-translocated papillary thyroid carcinoma. Am J Surg Pathol 2015; 39(5): 652-9.
[http://dx.doi.org/10.1097/PAS.0000000000000368] [PMID: 25501013]

[84] Zeng W, Chang H, Ma M, Li Y. CCL20/CCR6 promotes the invasion and migration of thyroid cancer cells *via* NF-kappa B signaling-induced MMP-3 production. Exp Mol Pathol 2014; 97(1): 184-90.
[http://dx.doi.org/10.1016/j.yexmp.2014.06.012] [PMID: 24984269]

[85] Bauerle KT, Schweppe RE, Lund G, *et al.* Nuclear factor κB-dependent regulation of angiogenesis, and metastasis in an *in vivo* model of thyroid cancer is associated with secreted interleukin-8. J Clin Endocrinol Metab 2014; 99(8): E1436-44.
[http://dx.doi.org/10.1210/jc.2013-3636] [PMID: 24758177]

[86] Bauerle KT, Schweppe RE, Haugen BR. Inhibition of nuclear factor-kappa B differentially affects thyroid cancer cell growth, apoptosis, and invasion. Mol Cancer 2010; 9: 117.
[http://dx.doi.org/10.1186/1476-4598-9-117] [PMID: 20492683]

[87] Nikiforova M N, Tseng G C, Steward D, Diorio D, Nikiforov Y E. MicroRNA expression profiling of thyroid tumors: biological significance and diagnostic utility. e Journal of Clinical Endocrinology & Metabolism 2008; 93(5): 1600-8.
[http://dx.doi.org/10.1210/jc.2007-2696]

[88] Chou C-K, Yang K D, Chou F-F, *et al.* Prognostic implications of miR-146b expression and its functional role in papillary thyroid carcinoma e Journal of Clinical Endocrinology & Metabolism

2013; 98(2): E196-205.
[http://dx.doi.org/10.1210/jc.2012-2666]

[89] Acibucu F, Dökmetaş HS, Tutar Y, Elagoz S, Kilicli F. Correlations between the expression levels of micro-RNA146b, 221, 222 and p27Kip1 protein mRNA and the clinicopathologic parameters in papillary thyroid cancers. Exp Clin Endocrinol Diabetes 2014; 122(3): 137-43.
[http://dx.doi.org/10.1055/s-0034-1367025] [PMID: 24643689]

[90] Liu L, Wang J, Li X, *et al.* MiR-204-5p suppresses cell proliferation by inhibiting IGFBP5 in papillary thyroid carcinoma. Biochem Biophys Res Commun 2015; 457(4): 621-6.
[http://dx.doi.org/10.1016/j.bbrc.2015.01.037] [PMID: 25603050]

[91] Santiago K, Chen Wongworawat Y, Khan S. Differential MicroRNA-signatures in thyroid cancer subtypes. J Oncol 2020; 2020: 2052396.
[http://dx.doi.org/10.1155/2020/2052396]

[92] Santos JE, Freitas M, Fonseca CP, *et al.* Iodine deficiency a persisting problem: assessment of iodine nutrition and evaluation of thyroid nodular pathology in Portugal. J Endocrinol Invest 2017; 40(2): 185-91.
[http://dx.doi.org/10.1007/s40618-016-0545-2] [PMID: 27619914]

[93] Jang EK, Song DE, Sim SY, *et al.* NRAS codon 61 mutation is associated with distant metastasis in patients with follicular thyroid carcinoma. Thyroid 2014; 24(8): 1275-81.
[http://dx.doi.org/10.1089/thy.2014.0053] [PMID: 24820091]

[94] Boufraqech M, Nilubol N. Multi-omics Signatures and Translational Potential to Improve Thyroid Cancer Patient Outcome. Cancers (Basel) 2019; 11(12): 10.
[http://dx.doi.org/10.3390/cancers11121988] [PMID: 31835496]

[95] Cinpolat O, Unal ZN, Ismi O, Gorur A, Unal M. Comparison of microRNA profiles between benign and malignant salivary gland tumors in tissue, blood and saliva samples: a prospective, case-control study. Rev Bras Otorrinolaringol (Engl Ed) 2017; 83(3): 276-84.
[http://dx.doi.org/10.1016/j.bjorl.2016.03.013] [PMID: 27184509]

[96] Miyashita T, Burford JL, Hong Y-K, *et al.* Localization and proliferation of lymphatic vessels in the tympanic membrane in normal state and regeneration. Biochem Biophys Res Commun 2013; 440(3): 371-3.
[http://dx.doi.org/10.1016/j.bbrc.2013.09.054] [PMID: 24055876]

[97] Urík M, Hurník P, Žiak D, *et al.* Immunohistochemical analysis of retraction pocket pars tensa of tympanic membrane in children. Int J Pediatr Otorhinolaryngol 2019; 122: 111-6.
[http://dx.doi.org/10.1016/j.ijporl.2019.04.008] [PMID: 30999159]

[98] Harabagiu OE, Cosgarea M, Mogoantă CA, Leucuţa DC, Maniu AA. Keratinocyte growth factor and its receptor expression in chronic otitis media with and without cholesteatoma. Rom J Morphol Embryol 2017; 58(4): 1333-8.
[PMID: 29556625]

[99] Thorup MB, Munk M, Poulsen SS, *et al.* Expression of the epidermal growth factor system in human middle ear cholesteatoma. Acta Otolaryngol 2014; 134(2): 124-34.
[http://dx.doi.org/10.3109/00016489.2013.847282] [PMID: 24256036]

[100] Xie S, Wang X, Ren H, Liu X, Ren J, Liu W. HB-EGF expression as a potential biomarker of acquired middle ear cholesteatoma. Acta Otolaryngol 2017; 137(8): 797-802.
[http://dx.doi.org/10.1080/00016489.2017.1284343] [PMID: 28498080]

[101] Koizumi H, Kawaguchi R, Ohkubo JI, *et al.* Expressions of isopeptide bonds and corneodesmosin in middle ear cholesteatoma. Clin Otolaryngol 2017; 42(2): 252-62.
[http://dx.doi.org/10.1111/coa.12703] [PMID: 27390311]

[102] Zang J, Hui L, Yang N, Yang B, Jiang X. Downregulation of MiR-203a Disinhibits Bmi1 and Promotes Growth and Proliferation of Keratinocytes in Cholesteatoma. Int J Med Sci 2018; 15(5):

447-55.
[http://dx.doi.org/10.7150/ijms.22410]

[103] Zhang W, Chen X, Qin Z. MicroRNA let-7a suppresses the growth and invasion of cholesteatoma keratinocytes. Mol Med Rep 2015; 11(3): 2097-103.
[http://dx.doi.org/10.3892/mmr.2014.2971] [PMID: 25405753]

[104] Song G, Ouyang G, Bao S. The activation of Akt/PKB signaling pathway and cell survival. J Cell Mol Med 2005; 9(1): 59-71.
[http://dx.doi.org/10.1111/j.1582-4934.2005.tb00337.x] [PMID: 15784165]

[105] Huisman MA, De Heer E, Grote JJ. Survival signaling and terminal differentiation in cholesteatoma epithelium. Acta Otolaryngol 2007; 127(4): 424-9.
[http://dx.doi.org/10.1080/00016480600868430] [PMID: 17453465]

[106] Yune TY, Byun JY. Expression of PTEN and phosphorylated Akt in human cholesteatoma epithelium. Acta Otolaryngol 2009; 129(5): 501-6.
[http://dx.doi.org/10.1080/00016480802258802] [PMID: 18615333]

[107] Hawkins PT, Anderson KE, Davidson K, Stephens LR. Signalling through Class I PI3Ks in mammalian cells. Biochem Soc Trans 2006; 34(Pt 5): 647-62.
[http://dx.doi.org/10.1042/BST0340647] [PMID: 17052169]

[108] Gao J, Tang Q, Xue R, *et al.* Comprehensive circular RNA expression profiling with associated ceRNA network reveals their therapeutic potential in cholesteatoma. Oncol Rep 2020; 43(4): 1234-44.
[http://dx.doi.org/10.3892/or.2020.7501] [PMID: 32323838]

[109] Hansdah K, Singh N, Bouzid A, *et al.* Evaluation of the Genetic Association and mRNA Expression of the *COL1A1*, *BMP2*, and *BMP4* Genes in the Development of Otosclerosis. Genet Test Mol Biomarkers 2020; 24(6): 343-51.
[http://dx.doi.org/10.1089/gtmb.2019.0235] [PMID: 32379989]

[110] Csomor P, Liktor B, Liktor B, Szekanecz Z, Sziklai I, Karosi T. Expression of bone morphogenetic protein 2, 4, 5, and 7 correlates with histological activity of otosclerotic foci. Acta Otolaryngol 2012; 132(6): 624-31.
[http://dx.doi.org/10.3109/00016489.2011.653669] [PMID: 22385409]

[111] Stankovic KM, Adachi O, Tsuji K, *et al.* Differences in gene expression between the otic capsule and other bones. Hear Res 2010; 265(1-2): 83-9.
[http://dx.doi.org/10.1016/j.heares.2010.02.006] [PMID: 20146935]

[112] Adams JC. Clinical implications of inflammatory cytokines in the cochlea: a technical note. Otol Neurotol 2002; 23(3): 316-22.
[http://dx.doi.org/10.1097/00129492-200205000-00015] [PMID: 11981388]

[113] Lang H, Schulte BA, Zhou D, Smythe N, Spicer SS, Schmiedt RA. Nuclear factor kappaB deficiency is associated with auditory nerve degeneration and increased noise-induced hearing loss. J Neurosci 2006; 26(13): 3541-50.
[http://dx.doi.org/10.1523/JNEUROSCI.2488-05.2006] [PMID: 16571762]

[114] Lin H-C, Ren Y, Lysaght AC, Kao S-Y, Stankovic KM. Proteome of normal human perilymph and perilymph from people with disabling vertigo. PLoS One 2019; 14(6): e0218292.
[http://dx.doi.org/10.1371/journal.pone.0218292] [PMID: 31185063]

[115] Jin Y, Wang J, Ye X, *et al.* Identification of GlcNAcylated alpha-1-antichymotrypsin as an early biomarker in human non-small-cell lung cancer by quantitative proteomic analysis with two lectins. Br J Cancer 2016; 114(5): 532-44.
[http://dx.doi.org/10.1038/bjc.2015.348] [PMID: 26908325]

[116] Hartman BH, Nelson BR, Reh TA, Bermingham-McDonogh O. Bermingham-McDonogh O. Delta/Notch-Like EGF-Related Receptor (DNER) is Expressed in Hair Cells and Neurons in the Developing and Adult Mouse Inner Ear. Journal of the Association for Research in Otolaryngology

2010; 11: 187-201.
[http://dx.doi.org/10.1007/s10162-009-0203-x]

[117] Eckhard A, Gleiser C, Arnold H, *et al.* Water channel proteins in the inner ear and their link to hearing impairment and deafness. Mol Aspects Med 2012; 33(5-6): 612-37.
[http://dx.doi.org/10.1016/j.mam.2012.06.004] [PMID: 22732097]

[118] Shew M, Wichova H, Bur A, *et al.* MicroRNA Profiling as a Methodology to Diagnose Ménière's Disease: Potential Application of Machine Learning. Otolaryngol Head Neck Surg 2020; 194599820940649.
[http://dx.doi.org/10.1177/0194599820940649] [PMID: 32663060]

[119] Korkmaz T, Bicer YO, Serin E, Seyhan S, Sanal SK. Salivary α-amylase levels in vertigo: Can it be an autonomic dysfunction? Ear Nose Throat J 2018; 97(9): 278-82.
[http://dx.doi.org/10.1177/014556131809700917] [PMID: 30273427]

[120] Takeda T, Kakigi A, Saito H. Antidiuretic hormone (ADH) and endolymphatic hydrops. Acta Otolaryngol Suppl 1995; 519: 219-22.
[http://dx.doi.org/10.3109/00016489509121909] [PMID: 7610873]

[121] Landegger LD, Vasilijic S, Fujita T, *et al.* Cytokine Levels in Inner Ear Fluid of Young and Aged Mice as Molecular Biomarkers of Noise-Induced Hearing Loss. Front Neurol 2019; 10: 977.
[http://dx.doi.org/10.3389/fneur.2019.00977]

[122] Eggermont JJ, Roberts LE. The neuroscience of tinnitus. Trends Neurosci 2004; 27(11): 676-82.
[http://dx.doi.org/10.1016/j.tins.2004.08.010] [PMID: 15474168]

[123] Chen XH, Zheng LL. Expression of pro-inflammatory cytokines in the auditory cortex of rats with salicylate-induced tinnitus. Mol Med Rep 2017; 16(4): 5643-8.
[http://dx.doi.org/10.3892/mmr.2017.7235] [PMID: 28849125]

[124] Stellwagen D, Malenka RC. Synaptic scaling mediated by glial TNF-alpha. Nature 2006; 440(7087): 1054-9.
[http://dx.doi.org/10.1038/nature04671] [PMID: 16547515]

[125] Lee RH, Mills EA, Schwartz N, *et al.* Neurodevelopmental effects of chronic exposure to elevated levels of pro-inflammatory cytokines in a developing visual system. Neural Dev 2010; 5: 2.
[http://dx.doi.org/10.1186/1749-8104-5-2] [PMID: 20067608]

[126] Marshall GD Jr, Agarwal SK, Lloyd C, Cohen L, Henninger EM, Morris GJ. Cytokine dysregulation associated with exam stress in healthy medical students. Brain Behav Immun 1998; 12(4): 297-307.
[http://dx.doi.org/10.1006/brbi.1998.0537] [PMID: 10080859]

[127] Seidman MD, Khan MJ, Tang WX, Quirk WS. Influence of lecithin on mitochondrial DNA and age-related hearing loss. Otolaryngol Head Neck Surg 2002; 127(3): 138-44.
[http://dx.doi.org/10.1067/mhn.2002.127627] [PMID: 12297801]

[128] Fujimoto C, Yamasoba T. Oxidative stresses and mitochondrial dysfunction in age-related hearing loss. Oxid Med Cell Longev 2014; 2014: 582849.
[http://dx.doi.org/10.1155/2014/582849] [PMID: 25110550]

[129] Haider HF, Flook M, Aparicio M, *et al.* Biomarkers of presbycusis and tinnitus in a portuguese older population. Front Aging Neurosci 2017; 9: 346.
[http://dx.doi.org/10.3389/fnagi.2017.00346] [PMID: 29163129]

CHAPTER 12

Pharmacogenomic Biomarkers

Zeynep Gizem Todurga Seven[1,*], **Deniz Özen**[1] and **Sibel Özyazgan**[1]

[1] *Department of Medical Pharmacology, Cerrahpasa Medical Faculty, Istanbul University-Cerrahpasa, Istanbul, Turkey*

Abstract: Why does the usual dose of medication work for a person while another individual cannot give the expected response to the same drug? On the other hand, how come half of the usual dose of an analgesic relieves an individual's pain immediately, as another man continue to suffer even after taking double dose? Although a treatment method has been successfully used in majority of the population for many years, why does the same therapy cause serious side effects in another region of the world? Most presently approved therapies are not effective in all patients. For example, 20-40% of patients with depression respond poorly or not at all to antidepressant drug therapy. Many patients are resistant to the effects of antiasthmatics and antiulcer drugs or drug treatment of hyperlipidemia and many other diseases. The reason for all those is basically interindividual differences in genomic structures of people, which are explained in this chapter in terms of the systems and the most frequently used drugs in clinical treatment.

Keywords: Adverse reactions, Carriers, Clinical pharmacology, Cytochrome P-450 enzyme family, Drug response, Enzyme activity, Extensive metabolizer, Intermediate metabolizer, Metabolism, Optimal dosage, Personalized medicine, Pharmacodynamics, Pharmacogenetics, Pharmacogenomic biomarkers, Pharmacokinetics, Polymorphism, Poor metabolizer, Targeted treatment, Transporter molecules, Ultra-rapid metabolizer.

INTRODUCTION

Personalized medicine has been an outstanding subject for overcoming this challenge and providing each individual with a tailored treatment rather than the traditional method of "trial-and-error" dosing. The primary goals of personalized medicine are to minimize the potential adverse drug reactions and obtain the optimal efficacy from the medication.

[*] **Corresponding author Zeynep Gizem Todurga Seven:** Department of Medical Pharmacology, Cerrahpasa Medical Faculty, Istanbul University-Cerrahpasa, Istanbul, Turkey; Tel: +90 (212) 414 30 00; E-mail: zeyneptodurga90@gmail.com

Pınar Atukeren and Hafize Uzun (Eds.)

Many drugs were withdrawn from the market because they caused severe adverse effects, so targeted drugadministration or dosing based on an individual's genetics has become important to avoid withdrawals. Moreover, the optimum efficacy can be reached when the drug is administered in the right dosage for each individual. For instance, 40-50% of patients with asthma, migraine, or diabetes cannot benefit from their medications, while the response rate of patients to cancer chemotherapy is only 25%.

The individual risk for drug inefficacy or drug toxicity is a product of the interaction between genes and the environment. The environmental variables can be nutritional factors, lifestyle influences such as smoking, barbequed meat or alcohol consumption, and also diseases or concomitantly administered drugs [1]. These factors act in concert with an individual's genes that code for pharmacokinetic and pharmacodynamic determinants of a drug, such as receptors, ion channels, drug-metabolizing enzymes, and drug transporters. In brief, the aim of personalized medicine is to give the right drug in the right dose at the right time to the right patient, having the right symptom for optimum efficacy while avoiding adverse effects [2].

Pharmacogenetics deals with the effect of individual gene variants on drug action. It investigates the simple relation between hereditary factors and the diversity of drug response. Simple mutations, which play important roles in the response of an individual to a specific drug, are determined by combining genetics and pharmacological methods. On the other hand, genomics involves the systematic identification of all human genes and gene products. It is the study of human genetic variations in health and disease, as well as changes in gene and protein expression over time. Pharmacogenetics has been evolved into pharmacogenomics to study the entire genomic structure that determines drug response, including the diversity of the human genome sequence and its clinical consequences. According to the definition of Food and Drug Administration (FDA), "Pharmacogenomics is the discipline for the estimation of drug response of an individual based on the genetic structure." Genetic information is being used to guide both drug and dosage selection for individual patients in clinical practice.

Pharmacogenomic biomarkers are variations, leading to person-to-person differences in the metabolism of drugs and thus, in drug kinetics and response. Single nucleotide polymorphisms (SNPs) are single-base differences that exist between individuals with a frequency of more than 1%, and they are the most common types of variations among pharmacogenomic biomarkers. An SNP may cause an amino acid substitution during transcription (non-synonymous/ missense), which affects protein activity; on the other hand, it may not result in the amino acid change (synonymous), but still, it may alter the gene expression and the amount of the product. Other types of variations include insertion, deletion, inversion of the gene, encoding the protein of interest, and also copy number variation, tandem duplication, dispersed duplication, mobile element insertion, or translocation (Fig. **1**).

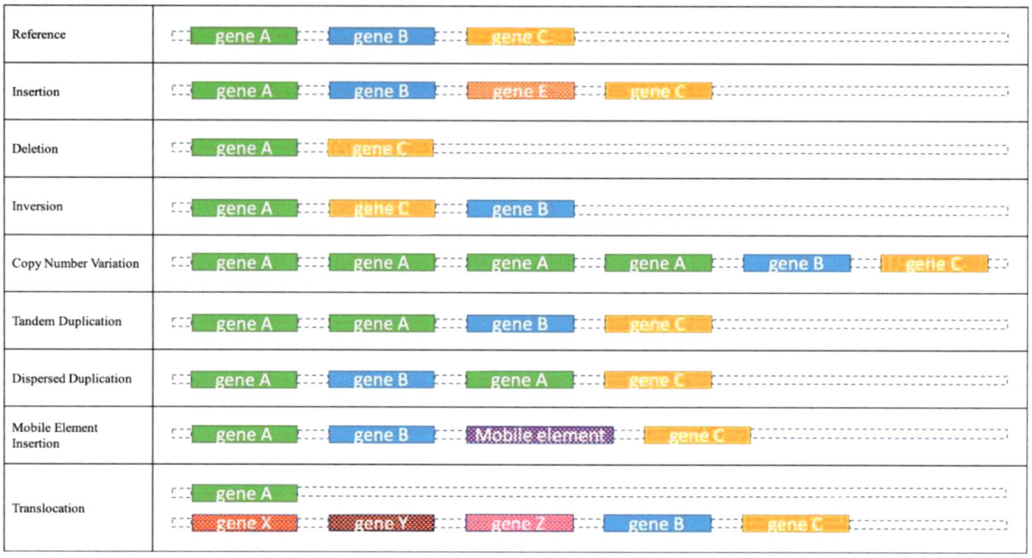

Fig. (1). Different types of genetic variations.

Polymorphisms and other genetic variations commonly occur for genes encoding proteins that function in drug metabolism, drug transporters, and drug target proteins. Drug metabolism and drug transporter genotypes can alter drug availability at the target site. In addition, drug target genotypes can affect a patient's sensitivity to a drug. If a large number of variations and their frequencies in different populations are known, they can be used to generate an individual's genetic fingerprint, indicating the individual's potential drug response to any given drug. Thus, those biomarkers can be used to provide a drug response profile, which is the resulting phenotype after all of the known affecting variations are contributed.

There are two main concepts, which are pharmacokinetics and pharmacodynamics, investigating the relationship between a drug and the human body after administration; there are specific proteins working in the process. If there is one of any aforementioned variations on the encoding gene of a protein, its function may be changed, which constitutes differences in the individual's drug response profile.

Pharmacokinetics explains drug movement in the body, how it gets into the system, distributes through the body, and gets out of the system. Pharmacokinetics mainly consists of four stages, including absorption, distribution, metabolism, and excretion. When a drug is taken orally, firstly, it enters the bloodstream through absorption, and this process involves transporter molecules and carriers. Once the drug enters the gut, it passes to the liver first for detoxification with the hepatic enzymes before it is distributed to other tissues and organs. During metabolism, the parent drug is chemically altered into a structure to be easily removed from the system by different enzymes. Metabolization can sometimes lead to inactivation of the drug, or sometimes a prodrug can be activated in the liver through enzymatic reactions. Finally, excretion is the process of the drug clearance from the body through the bowel or urine, in which transporter molecules and carriers are involved.

The cytochrome P-450 (CYP) superfamily of isoenzymes are mostly produced by hepatic cells, and they are grouped into 18 families and 44 subfamilies based on the similarity in their amino acid sequences. There are 57 different CYP enzymes, 15 of which are involved in the metabolism of drugs in humans (CYP1, CYP2, and CYP3 families). Main CYP enzymes are demonstrated based on the ratio of the drugs in Fig. (**2**) with some of their substrates. These enzymes are monooxygenases, which play a role in oxidation, reduction, and hydrolysis reactions during phase I metabolism of the drugs [3]. These reactions may convert inactive prodrugs into active metabolites or may convert active drugs into inactive metabolites for excretion.

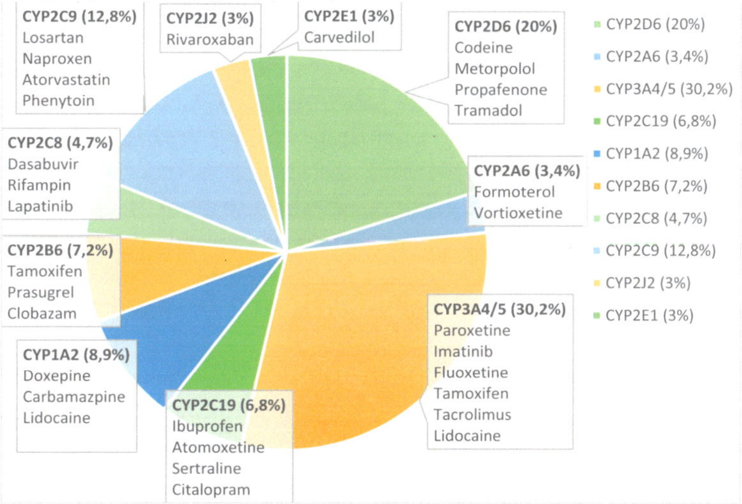

Fig. (2). Basic CYP enzymes and some of their drug substrates.

The genes encoding CYPs are highly polymorphic, resulting in variations in the expression or activity of the enzyme. CYP3A4/5, CYP2C9, CYP2D6, and CYP2C19 are the enzymes that mediate the oxidation of almost 79% of all registered drugs. CYP3A4, 2C9, 2C8, 2E1, and 1A2 are expressed in the highest amounts by hepatic cells, while CYP2A6, 2D6, 2C19, and 3A5 are the less abundant enzymes. Moreover, CYP2J2, 1A1, and 1B1 are expressed extrahepatically but still have roles in the pharmacokinetics of drugs.

The differences in the number of functional variants of alleles of the gene encoding a specific CYP determine the enzyme activity and expression rate. The individuals are classified into 4 general metabolic phenotypes based on their enzyme activity:

1. Extensive metabolizers: They have two functional alleles, so the enzyme works ideally with normal activity.

2. Intermediate metabolizers: They carry either one functional or two partially defective alleles. In both cases, the enzyme has reduced activity.

3. Ultra-rapid metabolizers: They carry a duplicated or amplified gene variant, resulting in two or more multiple copies of the functional allele. The enzyme activity is very high.

4. Poor metabolizers: They have defected or deleted alleles, resulting in abolished or no enzyme activity [2].

For the majority of the drugs, poor metabolizers exhibit a higher risk of adverse drug reactions, and ultra-rapid metabolizers experience lower efficacy when the standard dosage regimen of a drug is administered, which is mostly dependent on the polymorphic enzyme for elimination. Nevertheless, in the case of administration of a prodrug, ultra-rapid metabolizers have a higher risk of adverse drug reactions, while poor metabolizers obtain a lower efficacy from the drug.

Non-CYP drug-metabolizing enzymes are conjugating enzymes, which often mediate phase II metabolism of the drugs during the metabolism of elimination. They make small modifications to the drugs to make the drug soluble by the addition or subtraction of chemical groups, like methyl, sulfate, phosphate, uridine groups. The most abundant conjugating enzymes are UDP-glucuronosyltransferase (UGT), thiopurine-S-methyltransferase (TPMT), dihydropyrimidine dehydrogenase (DPD), N-acetyltransferase (NAT), and glutathione-S-transferase (GST). Individual drug response may be affected by variations in the genes encoding these enzymes. UGTs are involved in the inactivation and elimination of several nucleophilic substances by increasing the polarity of the molecule *via* glucuronidation reaction. They are localized to the internal membrane of the endoplasmic reticulum and take place in about 35% of phase II drug metabolism. TPMT mediated the inactivation of thiopurine drugs. DPDs catalyze the metabolism of 5-fluorouracil (5-FU). NATs participate in the transfer of an acetyl group from acetyl-CoA during the conjugation of metabolites for the excretion of the byproducts. GSTs catalyze the detoxification of drugs through the conjugation of reduced glutathione with different substrates.

During absorption and excretion, a variety of transporters and carriers mediate the passage of the drug molecule through the cell membrane. Drug transporters can be broadly classified into two groups: the efflux adenosine triphosphate binding cassette (ABC) family and the uptake solute carrier (SLC) family of transporters. ABCs are transmembrane proteins, which transport a wide variety of substrates across extra- and intracellular membranes, including the drugs. They are also involved in some inherited diseases in the means of resistance development to multiple drugs. There are several ABC transporter subtypes based on their genetic similarities, including ABCB1, ABCC1, ABCC2, and ABCG2. SLCs are solute carriers, which mediate the absorption of drugs. Organic anion transporting polypeptides (OATP), including OATP1B1, OATP1B3, OATP2B1, belong to the SLC family and have roles in hepatic uptake of drugs, especially in large hydrophobic organic anions. Organic cation transporter (OCT) is another subfamily of SLCs. OCT1 and OCT2 are expressed in the liver and kidney and facilitate the intracellular uptake of small organic cations. Organic anion transporters (OAT) mediate transport of low molecular weight substrates. While OAT1, OAT2, and OAT3 facilitate the uptake of drugs into the tubular cells, OAT4 takes place in the secretion of drugs into the renal tubule.

Secondly, pharmacodynamics investigates how the drug works in the body and how it changes the physiology of the patient. Pharmacodynamics is related to how the drug binds its target to create a pathway of effects leading to clinically measurable effects. It deals with drug-receptor interactions, dose-response relation and signal transduction. Genetic variations on the genes encoding a drug-related protein can affect how tightly the drug binds to the protein and may change the effectiveness of the drug. If the variation in a protein is involved in the signaling pathway or is interacting with the drug target; then, it may alter the response of the physiological system.

Differences in pharmacological response are one of key elements of morbidity and mortality of the patients. 5-13% of outpatients usually experience drug-related adverse events in the forms of adverse drug reactions and sub-therapeutic effects of drug treatment. In addition to genetic factors, a wide variety of other factors, like age, polypharmacy, concomitant diseases and diet, contribute to the genetic variations, which explain 20-30% of interindividual differences in drug response. Blood concentration of a drug might change when it is combined with another drug, which inhibits or induces the first drug's metabolizing enzyme activity. In that case, dosage adjustment or an alternative treatment method can be considered for both drugs. For example, if a CYP2C9 substrate drug is combined with a CYP2C9 inhibitor, it will be metabolized lower than usual and act as a poor metabolizer; thus, the prescribed dose of the drug needs to be adjusted. Basic metabolizing enzymes are specified with their inhibitors and inducers in Table **1**.

Table 1. Inhibitor and inducer drugs of main metabolizing enzymes.

Enzyme	Inhibitor	Inducer
CYP1A2	Acyclovir, allupurinol, cimetidine, ciprofloxacin, enoxacin, fluvoxamine, methoxsalen, mexiletine, oral contraceptives, peginterferon alpha-2a, piperine, zileuton	Phenytoin, rifampin, ritonavir, smoking, teriflunomide
CYP2B6	Clopidogrel, tenofovir, ticlopidine, voriconazole	Carbamazepine, efavirenz, nevirapine, rifampin, ritonavir
CYP2C8	Clopidogrel, defasirox, gemfibrozil, teriflunomide, trimethoprim	Rifampin
CYP2C9	Amiodarone, diosmin, fluconazole, fluvastatin, fluvoxamine, miconazole, piperine, voriconazole	Apalutamide, aprepitant, carbamazepine, enzalutamide, rifampin, ritonavir
CYP2C19	Felbamate, fluconazole, fluoxetine, fluvoxamine, omeprazole, ticlopidine, voriconazole	Apalutamide, efavirenz, enzalutamide, phenytoin, rifampin, ritonavir
CYP2D6	Abiraterone, amiodarone, bupropion, celecoxib, cimetidine, cinacalcet, clobazam, duloxetine, fluoxetine, fluvoxamine, labetalol, lorcaserin, mirabegron, paroxetine, ritonavir, quinidine, sertraline, terbinafine, vemurafenib	-
CYP3A	Clarithromycin, erythromycin, fluconazole, itraconazole, ketoconazole, troleandomycin, verapamil	Apalutamide, armodafinil, bosentan, carbamazepine, efavirenz, enzalutamide, etravirine, modafinil, mitotane, phenobarbital, phenytoin, primodone, rifampin, rufinamide, St. John's wort

(Table 1) cont.....

Enzyme	Inhibitor	Inducer
CYP3A4	Aprepitant, boceprevir, chlorzoxazone, cilostazol, cimetidine, ciprofloxacin, clotrimazole, conivaptan, cobicistat, crizotinib, cyclosporine, danoprevir, diltiazem, dronedarone, elvitegravir, erythromycin, fluconazole, fluvoxamine, fosaprepitant, grapefruit juice, imatinib, indinavir, istradefylline, itraconazole, ivacaftor, ketoconazole, lomitapide, lopinavir, pariteprevir, posaconazole, ranitidine, ranolazine, ritonavir, saquinavir, telaprevir, ticagrelor, tipranavir, telithromycin, tofisopam, troleandomycin, verapamil, voriconazole	Carbamazepin,e dexamethasone, ethosuximide, glucocorticoids, phenytoin, primodone, progesterone, rifabutin, nelfinavir, nevirapine, oxcarbazepine, phenobarbital, phenylbutazone, rofecxib, St. John's wort, sulfadimidine, sulfinpyrazone
P-gp	Amiodarone, carvedilol, clarithromycin, dronedarone, itraconazole, lapatinib, lopinavir, propafenone, quinidine, ranolazine, ritonavir, saquinavir, telaprevir, tipranavir, verapamil	Apalutamide, carbamazepine, fosphenytoin, lorlatinib, phenobarbital, phenytoin, rifampin, St. John's wort
UGT1A1	Atazanavir, indinavir, lopinavir, nilotinib, ritonavir, sorafenib	Phenobarbital, rifampin

AUTONOM AND CARDIOVASCULAR SYSTEM DRUGS

The drugs, which are commonly used in the treatment of autonomic nervous system and cardiovascular system diseases (Table **2**), and the pharmacogenomic biomarkers, which cause dosage adjustment for those drugs due to interindividual variability, are indicated below.

Table 2. Clinically used drugs in autonomic nervous system and cardiovascular system diseases.

Clinical Use	Drugs
Hypertension	Carvedilol, hydralazine, metoprolol, nebivolol, propranolol
Antiarrhythmic	Propafenone, quinidine
Dental (dry mouth)	Cevimelin
Gastrointestinal disorders	Darifenacin, fesoterodine, mirabegron, tamsulosin, tolterodine
Pulmonary	Formoterol, indacaterol, ivacaftor, umeclidinium
ADHD and narcolepsy	Amphetamine, modafinil

Carvedilol

Carvedilol is used in the treatment of mild and severe chronic heart failure and hypertension in left ventricular dysfunction after myocardial infarction (MI). CYP3A4, CYP2C19, CYP1A2, and CYP2E1 microsomal enzymes are involved in the metabolism of Carvedilol, but the enzymes responsible for their main metabolism are CYP2D6 and CYP2C9. CYP2D6 enzyme is responsible for the 4'- and 5'-hydroxylation of carvedilol. CYP2C9 enzyme plays a primary role in the O-methylation pathway of S (-)-carvedilol.

The plasma concentration of R (+) - carvedilol was 2-3 times higher in individuals with genetic polymorphisms in poor metabolizers of carvedilol, a CYP2D6 marker, compared to extensive metabolizers. However, plasma levels of S (-) - carvedilol have been reported to increase by -25 in poor metabolizers.

Interactions of CYP2D6 inhibitors with carvedilol have not been studied. When retrospective analysis of side effects seen in clinical studies was performed, it was shown that poor CYP2D6 metabolizers had a higher dizziness rate during up-titration [4].

Metoprolol

Metoprolol is used in the treatment of hypertension to lower blood pressure. By lowering blood pressure, it reduces fatal and non-fatal cardiovascular risks, primarily stroke and MI. It is also used in the treatment of angina pectoris and heart failure.

The main CYP450 enzyme responsible for the metabolism of metoprolol is CYP2D6, and about 8% of the Caucasians are weak metabolizers of this enzyme, and about 2% of other societies do not have this enzyme. The CYP2D6 enzyme is inhibited by many drugs. For this reason, the cardioselectivity of metoprolol will decrease as the drug level changes in the blood of weak metabolizers and extensive metabolizers that use metoprolol together with a drug that inhibits this enzyme.

Drugs that inhibit CYP2D6 are likely to increase the concentration of metoprolol. In healthy individuals with CYP2D6 extensive metabolizer phenotype, it was found that when used together with 100 mg quinidine and 200 mg immediate-release metoprolol, it increased the S-metoprolol level by 3 times and the elimination half-life of metoprolol was doubled. In 4 patients with cardiovascular disease using immediate-release 50 mg t.i.d metoprolol with 150 mg t.i.d propafenone, a 2 to 5-fold increase in metoprolol steady-state concentration was observed. This undesirable increase in plasma concentration will reduce the cardioselectivity of metoprolol [4].

Nebivolol

Nebivolol is used in the treatment of hypertension to lower blood pressure. Nebivolol is metabolized in several ways, including glucuronidation and hydroxylation by CYP2D6. The active isomer of nebivolol, d-nebivolol, showed an effect of approximately 12 hours in those with extensive metabolizers for CYP2D6 enzyme and 19 hours in poor metabolizers, and nebivolol exposure was significantly increased. However, this is not important for clinical use because nebivolol metabolites, including hydroxyl metabolites and glucuronides, contribute to the beta-blocking activity.

When nebivolol is used with CYP2D6 inhibitors, an increase in plasma levels of nebivolol is expected. If used with an inducer or inhibitor of the CYP2D6 enzyme, the dose of nebivolol should be adjusted according to the response observed in blood pressure. Also, *in vitro* studies have indicated that no cytochrome pathway is inhibited by d- and l-nebivolol at therapeutic concentrations.

CYP2D6 inhibition increases nebivolol exposure. Therefore, its dose should be reduced when used with CYP2D6 inhibitors [5].

Propranolol

Propranolol is used in the treatment of hypertension to lower blood pressure. Propranolol is metabolized by aromatic hydroxylation (42%), N-dealkylation followed by side-chain oxidation (41%), and direct glucuronidation (17%). However, the contribution of these reactions to the total metabolism varies among individuals. Aromatic hydroxylation and side-chain oxidation of propranolol are catalyzed by CYP2D6 and CYP1A2 enzymes, respectively. Besides, 4-hydroxy propranolol, one of its main metabolites, is a weak inhibitor of the CYP2D6 enzyme. Propranolol is also a substrate for CYP2C19 and the intestinal efflux transporter p-glycoprotein pump (P-gp). However, there is no dose of restriction in terms of P-gp.

There is no difference between poor metabolizers and extensive metabolizers of the CYP2D6 enzyme in terms of oral clearance and elimination half-life in healthy subjects. Partial clearance is quite high for 4-hydroxy propranolol in extensive metabolizers compared to poor metabolizers, whereas it is quite low for naphthyloxylactic acid, another metabolite of propranolol.

Exposure to propranolol increases when CYP2D6, CYP1A2, and CYP2C19 inhibitor drugs are used concurrently with propranolol. In this case, patients should be monitored for bradycardia and hypotension.

Smoking, as well as drugs that induce CYP1A2 enzyme, causes a decrease in the plasma level of propranolol. The same is true for use with CYP2C19 inducers. As a result, effectiveness is reduced in propranolol users [6].

Propafenone

Propafenone is a class IC antiarrhythmic drug. Approximately 90% of patients using propafenone metabolize the drug rapidly, and the elimination half-life of the drug is between 2-10 hours. In extensive metabolizers, propafenone is converted to 5-OH-propafenone by CYP2D6 and N-depropylpropafenone (norpropafenone) by CYP3A4 and CYP1A2. Propafenone metabolism is slower in the remaining 10% of patients. The reason for this is that the 5-hydroxypropafenone metabolite is not formed at all or very little in these patients.

When propafenone is used at a dose of 850 mg/day, its plasma concentration is approximately twice that of extensive metabolizers in poor metabolizers. In addition, the differences in plasma concentrations at low doses are greater, with slow metabolizers approximately 2-3 times higher than extensive metabolizers. Saturation of the hydroxylation step mediated by CYP2D6 in extensive metabolizers causes a nonlinear increase in the plasma levels of propafenone. However, propafenone pharmacokinetics are linear in poor metabolizers. The dose regimen applied in all patients is the same, both because of the lack of 5-OH metabolite in poor metabolizers and the decrease of the difference at high doses and the steady-state concentration reached 4-5 days after the use of propafenone. Due to the variability in blood levels between patients, the drug should be administered carefully.

Since propafenone has nonlinear pharmacokinetics in extensive metabolizers and is largely first-pass effect, its pharmacokinetics vary among individuals. This difference, which is seen more in extensive metabolizers, is thought to occur due to CYP2D6 polymorphism.

The R-isomer of propafenone is removed faster than the S-isomer by the 5-hydroxylation pathway mediated by CYP2D6. This causes a high rate of S-propafenone at a steady state. Both enantiomers of propafenone have equal blocking action to sodium channels, but its S-enantiomer is a stronger beta-blocker.

The use of propafenone in cases of CYP3A4 inhibition and deficiency/inhibition of the CYP2D6 enzyme results in a significant increase in propafenone plasma concentration, which increases the risk of proarrhythmia and other adverse events. Therefore, care should be taken not to use it together with inhibitors of either enzyme.

Combined use of fluoxetine and propafenone in extensive metabolizers causes an increase in S-propafenone C_{max} (39%) and AUC (50%) values, and R-propafenone C_{max} (71%) and AUC (50%). Quinidine completely inhibits the hydroxylation pathway mediated by CYP2D6 even at low doses, and consequently, all patients using quinidine become poor metabolizers in terms of CYP2D6 enzyme. In extensive metabolizers using 150 mg immediate-release propafenone 3 times a day and quinidine 50 mg 3 times a day, propafenone clearance was reduced by approximately 60%. In addition, a 100 mg dose of propafenone was found to increase the steady-state concentration 3-fold. Therefore, the use of these two drugs together should be avoided. When rifampin and propafenone were administered together in extensive metabolizers, the plasma concentration of propafenone decreased by 67% and consequently, 5-OH-propafenone plasma concentration decreased by 65%, while norpropafenone plasma concentration increased by 30%. In poor metabolizers, propafenone plasma concentration decreased by 50% and norpropafenone AUC (74%) and E_{max} (20%) values decreased. Propafenone and its metabolites were significantly reduced in urinary excretion.

Propafenone is metabolized by CYP1A2, CYP2C19 and CYP2D6 enzymes. CYP2D6, CYP3A4 and CYP1A2 inhibitors are expected to increase the plasma level of propafenone. Increased propafenone exposure causes cardiac arrhythmias and increased beta-blocker activity [7].

Quinidine

Quinidine is a class IA antiarrhythmic drug. Quinidine is eliminated hepatically by CYP3A4. Several different hydroxylated metabolites show antiarrhythmic activity. Its most important metabolite is 3-hydroxy-quinine, and its plasma concentration is almost the same in people taking normal doses of quinidine.

Although CYP2D6 does not contribute to the metabolism of quinidine, it inhibits this enzyme at therapeutic doses, converting extensive metabolizers into poor metabolizers. Therefore, caution is required when prescribing drugs metabolized by CYP2D6.

Grapefruit juice inhibits the CYP3A4-mediated metabolism of quinidine. For this reason, it is recommended that people using quinidine avoid the use of grapefruit juice. When administered with nifedipine, a calcium channel blocker, the metabolism of nifedipine slows down. The reason for this is metabolized by CYP3A4 in both drugs. This effect is also possible when used with other calcium channel blockers (nimodipine, nicardipine, etc.) metabolized by the same enzyme [6].

Cevimeline

It is a muscarinic agonist used in the treatment of dry mouth symptoms in patients with keratoconjunctivitis sicca (Sjogren's syndrome). Cevimeline is metabolized by CYP2D6, CYP3A3 and CYP3A4.

Drugs that inhibit CYP2D6 and CYP3A4 enzymes also inhibit the metabolism of cevimeline. Caution should be exercised in the use of cevimeline, as the risk of adverse effects will increase in people with or suspected CYP2D6 deficiency [8].

Darifenacin

Darifenacin is a cholinergic antagonist used in active bladder treatment. Darifenacin is metabolized by CYP3A4 and CYP2D6 enzymes. In CYP2D6 poor metabolizers, darifenacine metabolism is mediated by CYP3A4. In the steady-state, after 1.5 mg darifenacin once a day administration, C_{max} and AUC values were determined as 1.9 and 1.7 in extensive metabolizers, respectively, compared to poor metabolizers.

The daily dose of darifenacin should not exceed 7.5 mg in combination with CYP3A4 inhibitors or in patients with moderate hepatic impairment.

CYP3A4 inhibitors increase darifenacin exposure, so when used concomitantly, the daily dose of darifenacin should not exceed 7.5 mg. Darifenacin exposure increases when combined with CYP3A4 inhibitors.

Caution should be exercised when darifenacin is used with drugs such as tricyclic antidepressant that have a narrow therapeutic range and are predominantly metabolized by CYP2D6. For example, when used in combination with a 20 mg dose of paroxetine, a potent CYP2D6 inhibitor, darifenacin exposure was found 33% higher in steady-state at a dose of 30 mg once a day.

Concomitant use of darifenacin with other antimuscarinic agents may increase the frequency and severity of dry mouth, constipation, blurred vision, and other antimuscarinic effects. Since antimuscarinic drugs affect the motility of the gastrointestinal system, they may cause changes in the absorption rate and degree of the drugs they are used with [9].

Fesoterodine

Fesoterodine is a cholinergic antagonist used for the treatment of overactive bladder, with symptoms of urinary incontinence, urgency and frequency. After oral administration, fesoterodine is rapidly and intensely hydrolyzed into carboxy, carboxy-N-desisopropyl, and N-desisopropyl, which are active metabolites and do not contribute significantly to its antimuscarinic effect. Enzymes that play a major role in this pathway are CYP2D6 and CYP3A4 enzymes.

In CYP2D6 poor metabolizers, the C_{max} and AUC of active metabolites of fesoterodine increased 1.7 to 2-fold, respectively, compared to extensive metabolizers.

The active metabolites of fesoterodine do not inhibit CYP1A2, 2B6, 2C8, 2C9, 2C19, 2D6, 2E1 and 3A4 enzymes, nor do they induce CYP1A2, 2B6, 2C9, 2C19 and 3A4 enzymes at therapeutic concentrations.

Potent CYP3A4 inhibitors cause an approximately doubling of the C_{max} and AUC of the active metabolites of fesoterodine. In CYP2D6 extensive metabolizers, when ketoconazole, a potent CYP3A4 inhibitor, was given 200 mg twice a day for 5 days, the C_{max} and AUC of the active metabolites of fesoteradine increased 2.0 and 2.3-fold, respectively. C_{max} (2.1) and AUC (2.5) values also increased in CYP2D6 poor metabolizers. C_{max} and AUC values were found to be 4.5 and 5.7 times higher, respectively, when the CYP2D6 poor metabolizers and those using ketoconazole were compared with those who were CYP2D6 extensive metabolizers but not ketoconazole.

CYP3A4 inducers cause decreased plasma levels of active metabolites of fesoterodine. However, dose adjustment is not required when it comes to using with CYP3A4 inhibitors. In patients receiving potent CYP3A4 inhibitors, the daily dose of fesoterodine should not exceed 4 mg [10].

Mirabegron

Mirabegron is a beta-3 adrenergic agonist used for the treatment of overactive bladder with symptoms of urinary incontinence, urinary urgency, and frequency.

Dealkylation, oxidation, direct glucuronidation and amide hydrolysis are pathways that contribute to the metabolism of mirabegron. CYP3A4 and CYP2D6 enzymes are involved in the oxidation step of mirabegron metabolism. Healthy subjects with CYP2D6 poor metabolizers were found to have higher Cmax and AUC values compared to extensive metabolizers.

Mirabegron is a substrate for CYP3A4, CYP2D6, butyrylcholinesterase, UGT, efflux transporter P-gp and influx OCTs OCT1, OCT2 and OCT3. Mirabegron is a time-dependent moderate inhibitor of the CYP2D6 enzyme and a weak inhibitor of the CYP3A4 enzyme.

Mirabegron inhibits drug transport with high concentrations of P-gp. Concomitant use of mirabegron with ketoconazole, a potent CYP3A4 inhibitor, increases exposure to mirabegron. Also, systemic exposure to CYP2D6 substrates such as metoprolol and desipramine is increased with use with mirabegron, a moderated

CYP2D6 inhibitor. Besides, changes in tamsulosin pharmacokinetics (increase in C_{max} and AUC values) when mirabegron is used with tamsulosin are associated with inhibition of the CYP pathway.

Systemic exposure to CYP2D6 substrates is increased with use with mirabegron, a moderated CYP2D6 inhibitor. Therefore, drug monitoring and dose adjustment may be required, especially during use with drugs metabolized by CYP2D6 enzyme such as thioridazine, flecainide, and propafenone, which have a narrow therapeutic window [11].

Tamsulosin

Tamsulosin is used in the treatment of benign prostatic hyperplasia signs and symptoms. Tamsulosin is mainly metabolized by CYP3A4 and CYP2D6, but other CYP isozymes are also partially involved in its metabolism.

With the use of tamsulosin in combination with potent CYP3A4 inhibitors, its C_{max} and AUC increase approximately twice. When used with paroxetine, a strong CYP2D6 inhibitor, an increase in C_{max} (1.3) and AUC (1.6) values is also observed. An increase in tamsulosin exposure is expected in CYP2D6 poor metabolisers compared to extensive metabolizers. Concomitant use of 0.4 mg tamsulosin with CYP2D6 and CYP3A4 inhibitors significantly increases exposure to tamsulosin.

Since tamsulosin is mainly metabolized by CYP3A4 and CYP2D6, it should not be used in combination with strong CYP3A4 inhibitors in the form of 0.4 mg tablets. In addition, if it will be used at doses higher than 0.4 mg, it should be paid attention to use in combination with strong or moderate CYP2D6 inhibitors, especially in CYP2D6 poor metabolizers [12].

Tolterodine

Tolterodine is an antimuscarinic agent used for the treatment of overactive bladder, with symptoms of urinary incontinence, urgency and frequency. There is a relationship between the plasma concentration of tolterodine and its effect on the QT interval. In one study, after tolterodine treatment, there was a large increase in the QTc interval of CYP2D6 poor metabolizers compared to extensive metabolizers.

After tolterodine is taken orally, most of it is metabolized in the liver. As a result of the metabolism of tolterodine mainly through the CYP2D6 enzyme, the compound 5-hydroxymethyl tolterodine (5-HMT) is formed and this compound is pharmacologically active. The rest of the tolterodine is converted into 5-

carboxylic acid and N-dealkylated 5-carboxylic acid metabolites. The metabolic pathway to tolterodine in individuals who are poor metabolisers for the CYP2D6 enzyme is dealkylation with CYP3A4 resulting in N-dealkylated tolterodine.Since tolterodine is metabolized more slowly in poor metabolizers, it causes an increase in serum concentration and consequently the concentration of the active metabolite 5-HTM to remain at insignificant levels.

When a potent CYP2D6 inhibitor was used in combination with tolterodine, its metabolism was significantly inhibited in extensive metabolizers and there was a 4.8-fold increase in tolterodine AUC. Thus, for 5-HMT, AUC decreased by 20% and C_{max} by about 50%. However, no dose adjustment is required for tolterodine serum concentrations that occur during the concomitant use of these two agents.

A study in a small number of CYP2D6 poor metabolizers healthy volunteers increased the averages of tolterodine C_{max} and AUC approximately 2-2.5-fold after co-administration of tolterodine at a dose of 200 mg/day, a potent CYP3A4 inhibitor, ketoconazole. Therefore, the daily dose of tolterodine is recommended as 2 mg/day while using tolterodine with potent CYP3A4 inhibitors.

In a clinical study, the effect of tolterodine immediate-release tablets on the QT interval in CYP2D6 poor metabolizers was more pronounced than extensive metabolizers, especially for 8 mg/day dose. Therefore, caution should be exercised when prescribing tolterodine in patients receiving antiarrhythmics such as quinidine, procainamide or amiodarone, sotalol, or in people with known QT prolongation [10].

Formoterol

It is a beta-2 adrenergic receptor agonist used to prevent bronchoconstriction in patients with chronic obstructive pulmonary disease (COPD), chronic bronchitis and emphysema.

Two main pathways play a role in the metabolism of formoterol. It involves first direct glucuronidation to phenolic or aliphatic hydroxyl groups, followed by O-demethylation. While the glucuronidation step, which is the first of this two-step metabolism, is catalyzed by the enzymes UGT1A1, 1A8, 1A9, 2B7 and 2B15, O-demethylation is catalyzed by CYP2D6, CYP2C19, CYP2C9 and CYP2A6 enzymes. Formoterol does not inhibit CYP450 enzymes at therapeutic concentrations. However, in patients deficient in CYP2D6 and CYP2C19, or both, exposure to formoterol may be increased and systemic adverse effects may occur [13].

Indacaterol

It is a beta-2 adrenergic receptor agonist used to prevent bronchoconstriction in patients with COPD, chronic bronchitis and emphysema. Indacaterol is metabolized by UGT1 to phenolic o-glucuronide. CYP3A4 is the main isoenzyme responsible for indacaterol hydroxylation. Indacaterol pharmacokinetics are based on the results of a prospective study in patients with the *UGT1A1* (TA)7/(TA)7 genotype (low expression of UGT1A1; also called *28) and the (TA)6/(TA)6 genotype, indacaterol AUC and C_{max} values were found to increase 1.2 times in individuals with (TA)7/(TA)7 genotype. However, changes in systemic exposure due to interactions with concomitantly used drugs for indacaterol are negligible.

The results of a clinical trial for the concomitant use of indacaterol and potent and specific CYP3A4 inhibitors and P-gp inhibitors are as follows: As a result of the co-administration of each of these drugs with indacaterol, an increase in indacaterol AUC and C_{max} values were observed [14].

Ivacaftor

Ivacaftor is a cystic fibrosis transmembrane conductance regulator (CFTR) booster used for the treatment of cystic fibrosis (CF) in patients 6 months old and older with a mutation in the CFTR gene. Ivacaftor is a CFTR booster. CFTR protein is a chloride channel found on the surface of epithelial cells of many organs. Ivacaftor facilitates chloride transport by enhancing the opening of this channel.

CYP3A4 is the main enzyme responsible for ivacaftor metabolism. As a result of the metabolism, two metabolites called M1 and M2 are formed and M1, which has about 1 in 6 potency of ivacaftor, is thought to be pharmacologically active.

Ivacaftor and its pharmacologically active metabolite M1 are inhibitors of CYP3A4, CYP2C9 and P-gp. In clinical studies, concomitant use with CYP3A4 inducers decreased ivacaftor exposure, while exposure increased with CYP3A4 inhibitors.

Concomitant use of ivacaftor with CYP3A4 inhibitors the dose of ivacaftor in the form of a tablet or oral granule packs twice a week should be adjusted. When used with moderate CYP3A4 inhibitors such as fluconazole, it should be taken as a tablet/oral granule package once a day. Concomitant use of a drug that is a CYP3A4 substrate such as midazolam and ivacaftor increased midazolam exposure 1.5-fold, while exposure to digoxin increased 1.3-fold when used with a P-gp substrate such as digoxin. Caution and monitoring are recommended, as the concomitant use of ivacaftor and CYP3A4 and P-gp substrates will increase the systemic exposure of these drugs. Concomitant use with potent CYP3A4 inducers will reduce exposure to ivacaftor, resulting in decreased therapeutic efficacy. Therefore, use with potent CYP3A4 inducers is not recommended. Ivacaftor is not recommended for children younger than 6 months as its effectiveness and safety have not been established [15].

Umeclidinium

Umeclidinium is an antimuscarinic agent used for the treatment of COPD disease. Umeclidinium is mainly metabolized in the liver by the CYP2D6 enzyme. It is also a substrate for the P-gp transporter. Systemic exposure to metabolites resulting from oxidation and conjugation reactions is low.

When the moderate P-gp transporter inhibitor verapamil was used with a once-daily dose of 240 mg/kg, an approximately 1.4-fold increase in the AUC of umeclidinium occurred. There was no clinical change in systemic exposure to umeclidinium after repeated daily inhaled doses in normal and poor metabolizers of the CYP2D6 enzyme [11].

Amphetamine

Amphetamine is a central nervous system stimulant used in the treatment of attention deficit hyperactivity disorder. CYP enzymes involved in the metabolism of amphetamine are not fully known, but it is known that CYP2D6 enzyme is involved in the formation of 4-hydroxy-amphetamine, its active metabolite.

Serotonin syndrome can occur when amphetamine is used with drugs that affect the serotonergic system, such as monoamine oxidase inhibitors or selective serotonin re-uptake inhibitors. Thus, a drug that does not inhibit the CYP2D6 enzyme or is non-serotonergic should be considered as an alternative, as exposure will increase when amphetamine is used with CYP2D6 inhibitory drugs [16].

Modafinil

Modafinil is used in the treatment of excessive sleepiness caused by conditions such as narcolepsy and obstructive sleep apnea. Modafinil is metabolized by hydrolytic deamination, S-oxidation, hydroxylation of the aromatic ring and conjugation with glucuronide. Two of the metabolites formed as a result of metabolism (modafinil acid and modafinil sulfone) reach concentrations that can be considered important in plasma.

Modafinil induces CYP3A4 strongly and CYP1A2 and CYP2B6 enzymes weakly. It reversibly inhibits the activity of CYP2C19 enzyme. Because modafinil induces the CYP3A4 enzyme, when used with steroidal contraceptives, cyclosporine, and CYP3A4 substrates such as the short-acting benzodiazepines midazolam, triazolam, the blood concentration and effects of these drugs may be reduced. In the case of concomitant use for cyclosporine, blood concentrations of cyclosporine may be reduced by modafinil, and the concentration of cyclosporine should be monitored and appropriate dose adjustment should be made.

Modafinil suppresses the activity of the CYP2C9 enzyme and therefore has the potential to alter blood concentrations of these drugs when used with drugs such as warfarin and phenytoin, which are substrates for this enzyme. Modafinil and its metabolite, modafinil sulfone, can induce continuous partial inhibition of the CYP2C19 enzyme. In this case, exposure to these drugs can be increased by modafinil when used with CYP2C19 enzyme substrates such as phenytoin, diazepam, propranolol. CYP2C19 is also important in the metabolism of SSRIs (which are mainly metabolized by the CYP2D6 enzyme) and some tricyclic antidepressants. However, in patients with CYP2D6 enzyme deficiency, the concentrations of drugs will decrease, modafinil may increase the concentrations of these drugs, and dose adjustment should be considered in this situation. Since it is known that estradiol activity may decrease even 1 month after modafinil use is discontinued, an alternative method of contraception (other than steroidal contraceptives) should be recommended [16].

ENDOCRINE SYSTEM DRUGS

The drugs, which are commonly used in treatment of endocrine system diseases (Table **3**), and the pharmacogenomic biomarkers, which cause dosage adjustment for those drugs due to interindividual variability, are indicated below.

Table 3. Clinically used drugs in endocrine system diseases.

Clinical Use	Drugs
Diabetes	Chlorpropamide, glipizide, glyburide, tolazamide, glimepiride
Antilipidemic	Rosuvastatin
Gynecology	Drospirenone and ethinyl estradiol, elagolix, flibanserin, ospemifene

Chlorpropamide, Glipizide, Glyburide, Tolazamide

Chlorpropamide, glipizide, glyburide and tolazamide are used as an adjuvant to diet and exercise in patients with type 2 Diabetes Mellitus (T2DM) to improve glycemic control. Chlorpropamide, glipizide, glyburide and tolazamide belong to the sulfonylurea family and administration of sulfonylurea agents to the patients with glucose-6-phosphate dehydrogenase (G6PD) deficiency might cause hemolytic anemia [17].

Glimepiride

Glimepiride is used as an adjuvant to diet and exercise in patients with T2DM to improve glycemic control. There can be drug interactions between glimepiride and both inhibitors and inducers of CYP2C9. Fluconazole, a CYP2C9 inhibitor, inhibits glimepiride metabolism, leading to increase in plasma concentration and hypoglycemia accordingly.

Glimepiride belongs to the sulfonylurea family and administration of sulfonylurea agents to the patients with G6PD deficiency might cause hemolytic anemia [17].

Rosuvastatin

Rosuvastatin is an HMG-CoA reductase inhibitor that is used in adults with primer hyperlipidemia and mixed dyslipidemia to decrease the increased total-C, LDL-C, ApoB, nonHDL-C, TG levels and to increase HDL-C. It is used to decrease the increased total-C, LDL-C and ApoB levels when proper diet is not adequate in pediatric patients with age of 8 – 17 as monotherapy, in pediatric patients with age of 7 – 17 as monotherapy or combined with a lipid lowering therapy and in adults with homozygous familial hypercholesterolemia. It is used as an adjuvant to a proper diet in adult patients with hypertriglyceridemia and dsibetalipoproteinemia.

The plasma rosuvastatin concentration is higher in patients, having two low-functioning alleles of OATP1B1 encoding gene (*SLCO1B1* 521T>C). However, the effect of that polymorphism on rosuvastatin efficacy and/or safety has not been enlightened yet [4].

Drospirenone and Ethinyl Estradiol

Drospirenone and ethinyl estradiol, a combined oestrogen/progestin oral contraceptive, is used to prevent pregnancy in women. It is used to treat premenstrual dysphoric disease symptoms in women, using an oral contraceptive for birth control.

CYP3A4 is responsible for oxidative metabolism of drospirenone and 2-hydroxylation of ethinyl estradiol in the liver. Ethinyl estradiol and its oxidative metabolites are metabolized through glucuronide or sulfate conjugation.

CYP3A4 inducers, such as phenytoin, barbiturates, carbamazepine, rifampin and some herbal products can reduce efficacy of the combined oral contraceptive or breakthrough bleeding. CYP3A4 inhibitors, such as azole antifungals, macrolides, diltiazem and grapefruit juice can increase plasma concentrations of estrogen and/or progestin when combined with drospirenone and ethinyl estradiol.

When combined with a CYP3A4 inhibitor, such as azole antifungals and human immunodeficiency virus/ hepatitis C virus (HIV/HCV) protease inhibitors for a long term, serum potassium concentrations should be monitored in patients, using drospirenone and esthinyl estradiol [18].

Elagolix

Elagolix is used to relieve moderate to severe pain in patients with endometriosis. Elagolix is metabolized in the liver by CYP3A, in addition to minor roles of CYP2D6, CYP2C8 and UGTs. The plasma elagolix concentration is higher in patients, having two low-functioning alleles of OATP1B1 encoding gene (*SLCO1B1* 521T>C) [19].

Flibanserin

Flibanserin is used for treatment of hypoactive sexual desire disorder characterized by low sexual desire without any medical or psychiatric base, relationship problems, any treatment or medication in women in premenopausal period.

CYP3A4 is the primer enzyme eliminating flibanserin during metabolization of the drug, in which CYP2C19 also plays a minor role. Plasma flibanserin concentration is higher in CYP2C19 poor metabolizers than in extensive metabolizers. Drug concentration has been observed as lower in both CYP2D6 and CYP2C9 poor metabolizers.

When combined with CYP3A4 inhibitors, plasma flibanserin levels might increase leading to higher risk of hypotension and syncope or adverse reactions. When combined with CYP3A4 inducers, efficiency of flibanserin may decrease; thus, it is not recommended. CYP2C19 inhibitors might increase flibanserin efficiency, so their combination may cause hypotension, syncope and central nervous system depression [20].

Ospemifene

Ospemifene is an estrogen agonist/antagonist used to treat vaginal dryness and dyspareunia with vulvar and vaginal atrophy symptoms in menopausal period. CYP3A4 and CYP2C9 are primarily responsible for ospemifene metabolism and CYP2C19 has also minor roles.

When combined with fluconazole, a CYP3A4, CYP2C9 and CYP2C19 inhibitor, ospemifene efficiency might increase leading to adverse reactions. Since rifampin is a CYP3A4, CYP2C9 and CYP2C19 inducer, combination with ospemifene may decrease the clinical effect of the drug. When ospemifene is combined with a strong CYP3A4 inhibitor, the risk of adverse reactions may increase [21].

CENTRAL NERVOUS SYSTEM DRUGS

The drugs, which are commonly used in treatment of central nervous system diseases (Table **4**), and the pharmacogenomic biomarkers, which cause dosage adjustment for those drugs due to interindividual variability, are indicated below.

Table 4. Clinically used drugs in central nervous system diseases.

Clinical Use		Drugs
Antipsychotics		Risperidone, aripiprazole, clozapine, iloperidone, paliperidone, pimozide
Antidepressants		Amitriptyline, nortriptyline, desipramine, venlafaxine, duloxetine, vortioxetine, atomoxetine, doxepin, fluoxetine, paroxetine
Antiepileptics		Brivaracetam, fosphenytoin, oxcarbazepine, carbamazepine, clobazam, diazepam, lacosamide, phenytoin, valproic acid
Anesthesiology		Articaine and epinephrine, codeine, lidocaine and tetracaine, mepivacaine, oxymetazoline and tetracaine
Antitussive		Dextromethorphan and quinidine
Central nervous system degenerative disorders	Alzheimer's disease	Donepezil, galantamine
	Huntington disease	Deutetrabenazine, tetrabenazine
	Polyneuropathy	Inotersen, patisiran
Neuromuscular disorders	Duchene muscular dystrophy	Eteplirsen, nusinersen
	Lambert-Eaton myasthenic syndrome	Amifampridine, amifampridine phosphate
Motion sickness		Meclizine
Multiple sclerosis		Siponimod
Tardive dyskinesia		Valbenazine

Risperidone

Risperidone is an antipsychotic, which is solely used in schizophrenia, autism-related irritability. It is also used solely or combined with lithium or valproate in treatment of acute manic or mixed episodes related to bipolar disorder.

The main metabolization mechanism of risperidone is hydroxylation to 9-hydroxyrisperidone by CYP2D6. Extensive CYP2D6 metabolizers can convert risperidone into its metabolite rapidly, but this occurs slower in poor metabolizers. However, pharmacokinetics of the drug has been observed very similar in both extensive and poor metabolizers.

When prescribed with CYP2D6 inhibitors, risperidone dosage should be adjusted and daily dose cannot exceed 8 mg/day. When combined with carbamazepine, which is a CYP3A and P-gp inhibitor, a higher dose of risperidone should be prescribed; but the dosage cannot be more than 2-fold of the usual dose. On the other hand, dosage adjustment is not necessary when risperidone is combined with CYP3A inhibitors [22, 23].

Aripiprazole

Aripiprazole is used orally for acute treatment of Tourette's disorder, major depressive disorder, schizophrenia and bipolar-related manic and mixed episodes, and also used i.v. to treat schizophrenia and bipolar mania-related agitation.

Aripiprazole is eliminated by CYP2D6 and CYP3A4 catalysis. Poor CYP2D6 metabolizers should use half of the usual dosage of aripiprazole.

Poor CYP2D6 metabolizers should administer a quarter of the usual aripiprazole dosage when combined with a strong CYP3A4 inhibitor. Aripiprazole dosage should be half of the usual when combined with a strong CYP2D6 inhibitor or a CYP3A4 inhibitor. When combined with both a strong CYP2D6 inhibitor and CYP3A4 inhibitor, aripiprazole dosage should be quarter of the usual. When combined with a strong CYP3A4 inducer, aripiprazole should be administered in 2-fold of the usual dose for one or two weeks [22, 23].

Clozapine

Clozapine is used for treatment of schizophrenia in patients, who are resistant to treatment or to decrease suicidal tendency in patients with schizoaffective disorder.

CYP1A2, CYP2D6 and CYP3A4 are basically responsible from metabolization of clozapine. Since the plasma concentration of clozapine will be higher in poor CYP2D6 metabolizers, the prescribed dose should be less than usual dosage.

Clozapine should be given in one third of the usual dosage when combined with a strong CYP1A2 inhibitor. Combination of clozapine with a moderate or weak CYP1A2 inhibitor or a CYP2D6 inhibitor or a CYP3A4 inhibitor might cause clozapine-related adverse reactions, so given clozapine dosage should be less than usual if necessary. It is not recommended to use clozapine with a strong CYP3A4 inducer. However, if the CYP3A4 inducer is necessary, clozapine can be given in a higher dosage. Since the efficacy of clozapine will be reduced when combined with a moderate or strong CYP1A2 inducer or a CYP3A4 inducer, the dosage can be increased if necessary [23, 24].

Iloperidone

Iloperidone is an atypical antipsychotic used to treat schizophrenia in adults. Iloperidone is eliminated through hepatic metabolism by CYP3A4 and CYP2D6. CYP2D6 catalyzes hydroxylation and carbonyl reduction and CYP3A4 catalyzes O-demethylation of the drug. Since blood iloperidone concentration will be decreased in poor CYP2D6 metabolizers, iloperidone should be administered in half of the usual dosage in those patients.

When combined with a CYP3A4 inhibitor or a CYP2D6 inhibitor, iloperidone elimination will be inhibited, leading to increased plasma concentration. Thus; iloperidone should be prescribed in lower dosage than usual [25].

Pimozide

Pimozide is used for soothing motor and phonic twitches in patients of Tourette's disorder with no response to standard treatment.

N-dealkylation of pimozide in the liver is primarily mediated by CYP3A4. In addition, CYP1A2 and CYP2D6 also play minor roles in metabolization. Since drug concentration in circulation is higher in poor CYP2D6 metabolizers, dosage adjustment is required in those patients. Administered pimozide dose should not exceed 4 mg/day and the dosage should not be increased earlier than 14 days of use.

Pimozide is not recommended to be used with a strong CYP2D6 inhibitor. It is contraindicated to use pimozide with a CYP3A4 inhibitor [23].

Amitriptyline, Nortriptyline, Desipramine

Amitriptyline, nortriptyline and desipramine are tricyclic antidepressants, used in depression treatment. Amitriptyline is also used to treat neuropathic algia and fibromyalgia.

Metabolization of amitriptyline is primarily mediated by CYP2D6 and CYP2C19, and nortriptyline is also a metabolite of amitriptyline. Plasma concentration of the drug is higher in poor CYP2D6 metabolizers, so the prescribed dosage of amitriptyline, nortriptyline and desipramine should be decreased than usual.

The dosage of amitriptyline, nortriptyline and desipramine should be reduced when combined with a CYP2D6 inhibitor [26].

Venlafaxine

Venlafaxine is a serotonin norepinephrine reuptake inhibitor, used in treatment of major depressive disorder, generalized anxiety disorder, panic disorder and social anxiety disorder.

Venlafaxine is converted into its major metabolite by CYP2D6 in the liver. It has been observed that systemic venlafaxine level is higher and the concentration of its metabolite is lower in poor CYP2D6 metabolizers. No dosage adjustment is required for venlafaxine when combined with a CYP2D6 inhibitor.

Duloxetine

Duloxetine is a serotonin norepinephrine reuptake inhibitor, used to treat major depressive disorder. In addition, it is used to treat fibromyalgia in pediatric patients older than 13 years of age and adults, and for treatment of musculoskeletal pain, generalized anxiety disorder and diabetic peripheral neuropathic pain in adults.

Duloxetine is eliminated by CYP1A2 and CYP2D6 catalysis during hepatic metabolization. Duloxetine concentration in circulation might increase when combined with a CYP1A2 inhibitor or a CYP2D6 inhibitor [27].

Vortioxetine

Vortioxetine is used for treatment of major depressive disorder. Metabolization of vortioxetine is mediated by CYP2D6, CYP3A4/5, CYP2C19, CYP2A6, CYP2C8 and CYP2B6, and CYP2D6 is primarily responsible from conversion of the drug into its pharmacologically inactive metabolite. In poor CYP2D metabolizers, plasma vortioxetine concentration is almost twice that of extensive metabolizers. Thus, maximum vortioxetine dosage is suggested to be 10 mg/day in poor CYP2D6 metabolizers.

Vortioxetine should be prescribed in reduced dosage when combined with a CYP2D6 inhibitor. Vortioxetine dosage should be increased, but cannot exceed 3-fold of the usual dose when combined with a strong CYP inducer [28].

Atomoxetine

Atomoxetine is a selective norepinephrine reuptake inhibitor, used to treat attention-deficit/hyperactivity disorder.

Atomoxetine is eliminated *via* oxidative metabolization by CYP2D6. Since poor CYP2D6 metabolizers have deficit enzyme activity, plasma concentration of the drug might increase up to 5-fold in those individuals. Starting dose of atomoxetine should be 0.5 mg/kg/day in pediatric patients and adults under 70 kg of weight, who are poor CYP2D6 metabolizers and the administered dose can be increased up to the 1.2 mg/kg/day of usual dose after at least 4 weeks only if starting dose is tolerated and no response is given to the treatment. Starting dose of atomoxetine should be 40 mg/day in poor CYP2D6 metabolizer pediatric patients and adults over 70 kg of weight and the given dose can be increased up to 80 mg/day after at least 4 weeks only if starting dose is tolerated and no response is given to the treatment.

Plasma atomoxetine concentration was determined as the same level in both poor and extensive metabolizers when combined with a CYP2D6 inhibitor. Thus; if prescribed with a CYP2D6 inhibitor, the atomoxetine dosage should be adjusted, considering the patient as a poor CYP2D6 metabolizer.

Several adverse reactions, such as sleep problems, losing weight, constipation, depression, tremor, syncope, sedation and mydriasis, are observed significantly more often in more than 2% of pediatric poor CYP2D6 metabolizers than in extensive metabolizers. In addition to those adverse reactions, urinary retention, dry mouth, erectile dysfunction, ejaculation disorder and peripheral coldness are also observed more often in more than 2% of adult poor CYP2D6 metabolizers than in extensive metabolizers [27].

Doxepine

Doxepine is used to treat sleep-maintenance insomnia. Doxepine metabolization is mainly mediated by CYP2C19 and CYP2D6, and also CYP1A1 and CYP2C9 enzymes have minor roles. Poor CYP2C19 and CYP2D6 metabolizers have higher doxepin plasma levels than normal.

Efficacy of doxepine is doubled when combined with cimetidine, which is a nonspecific CYP1A2, CYP2C19, CYP2D6 and CYP3A4 inhibitor, so in that case, maximum dose of doxepine is suggested to be 3 mg/day in adults [26].

Fluoxetine

Fluoxetine is a psychotropic drug, used to treat major depressive disorder, obsessive compulsive disorder, panic disorder and bulimia nervosa. CYP2D6 is the main enzyme in metabolization of fluoxetine.

Because fluoxetine is a CYP2D6 inhibitor, it decreases CYP2D6 enzyme activity to a poor metabolizer level even in extensive CYP2D6 metabolizers. Therefore, dose adjustment is required when fluoxetine is combined with a drug, which is metabolized by CYP2D6.

When a drug, which is metabolized by CYP2D6, is combined with fluoxetine, dose adjustment is required for the aforementioned drug [29].

Paroxetine

Paroxetine is used in treatment of major depressive disorder, obsessive compulsive disorder, generalized anxiety disorder and panic disorder. CYP2D6 is the main enzyme in metabolization of paroxetine.

When combined with many CYP inhibitor cimetidine, plasma paroxetine concentration increases, so paroxetine dosage, following the 20 mg of starting dose, should be adjusted. When combined with many CYP inducer phenobarbital, plasma paroxetine concentration decreases, so paroxetine dosage, following the starting dose, should be adjusted based on clinical observations. Because paroxetine is a CYP2D6 inhibitor, it decreases CYP2D6 enzyme activity to a poor metabolizer level even in extensive CYP2D6 metabolizers. Therefore, dose adjustment is required when paroxetine is combined with a drug, which is metabolized by CYP2D6.

It has been observed that paroxetine reduces the efficacy of tamoxifen by irreversible CYP2D6 inhibition. Thus, an alternative antidepressant, which is not a CYP2D6 inhibitor, is suggested to be combined with tamoxifen [28].

Brivaracetam

Brivaracetam is used to treat partial-onset seizures in patients older than 4 years of age. The drug can be administered i.v. only to the patients older than 16 years of age.

Brivaracetam is hydrolyzed by CYP2C19 during its metabolization. Variations on *CYP2C19* gene may cause 2- or 10- fold of decrease in hydroxy metabolite of the drug. Prescribed brivaracetam dose should be less than usual dose in poor CYP2C19 metabolizers.

When combined with a CYP2C19 inducer, plasma brivaracetam concentration is decreased, so the dosage should be doubled. When combined with a CYP2C19 inhibitor, administered brivaracetam dose should be less than usual dosage.

Phenytoin, Fosphenytoin

Phenytoin is used to treat generalized tonic-clonic status epilepticus and to both prevent and treat seizures during neurosurgery. Parenteral phenytoin administration can be applied when the drug cannot be administered orally. Fosphenytoin is used as a substitution of phenytoin only if oral phenytoin administration is not possible.

Phenytoin is metabolized in the liver by CYP2C9 and CYP2C19 after it is administered orally or it is obtained by conversion of fosphenytoin in the body.

Drug interactions of phenytoin with CYP2C9 and CYP2C19 inhibitors might cause toxicity because of the increment in blood phenytoin concentration. Thus, dosage adjustment for fosphenytoin may be required.

Asian patients having *HLA-B**1502 genotype have higher risk of experiencing Stevens-Johnson Syndrome/toxic epidermal necrolysis after using fosphenytoin or phenytoin combined with carbamazepine. Therefore, those patients should not use fosphenytoin or phenytoin while using carbamazepine [30].

Oxcarbazepine

Oxcarbazepine is an antiepileptic, used as an adjunctive therapy for the treatment of partial seizures in adults and pediatric patients with age of 6-17. Both oxcarbazepine and its active metabolite, 10-monohydroxy derivative, induce CYP3A4 and CYP3A5 enzymes.

Combination of oxcarbazepine with any of CYP450 inducer antiepileptic drugs may lead to decrease in plasma concentration of both oxcarbazepine and its active metabolite. It is highly recommended that starting oxcarbazepine dose should be higher than usual when it is combined with carbamazepine. Dosage adjustment can be considered based on clinical observations during the treatment.

Patients having *HLA-B**1502 genotype have higher risk of experiencing Stevens-Johnson Syndrome/toxic epidermal necrolysis after using oxcarbazepine. Therefore, unless the benefits of the drug surpass the risks, oxcarbazepine administration is not recommended for those patients [31].

Carbamazepine

Carbamazepine is an anticonvulsant drug, used in treatment of partial and generalized tonic-clonic seizures and true trigeminal neuralgia-related pain. Carbamazepine is metabolized in the liver mainly by CYP3A4.

Carbamazepine metabolization will be inhibited when combined with CYP3A4 inhibitor, leading to higher plasma carbamazepine concentration. On the other hand, plasma carbamazepine concentration is reduced when combined with a CYP3A4 inducer.

Carbamazepine is an inducer of CYP1A2, CYP2B6, CYP2C19 and CYP3A4 enzymes. Thus; when carbamazepine is combined with a drug, which is metabolized by any of those enzymes, dosage adjustment is required for the aforementioned drug.

Patients, having *HLA-B**1502 genotype, have higher risk of experiencing Stevens-Johnson Syndrome/toxic epidermal necrolysis after using carbamazepine. Patients, having *HLA-A**3101 genotype, have higher risk of experiencing maculopapular eruption and Stevens-Johnson Syndrome/toxic epidermal necrolysis after using carbamazepine. In those cases, carbamazepine administration is not recommended for those patients unless the benefits of the drug surpass the risks [30].

Clobazam

Clobazam is a benzodiazepine, used as an adjunctive therapy for the treatment of seizures in Lennox-Gastaut syndrome patients with older than 2 years of age. N-demethylation of clobazam is mediated by CYP3A4, and the drug is metabolized by also CYP2B6 and CYP2C19. The active metabolite, N-desmethylclobazam, is mainly metabolized by CYP2C19. Plasma N-desmethylclobazam concentration increases up to 5-fold in poor CYP2C19 metabolizers. Thus, the starting dose should be 5 mg/day in those patients and the dosage can be increased based on the weight of the patient during treatment; however, it cannot exceed half of the usual dose.

When combined with a strong or a moderate CYP2C19 inhibitor, concentration of active metabolite of clobazam increases, leading to adverse reactions. Therefore, prescribed clobazam dose should be less than usual dosage.

Clobazam is a weak CYP3A4 inducer; thus, when combined with a CYP3A4 substrate, like hormonal contraceptives, efficacy of the aforementioned drug might decrease. In that case, a non-hormonal drug should be considered to administer. Clobazam is also a CYP2D6 inhibitor, so when it is combined with a drug, which is metabolized by CYP2D6, dosage adjustment is required for the aforementioned drug.

Diazepam

Diazepam is discontinuously used to keep under control the increased seizures in epilepsy patients, getting regular antiepileptic medication. Diazepam is demethylated by CYP2C19 and CYP3A4, and 3-hydroxylated by CYP3A4 during hepatic elimination.

When combined with a CYP2C19 inhibitor or a CYP3A4 inhibitor, elimination of diazepam is reduced. On the other hand, elimination of diazepam increases when combined with a CYP2C19 inducer or a CYP3A4 inducer. Dosage adjustment is recommended based on the combined drug [32].

Lacosamide

Lacosamide is used to treat partial-onset seizures in patients older than 4 years of age. The drug can be administered iv only to the patients older than 17 years of age.

Lacosamide is converted into its primary metabolite, O-desmethyl, by CYP3A4, CYP2C9 and CYP2C19. However; no significant difference between poor and extensive CYP2C19 metabolizers in the means of pharmacokinetic properties has been reported.

Lacosamide concentration gets higher when combined with a strong CYP3A4 and CYP2C9 inhibitor in patients with renal or hepatic impairment; thus, the prescribed dose should be less than the usual dosage [33].

Valproic Acid

Valproic acid is used as a monotherapy or an adjunctive therapy in treatment of complex partial seizures, simple and complex absence seizures. It is also used as an adjunctive therapy in treatment of multiple seizure types, including absence seizures.

Glucuronide conjugate formation and mitochondrial b-oxidation reactions occur during hepatic metabolization of valproic acid. Valproic acid should not be administered in patients with a mitochondrial disease, caused by a mutation on *POLG* gene, encoding mitochondrial DNA polymerase c. Clearance of valproic acid might increase when combined with a drug, inducing glucuronyltransferase activity [30].

Articaine and Epinephrine

Combination of local anesthetic articaine and vasoconstrictor epinephrine is used for anesthesia in simple and complex dental procedures in adults and pediatric patients, older than 4 years of age.

Administration of articaine and epinephrine to the patients with G6PD deficiency might cause methemoglobinemia; thus, it is not recommended to be used in those patients [11].

Lidocaine and Tetracaine

Combination of amide and ester local anesthetic lidocaine and tetracaine is used externally only on intact skin for local dermal anesthesia in excision and shave biopsy of skin lesions.

Lidocaine is converted by CYP1A2 and CYP3A4 into monoethylglycinexylidide and glycinexylidide, having lower potency than lidocaine. Administration of lidocaine and tetracaine to the patients with G6PD deficiency might cause methemoglobinemia; thus, it is not recommended to be used in those patients [34].

Mepivacaine

Mepivacaine is used for local analgesia and anesthesia through central neural techniques, peripheral nerve block techniques and local infiltration. Administration of mepivacaine to the patients with G6PD deficiency might cause methemoglobinemia; thus, it is not recommended to be used in those patients [11].

Oxymetazoline and Tetracaine

Combination of ester local anesthetic oxymetazoline and vasoconstrictor oxymetazoline is used for local anesthesia during restorative procedures on teeth 4-13 and A-J in adults and pediatric patients over 40 kg of weight.

Oxymetazoline is converted into its glucuronide conjugate by UGT1A9 during metabolization. Administration of oxymetazoline and tetracaine to the patients with G6PD deficiency might cause methemoglobinemia; thus, it is not recommended to be used in those patients [35].

Codeine

Codeine is used for mild to moderate pain control only in patients, who can tolerate the drug to avoid addiction and abuse risks, if and only if opioid administration is required and alternative therapy methods are not sufficient.

Codeine is converted into morphine by CYP2D6 during metabolization and CYP3A4 catalyzed norcodeine formation. Then, morphine and norcodeine are metabolized by glucuronic acid conjugation.

When combined with a CYP3A4 inhibitor, metabolization rate of codeine by CYP2D6 increases, leading to higher morphine levels, which may cause increased adverse reaction risks and fatal respiratory depression. Codeine dosage should be reduced if the CYP3A4 inhibitor must be administered to the patient. When combined with a CYP3A4 inducer, codeine level decreases as norcodeine level gets higher. As a result of that, since generated morphine concentration by CYP2D6 metabolization becomes lower, efficacy of the drug will be reduced and withdrawal syndrome may appear in patients with physical dependence. If the CYP3A4 inducer must be administered to the patient, increment in codeine dosage might be considered based on clinical observations. When combined with a CYP2D6 inhibitor, plasma codeine concentration rises and morphine level decreases. In that case, the required analgesic effect will be reduced and withdrawal syndrome symptoms may appear. If the CYP2D6 inhibitor must be administered to the patient, increment in codeine dosage might be considered based on clinical observations.

In ultra-rapid CYP2D6 metabolizers, plasma concentration of morphine, active metabolite of codeine, increases quickly, which may cause overdose symptoms like hypersomnia, confusion or shallow breathing and also fatal respiratory depression. Thus, codeine administration should not be considered in ultra-rapid metabolizers [36].

Dextromethorphan and Quinidine

Combination of quinidine and uncompetitive N-methyl-D-aspartate receptor antagonist and sigma-1 agonist dextromethorphan is used for treatment of pseudobulbar effect. Dextromethorphan is metabolized by CYP2D6 and quinidine metabolization is catalyzed by CYP3A4.

When combined with a moderate or strong CYP3A4 inhibitor, which prolongs the QT interval, response of the patient to dextromethorphan and quinidine therapy should be monitored.

Dextromethorphan and quinidine should not be combined with drugs, which both are metabolized by CYP2D6 and prolong QT interval. Since quinidine is a CYP2D6 inhibitor, when dextromethorphan and quinidine are combined with a CYP2D6 substrate, required response may not be received from the treatment. Therefore, alternative treatments should be considered. On the other hand, quinidine is also a P-gp inhibitor, leading to almost 2-fold increment in serum digoxin levels when two drugs are administered together. Thus, digoxin dosage should be lower than usual when combined with dextromethorphan and quinidine.

When a CYP2D6 inhibitor quinidine is combined with a drug, which is metabolized by CYP2D6, parent drug concentration may excessively increase and/or it may not be converted into its active metabolite, leading to reduced efficacy or toxicity. In poor CYP2D6 metabolizers, quinidine efficacy might not be observed and adverse reaction risks may develop [37].

Donepezil

Donepezil is an acetylcholinesterase inhibitor, used for treatment of dementia in patients with mild, moderate and severe Alzheimer's disease. Conversion of donepezil into its active metabolites is mediated by CYP2D6 and CYP3A4 catalysis. Donepezil clearance decreases down to one third of usual in poor CYP2D6 metabolizers. In ultra-rapid metabolizers, clearance of the drug is higher with almost a 20% of ratio when compared with extensive metabolizers. Therefore, dosage adjustment is recommended for those patients.

When combined with strong CYP3A and CYP2D6 inhibitors, donepezil metabolization is inhibited, leading to increased donepezil concentration. In addition, CYP3A inducers may elevate donepezil elimination rate.

Galantamine

Galantamine is an acetylcholinesterase inhibitor, used for treatment of patients with mild to moderate Alzheimer's disease. O-demethylation reaction of galantamine is catalyzed by CYP2D6 in the liver and the parent drug is N-oxidized by CYP3A4. Although enzyme activity is reduced in poor CYP2D6 metabolizers, plasma concentration of the drug or its metabolite is not affected. On the other hand, when the drug is administered as an extended-release capsule, drug exposure is higher than 50% in poor metabolizers when compared with extensive metabolizers.

When combined with a strong CYP2D6 and CYP3A4 inhibitor, galantamine concentration gets higher. Combined use of CYP2D6 inhibitor paroxetine increases bioavailability of galantamine up to a ratio of 40%. Thus; in those cases, prescribed galantamine dose should be less than usual dosage [38].

Deutetrabenazine

Deutetrabenazine is a vesicular monoamine transporter 2 inhibitor (vmat2), used for treatment of Huntington's disease and tardive dyskinesia. Deutetrabenazine metabolization is primarily mediated by CYP2D6 as CYP1A2 and CYP3A4/5 have minor roles. Single dose and total daily dose of deutetrabenazine should not exceed 18 and 36 mg, respectively, in poor CYP2D6 metabolizers.

When combined with a strong CYP2D6 inhibitor, active metabolite of deutetrabenazine increases up to 3-fold. In that case, the administered dose should be lower than usual dosage.

When combined with a CYP2D6 inhibitor or in poor CYP2D6 metabolizers, deutetrabenazine-related QT prolongation may be observed. Thus, single dose and total daily dose of deutetrabenazine should not exceed 18 and 36 mg, respectively, in those patients.

Tetrabenazine

Tetrabenazine is a vmat2, used for treatment of Huntington's disease. Tetrabenazine is rapidly metabolized in the liver following the absorption. The parent drug is converted into its active metabolite by CYP2D6 and also CYP1A2 have minor roles during metabolization. Recommended maximum daily dose of tetrabenazine is 25 mg. However, dosage should be adjusted based on CYP2D6 genotyping results if a patient must administer 50 mg or more of tetrabenazine daily. For extensive and intermediate metabolizers, the dose should be increased gradually, starting from 12.5 mg/day and single dose cannot exceed 37.5 mg as the total daily dose can be maximum of 100 mg. In poor metabolizers, the starting dose and dosage adjustment is similar with extensive and intermediate metabolizers. However, single dose and total daily dose should not exceed 25 ve 50 mg, respectively, for those patients.

When combined with a strong CYP2D6 inhibitor, concentration of the active metabolite increases, so in that case, single dose and total daily dose of tetrabenazine should not exceed 25 and 50 mg, respectively [39].

Inotersen

Inotersen is a transthyretin (TTR)-directed antisense oligonucleotide, used for treatment of polyneuropathy in patients with hereditary TTR-mediated amyloidosis. Inotersen causes mutant and wild-type *TTR* mRNA degradation by binding to *TTR* mRNA, leading to lower TTR protein levels in serum and tissues. Metabolization of inotersen is mediated by nucleases, which splice the oligonucleotide into different length-nucleotides.

Patisiran

Patisiran is a TTR-directed double-stranded small interfering RNA, used for treatment of polyneuropathy in patients with hereditary TTR-mediated amyloidosis.

Patisiran causes mutant and wild-type *TTR* mRNA degradation by RNA interference, leading to lower TTR protein levels in serum and tissues. Metabolization of patisiran is mediated by nucleases, which splice the oligonucleotide into different length-nucleotides [40].

Eteplirsen

Eteplirsen is an antisense oligonucleotide, used for treatment of patients with Duchenne muscular dystrophy (DMD), having a mutation, which blocks exon 51 skipping on *DMD* gene.

Eteplirsen binds to the exon 51 region on dystrophin pre-mRNA and leads to skipping the exon during transcription; thus, internally truncated dystrophin is produced.

Nusinersen

Nusinersen is a survival motor neuron-2 (SMN2)-directed antisense oligonucleotide, used for treatment of spinal muscular atrophy in adults and pediatric patients. Nusinersen causes involvement of exon 7 during SMN2 mRNA transcription; and thus, full-length SMN protein is produced. Nusinersen is hydrolyzed by exonucleases during metabolization [41].

Amifampridine

Amifampridine is a potassium channel blocker, used to treat Lambert-Eaton myasthenic syndrome in patients with age of 6-17. Amifampridine is rapidly converted into its inactive metabolite, 3-N-acetyl amifampridine during metabolization. Poor NAT2 metabolizers acetylizes amifampridine slower than normal, which causes increment in amifampridine concentration. Thus; the lowest starting dose should be administered and the patient should be monitored to prevent adverse reactions. Recommended starting dose for pediatric poor NAT2 metabolizers over 45 kg of weight is 15 mg in divided doses. Starting dose should not exceed 7,5 mg/day in divided doses for pediatric patients under 45 kg of weight.

Amifampridine Phosphate

Amifampridine phosphate is a potassium channel blocker, used to treat Lambert-Eaton myasthenic syndrome in adults. Amifampridine is rapidly converted into its inactive metabolite, 3-N-acetyl amifampridine during metabolization. Poor NAT2 metabolizers acetylizes amifampridine slower than normal, which causes increment in amifampridine concentration. Thus; the lowest starting dose should be administered and the patient should be monitored to prevent adverse reactions. Recommended starting dose for poor NAT2 metabolizers is 15 mg in 3 divided doses [42].

Meclizine

Meclizine is used to treat vertigo, which is caused by vestibular system-related diseases in adults. Meclizine is metabolized mainly by CYP2D6; thus, dosage adjustment is required for poor or ultra-rapid CYP2D6 metabolizers.

As a result of the interaction between CYP2D6 inhibitors and meclizine, the patient should be monitored to prevent adverse reaction risks and to observe clinical efficacy [43].

Siponimod

Siponimod is a sphingosine 1-phosphate receptor modulator, used in treatment relapsing forms of multiple sclerosis. Siponimod is mainly metabolized by CYP2C9 in the liver and CYP3A4 has also minor roles. Different CYP2C9 genotypes affect metabolism rate of siponimod. Drug concentration is 2- and 4-fold in patients, having CYP2C9*2/*3 and CYP2C9*3/*3 genotypes, respectively. Thus, dosage adjustment is highly recommended for those patients.

When combined with a CYP2C9 or CYP3A4 inhibitor or both of them, increased siponimod concentration may cause adverse reactions. On the other hand, when combined with a moderate CYP2C9 inducer or a strong CYP3A4 inducer or both of them, siponimod concentration will be reduced. Therefore, siponimod should not be administered with any of those drugs. Patients, having CYP2C9*1/*1 and *2/*3 genotypes should not use siponimod with a moderate or strong CYP3A4 inducer [44, 45].

IMMUNOMODULATION AND CHEMOTHERAPEUTIC DRUGS

The drugs, which are commonly used in immunomodulation and chemotherapy (Table **5**), and the pharmacogenomic biomarkers, which cause dosage adjustment for those drugs due to interindividual variability, are indicated below.

Table 5. Commonly used drugs in immunomodulation and chemotherapy.

Clinical use	Drugs
Infectious diseases	Abacavir, ceftriaxone, dasabuvir, ombitasvir, paritaprevir and ritonavir, dolutegravir, isoniazid, pyrazinamide and rifampin, ledipasvir and sofosbuvir, raltegravir, sulfamethoxazole and trimethoprim, voriconazole
Oncology	Abemaciclib, alectinib, anastrozole, capecitabine, cetuximab, crizotinib, erlotinib, everolimus, fluorouracil, imatinib, irinotecan, lapatinib, olaparib, panitumumab, pazopanib, tamoxifen

Abacavir

It is an agent recommended for use in combination with other antiretroviral drugs in the treatment of people infected with HIV-1. CYP450 enzymes do not play an important role in abacavir metabolism. Its metabolism is mainly the alcohol dehydrogenase pathway and it does not have antiviral activity in the formed metabolites. Therefore, no clinically significant interactions occur between CYP isoforms and other drugs metabolized.

During treatment with abacavir, potentially fatal hypersensitivity reactions (multiple organ failure, anaphylaxis that may occur during the first 6 weeks of treatment) have been reported. Patients with the *HLA-B**5701 allele for these hypersensitivity reactions are in the high-risk group. Hypersensitivity reactions in individuals to be treated with abacavir should be seen as a marker for clinical judgment and patients should be screened for the *HLA-B**5701 allele before starting treatment. Because abacavir is contraindicated in *HLA-B**5701-positive patients [46].

Ceftriaxone

It is an antibiotic of the cephalosporin group used in the treatment of infections caused by Gram-negative bacteria such as *E. aerogenes, E.cloacae, E. coli, H. influenzae, K. pneumoniae, N. gonorrhoeae, N.meningitidis, P. mirabilis*, Gram-positive bacteria such as *S. aureus, S.epidermidis, S.pyogenes* and anaerobes (Bacteroides and Clostridium species).

The risk of developing methemoglobinemia has been reported for ceftriaxone in combination use with local anesthetic agents such as lidocaine. The risk of developing methemoglobinemia appears to be higher in people with G6PD deficiency. If G6PD is used in combination with a local anesthetic in patients with enzyme deficiency, methemoglobinemia symptoms and signs should be monitored [11].

Dasabuvir, Ombitasvir, Paritaprevir and Ritonavir

They are combined antiviral drugs used in the treatment of adult patients infected with chronic HCV. Dasabuvir is predominantly metabolized by oxidative metabolism mediated by the CYP2C8 enzyme and by its contribution to the CYP3A4 enzyme. Ombitasvir is metabolized by hydrolysis reactions followed by amide hydrolysis in oxidative metabolism. CYP3A4 enzyme plays a leading role in the metabolism of paritaprevir and ritonavir, but CYP3A5 enzymes contribute to the metabolism of paritaprevir and CYP2D6 enzymes to metabolism of ritonavir.

Ombitasvir, paritaprevir, and dasabuvir are inhibitors of the *UGT1A1* gene (Gilbert's Syndrome) and ritonavir of the CYP3A4 enzyme. In addition, paritaprevir alone is an inhibitor of OATPs such as both OATP1B1 and OATP1B3, in combination with ritonavir and dasabuvir, it is an inhibitor of (BCRP). Therefore, if the combined preparation containing these 4 antivirals is used with drugs that are substrates of CYP3A, UGT1A1, BCRP, OATP1B1, or OATP1B3, an increase in plasma concentrations of these drugs may occur.

Since paritaprevir and ritonavir are predominantly metabolized by the CYP3A4 enzyme, increased plasma concentrations of these two drugs may occur when used in combination with potent CYP3A4 inhibitors. Likewise, when dasabuvir, which is metabolized by CYP2C8, is used in combination with these enzyme inhibitors, it is possible that the plasma concentration will increase [46, 47].

Dolutegravir

Dolutegravir is an antiviral agent used in combination therapy with other antiretroviral agents in the treatment of HIV1 infection. Dolutegravir is mainly metabolized by UGT1A1, but partially by the CYP3A enzyme.

Dolutegravir inhibits *in vitro* renal OCTs, OCT2 and multidrug and toxin extrusion transporter (MATE) 1 and actually inhibits creatinine tubular secretion *in vivo* due to this inhibition. Therefore, plasma concentrations of drugs such as dofetilide, dalfampridine, and metformin, in which OCT2 and MATE1 are responsible for elimination, may increase when they are used together with dolutegravir. It is also a substrate for dolutegravir UGT1A3, UGT1A9, BCRP, and P-gp, and both its clinical efficacy and plasma concentration may be reduced if these enzymes and transporters are used in conjunction with inducing agents [48].

Isoniazid, Pyrazinamide and Rifampin

Isoniazid, pyrazinamide and rifampin combination tablets, known under the trade name Rifater, are used in the treatment of tuberculosis. Isoniazid is generally metabolized in the liver by undergoing acetylation and dehydranization reaction. In terms of metabolizing this drug, almost half of African Americans and Caucasians are slow inactivators, while the majority of Eskimos and Asians are fast inactivators. The acetylation rate is not very important in terms of isoniazid activity, in the case of slow acetylation, the metabolism will slow down, so the drug may accumulate and reach a toxic dose.

Rifampin affects CYP1A2, CYP2B6, CYP2C8, CYP2C9, CYP2C19, and CYP3A4 enzymes, UGT, sulfotransferases, carboxylesterases and transporters such as P-gp and multidrug resistance-associated protein 2. Therefore, when these enzymes or many drugs that are substrates of transporters are used together with rifampin, drug interaction may occur, so it is possible that the effects of these drugs are reduced by rifampin and their metabolism accelerates. For example; when a drug whose metabolism is mainly dependent on CYP3A4 enzyme, such as toremifene, is used together, steady-state concentrations of toremifene in the blood decrease, exposure to warfarin, which is mainly a CYP2C9 enzyme, decreases, exposure to statins metabolized by CYP3A4 decreases, and plasma concentrations of antipsychotics decrease. When used together with saquinavir, an antiretroviral agent in which CYP3A4 enzyme is involved in metabolism, the AUC value of this drug decreases by 70% and severe hepatocellular toxicity occurs.

Isoniazid inhibits CYP1A2, CYP2C9, CYP2C19 and CYP3A4 enzymes. In other words, it inhibits the metabolism of antiepileptics such as carbamazepine, phenytoin, valproic acid and drugs such as diazepam, haloperidol, warfarin and ketoconazole, which are metabolized by these enzymes, and these drugs may accumulate in the body and cause toxic doses. In case of concomitant use, the dosage of rifater should be made according to therapeutic drug monitoring, taking into account the risk of toxicity [49].

Ledipasvir and Sofosbuvir

Ledipasvir and sofosbuvir combination tablets, known under the trade name Harvoni, are used in the treatment of chronic HCV in pediatric patients over 3 years of age and adult patients.

While *in vitro* studies on ledipasvir have found no evidence for a mediated metabolism of CYP enzymes, it has been stated that it is metabolized by a slow oxidative mechanism, the exact mechanism of which is unknown.

Sofosbuvir is metabolized in the liver into the nucleoside analog triphosphate GS-461203. This conversion pathway includes sequential hydrolysis of the carboxyl ester group followed by phosphorylation mediated by the pyrimidine nucleotide biosynthesis pathway. As a result of the dephosphorylation step, GS-331007, a nucleoside metabolite without *in vitro* anti-HCV activity, is formed.

Ledipasvir and sofosbuvir are substrates for P-gp and BCRP. When used together with drugs that are P-gp inducers such as rifampin, their pharmacological effectiveness decreases due to their plasma concentrations. For this reason, they should not be used with drugs that are P-gp inducers. On the contrary, P-gp or BCRP inhibitor drugs can be used together since they will increase the plasma concentration of ledipasvir and sofosbuvir. Ledipasvir inhibits drugs that are P-gp and BCRP transporters. Therefore, if ledipasvir and these drugs are used together, their intestinal absorption may increase.

When used with P-gp inducers, the plasma concentrations of ledipasvir and sofosbuvir may decrease and their therapeutic efficacy will decrease. Therefore, combined use is not recommended [50].

Raltegravir

It is indicated for use with other antiretroviral drugs in the treatment of HIV-1. Raltegravir is mainly metabolized by glucuronidation *via* the UGT1A1 enzyme and its elimination occurs in this way. Although there are no strong data that the *UGT1A1* polymorphism exerts a clinical effect on raltegravir pharmacokinetics, subjects with the *28/*28 genotype (associated with a decrease in UGT1A1 activity) were found to have a decrease in mean AUC compared to those with the wild-type genotype.

Since UGT1A1 plays the main role in raltegravir metabolism, drugs that inhibit this enzyme will increase plasma concentrations, while inducers (such as rifampin, etravirine) will decrease plasma concentrations. Therefore, combined use is not recommended [51].

Sulfamethoxazole and Trimethoprim

The combination of sulfamethoxazole/trimethoprim inhibits the growth of resistant bacteria and provides increased efficacy against bacterial infections through combined use. It should be used if infection by bacteria is strongly suspected. It is indicated for urinary system infections, acute otitis media, *Pneumocystis jiroveci* Pneumonia, shigellosis.

Sulfamethoxazole is metabolized to N4-hydroxy metabolites (N4-acetyl-, N4-hydroxy-, 5-methylhydroxy-, N4-acetyl-5-methylhydroxy-sulfamethoxazole) and N-glucuronide conjugates mediated by the CYP2C9 microsomal enzyme. Trimethoprim has 11 metabolites, five of which are conjugated with glutathione and six of them are oxidative metabolites (major metabolites: 1- and 3-oxides and the 3- and 4-hydroxy derivatives). The metabolites that are not bound in either drug are thought to be pharmacologically active metabolites.

Trimethoprim is an inhibitor for the CYP2C8 enzyme and the OCT2 transporter, while sulfamethoxazole is an inhibitor for the CYP2C9 microsomal enzyme. Therefore, caution should be exercised when the trimethoprim/sulfamethoxazole drug combination is used with drugs that are CYP2C8, CYP2C9 and OCT2 substrates. Prolongation of prothrombin time has been reported in patients when coadministered with warfarin, a CYP2C9 substrate. Therefore, when these drugs and warfarin, an anticoagulant, are used together, the coagulation time should be carefully evaluated.

In addition, these drugs can alter the metabolism of phenytoin, a CYP2C9 substrate. When sulfamethoxazole/trimethoprim is given at normal clinical doses, the half-life of phenytoin increases 9-fold and its metabolic clearance decreases. Therefore, caution should be exercised against unwanted phenytoin effects that may occur due to combined use.

The combination of sulfamethoxazole/trimethoprim potentiates the effects of hypoglycemics metabolized by CYP2C8 such as pioglitazone, repaglinide, and rosiglitazone and CYP2C9 enzymes such as glipizide, glyburide, in addition to metformin eliminated by OCT2.

The combination of sulfamethoxazole/trimethoprim may cause hemolysis in individuals with G6PD, depending on the dose frequency. Also, since slow acetylating agents may be more prone to idiosyncratic reactions to sulfonamides, adequate fluid intake and urine output should be provided during the treatment period against the risk of crystalluria development [52].

Voriconazole

Voriconazole is an antifungal drug used in the treatment of invasive aspergillosis, candida infections, Scedosporiosis, and Fusariosis. The metabolism of voriconazole is mediated by CYP2C9, CYP2C19, and CYP3A4 enzymes. Among these, CYP2C19 is important for voriconazole and this enzyme shows genetic polymorphism (approximately 15-20% of Asians, Caucasians, and Blacks 3-5% are poor metabolizers). In the studies conducted, voriconazole exposure of healthy people with poor metabolizers was found to be 4 times higher than homozygous extensive metabolizers and 2 times higher than heterozygous extensive metabolizers.

The affinity of voriconazole for CYP enzymes is from very low to low, respectively CYP2C19, CYP2C9 and very low CYP3A4. If used with drugs that inhibit or induce these enzymes, plasma concentration may decrease or increase.

Rifampin (600 mg/day), a strong CYP450 enzyme inducer, reduced the steady-state C_{max} and AUC values of voriconazole (200 mg/12h/7 days) by 93% and 96%, respectively. Therefore, the use of these two drugs together is contraindicated.

Ritonavir (400 mg/12h/9d), a strong CYP450 enzyme inducer and CYP3A4 inducer/substrate, reduced the steady-state C_{max} and AUC values of voriconazole (200 mg/12h/8 d) by 66% and 82%, respectively. With low dose ritonavir (100 mg/12h/9d) administration, these pharmacokinetic parameters decreased in smaller rates, while repeated high-dose voriconazole (400 mg every 12 hours) administration did not cause any change. Therefore, while the combined use of high dose ritonavir and voriconazole is contraindicated, it should be used considering the benefit/risk assessment in low dose applications.

St. John's Wort is an inducer of CYP450 and P-gp. Therefore, care should be taken in long-term combined use. Long-acting barbiturates such as carbamazepine and phenobarbital, which are potent CYP450 inducers, should also not be used together as they produce a significant reduction in voriconazole plasma concentration.

Although the combination of CYP2C19, CYP2C9 and CYP3A4 inhibitor fluconazole and voriconazole is not recommended, drug monitoring should be performed for adverse events related to voriconazole when it is considered.

Concomitant use of voriconazole with drugs that are CYP3A4 substrates such as sirolimus, cisapride, pimozide and quinidine is contraindicated, as systemic exposure to these drugs will increase significantly. Furthermore, combined use is not recommended since everolimus, a CYP3A4 and P-gp substrate, may increase the plasma concentration of everolimus when used with voriconazole.

In addition to all these drug interactions, CYP3A4 substrates (alfentanil, fentanyl, oxycodone, cyclosporine, methadone, tacrolimus, oral coumarin anticoagulants, statins, benzodiazepines, calcium channel blockers, vinca alkaloids), CYP2C19 substrates (methadone) and CYP2C19 substrates (methadone, warfarin, oral coumarin anticoagulants, sulfonylureas, ibuprofen, diclofenac) increase systemic exposure to these drugs when used in combination with voriconazole. Therefore, drug monitoring and dose adjustment may be required when used in combination, and these patients should be followed carefully for adverse events.

Concomitant use of voriconazole with CYP450 inducer/CYP3A4 inhibitor and its substrate, efavirenz, may cause toxic effects. These two drugs are contraindicated due to the significant changes in pharmacokinetic parameters (C_{max} and AUC) during concomitant use at standard doses (400 mg every 24 hours or higher). However, if the maintenance dose of voriconazole is 400 mg every 12 hours and the efavirenz dose is 300 mg every 24 hours, these two drugs can be used together.

Phenytoin is a CYP2C9 substrate and a potent CYP450 inducer. If these two drugs are to be used together, voriconazole maintenance dose should be changed: 5 mg/kg intravenously every 12 hours or 400 mg orally every 12 hours. Plasma phenytoin concentrations of patients should also be monitored in terms of phenytoin side effects due to concomitant use.

Voriconazole plasma exposure increases when CYP2C19 inhibitor oral contraceptives containing estradiol and norethindrone are used in combination with voriconazole. Therefore, drug monitoring should be performed due to toxic effects and adverse events that may occur during concomitant use.

Coadministration of cisapride, pimozide and quinidine with voriconazole may increase the plasma exposure of these drugs, leading to QT prolongation and rarely torsade de pointes [53].

Abemaciclib

Abemaciclib in combination with letrozole or anastrozole is used in the following situations:

– Hormone receptor (HR) -positive and human epidermal growth factor receptor 2 (HER2) negative postmenopausal metastatic breast cancer,
– Patients who have recurred the disease 12 months after adjuvant therapy or who have not previously received any hormonal therapy for advanced breast cancer,
– Also, in combination with fulvestrant in the treatment of women with HR-positive, HER2-negative or advanced metastatic breast cancer after endocrine therapy.

Abemaciclib is mainly metabolized by CYP3A4 enzyme. As a result of its metabolism, N-desethylabemaciclib (M2), hydroxyabemaciclib (M20), hydroxy-N-desethylabemaciclib (M18), and an oxidative metabolite (M1) are formed. The

M2, M18 and M20 metabolites have a potency similar to abemaciclib.

Potent CYP3A4 inhibitors increase the concentration of abemaciclib and its active metabolites. If patients stop using CYP3A4 inhibitors while using abemaciclib with strong CYP3A4 inhibitors, the dose of abemaciclib should be increased to the dose used before the combination.

When moderate CYP3A4 inhibitors are administered together with abemaciclib, it is expected that the active metabolites of abemaciclib will increase the AUC values 1.6 and 2.4 times, respectively. In such cases, adverse reactions related to abemaciclib should be monitored and the dose should be reduced if necessary.

Potent or moderate CYP3A4 inducers decrease the AUC value of the active metabolites of abemaciclib. Therefore, its use with these agents should be avoided against decreased activity and alternative therapies should be evaluated. Abemaciclib and its active metabolites inhibit the transporters OCT2, MATE1, and MATE2-K at recommended concentrations [54, 55].

Alectinib

Alectinib is used in the treatment of patients with anaplastic lymphoma kinase (ALK) -positive metastatic non-small cell lung cancer (NSCLC) after the disease is determined by FDA approved tests.

Alectinib is converted to M4 active metabolite after mainly metabolized by CYP3A4 enzyme and this metabolite is metabolized by CYP3A4 enzyme. Also, his active metabolite is a P-gp substrate. Patients to be treated with alectinib are tested to detect ALK positivity in pre-treatment tumor samples [56].

Anastrozole

Anastrozole is used in the following situations:

– As an adjuvant in the treatment of women with postmenopausal early-stage HR-positive breast cancer,

– First-line treatment of postmenopausal women with HR-positive or locally advanced or metastatic breast cancer of unknown origin

– In the treatment of postmenopausal women with advanced breast cancer who have disease progression after tamoxifen therapy

Anastrozole is metabolized as N-dealkylation, hydroxylation and glucuronidation, and triazole is the main metabolite (pharmacologically poorly active). Anastrozole strongly inhibits CYP1A2, CYP3A4, CYP2C8 and CYP2C9 enzymes catalyzed reactions [57].

Capecitabine

Capecitabine is indicated as a single agent for adjuvant therapy in Dukes'C colon cancer patients with complete resection of the primary tumor when fluoropyrimidine treatment is preferred and as a first-line treatment for patients with metastatic colorectal carcinoma, where fluoropyrimidine treatment alone is preferred.

Capecitabine is used in the treatment of patients with metastatic breast cancer after chemotherapy containing anthracycline. It is also used in the treatment of patients with metastatic breast cancer who are resistant to chemotherapy containing both anthracycline and paclitaxel or who are resistant to paclitaxel and who cannot receive more anthracycline.

Capecitabine is metabolized to 5-FU by enzymatic means, and most of this compound is converted to 5'-deoxy-5-fluorocytidine (5'-DFCR) by carboxylesterase mediated hydrolysis in the liver. The 5'-DFCR is then converted to 5'-DFUR by cytidine deaminase, an enzyme found in many tissues, including tumors.

Patients with homozygous or heterozygous deficient dihydropyrimidine dehydrogenase deficiency (DDP) gene mutations with complete or DDP activity due to the use of capecitabine are at high risk for serious or fatal adverse reactions. Likewise, patients with partial DPD activity may experience severe adverse reactions such as diarrhea, neutropenia and neurotoxicity. Patients with DPD deficiency should be asked to inform their healthcare professionals [58].

Cetuximab

Cetuximab is a monoclonal antibody that act by binding to epidermal growth factor receptors (EGFR) found on the surface of tumor cells. It is used in the treatment of metastatic colorectal cancers without Ras gene mutations. Cetuximab is combined with irinotecan and FU for first-line treatment. It is also used in combination with radiation therapy in the initial treatment of patients with locally or regionally advanced squamous cell head and neck cancer, and in combination with FU and platinum-based therapy for the first-line treatment of patients with a recurrent locoregional disease or metastatic squamous cell head and neck cancer.

EGFR is a transmembrane glycoprotein expressed in many epithelial tissues such as skin and hair follicles. EGFR expression can be detected in human tissues with cancer of the head and neck, rectum and colon. EGFR mediated signal transduction causes activation of the wild-type RAS protein. However, the mutant Ras proteins formed are continuously active regardless of EGFR regulation.

EGFR expression and the presence of ras mutation should also be determined using FDA approved tests before treatment with cetuximab. Because cetuximab does not provide clinical benefit in patients with ras mutations due to toxicity [59, 60].

Crizotinib

Crizotinib is used in the treatment of metastatic NSCLC with ALK tumors or ROS1-positive as determined by FDA-approved tests. Crizotinib is an inhibitor of tyrosine kinase receptors, including ALK, hepatocyte growth factor receptor (c-Met), ROS1 (c-ros), and recepteur d'origine nantais (RON). Oncogenic fusion proteins whose expression increases as a result of translocations that occur may affect the ALK gene. These ALK fusion proteins can contribute to gene expression and signal activation that can increase cell proliferation and survival in tumor cells in which they are expressed.

According to an ECG analysis performed in ALK-positive NSCLC patients, the maximum mean change in QTcF (corrected QT by the Fridericia method) according to baseline after oral administration of 250 mg crizotinib twice a day was calculated as 12.3 ms. Crizotinib increased the QTcF value depending on the concentration.

Crizotinib is mainly metabolized by CYP3A4 enzyme. Patients with metastatic NSCLC to be treated with crizotinib are selected based on ALK or ROS1 positivity in their tumor cells.

Concomitant use of crizotinib with strong CYP3A inhibitors increases crizotinib plasma concentrations, which may cause adverse effects. Therefore, these agents should not be used together, but the crizotinib dose should be reduced if it should be used. Also, caution should be exercised when used with moderate CYP3A inhibitors. In contrast, concomitant use of crizotinib with potent CYP3A4 inducers decreases crizotinib plasma concentration and thus its efficacy. Therefore, concomitant use of crizotinib with potent CYP3A4 inducers should be avoided. When CYP3A substrates are co-administered with crizotinib, there is an increase in crizotinib plasma concentration. Crizotinib should not be used in combination with CYP3A substrates, as changes in crizotinib concentrations can cause serious adverse reactions. However, if crizotinib and CYP3A substrates are to be used together, it is recommended to reduce the CYP3A4 substrate dose [59].

Erlotinib

Erlotinib is used in the treatment of NSCLC patients with EGFR exon 19 deletion or exon 21 (L858R) substitution mutations (determined by FDA approved tests) in their tumor cells. It is also used as a first-line treatment with gemcitabine for locally advanced, unresectable or metastatic pancreatic cancers.

EGFR kinase activity is reversibly inhibited by erlotinib. In the presence of EGFR exon 19 deletion and exon 21 mutations, erlotinib binding affinity is considerably higher than wild-type receptors. The most important enzyme in erlotinib metabolism is the CYP3A4 enzyme, but it is involved in the CYP1A2 enzyme with less contribution.

Before treating patients with metastatic NSCLC with erlotinib, EGFR exon 19 deletion or exon 21 (L858R) substitution mutations should be determined in plasma samples or tumor tissues with FDA approved tests. Patient selection should be made according to these test results.

Concomitant use of erlotinib with CYP3A4 and CYP1A2 inhibitors may cause serious adverse reactions because erlotinib exposure is increased. In this case, the erlotinib dose should be reduced to 50 mg. However, concomitant use of these agents should be avoided.

If erlotinib is used with CYP3A4 inducers, erlotinib dose should be increased by 50 mg at 2-week intervals to a maximum of 450 mg, because erlotinib exposure is reduced. However, concomitant use of these agents should be avoided.

In addition, exposure to erlotinib is reduced by smoking, which is a CYP1A2 inducer. Therefore, the use of erlotinib with cigarettes or drugs that cause induction of the CYP1A2 enzyme should be avoided [59, 60].

Everolimus

Everolimus is used in the following situations:

– In combination with exemestane in the treatment of women with HR-positive, HER2-negative advanced postmenopausal breast cancer.
– In the treatment of locally advanced or metastasized unresectable pancreatic progressive neuroendocrine tumors,
– In the treatment of locally advanced or metastasized unresectable gastrointestinal or lung-origin progressive, highly differentiated, non-functional neuroendocrine tumors,
– In the treatment of advanced renal cell carcinoma (RCC) in which treatment

with sunitinib or sorafenib has failed.

– In the treatment of Tuberous Sclerosis Complex (TSC) and renal angiomyolipoma.

Everolimus should be administered orally at 10 mg/day in patients with HR-positive, HER-2 negative breast cancer, until any toxicity is observed or disease progression.

Everolimus is a substrate CYP3A4 enzyme. After oral administration, it is converted to 6 main metabolites in the blood and their effect is lower than that of everolimus.

P-gp and CYP3A4 inhibitor drugs such as ketoconazole, erythromycin and verapamil increase everolimus C_{max} and AUC values, thus increasing everolimus exposure. However, P-gp and CYP3A4 inducers decrease these values, so everolimus exposure decreases. Everolimus dosage should be changed in the following cases: in patients with hepatic insufficiency, in the case of P-gp and CYP3A4 inhibitor or inducer drugs.

In indications of renal angiomyolipoma associated with breast cancer, neutrophil extracellular trap, RCC and TSC and when used with a P-gp or moderate CYP3A4 inhibitor drug, the dose of everolimus should be reduced to 2.5 mg per day. If patients can tolerate, the daily dose can be increased to 5 mg. Three days after discontinuation of the inhibitory drugs, the previous dose of these drugs is started. Concomitant use with P-gp or moderate CYP3A4 inducing drugs should be avoided, if this is not possible, the everolimus daily dose should be doubled in increments of 5mg or less. Everolimus dose may return to normal 5 days after discontinuation of P-gp or moderated CYP3A4 inducer drugs. These dose adjustments are also applicable when everolimus is used with P-gp or moderate CYP3A4 inducing drugs.

In cases of TSC-associated subependymal giant cell astrocytoma and TSC-Associated Partial-Onset Seizures, when everolimus and P-gp or a moderate CYP3A4 inhibitor are used together, the everolimus daily dose should be reduced by 50%. Three days after discontinuation of inhibitor drugs, the previous dose of these drugs are continued.

Concomitant use of everolimus with P-gp and potent CYP3A4 inhibitors should be avoided. If this cannot be achieved, the dose of everolimus should be reduced. The dose of everolimus should be increased when everolimus is used with P-gp and strong CYP3A4 inducer drugs.

In geriatric patients with HR-positive, HER2-negative breast cancer, patients should be carefully monitored against the risk of adverse reactions that may develop depending on the everolimus dose [61].

Fluorouracil

Fluorouracil is used in the following situations:

– In the topical treatment of multiple actinic or solar keratoses of the face and anterior scalp
– In the treatment of colon and rectum adenocarcinoma, breast adenocarcinoma, gastric adenocarcinoma, pancreatic adenocarcinoma

An increase in coagulation time has been reported in patients using warfarin in combination with fluorouracil. Therefore, it should be considered that the CYP2C9 enzyme may be inhibited by fluorouracil and its metabolites.

The majority of fluorouracil is catabolized by the DPD enzyme. This enzyme deficiency can cause various toxicities (stomatitis, diarrhea, neutropenia, and neurotoxicity). Therefore, fluorouracil should not be used in patients with DPD deficiency [62, 63].

Imatinib

Imatinib is used in the following situations:

– In newly diagnosed adult and pediatric patients with Philadelphia chromosome positive chronic myeloid leukemia (Ph+CML) and in patients with Ph+CML in blast crisis (BC), accelerated phase (AP) or in chronic phase (CP) after unsuccessful interferon alpha therapy.
– In adult and pediatric patients with Ph+ acute lymphoblastic leukemia (ALL)
– In adult patients with myelodysplastic or myeloproliferative diseases and aggressive systemic mastocytosis (ASM)
– In adult patients with hypereosinophilic syndrome and chronic eosinophilic leukemia
– In adult patients with dermatofibrosarcoma protuberans
– As normal and adjuvant therapy in patients with Kit (CD117) positive gastrointestinal stromal tumors.

Imatinib acts by inhibiting BCR-ABL tyrosine kinase. In the case of CML, a BCR-ABL chimeric gene is formed on the Philadelphia chromosome. Imatinib both inhibits this gene and causes apoptosis of BCR-ABL positive cells.

Although the metabolism of imatinib is predominantly provided by the CYP3A4 enzyme, CYP1A2, CYP2D6, CYP2C9, and CYP2C19 enzymes also contribute to its metabolism. As a result of metabolism by CYP3A4, N-demethylated, the main active metabolite is formed. Also, imatinib for CYP2C9, CYP2D6, and CYP3A4/5 enzymes is a potent competitive inhibitor.

The D816V c-Kit mutation must be identified before starting treatment with imatinib in adult ASM patients. In the absence of such mutation, the recommended dose is 400 mg/day for adult patients. If the mutation status is not detected or there is no response to other treatments, the imatinib dose is still the same.

In adult patients with Ph+CML CP, AP, or BC, the imatinib dose should be regulated as 400 mg/day for CML CP and 600 mg/day for AP or BC.

Imatinib should not be used in combination with potent CYP3A4 inhibitors. Potent CYP3A4 inhibitors may increase imatinib plasma concentration by approximately 50%. On the other hand, CYP3A4 inducers will reduce imatinib exposure. Therefore, the clinical effect should be carefully monitored when these agents are used in combination. Plasma concentration of drugs such as triazole-benzodiazepines, dihydropyridine calcium channel blockers, whose metabolism is mediated by CYP3A4 enzyme, will be increased by imatinib. Caution should be exercised when imatinib and CYP3A4 substrates are used together. Also, caution should be exercised when using imatinib with CYP2D6 substrates with a narrow therapeutic window.

CYP2C9 and CYP3A4 are two important enzymes involved in the metabolism of warfarin. When anticoagulation is required, heparin is recommended instead of warfarin [56].

Irinotecan

Irinotecan is used in combination with 5-FU and leucovorin in the first-line treatment of patients with metastatic carcinoma of the colon or rectum, or in the treatment of patients who have previously been treated with 5-FU and have worsened or relapsed.

Irinotecan and its active metabolite SN-38 are metabolized by CYP3A4 enzyme and UGT1A1, respectively. UGT1A1 shows genetic polymorphism among individuals, resulting in a decrease in enzyme activity.

When irinotecan is administered alone or in combination to patients homozygous for the UGT1A1 * 28 allele (*UGT1A1* 7/7 genotype), a reduction in the initial dose may be considered. However, there is no definite information on this subject.

Therefore, patients should be monitored for toxicity and dose adjustment should be made accordingly.

Exposure to irinotecan and its active metabolite SN38 in adult and pediatric patients is increased if irinotecan is used with CYP3A4 inducers and UGT1A1 inducers such as gemfibrozil, indinavir. Therefore, drugs that are not CYP3A4 enzyme inducers should be used at least 2 weeks before starting irinotecan treatment. Also, CYP3A4 inhibitors should be stopped at least one week ago against adverse reactions.

Persons homozygous for the *UGT1A1**28 allele early in treatment with irinotecan should be monitored for the risk of developing neutropenia. For this, there are tests that will determine the UGT1A1 genotype of the patients before starting treatment [58].

Lapatinib

It is used in combination with capecitabine in the treatment of metastatic or advanced breast cancers with HER2 overexpression, previously treated with anthracycline, taxane, and trastuzumab, and in combination with letrozole in the treatment of postmenopausal HR-positive women with metastatic breast cancer with HER2 overexpression for which hormone therapy is indicated. Lapatinib is a 4-anilinoquinazoline kinase inhibitor of the intracellular tyrosine kinase domains of both EGFR and HER2 receptors.

Although lapatinib is mainly metabolized by CYP3A4 and CYP3A5 enzymes, the CYP2C19 and CYP2C8 enzymes also play a minor role.

When administered in combination with 2000 mg/m^2/day capecitabine in the treatment of HER2-positive metastatic breast cancer, the recommended dose for lapatinib is 1250 mg orally once a day on days 1-21st.

When administered in combination with letrozole in the treatment of HR-pozitive, HER2-positive metastatic breast cancer, the recommended dose for lapatinib is 1500 mg orally once a day.

Hepatotoxicity reactions due to lapatinib in a monotherapy study were associated with HLA alleles *DQA1**02:01 and *DRB1**07:01. Serious liver damage has occurred in people with these alleles. These HLA alleles are found in approximately 15-25% of Caucasian, Asian, African, and Hispanic populations and 1% of Japanese. Patients receiving lapatinib treatment without genotype determination should be followed up for liver damage.

Lapatinib is an inhibitor of CYP3A4, CYP2C8, and P-gp. Therefore, in combination therapy should be exercised and CYP3A4, CYP2C8, and P-gp substrate doses should be reduced. For example, if lapatinib is used in combination with CYP3A4 inhibitors, the dose of lapatinib should be reduced to 500 mg/day. After stopping the CYP3A4 inhibitor, one week should wait before increasing the lapatinib dose.

The use of Lapatinib with strong CYP3A4 inducers is not recommended. However, in case of concomitant use, the dose of lapatinib should be changed as follows: 4500 mg/day in HER2-positive metastatic breast cancer cases, and 5500 mg/day in HR-positive, HER2-positive metastatic breast cancer cases [64].

Olaparib

Olaparib is a poly (ADP-ribose) polymerase (PARP) inhibitor used in the first-line treatment of ovarian cancers with advanced *BRCA* mutation, in the treatment of ovarian cancers with germline *BRCA* mutation (gBRCAm), in the treatment of *gBRCAm* HER2-negative breast cancers, *gBRCAm* in the treatment of metastatic pancreatic adenosarcomas, homologous recombination deficiency (HRD) - positive progressive ovarian cancers.

Olaparib is mainly metabolized by CYP3A enzyme. Olaparib is both an inhibitor and an inducer of CYP3A enzyme, but olaparib is only an inducer of CYP2D6 enzyme. It is also an inhibitor of the UGT1A1 enzyme. Apart from these, it is an inhibitor for some transporters: BCRP, OATP1B1, OCT1, OCT2, OAT3, MATE1, MATE2K and P-gp.

The recommended oral dose for olaparib is 300 mg twice a day. Although its use with strong or moderated CYP3A inhibitors is not recommended, dose adjustment should be made as follows if it will still be used: 100 mg twice daily for use with potent inhibitors and 150 mg twice daily for use with moderate inhibitors. The usual administration dose for olaparib can be resumed 3-4 half-life after CYP3A inhibitor cessation.

Before starting treatment with Olaparip, genetic mutations should be detected with tests approved by the FDA. Patients should be evaluated for the following biomarkers: *BRCA1* and *BRCA2* mutations for ovarian cancer patients who develop germline or somatic *BRCA* mutations, *BRCA1* and *BRCA2* mutations genomic instability and ovarian cancer patients who advenced HRD-positive, advanced *gBRCA* mutation in ovarian cancer, gBRCA1 ve gBRCA2 mutations for HER2-negative metastatic breast cancer with *gBRCA* mutation and for metastatic pancreatic adenocarcinoma with *gBRCAm.* In addition, before starting treatment with olaparib for germline or somatic HRR gene-mutated metastatic castration-resistant prostate cancer, patients should be screened in blood or tumor samples for *ATM, BRCA1, BRCA2, BARD1, BRIP1, CDK12, CHEK1, CHEK2, FANCL, PALB2, RAD51B, RAD51C, RAD51D, RAD54L, gBRCA1* and *gBRCA2* mutations.

When used with CYP3A inhibitors, the concentration of olaparib may increase and adverse reactions may occur. Therefore, concomitant use with drugs that are CYP3A inhibitors should be avoided or the dose of olaparib should be reduced when used together. Conversely, when used with CYP3A inducers, olaparib concentration decreases. Consequently, it is not recommended to be used together as the effectiveness of the drug will be reduced [65].

Panitumumab

Panitumumab is an EGFR antagonist human IgG_2 kappa monoclonal antibody used in the treatment of metastatic colorectal cancer with a wild-type RAS mutation. If the RAS mutation status is unknown in patients, it is not indicated.

Before starting treatment with panitumumab, the RAS mutation status should be determined in patients with tests approved by the FDA. RAS mutations in both KRAS and NRAS in exon 2 (codons 12 and 13), exon 3 (codons 59 and 61), and exon 4 (codons 117 and 146) should be determined with the help of these tests.

Panitumumab, an anti-EGFR antibody, caused adverse reactions associated with anti-EGFR without clinical benefit in patients with tumors with many RAS mutations [66].

Pazopanib

Pazopanib is a drug used in the treatment of patients with advanced RCC and advanced soft tissue sarcoma patients who previously received chemotherapy.

When used with CYP3A4 inhibitors, it should be avoided due to the increase in pazopanib concentration. If they are to be used together, then the dose of pazopanib should be reduced to 400 mg. On the other hand, potent CYP3A4 inducers will cause a decrease in pazopanib concentration. Therefore, concomitant use of these agents should be avoided.

Pazopanib is mainly metabolized by the CYP3A4 enzyme, but CYP1A2 and CYP2C8 enzymes also mediate its metabolism. Therefore, drugs that are inhibitors or inducers of these enzymes may cause changes in pazopanib metabolism.

Pazopanib inhibits the UGT1A1 enzyme. Therefore, pazopanib treatment can increase the total bilirubin level. As a result of the studies, a significant increase in the incidence of hyperbilirubinemia was found in individuals with *(TA)7/(TA)7* (*UGT1A1**28/*28) genotype compared to *(TA)6/(TA)6* and *(TA)6/(TA)7* genotypes.

Pazopanib is a P-gp and BCRP substrate. Drugs affecting these transporters may cause changes in pazopanib absorption and elimination. Therefore, concomitant use of pazopanib with potent P-gp and BCRP inhibitors should be avoided [67].

Tamoxifen

Tamoxifen is an estrogen agonist/antagonist drug used in the treatment of estrogen receptor positive breast cancer. Tamoxifen is metabolized by CYP3A, CYP2D6, CYP2C9, CYP2C19, and CYP2B6 enzymes. It is converted by the CYP3A enzyme to N-desmethyltamoxifen, its major metabolite having the same effect as tamoxifen. Endoxifen and 4-hydroxytamoxifen are minor metabolites that have an affinity for estrogen receptors approximately 100 times more than tamoxifen. This minor metabolite is formed by CYP2D6. Changes occur in endoxifene concentrations due to the CYP2D6 polymorphism. Apart from microsomal enzymes, phase 2 enzymes such as SULT1A1, UGT2B7, and UGT1A4 are also involved in the clearance of tamoxifen from plasma.

When tamoxifen is used with strong CYP3A4 inducers, AUC and C_{max} values decrease. Therefore, these agents should not be used together. Potent CYP2D6 inhibitors reduce the active metabolites of tamoxifen. Accordingly, the effectiveness of tamoxifen decreases. Therefore, the use of tamoxifen in combination with strong CYP2D6 inhibitor drugs should be avoided [64].

DRUGS WITH IMPORTANT ACTIONS IN BLOOD, INFLAMMATION, AND GOUT

The drugs, which are commonly used in treatment of blood- and inflammation-related diseases (Table **6**), and the pharmacogenomic biomarkers, which cause dosage adjustment for those drugs due to interindividual variability, are indicated below.

Table 6. Commonly used drugs in blood- and inflammation-related diseases.

Clinical use	Drugs
Anticoagulant drugs	Clopidogrel, rivaroxaban, warfarin, prasugrel, ticagrelor
Rheumatology	Azathioprine, Carisoprodol, Flurbiprofen, Lesinurad, Pegloticase, Probenecid

Clopidogrel

Clopidogrel is a P2Y12 platelet inhibitor, used for treatment of unstable angina/non-ST-elevation, non-ST-segment elevation characterized with MI, acute coronary syndrome and ST-elevation MI or stroke, and also patients with peripheral arterial disease.

Clopidogrel is converted into its active metabolite, 2-oxo-clopidogrel by CYP2C19, and also CYP1A2, CYP2B6 and CYP3A play roles in the metabolization pathway. Concentration of active metabolite is reduced in poor CYP2C19 metabolizers.

When combined with CYP2C19 inhibitors, the efficacy of clopidogrel decreases; thus, it is not recommended to prescribe those drugs together. Since acyl--glucuronide metabolite of clopidogrel is a strong CYP2C8 inhibitor, when clopidogrel is combined with a CYP2C8 substrate, dosage adjustment is recommended for the aforementioned drug.

Clopidogrel must be metabolized by CYP2C19 to be effective. It is highly recommended to use another platelet P2Y12 inhibitor for poor CYP2C19 metabolizers [4].

Rivaroxaban

Rivaroxaban is a factor Xa inhibitor, used to decrease the risk of stroke and systemic embolism in nonvalvular atrial fibrillation, to treat and to prevent recurrence of deep vein thrombosis and pulmonary embolism, and also to prevent major cardiovascular events risk in patients with coronary artery disease and peripheral artery disease.

Oxidative degradation of rivaroxaban is mediated by CYP3A4/5 and CYP2J2 during metabolization. The drug is excreted by P-gp and ABCG2. Since mutations on the *F5* gene may cause risk factors for recurrence of deep vein thrombosis and pulmonary embolism, those mutations may change the efficacy of rivaroxaban. Thus; dosage adjustment is required in patients having those mutations.

When combined with a P-gp or a CYP3A inhibitor, efficacy of rivaroxaban might increase, leading to bleeding. When combined with a P-gp or a strong CYP3A inducer, efficacy of rivaroxaban will be reduced, which causes increased thromboembolic event risk. Therefore; it is not recommended to prescribe those drugs with rivaroxaban.

Rivaroxaban should not be combined with P-gp and strong CYP3A inhibitors and inducers [68].

Warfarin

Warfarin is a vitamin K antagonist, used to treat and prevent venous thrombosis and related pulmonary embolism, atrial fibrillation and/or cardiac valve replacement-related thromboembolic events, post-MI stroke or recurrent MI.

Warfarin is eliminated primarily by CYP2C9, and also CYP2C19, CYP2C8, CYP2C18, CYP1A2 and CYP3A4 play roles in metabolization. Patients, having CYP2C9*2 and CYP2C9*3 alleles, have reduced enzyme activity, leading to lower hydroxylation rate. Efficiency of warfarin is altered in patients with -1639G>A polymorphism on *VKORC1* gene, so dosage adjustment is recommended for those patients.

When combined with a CYP2C9, CYP1A2 and/or CYP3A4 inhibitor, clinical efficacy of warfarin may increase. When combined with a CYP2C9, CYP1A2 and/or CYP3A4 inducer, the efficacy of warfarin may decrease. In those cases, dosage adjustment is recommended and the patient should be monitored carefully [6].

Prasugrel

Prasugrel is a platelet P2Y12 inhibitor, used to prevent risk of thrombotic cardiovascular events, like stent thrombosis in patients with acute coronary syndrome. Prasugrel is metabolized by CYP3A4 and CYP2B6 in the liver, and also CYP2C9 and CYP2C19 have minor roles in rapid metabolization of the drug [69].

Ticagrelor

Ticagrelor is a platelet P2Y12 inhibitor, used to prevent stroke or MI and cardiovascular death risk in patients with acute coronary syndrome or MI history. Ticagrelor is converted into its active metabolite in the liver by CYP3A4.

Blood level of ticagrelor gets higher when combined with a CYP3A inhibitor, while ticagrelor exposure of the patient decreases when combined with a CYP3A inducer. Therefore, it is not recommended to prescribe ticagrelor with those drugs [69].

Azathioprine

Azathioprine is used as an adjunctive therapy to prevent rejection after renal homotransplantation, and to control active rheumatoid arthritis. Azathioprine is excreted by TPMT and nudix hydrolase 15 (NUDT15). For patients with reduced TPMT and NUDT15 enzyme activity, azathioprine accumulates inside of the cells, leading to adverse reactions. Thus; prescribed dose of azathioprine should be lower than usual dosage for those patients or an alternative therapy method should be considered.

Genotyping is required to determine TPMT and NUDT15 deficiency in patients with severe myelosuppression and azathioprine dose should be adjusted based on the results [70].

Carisoprodol

Carisoprodol is a muscle relaxant, used to relieve acute painful musculoskeletal conditions-related discomfort in adults for only acute treatment periods of 2-3 weeks. Carisoprodol is converted into its active metabolite, meprobamate, by CYP2C19. Carisoprodol concentration is increased up to 4-fold in patients with low CYP2C19 activity, so dosage adjustment is required.

When combined with a CYP2C19 inhibitor, carisoprodol concentration gets higher than normal and concentration of its active metabolite is reduced. On the other hand, the alterations in concentrations are the exact opposite of that when carisoprodol is combined with a CYP2C19 inducer [71].

Flurbiprofen

Flurbiprofen is a non-steroidal anti-inflammatory, used to relieve symptoms of rheumatoid arthritis and osteoarthritis.

Flurbiprofen is converted into its major metabolite, 4-hydroxy-flurbiprofen, in the liver by CYP2C9. Because flurbiprofen clearance is reduced in poor CYP2C9 metabolizers, the prescribed dose should be lower than usual dosage.

When combined with another CYP2C9 substrate, flurbiprofen clearance will be reduced, so prescribed dose should be lower than usual dosage [72].

Lesinurad

Lesinurad is a URAT1 inhibitor, used for treatment of hyperuricemia in patients with gout disease when required serum uric acid levels cannot be obtained only with a xanthine oxidase inhibitor.

Lesinurad is metabolized mainly by CYP2C9 enzyme. Poor CYP2C9 metabolizers with lower enzyme activity have higher lesinurad concentration than normal, so prescribed dose should be reduced.

When combined with a CYP2C9 inhibitor, lesinurad concentration elevates; thus, administered lesinurad dose should be lower than usual dosage. When combined with a CYP3A substrate, plasma concentration will be reduced, so dosage adjustment should be considered [73].

Pegloticase, Probenecide

Pegloticase is a PEGylated uric acid-specific enzyme, used to treat chronic gout in adults and should not be used for treatment of asymptomatic hyperuricemia. Probenecide is used to treat chronic gouty arthritis.

Administration of pegloticase and probenecide to the patients with G6PD deficiency might cause hemolysis and methemoglobinemia; thus, it is not recommended to be used in those patients [11].

SPECIAL TOPICS

The drugs, which are commonly used in treatment of diseases in dermatology and gastroenterology (Table 7), and the pharmacogenomic biomarkers, which cause dosage adjustment for those drugs due to interindividual variability, are indicated below.

Table 7. Clinically used drugs in dermatology and gastroenterology.

Clinical use	Drugs
Dermatology	Dapsone, Ustekinumab
Gastroenterology	Dronabinol, Esomeprazole, Lansoprazole, Metoclopramide, Omeprazole, Ondansetron, Pantoprazole, Sulfasalazine

Dapsone

Dapsone is used to treat acne vulgaris. Administration of dapsone to the patients with G6PD deficiency might cause methemoglobinemia; thus, it is not recommended to be used in those patients [74].

Ustekinumab

Ustekinumab is a human IL-12 and -23 antagonist, used to treat plaque psoriasis, active psoriatic arthritis, ulcerative colitis and Crohn's disease in adults. It is used for treatment of plaque psoriasis in adolescent patients with older than 12 years of age.

Increment in certain cytokine levels during inflammation might block formation of CYP450 enzymes. Since administration of IL-12 ve IL-23 antagonist ustekinumab will alter that balance, when ustekinumab is combined with a CYP substrate with a narrow therapeutic index, clinical efficiency of the aforementioned drug should be monitored carefully. Dosage adjustment should be considered if necessary [75].

Dronabinol

Dronabinol is a cannabinoid, used for treatment of weight loss-related anorexia in AIDS and to treat cancer chemotherapy-related nausea and vomiting when conventional antiemetic treatments are not necessary.

Dronabinol is converted into its active metabolite primarily by CYP29 and CYP3A4 have also minor roles in metabolization. Dornabinol clearance may increase or decrease because of the polymorphisms on CYP2C9 encoding gene. Dronabinol concentration is higher up to 2-3-fold in patients with reduced CYP2C9 activity, so those patients should be monitored carefully to prevent adverse reaction risk.

When combined with a CYP2C9 or a CYP3A4 inhibitor, dronabinol concentration may elevate. Thus, the patient should be monitored carefully to prevent adverse reactions [76].

Esomeprazole

Esomeprazole is a proton pump inhibitor, used to treat gastroesophageal reflux and pathological hypersecretory conditions, and also to prevent duodenal ulcer recurrence risk in *H. pylori* eradication and non-steroidal anti-inflammatory (NSAID)-related gastric ulcer risk.

Esomeprazole is metabolized by CYP2C19 and CYP3A4 in the liver. Drug concentration is almost doubled in poor CYP2C19 metabolizers than in extensive metabolizers. Thus, prescribed esomeprazole dose should be lower than usual dosage in those patients.

When combined with a CYP2C19 and/or a CYP3A4 inducer, serum esomeprazole concentration decreases. Therefore, combined administration is not recommended.

When esomeprazole is combined with a CYP2C19 substrate, interaction between the enzyme and the aforementioned drug may be suppressed, leading to altered efficacy. When esomeprazole is combined with clopidogrel, plasma concentration of the active metabolite of clopidogrel is reduced. When esomeprazole is combined with diazepam, diazepam clearance is decreased almost 45%. In those cases, use of alternative therapy methods or dosage adjustment should be considered.

Esomeprazole, a CYP2C19 inhibitor, should not be combined with clopidogrel, which is converted to its active metabolite by CYP2C19. An alternative anti-platelet treatment method should be considered.

Lansoprazole

Lansoprazole is a proton pump inhibitor, used to treat and to maintain the treatment of active duodenal ulcer, to prevent and to treat NSAID-associated gastric ulcer formation, to treat pathological hypersecretory conditions, active benign gastric ulcer, to maintain the treatment of erosive esophagitis, to prevent duodenal ulcer recurrence risk in *H. pylori* eradication in adults. It is also used to treat symptomatic gastroesophageal reflux and erosive esophagitis in both adults and pediatric patients with older than 1 of age.

Lansoprazole is metabolized by CYP2C19 and CYP3A4 in the liver. Dosage adjustment is recommended for poor CYP2C19 metabolizers.

When combined with a strong CYP2C19 or a CYP3A4 inducer, lansoprazole concentration is reduced, so combined administration is not recommended. When combined with a strong CYP2C19 and a CYP3A4 inhibitor, lansoprazole concentration elevates. In that case, dosage adjustment is required.

Lansoprazole, a CYP2C19 inhibitor, should not be combined with clopidogrel, which is converted into its active metabolite by CYP2C19. An alternative anti-platelet treatment method should be considered.

Omeprazole

Omeprazole is a proton pump inhibitor, used to treat active duodenal ulcer, active benign gastric ulcer and pathologic hypersecretory conditions and to prevent duodenal ulcer recurrence risk in *H. pylori* eradication in adults. It is also used to treat and to maintain the treatment of symptomatic gastroesophageal reflux disease and related erosive esophagitis in both adults and pediatric patients with older than 1 of age. Omeprazole is metabolized by CYP2C19 and CYP3A4 in the liver. Dosage adjustment is recommended for poor CYP2C19 metabolizers.

When combined with a CYP2C19 or a CYP3A4 inhibitor or inducer, omeprazole concentration will be altered. Thus, dosage adjustment or an alternative therapy method should be considered.

Omeprazole and CYP2C19 substrates should be combined carefully. Since omeprazole reduces active metabolite concentration of clopidogrel, those drugs should not be administered concomitantly. Since omeprazole increases citalopram exposure, leading to QT prolongation risk, maximum citalopram dose should not exceed 20 mg/day when combined with omeprazole. Omeprazole may elevate concentrations of diazepam, phenytoin and digoxin, so the patients should be monitored carefully and dosage adjustment might be considered. When combined with omeprazole in poor CYP2C19 metabolizers, prescribed tacrolimus dose should be lower than usual dosage.

Omeprazole, a CYP2C19 inhibitor, should not be combined with clopidogrel, which is converted to its active metabolite by CYP2C19. An alternative antiplatelet therapy method should be considered [77].

Pantoprazole

Pantoprazole is a proton pump inhibitor, used for a short-term in treatment and maintenance of the treatment of gastroesophageal reflux-related erosive esophagitis and in treatment of pathological hypersecretory conditions.

Pantoprazole is metabolized by CYP2C19 and CYP3A4 in the liver. Oral clearance of the drug is 10-fold reduced in pediatric poor CYP2C19 metabolizers. Thus, prescribed pantoprazole dose should be lower than usual dosage [76].

Metoclopramide

Metoclopramide is used for 4-12 weeks to relieve the symptoms in patients with acute and recurrent diabetic gastroparesis and to treat symptomatic gastroesophageal reflux in patients that do not give any response to the conventional therapy.

Metoclopramide is oxidized by CYP2D6 during hepatic metabolization. Elimination rate of metoclopramide is reduced in poor CYP2D6 metabolizers, leading to higher adverse reaction risk. Thus, prescribed metoclopramide dose should be lower than usual dosage in those patients.

When combined with a strong CYP2D6 inhibitor, plasma metoclopramide concentration elevates, which may increase adverse reaction risk; therefore, prescribed metoclopramide dose should be lower than usual dosage.

Administration of metoclopramide to the patients with G6PD and NADH-cytochrome b5 reductase deficiency might cause methemoglobinemia; thus, it is not recommended to be used in those patients [78].

Ondansetron

Ondansetron is a 5-HT_3 receptor antagonist, used to prevent nausea and vomiting during emetogenic cancer chemotherapy and in the postoperative period. Ondansetron is primarily metabolized by CYP1A2 and CYP3A4, and also CYP2D6 has minor roles [76].

Sulfasalazine

Sulfasalazine is an anti-inflammatory agent, used to treat ulcerative colitis and to prolong the remission period between acute attacks of ulcerative colitis. Sulfasalazine is metabolized into sulfapyridine and 5-aminosalicylic acid by intestinal bacteria. Sulfapyridine is eliminated by acetylation reaction. Sulfapyridine half-life is longer in slow acetylators, leading to higher risk of adverse reactions. Thus, dosage adjustment should be considered.

Administration of sulfasalazine to the patients with G6PD deficiency might cause hemolytic anemia; thus, it is not recommended to be used in those patients [79].

CONCLUSION

Interindividual variances have come into prominence since they are effective on both pharmacokinetics and pharmacodynamics of certain drugs. Thus, indicated pharmacogenomic biomarkers are important determinants, which should be considered in clinics to be able to prevent improper dosage-related-adverse drug reactions and to obtain optimal efficacy from medications.

CONSENT FOR PUBLICATION

Not applicable.

CONFLICT OF INTEREST

The authors declare no conflict of interest, financial or otherwise.

ACKNOWLEDGEMENT

Declared none.

LIST OF ABBREVIATIONS

5-FU	5- fluorouracil
5-HMT	5-hydroxymethyl tolterodine
5'-DFCR	5'-deoxy-5-fluorocytidine
ABC	Adenosine triphosphate binding cassette
ALK	Anaplastic lymphoma kinase
ALL	Acute lymphoblastic leukemia
AP	Accelerated phase
ASM	Aggressive systemic mastocytosis
BC	Blast crisis
BCRP	Breast cancer resistance protein
CF	Cystic fibrosis
CFTR	Cystic fibrosis transmembrane conductance regulator
COPD	Chronic obstructive pulmonary disease
CP	Chronic phase
CYP	Cytochrome P450
DDP	Dihydropyrimidine dehydrogenase deficiency
DMD	Duchenne muscular dystrophy
DPD	Dihydropyrimidine dehydrogenase
EGFR	Epidermal growth factor receptors
FDA	Food and Drug Administration
G6PD	Glucose-6-phosphate dehydrogenase
gBRCAm	Germline BRCA mutation
GST	Glutathione-S-transferase
HCV	Hepatitis C virus
HER2	Human epidermal growth factor receptor 2
HIV	Human immunodeficiency virus
HR	Hormone receptor
HRD	Homologous recombination deficiency
MATE	Multidrug and toxin extrusion transporter
MI	Myocardial infarction
NAT	N-acetyltransferase
NSAID	Non-steroidal anti-inflammatory
NSCLC	Non-small cell lung cancer
NUDT15	Nudix hydrolase 15

OAT	Organic anion transporter
OATP	Organic anion transporting polypeptide
OCT	Organic cation transporter
P-gp	P-glycoprotein pump
Ph+CML	Philadelphia chromosome positive chronic myeloid leukemia
QTcF	Corrected QT by the Fridericia method
RCC	Renal cell carcinoma
SLC	Solute carrier
SMN	Survival motor neuron
SNP	Single nucleotide polymorphism
T2DM	Type 2 Diabetes Mellitus
TPMT	Thiopurine-S-methyltransferase
TSC	Tuberous sclerosis complex
TTR	Transthyretin
UGT	UDP-glucuronosyltransferase
Vmat2	Vesicular monoamine transporter 2 inhibitor

REFERENCES

[1] Giacomini KM, Brett CM, Altman RB, *et al.* The pharmacogenetics research network: from SNP discovery to clinical drug response. Clin Pharmacol Ther 2007; 81(3): 328-45. [http://dx.doi.org/10.1038/sj.clpt.6100087] [PMID: 17339863]

[2] Ji Y, Skierka JM, Blommel JH, *et al.* Preemptive Pharmacogenomic Testing for Precision Medicine: A Comprehensive Analysis of Five Actionable Pharmacogenomic Genes Using Next-Generation DNA Sequencing and a Customized CYP2D6 Genotyping Cascade. J Mol Diagn 2016; 18(3): 438-45. [http://dx.doi.org/10.1016/j.jmoldx.2016.01.003] [PMID: 26947514]

[3] Elfaki I, Mir R, Almutairi FM, Duhier FMA. Cytochrome P450: Polymorphisms and Roles in Cancer, Diabetes and Atherosclerosis. Asian Pac J Cancer Prev 2018; 19(8): 2057-70. [PMID: 30139042]

[4] Johnson JA, Cavallari LH. Pharmacogenetics and cardiovascular disease--implications for personalized medicine. Pharmacol Rev 2013; 65(3): 987-1009. [http://dx.doi.org/10.1124/pr.112.007252] [PMID: 23686351]

[5] De T, Park CS, Perera MA. Cardiovascular Pharmacogenomics: Does It Matter If You're Black or White? Annu Rev Pharmacol Toxicol 2019; 59(1): 577-603. [http://dx.doi.org/10.1146/annurev-pharmtox-010818-021154] [PMID: 30296897]

[6] Roden DM, Johnson JA, Kimmel SE, *et al.* Cardiovascular pharmacogenomics. Circ Res 2011; 109(7): 807-20. [http://dx.doi.org/10.1161/CIRCRESAHA.110.230995] [PMID: 21921273]

[7] Dean L. Propafenone Therapy and CYP2D6 Genotype. In: Pratt VM, McLeod HL, Rubinstein WS, Scott SA, Dean LC, Kattman BL, Eds. Medical Genetics Summaries. Bethesda, MD 2012.

[8] Weinshilboum RM, Wang L. Pharmacogenomics: Precision Medicine and Drug Response. Mayo Clin Proc 2017; 92(11): 1711-22. [http://dx.doi.org/10.1016/j.mayocp.2017.09.001] [PMID: 29101939]

[9] Chancellor MB. New frontiers in the treatment of overactive bladder and incontinence. Rev Urol 2002; 4 (Suppl. 4): S50-6.
[PMID: 16986022]

[10] Malhotra B, Darsey E, Crownover P, Fang J, Glue P. Comparison of pharmacokinetic variability of fesoterodine vs. tolterodine extended release in cytochrome P450 2D6 extensive and poor metabolizers. Br J Clin Pharmacol 2011; 72(2): 226-34.
[http://dx.doi.org/10.1111/j.1365-2125.2011.03948.x] [PMID: 21352267]

[11] Whirl-Carrillo M, McDonagh EM, Hebert JM, *et al.* Pharmacogenomics knowledge for personalized medicine. Clin Pharmacol Ther 2012; 92(4): 414-7.
[http://dx.doi.org/10.1038/clpt.2012.96] [PMID: 22992668]

[12] Jessurun NT, Wijnen PA, Bast A, van Puijenbroek EP, Bekers O, Drent M. Tamsulosin associated with interstitial lung damage in CYP2D6 variant alleles carriers. Int J Mol Sci 2020; 21(8): E2770.
[http://dx.doi.org/10.3390/ijms21082770] [PMID: 32316326]

[13] Blake K, Lima J. Pharmacogenomics of long-acting β2-agonists. Expert Opin Drug Metab Toxicol 2015; 11(11): 1733-51.
[http://dx.doi.org/10.1517/17425255.2015.1071793] [PMID: 26235677]

[14] PubChem Compound Summary for CID 6918554, Indicaterol: National Center for Biotechnology Information. 2020. Available from: https://pubchem.ncbi.nlm.nih.gov/compound/Indacaterol

[15] Clancy JP, Johnson SG, Yee SW, *et al.* Clinical Pharmacogenetics Implementation Consortium (CPIC) guidelines for ivacaftor therapy in the context of CFTR genotype. Clin Pharmacol Ther 2014; 95(6): 592-7.
[http://dx.doi.org/10.1038/clpt.2014.54] [PMID: 24598717]

[16] Tafti M, Dauvilliers Y. Pharmacogenomics in the treatment of narcolepsy. Pharmacogenomics 2003; 4(1): 23-33.
[http://dx.doi.org/10.1517/phgs.4.1.23.22590] [PMID: 12517283]

[17] Aquilante CL. Sulfonylurea pharmacogenomics in Type 2 diabetes: the influence of drug target and diabetes risk polymorphisms. Expert Rev Cardiovasc Ther 2010; 8(3): 359-72.
[http://dx.doi.org/10.1586/erc.09.154] [PMID: 20222815]

[18] Wang B, Sanchez RI, Franklin RB, Evans DC, Huskey SE. The involvement of CYP3A4 and CYP2C9 in the metabolism of 17 alpha-ethinylestradiol. Drug Metab Dispos 2004; 32(11): 1209-12.
[http://dx.doi.org/10.1124/dmd.104.000182] [PMID: 15304426]

[19] Shebley M, Polepally AR, Nader A, *et al.* Clinical Pharmacology of Elagolix: An Oral Gonadotropin-Releasing Hormone Receptor Antagonist for Endometriosis. Clin Pharmacokinet 2020; 59(3): 297-309.
[http://dx.doi.org/10.1007/s40262-019-00840-7] [PMID: 31749075]

[20] Dean L. Flibanserin Therapy and CYP2C19 Genotype. In: Pratt VM, McLeod HL, Rubinstein WS, Scott SA, Dean LC, Kattman BL, Eds. Medical Genetics Summaries. Bethesda, MD 2012.

[21] Tolonen A, Koskimies P, Turpeinen M, Uusitalo J, Lammintausta R, Pelkonen O. Ospemifene metabolism in humans *in vitro* and *in vivo*: metabolite identification, quantitation, and CYP assignment of major hydroxylations. Drug Metabol Drug Interact 2013; 28(3): 153-61.
[http://dx.doi.org/10.1515/dmdi-2013-0016] [PMID: 23729558]

[22] Yoshida K, Müller DJ. Pharmacogenetics of Antipsychotic Drug Treatment: Update and Clinical Implications. Mol Neuropsychiatry 2020; 5 (Suppl. 1): 1-26.
[http://dx.doi.org/10.1159/000492332] [PMID: 32399466]

[23] Arranz MJ, Gonzalez-Rodriguez A, Perez-Blanco J, *et al.* A pharmacogenetic intervention for the improvement of the safety profile of antipsychotic treatments. Transl Psychiatry 2019; 9(1): 177.
[http://dx.doi.org/10.1038/s41398-019-0511-9] [PMID: 31346157]

[24] Li KJ, Solomon HV, DeLisi LE. Clozapine pharmacogenomics: a review of efficacy, pharmacokinetics, and agranulocytosis. Curr Opin Psychiatry 2018; 31(5): 403-8. [http://dx.doi.org/10.1097/YCO.0000000000000440] [PMID: 29985178]

[25] Nnadi CU, Malhotra AK. Clinical and pharmacogenetic studies of iloperidone. Per Med 2008; 5(4): 367-75. [http://dx.doi.org/10.2217/17410541.5.4.367] [PMID: 29783461]

[26] Hicks JK, Sangkuhl K, Swen JJ, *et al.* Clinical pharmacogenetics implementation consortium guideline (CPIC) for CYP2D6 and CYP2C19 genotypes and dosing of tricyclic antidepressants: 2016 update. Clin Pharmacol Ther 2017; 102(1): 37-44. [http://dx.doi.org/10.1002/cpt.597] [PMID: 27997040]

[27] Namerow LB, Walker SA, Loftus M, Bishop JR, Ruaño G, Malik S. Pharmacogenomics: an Update for Child and Adolescent Psychiatry. Curr Psychiatry Rep 2020; 22(5): 26. [http://dx.doi.org/10.1007/s11920-020-01145-4] [PMID: 32377970]

[28] Smith T, Sharp S, Manzardo AM, Butler MG. Pharmacogenetics informed decision making in adolescent psychiatric treatment: a clinical case report. Int J Mol Sci 2015; 16(3): 4416-28. [http://dx.doi.org/10.3390/ijms16034416] [PMID: 25710722]

[29] Peñas-Lledó EM, Trejo HD, Dorado P, *et al.* CYP2D6 ultrarapid metabolism and early dropout from fluoxetine or amitriptyline monotherapy treatment in major depressive patients. Mol Psychiatry 2013; 18(1): 8-9. [http://dx.doi.org/10.1038/mp.2012.91] [PMID: 22733128]

[30] Franco V, Perucca E. The pharmacogenomics of epilepsy. Expert Rev Neurother 2015; 15(10): 1161-70. [http://dx.doi.org/10.1586/14737175.2015.1083424] [PMID: 26327193]

[31] Phillips EJ, Sukasem C, Whirl-Carrillo M, *et al.* Clinical Pharmacogenetics Implementation Consortium Guideline for HLA Genotype and Use of Carbamazepine and Oxcarbazepine: 2017 Update. Clin Pharmacol Ther 2018; 103(4): 574-81. [http://dx.doi.org/10.1002/cpt.1004] [PMID: 29392710]

[32] Tiwari AK, Souza RP, Müller DJ. Pharmacogenetics of anxiolytic drugs. J Neural Transm (Vienna) 2009; 116(6): 667-77. [http://dx.doi.org/10.1007/s00702-009-0229-6] [PMID: 19434367]

[33] Heavin SB, McCormack M, Wolking S, *et al.* Genomic and clinical predictors of lacosamide response in refractory epilepsies. Epilepsia Open 2019; 4(4): 563-71. [http://dx.doi.org/10.1002/epi4.12360] [PMID: 31819912]

[34] Bach-Rojecky L, Vađunec D, Lozić M, *et al.* Challenges in anesthesia personalization: resolving the pharmacogenomic puzzle. Per Med 2019; 16(6): 511-25. [http://dx.doi.org/10.2217/pme-2019-0056] [PMID: 31692418]

[35] Mahajan MK, Uttamsingh V, Gan LS, Leduc B, Williams DA. Identification and characterization of oxymetazoline glucuronidation in human liver microsomes: evidence for the involvement of UGT1A9. J Pharm Sci 2011; 100(2): 784-93. [http://dx.doi.org/10.1002/jps.22303] [PMID: 20669329]

[36] Jannetto PJ, Bratanow NC. Pharmacogenomic considerations in the opioid management of pain. Genome Med 2010; 2(9): 66. [http://dx.doi.org/10.1186/gm187] [PMID: 20854646]

[37] Funck-Brentano C, Boëlle PY, Verstuyft C, Bornert C, Becquemont L, Poirier JM. Measurement of CYP2D6 and CYP3A4 activity *in vivo* with dextromethorphan: sources of variability and predictors of adverse effects in 419 healthy subjects. Eur J Clin Pharmacol 2005; 61(11): 821-9. [http://dx.doi.org/10.1007/s00228-005-0051-5] [PMID: 16315033]

[38] Noetzli M, Eap CB. Pharmacodynamic, pharmacokinetic and pharmacogenetic aspects of drugs used

in the treatment of Alzheimer's disease. Clin Pharmacokinet 2013; 52(4): 225-41.
[http://dx.doi.org/10.1007/s40262-013-0038-9] [PMID: 23408070]

[39] Mehanna R, Hunter C, Davidson A, Jimenez-Shahed J, Jankovic J. Analysis of CYP2D6 genotype and response to tetrabenazine. Mov Disord 2013; 28(2): 210-5.
[http://dx.doi.org/10.1002/mds.25278] [PMID: 23280482]

[40] Minamisawa M, Claggett B, Adams D, *et al.* Association of patisiran, an RNA interference therapeutic, with regional left ventricular myocardial strain in hereditary transthyretin amyloidosis: the apollo study. JAMA Cardiol 2019; 4(5): 466-72.
[http://dx.doi.org/10.1001/jamacardio.2019.0849] [PMID: 30878017]

[41] Bilkey GA, Burns BL, Coles EP, Mahede T, Baynam G, Nowak KJ. Optimizing precision medicine for public health. Front Public Health 2019; 7: 42.
[http://dx.doi.org/10.3389/fpubh.2019.00042] [PMID: 30899755]

[42] Mantegazza R. Amifampridine tablets for the treatment of Lambert-Eaton myasthenic syndrome. Expert Rev Clin Pharmacol 2019; 12(11): 1013-8.
[http://dx.doi.org/10.1080/17512433.2019.1681972] [PMID: 31639317]

[43] Wang Z, Lee B, Pearce D, *et al.* Meclizine metabolism and pharmacokinetics: formulation on its absorption. J Clin Pharmacol 2012; 52(9): 1343-9.
[http://dx.doi.org/10.1177/0091270011414575] [PMID: 21903894]

[44] Gardin A, Ufer M, Legangneux E, *et al.* Effect of Fluconazole Coadministration and CYP2C9 Genetic Polymorphism on Siponimod Pharmacokinetics in Healthy Subjects. Clin Pharmacokinet 2019; 58(3): 349-61.
[http://dx.doi.org/10.1007/s40262-018-0700-3] [PMID: 30088221]

[45] Gardin A, Gray C, Neelakantham S, *et al.* Siponimod pharmacokinetics, safety, and tolerability in combination with rifampin, a CYP2C9/3A4 inducer, in healthy subjects. Eur J Clin Pharmacol 2018; 74(12): 1593-604.
[http://dx.doi.org/10.1007/s00228-018-2533-2] [PMID: 30105453]

[46] Mahungu T, Owen A. Current progress in the pharmacogenetics of infectious disease therapy. Elsevier 2011; pp. 555-78.

[47] King JR, Dutta S, Cohen D, *et al.* Drug-drug interactions between sofosbuvir and ombitasvir-paritaprevir-ritonavir with or without dasabuvir. Antimicrob Agents Chemother 2015; 60(2): 855-61.
[http://dx.doi.org/10.1128/AAC.01913-15] [PMID: 26596948]

[48] Lepist EI, Zhang X, Hao J, *et al.* Contribution of the organic anion transporter OAT2 to the renal active tubular secretion of creatinine and mechanism for serum creatinine elevations caused by cobicistat. Kidney Int 2014; 86(2): 350-7.
[http://dx.doi.org/10.1038/ki.2014.66] [PMID: 24646860]

[49] Yew WW. Clinically significant interactions with drugs used in the treatment of tuberculosis. Drug Saf 2002; 25(2): 111-33.
[http://dx.doi.org/10.2165/00002018-200225020-00005] [PMID: 11888353]

[50] Cusato J, De Nicolò A, Boglione L, *et al.* Pharmacogenetics of the anti-HCV drug sofosbuvir: a preliminary study. J Antimicrob Chemother 2018; 73(6): 1659-64.
[http://dx.doi.org/10.1093/jac/dky053] [PMID: 29509884]

[51] Arab-Alameddine M, Fayet-Mello A, Lubomirov R, *et al.* Population pharmacokinetic analysis and pharmacogenetics of raltegravir in HIV-positive and healthy individuals. Antimicrob Agents Chemother 2012; 56(6): 2959-66.
[http://dx.doi.org/10.1128/AAC.05424-11] [PMID: 22371894]

[52] Ho JM, Juurlink DN. Considerations when prescribing trimethoprim-sulfamethoxazole. CMAJ 2011; 183(16): 1851-8.
[http://dx.doi.org/10.1503/cmaj.111152] [PMID: 21989472]

[53] Dean L. Voriconazole Therapy and CYP2C19 Genotype. In: Pratt VM, McLeod HL, Rubinstein WS, Scott SA, Dean LC, Kattman BL, Eds. Medical Genetics Summaries. Bethesda, MD 2012.

[54] Chappell JC, Turner PK, Pak YA, *et al.* Abemaciclib Inhibits Renal Tubular Secretion Without Changing Glomerular Filtration Rate. Clin Pharmacol Ther 2019; 105(5): 1187-95. [http://dx.doi.org/10.1002/cpt.1296] [PMID: 30449032]

[55] Posada MM, Morse BL, Turner PK, Kulanthaivel P, Hall SD, Dickinson GL. Predicting clinical effects of cyp3a4 modulators on abemaciclib and active metabolites exposure using physiologically based pharmacokinetic modeling. J Clin Pharmacol 2020; 60(7): 915-30. [http://dx.doi.org/10.1002/jcph.1584] [PMID: 32080863]

[56] Touma JA, McLachlan AJ, Gross AS. The role of ethnicity in personalized dosing of small molecule tyrosine kinase inhibitors used in oncology. Transl Cancer Res 2017; 6(S10): S1558-91. [http://dx.doi.org/10.21037/tcr.2017.09.09]

[57] Hertz DL, Henry NL, Rae JM. Germline genetic predictors of aromatase inhibitor concentrations, estrogen suppression and drug efficacy and toxicity in breast cancer patients. Pharmacogenomics 2017; 18(5): 481-99. [http://dx.doi.org/10.2217/pgs-2016-0205] [PMID: 28346074]

[58] Ruwali M. Pharmacogenetics and cancer treatment: progress and prospects. In: Nalbantoglu S, Amri H, Eds. Molecular Medicine. London: IntechOpen 2019. [http://dx.doi.org/10.5772/intechopen.83424]

[59] Santarpia M, Rolfo C, Peters GJ, Leon LG, Giovannetti E. On the pharmacogenetics of non-small cell lung cancer treatment. Expert Opin Drug Metab Toxicol 2016; 12(3): 307-17. [http://dx.doi.org/10.1517/17425255.2016.1141894] [PMID: 26761638]

[60] Delaney C, Frank S, Huang RS. Pharmacogenomics of EGFR-targeted therapies in non-small cell lung cancer: EGFR and beyond. Chin J Cancer 2015; 34(4): 149-60. [http://dx.doi.org/10.1186/s40880-015-0007-9] [PMID: 25962919]

[61] Moes DJ, Press RR, den Hartigh J, van der Straaten T, de Fijter JW, Guchelaar HJ. Population pharmacokinetics and pharmacogenetics of everolimus in renal transplant patients. Clin Pharmacokinet 2012; 51(7): 467-80. [http://dx.doi.org/10.2165/11599710-000000000-00000] [PMID: 22624503]

[62] Ma WW, Saif MW, El-Rayes BF, *et al.* Emergency use of uridine triacetate for the prevention and treatment of life-threatening 5-fluorouracil and capecitabine toxicity. Cancer 2017; 123(2): 345-56. [http://dx.doi.org/10.1002/cncr.30321] [PMID: 27622829]

[63] Baek JH, Kim JG, Kim SN, *et al.* Unpredicted severe toxicity after 5-fluorouracil treatment due to dihydropyrimidine dehydrogenase deficiency. Korean J Intern Med (Korean Assoc Intern Med) 2006; 21(1): 43-5. [http://dx.doi.org/10.3904/kjim.2006.21.1.43] [PMID: 16646564]

[64] Hertz DL, Rae J. Pharmacogenetics of cancer drugs. Annu Rev Med 2015; 66(1): 65-81. [http://dx.doi.org/10.1146/annurev-med-053013-053944] [PMID: 25386932]

[65] Sugarman R, Patel R, Sharma S, Plenker D, Tuveson D, Saif MW. Pharmacokinetics and pharmacodynamics of new drugs for pancreatic cancer. Expert Opin Drug Metab Toxicol 2019; 15(7): 541-52. [http://dx.doi.org/10.1080/17425255.2019.1637417] [PMID: 31241371]

[66] Ketzer S, Schimmel K, Koopman M, Guchelaar HJ. Clinical Pharmacokinetics and Pharmacodynamics of the Epidermal Growth Factor Receptor Inhibitor Panitumumab in the Treatment of Colorectal Cancer. Clin Pharmacokinet 2018; 57(4): 455-73. [http://dx.doi.org/10.1007/s40262-017-0590-9] [PMID: 28853050]

[67] Thorn CF, Sharma MR, Altman RB, Klein TE. PharmGKB summary: pazopanib pathway, pharmacokinetics. Pharmacogenet Genomics 2017; 27(8): 307-12.

[http://dx.doi.org/10.1097/FPC.0000000000000292] [PMID: 28678138]

[68] Kanuri SH, Kreutz RP. Pharmacogenomics of novel direct oral anticoagulants: newly identified genes and genetic variants. J Pers Med 2019; 9(1): E7. [http://dx.doi.org/10.3390/jpm9010007] [PMID: 30658513]

[69] Cavallari LH, Obeng AO. Genetic Determinants of $P2Y_{12}$ Inhibitors and Clinical Implications. Interv Cardiol Clin 2017; 6(1): 141-9. [http://dx.doi.org/10.1016/j.iccl.2016.08.010] [PMID: 27886818]

[70] Dean L. Azathioprine Therapy and TPMT and NUDT15 Genotype. In: Pratt VM, McLeod HL, Rubinstein WS, Scott SA, Dean LC, Kattman BL, Eds. Medical Genetics Summaries. Bethesda, MD 2012.

[71] Dean L. Carisoprodol Therapy and CYP2C19 Genotype. In: Pratt VM, McLeod HL, Rubinstein WS, Scott SA, Dean LC, Kattman BL, Eds. Medical Genetics Summaries. Bethesda, MD 2012.

[72] Dean L. Flurbiprofen Therapy and CYP2C9 Genotype. In: Pratt VM, McLeod HL, Rubinstein WS, Scott SA, Dean LC, Kattman BL, Eds. Medical Genetics Summaries. Bethesda, MD 2012.

[73] Dean L. Lesinurad Therapy and CYP2C9 Genotype. In: Pratt VM, McLeod HL, Rubinstein WS, Scott SA, Dean LC, Kattman BL, Eds. Medical Genetics Summaries. Bethesda, MD 2012.

[74] Dapsone. Drugs and Lactation Database (LactMed). Bethesda (MD) 2006.

[75] Dobbin-Sears I, Roberts J, O'Rielly DD, Rahman P. Ustekinumab in psoriatic arthritis and related phenotypes. Ther Adv Chronic Dis 2018; 9(10): 191-8. [http://dx.doi.org/10.1177/2040622318781760] [PMID: 30263103]

[76] Halawi H, Camilleri M. Pharmacogenetics and the treatment of functional gastrointestinal disorders. Pharmacogenomics 2017; 18(11): 1085-94. [http://dx.doi.org/10.2217/pgs-2017-0049] [PMID: 28686075]

[77] Zheng CL, Alzghari SK. Putting out the fire: the relationship of pharmacogenetics and proton pump inhibitors for the treatment of gastroesophageal reflux disease. Cureus 2018; 10(5): e2687. [http://dx.doi.org/10.7759/cureus.2687] [PMID: 30050742]

[78] Parkman HP, Mishra A, Jacobs M, *et al.* Clinical response and side effects of metoclopramide: associations with clinical, demographic, and pharmacogenetic parameters. J Clin Gastroenterol 2012; 46(6): 494-503. [http://dx.doi.org/10.1097/MCG.0b013e3182522624] [PMID: 22688145]

[79] Weber WW. Populations and genetic polymorphisms. Mol Diagn 1999; 4(4): 299-307. [http://dx.doi.org/10.1016/S1084-8592(99)80006-X] [PMID: 10671640]

CHAPTER 13

Clinical Application of Circulating MicroRNAs as Novel Biomarkers for Different Diseases

Volkan Sozer[1], **Pınar Atukeren**[2,3] and **Hafize Uzun**[4,*]

[1] *Department of Biochemistry, Yildiz Technical University, Istanbul, Turkey*

[2] *Department of Medical Biochemistry, Cerrahpaşa Faculty of Medicine, Istanbul University-Cerrahpasa, Istanbul, Turkey*

[3] *Faculty of Pharmacy, Istanbul University-Cerrahpasa, Istanbul, Turkey*

[4] *Department of Biochemistry, Faculty of Medicine, Istanbul Atlas University, Istanbul, Turkey*

Abstract: Biomarker research has become increasingly interesting in many areas nowadays. Biomarkers are indicators of the biological process that can show changes in disease and health status and reveal pathological conditions. There is always a need for markers that divide patients into risk categories that can help in early diagnosis, detect complications ahead of time, guide treatment, and predict adverse outcomes in a chronic complex and certain diseases such as cancer. microRNAs (miRNAs) are ~ 22 nt long npcRNAs involved in post-transcriptional arrangements. miRNAs regulate messenger RNAs (mRNA), especially through negative regulation of gene expression. The fact that miRNAs have come to the fore in many disease mechanisms brings up their use as biomarkers in the early stage. The purpose of this review is to gather the latest information on this subject by bringing together recent articles and reviews to contribute to understanding the role of miRNAs, which act as biomarkers in different ways in vital processes, in the formation, early diagnosis, and treatment of diseases. miRNAs have an important potential to become a next-generation biomarker and therapeutic. But, each miRNA molecule can bind to a large number of different mRNAs, and different miRNAs in each mRNA. Therefore, new findings are needed to determine the expression activities and targets of miRNAs.

Keywords: Autoimmune Diseases, Biomarkers, Cancer, Coronary Artery Disease, Genital Diseases, Female, MicroRNAs, Nervous System Diseases, SARS-CoV-2.

INTRODUCTION

A feature that can be objectively measured and evaluated, which can be an indicator of pharmacological response to normal biological processes, patho-

* **Corresponding author Hafize Uzun**: Department of Biochemistry, Faculty of Medicine, Istanbul Atlas University, Istanbul, Turkey; Tel: +90 (535) 542 11 47; E-mail: huzun59@hotmail.com

Pınar Atukeren and Hafize Uzun (Eds.)

logical processes, or a therapeutic process, is called a "biomarker." This definition was developed in 2001 by the "Biomarkers Definitions Working Group" [1].

Biomarkers are biological signals originating from the human body. In this case, virtually any biological sign, if measurable, *e.g.*, biological material analysis, electrocardiography (ECG), blood pressure, etc., are all biomarkers.

The Purpose of Biomarkers

1. Diagnosis
2. Disease staging
3. Risk identification
4. Disease prognosis
5. Treatment evaluation
6. Treatment follow-up
7. Population screening

An ideal biomarker [1] should be biochemically stable and able to be studied in a small blood volume [2]; sampling time should be wide (> 48 hours); It should be detected by fast, cheap, simple, and automated analysis method [3]; results should be comparable between laboratories, and the threshold value should be well established [4]; sensitivity/negative predictive value should be 100%, and specificity/positive predictive value> 85% [5]; should be able to predict a complicated course in determining prognosis, and [6] should have features that can demonstrate the efficacy of treatment.

MicroRNAs (miRNAs)

miRNAs are a class of endogenous small RNAs and are functional RNA molecules that are transcribed from the intron or exon regions encoding protein on the genome and RNA genes in regions that do not encode a protein and are not translated into protein [2]. miRNAs consist of an average of 20-22 nucleotides. They regulate post-transcriptional gene expression by targeting messenger RNA (mRNA). Suppression of translation or fragmentation of the target mRNA with the relationship between miRNA and mRNA are possible consequences that may occur. But what if miRNAs are more than gene regulation? Many studies have shown that the miRNA profile in the blood of patients changes compared to healthy people. These discoveries have been of great interest in the use of the level of extracellular (extracellular) miRNAs as a novel biomarker in various diseases [3, 4].

miRNAs have the ability to leave the cell. Extracellular miRNAs were first discovered in a cell culture medium. Extracellular miRNAs encapsulated in exosomes and secreted from donor cells could be transferred to other recipient cells. The presence of extracellular miRNAs in human body fluids has been mentioned in several studies. Extracellular miRNAs have been discovered in various macromolecules in blood and other body fluids. As a result of disease or other necrotic events, cells die, and miRNAs are released passively from the cytoplasm, which can be detected in the blood. Exosomal miRNAs from miRNAs released outside the cell are transported in the vesicle covered with a lipid bilayer membrane. These are more stable than free circulating miRNAs [5].

A biomarker based on miRNA levels has several advantages over biomarkers currently used at the levels of specific proteins in the blood [5, 6].

1. The strongest advantage is that the global miRNA level can be measured quickly and accurately.

2. Because of the high precision and efficiency of transcriptome profiling techniques, it is much easier to profile global miRNA levels than proteins or metabolites.

3. The miRNA population consisting of different sequences in humans can be rapidly measured with high-throughput sequencing technology because many miRNAs are expressed differently in different tissues and cellular situations. It can help more with the combination of miRNA levels.

4. The origin of cancer tissues can be classified at the miRNA level.

5. If we compare it with protein or other types of molecules where no amplification method can be applied, polymerase chain reaction (PCR) based amplification enables measurement of the low amount of miRNA in body fluids, even in small volumes.

The fact that miRNAs have come to the fore in many disease mechanisms brings to the agenda their use as biomarkers in the early stage. Much research has been done to find extracellular miRNAs that can be used as diagnostic or prognostic markers for specific diseases. After the first promising results of the studies, it turned out that the exact quantification of circulating miRNAs is more complex than expected. One of the uncertainties it brings is why the expression profiles of certain circulating miRNAs are not always specific in a particular pathology, and the other is that they are effective biomarkers. It is possible to project miRNAs to the clinic [7]. With the completion of experimental techniques and our knowledge of extracellular miRNAs, it will be inevitable to use miRNAs as non-invasive biomarkers.

With miRNA studies, it can be understood that at a translational level which miRNAs contribute to which disease pathogenesis, which miRNAs can be used as biomarkers in the diagnosis of which diseases, and what kind of contribution miRNAs contribute to the clinical course of diseases. In this way, potential new drug targets are revealed and new potential targets are determined in the treatment of aging, neurological diseases, heart diseases, cancer diseases as well as rheumatological and immunological diseases. At the same time, contributing to understanding the action mechanisms and drug resistance of drugs currently used in the treatment of diseases is one of the goals of translational studies in this field.

In general, miRNA profiling studies related to any disease model are carried out and where and when miRNAs are expressed with the help of bioinformatics studies are tried to be revealed. In this sense, web-based tools have become necessary resources for gathering information and creating new hypotheses. With the increase in studies on miRNAs in recent years, a large number of bioinformatics databases have been developed for miRNA-related data. These databases are concerned with the detection of individual miRNAs, miRNA sequences, target estimation of miRNAs, verification of miRNA targets, miRNA expression, search for miRNA regulatory and metabolic pathways, investigation of transcription factor (TF) and miRNA relationship, and the relationship of miRNAs to diseases. they are divided into separate groups. These databases examine the detection of individual miRNAs, miRNA sequences, target estimation of miRNAs, verification of miRNA targets, miRNA expression, search for miRNA regulatory and metabolic pathways, investigation of transcription factor (TF) and miRNA relationship. It is also divided into separate groups regarding the relationship of miRNAs to diseases. miRNAPath are bioinformatics data banks where miRTar and miTHALOS miRNAs interact with pathways and store data [8].

MiRNAs, Signaling Pathways and Related Diseases

Immune System and miRNAs

Studies in recent years have focused on the regulatory role of miRNAs in the immune system. A healthy immune system can detect and destroy pathogens. However, excessive or inappropriate responses to pathogens, toxins, damaged cells, allergens, irritants and the body's own antigens lead to severe inflammation, cell damage, autoimmune reactions or allergic diseases [9].

Data in the literature have shown that miRNAs play a role in the hereditary and adaptive immune system. miR-21, miR-146a and miR-155 are miRNAs that play the most active role in the immune system. miR-21 is highly expressed in many cells and is involved in many immunological processes. Expression of miR-21 is regulated by the signal transducer and activator of transcription (STAT)-3 and nuclear factor kappa-B (NF-κB) [10 - 12]. Similar to miR-21, miR-146a acts as an anti-inflammatory miRNA and has been shown in many studies to suppress the activation of NF-κB [13 - 18]. Unlike miR-21 and miR-146a, miR-155 acts as a proinflammatory factor. It has been shown that miR-155 expression is stimulated by vesicular stomatitis virus infection in macrophages [19].

Apart from miR-21, miR-146a and miR-155, many other miRNAs are also involved in the regulation of the immune response. miR-301a, miR-9, miR-147b and miR-125a in the regulation of NF-κB, miR-10a, miR-17-92, miR-181a, miR-182, miR-29a/b development of T-cells and miR-17-92, miR-34a, miR-150, miR-181b, miR-125b and miR-217 are effective in the development and functions of B cells [20, 21].

Immunological Originated Diseases and miRNAs

Cancer

miRNAs function as tumor suppressors or oncogenes in cancer initiation and progression depending on the character of the gene they target. Some miRNAs, on the other hand, can exhibit both tumor suppressor and oncogene characteristics depending on their conditions. For example, miR-29a functions as a tumor suppressor in lymphocytic leukemia and lung cancer and as an oncogene in breast cancer [22].

Changes in protein-encoding oncogene and/or tumor suppressor genes are known to cause cancer. In recent years, it has been reported that the genetic causes of cancer are more complex, with the demonstration that miRNAs are also effective in tumor formation [23]. The genomic domains or fragile regions associated with cancer consisted of genes encoding miRNA in more than 50%, suggesting that miRNAs are important in cancer pathogenesis [24].

Various miRNAs have been shown to play specific roles in the differentiation of cancer cells, tumor progression and metastasis [25]. The first study revealing the relationship between miRNA genes and cancer was done by Calin *et al.* [26]. In a study with CLL patients, they showed that the miR-15 and miR-16 genes might be responsible for the 13q14 deletion seen in more than half of the patients, rather than targeting a tumor suppressor gene as previously thought. Cimmino *et al.* [27] reported that the BCL2 antiapoptotic gene is manipulated by two different miRNAs and can be used therapeutically in leukemia cell lines in which BCL2 is overexpressed.

Studies have found that approximately 200 miRNAs are associated with cancer, and some difficulties encountered in early diagnosis of colon and other occult cancers can be overcome with miRNA profiles obtained from patients' serum, plasma, saliva and tissues. The most common increase in miRNA in human cancers is miR-21 [28]. It has been shown in studies that overexpression of this miRNA increases cell proliferation and invasion in breast, lung, prostate and other cancers (Table **1**).

Table 1. Potential miRNA biomarkers in tissue and blood samples in cancer.

Cancer Type	Tissue Samples	Blood Samples
Lung cancer	miR-21, miR-205; miR-146b, miR-205, miR-29c, miR-30b; miR-145 [29 - 31]. ,	**Serum:** miR-146b; miR-10b, miR-155; miR-141, miR-193b, miR-200b, miR-301a; (miR-130a, miR-25, miR-191; miR-21-5p, miR-141-3p, miR-222-3p, miR-486-5p; miR-145, miR-20a [32 - 37].
Breast cancer	miR-21, miR-155, miR-191, miR-196a, miR125b, miR-221, let-7a, miR-145, miR-205, miR-18a, miR-210 [38 - 40].	**Serum:** miR-10b, miR-21, miR-23a, miR-34a, miR-155, miR-21, miR-106a, miR-155, miR-126, miRNA-195, miR-199a, miR-335, miR-1246, miR-1307-3p, miR-4634, miR-6861-5p, miR-6875-5p, miR-518b, 4719, 6757-3p, miR-155, let-7c. **Whole blood:** miR-19a, miR-21, miR-22, miR-31, miR-127, miR-143, miR-145, miR-195. **Urine:** miR10b-5p [41 - 50].
Gastric cancer	miR-31, miR-34a, miR-106a, miR-532 [51 - 54].	**Serum**: miR-10a, miR-17, miR-22, miR-25, miR-100, miR-127-3p, miR-133a, miR-133b, miR-148b, miR-223, miR-1, miR-20a, miR-27a, miR-34, miR-331, miR-423-5p, miR-RNA-381, miR-1246 [55 - 60].
Pancreatic cancer	miR-125a-3p, miR-452, miR-105, miR-127, miR-518a-2, miR-187, miR-30a-3p, miR-21, miR-155, miR-221, miR-222, let-7a [61 - 65].	**Serum:** miR-**7**, miR-125a-3p, miR-6893-5p, miR-125b-1-3p, miR-6075, miR-6836-3p, miR-1469, miR-6729-5p, miR-575, miR-204-3p, miR-6820-5p, miR-4294, miR-4476, and miR-4792, miR-21, miR-155, miR-196a, miR-133a Plasma: miR-21, miR-155, miR-125b-5p, miR-196a, miR-210 [66 - 73].

(Table 1) cont.....

Cancer Type	Tissue Samples	Blood Samples
Colon cancer	miR-21, miR-195, miR-17, miR-20a, miR-145, miR-224, miR-139, miR-423-5p, miR-1228-3p, miR-21, miR-17 [74 - 78].	**Serum:** miR-21, miR-17, miR-20a, miR-32, miR-18a, miR-21, and miR-92a, **Plasma:** miR-423-5p, miR-1224, miR-205-5p, miR-124, miR-127-3p, miR-138, miR-143, miR146a, miR-222 **Whole blood:** mir21, mir143, mir145 **Fecal:** miR-421, miR-27a-3p, miR-106a, miR-130--3p, miR-20a, miR-32, miR-421, miR-27a3p [47, 78 - 86].
Endometrial cancer	miR-23a, miR-144-3p, miR-29c, miR-101, miR-139-5p, miR-182, miR-183, miR-153, miR-27a, miR-96, miR-29c, miR-31, miR-185, miR-652, miR-216b [87 - 94].	**Serum/plasma:** miR-484, miR-23a, miR-222, miR-223, miR-186, miR206, miR-887-5p **Urine:** miR-3973, miR-4426, miR-5089-5p, miR-6841 [50, 95, 96].
Ovarian cancer	miR-21, miR-205, miR-135a-3p, miR-200c, miR-216a, miR-340, miR-665 [97 - 100].	**Serum:** miR-21, miR-484, miR-23a, miR-34a, mir-200a, mir-200c, miR-145, miR-200c-3p, miR-346, miR-127-3p, miR-141, miR-1181, miR-4314 **Urine:** miR-3973, miR-4426, miR-5089-5p, miR-6841 [50, 101 - 108].
Prostate cancer	miR-21, miR-17, miR-20a, miR-32, miR155 [78, 108, 109].	**Serum:** miR-100, miR-125b, miR -141, miR-143, miR-205, miR-296, let-7a, miR-103, -451, -24, -26b, -30c, -93, -106a, -223, -874, -146a, -125b, -100, -107, miR-130b, miR-106a, miR-1274, miR-93, miR-223, miR-874, miR-1207, miR-24, miR-93, miR-106a, miR-24 **Plasma:** miR-622, -1285, -30c, let-7e, miR-let-7c **Urinary:** miR-21, miR-141, miR-574 [109 - 114].

It has become important to regulate up or down miRNA expression in order to study the functions of miRNA in cancer pathology. For this purpose, while the function of miRNA was working, various methods such as antisense inhibitors, transgenics, specific promoters, real time PCR and miRNA microarray have been used [115, 116].

Recognition of miRNAs expressed at different levels in tumor tissues and normal tissues, determining the roles of these miRNAs in human cancers can be useful to elucidate their role in various cancers. The miRNA expression profile is a good starting point for investigating the function of miRNAs in cancer pathogenesis. The absence or overexpression of a particular miRNA in the cancerous cell allows studying the specific role of miRNA in cancer initiation and development. miRNAs appear to be candidates to answer questions about the molecular pathology of cancer. In order for miRNAs to be used in diagnosis and treatment in various cancers, the data must be brought to certain standards and their targets must be defined correctly. Each miRNA molecule can bind to many different mRNAs, different miRNAs in each mRNA. Therefore, new findings are needed to determine the expression activities and targets of miRNAs.

MiRNAs in Infectious Diseases

MiRNAs and Viral Infections

It has been observed that pathogens can induce up / down regulation of various host miRNAs to evade the host immune system. However, some miRNAs studied have been shown to have an antiviral effect and contribute to defense mechanisms to fight infection. It has also been shown that miRNAs play a role in inducing the transition of some pathogens to the latent state. Also, some viruses encode their own miRNAs that modulate the host's signaling pathways and thus support the survival and replication of the virus. Studies have shown that certain miRNA families are effective on the activities of Wnt, MAPK, PI3K / akt, Notch signaling pathways in relation to viral infections [117].

Host miRNAs could control viral infection and replication. Zhang *et al.* [118] suggest that up-regulation of miR-199a-3p and miR-210 in HepG2 2.2.15 cells may play a role in regulating hepatitis B virus (HBV) replication and maintenance of a suitable level of virion production in persistent infection by targeting crucial HBV genes. Serum miR-146a-5p, miR-122-3p and miR-328-3p positively correlate with the severity of liver inflammation in patients with acute-on-chronic liver failure (ACLF) and may be useful to predict HBV-ACLF severity [119]. Circulating miR-223-3p was differentially expressed in cancerous compared with paired adjacent non-cancerous tissues and it could represent a novel diagnostic and prognostic marker for patients with HBV-related hepatocellular carcinoma (HCC) [120]. Serum miR-122 and miR-185 expressions are increased, while miR-29, -143, -21 and miR-223 expressions are decreased during fibrosis progression in patients with HBV chronic infection (CHB). Serum miR-143 and miR-223 expressions are increased, while miR-122 expression is decreased during fibrosis progression in patients with HCV chronic infection (CHC) [121]. Serum miR-21 correlated with the histological stage of liver fibrosis in patients with CHB [122]. Serum miR-26a and miR-21 are closely related to the extent of liver decompensation, and the decreased levels are capable of discriminating patients with cirrhosis at Child-Pugh class C from the whole cirrhosis cases [123]. Circulating miR-122 has a relatively high diagnostic value for chronic viral hepatitis detection, especially in patients with HCV-associated chronic viral hepatitis [124].

At the same time, miRNAs play a role in the defense mechanism against viruses such as Epstein-Barr virus, cytomegalovirus, and herpes simplex virus and infections due to Mycobacterium tuberculosis [125, 126].

Knowing which miRNAs are involved in which viral infections and the role they play can help develop new antiviral therapeutic agents for some diseases for which there is no definitive cure available. Inhibition of targeted miRNAs in clinical applications can have significant therapeutic effects [117].

Especially three important miRNAs (miR-150, miR-182, miR-342-5p) were found to be higher in the blood of patients with septic shock compared to the control group, and it was suggested that they could be a potential marker in early diagnosis [127].

MiRNAs and SARS-CoV-2

COVID-19 pandemic caused by acute respiratory syndrome Coronavirus-2 (SARS-CoV-2) from the Coronavirus family is a highly contagious disease that first appeared in the People's Republic of China in late 2019. COVID-19 affected almost all countries within weeks and was declared a pandemic by the World Health Organization [128].

It is thought that miRNAs can function as proviral determinants. Both SARS-CoV-1 and SARS-CoV-2 viral miRNAs can target a wide variety of immune system-related signaling pathways, and some SARS-CoV-2 miRNAs can target a range of immune-related signaling pathways including autophagy and IFN-I signaling pathways have been shown to target. It is thought that miRNAs can be considered critical epigenetic modulators in the development of complications in COVID-19 cases. Studies have supported the fact that miRNAs from both host and virus (SARS-CoV-2) can contribute to pathological events. These results support that the development of miRNA-based therapeutic agents could be beneficial to alleviate the complications of COVID-19 [129].

The mRNA region of the spike (S) protein in the viral genome represents a good target for miRNA development using exosome against spike mRNA of SARS-CoV-2. In this case, viral spike mRNA translation would be blocked after hybridization with selected complementary miRNAs. A synthetic miRNA can be used to inhibit viral genomic mRNA replication by reducing RNA-dependent RNA polymerase expression. However, miRNAs can be used as a biomarker to evaluate viral replications before and after vaccination in SARS-CoV-2 vaccine development studies [130]. In addition, it is aimed to develop vaccines that will affect the RISC mechanism of the host with miRNAs to be designed specifically for the functional proteins of SARS-CoV-2 [131].

miRNAs in Noninfectious Diseases

miRNAs and Coronary Artery Disease (CAD)

Heart diseases are still the main cause of disease-related deaths. The expression of miRNAs, which have an important role in cardiovascular diseases, changes in diseases such as CAD, acute myocardial infarction (AMI), cardiac hypertrophy and cardiomyopathy. miRNAs, which can play a role in tissue-specific developments and regulate target genes, are at different levels in cardiovascular diseases and provide more sensitive information than important cardiac markers such as cardiac troponin I [132]. As a result of a study conducted by evaluating 259 miRNAs in the circulation of AMI patients, miRNAs such as miR-1, miR-133a and miR-133b peaked before cardiac troponin I (cTnI). Therefore, miRNAs are used as biomarkers in the diagnosis of cardiovascular diseases [133]. In the study of Wang *et al.* [134] while the diagnostic accuracy value for miR-133a in CAD was 0.918, the accuracy value for cTnI was 0.741. These miRNAs are expressed not only in the heart but also in skeletal muscle, this situation negatively affects the specificity of miRNAs.

In patients with abnormal coronary endothelial function, miR-92a and miR-133 were found to be significantly higher. Thus, it has been determined that these miRNAs, which are high in the coronary circulation, play a role in atherosclerosis by causing endothelial dysfunction [135]. In another study, the expression level of miR-487a, miR-508, miR-208 and miR-215 was found to be very high in CAD patients, and the expression level of miR-29b was found to be low. In the same study, cTnI levels were also measured and it was determined that miRs give more sensitive results than cTnI. As a result of the study, it was determined that miR-487a, miR-508, miR-208, miR-215 and miR-29b are independent biomarkers for disease diagnosis in CAD patients [132].

In the plasma of patients with over 70% stenosis, 8 miRNAs (has-miR-494, has-miR-1183, has-miR-490, has-miR-1273, has-miR-585, has-miR-769-3p, has-miR-150, has-miR-7-2) was found to be expressed twice more than patients with stenosis below 70%, and has-miR-769-3p was associated with coronary atherosclerosis [136].

CAD has a relationship with lipid metabolism. There are many studies showing that miRNAs are the main regulators of lipid metabolism. It was first shown in a study conducted in Drosophila in 2003 that miR-14 has an important function in lipid metabolism. In this study, it was revealed that the amount of total body triacylglycerol (TAG) doubled as a result of the inhibition of miR-14 [137]. According to other studies, it has been determined that miRNAs play an active role in lipid and amino acid metabolism and glucose regulation [137 - 139].

In addition, miRNAs play a role in the development of congenital heart diseases such as atrial septal defects, ventricular septal defects, patent ductus arteriosus, and Fallot tetralogy. Emryogenesis of the heart is regulated by signaling pathways. Dysregulations in miRNAs cause structural anomalies in heart and valve structures by affecting multiple signal pathways. In this context, miRNAs can be used as biomarkers in congenital heart diseases and are also a therapeutic target [140].

miRNAs and Gynecological Diseases

Regarding gynecological malignancies, miRNAs such as miR-145 have been identified as central players in cervical carcinogenesis, while miR-125b, miR-145, miR-21 and miR-155 have been shown to play very important roles in breast malignancies. miR-200 and let-7 have been identified as key modulators in ovarian neoplasms, while miR-185, miR-210, miR-423, let-7c, miR-205 and miR-429 have been associated with oncogenesis, invasion, and metastasis in endometrial carcinomas. In addition to its use as prognostic or diagnostic biomarkers, therapeutic target strategies for the inhibition of miRNAs continue to be developed [141]. Another disease to be mentioned about the diagnostic and therapeutic importance of miRNAs is polycystic ovary syndrome (PCOS). It has been shown that miR-29a-5p is a biomarker that can be used in the diagnosis of PCOS [142].

miRNAs and Autoimmune Diseases

In vivo and *in vitro* studies have shown that miRNA expression changes in various autoimmune diseases as well as in cancer. As a result of the research conducted with rheumatoid arthritis patients, it was observed that expressions of miR-155 and miR-146a were different in synovial fluid samples, and expression of miR-146a, miR-155, and miR-16 was increased in isolated T cells [143].

In another study, it was revealed that expressions of miR-18b, miR-599 and miR-96 in peripheral mononuclear cells of multiple sclerosis patients were different compared to the control [144].

In a study conducted with mice, it has been shown that miR-326 induces the progression of the disease by stimulating the production of Th-17 cells in the development of experimental autoimmune encephalomyelitis [145].

Serum miR-448, miR-124, and miR-551b could serve as valuable specific biomarkers for distinguishing patients with autoimmune diseases [rheumatoid arthritis (RA), systemic lupus erythematosus (SLE), Sjogren's syndrome (SS), and ulcerative colitis (UC)] from healthy controls [146].

MiRNAs in autoimmune diseases is shown in Table **2**.

Table 2. MiRNAs in autoimmune diseases.

Disease	Associated microRNAs
Diabetes mellitus, Type 1	miR-21-5p, miR-24-3p, miR-34a, miR-100-5p, miR-146a-5p, miR-148a-3p, miR-150-5p, miR-181a-5p, miR-210-5p, miR-342-3p, miR-375 and miR-1275, miR-204, miR-126, miR- 101-3p, miR- 181, miR- 210 [147 - 152].
Rheumatoid arthritis (RA),	miR-448, miR-124, miR-551b, miR-223, miR-340-5p, miR-181a, miR-145, miR-29a, miR-132, miR-126, miR-155 [146, 153 - 155].
Systemic lupus erythematosus (SLE),	miR-21, miR-155, miRNA-146, miR-132, miR-143, miR-29a [155 - 157].
Sjogren's syndrome (SS)	miR-127-3p, miR-409-3p, miR-410-3p, miR-541-5p, miR-540-5p. miR-127-3p, 541-5p [158].
Ulcerative colitis (UC)	miR-448, miR-124, miR-551b, miR-29b-3p, miR-122-5p, miR-140-5p, miR-14--3p, miR-146a-3p,miR-148a-3p, miR-150-5p, miR-152-3p, miR-192-5p, miR-194-5p, miR-195a-5p, miR-199a-3p, miR-335-5p, miR-375-3p, miR-146a, miR-219-5p [146, 159 - 161].
Inflammatory bowel diseases (IBD)	miR-126, let-7c, miR-146a, miR-146b, miR-320a [162].
Psoriasis	miR-1910-3p, miR-155, miR-210, miR-20b, miR-203, miR-31, miR-146a, miR-21, miR-424, miR-151a [163 - 169].

miRNAs and the Nervous System Diseases

It has been demonstrated that miRNAs are quite abundant in the brain and play important roles in various biological processes such as neuronal differentiation, brain development, synapse formation and plasticity, neuronal survival, and neurodegenerative diseases [170].

Studies have shown 102 different miRNAs associated with neuronal differentiation. Most of the identified miRNAs act similarly on the cell cycle using different pathways. MiR-125a-5p, miR-423-5p, miR-320 are the most common [171].

Based on extensive fluorescent miRNA sequencing and RNA sequencing analysis, significant increases were noted in the amounts of miRNA-9, miRNA-34a, miRNA-125b, miRNA-146a, miRNA-155 in common in brain tissues, extracellular fluids, and cerebrospinal fluid from Alzheimer's patients. However, it is known that the transcription of the mentioned miRNA genes is related to the

NF-kB signaling pathway [172]. MiRNAs related to different neurological pathologies are shown in Table **3**.

Table 3. MiRNAs related to different neurological pathologies.

Disease	Associated microRNAs
Huntington Disease	miR-520f-3p, miR-135b-3p, miR-4317, miR-3928-5p, miR-8082, miR-140-5p, miR-124 [173, 174].
Spinal muscular atrophy (SMA)	miR-9, miR-132, miR-133a [175, 176].
Alzheimer Disease	miR-135a, miR-193b, miR-384, miR-125, miR-181c, miR-191-5p, miR-146a, miR-9, miR-455-3p, miR-222, hsa-miR-101-3p, hsa-miR-20a-5p, hsa-mi--106b5p, hsa-miR-15a-5p, hsa-miR-1306-5p, hsa-miR-15b-3p, 106b-5p, hsa-mi--106a-5p, hsa-miR-424-5p, miR-195, miR-155, miR-34a, miR-9, miR-206, miR-125b, miR-29, miR-22, miR-106b [177 - 183].
Schizophrenia	miR-181b, miR-219-2-3p, miR-346, miR-195, miR-1308, miR-92a, miR-17, miR-103, let-7g, hsa-miR-146a, miR-137 [184 - 188].
Bipolar disorder	miR-134, miR-7-5p, miR-23b-3p, miR-142-3p, miR-221-5p, miR-370-3p [189, 190].
Epilepsy	miR-134, miR- 146a, miR- 124, miR-199a, miR-128, miR-15a-5p, miR-194-5p, miR-34c-5p, miR-106b, miR-139-5p, miR-15a-5p, miR-146a, miR-155, miR-132, miR-654-3p, miR-328-3p, hsa-miR-1275, miR-21-5p [191 - 200].
HIV encephalitis	miR-1296, miR-134, miR-16, miR-495, miR-1296, miR-937, miR-595, miR-300, miR-598, miR-555, miR-222, miR-105 [201].
Rett Sendromu	miR-16, miR-27a, miR-132, miR-335-5p, miR-323-3p [202 - 204].
Multiple sclerosis	miR-660, miR-939, miR-22, miR-30e, miR-140-3p, miR-210, miR-500, miR-57--3p, miR-92a-1, miR-135a, miR-454, miR-15b, miR-23a, miR-223, miR-181c, miR-633, miR-128-3p, miR-146a, miR-155, miR-300, miR-450b-5p [205 - 211].
Parkinson's disease	miR-29c, miR-29s, miR-22, miR-29a, let-7d, miR-15b, miR-24, miR-142-3p, miR-181c, miR-222, miR-150, miR-214, miR-31, miR-141, miR-181c, miR-193--3p [212 - 216].

CONCLUSION

microRNAs are from the class of non-coding RNAs and they play an important role in the regulation of gene expression. While miRNAs regulate transcription and translation at the cell level along with other non-coding RNAs, they also act as chemical messengers that are released outside the cell and mediate cell-cell communication. It also functions in the homeostasis of signaling pathways and plays a role in disease development through related signal pathways, especially cancers. The fact that miRNAs remain intact in body fluids suggests that they can be good biomarkers. miRNAs can serve as oncogenes or tumor suppressors. They take part in pathological and physiological pathways since they regulate gene expression by matching their targets with base. They can be obtained from a wide variety of biological specimens such as blood plasma, urine, stool or biopsy specimen and are generally resistant to a variety of storage conditions. Changes in miRNAs can be detected noninvasively and measured in body fluids using highly sensitive and standardized techniques. In the light of this information, it is possible to state that miRNAs are good candidates for molecular markers in terms of diagnosis, prognosis and clinical follow-up. Undoubtedly, the evidences that we have very little information about miRNA studies, which have accelerated in recent years, have been and will continue to be added to the scientific world with many innovative studies. Before these new molecules can be used in diagnosis, more clinical and experimental studies are needed to elucidate how biochemical and immunological processes in humans progress.

CONSENT FOR PUBLICATION

Not applicable.

CONFLICT OF INTEREST

The authors declare no conflict of interest, financial or otherwise

ACKNOWLEDGEMENT

Declared none.

REFERENCES

[1] Biomarkers Definitions Working Group.. Biomarkers and surrogate endpoints: preferred definitions and conceptual framework. Clin Pharmacol Ther 2001; 69(3): 89-95.
[http://dx.doi.org/10.1067/mcp.2001.113989] [PMID: 11240971]

[2] Lu TX, Rothenberg ME. MicroRNA. J Allergy Clin Immunol 2018; 141(4): 1202-7.
[http://dx.doi.org/10.1016/j.jaci.2017.08.034] [PMID: 29074454]

[3] Bartel DP. MicroRNAs: genomics, biogenesis, mechanism, and function. Cell 2004; 116(2): 281-97.
[http://dx.doi.org/10.1016/S0092-8674(04)00045-5] [PMID: 14744438]

[4] Dutta RK, Chinnapaiyan S, Unwalla H. Aberrant MicroRNAomics in Pulmonary Complications: Implications in Lung Health and Diseases. Mol Ther Nucleic Acids 2019; 18: 413-31.
[http://dx.doi.org/10.1016/j.omtn.2019.09.007] [PMID: 31655261]

[5] Kim YK. Extracellular microRNAs as Biomarkers in Human Disease. Chonnam Med J 2015; 51(2): 51-7.
[http://dx.doi.org/10.4068/cmj.2015.51.2.51] [PMID: 26306299]

[6] Rosenfeld N, Aharonov R, Meiri E, *et al.* MicroRNAs accurately identify cancer tissue origin. Nat Biotechnol 2008; 26(4): 462-9.
[http://dx.doi.org/10.1038/nbt1392] [PMID: 18362881]

[7] Gareev I, Beylerli O, Yang G, *et al.* The current state of MiRNAs as biomarkers and therapeutic tools. Clin Exp Med 2020; 20(3): 349-59.
[http://dx.doi.org/10.1007/s10238-020-00627-2] [PMID: 32399814]

[8] Shukla V, Varghese VK, Kabekkodu SP, Mallya S, Satyamoorthy K. A compilation of Web-based research tools for miRNA analysis. Brief Funct Genomics 2017; 16(5): 249-73. [http://dx.doi.org/10.1093/bfgp/elw042] [PMID: 28334134]

[9] Akdis CA. Therapies for allergic inflammation: refining strategies to induce tolerance. Nat Med 2012; 18(5): 736-49. [http://dx.doi.org/10.1038/nm.2754] [PMID: 22561837]

[10] Kumarswamy R, Volkmann I, Thum T. Regulation and function of miRNA-21 in health and disease. RNA Biol 2011; 8(5): 706-13. [http://dx.doi.org/10.4161/rna.8.5.16154] [PMID: 21712654]

[11] Iliopoulos D, Jaeger SA, Hirsch HA, Bulyk ML, Struhl K. STAT3 activation of miR-21 and miR-181b-1 via PTEN and CYLD are part of the epigenetic switch linking inflammation to cancer. Mol Cell 2010; 39(4): 493-506. [http://dx.doi.org/10.1016/j.molcel.2010.07.023] [PMID: 20797623]

[12] Sheedy FJ, Palsson-McDermott E, Hennessy EJ, *et al.* Negative regulation of TLR4 via targeting of the proinflammatory tumor suppressor PDCD4 by the microRNA miR-21. Nat Immunol 2010; 11(2): 141-7. [http://dx.doi.org/10.1038/ni.1828] [PMID: 19946272]

[13] Rebane A, Akdis CA. MicroRNAs in allergy and asthma. Curr Allergy Asthma Rep 2014; 14(4): 424. [http://dx.doi.org/10.1007/s11882-014-0424-x] [PMID: 24504527]

[14] Taganov KD, Boldin MP, Chang KJ, Baltimore D. NF-kappaB-dependent induction of microRNA miR-146, an inhibitor targeted to signaling proteins of innate immune responses. Proc Natl Acad Sci USA 2006; 103(33): 12481-6. [http://dx.doi.org/10.1073/pnas.0605298103] [PMID: 16885212]

[15] Zhao JL, Rao DS, Boldin MP, Taganov KD, O'Connell RM, Baltimore D. NF-kappaB dysregulation in microRNA-146a-deficient mice drives the development of myeloid malignancies. Proc Natl Acad Sci USA 2011; 108(22): 9184-9. [http://dx.doi.org/10.1073/pnas.1105398108] [PMID: 21576471]

[16] Boldin MP, Taganov KD, Rao DS, *et al.* miR-146a is a significant brake on autoimmunity, myeloproliferation, and cancer in mice. J Exp Med 2011; 208(6): 1189-201. [http://dx.doi.org/10.1084/jem.20101823] [PMID: 21555486]

[17] Lu LF, Boldin MP, Chaudhry A, *et al.* Function of miR-146a in controlling Treg cell-mediated regulation of Th1 responses. Cell 2010; 142(6): 914-29. [http://dx.doi.org/10.1016/j.cell.2010.08.012] [PMID: 20850013]

[18] Guo Q, Zhang J, Li J, *et al.* Forced miR-146a expression causes autoimmune lymphoproliferative syndrome in mice via downregulation of Fas in germinal center B cells. Blood 2013; 121(24): 4875-83. [http://dx.doi.org/10.1182/blood-2012-08-452425] [PMID: 23645835]

[19] Wang P, Hou J, Lin L, *et al.* Inducible microRNA-155 feedback promotes type I IFN signaling in antiviral innate immunity by targeting suppressor of cytokine signaling 1. J Immunol 2010; 185(10): 6226-33. [http://dx.doi.org/10.4049/jimmunol.1000491] [PMID: 20937844]

[20] Baumjohann D, Ansel KM. MicroRNA-mediated regulation of T helper cell differentiation and plasticity. Nat Rev Immunol 2013; 13(9): 666-78. [http://dx.doi.org/10.1038/nri3494] [PMID: 23907446]

[21] de Yébenes VG, Bartolomé-Izquierdo N, Ramiro AR. Regulation of B-cell development and function by microRNAs. Immunol Rev 2013; 253(1): 25-39. [http://dx.doi.org/10.1111/imr.12046] [PMID: 23550636]

[22] Gebeshuber CA, Zatloukal K, Martinez J. miR-29a suppresses tristetraprolin, which is a regulator of

epithelial polarity and metastasis. EMBO Rep 2009; 10(4): 400-5.
[http://dx.doi.org/10.1038/embor.2009.9] [PMID: 19247375]

[23] Calin GA, Croce CM. MicroRNA signatures in human cancers. Nat Rev Cancer 2006; 6(11): 857-66.
[http://dx.doi.org/10.1038/nrc1997] [PMID: 17060945]

[24] Calin GA, Sevignani C, Dumitru CD, *et al.* Human microRNA genes are frequently located at fragile sites and genomic regions involved in cancers. Proc Natl Acad Sci USA 2004; 101(9): 2999-3004.
[http://dx.doi.org/10.1073/pnas.0307323101] [PMID: 14973191]

[25] Kim YK, Yeo J, Ha M, Kim B, Kim VN. Cell adhesion-dependent control of microRNA decay. Mol Cell 2011; 43(6): 1005-14.
[http://dx.doi.org/10.1016/j.molcel.2011.07.031] [PMID: 21925388]

[26] Calin GA, Dumitru CD, Shimizu M, *et al.* Frequent deletions and down-regulation of micro- RNA genes miR15 and miR16 at 13q14 in chronic lymphocytic leukemia. Proc Natl Acad Sci USA 2002; 99(24): 15524-9.
[http://dx.doi.org/10.1073/pnas.242606799] [PMID: 12434020]

[27] Cimmino A, Calin GA, Fabbri M, *et al.* miR-15 and miR-16 induce apoptosis by targeting BCL2. Proc Natl Acad Sci USA 2005; 102(39): 13944-9.
[http://dx.doi.org/10.1073/pnas.0506654102] [PMID: 16166262]

[28] Hydbring P, Badalian-Very G. Clinical applications of microRNAs. F1000 Res 2013; 2: 136.
[http://dx.doi.org/10.12688/f1000research.2-136.v1] [PMID: 24627783]

[29] Lebanony D, Benjamin H, Gilad S, *et al.* Diagnostic assay based on hsa-miR-205 expression distinguishes squamous from nonsquamous non-small-cell lung carcinoma. J Clin Oncol 2009; 27(12): 2030-7.
[http://dx.doi.org/10.1200/JCO.2008.19.4134] [PMID: 19273703]

[30] Yang X, Zhang Q, Zhang M, *et al.* Serum microRNA Signature Is Capable of Early Diagnosis for Non-Small Cell Lung Cancer. Int J Biol Sci 2019; 15(8): 1712-22.
[http://dx.doi.org/10.7150/ijbs.33986] [PMID: 31360113]

[31] Hu J, Qiu M, Jiang F, *et al.* MiR-145 regulates cancer stem-like properties and epithelial-t--mesenchymal transition in lung adenocarcinoma-initiating cells. Tumour Biol 2014; 35(9): 8953-61.
[http://dx.doi.org/10.1007/s13277-014-2158-8] [PMID: 24903381]

[32] Roth C, Kasimir-Bauer S, Pantel K, Schwarzenbach H. Screening for circulating nucleic acids and caspase activity in the peripheral blood as potential diagnostic tools in lung cancer. Mol Oncol 2011; 5(3): 281-91.
[http://dx.doi.org/10.1016/j.molonc.2011.02.002] [PMID: 21398193]

[33] Nadal E, Truini A, Nakata A, *et al.* A Novel Serum 4-microRNA Signature for Lung Cancer Detection. Sci Rep 2015; 5: 12464.
[http://dx.doi.org/10.1038/srep12464] [PMID: 26202143]

[34] Yang X, Su W, Chen X, *et al.* Validation of a serum 4-microRNA signature for the detection of lung cancer. Transl Lung Cancer Res 2019; 8(5): 636-48.
[http://dx.doi.org/10.21037/tlcr.2019.09.11] [PMID: 31737499]

[35] Lv J, An J, Zhang YD, *et al.* A three serum miRNA panel as diagnostic biomarkers of radiotherapy-related metastasis in non-small cell lung cancer. Oncol Lett 2020; 20(5): 236.
[http://dx.doi.org/10.3892/ol.2020.12099] [PMID: 32968458]

[36] Wu Q, Yu L, Lin X, *et al.* Combination of Serum miRNAs with Serum Exosomal miRNAs in Early Diagnosis for Non-Small-Cell Lung Cancer. Cancer Manag Res 2020; 12: 485-95.
[http://dx.doi.org/10.2147/CMAR.S232383] [PMID: 32021461]

[37] Chaniad P, Trakunran K, Geater SL, Keeratichananont W, Thongsuksai P, Raungrut P. Serum miRNAs associated with tumor-promoting cytokines in non-small cell lung cancer. PLoS One 2020; 15(10): e0241593.

[http://dx.doi.org/10.1371/journal.pone.0241593] [PMID: 33125430]

[38] Hui AB, Shi W, Boutros PC, *et al.* Robust global micro-RNA profiling with formalin-fixed paraffin-embedded breast cancer tissues. Lab Invest 2009; 89(5): 597-606. [http://dx.doi.org/10.1038/labinvest.2009.12] [PMID: 19290006]

[39] Sempere LF, Christensen M, Silahtaroglu A, *et al.* Altered MicroRNA expression confined to specific epithelial cell subpopulations in breast cancer. Cancer Res 2007; 67(24): 11612-20. [http://dx.doi.org/10.1158/0008-5472.CAN-07-5019] [PMID: 18089790]

[40] Rohan T, Ye K, Wang Y, Glass AG, Ginsberg M, Loudig O. MicroRNA expression in benign breast tissue and risk of subsequent invasive breast cancer. PLoS One 2018; 13(2): e0191814. [http://dx.doi.org/10.1371/journal.pone.0191814] [PMID: 29432432]

[41] Roth C, Rack B, Müller V, Janni W, Pantel K, Schwarzenbach H. Circulating microRNAs as blood-based markers for patients with primary and metastatic breast cancer. Breast Cancer Res 2010; 12(6): R90. [http://dx.doi.org/10.1186/bcr2766] [PMID: 21047409]

[42] Wang F, Zheng Z, Guo J, Ding X. Correlation and quantitation of microRNA aberrant expression in tissues and sera from patients with breast tumor. Gynecol Oncol 2010; 119(3): 586-93. [http://dx.doi.org/10.1016/j.ygyno.2010.07.021] [PMID: 20801493]

[43] Heneghan HM, Miller N, Kelly R, Newell J, Kerin MJ. Systemic miRNA-195 differentiates breast cancer from other malignancies and is a potential biomarker for detecting noninvasive and early stage disease. Oncologist 2010; 15(7): 673-82. [http://dx.doi.org/10.1634/theoncologist.2010-0103] [PMID: 20576643]

[44] Shimomura A, Shiino S, Kawauchi J, *et al.* Novel combination of serum microRNA for detecting breast cancer in the early stage. Cancer Sci 2016; 107(3): 326-34. [http://dx.doi.org/10.1111/cas.12880] [PMID: 26749252]

[45] Liu Y, Tang D, Zheng S, Su R, Tang Y. Serum microRNA-195 as a potential diagnostic biomarker for breast cancer: a systematic review and meta-analysis. Int J Clin Exp Pathol 2019; 12(11): 3982-91. [PMID: 31933794]

[46] Alunni-Fabbroni M, Majunke L, Trapp EK, *et al.* Whole blood microRNAs as potential biomarkers in post-operative early breast cancer patients. BMC Cancer 2018; 18(1): 141. [http://dx.doi.org/10.1186/s12885-018-4020-7] [PMID: 29409452]

[47] Kundaktepe BP, Sozer V, Papila C, *et al.* Associations Between miRNAs and Two Different Cancers: Breast and Colon. Cancer Manag Res 2020; 12: 871-9. [http://dx.doi.org/10.2147/CMAR.S227628] [PMID: 32104069]

[48] Savari B, Boozarpour S, Tahmasebi-Birgani M, Sabouri H, Hosseini SM. Overexpression of *microRNA-21* in the Serum of Breast Cancer Patients. MicroRNA 2020; 9(1): 58-63. [http://dx.doi.org/10.2174/2211536608666190318105757] [PMID: 30887933]

[49] Aksan H, Kundaktepe BP, Sayili U, *et al.* Circulating miR-155, let-7c, miR-21, and PTEN levels in differential diagnosis and prognosis of idiopathic granulomatous mastitis and breast cancer. Biofactors 2020; 46(6): 955-62. [http://dx.doi.org/10.1002/biof.1676] [PMID: 32941675]

[50] Ritter A, Hirschfeld M, Berner K, *et al.* Discovery of potential serum and urine-based microRNA as minimally-invasive biomarkers for breast and gynecological cancer. Cancer Biomark 2020; 27(2): 225-42. [http://dx.doi.org/10.3233/CBM-190575] [PMID: 32083575]

[51] Xiao B, Guo J, Miao Y, *et al.* Detection of miR-106a in gastric carcinoma and its clinical significance. Clin Chim Acta 2009; 400(1-2): 97-102. [http://dx.doi.org/10.1016/j.cca.2008.10.021] [PMID: 18996365]

[52] Zhang Y, Guo J, Li D, *et al.* Down-regulation of miR-31 expression in gastric cancer tissues and its

clinical significance. Med Oncol 2010; 27(3): 685-9.
[http://dx.doi.org/10.1007/s12032-009-9269-x] [PMID: 19598010]

[53] Li Z, Liu ZM, Xu BH. A meta-analysis of the effect of microRNA-34a on the progression and prognosis of gastric cancer. Eur Rev Med Pharmacol Sci 2018; 22(23): 8281-7.
[PMID: 30556868]

[54] Hu S, Zheng Q, Wu H, Wang C, Liu T, Zhou W. miR-532 promoted gastric cancer migration and invasion by targeting NKD1. Life Sci 2017; 177: 15-9.
[http://dx.doi.org/10.1016/j.lfs.2017.03.019] [PMID: 28356225]

[55] Zhang C, Wang C, Chen X, *et al.* Expression profile of microRNAs in serum: a fingerprint for esophageal squamous cell carcinoma. Clin Chem 2010; 56(12): 1871-9.
[http://dx.doi.org/10.1373/clinchem.2010.147553] [PMID: 20943850]

[56] Liu R, Zhang C, Hu Z, *et al.* A five-microRNA signature identified from genome-wide serum microRNA expression profiling serves as a fingerprint for gastric cancer diagnosis. Eur J Cancer 2011; 47(5): 784-91.
[http://dx.doi.org/10.1016/j.ejca.2010.10.025] [PMID: 21112772]

[57] ZiaSarabi P, Sorayayi S, Hesari A, Ghasemi F. Circulating microRNA-133, microRNA-17 and microRNA-25 in serum and its potential diagnostic value in gastric cancer. J Cell Biochem 2019; 120(8): 12376-81.
[http://dx.doi.org/10.1002/jcb.28503] [PMID: 30861177]

[58] Li Y, Sun H, Guan J, Ji T, Wang X. Serum microRNA-381: A Potential Marker for Early Diagnosis of Gastric Cancer. Yonsei Med J 2019; 60(8): 720-6.
[http://dx.doi.org/10.3349/ymj.2019.60.8.720] [PMID: 31347326]

[59] Shi Y, Wang Z, Zhu X, *et al.* Exosomal miR-1246 in serum as a potential biomarker for early diagnosis of gastric cancer. Int J Clin Oncol 2020; 25(1): 89-99.
[http://dx.doi.org/10.1007/s10147-019-01532-9] [PMID: 31506750]

[60] Sierzega M, Kaczor M, Kolodziejczyk P, Kulig J, Sanak M, Richter P. Evaluation of serum microRNA biomarkers for gastric cancer based on blood and tissue pools profiling: the importance of miR-21 and miR-331. Br J Cancer 2017; 117(2): 266-73.
[http://dx.doi.org/10.1038/bjc.2017.190] [PMID: 28641313]

[61] Bloomston M, Frankel WL, Petrocca F, *et al.* MicroRNA expression patterns to differentiate pancreatic adenocarcinoma from normal pancreas and chronic pancreatitis. JAMA 2007; 297(17): 1901-8.
[http://dx.doi.org/10.1001/jama.297.17.1901] [PMID: 17473300]

[62] Habbe N, Koorstra JB, Mendell JT, *et al.* MicroRNA miR-155 is a biomarker of early pancreatic neoplasia. Cancer Biol Ther 2009; 8(4): 340-6.
[http://dx.doi.org/10.4161/cbt.8.4.7338] [PMID: 19106647]

[63] du Rieu MC, Torrisani J, Selves J, *et al.* MicroRNA-21 is induced early in pancreatic ductal adenocarcinoma precursor lesions. Clin Chem 2010; 56(4): 603-12.
[http://dx.doi.org/10.1373/clinchem.2009.137364] [PMID: 20093556]

[64] Moriya C, Taniguchi H, Miyata K, Nishiyama N, Kataoka K, Imai K. Inhibition of PRDM14 expression in pancreatic cancer suppresses cancer stem-like properties and liver metastasis in mice. Carcinogenesis 2017; 38(6): 638-48.
[http://dx.doi.org/10.1093/carcin/bgx040] [PMID: 28498896]

[65] Liu G, Ji L, Ke M, Ou Z, Tang N, Li Y. miR-125a-3p is responsible for chemosensitivity in PDAC by inhibiting epithelial-mesenchymal transition via Fyn. Biomed Pharmacother 2018; 106: 523-31.
[http://dx.doi.org/10.1016/j.biopha.2018.06.114] [PMID: 29990840]

[66] Wang J, Chen J, Chang P, *et al.* MicroRNAs in plasma of pancreatic ductal adenocarcinoma patients as novel blood-based biomarkers of disease. Cancer Prev Res (Phila) 2009; 2(9): 807-13.

[http://dx.doi.org/10.1158/1940-6207.CAPR-09-0094] [PMID: 19723895]

[67] Ho AS, Huang X, Cao H, *et al.* Circulating miR-210 as a Novel Hypoxia Marker in Pancreatic Cancer. Transl Oncol 2010; 3(2): 109-13.
[http://dx.doi.org/10.1593/tlo.09256] [PMID: 20360935]

[68] Kong X, Du Y, Wang G, *et al.* Detection of differentially expressed microRNAs in serum of pancreatic ductal adenocarcinoma patients: miR-196a could be a potential marker for poor prognosis. Dig Dis Sci 2011; 56(2): 602-9.
[http://dx.doi.org/10.1007/s10620-010-1285-3] [PMID: 20614181]

[69] Yan Q, Hu D, Li M, *et al.* The Serum MicroRNA Signatures for Pancreatic Cancer Detection and Operability Evaluation. Front Bioeng Biotechnol 2020; 8: 379.
[http://dx.doi.org/10.3389/fbioe.2020.00379] [PMID: 32411694]

[70] Zou X, Wei J, Huang Z, *et al.* Identification of a six-miRNA panel in serum benefiting pancreatic cancer diagnosis. Cancer Med 2019; 8(6): 2810-22.
[http://dx.doi.org/10.1002/cam4.2145] [PMID: 31006985]

[71] Ye ZQ, Zou CL, Chen HB, Jiang MJ, Mei Z, Gu DN. MicroRNA-7 as a potential biomarker for prognosis in pancreatic cancer. Dis Markers 2020; 2020: 2782101.
[http://dx.doi.org/10.1155/2020/2782101] [PMID: 32566037]

[72] Wang Z. Diagnostic performance for declined microRNA-133a in pancreatic cancer. J Cell Biochem 2019.
[http://dx.doi.org/10.1002/jcb.29547]

[73] Zhou X, Lu Z, Wang T, Huang Z, Zhu W, Miao Y. Plasma miRNAs in diagnosis and prognosis of pancreatic cancer: A miRNA expression analysis. Gene 2018; 673: 181-93.
[http://dx.doi.org/10.1016/j.gene.2018.06.037] [PMID: 29913239]

[74] Jia W, Yu T, An Q, Cao X, Pan H. MicroRNA-423-5p inhibits colon cancer growth by promoting caspase-dependent apoptosis. Exp Ther Med 2018; 16(2): 1225-31.
[http://dx.doi.org/10.3892/etm.2018.6288] [PMID: 30116373]

[75] Cojocneanu R, Braicu C, Raduly L, *et al.* Plasma and Tissue Specific miRNA Expression Pattern and Functional Analysis Associated to Colorectal Cancer Patients. Cancers (Basel) 2020; 12(4): E843.
[http://dx.doi.org/10.3390/cancers12040843] [PMID: 32244548]

[76] Volinia S, Calin GA, Liu CG, *et al.* A microRNA expression signature of human solid tumors defines cancer gene targets. Proc Natl Acad Sci USA 2006; 103(7): 2257-61.
[http://dx.doi.org/10.1073/pnas.0510565103] [PMID: 16461460]

[77] Masuda T, Hayashi N, Kuroda Y, Ito S, Eguchi H, Mimori K. MicroRNAs as Biomarkers in Colorectal Cancer. Cancers (Basel) 2017; 9(9): E124.
[http://dx.doi.org/10.3390/cancers9090124] [PMID: 28902152]

[78] Gmerek L, Martyniak K, Horbacka K, *et al.* MicroRNA regulation in colorectal cancer tissue and serum. PLoS One 2019; 14(8): e0222013.
[http://dx.doi.org/10.1371/journal.pone.0222013] [PMID: 31469874]

[79] Elaguizy M, Sheta M, Ibrahim N, Eltaweel A, Mostafa A. Serum microRNA-18a, MicroRNA-21 and microRNA-92a as diagnostic markers in colorectal cancer patients. J BUON 2020; 25(3): 1443-8.
[PMID: 32862588]

[80] Zanutto S, Pizzamiglio S, Ghilotti M, *et al.* Circulating miR-378 in plasma: a reliable, haemolysis-independent biomarker for colorectal cancer. Br J Cancer 2014; 110(4): 1001-7.
[http://dx.doi.org/10.1038/bjc.2013.819] [PMID: 24423916]

[81] Gulei D, Magdo L, Jurj A, *et al.* The silent healer: miR-205-5p up-regulation inhibits epithelial to mesenchymal transition in colon cancer cells by indirectly up-regulating E-cadherin expression. Cell Death Dis 2018; 9(2): 66.
[http://dx.doi.org/10.1038/s41419-017-0102-8] [PMID: 29352232]

[82] Gulei D, Irimie AI, Cojocneanu-Petric R, Schultze JL, Berindan-Neagoe I. Exosomes-Small Players, Big Sound. Bioconjug Chem 2018; 29(3): 635-48.
[http://dx.doi.org/10.1021/acs.bioconjchem.8b00003] [PMID: 29369619]

[83] Link A, Balaguer F, Shen Y, *et al.* Fecal MicroRNAs as novel biomarkers for colon cancer screening. Cancer Epidemiol Biomarkers Prev 2010; 19(7): 1766-74.
[http://dx.doi.org/10.1158/1055-9965.EPI-10-0027] [PMID: 20551304]

[84] Sobin LH, Wittekind CH, Eds. UICC TNM Classification of Malignant Tumors. 6th ed. New York: John Wiley 2002; pp. 170-3.

[85] Greene FL, Page DL, Fleming ID, Eds. AJCC Cancer Staging Manual. 6th ed., New York: Springer-Verlag 2002.
[http://dx.doi.org/10.1007/978-1-4757-3656-4]

[86] Duran-Sanchon S, Moreno L, Augé JM, *et al.* Identification and Validation of MicroRNA Profiles in Fecal Samples for Detection of Colorectal Cancer. Gastroenterology 2020; 158(4): 947-957.e4.
[http://dx.doi.org/10.1053/j.gastro.2019.10.005] [PMID: 31622624]

[87] Li HL, Sun JJ, Ma H, *et al.* MicroRNA-23a inhibits endometrial cancer cell development by targeting SIX1. Oncol Lett 2019; 18(4): 3792-802.
[http://dx.doi.org/10.3892/ol.2019.10694] [PMID: 31579409]

[88] Wang W, Ge L, Xu XJ, *et al.* LncRNA NEAT1 promotes endometrial cancer cell proliferation, migration and invasion by regulating the miR-144-3p/EZH2 axis. Radiol Oncol 2019; 53(4): 434-42.
[http://dx.doi.org/10.2478/raon-2019-0051] [PMID: 31747378]

[89] Van Sinderen M, Griffiths M, Menkhorst E, Niven K, Dimitriadis E. Restoration of microRNA-29c in type I endometrioid cancer reduced endometrial cancer cell growth. Oncol Lett 2019; 18(3): 2684-93.
[http://dx.doi.org/10.3892/ol.2019.10588] [PMID: 31404303]

[90] Liu Y, Li H, Zhao C, Jia H. MicroRNA-101 inhibits angiogenesis via COX-2 in endometrial carcinoma. Mol Cell Biochem 2018; 448(1-2): 61-9.
[http://dx.doi.org/10.1007/s11010-018-3313-0] [PMID: 29404887]

[91] Liu J, Li C, Jiang Y, Wan Y, Zhou S, Cheng W. Tumor-suppressor role of miR-139-5p in endometrial cancer. Cancer Cell Int 2018; 18: 51.
[http://dx.doi.org/10.1186/s12935-018-0545-8] [PMID: 29618950]

[92] Fang Q, Sang L, Du S. Long noncoding RNA LINC00261 regulates endometrial carcinoma progression by modulating miRNA/FOXO1 expression. Cell Biochem Funct 2018; 36(6): 323-30.
[http://dx.doi.org/10.1002/cbf.3352] [PMID: 30019459]

[93] Ushakov DS, Dorozhkova AS, Babayants EV, *et al.* Expression of microRNA Potentially Regulated by AhR and CAR in Malignant Tumors of the Endometrium. Bull Exp Biol Med 2018; 165(5): 688-91.
[http://dx.doi.org/10.1007/s10517-018-4242-3] [PMID: 30225717]

[94] Xie P, Cao H, Li Y, Wang J, Cui Z. Knockdown of lncRNA CCAT2 inhibits endometrial cancer cells growth and metastasis via sponging miR-216b. Cancer Biomark 2017; 21(1): 123-33.
[http://dx.doi.org/10.3233/CBM-170388] [PMID: 29036788]

[95] Rižner TL. Discovery of biomarkers for endometrial cancer: current status and prospects. Expert Rev Mol Diagn 2016; 16(12): 1315-36.
[http://dx.doi.org/10.1080/14737159.2016.1258302] [PMID: 27817223]

[96] Jiang Y, Wang N, Yin D, *et al.* Changes in the Expression of Serum MiR-887-5p in Patients With Endometrial Cancer. Int J Gynecol Cancer 2016; 26(6): 1143-7.
[http://dx.doi.org/10.1097/IGC.0000000000000730] [PMID: 27177284]

[97] Chan JK, Blansit K, Kiet T, *et al.* The inhibition of miR-21 promotes apoptosis and chemosensitivity in ovarian cancer. Gynecol Oncol 2014; 132(3): 739-44.

[http://dx.doi.org/10.1016/j.ygyno.2014.01.034] [PMID: 24472409]

[98] He L, Zhu W, Chen Q, *et al.* Ovarian cancer cell-secreted exosomal miR-205 promotes metastasis by inducing angiogenesis. Theranostics 2019; 9(26): 8206-20. [http://dx.doi.org/10.7150/thno.37455] [PMID: 31754391]

[99] Ghafouri-Fard S, Shoorei H, Taheri M. miRNA profile in ovarian cancer. Exp Mol Pathol 2020; 113: 104381. [http://dx.doi.org/10.1016/j.yexmp.2020.104381] [PMID: 31954715]

[100] Zhou P, Xiong T, Yao L, Yuan J. MicroRNA-665 promotes the proliferation of ovarian cancer cells by targeting SRCIN1. Exp Ther Med 2020; 19(2): 1112-20. [PMID: 32010277]

[101] Maeda K, Sasaki H, Ueda S, *et al.* Serum exosomal microRNA-34a as a potential biomarker in epithelial ovarian cancer. J Ovarian Res 2020; 13(1): 47. [http://dx.doi.org/10.1186/s13048-020-00648-1] [PMID: 32336272]

[102] Elias KM, Fendler W, Stawiski K, *et al.* Diagnostic potential for a serum miRNA neural network for detection of ovarian cancer. eLife 2017; 6: e28932. [http://dx.doi.org/10.7554/eLife.28932] [PMID: 29087294]

[103] Mahmoud EH, Fawzy A, A Elshimy RA. Serum MicroRNA-21 Negatively Relates to Expression of Programmed Cell Death-4 in Patients with Epithelial Ovarian Cancer. Asian Pac J Cancer Prev 2018; 19(1): 33-8. [PMID: 29373877]

[104] Liang H, Jiang Z, Xie G, Lu Y. Serum microRNA-145 as a novel biomarker in human ovarian cancer. Tumour Biol 2015; 36(7): 5305-13. [http://dx.doi.org/10.1007/s13277-015-3191-y] [PMID: 25722112]

[105] Wang W, Wu L-R, Li C, *et al.* Five serum microRNAs for detection and predicting of ovarian cancer. Eur J Obstet Gynecol Reprod Biol X 2019; 3(100017): 100017. [http://dx.doi.org/10.1016/j.eurox.2019.100017] [PMID: 31404211]

[106] Yoshimura A, Sawada K, Nakamura K, *et al.* Exosomal miR-99a-5p is elevated in sera of ovarian cancer patients and promotes cancer cell invasion by increasing fibronectin and vitronectin expression in neighboring peritoneal mesothelial cells. BMC Cancer 2018; 18(1): 1065. [http://dx.doi.org/10.1186/s12885-018-4974-5] [PMID: 30396333]

[107] Gao Y-C, Wu J. MicroRNA-200c and microRNA-141 as potential diagnostic and prognostic biomarkers for ovarian cancer. Tumour Biol 2015; 36(6): 4843-50. [http://dx.doi.org/10.1007/s13277-015-3138-3] [PMID: 25636451]

[108] Ruan L, Xie Y, Liu F, Chen X. Serum miR-1181 and miR-4314 associated with ovarian cancer: MiRNA microarray data analysis for a pilot study. Eur J Obstet Gynecol Reprod Biol 2018; 222: 31-8. [http://dx.doi.org/10.1016/j.ejogrb.2018.01.006] [PMID: 29353130]

[109] Yang Y, Guo J-X, Shao Z-Q. miR-21 targets and inhibits tumor suppressor gene PTEN to promote prostate cancer cell proliferation and invasion: An experimental study. Asian Pac J Trop Med 2017; 10(1): 87-91. [http://dx.doi.org/10.1016/j.apjtm.2016.09.011] [PMID: 28107872]

[110] Mitchell PS, Parkin RK, Kroh EM, *et al.* Circulating microRNAs as stable blood-based markers for cancer detection. Proc Natl Acad Sci USA 2008; 105(30): 10513-8. [http://dx.doi.org/10.1073/pnas.0804549105] [PMID: 18663219]

[111] Mihelich BL, Maranville JC, Nolley R, Peehl DM, Nonn L. Elevated serum microRNA levels associate with absence of high-grade prostate cancer in a retrospective cohort. PLoS One 2015; 10(4): e0124245. [http://dx.doi.org/10.1371/journal.pone.0124245] [PMID: 25874774]

[112] Chen Z-H, Zhang G-L, Li H-R, *et al.* A panel of five circulating microRNAs as potential biomarkers

for prostate cancer. Prostate 2012; 72(13): 1443-52.
[http://dx.doi.org/10.1002/pros.22495] [PMID: 22298030]

[113] Moltzahn F, Olshen AB, Baehner L, *et al.* Microfluidic-based multiplex qRT-PCR identifies diagnostic and prognostic microRNA signatures in the sera of prostate cancer patients. Cancer Res 2011; 71(2): 550-60.
[http://dx.doi.org/10.1158/0008-5472.CAN-10-1229] [PMID: 21098088]

[114] Samsonov R, Shtam T, Burdakov V, Glotov A, Tsyrlina E, Berstein L, *et al.* Lectin-induced agglutination method of urinary exosomes isolation followed by miRNA analysis: Application for prostate cancer diagnostic: Agglutination of Urinary Exosomes by Lectins. Prostate 2016; 76(1): 68-79.
[http://dx.doi.org/10.1002/pros.23101] [PMID: 26417675]

[115] Zhang L, Huang J, Yang N, *et al.* microRNAs exhibit high frequency genomic alterations in human cancer. Proc Natl Acad Sci USA 2006; 103(24): 9136-41.
[http://dx.doi.org/10.1073/pnas.0508889103] [PMID: 16754881]

[116] Jiang J, Lee EJ, Gusev Y, Schmittgen TD. Real-time expression profiling of microRNA precursors in human cancer cell lines. Nucleic Acids Res 2005; 33(17): 5394-403.
[http://dx.doi.org/10.1093/nar/gki863] [PMID: 16192569]

[117] Barbu MG, Condrat CE, Thompson DC, *et al.* MicroRNA involvement in signaling pathways during viral infection. Front Cell Dev Biol 2020; 8: 143.
[http://dx.doi.org/10.3389/fcell.2020.00143] [PMID: 32211411]

[118] Zhang GL, Li YX, Zheng SQ, Liu M, Li X, Tang H. Suppression of hepatitis B virus replication by microRNA-199a-3p and microRNA-210. Antiviral Res 2010; 88(2): 169-75.
[http://dx.doi.org/10.1016/j.antiviral.2010.08.008] [PMID: 20728471]

[119] Wen Y, Peng SF, Fu L, *et al.* Serum levels of miRNA in patients with hepatitis B virus-associated acute-on-chronic liver failure. Hepatobiliary Pancreat Dis Int 2018; 17(2): 126-32.
[http://dx.doi.org/10.1016/j.hbpd.2018.03.004] [PMID: 29602672]

[120] Pratedrat P, Chuaypen N, Nimsamer P, *et al.* Diagnostic and prognostic roles of circulating miRNA-223-3p in hepatitis B virus-related hepatocellular carcinoma. PLoS One 2020; 15(4): e0232211.
[http://dx.doi.org/10.1371/journal.pone.0232211] [PMID: 32330203]

[121] Loureiro D, Tout I, Narguet S, Benazzouz SM, Mansouri A, Asselah T. MiRNAs as potential biomarkers for viral hepatitis B and C. Viruses 2020; 12(12): 1440.
[http://dx.doi.org/10.3390/v12121440] [PMID: 33327640]

[122] Wu H-C, Huang C-L, Wang H-W, *et al.* Serum miR-21 correlates with the histological stage of chronic hepatitis B-associated liver fibrosis. Int J Clin Exp Pathol 2019; 12(10): 3819-29.
[PMID: 31933770]

[123] Jiang S, Jiang W, Xu Y, Wang X, Mu Y, Liu P. Serum miR-21 and miR-26a levels negatively correlate with severity of cirrhosis in patients with chronic hepatitis B. MicroRNA 2019; 8(1): 86-92.
[http://dx.doi.org/10.2174/2211536607666180821162850] [PMID: 30147020]

[124] Zhou X, Fang S, Wang M, *et al.* Diagnostic value of circulating miRNA-122 for hepatitis B virus and/or hepatitis C virus-associated chronic viral hepatitis. Biosci Rep 2019; 39(9): BSR20190900.
[http://dx.doi.org/10.1042/BSR20190900] [PMID: 31427483]

[125] Xiao B, Liu Z, Li B-S, *et al.* Induction of microRNA-155 during Helicobacter pylori infection and its negative regulatory role in the inflammatory response. J Infect Dis 2009; 200(6): 916-25.
[http://dx.doi.org/10.1086/605443] [PMID: 19650740]

[126] Vigorito E, Kohlhaas S, Lu D, Leyland R. miR-155: an ancient regulator of the immune system. Immunol Rev 2013; 253(1): 146-57.
[http://dx.doi.org/10.1111/imr.12057] [PMID: 23550644]

[127] Vasilescu C, Rossi S, Shimizu M, *et al.* MicroRNA fingerprints identify miR-150 as a plasma

prognostic marker in patients with sepsis. PLoS One 2009; 4(10): e7405.
[http://dx.doi.org/10.1371/journal.pone.0007405] [PMID: 19823581]

[128] World Health Organization Coronavirus (COVID-19) Situtiation Report-114. 13 May 2020. 2020. https://apps.who.int/iris/handle/10665/332089

[129] Mirzaei R, Mahdavi F, Badrzadeh F, *et al.* The emerging role of microRNAs in the severe acute respiratory syndrome coronavirus 2 (SARS-CoV-2) infection. Int Immunopharmacol 2021; 90: 107204.
[http://dx.doi.org/10.1016/j.intimp.2020.107204] [PMID: 33221169]

[130] Cruz-Rodrıguez L, Dilsiz N, Zıaratı P, Lambert BD, Hochwımmer B. Cruz-Rodrıguez L, Dilsiz N, Zıaratı P, Lambert BD, Hochwımmer B. A miRNA-peptıde fusıon as a vaccıne candıdate agaınst the novel coronavırus (covıd-19). Exosomes as potentıal bıomarkers of sars-cov-2 ın lung: after and before vaccınatıon LCR2020B008-13. Journal of Bioscience &p Biomedical Engineering 2020; 1-11.

[131] Abdel-Ghany S, Sabit H. microRNABased Vaccination and Treatment for COVID-19. Curr Trends Vaccine Vaccinol 2020; 3(1): 109.

[132] Wang J, Pei Y, Zhong Y, Jiang S, Shao J, Gong J. Altered serum microRNAs as novel diagnostic biomarkers for atypical coronary artery disease. PLoS One 2014; 9(9): e107012.
[http://dx.doi.org/10.1371/journal.pone.0107012] [PMID: 25198728]

[133] Rotllan N, Fernández-Hernando C. MicroRNA Regulation of Cholesterol Metabolism. Cholesterol 2012; 2012: 847849.
[http://dx.doi.org/10.1155/2012/847849] [PMID: 22919472]

[134] Wang F, Chen C, Wang D. Circulating microRNAs in cardiovascular diseases: from biomarkers to therapeutic targets. Front Med 2014; 8(4): 404-18.
[http://dx.doi.org/10.1007/s11684-014-0379-2] [PMID: 25445171]

[135] Widmer RJ, Chung WY, Herrmann J, Jordan KL, Lerman LO, Lerman A. The association between circulating microRNA levels and coronary endothelial function. PLoS One 2014; 9(10): e109650.
[http://dx.doi.org/10.1371/journal.pone.0109650] [PMID: 25310838]

[136] Freedman JE, Ercan B, Morin KM, *et al.* The distribution of circulating microRNA and their relation to coronary disease. F1000 Res 2012; 1: 50.
[http://dx.doi.org/10.12688/f1000research.1-50.v1] [PMID: 24358814]

[137] Krützfeldt J, Stoffel M. MicroRNAs: a new class of regulatory genes affecting metabolism. Cell Metab 2006; 4(1): 9-12.
[http://dx.doi.org/10.1016/j.cmet.2006.05.009] [PMID: 16814728]

[138] Rottiers V, Näär AM. MicroRNAs in metabolism and metabolic disorders. Nat Rev Mol Cell Biol 2012; 13(4): 239-50.
[http://dx.doi.org/10.1038/nrm3313] [PMID: 22436747]

[139] Güzelgül F, Aksoy K. Bir gen ifade düzenleyicisi miRNA. Arşiv Kaynak Tarama Dergisi 2015; 24(4): 472-93.
[http://dx.doi.org/10.17827/aktd.95263]

[140] Colpaert RMW, Calore M. MicroRNAs in Cardiac Diseases. Cells 2019; 8(7): E737.
[http://dx.doi.org/10.3390/cells8070737] [PMID: 31323768]

[141] Duică F, Condrat CE, Dănila CA, *et al.* MiRNAs: A Powerful Tool in Deciphering Gynecological Malignancies. Front Oncol 2020; 10: 591181.
[http://dx.doi.org/10.3389/fonc.2020.591181] [PMID: 33194751]

[142] Deswal R, Dang AS. Dissecting the role of micro-RNAs as a diagnostic marker for polycystic ovary syndrome: a systematic review and meta-analysis. Fertil Steril 2020; 113(3): 661-669.e2.
[http://dx.doi.org/10.1016/j.fertnstert.2019.11.001] [PMID: 32192599]

[143] Lewohl JM, Nunez YO, Dodd PR, Tiwari GR, Harris RA, Mayfield RD. Up-regulation of microRNAs

in brain of human alcoholics. Alcohol Clin Exp Res 2011; 35(11): 1928-37.
[http://dx.doi.org/10.1111/j.1530-0277.2011.01544.x] [PMID: 21651580]

[144] Giorgi Silveira R, Perelló Ferrúa C, do Amaral CC, Fernandez Garcia T, de Souza KB, Nedel F. MicroRNAs expressed in neuronal differentiation and their associated pathways: Systematic review and bioinformatics analysis. Brain Res Bull 2020; 157: 140-8.
[http://dx.doi.org/10.1016/j.brainresbull.2020.01.009] [PMID: 31945407]

[145] Zhu B, Ju S, Chu H, *et al.* The potential function of microRNAs as biomarkers and therapeutic targets in multiple myeloma. Oncol Lett 2018; 15(5): 6094-106.
[http://dx.doi.org/10.3892/ol.2018.8157] [PMID: 29731841]

[146] Jin F, Hu H, Xu M, *et al.* Serum microRNA Profiles Serve as Novel Biomarkers for Autoimmune Diseases. Front Immunol 2018; 9: 2381.
[http://dx.doi.org/10.3389/fimmu.2018.02381] [PMID: 30459760]

[147] Assmann TS, Recamonde-Mendoza M, De Souza BM, Crispim D. MicroRNA expression profiles and type 1 diabetes mellitus: systematic review and bioinformatic analysis. Endocr Connect 2017; 6(8): 773-90.
[http://dx.doi.org/10.1530/EC-17-0248] [PMID: 28986402]

[148] Xu G, Thielen LA, Chen J, *et al.* Serum miR-204 is an early biomarker of type 1 diabetes-associated pancreatic beta-cell loss. Am J Physiol Endocrinol Metab 2019; 317(4): E723-30.
[http://dx.doi.org/10.1152/ajpendo.00122.2019] [PMID: 31408375]

[149] Barutta F, Bruno G, Matullo G, *et al.* MicroRNA-126 and micro-/macrovascular complications of type 1 diabetes in the EURODIAB Prospective Complications Study. Acta Diabetol 2017; 54(2): 133-9.
[http://dx.doi.org/10.1007/s00592-016-0915-4] [PMID: 27696070]

[150] Santos AS, Cunha Neto E, Fukui RT, Ferreira LRP, Silva MER. Increased Expression of Circulating microRNA 101-3p in Type 1 Diabetes Patients: New Insights Into miRNA-Regulated Pathophysiological Pathways for Type 1 Diabetes. Front Immunol 2019; 10: 1637.
[http://dx.doi.org/10.3389/fimmu.2019.01637] [PMID: 31396209]

[151] Ibrahim AA, Wahby AA, Ashmawy I, Saleh RM, Soliman H. Association of Exosomal miR-34a with Markers of Dyslipidemia and Endothelial Dysfunction in Children and Adolescents with T1DM. J Clin Res Pediatr Endocrinol 2020; 12(4): 401-9.
[http://dx.doi.org/10.4274/jcrpe.galenos.2020.2020.0134] [PMID: 32654473]

[152] Margaritis K, Margioula-Siarkou G, Margioula-Siarkou C, *et al.* Circulating serum and plasma levels of micro-RNA in type-1 diabetes in children and adolescents: A systematic review and meta-analysis. Eur J Clin Invest 2021; 51(7): e13510.
[http://dx.doi.org/10.1111/eci.13510] [PMID: 33565089]

[153] Taha M, Shaker OG, Abdelsalam E, Taha N. Serum a proliferation-inducing ligand and MicroRNA-223 are associated with rheumatoid arthritis: diagnostic and prognostic implications. Mol Med 2020; 26(1): 92.
[http://dx.doi.org/10.1186/s10020-020-00199-7] [PMID: 32998682]

[154] Zhang S, Meng T, Tang C, *et al.* MicroRNA-340-5p suppressed rheumatoid arthritis synovial fibroblast proliferation and induces apoptotic cell number by targeting signal transducers and activators of transcription 3. Autoimmunity 2020; 53(6): 314-22.
[http://dx.doi.org/10.1080/08916934.2020.1793134] [PMID: 32706318]

[155] Stypinska B, Wajda A, Walczuk E, *et al.* The Serum Cell-Free microRNA Expression Profile in MCTD, SLE, SSc, and RA Patients. J Clin Med 2020; 9(1): E161.
[http://dx.doi.org/10.3390/jcm9010161] [PMID: 31936082]

[156] Li W, Liu S, Chen Y, *et al.* Circulating Exosomal microRNAs as Biomarkers of Systemic Lupus Erythematosus. Clinics (São Paulo) 2020; 75: e1528.
[http://dx.doi.org/10.6061/clinics/2020/e1528] [PMID: 32876110]

[157] Fan Y, Ji Y, Wang X, *et al.* Relationship of miRNA-146a to systemic lupus erythematosus: A PRISMA-compliant meta-analysis. Medicine (Baltimore) 2020; 99(40): e22444.
[http://dx.doi.org/10.1097/MD.0000000000022444] [PMID: 33019429]

[158] Kakan SS, Janga SR, Cooperman B, *et al.* Small RNA Deep Sequencing Identifies a Unique miRNA Signature Released in Serum Exosomes in a Mouse Model of Sjögren's Syndrome. Front Immunol 2020; 11: 1475.
[http://dx.doi.org/10.3389/fimmu.2020.01475] [PMID: 32849505]

[159] Viennois E, Zhao Y, Han MK, *et al.* Serum miRNA signature diagnoses and discriminates murine colitis subtypes and predicts ulcerative colitis in humans. Sci Rep 2017; 7(1): 2520.
[http://dx.doi.org/10.1038/s41598-017-02782-1] [PMID: 28566745]

[160] Feng J, Zhu Y, Chen L, Wang M. Clinical Significance of microRNA-146a in Patients with Ulcerative Colitis. Ann Clin Lab Sci 2020; 50(4): 463-7.
[PMID: 32826242]

[161] Li X, Sun L, Chen L, Xu Y, Kong X. Upregulation of microRNA-219-5p relieves ulcerative colitis through balancing the differentiation of Treg/Th17 cells. Eur J Gastroenterol Hepatol 2020; 32(7): 813-20.
[http://dx.doi.org/10.1097/MEG.0000000000001712] [PMID: 32175983]

[162] Batra SK, Heier CR, Diaz-Calderon L, *et al.* Serum miRNAs Are Pharmacodynamic Biomarkers Associated With Therapeutic Response in Pediatric Inflammatory Bowel Disease. Inflamm Bowel Dis 2020; 26(10): 1597-606.
[http://dx.doi.org/10.1093/ibd/izaa209] [PMID: 32793975]

[163] Karabacak M, Erturan İ, Hekİmler ÖztÜrk K, Ayvaz HH, Korkmaz S, Yildirim M. Is microRNA 1910-3p (miR-1910-3p) a really distinctive marker for psoriasis? Turk J Med Sci 2021; 51(3): 1098-105.
[PMID: 33356031]

[164] El-Komy M, Amin I, El-Hawary MS, Saadi D, Shaker O. Upregulation of the miRNA-155, miRNA-210, and miRNA-20b in psoriasis patients and their relation to IL-17. Int J Immunopathol Pharmacol 2020; 34: 2058738420933742.
[http://dx.doi.org/10.1177/2058738420933742] [PMID: 32602388]

[165] Sonkoly E, Wei T, Janson PC, *et al.* MicroRNAs: novel regulators involved in the pathogenesis of psoriasis? PLoS One 2007; 2(7): e610.
[http://dx.doi.org/10.1371/journal.pone.0000610] [PMID: 17622355]

[166] Xu Y, Ji Y, Lan X, Gao X, Chen HD, Geng L. miR□203 contributes to IL□17□induced VEGF secretion by targeting SOCS3 in keratinocytes. Mol Med Rep 2017; 16(6): 8989-96.
[http://dx.doi.org/10.3892/mmr.2017.7759] [PMID: 29039484]

[167] Solé C, Domingo S, Ferrer B, Moliné T, Ordi-Ros J, Cortés-Hernández J. MicroRNA Expression Profiling Identifies miR-31 and miR-485-3p as Regulators in the Pathogenesis of Discoid Cutaneous Lupus. J Invest Dermatol 2019; 01;139(1): 51-61.

[168] Ichihara A, Jinnin M, Yamane K, *et al.* microRNA-mediated keratinocyte hyperproliferation in psoriasis vulgaris. Br J Dermatol 2011; 165(5): 1003-10.
[http://dx.doi.org/10.1111/j.1365-2133.2011.10497.x] [PMID: 21711342]

[169] Chen XF, Zhang LJ, Zhang J, *et al.* MiR-151a is involved in the pathogenesis of atopic dermatitis by regulating interleukin-12 receptor β2. Exp Dermatol 2018; 27(4): 427-32.
[http://dx.doi.org/10.1111/exd.13276] [PMID: 27992076]

[170] Garzon R, Liu S, Fabbri M, *et al.* MicroRNA-29b induces global DNA hypomethylation and tumor suppressor gene reexpression in acute myeloid leukemia by targeting directly DNMT3A and 3B and indirectly DNMT1. Blood 2009; 113(25): 6411-8.
[http://dx.doi.org/10.1182/blood-2008-07-170589] [PMID: 19211935]

[171] Otaegui D, Baranzini SE, Armañanzas R, *et al.* Differential micro RNA expression in PBMC from multiple sclerosis patients. PLoS One 2009; 4(7): e6309. [http://dx.doi.org/10.1371/journal.pone.0006309] [PMID: 19617918]

[172] Du C, Liu C, Kang J, *et al.* MicroRNA miR-326 regulates TH-17 differentiation and is associated with the pathogenesis of multiple sclerosis. Nat Immunol 2009; 10(12): 1252-9. [http://dx.doi.org/10.1038/ni.1798] [PMID: 19838199]

[173] Reed ER, Latourelle JC, Bockholt JH, *et al.* MicroRNAs in CSF as prodromal biomarkers for Huntington disease in the PREDICT-HD study. Neurology 2018; 90(4): e264-72. [http://dx.doi.org/10.1212/WNL.0000000000004844] [PMID: 29282329]

[174] Liu T, Im W, Mook-Jung I, Kim M. MicroRNA-124 slows down the progression of Huntington's disease by promoting neurogenesis in the striatum. Neural Regen Res 2015; 10(5): 786-91. [http://dx.doi.org/10.4103/1673-5374.156978] [PMID: 26109954]

[175] Catapano F, Zaharieva I, Scoto M, *et al.* Altered Levels of MicroRNA-9, -206, and -132 in Spinal Muscular Atrophy and Their Response to Antisense Oligonucleotide Therapy. Mol Ther Nucleic Acids 2016; 5(7): e331. [http://dx.doi.org/10.1038/mtna.2016.47] [PMID: 27377135]

[176] Bonanno S, Marcuzzo S, Malacarne C, *et al.* Circulating MyomiRs as Potential Biomarkers to Monitor Response to Nusinersen in Pediatric SMA Patients. Biomedicines 2020; 8(2): E21. [http://dx.doi.org/10.3390/biomedicines8020021] [PMID: 31991852]

[177] Kumar S, Reddy PH. Are circulating microRNAs peripheral biomarkers for Alzheimer's disease? Biochim Biophys Acta 2016; 1862(9): 1617-27. [http://dx.doi.org/10.1016/j.bbadis.2016.06.001] [PMID: 27264337]

[178] Kumar S, Vijayan M, Reddy PH. MicroRNA-455-3p as a potential peripheral biomarker for Alzheimer's disease. Hum Mol Genet 2017; 26(19): 3808-22. [http://dx.doi.org/10.1093/hmg/ddx267] [PMID: 28934394]

[179] Zeng Q, Zou L, Qian L, *et al.* Expression of microRNA□222 in serum of patients with Alzheimer's disease. Mol Med Rep 2017; 16(4): 5575-9. [http://dx.doi.org/10.3892/mmr.2017.7301] [PMID: 28849039]

[180] Cheng L, Doecke JD, Sharples RA, *et al.* Prognostic serum miRNA biomarkers associated with Alzheimer's disease shows concordance with neuropsychological and neuroimaging assessment. Mol Psychiatry 2015; 20(10): 1188-96. [http://dx.doi.org/10.1038/mp.2014.127] [PMID: 25349172]

[181] Zhao Y, Zhang Y, Zhang L, Dong Y, Ji H, Shen L. The Potential Markers of Circulating microRNAs and long non-coding RNAs in Alzheimer's Disease. Aging Dis 2019; 10(6): 1293-301. [http://dx.doi.org/10.14336/AD.2018.1105] [PMID: 31788340]

[182] Han C, Guo L, Yang Y, *et al.* Mechanism of microRNA-22 in regulating neuroinflammation in Alzheimer's disease. Brain Behav 2020; 10(6): e01627. [http://dx.doi.org/10.1002/brb3.1627] [PMID: 32307887]

[183] Madadi S, Saidijam M, Yavari B, Soleimani M. Downregulation of serum miR-106b: a potential biomarker for Alzheimer disease. Arch Physiol Biochem 2020; 1-5. [http://dx.doi.org/10.1080/13813455.2020.1734842] [PMID: 32141790]

[184] Perkins DO, Jeffries CD, Jarskog LF, *et al.* microRNA expression in the prefrontal cortex of individuals with schizophrenia and schizoaffective disorder. Genome Biol 2007; 8(2): R27. [http://dx.doi.org/10.1186/gb-2007-8-2-r27] [PMID: 17326821]

[185] Shi W, Du J, Qi Y, *et al.* Aberrant expression of serum miRNAs in schizophrenia. J Psychiatr Res 2012; 46(2): 198-204. [http://dx.doi.org/10.1016/j.jpsychires.2011.09.010] [PMID: 22094284]

[186] Beveridge NJ, Cairns MJ. MicroRNA dysregulation in schizophrenia. Neurobiol Dis 2012; 46(2): 263-71.
[http://dx.doi.org/10.1016/j.nbd.2011.12.029] [PMID: 22207190]

[187] Lett TA, Chakravarty MM, Felsky D, *et al.* The genome-wide supported microRNA-137 variant predicts phenotypic heterogeneity within schizophrenia. Mol Psychiatry 2013; 18(4): 443-50.
[http://dx.doi.org/10.1038/mp.2013.17] [PMID: 23459466]

[188] Ibrahim RR, Amer RA, Abozeid AA, Elsharaby RM, Shafik NM. Micro RNA 146a gene variant / TNF-α / IL-6 / IL-1 β; A cross-link axis inbetween oxidative stress, endothelial dysfunction and neuro-inflammation in acute ischemic stroke and chronic schizophrenic patients. Arch Biochem Biophys 2020; 679: 108193.
[http://dx.doi.org/10.1016/j.abb.2019.108193] [PMID: 31738890]

[189] Rong H, Liu TB, Yang KJ, *et al.* MicroRNA-134 plasma levels before and after treatment for bipolar mania. J Psychiatr Res 2011; 45(1): 92-5.
[http://dx.doi.org/10.1016/j.jpsychires.2010.04.028] [PMID: 20546789]

[190] Lee SY, Lu RB, Wang LJ, *et al.* Serum miRNA as a possible biomarker in the diagnosis of bipolar II disorder. Sci Rep 2020; 10(1): 1131.
[http://dx.doi.org/10.1038/s41598-020-58195-0] [PMID: 31980721]

[191] Ma Y. The Challenge of microRNA as a Biomarker of Epilepsy. Curr Neuropharmacol 2018; 16(1): 37-42.
[PMID: 28676013]

[192] Wang X, Sun Y, Tan Z, *et al.* Serum MicroRNA-4521 is a Potential Biomarker for Focal Cortical Dysplasia with Refractory Epilepsy. Neurochem Res 2016; 41(4): 905-12.
[http://dx.doi.org/10.1007/s11064-015-1773-0] [PMID: 26645999]

[193] An N, Zhao W, Liu Y, Yang X, Chen P. Elevated serum miR-106b and miR-146a in patients with focal and generalized epilepsy. Epilepsy Res 2016; 127: 311-6.
[http://dx.doi.org/10.1016/j.eplepsyres.2016.09.019] [PMID: 27694013]

[194] Wang L, Song L, Chen X, *et al.* microRNA-139-5p confers sensitivity to antiepileptic drugs in refractory epilepsy by inhibition of MRP1. CNS Neurosci Ther 2020; 26(4): 465-74.
[http://dx.doi.org/10.1111/cns.13268] [PMID: 31750616]

[195] Li N, Pan J, Liu W, Li Y, Li F, Liu M. MicroRNA-15a-5p serves as a potential biomarker and regulates the viability and apoptosis of hippocampus neuron in children with temporal lobe epilepsy. Diagn Pathol 2020; 15(1): 46.
[http://dx.doi.org/10.1186/s13000-020-00944-w] [PMID: 32384924]

[196] Leontariti M, Avgeris M, Katsarou MS, *et al.* Circulating miR-146a and miR-134 in predicting drug-resistant epilepsy in patients with focal impaired awareness seizures. Epilepsia 2020; 61(5): 959-70.
[http://dx.doi.org/10.1111/epi.16502] [PMID: 32314378]

[197] Martins-Ferreira R, Chaves J, Carvalho C, *et al.* Circulating microRNAs as potential biomarkers for genetic generalized epilepsies: a three microRNA panel. Eur J Neurol 2020; 27(4): 660-6.
[http://dx.doi.org/10.1111/ene.14129] [PMID: 31746515]

[198] Ioriatti ES, Cirino MLA, Lizarte Neto FS, *et al.* Expression of circulating microRNAs as predictors of diagnosis and surgical outcome in patients with mesial temporal lobe epilepsy with hippocampal sclerosis. Epilepsy Res 2020; 166: 106373.
[http://dx.doi.org/10.1016/j.eplepsyres.2020.106373] [PMID: 32682279]

[199] Zhao Y, Lu C, Wang H, *et al.* Identification of hsa-miR-1275 as a Novel Biomarker Targeting *MECP2* for Human Epilepsy of Unknown Etiology. Mol Ther Methods Clin Dev 2020; 19: 398-410.
[http://dx.doi.org/10.1016/j.omtm.2020.10.005] [PMID: 33251277]

[200] Zhang X, Li X, Li B, Sun C, Zhang P. miR-21-5p protects hippocampal neurons of epileptic rats via inhibiting STAT3 expression. Adv Clin Exp Med 2020; 29(7): 793-801.

[http://dx.doi.org/10.17219/acem/121929] [PMID: 32745381]

[201] Pacifici M, Delbue S, Ferrante P, *et al.* Cerebrospinal fluid miRNA profile in HIV-encephalitis. J Cell Physiol 2013; 228(5): 1070-5.
[http://dx.doi.org/10.1002/jcp.24254] [PMID: 23042033]

[202] Sheinerman K, Djukic A, Tsivinsky VG, Umansky SR. Brain-enriched microRNAs circulating in plasma as novel biomarkers for Rett syndrome. PLoS One 2019; 14(7): e0218623.
[http://dx.doi.org/10.1371/journal.pone.0218623] [PMID: 31291284]

[203] Sheinerman KS, Umansky SR. Early detection of neurodegenerative diseases: circulating brain-enriched microRNA. Cell Cycle 2013; 12(1): 1-2.
[http://dx.doi.org/10.4161/cc.23067] [PMID: 23255116]

[204] Pejhan S, Del Bigio MR, Rastegar M. The MeCP2E1/E2-BDNF-*miR132* Homeostasis Regulatory Network Is Region-Dependent in the Human Brain and Is Impaired in Rett Syndrome Patients. Front Cell Dev Biol 2020; 8: 763.
[http://dx.doi.org/10.3389/fcell.2020.00763] [PMID: 32974336]

[205] Søndergaard HB, Hesse D, Krakauer M, Sørensen PS, Sellebjerg F. Differential microRNA expression in blood in multiple sclerosis. Mult Scler 2013; 19(14): 1849-57.
[http://dx.doi.org/10.1177/1352458513490542] [PMID: 23773985]

[206] Gandhi R, Healy B, Gholipour T, *et al.* Circulating microRNAs as biomarkers for disease staging in multiple sclerosis. Ann Neurol 2013; 73(6): 729-40.
[http://dx.doi.org/10.1002/ana.23880] [PMID: 23494648]

[207] Fenoglio C, Ridolfi E, Cantoni C, *et al.* Decreased circulating miRNA levels in patients with primary progressive multiple sclerosis. Mult Scler 2013; 19(14): 1938-42.
[http://dx.doi.org/10.1177/1352458513485654] [PMID: 24277735]

[208] Haghikia A, Haghikia A, Hellwig K, *et al.* Regulated microRNAs in the CSF of patients with multiple sclerosis: a case-control study. Neurology 2012; 79(22): 2166-70.
[http://dx.doi.org/10.1212/WNL.0b013e3182759621] [PMID: 23077021]

[209] Zanoni M, Orlandi E, Rossetti G, *et al.* Upregulated serum miR-128-3p in progressive and relapse-free multiple sclerosis patients. Acta Neurol Scand 2020; 142(5): 511-6.
[http://dx.doi.org/10.1111/ane.13288] [PMID: 32432792]

[210] Shademan B, Nourazarian A, Nikanfar M, Biray Avci C, Hasanpour M, Isazadeh A. Investigation of the miRNA146a and miRNA155 gene expression levels in patients with multiple sclerosis. J Clin Neurosci 2020; 78: 189-93.
[http://dx.doi.org/10.1016/j.jocn.2020.04.071] [PMID: 32331943]

[211] Ibrahim SH, El-Mehdawy KM, Seleem M, El-Sawalhi MM, Shaheen AA. Serum ROCK2, miR-300 and miR-450b-5p levels in two different clinical phenotypes of multiple sclerosis: Relation to patient disability and disease progression. J Neuroimmunol 2020; 347: 577356.
[http://dx.doi.org/10.1016/j.jneuroim.2020.577356] [PMID: 32781341]

[212] Ozdilek B, Demircan B. Serum microRNA expression levels in Turkish patients with Parkinson's disease. Int J Neurosci 2021; 131(12): 1181-9.
[http://dx.doi.org/10.1080/00207454.2020.1784165] [PMID: 32546033]

[213] Han L, Tang Y, Bai X, *et al.* Association of the serum microRNA-29 family with cognitive impairment in Parkinson's disease. Aging (Albany NY) 2020; 12(13): 13518-28.
[http://dx.doi.org/10.18632/aging.103458] [PMID: 32649312]

[214] Barbagallo C, Mostile G, Baglieri G, *et al.* Specific Signatures of Serum miRNAs as Potential Biomarkers to Discriminate Clinically Similar Neurodegenerative and Vascular-Related Diseases. Cell Mol Neurobiol 2020; 40(4): 531-46.
[http://dx.doi.org/10.1007/s10571-019-00751-y] [PMID: 31691877]

[215] Li H, Yu L, Li M, *et al.* MicroRNA-150 serves as a diagnostic biomarker and is involved in the

inflammatory pathogenesis of Parkinson's disease. Mol Genet Genomic Med 2020; 8(4): e1189. [http://dx.doi.org/10.1002/mgg3.1189] [PMID: 32077254]

[216] Yan JH, Hua P, Chen Y, *et al.* Identification of microRNAs for the early diagnosis of Parkinson's disease and multiple system atrophy. J Integr Neurosci 2020; 19(3): 429-36. [http://dx.doi.org/10.31083/j.jin.2020.03.163] [PMID: 33070521]

CHAPTER 14

Advancements in Gastrointestinal System Biomarkers

Suleyman Demiryas[1,*] and **Anıl Orhan**[2]

[1] *Department of General Surgery, Cerrahpaşa School of Medicine, IIstanbul University – Cerrahpaşa, Istanbul, Turkey*

[2] *Department of General Surgery, Haseki Training and Research Hospital, University of Health Sciences, Istanbul, Turkey*

Abstract: The requirement for diagnostic surgical operations of gastrointestinal system diseases significantly decreases with the help of proper diagnostic tools. These modalities are also beneficial for identifying postoperative complications, allowing us to diagnose them in earlier stages, and increasing the postoperative survival rates. Biomarkers are considered an integral part of diagnostic examinations. Developments of biomarkers used for diagnosing and treating abdominal diseases are essential for improving our capabilities in non-invasive monitoring. In this chapter, we review both the novel and the routine biomarkers in the diagnosis and follow-up of gastrointestinal system diseases.

Keywords: Benign Gastrointestinal System Diseases, Biomarker, Diagnostic biomarkers, Malignant Gastrointestinal System Diseases, Surgery.

INTRODUCTION

In the last century, great advancements have been made in the preoperative preparation, intraoperative technique, postoperative follow-up, and treatment stages in surgery. While these advancements facilitate our routine surgical procedures, they also increase disease-free survival rates after surgery. Especially in the last thirty years, surgery has enormously evolved in an attempt to ensure minimally invasive operations.

With the advancements in surgery, the routine practices we use in the screening and diagnosis of diseases have also advanced. These advances reached their pinnacle in biochemical and radiological examinations. Thus, differential diagnosis of acute abdominal diseases can be performed much easily, and the

* **Corresponding author Suleyman Demiryas**: Department of General Surgery, Cerrahpaşa School of Medicine, IIstanbul University – Cerrahpaşa, Istanbul, Turkey; Tel: +90 (542) 366 72 32; E-mail: suleyman.demiryas@iuc.edu.tr

Pınar Atukeren and Hafize Uzun (Eds.)

need for diagnostic laparotomy has decreased significantly. Brilliant advances in endoscopic methods have paved the way for minimally invasive interventions, especially in the screening, diagnosis, and treatment of gastrointestinal diseases.

These advancements allowed surgeons to perform cost-effective, easily accessible procedures which cause the least trauma for the patient. Although previous studies have focused on the structure and functions of these biomarkers, recent studies have also focused on their clinical usage. Therefore, the clinical usage of biochemical markers as diagnostic tests and patient follow-ups has gradually increased. This chapter focuses on biomarkers that we are now using or incorporating into our clinical practice, which do not require excessive surgical procedures or post-operative pathological evaluation.

Diagnostic Biomarkers of Upper Gastrointestinal System Diseases

Upper gastrointestinal system diseases are one of the most common causes of global hospital admissions. More than 10 million applications for upper gastrointestinal system complaints were made in the United States in 2018 [1]. According to 2019 clinical data, out of 1,806,590 cancer cases in the United States, 333,680 were diagnosed in the gastrointestinal system and approximately 46,000 in the upper gastrointestinal tract (stomach and esophagus) [2]. Many methods have been developed for the diagnosis and treatment of these disorders, and undoubtedly endoscopic interventions and radiological examinations account for a large percentage. According to the clinical data of 2013, 6.1 million upper gastrointestinal system endoscopies were performed in the United States [3]. However, the fact that endoscopic interventions are considered invasive, which may also require anesthesia for the procedure, and their occasional inappropriate usage, researchers are looking for easier methods for diagnosing these disorders.

Diagnostic Biomarkers of Benign Upper Gastrointestinal System Diseases

Pepsinogen I & II

Pepsinogen (PG) is the zymogenic form of pepsin and it is classified as an endoproteinase with two different isoforms. While Pepsinogen I is released from the chief cells and neck cells in the stomach fundus and corpus, Pepsinogen II is secreted from Brunner glands in cardia, antrum, pylorus, and duodenum [1, 4]. It turns into its activated form pepsin in acidic conditions and takes part in the hydrolysis of nutritional proteins. Pepsinogens I & II are useful markers that can reveal gastritis. Especially in *Helicobacter pylori*-associated gastritis, both PG values elevate in serum, predominantly PG II [5, 6]. A rapid decrease in PG II after the treatment of H.pylori infection makes serum PGs useful in the control after eradication treatment [5]. Furthermore, PG tests can be used as a confirmation test for other non-invasive tests, as a high false positivity rate can be seen in other non-invasive tests such as urea breath test and *H. pylori* stool antigen test, especially in acute cases with a low prevalence of *H. pylori* [6].

PG values can also be used to diagnose atrophic gastritis. In chronic atrophic gastritis cases in which the corpus is affected, a decrease in PG values is observed. It has been reported that serum PG I - PG II values and PG I / II ratio may be useful for the early diagnosis of gastric atrophy [7, 8].

The use of serum PGs in gastric cancer screening is also increasing day by day. Recent studies have shown that there is a possible link between low serum PG I value and the risk of stomach cancer due to atrophic gastritis [9, 10]. In the study conducted in asymptomatic patient groups, higher rates of gastric cancer and intestinal-type intramucosal cancer were observed in upper GIS endoscopy in patients with low PG values (PG test positive) compared to the patients with normal PG values (PG test negative) [11]. Thus, it is thought that serum PG values may be a useful biomarker in terms of screening for gastric cancer in the patient population with high risk.

Gastrin 17

Gastrin 17 (G-17) is a peptide hormone consisting of 17 amino acids produced by antral G cells, and it provides the control of gastric acid secretion with the help of a negative feedback mechanism. G-17 release is directly related to intragastric acidity, so it is a significant indicator of gastric acidity [12].

Gastroesophageal reflux (GERD) and reflux-related diseases are conditions in which gastric acidity increases significantly. Thus, G-17 values are found to be decreased in GERD and reflux-related diseases. A study in which patients with retrosternal burning complaints were evaluated showed a correlation between different reflux patterns and G - 17 values, and no additional invasive procedures were required [13]. Barrett's esophagitis (BE) is another reflux-related disorder that occurs due to high gastric acidity. BE patients have lower G - 17 levels compared to normal patients, suggesting that it may help the diagnosis for early detection of esophageal cancer [12].

Another condition which both gastric acidity and G - 17 affected is atrophic gastritis. The stomach can be evaluated in two different segments from the corpus-antrum junction; the proximal segment mainly produces exocrine secretion, and the distal segment is widely responsible for the endocrine secretion

[14]. Consequently, differences in G - 17 values may occur depending on the part of the stomach that is atrophied. In atrophic gastritis where the corpus is affected, G - 17 levels are expected to increase due to the decrease in gastric acidity; and in atrophic gastritis, where the antrum or the entire stomach is affected, a decrease in G - 17 is expected due to the loss of antral G cells. Although there is controversy in the specificity and sensitivity of the relationship between G - 17 values and gastric atrophy, the general consensus in current literature is that in cases of atrophy where the corpus and antrum are affected, a decrease in all PG I, PG II and G - 17 values are observed [8, 15].

Helicobacter pylori IgA - IgG and Stool Antigen Tests

Helicobacter pylori is a gram-negative pathogenic bacterium detected in patients with gastric complaints, and it was first described back in the early 1980s [16]. Apart from gastritis and ulcers, it may cause mucosa-associated lymphoid tissue (MALT) lymphoma and gastric cancer. *H. pylori* generates a significant immune response, and immunomodulators resulting from this immunity are useful for evaluating *H. pylori* infection. Some of the antibodies, such as serum immunoglobulin IgG and IgA, which are the main products of the humoral immune response, are known as highly accurate and easily applicable non-invasive biomarkers in terms of showing the status of *H. pylori* infection. There is a significant relationship between the severity of the disease and antibody levels in patients with chronic gastritis due to *H. pylori* [17].

In addition to serum antibodies, a stool antigen test is another diagnostic method that can reveal the presence of *H. pylori*. Enzyme immunoassays using monoclonal or polyclonal antibodies and immunochromatography are used as stool antigen tests in current clinical utilization. Tests utilizing monoclonal antibodies have a higher specificity ratio compared to the polyclonal ones. Although these tests give reliable results for the diagnosis of pylori, false negativity might be seen in the post-eradication control [18]. Moreover, stool antigen tests are more widely practiced in the pediatric patient group compared to adults, due to their lower exposure to *H. pylori* compared to the adult patient group [19].

Urea Breath Test – Rapid Test

Urease is a metalloenzyme that hydrolyzes urea and breaks it down into ammonia and carbon dioxide. It uses nickel ions as a co-factor. Urease enzyme can be found in many bacteria, and among these bacteria, *Helicobacter pylori* is by far the most common pathogen seen in our clinical routine. The urease activity of *H. pylori* enables the bacteria to maintain its viability by neutralizing the acidic environment in the stomach submucosa. Bacteria colonizing in the stomach triggers immune reactions which take place in the submucosal region and induce an inflammatory response. Some of the diagnostic tests performed to detect the presence of *H. pylori* are based on its urease activity. Among these tests, the urea breath test and rapid test are the most widely used for clinical purposes.

The urea breath test is a useful non-invasive test for high risk comorbid patients or patients who cannot tolerate the procedure to undergo upper gastrointestinal endoscopy [20]. Carbon dioxide levels are determined with a measurement after the patient consumes radioactive urea labeled C_{13} or C_{14}. C_{13} can be detected by spectrometry, and C_{14} can be detected by scintillation. The test has high sensitivity and specificity rates, making it a useful alternative for the diagnosis and post-eradication control of *H. pylori* [21]. There are breath tests based on similar principles to the urea breath test and can be used to diagnose different conditions (*e.g.*, intestinal bacterial overgrowth, gastric emptying, carbohydrate absorption disorders) [21].

Rapid Urease Test is performed by inoculating gastric biopsies taken during upper gastrointestinal system endoscopy into test kits previously prepared and impregnated with phenol indicator. Ammonia, which appears in the presence of urease activity, causes a change in the test kit with the increase of pH values (The test kit, which is yellow at pH 6.8, turns red when pH 8.4) [22]. Thanks to the rapid urease test, *H. pylori* treatment can be initiated in patients without waiting for histopathological evaluation results; however, it has handicaps such as invasive intervention, the size and depth of the biopsy and the variability of the test result. The number of biopsies taken may increase the sensitivity of the test, but in this case, the possibility of prolonging the duration of the procedure should also need to be considered [22].

Citrulline

Citrulline is a nonproteinic amino acid, which is an intermediate product of the urea cycle. Citrulline is synthesized in kidneys, liver, and enterocytes, which plays a role in nitric oxide formation [23]. Although it can be assessed even in low sample volumes, it has no diagnostic value in patients with renal dysfunction (creatinine clearance value is less than 50 ml/min), which have increased blood citrulline concentrations [24, 25].

Lower plasma citrulline levels are observed in cases where the integrity of the gastrointestinal mucosa is impaired (*e.g.*, enterocolitis due to radiotherapy, mucosal damage due to non-steroid inflammatory drugs) or when the concentration of healthy intestinal tissue is getting reduced (*e.g.*, celiac disease). Thus, plasma citrulline levels are considered to be a useful indicator in terms of determining the intestinal absorption capacity and enterocyte mass [23, 25].

Plasma citrulline levels are usually assessed by commercial enzymatic immunoassay kits, but its sensitivity is relatively low. However, the use of highly sensitive and specific analytical methods such as high-performance liquid chromatographs and tandem mass spectrophotometric indicate that citrulline might be a potential biomarker [25].

Biomarkers for Malignant Upper Gastrointestinal System Diseases

Carbohydrate Antigens (CA 19-9 & CA 72-4)

CA 19-9 is a high molecular weight (200-1000 kDa) glycoprotein that was first identified in the late 1970s. It is synthesized in pancreatic cells, biliary ductal cells, gastric, colonic, endometrial, and salivary gland epithelium. Although, its serum levels might increase in some benign conditions (*e.g.*, acute and chronic pancreatitis, liver cirrhosis, cholangitis and mechanical icterus), CA 19-9 levels predominantly increase in the stomach, colorectal, hepatocellular, biliary tract and especially in pancreatic neoplasms [26, 27].

The usage of CA 19-9 in gastric cancer depends heavily on the American Joint Committee on Cancer Tumor-Node-Metastasis (TNM) [28] status of the tumor. Studies have shown that its use is fairly limited in the early stage of gastric cancer due to its sensitivity and specificity [29 - 31]. On the other hand, it is widely accepted that CA 19-9 is more useful in the prognostic evaluation of advanced gastric cancer, metastatic diseases, and recurrent diseases [32, 33]. Therefore, CA 19-9 is more helpful in evaluating the metastatic status and recurrence after surgical resection or chemotherapy in advanced stage patients rather than gastric cancer screening [34].

Another glycoprotein antigen is CA 72-4 which is not physiologically synthesized in non-malignant tissues, and its levels may increase in gastrointestinal or gynecological epithelial malignancies. CA 72-4 levels may be detected higher like CA 19-9 in some benign conditions such as pancreatitis, biliary cirrhosis, pneumonia, rheumatological diseases, and ovarian cysts (34). CA 72-4 has a higher sensitivity for diagnosing and following advanced, metastatic, and recurrent disease of gastric cancer compared to CA 19-9 and carcinoembryonic antigens. The prognosis is supposed to be worse for the patients with positive serum CA 72-4, and it is generally correlated with lymphatic involvement and advanced-stage disease. On the other hand, it has been shown that the decrease of

CA 72-4 values in post-operative period is associated with curative surgery and longer term survival rate [34, 35].

Circular RNA (circRNA) - Use in Gastric Cancer

Non-coding RNA fragments (ncRNA) are the functional RNA fragments that do not encode any protein and display transcriptional and post-transcriptional gene expressions. Circular RNA (circRNA), one of the ncRNAs, are RNA fragments that have a more stable structure compared to linear RNAs and can be found in significant amounts in the cell. When cell death occurs, ncRNAs pass from cytosol to the surrounding tissue and remain stable for a certain period of time in the circulation. Therefore, they may be useful biomarkers in the evaluation of gastric and intestinal mucosal damage [36, 37].

Recently, it has been established that there is a casual relationship between circRNA and gastric cancer progression. It was observed that the expression of hsa_circ_002059 from the circRNA family was significantly decreased in tumors with advanced TNM stages and patients with distant metastases. It has been determined that circR-7, another circRNA, plays a role in the downregulation of miR-7 expression. With circR-7 downregulating miR-7, E calcium mucin synthesis decreases and the proliferation, invasion and migration capacity of gastric cancer cells increase [36]. Another RNA fragment belonging to the circRNA family is hsa_circ_0001649, its downregulation is more pronounced in gastric cancer cells compared to surrounding cells, it undergoes upregulation after surgical excision and correlates with the pathological difference of the tumor [36, 38].

Diagnostic Biomarkers of Lower Gastrointestinal System Diseases

Ligamentum Suspensorium Duodeni, (The Ligamentum of Treitz), which divides the fourth segment of the duodenum from proximal jejunum, is accepted as a landmark to divide upper and lower gastrointestinal tracts. The lower gastrointestinal tract is a 5-11-meter-long segment from the jejunum to the anal verge, and it is basically responsible for chemical digestion, absorption, and several immunomodulatory functions. Like upper gastrointestinal system disorders, annually, over 10 million hospital visits occur in the United States of America due to lower gastrointestinal complaints [3]. The lower gastrointestinal system disorders display a broad spectrum; clinicians diagnose over 150.000 lower gastrointestinal tract related malignancies, and over 1.500.000 patients suffering from inflammatory bowel diseases apply to hospitals annually [2, 3]. According to 2013 data, over 11 million lower gastrointestinal system endoscopies were performed by clinicians [2, 3]. Radiological techniques remain useful as a diagnostic tool only for the determination of macroscopic pathologies. Due to the absence of low-cost non-invasive diagnostic surgical methods, biomarker usage emerged as a useful alternative for diagnosing and screening patients suffering from lower gastrointestinal diseases.

Biomarkers for Benign Lower Gastrointestinal Disorders

Calprotectin

Calprotectin is a 36.5 kDa non-glycosylated calcium and zinc binding protein complex, which can be found in neutrophilic cytosol [37]. It can be detected in monocytes & macrophages and released into the extracellular area as a result of the immune response [39, 40]. Therefore, it is possible to assess calprotectin in serum, faeces, synovium, and saliva according to the severity of the inflammation [41]. Due to its relatively stable molecular structure and high sensitivity, we can use calprotectin for the diagnosis or follow-up of several disorders [37].

The widespread use of calprotectin has entered into our clinical practice with its ability to diagnose and follow the status of patients suffering from inflammatory bowel diseases (IBD) (*e.g.*, Crohn's disease and ulcerative colitis). Despite patients might have different genetic predispositions, clinical features, and disease activation sites; inflammatory bowel diseases are considered as progressive and relapsing disorders [42, 43]. Fecal calprotectin is very useful to determine whether the disease is at an active phase or not. Higher fecal calprotectin levels can be present if the patient is suffering from an inflammation-related attack; on the other hand, its levels decrease if the patient enter a remission period and the intestinal mucosa is healing [44].

Surgical intervention for inflammatory bowel diseases is the last resort countermeasure, and it is only used when the patient suffers from increased symptoms despite medical treatment or a complication occurs. (*e.g.*, perforation, enteroenteric fistulas, toxic megacolon). Fecal calprotectin is suggested as a useful biomarker for the follow-ups of operated patients; its assessment is much more reliable compared to Crohn Disease Activation Index (CDAI) and C Reactive Protein (CRP) [45]. If an asymptomatic patient has 100 μg/g or above levels of calprotectin, a colonoscopy should be scheduled in order to evaluate whether the disease is reactivating or not [46].

Apart from inflammatory bowel diseases, fecal calprotectin levels may increase in medical conditions such as mucosal damage due to anti-inflammatory drugs, infectious colitis, microscopic colitis, necrotizing enterocolitis, Celiac disease, gastrointestinal involvement of spondyloarthropathies, and colonic cancer [39, 47].

Lactoferrin

Lactoferrin is a 76-80 kDa iron-binding glycosylated protein composed of 690 amino acids which exhibits a similar protein architecture like transferrin. Generally, lactoferrin is found in secretions or located in the granules of neutrophils. Primarily, lactoferrin inhibits the adhesion of microorganisms to the cell membrane, and prevents their transportation into cytosolic space. Too add to this, it also has antimicrobial activity itself due to its iron-binding capacity. All these features make lactoferrin an important immunological molecule, especially in the infant immune system. Apart from its antimicrobial properties, lactoferrin also takes place in immune modulation and inflammatory response [48, 49].

Lactoferrin, like calprotectin, is resistant to proteolysis, and due to its stable molecular structure, it can be a beneficial biomarker to determine gastrointestinal luminal inflammation. As a biomarker, lactoferrin is mostly used for the diagnosis and routine follow-ups of patients suffering from inflammatory bowel diseases. Being a cheap, safe to conduct, and reliable test makes lactoferrin a useful biomarker for the assessment of treatment response of patients. Fecal calprotectin levels may increase when inflammation progresses parallel to disease activation and decrease during the recovery period when the inflammation slows down [50].

Fatty Acid Binding Proteins (FABP)

Fatty acid binding proteins (FABPs) are intracellular proteins responsible for the absorption of long chain fatty acids from the intestinal lumen into the cytosolic part of the enterocyte [51]. Later studies revealed FABPs are also present in other tissues such as brain, heart, liver, and nine different tissue-specific FABPs have been identified [52]. Among these FABPs, liver FABPs (L-FABP, Fabp1), intestinal FABPs (I-FABP, Fabp2), and ileal FABPs (Il-FABP, Fabp6) are extensively present in the gastrointestinal tract. FABPs are emerged as a potential biomarker to determine intestinal mucosa integrity, intestinal cell mass, and intestinal mucosal inflammation. After the mucosal damage and enterocyte death initiates, FABPs are released into the peripheral circulation, and increasing levels of FABPs correlate with the severity inflammation & cell death. Increasing levels of plasma FABPs may help to diagnose and follow up the treatment responses of clinical conditions like mesenteric ischemia, necrotizing enterocolitis, inflammatory bowel diseases and Celiac disease [52].

Acute mesenteric ischemia is a circulatory disorder that originates from the occlusion of mesenteric arteries and veins or due to insufficient perfusion without any significant vascular pathology (non-occlusive mesenteric ischemia, NOMI). Even though conventional diagnostic tools (*e.g.*, physical examination, laboratory findings, radiological examinations) are able to differentiate occlusive mesenteric ischemia, these tests may be insufficient for patients with non-specific clinical findings and perfusion related ischemia [53]. D-Dimer and lactate are the most preferred biomarkers in routine clinical use. However, despite having a high sensitivity, D-Dimer has fairly low specificity, and blood lactate levels might also increase due to other ischemic conditions, which makes the use of these two biomarkers not so practical [54, 55]. Recent studies have shown I-FABP might be a relatively sensitive biomarker to determine intestinal ischemia and reperfusion damage, but it still lacks an established standard cut-off value for clinical utility [52].

Necrotizing enterocolitis is known as a highly mortal condition for premature infants with low birth weight, and the etiology of the disease still remains unclear [56]. While 90% of the cases have a history of enteral feeding, infants who received breastfeeding have a three to tenfold reduced risk for the disease [57, 58]. In recent years, FABPs emerged as a promising biomarker for diagnosing necrotizing enterocolitis due to their high specificity. While L-FABPs tend to be increased during the early stages (stage I) of the disease, I-FABP levels elevate as the disease progresses (stage III) [59]. Predictive values of the FABPs increase when it is used together with fecal calprotectin, interleukin-1 receptor antagonist (IL-1ra), interleukin-8 (IL-8), interleukin-10 (IL-10), and monocyte chemoattractant protein-1(MCP-1) [52].

Celiac disease is an auto-immune disorder due to an immune reaction in the intestinal lumen caused by inadequate digestion of gluten in individuals with genetic predisposition [60]. Even though duodenal biopsies revealing duodenal morphology is the golden standard for diagnosis, routine clinical use of more practical and non-invasive tests are rapidly increasing [52, 60]. Serum and urine I-FABP levels exhibit a positive correlation with villous atrophy and tissue transglutaminase IgA (tTG-IgA) levels [51]. Moreover, when the mucosal healing period starts in patients on a gluten-free diet, I-FABP levels significantly decrease to their reference range values [52].

Anti-gliadine, Anti-endomysial and Tissue Transglutaminase Autoantibodies

Autoantibodies are another group of biomarkers for Celiac disease diagnosis. Patients suffering from Celiac disease produce these antibodies in the affected intestinal mucosa due to gluten uptake, and these antibodies appear in the diseased intestinal mucosa, saliva, and blood [61]. Anti-gliadin antibodies develop against dietary gliadin, anti-endomysial antibodies develop against smooth muscle cell endomysium (connective tissue around smooth muscle cells), and tissue transglutaminase antibodies also appear against transglutaminase 2 protein [61, 62].

Anti-gliadin autoantibody (AGA) was the first identified disease related antibody to enter into widespread clinical use. However, due to its low sensitivity & specificity, AGA quickly began to fall out of favor. Later, anti-endomysial autoantibodies, which were shown to be related to dermatitis herpetiformis and Celiac disease, came into our clinical practice. Indirect immunofluorescence method is the preferred method to identify endomysial antibodies, and primate (endomysial autoantibody – EMA) or rodent (IgA R1 type reticulin – ARA) tissues are preferred as antigens [63, 64]. Even though EMA is a reliable test, it is an expensive time consuming work-up, and results may differ according to the performer [63, 65]. The diagnosis and follow-up of Celiac disease evolved after the demonstration of tissue transglutaminase antibodies (tTGA) and deaminated gliadin peptide antibodies (DGPA) [63].

EMA, tTG2A, and DGPA tests have high sensitivity and specificity. Among these tests, EMA has the highest specificity, and anti-tTG IgA has the highest sensitivity [65]. Anti-tTG IgA remains the most useful test for screening patients, and EMA can be used as a confirmatory test for the patients whose tTGA IgA results are positive. Since tTGA IgA cannot be detected in patients with selective IgA deficiency, tTGA IgG is a useful antibody in this group, and it generally remains positive [65]. However, all these antibodies are significantly dependent on dietary gluten, and their blood levels may reduce within one year after a strict gluten-free diet [63].

Biomarkers for Malign Lower Gastrointestinal Disorders

Carbohydrate Antigen (CA 19-9) and Carcinoembryonic Antigen (CEA)

As mentioned above, carbohydrate antigen 19-9 (CA 19-9) is still considered to be a useful biomarker for gastric, pancreatic, and colonic cancers. CA 19-9 is not a glycoprotein found in healthy tissues, and it tends to elevate if a neoplastic process occurs [66]. With its tetrasaccharide structure, CA 19-9 is a fairly large molecule (200 – 1000 kDa) in the plasma [27]. However, despite being an easy biomarker to analytically detect, CA 19-9 does not have high levels of sensitivity since it does not only exceed the reference range with malignant conditions [27]. To add to this, CA 19-9 is not specific to the histological subtype of the tumor, so it does not differentiate the origin of the malignancy [67]. The primary role of CA

19-9 in clinical use remains to be supportive. Due its low sensitivity (~30%), CA 19-9 is not a useful biomarker to determine the disease on its own [68]. The practical utility of CA 19-9 increases with the size of the tumor, and in combination with other biomarkers. The sensitivity of CA 19-9 increases with the higher Dukes stage colorectal cancers; but it does not show any correlation with the tumor location and the number of positive lymph nodes [67, 69].

Carcinoembryonic antigen (CEA) is a membrane glycoprotein that can be located in the intestinal epithelium, colonic polyps, and tumors [68, 70]. It was first described in colonic cancer tissue extracts, and it has a molecular weight of 150-180 kDa [70, 71]. CEA has a cut-off value in healthy patients up to 5 ng/mL, and 10 ng/mL in smokers, patients suffering from inflammatory bowel diseases, and liver cirrhosis [67]. Like CA 19-9, CEA might increase in benign gastrointestinal conditions such as inflammatory bowel diseases, pancreatitis, and hepatitis [68]. Since CEA is a more commonly found protein in the intestinal epithelium, it has the privilege of being a more sensitive and specific biomarker compared to CA 19-9. Even though the suggested use of CEA prior to the surgery, the most beneficial use of CEA is for disease recurrence after the resection [67, 72]. Up to this date, CEA is considered to be one of the most feasible biomarkers in terms of detecting and predicting the prognosis in patients suffering from colorectal cancer recurrence [72, 73].

In the current clinical practice, the primary functions of CA 19-9 and CEA are to evaluate the post-operative status of the patient for disease recurrence and to assist the clinician for overseeing the prognosis.

Circular RNA – Use in Colorectal Cancer

Circulating RNA fragments (circRNA) are non-coding RNA fragments, which are endorsed as potential novel biomarkers for colorectal malignancies. Due to the lack of 5' terminal caps and 3' terminal tails, circRNAs form a closed-loop structure, which makes it more stable compared to mRNAs [74, 75]. circRNA overexpression seems to be related to tumor differentiation, tumor size, distant metastasis, lymphatic metastasis, depth of tumor invasion, and higher TNM staged colorectal cancers [72]. All these factors have an impact on the prognosis of the patient, so it is fair to say circRNA levels exhibit a significant correlation with overall survival for the patient. On the other hand, circRNA level shows no correlation with the levels of other biomarkers, tumor location, and demographic factors such as age & gender [75]. A recent study showed that hsa_circ_001569 exhibits higher expression in colorectal malignancies [76].

The prospect of circRNA use has entered a new era after the recent studies. Previously, it was accepted as circRNAs were products of transcription errors.

However, it has been shown that circRNAs can encode proteins, and they are closely related to many biochemical processes. Influencing the behaviors of malignant cells is one of its many functions, and a better understanding of circRNAs will certainly increase our ability to diagnose colorectal cancers [77].

CONCLUSION

The clinical approach for treating gastrointestinal disorders dramatically changed in recent years, and biomarkers proved to be useful diagnostic materials in the process. These biomarkers not only prevent unnecessary surgical interventions, but they are also helpful for shortening the duration to start necessary treatment for patients. Even though we revealed most of the biomolecular properties for some of these biomarkers, they are still far away from widespread clinical use. Especially, genomic biomarkers show promising potential to increase our ability to diagnose and subtyping malignant disorders; but they are still far from our laboratory routine. We believe gastrointestinal biomarkers will steadily continue to emerge into our routine in the near future, not only with increased efficiency but also with greater ease in utilization.

CONSENT FOR PUBLICATION

Not applicable.

CONFLICT OF INTEREST

The authors declare no conflict of interest, financial or otherwise.

ACKNOWLEDGEMENT

Declared none.

REFERENCES

[1] Barchi A, Miraglia C, Violi A, *et al.* A non-invasive method for the diagnosis of upper GI diseases. Acta Biomed 2018; 89(8-S): 40-3.
[PMID: 30561417]

[2] Siegel RL, Miller KD, Jemal A. Cancer statistics, 2020. CA Cancer J Clin 2020; 70(1): 7-30.
[http://dx.doi.org/10.3322/caac.21590] [PMID: 31912902]

[3] Peery AF, Crockett SD, Murphy CC, *et al.* Burden and Cost of Gastrointestinal, Liver, and Pancreatic Diseases in the United States: Update 2018. Gastroenterology 2019; 156(1): 254-272.e11.
[http://dx.doi.org/10.1053/j.gastro.2018.08.063] [PMID: 30315778]

[4] Rotter JI, Wong FL, Samloff IM, *et al.* Evidence for a major dominance component in the variation of serum pepsinogen I levels. Am J Hum Genet 1982; 34(3): 395-401.
[PMID: 7081219]

[5] Shan JH, Bai XJ, Han LL, Yuan Y, Sun XF. Changes with aging in gastric biomarkers levels and in biochemical factors associated with *Helicobacter pylori* infection in asymptomatic Chinese

population. World J Gastroenterol 2017; 23(32): 5945-53.
[http://dx.doi.org/10.3748/wjg.v23.i32.5945] [PMID: 28932086]

[6] Nurgalieva ZZ, Opekun AR, Graham DY. Problem of distinguishing false-positive tests from acute or transient *Helicobacter pylori* infections. Helicobacter 2006; 11(2): 69-74.
[http://dx.doi.org/10.1111/j.1523-5378.2006.00380.x] [PMID: 16579835]

[7] Storskrubb T, Aro P, Ronkainen J, *et al.* Serum biomarkers provide an accurate method for diagnosis of atrophic gastritis in a general population: The Kalixanda study. Scand J Gastroenterol 2008; 43(12): 1448-55.
[http://dx.doi.org/10.1080/00365520802273025] [PMID: 18663663]

[8] Germaná B, Di Mario F, Cavallaro LG, *et al.* Clinical usefulness of serum pepsinogens I and II, gastrin-17 and anti-Helicobacterpylori antibodies in the management of dyspeptic patients in primary care. Dig Liver Dis 2005; 37(7): 501-8.
[http://dx.doi.org/10.1016/j.dld.2005.01.016] [PMID: 15975537]

[9] Miki K. Gastric cancer screening using the serum pepsinogen test method. Gastric Cancer 2006; 9(4): 245-53.
[http://dx.doi.org/10.1007/s10120-006-0397-0] [PMID: 17235625]

[10] Kikuchi R, Abe Y, Iijima K, *et al.* Low serum levels of pepsinogen and gastrin 17 are predictive of extensive gastric atrophy with high-risk of early gastric cancer. Tohoku J Exp Med 2011; 223(1): 35-44.
[http://dx.doi.org/10.1620/tjem.223.35] [PMID: 21222340]

[11] Miki K, Fujishiro M, Kodashima S, Yahagi N. Long-term results of gastric cancer screening using the serum pepsinogen test method among an asymptomatic middle-aged Japanese population. Dig Endosc 2009; 21(2): 78-81.
[http://dx.doi.org/10.1111/j.1443-1661.2009.00839.x] [PMID: 19691778]

[12] Sipponen P, Vauhkonen M, Helske T, Kaariainen I, Harkonen M. Low circulating levels of gastrin-17 in patients with Barrett's esophagus. World J Gastroenterol 2005; 11(38): 5988-92.
[http://dx.doi.org/10.3748/wjg.v11.i38.5988] [PMID: 16273611]

[13] Savarino V, Bortoli ND, Coletta MD, Speroni S. Gastrin 17 in Singling Out Patients with Different Patterns of Refluxate: A Plot Study Using Impedance-pH as Reference Standard. Gastroenterology 2017; 152(5): 653.
[http://dx.doi.org/10.1016/S0016-5085(17)32302-8]

[14] Chang, EB ve Leung PS Gastric Physiology Leung PS The Gastrointestinal System. Dordrecht: Springer 2014; pp. 63-85.

[15] Sipponen P, Ranta P, Helske T, *et al.* Serum levels of amidated gastrin-17 and pepsinogen I in atrophic gastritis: an observational case-control study. Scand J Gastroenterol 2002; 37(7): 785-91.
[http://dx.doi.org/10.1080/gas.37.7.785.791] [PMID: 12190091]

[16] Marshall, BJ ve Warren, JR. Unidentified curved bacilli in the stomach of patients with gastritis and peptic ulceration. Lancet 1984; 16: 1311-5.

[17] Li S, Lu AP, Zhang L, Li YD. Anti-*Helicobacter pylori* immunoglobulin G (IgG) and IgA antibody responses and the value of clinical presentations in diagnosis of *H. pylori* infection in patients with precancerous lesions. World J Gastroenterol 2003; 9(4): 755-8.
[http://dx.doi.org/10.3748/wjg.v9.i4.755] [PMID: 12679926]

[18] Shimoyama T. Stool antigen tests for the management of *Helicobacter pylori* infection. World J Gastroenterol 2013; 19(45): 8188-91.
[http://dx.doi.org/10.3748/wjg.v19.i45.8188] [PMID: 24363508]

[19] She RC, Wilson AR, Litwin CM. Evaluation of *Helicobacter pylori* Immunoglobulin G (IgG), IgA, and IgM serologic testing compared to stool antigen testing. Clin Vaccine Immunol 2009; 16(8): 1253-5.

[http://dx.doi.org/10.1128/CVI.00149-09] [PMID: 19515865]

[20] Ferwana M, Abdulmajeed I, Alhajiahmed A, *et al.* Accuracy of urea breath test in *Helicobacter pylori* infection: meta-analysis. World J Gastroenterol 2015; 21(4): 1305-14. [http://dx.doi.org/10.3748/wjg.v21.i4.1305] [PMID: 25632206]

[21] Siddiqui I, Ahmed S, Abid S. Update on diagnostic value of breath test in gastrointestinal and liver diseases. World J Gastrointest Pathophysiol 2016; 7(3): 256-65. [http://dx.doi.org/10.4291/wjgp.v7.i3.256] [PMID: 27574563]

[22] Lim LL, Ho K. YuHo HB, Salto-Terrez M. Effects of biopsies on sensitivityandspecificity of ultra-rapidurease test fordetection of *Helicobacter pylori* infection: A prospective evaluation. World J Gastroenterol 2004; 10: 1907-10. [http://dx.doi.org/10.3748/wjg.v10.i13.1907] [PMID: 15222034]

[23] Shin S, Jeong HM, Chung SE, *et al.* Simultaneous analysis of acetylcarnitine, proline, hydroxyproline, citrulline, and arginine as potential plasma biomarkers to evaluate NSAIDs-induced gastric injury by liquid chromatography-tandem mass spectrometry. J Pharm Biomed Anal 2019; 165: 101-11. [http://dx.doi.org/10.1016/j.jpba.2018.11.051] [PMID: 30522064]

[24] Kuiken NSS, Rings EHHM, Blijlevens NMA, Tissing WJE. Biomarkers and non-invasive tests for gastrointestinal mucositis. Support Care Cancer 2017; 25(9): 2933-41. [http://dx.doi.org/10.1007/s00520-017-3752-2] [PMID: 28536886]

[25] Crenn P, Messing B, Cynober L. Citrulline as a biomarker of intestinal failure due to enterocyte mass reduction. Clin Nutr 2008; 27(3): 328-39. [http://dx.doi.org/10.1016/j.clnu.2008.02.005] [PMID: 18440672]

[26] Koprowski H, Steplewski Z, Mitchell K, Herlyn M, Herlyn D, Fuhrer P. Colorectal carcinoma antigens detected by hybridoma antibodies. Somatic Cell Genet 1979; 5(6): 957-71. [http://dx.doi.org/10.1007/BF01542654] [PMID: 94699]

[27] Scarà S, Bottoni P, Scatena R. CA 19-9: Biochemical and Clinical Aspects. Adv Exp Med Biol 2015; 867: 247-60. [http://dx.doi.org/10.1007/978-94-017-7215-0_15] [PMID: 26530370]

[28] Edge SB, Byrd DR, Compton CC, Fritz AG, Greene FL, Trotti A. American Joint Committee on Cancer Cancer Staging Manual. 7th ed., Chicago: Springer 2015.

[29] Feng F, Tian Y, Xu G, *et al.* Diagnostic and prognostic value of CEA, CA19-9, AFP and CA125 for early gastric cancer. BMC Cancer 2017; 17(1): 737. [http://dx.doi.org/10.1186/s12885-017-3738-y] [PMID: 29121872]

[30] Liang Y, Wang W, Fang C, *et al.* Clinical significance and diagnostic value of serum CEA, CA19-9 and CA72-4 in patients with gastric cancer. Oncotarget 2016; 7(31): 49565-73. [http://dx.doi.org/10.18632/oncotarget.10391] [PMID: 27385101]

[31] Zheng TH, Zhao JL, Guleng B. Advances in Molecular Biomarkers for Gastric Cancer. Crit Rev Eukaryot Gene Expr 2015; 25(4): 299-305. [http://dx.doi.org/10.1615/CritRevEukaryotGeneExpr.2015014360] [PMID: 26559090]

[32] Takahashi Y, Takeuchi T, Sakamoto J, *et al.* Tumar Marker Commitee. The usefulness of CEA and/or CA 19-9 in monitoring for recurrence in gastric cancer patients: a prospective study. Gastric Cancer 2003; 6(3): 142-5. [http://dx.doi.org/10.1007/s10120-003-0240-9] [PMID: 14520526]

[33] Choi SR, Jang JS, Lee JH, *et al.* Role of serum tumor markers in monitoring for recurrence of gastric cancer following radical gastrectomy. Dig Dis Sci 2006; 51(11): 2081-6. [http://dx.doi.org/10.1007/s10620-006-9166-5] [PMID: 17009116]

[34] Kotzev AI, Draganov PV. Carbohydrate Antigen 19-9, Carcinoembryonic Antigen, and Carbohydrate Antigen 72-4 in Gastric Cancer: Is the Old Band Still Playing? Gastrointest Tumors 2018; 5(1-2): 1-13.

[http://dx.doi.org/10.1159/000488240] [PMID: 30574476]

[35] Spila A, Roselli M, Cosimelli M, *et al.* Clinical utility of CA 72-4 serum marker in the staging and immediate post-surgical management of gastric cancer patients. Anticancer Res 1996; 16(4B): 2241-7. [PMID: 8694550]

[36] Luo B, Tang CM, Chen JS. circRNA and gastrointestinal cancer. J Cell Biochem 2019; 120(7): 10956-63. [http://dx.doi.org/10.1002/jcb.28610] [PMID: 30945338]

[37] Banerjee A. Gastrointestinal toxicity biomarkers Ramesh CG Biomarkers in Toxicology. Elsevier 2014; pp. 269-77.

[38] Li WH, Song YC, Zhang H, *et al.* Decreased Expression of Hsa_circ_00001649 in Gastric Cancer and Its Clinical Significance. Dis Markers 2017; 2017: 4587698. [http://dx.doi.org/10.1155/2017/4587698] [PMID: 28167847]

[39] Esatoğlu SN, Hatemi I, Özgüler Y, *et al.* Faecal but not serum calprotectin levels look promising in predicting active disease in Behçet's syndrome patients with gastrointestinal involvement. Clin Exp Rheumatol 2018; 36(6) (Suppl. 115): 90-6. [PMID: 30582504]

[40] Hale MF, Drew K, McAlindon ME, Sidhu R. The diagnostic accuracy of faecal calprotectin and small bowel capsule endoscopy and their correlation in suspected isolated small bowel Crohn's disease. Eur J Gastroenterol Hepatol 2016; 28(10): 1145-50. [http://dx.doi.org/10.1097/MEG.0000000000000696] [PMID: 27384306]

[41] Johne B, Fagerhol MK, Lyberg T, *et al.* Functional and clinical aspects of the myelomonocyte protein calprotectin. Mol Pathol 1997; 50(3): 113-23. [http://dx.doi.org/10.1136/mp.50.3.113] [PMID: 9292145]

[42] Baumgart DC, Sandborn WJ. Crohn's disease Lancet 202; 380(9853): 1590-605. [http://dx.doi.org/10.1016/S0140-6736(12)60026-9]

[43] Ordás I, Eckmann L, Talamini M, Baumgart DC, Sandborn WJ. Ulcerative colitis. Lancet 2012; 380(9853): 1606-19. [http://dx.doi.org/10.1016/S0140-6736(12)60150-0] [PMID: 22914296]

[44] D'Haens G, Ferrante M, Vermeire S, *et al.* Fecal calprotectin is a surrogate marker for endoscopic lesions in inflammatory bowel disease. Inflamm Bowel Dis 2012; 18(12): 2218-24. [http://dx.doi.org/10.1002/ibd.22917] [PMID: 22344983]

[45] Lynn AM, Loftus EV Jr. Luminal Crohn's Disease Ding NS & De Cruz P Biomarkers in Inflammatory Bowel Diseases. New York: Springer 2019; pp. 57-71.

[46] Patwala K, De Cruz P. Chapter: Postoperative Crohn's Disease Ding NS & De Cruz P Biomarkers in Inflammatory Bowel Diseases. New York: Springer 2019; pp. 89-97.

[47] Lawrance IC. Chapter: Refractory Proctitis Ding NS & De Cruz P Biomarkers in Inflammatory Bowel Diseases. New York: Springer 2019; pp. 143-52.

[48] Brock J. Lactoferrin: a multifunctional immunoregulatory protein? Immunol Today 1995; 16(9): 417-9. [http://dx.doi.org/10.1016/0167-5699(95)80016-6] [PMID: 7546203]

[49] Berlutti F, Pantanella F, Natalizi T, *et al.* Antiviral properties of lactoferrin--a natural immunity molecule. Molecules 2011; 16(8): 6992-7018. [http://dx.doi.org/10.3390/molecules16086992] [PMID: 21847071]

[50] Sipponen T, Björkesten CG A F, Färkkilä M, Nuutinen H, Savilahti E, Kolho KL. Faecal calprotectin and lactoferrin are reliable surrogate markers of endoscopic response during Crohn's disease treatment. Scand J Gastroenterol 2010; 45(3): 325-31. [http://dx.doi.org/10.3109/00365520903483650] [PMID: 20034360]

[51] Ockner RK, Manning JA, Poppenhausen RB, Ho WK. A binding protein for fatty acids in cytosol of intestinal mucosa, liver, myocardium, and other tissues. Science 1972; 177(4043): 56-8. [http://dx.doi.org/10.1126/science.177.4043.56] [PMID: 5041774]

[52] Ho SSC, Keenan JI, Day AS. The Role of Gastrointestinal-Related Fatty Acid-Binding Proteins as Biomarkers in Gastrointestinal Diseases. Dig Dis Sci 2020; 65(2): 376-90. [http://dx.doi.org/10.1007/s10620-019-05841-x] [PMID: 31529416]

[53] Treskes N, Persoon AM, van Zanten ARH. Diagnostic accuracy of novel serological biomarkers to detect acute mesenteric ischemia: a systematic review and meta-analysis. Intern Emerg Med 2017; 12(6): 821-36. [http://dx.doi.org/10.1007/s11739-017-1668-y] [PMID: 28478489]

[54] Cudnik MT, Darbha S, Jones J, Macedo J, Stockton SW, Hiestand BC. The diagnosis of acute mesenteric ischemia: A systematic review and meta-analysis. Acad Emerg Med 2013; 20(11): 1087-100. [http://dx.doi.org/10.1111/acem.12254] [PMID: 24238311]

[55] Memet O, Zhang L, Shen J. Serological biomarkers for acute mesenteric ischemia. Ann Transl Med 2019; 7(16): 394. [http://dx.doi.org/10.21037/atm.2019.07.51] [PMID: 31555708]

[56] Lin PW, Stoll BJ. Necrotising enterocolitis. Lancet 2006; 368(9543): 1271-83. [http://dx.doi.org/10.1016/S0140-6736(06)69525-1] [PMID: 17027734]

[57] Berseth CL. Feeding strategies and necrotizing enterocolitis. Curr Opin Pediatr 2005; 17(2): 170-3. [http://dx.doi.org/10.1097/01.mop.0000150566.50580.26] [PMID: 15800406]

[58] McGuire W, Anthony MY. Donor human milk *versus* formula for preventing necrotising enterocolitis in preterm infants: systematic review. Arch Dis Child Fetal Neonatal Ed 2003; 88(1): F11-4. [http://dx.doi.org/10.1136/fn.88.1.F11] [PMID: 12496220]

[59] Edelson MB, Sonnino RE, Bagwell CE, Lieberman JM, Marks WH, Rozycki HJ. Plasma intestinal fatty acid binding protein in neonates with necrotizing enterocolitis: a pilot study. J Pediatr Surg 1999; 34(10): 1453-7. [http://dx.doi.org/10.1016/S0022-3468(99)90102-1] [PMID: 10549746]

[60] Lebwohl B, Sanders DS, Green PHR. Coeliac disease. Lancet 2018; 391(10115): 70-81. [http://dx.doi.org/10.1016/S0140-6736(17)31796-8] [PMID: 28760445]

[61] Pasinszki T, Krebsz M. Biosensors for Non-Invasive Detection of Celiac Disease Biomarkers in Body Fluids. Biosensors (Basel) 2018; 8(2): 55. [http://dx.doi.org/10.3390/bios8020055] [PMID: 29914179]

[62] Rauhavirta T, Hietikko M, Salmi T, Lindfors K. Transglutaminase 2 and Transglutaminase 2 Autoantibodies in Celiac Disease: a Review. Clin Rev Allergy Immunol 2019; 57(1): 23-38. [http://dx.doi.org/10.1007/s12016-016-8557-4] [PMID: 27263022]

[63] Caja S, Mäki M, Kaukinen K, Lindfors K. Antibodies in celiac disease: implications beyond diagnostics. Cell Mol Immunol 2011; 8(2): 103-9. [http://dx.doi.org/10.1038/cmi.2010.65] [PMID: 21278768]

[64] Ortiz G, Messere G, Toca MDC, *et al.* IgA anti-tissue transglutaminase antibodies and IgG antibodies against deamidated gliadin peptides as predictors of celiac disease. Arch Argent Pediatr 2019; 117(1): 52-5. [PMID: 30652447]

[65] Brusca I. Overview of biomarkers for diagnosis and monitoring of celiac disease. Adv Clin Chem 2015; 68: 1-55. [http://dx.doi.org/10.1016/bs.acc.2014.12.006] [PMID: 25858867]

[66] Goonetilleke KS, Siriwardena AK. Systematic review of carbohydrate antigen (CA 19-9) as a

biochemical marker in the diagnosis of pancreatic cancer. Eur J Surg Oncol 2007; 33(3): 266-70. [http://dx.doi.org/10.1016/j.ejso.2006.10.004] [PMID: 17097848]

[67] Jelski W, Mroczko B. Biochemical Markers of Colorectal Cancer - Present and Future. Cancer Manag Res 2020; 12: 4789-97. [http://dx.doi.org/10.2147/CMAR.S253369] [PMID: 32606968]

[68] Troup S. Analyte monographs alongside the National Laboratory Medicine Catalogue. London: The Association for Clinical Biochemistry and Laboratory Medicine 2012.

[69] Filella X, Molina R, Grau JJ, *et al.* Prognostic value of CA 19.9 levels in colorectal cancer. Ann Surg 1992; 216(1): 55-9. [http://dx.doi.org/10.1097/00000658-199207000-00008] [PMID: 1632702]

[70] Garcia M, Seigner C, Bastid C, Choux R, Payan MJ, Reggio H. Carcinoembryonic antigen has a different molecular weight in normal colon and in cancer cells due to N-glycosylation differences. Cancer Res 1991; 51(20): 5679-86. [PMID: 1717152]

[71] Thomas P. Chapter: Carcinoembryonic Antigen Manfred Schwab Encyclopedia of Cancer. Heidelberg, Berlin: Springer 2014.

[72] Tong G, Xu W, Zhang G, *et al.* The role of tissue and serum carcinoembryonic antigen in stages I to III of colorectal cancer-A retrospective cohort study. Cancer Med 2018; 7(11): 5327-38. [http://dx.doi.org/10.1002/cam4.1814] [PMID: 30302946]

[73] Yang KM, Park IJ, Kim CW, Roh SA, Cho DH, Kim JC. The prognostic significance and treatment modality for elevated pre- and postoperative serum CEA in colorectal cancer patients. Ann Surg Treat Res 2016; 91(4): 165-71. [http://dx.doi.org/10.4174/astr.2016.91.4.165] [PMID: 27757393]

[74] Li C, He X, Zhang L, Li L, Zhao W. A pair-wise meta-analysis highlights circular RNAs as potential biomarkers for colorectal cancer. BMC Cancer 2019; 19(1): 957. [http://dx.doi.org/10.1186/s12885-019-6136-9] [PMID: 31615475]

[75] Yin Y, Long J, He Q, *et al.* Emerging roles of circRNA in formation and progression of cancer. J Cancer 2019; 10(21): 5015-21. [http://dx.doi.org/10.7150/jca.30828] [PMID: 31602252]

[76] Xie H, Ren X, Xin S, *et al.* Emerging roles of circRNA_001569 targeting miR-145 in the proliferation and invasion of colorectal cancer. Oncotarget 2016; 7(18): 26680-91. [http://dx.doi.org/10.18632/oncotarget.8589] [PMID: 27058418]

[77] Hao S, Cong L, Qu R, Liu R, Zhang G, Li Y. Emerging roles of circular RNAs in colorectal cancer. OncoTargets Ther 2019; 12: 4765-77. [http://dx.doi.org/10.2147/OTT.S208235] [PMID: 31354303]

CHAPTER 15

Novel Applications of Biomarkers in Chronic Obstructive Pulmonary Disease

Pelin Uysal[1,*]

[1] *Department of Chest Diseases, Faculty of Medicine, Mehmet Ali Aydınlar University, Atakent Hospital, Istanbul, Turkey*

Abstract: Chronic obstructive pulmonary disease (COPD) is an important health problem and an increasing cause of morbidity and mortality worldwide. Currently, COPD is considered a multisystem disease. Although it primarily affects the lungs, structural and functional changes occur in other organs due to systemic inflammation. It is stated that in patients with COPD, airway and systemic inflammatory markers are increased and that these markers are high are associated with a faster decline in lung functions. In recent years, numerous articles have been published on the discovery and evaluation of biomarkers in COPD. Many markers have also been studied to accurately assess COPD exacerbations and provide effective treatment. However, based on the evidence from published studies, a single molecule has not been adequately validated for broad clinical use.

Keywords: Chronic obstructive pulmonary disease, Biomarkers, Exacerbation, Therapy.

INTRODUCTION

Chronic obstructive pulmonary disease (COPD) was previously defined only by the presence of airflow limitation, is now defined as an inflammatory disease. It is also stated that COPD is a systemic disease, and comorbid and common comorbidities contribute to the severity and mortality of COPD. In COPD, the musculoskeletal system (muscle dysfunction and osteoporosis), cardiovascular system (atherosclerosis), endocrine, and nervous system are mainly affected [1]. Erythrocyte sedimentation rate (ESR), C-reactive protein (CRP), interleukin-8 (IL-8), IL-6, tumor necrosis factor-alpha (TNF-α), and leukotriene B4 (LTB4) are the major inflammatory markers that may cause COPD. The importance of changes in these markers during exacerbation and after treatment has been emphasized in various studies [2].

* **Corresponding author Pelin Uysal:** Department of Chest Diseases, Faculty of Medicine, Mehmet Ali Aydınlar University, Atakent Hospital, Istanbul, Turkey; Tel:+90 505 453 56 23; E-mail: pelinuysal@gmail.com

Pınar Atukeren and Hafize Uzun (Eds.)

With the demonstration that many inflammatory cells, cytokines and enzymes are involved in different stages of COPD, the place and importance of inflammation in the pathophysiology of the disease has been better understood in recent years. Many inflammatory cells and markers are involved in the pathogenesis of COPD and this inflammation has been demonstrated by different methods. Developments in this area will also help develop new approaches to the treatment of the disease [3].

Biomarkers are measurable indicator molecules that show biological developments, pathological changes, and responses to pharmacological treatment. A good biomarker should be disease-related, sensitive, reliable, consistent, reproducible, practically applicable, and should also be generalizable and cost-effective; these can be cells, cytokines, proteins, genes, and products, *etc* [4].

To show inflammation in COPD and other respiratory system diseases, samples of bronchial biopsy, induced sputum, bronchoalveolar lavage (BAL) fluid, and expiratory air can be taken. Many inflammatory markers can be measured in BAL fluid. Eosinophil cationic protein (ECP), myeloperoxidase and IL-8 levels were found to be higher in COPD and smokers than in healthy non-smokers. Protease and antiproteases can be detected in BAL fluid. In COPD, there is an increase in total elastase activity and a decrease in anti-elastase activity, which reveals the protease-antiprotease imbalance compared to normal smokers. It has been reported that the ratio of neutrophils and macrophages is increased in BAL fluid and sputum in patients with COPD. Neutrophils increase the release of proteases, such as neutrophil elastase and matrix metalloproteinase (MMP), and toxic peptides, such as oxidants and defensins in COPD. While neutrophil ratios in bronchial biopsies and induced sputum correlate with the severity of the disease, they also show parallelism with the rate of decline in pulmonary function tests. Macrophages are significantly increased in the airways, lung parenchyma, BAL fluid, and sputum. Macrophages increase the release of TNF-α and chemokines sensitive to some monocytes, especially with the effect of smoking in COPD. Macrophages also cause the release of many MMPs such as MMP-1, MMP-9, and MMP-12. In stable COPD patients, activated T-lymphocytes such as macrophages and especially CD-8 T-lymphocytes, interferon (IFN) -γ, CXCL10 (IFN--induced protein [IP-10]), IL-4 and IL- 9 oscillate. CD-8 T-cells cause a variety of inflammatory changes. CD-8 T-cells have been demonstrated to increase the airways of patients with COPD, and it is known that the CD-4/CD-8 ratio reverses in COPD. Natural killer (NK) cells increase numerically at a low rate in terms of phagocytosis ability in patients with COPD. However, more pronounced increases in NK cell counts were found in the sputum of COPD patients, especially during acute attack periods [5 - 8].

Increases in the number of neutrophils, CD-8 T cells, and sometimes eosinophils are observed in the induced sputum of patients with COPD. The increase in eosinophils helps identify coexisting asthma or higher bronchodilator response and subgroups of patients who will benefit from corticosteroids [3, 6, 9]. Increased levels of IL-8 in the sputum of patients with COPD correlate with bacterial load in the airways and myeloperoxidase released from activated neutrophils. IL-6, IL-1β, TNF-α, and IL-8 were increased in induced sputum; monocyte chemotactic factor-1 (MCF-1) increases, causing IL-8 release in the bronchial epithelium. Cytokines are proteins that are generally smaller than 80 kilodaltons and enable communication between cells. Cytokines have important roles in inflammation, and they function with surface receptors with high affinity for them. TNF-α increases ICAM-1 level in patients with COPD. The activation of macrophages results in the formation of MMP. In bronchial epithelial cells, extracellular TNF-α plays an important role in the production of tenascin, a matrix glycoprotein. Increased serum TNF-α concentration has been shown to be associated with weight loss in COPD. TNF-α and receptors in COPD are higher than non-smokers. IL-8 is a neutrophil-specific chemotactic factor and a CXC chemokine that activates LTB4 and neutrophil 5-lipoxygenase. LTB4 is known to provide chemotactic activity in the sputum of patients with COPD. IL-8 also functions as a chemotactic for T-cells. IL-8 was found to be higher in induced sputum in COPD patients compared to smokers without COPD and the control group. The value of IL-8 increases even more with the severity of the decrease in FEV1 values, and this becomes more pronounced in attacks. Leptin, another important cytokine, is detected in the induced sputum of patients with COPD and correlated with TNF-α and other inflammatory markers such as CRP [10].

Measurement of markers in exhaled air is preferred due to its noninvasiveness and easy reproducibility. Nitric oxide (NO) in the expiratory air (eNO, FeNO) has been extensively studied in asthma. However, the clinical benefit of measuring eNO in COPD is limited and normal or slight elevations can be seen. The increase in eNO in COPD correlates with the increase in bronchodilator and corticosteroid response, just like the increase in eosinophils [11, 12]. It is not useful for practical use as it is highly variable due to environmental CO levels and the effects of passive smoking Concentrated exhaled breath condensate (EBC) is easy to make It has been preferred frequently in recent years due to its nature and non-invasiveness, protein biomarkers in exhaled air could not be measured reliably by conventional methods [13, 14]. Hydrogen peroxide (H_2O_2) and 8-isoprostane, which are among oxidative stress markers, are found to be high in the EBCs of patients with COPD. pH is lower. While pro-inflammatory cytokines, including IL-1, IL-6, and TNF-α, are detected higher in exhalation air during COPD exacerbation, some important markers are fully observed in EBCs [6, 14].

During acute attacks of COPD, high rates of eosinophils are detected in sputum and bronchial biopsies. IL-6 and IL-8 levels in induced sputum were higher in patients with three or more attacks compared to patients with two or fewer attacks. IL-1 increases the release of neutrophils from the bone marrow. The release of IL-6, IL-8, GM-CSF, and TNF-α and T-cells -Chemokines (regulated on activation, normal T-cells expressed and secreted (RANTES) increases its production. RANTES mRNA expression is increased in the surface epithelium and subepithelial lympho mononuclear cells in patients with COPD attack. Increased release of CC chemokines such as eotacsin and RANTES increases eosinophilia in COPD attacks [6, 7, 15].

There are differences in the distribution of inflammatory cells and markers in COPD and asthma. Eosinophils and CD-4 T (Th2) cells in asthma, neutrophils and CD-8 T (Th1) cells are prominent in COPD. Cytokines such as IL-4, IL-5, IL-10 and IL-13 are released in asthma, and IL-8, IL-1, IL-2, IFN, and TNF-α in COPD. However, eosinophilic inflammation can also be observed in some COPD patients [6, 7, 16].

In terms of the general evaluation of biomarkers, there are large observational studies that increase our understanding of COPD by creating a very valuable biological, genetic, and clinical data set in large cohorts. SPIROMICS (the Subpopulations and Intermediate Outcomes in COPD Study) COPECLIPSE (Evaluation of COPD Longitudinally to Identify Predictive Surrogate End-points) DGene GWAS (Genom-wild association studies) are important examples [5, 17, 18]. The circulating biomarkers are validated with regards to sensitivity, accuracy, precision and reproducibility are shown in Table **1**.

Although there are several comments regarding the eosinophilic cut-off, a promising biomarker of eosinophilia in COPD, it has been shown to be useful in predicting the response of acute attack treatments. In stable COPD there is a positive relationship between blood and sputum eosinophilia [85, 86].

In the ECLIPSE cohort (1755 COPD, 297 SS, 202 NSS) 6 systemic inflammation markers CRP, fibrinogen, blood leukocyte, IL6, IL-8, TNFα combined were evaluated [17]; only fibrinogen and CRP clinical validation were provided. Plasma fibrinogen is promising as the most stable biomarker studied, it has been associated with symptoms, exercise capacity, exacerbation, BODE index and mortality, its regular follow-up may provide prognostic information, but further clinical studies are needed [18, 40, 87 - 92]. Biomarkers associated with exacerbation in COPD are shown in Table **2**.

Table 1. Circulating biomarkers in COPD patients.

Biomarker	Main finding
Serum Adiponectin	The adipose tissue, the pathogenesis and mortality [19 - 24].
Club cell protein 16 (CC-16, previously named Clara cell protein)	Pathogenesis and disease progression [25 - 29].
Monocyte chemoattractant protein-1 (MCP-1, CCL2)	The inflammatory processes, disease pathophysiology, and severity [30 - 33].
Chemokines (CC chemokine) as CXCL8 (IL-8) -CXCR1/2, CXCL9/10/11-CXCR3, CCL2-CCR2, and CXCL12/CXCR4	The pathophysiology, potential COPD treatment [34].
Pulmonary and activation-regulated chemokine= CC chemokine ligand-18 (PARC/CCL-18)	Diagnostic, prognostic, and survival [29, 35, 36]. Cardiovascular disease, cardiovascular events observed during the treatment period, morbidity and mortality [37 - 45].
Tumor necrosis factor (TNF-α)	The inflammatory processes, Anti-TNF therapies [46, 47].
Interleukin-1 beta (IL-1β)	Activates macrophages, pathogenesis and disease progression [48, 49].
Interleukin-6 (IL-6)	Innate and acquired immunity, treatment [50 - 52].
Interleukin-32 (IL-32)	Disease severity [53].
Interleukin-18 (IL-18)	Disease severity [54, 55], systemic manifestations of this disease [56].
Granulocyte-macrophage colony-stimulating factor (GM-CSF)	In the differentiation and survival of neutrophils, eosinophils, and macrophages in COPD [48].
Transforming growth factor-beta (TGF-β)	A multifunctional growth factor, which induces proliferation of fibroblasts and airway smooth muscle cells, deposition of extracellular matrix, and epithelial repair [57 - 59] and the fibrotic effects [59, 60].
Vascular-endothelial growth factor (VEGF)	In regulating the growth of new vessels and vascular leakage [59, 60].
Epidermal growth factor (EGF)	In regulating mucus secretion [62], in airway epithelial cells [61, 62].
IL-10	Inhibits the synthesis of many inflammatory proteins [63].
C-reactive protein (CRP)	Systemic inflammatory marker, disease severity, progression and mortality [64].
Fibrinogen	Prognosis, mortality [27, 65 - 67].
Surfactant Protein D	Pathogenesis, risk and progression, COPD treatment [68 - 76].
Plasma pro-surfactant protein B (pro-SFTPB)	Severity and progression, response to therapy and prognosis [77, 78].

(Table 1) cont.....

Biomarker	Main finding
Matrix metalloproteinase-9	Severity and progression, response to therapy and prognosis, the presence of osteoporosis [79 - 84].

Table 2. Biomarkers associated with exacerbation in COPD.

Biomarker	Function
Serum adiponectin	Exacerbation period [100 - 102].
Serum C-reactive protein (CRP)	Re-hospitalization, short-term mortality, bacterial exacerbation [103 - 106].
β-Defensin-2	Severe exacerbations these patients experience in response to viral-bacterial co-infections [107].
Monocyte chemoattractant protein-1 (MCP-1, CCL2)	During exacerbations [108].
Serum interferon-gamma-inducible protein 10 (IP-10)	Human Rhinovirus infection [109].
Pulmonary and activation-regulated chemokine= CC chemokine ligand-18 (PARC/CCL-18)	During exacerbations [110, 111].
Serum IL-1 Beta	Persistent neutrophilic airway inflammation and potential ongoing exacerbation [112].
Serum chemokine ligand 10 (CXCL10)	Frequent exacerbations [113].
Chemokines as CXCL8-CXCR1/2, CXCL9/10/11-CXCR3, and CCL2-CCR2	During exacerbations [34].
Peripheral blood eosinophil	Frequent and severe exacerbations [114].
Proadrenomedullin	Length of stay in the hospital [115, 116].
Prokalsitonin	Short term mortality [117 - 119].
Surfactant protein D	Exacerbation severity and mortality [120].
Matrix metalloproteinase-9	During exacerbations [121].

In recent years, studies on genetic biomarkers in COPD have been extremely important. Genome-wild association studies (Gwas) studies analyzed smoking-related genomic regions in 4 independent cohorts; smoking behavior, susceptibility to COPD development and genes associated with different COPD subtypes were examined Histone methylation dysregulation gene in COPD patients compared to healthy (smoker / nonsmoker) promoters have been shown to be hypo-methylated, and specific miRNA expressions that are effective in activation of inflammatory mediators in COPD patients have also been shown to change [17, 93 - 96]. There is no biomarker yet to distinguish exacerbation from stable disease. Serum biomarkers (IP-10 and IL-1β) may be useful in distinguishing bacterial exacerbations from viral causes. Procalcitonin can be used to prevent the unnecessary use of antibiotics [97 - 99].

CONCLUSION

Among the inflammatory markers that play a major role in inflammation in COPD is ESR, CRP, IL-8, IL-1, IL-6, TNF-α and leukotriene B4. The importance of changes in these markers during exacerbation and after treatment has been emphasized in various studies. As a result of these studies, only a few markers of inflammation (neutrophil ratio in sputum, IL-8 level in sputum and TNF-α) were determined by the development, severity, and progression of the disease. Relationship has been determined and information has been obtained that it can be used in clinical use. In order for inflammatory markers to be applicable in clinical studies, multi-center studies involving many patients should be conducted. Further validation is required to test the clinical use of these circulating markers as biomarkers in COPD progression.

CONSENT FOR PUBLICATION

Not applicable.

CONFLICT OF INTEREST

The authors declare no conflict of interest, financial or otherwise.

ACKNOWLEDGEMENT

Declared none.

REFERENCES

[1] Vestbo J, Hurd SS, Agustí AG, *et al.* Global strategy for the diagnosis, management, and prevention of chronic obstructive pulmonary disease: GOLD executive summary. Am J Respir Crit Care Med 2013; 187(4): 347-65. [http://dx.doi.org/10.1164/rccm.201204-0596PP] [PMID: 22878278]

[2] Barnes PJ, Shapiro SD, Pauwels RA. Chronic obstructive pulmonary disease: molecular and cellular mechanisms. Eur Respir J 2003; 22(4): 672-88. [http://dx.doi.org/10.1183/09031936.03.00040703] [PMID: 14582923]

[3] Fermont JM, Masconi KL, Jensen MT, *et al.* Biomarkers and clinical outcomes in COPD: a systematic review and meta-analysis. Thorax 2019; 74(5): 439-46. [http://dx.doi.org/10.1136/thoraxjnl-2018-211855] [PMID: 30617161]

[4] Vincent JL, Bogossian E, Menozzi M. The Future of Biomarkers. Crit Care Clin 2020; 36(1): 177-87. [http://dx.doi.org/10.1016/j.ccc.2019.08.014] [PMID: 31733679]

[5] Couper D, LaVange LM, Han M, *et al.* Design of the Subpopulations and Intermediate Outcomes in COPD Study (SPIROMICS). Thorax 2014; 69(5): 491-4. [http://dx.doi.org/10.1136/thoraxjnl-2013-203897] [PMID: 24029743]

[6] Barnes PJ, Chowdhury B, Kharitonov SA, *et al.* Pulmonary biomarkers in chronic obstructive pulmonary disease. Am J Respir Crit Care Med 2006; 174(1): 6-14. [http://dx.doi.org/10.1164/rccm.200510-1659PP] [PMID: 16556692]

[7] Chung KF. Cytokines in chronic obstructive pulmonary disease. Eur Respir J Suppl 2001; 34(1): 50s-9s.
[http://dx.doi.org/10.1183/09031936.01.00229701] [PMID: 12392035]

[8] Hattotuwa K, Gamble EA, O'Shaughnessy T, Jeffery PK, Barnes NC. Safety of bronchoscopy, biopsy, and BAL in research patients with COPD. Chest 2002; 122(6): 1909-12.
[http://dx.doi.org/10.1378/chest.122.6.1909] [PMID: 12475825]

[9] Brightling CE, Monteiro W, Ward R, *et al.* Sputum eosinophilia and short-term response to prednisolone in chronic obstructive pulmonary disease: a randomised controlled trial. Lancet 2000; 356(9240): 1480-5.
[http://dx.doi.org/10.1016/S0140-6736(00)02872-5] [PMID: 11081531]

[10] Riise GC, Larsson S, Löfdahl CG, Andersson BA. Circulating cell adhesion molecules in bronchial lavage and serum in COPD patients with chronic bronchitis. Eur Respir J 1994; 7(9): 1673-7.
[http://dx.doi.org/10.1183/09031936.94.07091673] [PMID: 7527787]

[11] Kharitonov SA, Barnes PJ. Exhaled markers of pulmonary disease. Am J Respir Crit Care Med 2001; 163(7): 1693-722.
[http://dx.doi.org/10.1164/ajrccm.163.7.2009041] [PMID: 11401895]

[12] ATS/ERS recommendations for standardized procedures for the online and offline measurement of exhaled lower respiratory nitric oxide and nasal nitric oxide, 2005. Am J Respir Crit Care Med 2005; 171(8): 912-30.
[http://dx.doi.org/10.1164/rccm.200406-710ST] [PMID: 15817806]

[13] Horváth I, Hunt J, Barnes PJ, *et al.* Exhaled breath condensate: methodological recommendations and unresolved questions. Eur Respir J 2005; 26(3): 523-48.
[http://dx.doi.org/10.1183/09031936.05.00029705] [PMID: 16135737]

[14] Gessner C, Scheibe R, Wötzel M, *et al.* Exhaled breath condensate cytokine patterns in chronic obstructive pulmonary disease. Respir Med 2005; 99(10): 1229-40.
[http://dx.doi.org/10.1016/j.rmed.2005.02.041] [PMID: 16140223]

[15] Wedzicha JA, Seemungal TA, MacCallum PK, *et al.* Acute exacerbations of chronic obstructive pulmonary disease are accompanied by elevations of plasma fibrinogen and serum IL-6 levels. Thromb Haemost 2000; 84(2): 210-5.
[PMID: 10959691]

[16] Jeffery PK. Structural and inflammatory changes in COPD: a comparison with asthma. Thorax 1998; 53(2): 129-36.
[http://dx.doi.org/10.1136/thx.53.2.129] [PMID: 9624299]

[17] Agusti A, Gea J, Faner R. Biomarkers, the control panel and personalized COPD medicine. Respirology 2016; 21(1): 24-33.
[http://dx.doi.org/10.1111/resp.12585] [PMID: 26172306]

[18] Faner R, Tal-Singer R, Riley JH, *et al.* Lessons from ECLIPSE: a review of COPD biomarkers. Thorax 2014; 69(7): 666-72.
[http://dx.doi.org/10.1136/thoraxjnl-2013-204778] [PMID: 24310110]

[19] Bianco A, Nigro E, Monaco ML, *et al.* The burden of obesity in asthma and COPD: Role of adiponectin. Pulm Pharmacol Ther 2017; 43: 20-5.
[http://dx.doi.org/10.1016/j.pupt.2017.01.004] [PMID: 28115224]

[20] Mokari-Yamchi A, Sharifi A, Kheirouri S. Increased serum levels of S100A1, ZAG, and adiponectin in cachectic patients with COPD. Int J Chron Obstruct Pulmon Dis 2018; 13: 3157-63.
[http://dx.doi.org/10.2147/COPD.S172996] [PMID: 30349224]

[21] Breyer MK, Rutten EP, Locantore NW, Watkins ML, Miller BE, Wouters EF. Dysregulated adipokine metabolism in chronic obstructive pulmonary disease. Eur J Clin Invest 2012; 42(9): 983-91.
[http://dx.doi.org/10.1111/j.1365-2362.2012.02686.x] [PMID: 22703238]

[22] Breyer MK, Rutten EPA, Vernooy JHJ, *et al.* Gender differences in the adipose secretome system in chronic obstructive pulmonary disease (COPD): a pivotal role of leptin. Respir Med 2011; 105(7): 1046-53.
[http://dx.doi.org/10.1016/j.rmed.2011.02.001] [PMID: 21367591]

[23] Tomoda K, Yoshikawa M, Itoh T, *et al.* Elevated circulating plasma adiponectin in underweight patients with COPD. Chest 2007; 132(1): 135-40.
[http://dx.doi.org/10.1378/chest.07-0227] [PMID: 17625082]

[24] Mohamed NA, Fawzy MA, Elgamry R, Gad DM, Ibraheem HA. Role of adiponectin and other inflammatory biomarkers in COPD patients. Egypt J Chest Dis Tuberc 2013; 62(1): 45-50.
[http://dx.doi.org/10.1016/j.ejcdt.2013.03.005]

[25] Milne S, Li X, Hernandez Cordero AI, *et al.* Protective effect of club cell secretory protein (CC-16) on COPD risk and progression: a Mendelian randomisation study. Thorax 2020; 75(11): 934-43.
[http://dx.doi.org/10.1136/thoraxjnl-2019-214487] [PMID: 32839289]

[26] Park HY, Churg A, Wright JL, *et al.* Club cell protein 16 and disease progression in chronic obstructive pulmonary disease. Am J Respir Crit Care Med 2013; 188(12): 1413-9.
[http://dx.doi.org/10.1164/rccm.201305-0892OC] [PMID: 24245748]

[27] Celli BR, Anderson JA, Brook R, *et al.* Serum biomarkers and outcomes in patients with moderate COPD: a substudy of the randomised SUMMIT trial. BMJ Open Respir Res 2019; 6(1): e000431.
[http://dx.doi.org/10.1136/bmjresp-2019-000431] [PMID: 31258919]

[28] Oh JY, Lee YS, Min KH, *et al.* Decreased serum club cell secretory protein in asthma and chronic obstructive pulmonary disease overlap: a pilot study. Int J Chron Obstruct Pulmon Dis 2018; 13: 3411-7.
[http://dx.doi.org/10.2147/COPD.S174545] [PMID: 30425470]

[29] Lock-Johansson S, Vestbo J, Sorensen GL. Surfactant protein D, Club cell protein 16, Pulmonary and activation-regulated chemokine, C-reactive protein, and Fibrinogen biomarker variation in chronic obstructive lung disease. Respir Res 2014; 15(1): 147.
[http://dx.doi.org/10.1186/s12931-014-0147-5] [PMID: 25425298]

[30] Di Stefano A, Coccini T, Roda E, *et al.* Blood MCP-1 levels are increased in chronic obstructive pulmonary disease patients with prevalent emphysema. Int J Chron Obstruct Pulmon Dis 2018; 13: 1691-700.
[http://dx.doi.org/10.2147/COPD.S159915] [PMID: 29872287]

[31] Wang Y, Zheng Y, Zhai YL, Liu FQ, Ding N. Comparative analysis of MCP-1 and TF in elderly patients with acute exacerbations of COPD and its clinical significance. Eur Rev Med Pharmacol Sci 2015; 19(2): 215-9.
[PMID: 25683933]

[32] de Boer WI, Sont JK, van Schadewijk A, Stolk J, van Krieken JH, Hiemstra PS. Monocyte chemoattractant protein 1, interleukin 8, and chronic airways inflammation in COPD. J Pathol 2000; 190(5): 619-26.
[http://dx.doi.org/10.1002/(SICI)1096-9896(200004)190:5<619::AID-PATH555>3.0.CO;2-6] [PMID: 10727989]

[33] Liu SF, Chin CH, Wang CC, Lin MC. Correlation between serum biomarkers and BODE index in patients with stable COPD. Respirology 2009; 14(7): 999-1004.
[http://dx.doi.org/10.1111/j.1440-1843.2009.01608.x] [PMID: 19740260]

[34] Henrot P, Prevel R, Berger P, Dupin I. Chemokines in COPD: From Implication to Therapeutic Use. Int J Mol Sci 2019; 20(11): 2785.
[http://dx.doi.org/10.3390/ijms20112785] [PMID: 31174392]

[35] Rosenberg SR, Kalhan R. Biomarkers in chronic obstructive pulmonary disease. Transl Res 2012; 159(4): 228-37.

[http://dx.doi.org/10.1016/j.trsl.2012.01.019] [PMID: 22424427]

[36] Pinto-Plata V, Toso J, Lee K, *et al.* Profiling serum biomarkers in patients with COPD: associations with clinical parameters. Thorax 2007; 62(7): 595-601. [http://dx.doi.org/10.1136/thx.2006.064428] [PMID: 17356059]

[37] Yoon HI, Sin DD. Biomarkers of therapeutic response in patients with chronic obstructive pulmonary disease: a critical review of the literature. Drugs 2011; 71(14): 1821-37. [http://dx.doi.org/10.2165/11595180-000000000-00000] [PMID: 21942975]

[38] Sin DD, Miller BE, Duvoix A, *et al.* Serum PARC/CCL-18 concentrations and health outcomes in chronic obstructive pulmonary disease. Am J Respir Crit Care Med 2011; 183(9): 1187-92. [http://dx.doi.org/10.1164/rccm.201008-1220OC] [PMID: 21216880]

[39] Pinto-Plata V, Casanova C, Müllerova H, *et al.* Inflammatory and repair serum biomarker pattern: association to clinical outcomes in COPD. Respir Res 2012; 13(1): 71. [http://dx.doi.org/10.1186/1465-9921-13-71] [PMID: 22906131]

[40] Lomas DA, Silverman EK, Edwards LD, Miller BE, Coxson HO, Tal-Singer R. Evaluation of serum CC-16 as a biomarker for COPD in the ECLIPSE cohort. Thorax 2008; 63(12): 1058-63. [http://dx.doi.org/10.1136/thx.2008.102574] [PMID: 18757456]

[41] Kraaijeveld AO, de Jager SC, de Jager WJ, *et al.* CC chemokine ligand-5 (CCL5/RANTES) and CC chemokine ligand-18 (CCL18/PARC) are specific markers of refractory unstable angina pectoris and are transiently raised during severe ischemic symptoms. Circulation 2007; 116(17): 1931-41. [http://dx.doi.org/10.1161/CIRCULATIONAHA.107.706986] [PMID: 17909104]

[42] Cazzola M, MacNee W, Martinez FJ, *et al.* Outcomes for COPD pharmacological trials: from lung function to biomarkers. Eur Respir J 2008; 31(2): 416-69. [http://dx.doi.org/10.1183/09031936.00099306] [PMID: 18238951]

[43] Muñoz-Esquerre M, Aliagas E, López-Sánchez M, *et al.* Vascular disease in COPD: Systemic and pulmonary expression of PARC (Pulmonary and Activation-Regulated Chemokine). PLoS One 2017; 12(5): e0177218. [http://dx.doi.org/10.1371/journal.pone.0177218] [PMID: 28545096]

[44] Pinto-Plata V, Casanova C, Müllerova H, *et al.* Inflammatory and repair serum biomarker pattern: association to clinical outcomes in COPD. Respir Res 2012; 13: 71. [http://dx.doi.org/10.1186/1465-9921-13-71]

[45] Sin DD, Man SF, Connett JE, *et al.* Pulmonary and activation-regulated chemokine (PARC): new lung-specific blood biomarker for COPD? Am J Respir Crit Care Med 2009; 179: A3763.

[46] Dentener MA, Creutzberg EC, Pennings HJ, Rijkers GT, Mercken E, Wouters EF. Effect of infliximab on local and systemic inflammation in chronic obstructive pulmonary disease: a pilot study. Respiration 2008; 76(3): 275-82. [http://dx.doi.org/10.1159/000117386] [PMID: 18277064]

[47] Suissa S, Ernst P, Hudson M. TNF-alpha antagonists and the prevention of hospitalisation for chronic obstructive pulmonary disease. Pulm Pharmacol Ther 2008; 21(1): 234-8. [http://dx.doi.org/10.1016/j.pupt.2007.03.003] [PMID: 17517530]

[48] Culpitt SV, Rogers DF, Shah P, *et al.* Impaired inhibition by dexamethasone of cytokine release by alveolar macrophages from patients with chronic obstructive pulmonary disease. Am J Respir Crit Care Med 2003; 167(1): 24-31. [http://dx.doi.org/10.1164/rccm.200204-298OC] [PMID: 12406856]

[49] Sapey E, Ahmad A, Bayley D, *et al.* Imbalances between interleukin-1 and tumor necrosis factor agonists and antagonists in stable COPD. J Clin Immunol 2009; 29(4): 508-16. [http://dx.doi.org/10.1007/s10875-009-9286-8] [PMID: 19291375]

[50] Bhowmik A, Seemungal TA, Sapsford RJ, Wedzicha JA. Relation of sputum inflammatory markers to symptoms and lung function changes in COPD exacerbations. Thorax 2000; 55(2): 114-20.

[http://dx.doi.org/10.1136/thorax.55.2.114] [PMID: 10639527]

[51] Bucchioni E, Kharitonov SA, Allegra L, Barnes PJ. High levels of interleukin-6 in the exhaled breath condensate of patients with COPD. Respir Med 2003; 97(12): 1299-302. [http://dx.doi.org/10.1016/j.rmed.2003.07.008] [PMID: 14682411]

[52] Barnes PJ, Celli BR. Systemic manifestations and comorbidities of COPD. Eur Respir J 2009; 33(5): 1165-85. [http://dx.doi.org/10.1183/09031936.00128008] [PMID: 19407051]

[53] Calabrese F, Baraldo S, Bazzan E, *et al.* IL-32, a novel proinflammatory cytokine in chronic obstructive pulmonary disease. Am J Respir Crit Care Med 2008; 178(9): 894-901. [http://dx.doi.org/10.1164/rccm.200804-646OC] [PMID: 18703789]

[54] Imaoka H, Hoshino T, Takei S, *et al.* Interleukin-18 production and pulmonary function in COPD. Eur Respir J 2008; 31(2): 287-97. [http://dx.doi.org/10.1183/09031936.00019207] [PMID: 17989120]

[55] Rovina N, Dima E, Gerassimou C, Kollintza A, Gratziou C, Roussos C. Interleukin-18 in induced sputum: association with lung function in chronic obstructive pulmonary disease. Respir Med 2009; 103(7): 1056-62. [http://dx.doi.org/10.1016/j.rmed.2009.01.011] [PMID: 19208460]

[56] Petersen AM, Penkowa M, Iversen M, *et al.* Elevated levels of IL-18 in plasma and skeletal muscle in chronic obstructive pulmonary disease. Lung 2007; 185(3): 161-71. [http://dx.doi.org/10.1007/s00408-007-9000-7] [PMID: 17436040]

[57] Wan YY, Flavell RA. Regulatory T cells, transforming growth factor-beta, and immune suppression. Proc Am Thorac Soc 2007; 4(3): 271-6. [http://dx.doi.org/10.1513/pats.200701-020AW] [PMID: 17607012]

[58] de Boer WI, van Schadewijk A, Sont JK, *et al.* Transforming growth factor beta1 and recruitment of macrophages and mast cells in airways in chronic obstructive pulmonary disease. Am J Respir Crit Care Med 1998; 158(6): 1951-7. [http://dx.doi.org/10.1164/ajrccm.158.6.9803053] [PMID: 9847291]

[59] Takizawa H, Tanaka M, Takami K, *et al.* Increased expression of transforming growth factor-beta1 in small airway epithelium from tobacco smokers and patients with chronic obstructive pulmonary disease (COPD). Am J Respir Crit Care Med 2001; 163(6): 1476-83. [http://dx.doi.org/10.1164/ajrccm.163.6.9908135] [PMID: 11371421]

[60] Ning W, Li CJ, Kaminski N, *et al.* Comprehensive gene expression profiles reveal pathways related to the pathogenesis of chronic obstructive pulmonary disease. Proc Natl Acad Sci USA 2004; 101(41): 14895-900. [http://dx.doi.org/10.1073/pnas.0401168101] [PMID: 15469929]

[61] Shao MX, Nadel JA. Neutrophil elastase induces MUC5AC mucin production in human airway epithelial cells *via* a cascade involving protein kinase C, reactive oxygen species, and TNF-alph--converting enzyme. J Immunol 2005; 175(6): 4009-16. [http://dx.doi.org/10.4049/jimmunol.175.6.4009] [PMID: 16148149]

[62] Takeyama K, Fahy JV, Nadel JA. Relationship of epidermal growth factor receptors to goblet cell production in human bronchi. Am J Respir Crit Care Med 2001; 163(2): 511-6. [http://dx.doi.org/10.1164/ajrccm.163.2.2001038] [PMID: 11179132]

[63] Takanashi S, Hasegawa Y, Kanehira Y, *et al.* Interleukin-10 level in sputum is reduced in bronchial asthma, COPD and in smokers. Eur Respir J 1999; 14(2): 309-14. [http://dx.doi.org/10.1183/09031936.99.14230999] [PMID: 10515406]

[64] Man SF, Xing L, Connett JE, *et al.* Circulating fibronectin to C-reactive protein ratio and mortality: a biomarker in COPD? Eur Respir J 2008; 32(6): 1451-7. [http://dx.doi.org/10.1183/09031936.00153207] [PMID: 18799503]

[65] Zemans RL, Jacobson S, Keene J, *et al.* Multiple biomarkers predict disease severity, progression and mortality in COPD. Respir Res 2017; 18(1): 117. [http://dx.doi.org/10.1186/s12931-017-0597-7] [PMID: 28610627]

[66] Dong T, Santos S, Yang Z, Yang S, Kirkhus NE. Sputum and salivary protein biomarkers and point-of-care biosensors for the management of COPD. Analyst (Lond) 2020; 145(5): 1583-604. [http://dx.doi.org/10.1039/C9AN01704F] [PMID: 31915768]

[67] Leuzzi G, Galeone C, Taverna F, Suatoni P, Morelli D, Pastorino U. C-reactive protein level predicts mortality in COPD: a systematic review and meta-analysis. Eur Respir Rev 2017; 26(143): 160070. [http://dx.doi.org/10.1183/16000617.0070-2016] [PMID: 28143876]

[68] Khan RA, Waseem HM, Ahmad HU, Lone KP. Plasma surfactant Protein-D levels in healthy subjects and COPD patients. J Pak Med Assoc 2019; 69(4): 494-8. [PMID: 31000851]

[69] Fakih D, Akiki Z, Junker K, *et al.* Surfactant protein D multimerization and gene polymorphism in COPD and asthma. Respirology 2018; 23(3): 298-305. [http://dx.doi.org/10.1111/resp.13193] [PMID: 28960651]

[70] Obeidat M, Li X, Burgess S, *et al.* Surfactant protein D is a causal risk factor for COPD: results of Mendelian randomisation. Eur Respir J 2017; 50(5): 1700657. [http://dx.doi.org/10.1183/13993003.00657-2017] [PMID: 29191953]

[71] Moreno D, Garcia A, Lema D, De Sanctis JB. Surfactant protein D in chornic obstructive pulmonary disease (COPD). Recent Pat Endocr Metab Immune Drug Discov 2014; 8(1): 42-7. [http://dx.doi.org/10.2174/1872214808666140209142640] [PMID: 24506680]

[72] Liao Y, Huang C, Wang J, Fan X. Association of Surfactant-Associated Protein D Gene Polymorphisms with the Risk of COPD: a Meta-Analysis. Clinics (São Paulo) 2019; 74: e855. [http://dx.doi.org/10.6061/clinics/2019/e855] [PMID: 31116231]

[73] Sin DD, Pahlavan PS, Man SF. Surfactant protein D: a lung specific biomarker in COPD? Ther Adv Respir Dis 2008; 2(2): 65-74. [http://dx.doi.org/10.1177/1753465808088903] [PMID: 19124360]

[74] Wang H, Li F, Huang H, *et al.* Serum Surfactant Protein D is a Potential Biomarker for Chronic Obstructive Pulmonary Disease: a Systematic Review and Meta-analysis. Clin Lab 2019; 65(12). [http://dx.doi.org/10.7754/Clin.Lab.2019.190539] [PMID: 31850711]

[75] Bowler RP. Surfactant protein D as a biomarker for chronic obstructive pulmonary disease. COPD 2012; 9(6): 651-3. [http://dx.doi.org/10.3109/15412555.2012.736317] [PMID: 23244171]

[76] Akiki Z, Fakih D, Jounblat R, *et al.* Surfactant protein D, a clinical biomarker for chronic obstructive pulmonary disease with excellent discriminant values. Exp Ther Med 2016; 11(3): 723-30. [http://dx.doi.org/10.3892/etm.2016.2986] [PMID: 26997985]

[77] Leung JM, Mayo J, Tan W, *et al.* Plasma pro-surfactant protein B and lung function decline in smokers. Eur Respir J 2015; 45(4): 1037-45. [http://dx.doi.org/10.1183/09031936.00184214] [PMID: 25614175]

[78] Um SJ, Lam S, Coxson H, Man SF, Sin DD. Budesonide/formoterol enhances the expression of pro Surfactant Protein-B in lungs of COPD patients. PLoS One 2013; 8(12): e83881. [http://dx.doi.org/10.1371/journal.pone.0083881] [PMID: 24386300]

[79] Gilowska I, Kasper Ł, Bogacz K, *et al.* Impact of Matrix Metalloproteinase 9 on COPD Development in Polish Patients: Genetic Polymorphism, Protein Level, and Their Relationship with Lung Function. BioMed Res Int 2018; 2018: 6417415. [http://dx.doi.org/10.1155/2018/6417415] [PMID: 30643813]

[80] Maskey-Warzęchowska M, Górska K, Nejman-Gryz P, *et al.* Matrix metalloproteinase 9 in exhaled

breath condensate in patients with stable chronic obstructive pulmonary disease: an observational study. Pol Arch Intern Med 2018; 128(7-8): 427-33.
[PMID: 30057384]

[81] Uysal P, Uzun H. Relationship Between Circulating Serpina3g, Matrix Metalloproteinase-9, and Tissue Inhibitor of Metalloproteinase-1 and -2 with Chronic Obstructive Pulmonary Disease Severity. Biomolecules 2019; 9(2): 62.
[http://dx.doi.org/10.3390/biom9020062] [PMID: 30781876]

[82] Linder R, Rönmark E, Pourazar J, Behndig A, Blomberg A, Lindberg A. Serum metalloproteinase-9 is related to COPD severity and symptoms - cross-sectional data from a population based cohort-study. Respir Res 2015; 16(1): 28.
[http://dx.doi.org/10.1186/s12931-015-0188-4] [PMID: 25849664]

[83] Bolton CE, Stone MD, Edwards PH, Duckers JM, Evans WD, Shale DJ. Circulating matrix metalloproteinase-9 and osteoporosis in patients with chronic obstructive pulmonary disease. Chron Respir Dis 2009; 6(2): 81-7.
[http://dx.doi.org/10.1177/1479972309103131] [PMID: 19411568]

[84] Wells JM, Parker MM, Oster RA, *et al.* Elevated circulating MMP-9 is linked to increased COPD exacerbation risk in SPIROMICS and COPDGene. JCI Insight 2018; 3(22): e123614.
[http://dx.doi.org/10.1172/jci.insight.123614] [PMID: 30429371]

[85] Pascoe S, Locantore N, Dransfield MT, Barnes NC, Pavord ID. Blood eosinophil counts, exacerbations, and response to the addition of inhaled fluticasone furoate to vilanterol in patients with chronic obstructive pulmonary disease: a secondary analysis of data from two parallel randomised controlled trials. Lancet Respir Med 2015; 3(6): 435-42.
[http://dx.doi.org/10.1016/S2213-2600(15)00106-X] [PMID: 25878028]

[86] Negewo N, McDonald V, Baines K, *et al.* Peripheral blood eosinophils: a surrogate marker for airway eosinophilia in stable COPD. Negewo NA, McDonald VM, Baines KJ, Wark PA, Simpson JL, Jones PW, Gibson PG. Int J Chron Obstruct Pulmon Dis 2016; 11: 1495-504.
[http://dx.doi.org/10.2147/COPD.S100338]

[87] Dickens JA, Miller BE, Edwards LD, Silverman EK, Lomas DA, Tal-Singer R. COPD association and repeatability of blood biomarkers in the ECLIPSE cohort. Respir Res 2011; 12(1): 146.
[http://dx.doi.org/10.1186/1465-9921-12-146] [PMID: 22054035]

[88] Celli BR, Locantore N, Yates J, *et al.* Inflammatory biomarkers improve clinical prediction of mortality in chronic obstructive pulmonary disease. Am J Respir Crit Care Med 2012; 185(10): 1065-72.
[http://dx.doi.org/10.1164/rccm.201110-1792OC] [PMID: 22427534]

[89] Duvoix A, Dickens J, Haq I, *et al.* Blood fibrinogen as a biomarker of chronic obstructive pulmonary disease. Thorax 2013; 68(7): 670-6.
[http://dx.doi.org/10.1136/thoraxjnl-2012-201871] [PMID: 22744884]

[90] Casaburi R, Celli B, Crapo J, *et al.* The COPD Biomarker Qualification Consortium (CBQC). COPD 2013; 10(3): 367-77.
[http://dx.doi.org/10.3109/15412555.2012.752807] [PMID: 23713597]

[91] Vestbo J, Edwards LD, Scanlon PD, *et al.* Changes in forced expiratory volume in 1 second over time in COPD. N Engl J Med 2011; 365(13): 1184-92.
[http://dx.doi.org/10.1056/NEJMoa1105482] [PMID: 21991892]

[92] Dahl M, Vestbo J, Lange P, Bojesen SE, Tybjaerg-Hansen A, Nordestgaard BG. C-reactive protein as a predictor of prognosis in chronic obstructive pulmonary disease. Am J Respir Crit Care Med 2007; 175(3): 250-5.
[http://dx.doi.org/10.1164/rccm.200605-713OC] [PMID: 17053205]

[93] Qiu W, Baccarelli A, Carey VJ, *et al.* Variable DNA methylation is associated with chronic obstructive pulmonary disease and lung function. Am J Respir Crit Care Med 2012; 185(4): 373-81.

[http://dx.doi.org/10.1164/rccm.201108-1382OC] [PMID: 22161163]

[94] Wan ES, Qiu W, Baccarelli A, *et al.* Systemic steroid exposure is associated with differential methylation in chronic obstructive pulmonary disease. Am J Respir Crit Care Med 2012; 186(12): 1248-55.
[http://dx.doi.org/10.1164/rccm.201207-1280OC] [PMID: 23065012]

[95] Ezzie ME, Crawford M, Cho JH, *et al.* Gene expression networks in COPD: microRNA and mRNA regulation. Thorax 2012; 67(2): 122-31.
[http://dx.doi.org/10.1136/thoraxjnl-2011-200089] [PMID: 21940491]

[96] Wang M, Huang Y, Liang Z, *et al.* Plasma miRNAs might be promising biomarkers of chronic obstructive pulmonary disease. Clin Respir J 2016; 10(1): 104-11.
[http://dx.doi.org/10.1111/crj.12194] [PMID: 25102970]

[97] Quint JK, Donaldson GC, Goldring JJP, Baghai-Ravary R, Hurst JR, Wedzicha JA. Serum IP-10 as a biomarker of human rhinovirus infection at exacerbation of COPD. Chest 2010; 137(4): 812-22.
[http://dx.doi.org/10.1378/chest.09-1541] [PMID: 19837822]

[98] Damera G, Pham TH, Zhang J, *et al.* A Sputum Proteomic Signature That Associates with Increased IL-1β Levels and Bacterial Exacerbations of COPD. Lung 2016; 194(3): 363-9.
[http://dx.doi.org/10.1007/s00408-016-9877-0] [PMID: 27083436]

[99] Corti C, Fally M, Fabricius-Bjerre A, *et al.* Point-of-care procalcitonin test to reduce antibiotic exposure in patients hospitalized with acute exacerbation of COPD. Int J Chron Obstruct Pulmon Dis 2016; 11: 1381-9.
[http://dx.doi.org/10.2147/COPD.S104051] [PMID: 27382274]

[100] Liu H, Liu JS, Huang J, Zhong LW, Xu JY. Unique association of adiponectin isoforms with serum cytokines and redox molecules in patients with chronic obstructive pulmonary disease. Chin Med J (Engl) 2013; 126(17): 3383-4.
[PMID: 24033970]

[101] Xie J, Yang XY, Shi JD, Deng XQ, Long W. A new inflammation marker of chronic obstructive pulmonary disease-adiponectin. World J Emerg Med 2010; 1(3): 190-5.
[PMID: 25214966]

[102] Kirdar S, Serter M, Ceylan E, Sener AG, Kavak T, Karadağ F. Adiponectin as a biomarker of systemic inflammatory response in smoker patients with stable and exacerbation phases of chronic obstructive pulmonary disease. Scand J Clin Lab Invest 2009; 69(2): 219-24.
[http://dx.doi.org/10.1080/00365510802474400] [PMID: 18946779]

[103] Prins HJ, Duijkers R, van der Valk P, *et al.* CRP-guided antibiotic treatment in acute exacerbations of COPD in hospital admissions. Eur Respir J 2019; 53(5): 1802014.
[http://dx.doi.org/10.1183/13993003.02014-2018] [PMID: 30880285]

[104] Butler CC, Gillespie D, White P, *et al.* C-Reactive Protein Testing to Guide Antibiotic Prescribing for COPD Exacerbations. N Engl J Med 2019; 381(2): 111-20.
[http://dx.doi.org/10.1056/NEJMoa1803185] [PMID: 31291514]

[105] Sacks HS. In primary care, CRP testing reduced antibiotic use in COPD exacerbations without worsening health. Ann Intern Med 2019; 171(10): JC51.
[http://dx.doi.org/10.7326/ACPJ201911190-051] [PMID: 31739334]

[106] Nikolakopoulou S, Hillas G, Perrea D, *et al.* Serum angiopoietin-2 and CRP levels during COPD exacerbations. COPD 2014; 11(1): 46-51.
[http://dx.doi.org/10.3109/15412555.2013.829436] [PMID: 24102428]

[107] Arnason JW, Murphy JC, Kooi C, *et al.* Human β-defensin-2 production upon viral and bacterial co-infection is attenuated in COPD. PLoS One 2017; 12(5): e0175963.
[http://dx.doi.org/10.1371/journal.pone.0175963] [PMID: 28489911]

[108] Zheng J, Shi Y, Xiong L, *et al.* The Expression of IL-6, TNF-α, and MCP-1 in Respiratory Viral

Infection in Acute Exacerbations of Chronic Obstructive Pulmonary Disease. J Immunol Res 2017; 2017: 8539294.
[http://dx.doi.org/10.1155/2017/8539294] [PMID: 28352642]

[109] Quint JK, Donaldson GC, Goldring JJ, Baghai-Ravary R, Hurst JR, Wedzicha JA. Serum IP-10 as a biomarker of human rhinovirus infection at exacerbation of COPD. Chest 2010; 137(4): 812-22.
[http://dx.doi.org/10.1378/chest.09-1541] [PMID: 19837822]

[110] Hurst JR, Donaldson GC, Perera WR, *et al.* Use of plasma biomarkers at exacerbation of chronic obstructive pulmonary disease. Am J Respir Crit Care Med 2006; 174(8): 867-74.
[http://dx.doi.org/10.1164/rccm.200604-506OC] [PMID: 16799074]

[111] Dilektasli AG, Demirdogen Cetinoglu E, Uzaslan E, *et al.* Serum CCL-18 level is a risk factor for COPD exacerbations requiring hospitalization. Int J Chron Obstruct Pulmon Dis 2017; 12: 199-208.
[http://dx.doi.org/10.2147/COPD.S118424] [PMID: 28115842]

[112] Zou Y, Chen X, Liu J, *et al.* Serum IL-1β and IL-17 levels in patients with COPD: associations with clinical parameters. Int J Chron Obstruct Pulmon Dis 2017; 12: 1247-54.
[http://dx.doi.org/10.2147/COPD.S131877] [PMID: 28490868]

[113] Rong B, Fu T, Rong C, Liu W, Li K, Liu H. Association between serum CCL-18 and IL-23 concentrations and disease progression of chronic obstructive pulmonary disease. Sci Rep 2020; 10(1): 17756.
[http://dx.doi.org/10.1038/s41598-020-73903-6] [PMID: 33082377]

[114] Wu HX, Zhuo KQ, Cheng DY. Peripheral Blood Eosinophil as a Biomarker in Outcomes of Acute Exacerbation of Chronic Obstructive Pulmonary Disease. Int J Chron Obstruct Pulmon Dis 2019; 14: 3003-15.
[http://dx.doi.org/10.2147/COPD.S226783] [PMID: 31920297]

[115] Citgez E, Zuur-Telgen M, van der Palen J, van der Valk P, Stolz D, Brusse-Keizer M. Stable-State Midrange Proadrenomedullin Is Associated With Severe Exacerbations in COPD. Chest 2018; 154(1): 51-7.
[http://dx.doi.org/10.1016/j.chest.2018.02.006] [PMID: 29475034]

[116] Grolimund E, Kutz A, Marlowe RJ, *et al.* Long-term Prognosis in COPD Exacerbation: Role of Biomarkers, Clinical Variables and Exacerbation Type. COPD 2015; 12(3): 295-305.
[http://dx.doi.org/10.3109/15412555.2014.949002] [PMID: 25230352]

[117] Abedini A, Kiani A, Emami H, Touhidi MH. Serum Procalcitonin Level as a Predictor of Bacterial Infection in Patients with COPD Exacerbation. Tanaffos 2019; 18(2): 112-7.
[PMID: 32440298]

[118] Mathioudakis AG, Chatzimavridou-Grigoriadou V, Corlateanu A, Vestbo J. Procalcitonin to guide antibiotic administration in COPD exacerbations: a meta-analysis. Eur Respir Rev 2017; 26(143): 160073.
[http://dx.doi.org/10.1183/16000617.0073-2016] [PMID: 28143877]

[119] Ni W, Bao J, Yang D, *et al.* Potential of serum procalcitonin in predicting bacterial exacerbation and guiding antibiotic administration in severe COPD exacerbations: a systematic review and meta-analysis. Infect Dis (Lond) 2019; 51(9): 639-50.
[http://dx.doi.org/10.1080/23744235.2019.1644456] [PMID: 31355690]

[120] Günaydın FE, Günlüoğlu G, Kalkan N, Aktepe EN, Demirkol B, Altın S. The relationship between serum levels of surfactant protein D in COPD exacerbation severity and mortality. Turk J Med Sci 2019; 49(3): 888-93.
[http://dx.doi.org/10.3906/sag-1809-6] [PMID: 31083857]

[121] Kwiatkowska S, Noweta K, Zieba M, Nowak D, Bialasiewicz P. Enhanced exhalation of matrix metalloproteinase-9 and tissue inhibitor of metalloproteinase-1 in patients with COPD exacerbation: a prospective study. Respiration 2012; 84(3): 231-41.
[http://dx.doi.org/10.1159/000339417] [PMID: 22832426]

CHAPTER 16

Biomarkers in Diabetes Mellitus

Fatih Orkun Kundaktepe[1,*]

[1] *Department of Internal Medicine, Istanbul Taksim Research and Training Hospital, Istanbul, Turkey*

Abstract: Diabetes mellitus (DM) is a chronic metabolic disease characterized by hyperglycemia that occurs as a result of impaired insulin secretion and/or insulin effect, or both of these factors. The fact that the disease has both individual and social dimensions makes it important to detect as early as possible and make the necessary lifestyle changes. For this purpose, it becomes necessary to develop fast, effortless, cheap, and reliable methods for diagnosis. We discussed which biochemical markers should enter routine use according to their sensitivity and specificity among the biochemical markers that have been used and are still being studied. In this chapter, we explored some methods that may be used as biomarkers and discussed advantages and pitfalls for each.

Keywords: Advanced glycation end-products, Biomarkers, Diabetes mellitus, Glycated albumin, HbA1c, microRNAs, Netrin, Salivary glucose.

INTRODUCTION

Diabetes mellitus (DM) is a chronic metabolic disease presents with hyperglycemia caused by impaired insulin secretion or insulin resistance and/or destruction of pancreatic β cells. By 2019, IDF declared that approximately 463 million adults were suffering from DM, and by 2045, this number is projected as 700 M. DM also caused 4.2 M deaths in the aforementioned year [1].

DM is classified into three major subgroups; Type I (5-10%), Type II (most common, 90-95%), gestational DM, and some other specific types. Type I diabetes mostly occurring in juveniles is immune-mediated and insulin-dependent diabetes. Patients with Type I DM have to use subcutaneous insulin injections several times a day to survive. Type II diabetes, also known as insulin-independent diabetes, is a result of impaired cellular insulin response, which may be called insulin resistance. In patients with type II DM, there is also a phase called pre-diabetes, which presents insulin resistance and impaired fasting plasma

* **Corresponding author Fatih Orkun Kundaktep:** Department of Internal Medicine, Istanbul Taksim Research and Training Hospital, Istanbul, Turkey; Tel: +90 (530) 826 00 99; E-mail: doktorkun@gmail.com

Pınar Atukeren and Hafize Uzun (Eds.)

glucose (FPG) levels. Insulin resistance and Type II DM are associated with obesity, age, and family history [1].

With novel approaches in DM treatment, it is possible to prevent DM progression before causing microvascular (retinopathy, nephropathy, and neuropathy) and macrovascular (coronary artery disease, peripheral arterial disease, and stroke) complications. In this context, detecting DM at the pre-diabetes phase provides better control of this disease. Thus, new diagnostic techniques and tools will help us to detect DM as earlier as possible so we can start treatment in an early phase of diabetes.

Currently, FPG and hemoglobin A1c (glycated hemoglobin, HbA1c) measurements are in use as a biomarker at DM diagnosis. There is also an oral glucose tolerance test (OGTT) for detecting pre-diabetes and gestational diabetes. DM is diagnosed with;

- An FPG level of 126 mg/dL (7.0 mmol/L) or higher, *or*
- A 2-hour plasma glucose level of 200 mg/dL (11.1 mmol/L) or higher during a 75-g OGTT, *or*
- Random plasma glucose of 200 mg/dL (11.1 mmol/L) or higher in a patient with classic symptoms of hyperglycemia or hyperglycemic crisis, *or*
- An HbA1c level of 6.5% (48 mmol/mol) or higher [2].

HbA1c is currently a reference test being used for diagnosis, screening, and follow-up in patients with DM. HbA1c reflects blood glucose levels over a 2-3 months period and is insufficient with rapid glycemic changes. Furthermore, it is not recommended in patients with uremia, hemoglobinopathy, anemia, and pregnancy. Patients with those situations are very common in daily clinical practice. Thus, alternative biomarkers would help clinicians for better diagnosis and follow-up in patients with DM.

In this chapter, we will explore some other methods may be used as biomarkers.

GLYCATED ALBUMIN (GA)

Albumin is the most abundant protein in plasma. With a molecular weight of 66.7 kDa, it contains a single polypeptide chain with 585 amino acids and 17 disulfide chains. Albumin makes up approximately 60% of total serum protein with concentrations in a range of 30-50 g/L and a half-life of 14 to 20 days [3].

Glycation is defined as a non-enzymatic spontaneous reaction that occurs in situations with high plasma glucose, initiated with the addition of reducing sugars and/or their metabolites to amine groups (typically lysine and arginine) within proteins. With forming a Schiff base, an intermediate product can undergo a rearrangement to create a more stable Amadori product. After dehydration and oxidation, reactive dicarbonyl compounds transform in to precursors of advanced glycation end-products (AGEs). This incident is known as the Maillard reaction [3].

Albumin glycation is affected by glucose concentration and the time of exposure between glucose and protein [4]. Compared to hemoglobin, albumin glycation rates are about 9 to 10 times greater [5]. Ueda and Matsumoto conducted an *in vitro* experiment, and they showed that GA production was 4.5 times greater than HbA1c production after adding the same glucose concentrations. That showed GA production is faster in the same *in vitro* conditions [6].

There are several ways to measure GA like ion-exchange high-performance liquid chromatography (HPLC), boronate affinity chromatography, immunoassays, and colorimetric method with thiobarbituric acid and enzymatic methods using proteinase and ketamine oxidase [7]. Enzymatic methods result in a shorter time and are easy to perform compared to other methods [8]. In all methods, GA levels increase 2 to 5 fold in patients with DM compared to healthy individuals [3].

GA is being considered as a biomarker for DM diagnosis since Tomigana and colleagues conducted a study to determine a reference interval for GA in 2006, which resulted in a range of 12.3% to 16.9% in 699 patients [8]. In 2011, a study with a larger patient number (N=1575) reported a similar interval with a cut-off point for DM diagnosis as 15.5% shows 83.3% sensitivity and specificity by using fasting glucose (≥126 mg/dL) and Hba1c (≥6.5%) as reference tests [9]. In years following, studies showed similar cut-off points with similar sensitivity and specificity rates.

After diagnosis, GA has also been considered as a consistent follow-up biomarker. As mentioned before, GA has a half-life of 14-20 days, making it suitable for detecting rapid glucose changes in a shorter time than Hba1c with a mean life of 120 days. GA can show the efficacy of drug therapy [10] and insulin treatment with a better correlation with fasting glucose than HbA1c [11]. Thus, GA is more useful for managing the medication or dosage of oral antidiabetics and titration of insulin therapy.

Long-term complications of DM are another challenge to predict by biomarkers. A study showed GA and HbA1c had a similar association with microvascular complications like retinopathy and nephropathy. Same study showed cardiovascular complications were predicted by only HbA1c [12].

Besides those advantages, GA also has some limitations. First of all, hypoalbuminemia causes increasing glycation of albumin. In any situation with impaired protein metabolism GA should not be the first test that comes to mind [13]. Thyroid hormone disorders, massive proteinuria, and liver cirrhosis, are other conditions to consider other biomarkers but GA [14].

In conclusion, GA is a useful biomarker for diagnosis and follow-up in DM. With showing short-term glucose control and variability, it is more reliable for treatment management than HbA1c. GA is the most promising biomarker that can be used in daily clinical practice. In the future, the importance and usefulness of this test will be better understood.

NETRİN

Netrins are described as a family of laminin-related extracellular proteins that promote cell and axon migration in embryogenesis. Netrins 1,3 and 4(secreted netrins) and netrins G1,G2 (glycosyl phosphatidyl inositol (GPI) -attached membrane proteins) have been identified in mammals [15].

Netrins are known for their tissue regenerative, angiogenic, and inflammation suppressive functions. Netrin-1 mRNA expression was detected in the fetal pancreas and post-ligation in the adult pancreatic duct in endocrine and exocrine cells. Especially a significant Netrin-4 expression was found in pancreatic ductal cells and vascular endothelium [16]. In hyperglycemic situations, exogenous Netrin -1 and -4 caused caspase-3 downregulation. Caspase-3 reduction was associated with the decrease of neogenin and UNC5-A receptors. Neogenin and UNC5-A receptors cause apoptosis without Netrin while they prevent apoptosis with Netrin interaction [17]. Those findings showed Netrin has a prosurvival role in β-cells. Conversely Netrin-1 and UNC5B expression in adipose tissue was found higher in high-fat diet-fed obese mice compared to mice who had a regular diet. High expression of Netrin-1 shows chronic inflammation, insulin resistance, and metabolic dysfunction [18].

Another study suggested that Netrin-1 also has a hypoglycemic effect possibly related to improved β-cell function and amplified insulin levels and pre-proinsulin mRNA expression [19].

Recently, a clinical study look upon Netrin as a biomarker for early detection of impaired fasting glucose (IFG) and/or Type 2 DM. They found significantly increased levels of Netrin-1 in patients with IFG and Type 2 DM compared to the control group. Netrin-1 was also found in a significant positive correlation with fasting glucose, HbA1c and HOMA-IR and in a negative correlation with HDL cholesterol and eGFR levels [20].

On the contrary, another study conducted by Lui *et al.* suggested significantly decreased Netrin-1 levels in patients with Type 2 DM and found a negative correlation with HbA1c, HOMA-IR, fasting and postprandial blood glucose and fasting insulin [21].

The role of Netrin, especially -1 and -4 are being investigated in DM and micro- and macrovascular complications of DM. Hence, in the future direction, specifically for diagnosis and follow-up of DM, new studies will show whether Netrins can be used as a biomarker.

SALIVARY GLUCOSE

As known, DM is a worldwide health problem with rapidly growing numbers of patients. Most of the patients can be diagnosed incidentally by blood tests while they applied to a clinician with irrelevant symptoms. Therefore a non-invasive screening test would provide early diagnosis so patients can make changes in their lifestyle and physicians can start treatment as soon as possible to prevent micro- and macro-vascular complications.

A study in 2016 was conducted with 180 patients divided into three equal groups, including patients with DM already on medication, patients newly-diagnosed with DM and nondiabetic individuals. Komal S. *et al.* found a significant difference in the mean salivary glucose levels among the three study groups. A significant positive, strong correlation was found between fasting blood glucose and fasting salivary glucose in all study groups. This study also suggested an optimal cutoff point as 7.05 mg/dl with high sensitivity and specificity (99.1% and 93,7%, respectively) [22]

Another study in the same year, Dhanya and Hegde compared 100 diabetic patients with 100 non-diabetic patients older than age 35 salivary glucose levels were significantly higher in the diabetic group.

In diabetics fasting serum glucose level was 136.30±28.58 mg/dL and fasting salivary glucose level was 8.47±4.20 mg/dl comparing levels of 97.78±6.39 md/dl and 1.20±0.86 mg/dl in non-diabetics, respectively. Also, fasting salivary glucose levels were in positive correlation with fasting serum glucose levels in diabetics. Diabetic patients are divided into three subgroups according to their fasting serum glucose levels: Patients with glucose levels below 130 mg/dL, 130 to 200 mg/dL and above 200 mg/dL. Median fasting salivary glucose levels are evaluated as 5.43 mg/dl, 10.99 mg/dL and 21.00 mg/dL, respectively [23].

Fasting salivary glucose may be a novel diagnostic biomarker for DM. Noninvasive, painless, and simple measurement is the biggest advantage. Positive correlation with fasting serum glucose shows it can be used for screening, diagnosis, and monitorization. Further studies in larger populations will show if it is suitable to use in clinical practice.

MicroRNAs (miRNAs)

Type 2 DM is a disease that can be triggered by genetic and environmental factors and lifestyle. Biomarkers in current clinical practice are not usable for prevention. Similarly DM medications on the market are in use for treatment and preventing complications, not the disease itself. Thus, knowing the genetic roots of DM will help us, physicians, to foresee the development of disease and modify the patients' lifestyle and work on new medications to prevent it.

The human genome sequencing project uncovered a detailed definition of the polymorphisms which can lead to specific phenotypes of monogenic disorders. More complex diseases like DM, prediction is not simple as that. In different populations minimum of 50 genome-wide associations analysis identified more than 60 loci (specific, fixed position on a chromosome where a particular gene or genetic marker is located) related to DM [24].

Variants in *TCF7L2* gene showed the strongest risk of type 2 DM. This gene is known for it's major role in *Wnt* signaling, which is related to β-cell proliferation and insulin exocytosis. *HMGA1, ADAMTS9, TSPAN8, LGR5, ADCY5, KCNJ11, JAZF1, NOTCH2, GCKR, CDC123/CAMK1D, GIPR, THADA,* etc. are other important loci associated with DM [25].

miRNAs are a group of non-coding, small, single-stranded, and sequence-specific RNAs that can bind to messenger RNAs (mRNAs) to inhibit protein synthesis [25].

A study in 2012 suggested that miRNAs have a major impact on pancreatic β-cell discrimination and islet maturation. Thereby they play a role in glucose metabolism and diabetes [26].

Zampetaki *et al.* conducted a randomized controlled trial in 2010. They found out that some miRNAs can be used for the prediction of type 2 DM. With a sensitivity of 70%, expression of five miRNAs(miR-15a, miR- 28-3p, miR-29b, miR-126 and miR-223) was dysregulated [27].

miR-375, an islet-specific miRNA, is one of the first which has a major role on β-cell function and insulin secretion. Decreased levels of miRNA-375 may lead to DM by lessening insulin levels. Reduced levels of miR-29b, miR-223, miR-15a and elevated levels of miR-29a, miR-9, miR-30d and miR-27a have been shown a relation with DM, suggested as potential biomarkers [25].

In the future direction, detection of miRNAs may get easier and more common in clinical practice. Hence, DM can be predictable years before diagnosis.

3-Carboxy-4-Methyl-5-Propyl-2-Furanopropanoic Acid (CMPF)

CMPF is a major endogenous metabolite of Furan Fatty Acids, which may lead to oxidative stress and disrupt insulin secretion [28].

Lui *et al.* conducted a prospective human cohort study and showed individuals who developed Type 2 DM with a progression of prediabetes had higher levels of CMPF than prediabetics who sustained chronically high levels [29].

In conclusion, CMPF increases oxidative stress and advanced glycation products (AGEs) and impairs insulin secretion. Thus, CMPF may be a leading factor for T2DM. Utilization of CMPF as a biomarker for diagnoses and screening for T2DM will remain unclear without new clinical studies with larger cohorts.

OTHER POTENTIAL BIOMARKERS

Glycated complement regulatory protein CD59 (GCD59) was found to be in a positive correlation with HbA1c and significantly discriminate DM with 93% sensitivity and 100% specificity [30].

8-hydroxyguanine (8-OH-Gua) detected an almost 3-fold higher urinary concentration in patients with DM than the control group. There was no statistical analysis for significance in the study. However, the study showed oxidative DNA damage might lead DM [31].

F_2-Isoprostanes (iPF2α-III, 2, 3-dinor-Ipf2a-III, Ipf2a-VI) were found to be significantly lower in patients with DM. Study suggested urinary concentration of F_2-Isoprostanes could be used as a biomarker for DM, but no further studies could show that yet [32].

2-aminoadipic acid (2-AAA) concentration showed the strongest association with future development of DM among 70 intermediary plasma metabolites in a study Wang *et al.* conducted in 2013. Curiously 2-AAA did not show a positive correlation with HbA1c but showed a significant correlation with fasting insulin and other insulin resistance indicators. In brief, 2-AAA levels may predict the development of DM even in years [33].

Plasma apelin concentrations were evaluated as a biomarker in DM and found higher in female patients but also associated with increased risk of DM in men. Thus, apelin looks useful for predicting DM risk but only in men [34].

CONCLUSION

Biomarker investigation for DM has reached exciting levels in the last decade. Optimal biomarker should have some features. First of all, it should be rapidly detectable in blood or urine by a common measurement method. Accessibility and affordability are other key points. Diagnosis of DM is possible with simple and cheap methods like measuring fasting plasma glucose and HbA1c. But especially following up those patients has some limitations like lack of detecting plasma glucose variability and response to treatment in the short term. Hopefully, mainly glycated albumin and other biomarkers mentioned above will be available in our daily clinical practice. Thus, patients with DM would be diagnosed in an early stage which provides better management of this disease and its complications.

CONSENT FOR PUBLICATION

Not applicable.

CONFLICT OF INTEREST

The author declares no conflict of interest, financial or otherwise

ACKNOWLEDGEMENT

Declared none.

REFERENCES

[1] IDF Diabetes Atlas. 9th ed., 2019.

[2] American diabetes association. classification and diagnosis of diabetes: standards of medical care in diabetes-2020. Diabetes Care 2020; 43 (Suppl. 1): S14-31. [http://dx.doi.org/10.2337/dc20-S002] [PMID: 31862745]

[3] Anguizola J, Matsuda R, Barnaby OS, *et al.* Review: Glycation of human serum albumin. Clin Chim Acta 2013; 425: 64-76. Intervention strategies to prevent pathogenetic effects of glycated albumin. Arch Biochem Biophys. 2003;419(1):25-30. [http://dx.doi.org/10.1016/j.cca.2013.07.013Cohen]

[4] Rondeau P, Bourdon E. The glycation of albumin: structural and functional impacts. Biochimie 2011; 93(4): 645-58. [http://dx.doi.org/10.1016/j.biochi.2010.12.003] [PMID: 21167901]

[5] Ueda Y, Matsumoto H. Recent topics in chemical and clinical research on glycated albumin. J Diabetes Sci Technol 2015; 9(2): 177-82. [http://dx.doi.org/10.1177/1932296814567225] [PMID: 25614014]

[6] Freitas PAC, Ehlert LR, Camargo JL. Glycated albumin: a potential biomarker in diabetes. Arch

Endocrinol Metab 2017; 61(3): 296-304.
[http://dx.doi.org/10.1590/2359-3997000000272] [PMID: 28699985]

[7] Kouzuma T, Usami T, Yamakoshi M, Takahashi M, Imamura S. An enzymatic method for the measurement of glycated albumin in biological samples. Clin Chim Acta 2002; 324(1-2): 61-71.
[http://dx.doi.org/10.1016/S0009-8981(02)00207-3] [PMID: 12204426]

[8] Tominaga M, Makino E, Yoshino G, *et al.* Report of the committee on standardization of laboratory testing related to diabetes mellitus of Japan Diabetes Society: determination of reference intervals of hemoglobin A1c (IFCC) and glycoalbumin in the Japanese population. J Japan Diab Soc 2006; 49: 825-33.

[9] Furusyo N, Koga T, Ai M, *et al.* Utility of glycated albumin for the diagnosis of diabetes mellitus in a Japanese population study: results from the Kyushu and Okinawa Population Study (KOPS). Diabetologia 2011; 54(12): 3028-36.
[http://dx.doi.org/10.1007/s00125-011-2310-6] [PMID: 21947435]

[10] Yoshiuchi K, Matsuhisa M, Katakami N, *et al.* Glycated albumin is a better indicator for glucose excursion than glycated hemoglobin in type 1 and type 2 diabetes. Endocr J 2008; 55(3): 503-7.
[http://dx.doi.org/10.1507/endocrj.K07E-089] [PMID: 18445997]

[11] Paroni R, Ceriotti F, Galanello R, *et al.* Performance characteristics and clinical utility of an enzymatic method for the measurement of glycated albumin in plasma. Clin Biochem 2007; 40(18): 1398-405.
[http://dx.doi.org/10.1016/j.clinbiochem.2007.08.001] [PMID: 17919531]

[12] Nathan DM, McGee P, Steffes MW, Lachin JM. Relationship of glycated albumin to blood glucose and HbA1c values and to retinopathy, nephropathy, and cardiovascular outcomes in the DCCT/EDIC study. Diabetes 2014; 63(1): 282-90.
[http://dx.doi.org/10.2337/db13-0782] [PMID: 23990364]

[13] Bhonsle HS, Korwar AM, Kote SS, *et al.* Low plasma albumin levels are associated with increased plasma protein glycation and HbA1c in diabetes. J Proteome Res 2012; 11(2): 1391-6.
[http://dx.doi.org/10.1021/pr201030m] [PMID: 22181049]

[14] Furusyo N, Hayashi J. Glycated albumin and diabetes mellitus. Biochim Biophys Acta 2013; 1830(12): 5509-14.
[http://dx.doi.org/10.1016/j.bbagen.2013.05.010] [PMID: 23673238]

[15] Rajasekharan S, Kennedy TE. The netrin protein family. Genome Biol 2009; 10(9): 239.
[http://dx.doi.org/10.1186/gb-2009-10-9-239] [PMID: 19785719]

[16] Yimer EM, Zewdie KA, Hishe HZ. Netrin as a novel biomarker and its therapeutic implications in diabetes mellitus and diabetes-associated complications. J Diabetes Res 2018; 8250521.
[http://dx.doi.org/10.1155/2018/8250521]

[17] Yang YHC, Szabat M, Bragagnini C, *et al.* Paracrine signalling loops in adult human and mouse pancreatic islets: netrins modulate beta cell apoptosis signalling *via* dependence receptors. Diabetologia 2011; 54(4): 828-42.
[http://dx.doi.org/10.1007/s00125-010-2012-5] [PMID: 21212933]

[18] Ramkhelawon B, Hennessy EJ, Ménager M, *et al.* Netrin-1 promotes adipose tissue macrophage retention and insulin resistance in obesity. Nat Med 2014; 20(4): 377-84.
[http://dx.doi.org/10.1038/nm.3467] [PMID: 24584118]

[19] Gao S, Zhang X, Qin Y, *et al.* Dual actions of Netrin-1 on islet insulin secretion and immune modulation. Clin Sci (Lond) 2016; 130(21): 1901-11.
[http://dx.doi.org/10.1042/CS20160133] [PMID: 27520508]

[20] Jung H-I, Bae J, Han E, *et al.* "Circulating Netrin-1 as a novel biomarker for impaired fasting glucose and newly diagnosed type 2 diabetes mellitus," Diabetes, vol. 67. Supplement 2018; 1: 1535-P.
[http://dx.doi.org/10.2337/db18-1535-P]

[21] Liu C, Ke X, Wang Y, *et al.* The level of netrin-1 is decreased in newly diagnosed type 2 diabetes

mellitus patients. BMC Endocr Disord 2016; 16(1): 33-5. [http://dx.doi.org/10.1186/s12902-016-0112-z] [PMID: 27255377]

[22] Smriti K, Pai KM, Ravindranath V, Gadicherla S, Pentapati KC. Salivary Glucose as a Diagnostic Marker for Diabetes Mellitus. J Diabetes Sci Technol 2016; 10(4): 991-2. [http://dx.doi.org/10.1177/1932296816637619] [PMID: 26961976]

[23] Dhanya M, Hegde S. Salivary glucose as a diagnostic tool in Type II diabetes mellitus: A case-control study. Niger J Clin Pract 2016; 19(4): 486-90. [http://dx.doi.org/10.4103/1119-3077.183314] [PMID: 27251965]

[24] Zhou Y, Park SY, Su J, *et al.* TCF7L2 is a master regulator of insulin production and processing. Hum Mol Genet 2014; 23(24): 6419-31. [http://dx.doi.org/10.1093/hmg/ddu359] [PMID: 25015099]

[25] Bhatia P, Raina S, Chugh J, Sharma S. miRNAs: early prognostic biomarkers for Type 2 diabetes mellitus? Biomarkers Med 2015; 9(10): 1025-40. [http://dx.doi.org/10.2217/bmm.15.69] [PMID: 26441333]

[26] Kumar M, Nath S, Prasad HK, Sharma GD, Li Y. MicroRNAs: a new ray of hope for diabetes mellitus. Protein Cell 2012; 3(10): 726-38. [http://dx.doi.org/10.1007/s13238-012-2055-0] [PMID: 23055040]

[27] Zampetaki A, Kiechl S, Drozdov I, *et al.* Plasma microRNA profiling reveals loss of endothelial miR-126 and other microRNAs in type 2 diabetes. Circ Res 2010; 107(6): 810-7. [http://dx.doi.org/10.1161/CIRCRESAHA.110.226357] [PMID: 20651284]

[28] Koppe L, Poitout V. CMPF: A Biomarker for Type 2 Diabetes Mellitus Progression? Trends Endocrinol Metab 2016; 27(7): 439-40. [http://dx.doi.org/10.1016/j.tem.2016.04.010] [PMID: 27132217]

[29] Liu Y, Prentice KJ, Eversley JA, *et al.* Rapid elevation inCMPF may act as a tipping point in diabetes development. Cell Rep 2016; 14(12): 2889-900. [http://dx.doi.org/10.1016/j.celrep.2016.02.079] [PMID: 26997281]

[30] Ghosh P, Sahoo R, Vaidya A, *et al.* A specific and sensitive assay for blood levels of glycated CD59: a novel biomarker for diabetes. Am J Hematol 2013; 88(8): 670-6. [http://dx.doi.org/10.1002/ajh.23478] [PMID: 23670858]

[31] Bishnoi S, Goyal RN, Shim YB. A novel nanogold-single wall carbon nanotube modified sensor for the electrochemical determination of 8-hydroxyguanine, a diabetes risk biomarker. Bioelectrochemistry 2014; 99: 24-9. [http://dx.doi.org/10.1016/j.bioelechem.2014.06.003] [PMID: 24976140]

[32] Il'yasova D, Spasojevic I, Base K, *et al.* Urinary F2-isoprostanes as a biomarker of reduced risk of type 2 diabetes. Diabetes Care 2012; 35(1): 173-4. [http://dx.doi.org/10.2337/dc11-1502] [PMID: 22100959]

[33] Wang TJ, Ngo D, Psychogios N, *et al.* 2-Aminoadipic acid is a biomarker for diabetes risk. J Clin Invest 2013; 123(10): 4309-17. [http://dx.doi.org/10.1172/JCI64801] [PMID: 24091325]

[34] Ma WY, Yu TY, Wei JN, *et al.* Plasma apelin: A novel biomarker for predicting diabetes. Clin Chim Acta 2014; 435: 18-23. [http://dx.doi.org/10.1016/j.cca.2014.03.030] [PMID: 24721640]

CHAPTER 17

Biomarkers and their Clinical Applications in Pediatrics

Yusuf Elgormus[1,*]

[1] *Department of Pediatric Health and Diseases, Medicine Hospital, Istanbul, Turkey*

Abstract: Biomarker studies are becoming increasingly interesting for many fields of medicine. The use of biomarkers in medicine is involved in detecting diseases and supporting diagnosis and treatment decisions. New research and new discoveries on the molecular basis of the disease show that there may be a number of promising new biomarkers for use in daily clinical practice. Clinical trials in children lag behind adult research both in quality and quantity. The number of biomarkers validated to optimize pediatric patient management is limited. In the pathogenesis of many diseases, it should not be extrapolated to the pediatric clinical setting, taking into account that biomarkers that are effective in adults are clearly different in children and that ontogeny directly affects disease development and therapeutic response in children. The search for ideal biomarkers or markers that can make an early and definitive diagnosis in neonatal sepsis is still ongoing. The ideal biomarker for pediatric diseases should be cost-effective, noninvasive, applicable to pediatric specific diseases, and its results should correspond to age-related physiological changes. Lactate, troponin and B-type natriuretic peptide are valuable biomarkers in the evaluation and management of critically ill children with cardiac disease. Tumor markers in children are biochemical substances used in the clinical treatment of pediatric tumors and to detect the presence of cancer (regression or progression). In this chapter, current and brief information about biomarkers and their clinical applications used in the diagnosis and monitoring of pediatric diseases is presented.

Keywords: Biomarker, Early predictor, Management, Monitoring, Premature, Tumor markers.

INTRODUCTION

Biomarker research has become increasingly interesting for many areas of medicine today. Biomarkers (biological markers) are indicators of the biological process that can show changes in disease and health status and pathological conditions. Biomarkers may include biofluids (urine, blood, saliva, stool, amniotic

* **Corresponding author Yusuf Elgormus:** Department of Pediatric Health and Diseases, Medicine Hospital, Istanbul, Turkey; Tel:+905322939282; E-mail: yusuf.elgormus@medicinehospital.com.tr

Pınar Atukeren and Hafize Uzun (Eds.)

fluid, bronchoalveolar lavage, cerebral spinal fluid), other tissue (including skin, hair or nails), radiographic, genetic, or any other physiologic testing. While biomarker provides important information in the diagnosis of a disease, they may be ineffective in showing the course of the disease. However, even if a biomarker may not be a good diagnostic biomarker, there are numerous other types of potential applications for newly discovered markers. For example, prostate-specific antigen (PSA) used in prostate cancer is an important parameter for diagnosis, but it is not a suitable marker for monitoring the disease. Similarly, carcinoembryonic antigen (CEA) used in ovarian carcinomas is not an important indicator for diagnosis, but it is a good biomarker in the follow-up of the disease [1, 2]. The most important characteristics of an ideal biomarker are summarized in Table **1**.

Table 1. The most important characteristics of an ideal biomarker.

• It should be specific for the tumor type identified.
• Analysis should not require an expensive tool.
• The results of the biomarker should be easily interpreted.
• It must have analytical and clinical validity.
• Its quantity should not show a wide variation in the general population.
• It must be precise, sensitive and specific for the condition of the disease.
• Its measurement is practical and easy; its levels should be compatible with the stages of the disease.
• Adequate and reliable amounts of body tissues and fluids should be obtained.

Despite the lack of knowledge on child-specific biomarkers, their use is beneficial as it is integrated into the pediatrician's daily practice. When a complete blood count (leukocyte) and C-reactive protein (CRP) are measured for the initial evaluation of bacterial infection in a febrile child or HbA1c to assess glycemic control in a diabetic patient, biological surrogates are used to assess pediatric disease processes and direct the care plan. There are many approaches when considering pediatric biomarkers. A more traditional approach uses well-established adult biomarkers and applies data directly to children. This approach is similar to the drug development process where a drug is developed for adult disease, studied in adult patients, and then investigated for potential use in children [3]. This chapter reviews and summarizes the different types of biomarkers and their clinical applications in many childhood diseases. Although many diseases have overlapping biological system involvement, they were broadly categorized as noncommunicable diseases, metabolic diseases, inflammatory diseases, liver diseases, urological diseases, autoimmune diseases, genetic diseases, and rare diseases.

CHILDHOOD NONCOMMUNICABLE DISEASES

Noncommunicable diseases (NCDs), such as cardiovascular diseases, cancer, diabetes and chronic respiratory diseases, are the leading global cause of death and are responsible for just over 70% of deaths worldwide [4].

Biomarkers in Patients with NCDs in Childhood

Allergy, Asthma, Atopic Dermatitis (AD)

Allergy is not one disease but a collection of a number of allergic conditions. The frequency of allergic diseases is increasing day by day. Asthma, atopic dermatitis (AD), hay fever, food allergy, eosinophilic esophagitis (EoE), and allergic rhinitis are among the most prevalent chronic NCDs in children from 0-18 years of age [5, 6]. If food allergy accompanies AD in young children, asthma and allergic rhinitis (AR) may develop in these children at later ages. This clinical course is known as the "**atopic march**". Epidemiologic studies indicate that asthma begins in the preschool years even when chronic symptoms do not appear until early adulthood [7, 8]. The lack of definitive diagnostic criteria in differentiating children with recurrent bronchiolitis/bronchitis patients who have progressed to asthma and those who have had one or two non-recurring bronchiolitis/bronchitis attacks prevents predicting the progression of these children. The genes containing differential methylation robustly associated with asthma and allergy are likely to play a role in allergic disease pathogenesis, and understanding the biological implications of these epigenetic signatures may help elucidate mechanistic pathways and define disease endotypes [9].

The diagnosis of allergic disorders is most often made on clinical history, supported by the results of skin tests and diagnostic markers *e.g.* immunoglobulin E (IgE). Although skin testing has advantages of relative sensitivity and specificity, rapid results, and good tolerability, it is subject to some operator, observer and interpretation variability. IgE is also known to be elevated in non-allergic conditions and many allergic patients have IgE levels within the normal range [10]. Schools *et al.* [11] demonstrated that there is a substantial disagreement between skin prick test (SPT) and specific IgE (sIgE) for diagnosing allergic sensitization in young children, which increases with age for food sensitization. sIgE and SPT cannot be used interchangeably as the choice of method significantly influences the diagnosis of IgE - mediated allergy in young children. The lack of agreement between tests and a poor correlation to clinical disease underscores that allergy testing should not be used as a screening tool in children [11].

AD, whose frequency increases with other allergic diseases, is the most common chronic inflammatory skin disease in children and usually begins in the first year of life [12]. It is known that food allergies are effective in the formation of AD in the first year. Foods are the leading allergens among the allergens held responsible in AD, and therefore food restrictions are frequently applied in patients. Cow's milk allergy is an important allergen in the development of AD [13]. In addition, egg and wheat allergies are among the factors that cause AD in children [14]. Food allergies can be either IgE-mediated or non-IgE-mediated (T-cell-mediated) [15]. There is no specific laboratory test. Allergy skin tests and allergen-sIgE tests can only give positive results in IgE-mediated food allergies. Simon *et al.* [16] stated that eosinophils are a strong marker of AD. Eosinophils, eosinophil-derived products basic proteins as eosinophil cationic protein (ECP), eosinophil-derived neurotoxin (EDN, EPX), and major basic protein (MBP) are present in increased amounts in the blood and the skin of AD patients. The serum ECP is a frequently used tool in monitoring AD activity [16, 17]. However, serum ECP levels do not correlate with total IgE levels [18]. sIgE can be used as a sensitization or predictive marker. The serum EDN [19], serum MBP [20], and urine EPX levels [21 - 23] have also been used as markers for monitoring AD activity.

Other immunoglobulin isotypes (IgG, IgA, IgM, IgD, and IgE), such as IgE, may also play a role in allergic reactions. Serum IgG, IgD and IgE levels were elevated significantly in asthma patients compared to controls while IgA and IgM levels were normal [24]. Allergy was associated with more intense IgA and IgG responses to ovalbumin (OVA). Furthermore, OVA-specific IgA and IgG antibodies may help in assessing the risk for atopy [25].

IgA deficiency, one of the B cell deficiencies, is the most common primary immune deficiency. Serum and secretory IgA levels have been suggested to be inversely associated with allergic symptoms in children. The role of IgA in defense of the body's immune system against external factors is great. It has been reported that IgA prevents the colonization of bacteria the effects of their toxins and prevents food allergies [26]. In cases diagnosed with IgA deficiency, secretory IgA deficiency is also common. The absence of a protective barrier on respiratory and gastrointestinal mucosal surfaces in these cases explains the predisposition to atopic diseases due to aeroallergens and food allergens [27]. Increased serum IgA concentration at age 2 months was associated with the development of subsequent allergic symptoms and sensitization in childhood and adolescence [28]. It has been reported that IgA antibodies have a suppressive role in mediator release and its disappearance in its deficiency leads to mediator release [29].

IgD was found to bind to basophils and mast cells may also have some role in allergic reactions. A significant increase of IgD levels was observed in children at first signs of asthma, and the following normalization after 18 months may represent a non-specific response or an attempt of the organism to block asthma, therefore favoring immunologic tolerance [30]. Serum IgD and IgE were significantly higher in atopic subjects than that in normal subjects. However, total serum IgD was not significantly correlated with total serum IgE, indicating that immunoregulatory control of the basal levels of the two isotypes is not linked [31]. Both studies were conducted >15 years ago and recent data on IgD is therefore lacking.

The immune system consists of 4 main parts: humoral, cellular, phagocyte, and complement systems. Primary immunodeficiency (PID) diseases are a group of diseases with chronic and / or recurrent bacterial, fungal, protozoal and viral infections that develop as a result of primary or congenital immunodeficiency. Secondary immune deficiency is the deterioration of the host's defense system as a result of another disease. It can develop due to hereditary, metabolic, hematological or infectious disease and is more common than PID. Rational laboratory analyses provide helpful information in differential diagnosis [32]. The first examinations to be performed in immune deficiencies are complete blood count (CBC) and immunoglobulins. With a CBC, the number of leukocytes and their formula (lymphocyte, neutrophil and eosinophil), erythrocyte and platelet count is evaluated. Lymphopenia, combined immunodeficiency; leukocytosis, leukocyte adhesion defect (LAD); thrombocytopenia, Wiskott-Aldrich syndrome and hyper IgM syndrome; neutropenia, some combined immunodeficiencies, X-linked agammaglobulinemia and congenital neutropenias; Eosinophilia may be a clue to hyper-IgE syndrome. The appearance of Howell-Jolly bodies on peripheral blood smear indicates splenic dysfunction or asplenia [33].

It may be due to decreased globulin fraction, hypogammaglobulinemia, complement deficiencies, malnutrition or protein loss. High globulin levels accompany HIV, some autoimmune diseases, hepatitis, myeloma or chronic infections. Immunoglobulin levels can be low in humoral and combined immunodeficiencies to varying degrees, while normal immunoglobulin levels do not exclude antibody production disorders. If the immunoglobulin levels are significantly low or there is a lack of antibody production, the number of T and B cells should be evaluated to evaluate the B cell number and whether there is a T cell failure. B cell functions can be evaluated by measuring antibodies against polysaccharide antigens against diphtheria, pertussis, tetanus and H. influenza. Isohemagglutinins are another test used to evaluate antibodies against group A and B antigens in erythrocytes [33, 34].

In the evaluation of cellular immunodeficiency; Evaluation of proliferative response with mitogen / antigens (candida skin tests), detailed T cell subgroups (used CD3 +; T lymphocyte, CD3 + / CD4 +; helper T lymphocyte, CD3 + / CD8 +; cytotoxic T lymphocyte, CD3 + / HLA-DR +; activated T lymphocyte). NK cell cytotoxicity, cytokine production, chimerism analysis, mutation analysis are used [33, 35].

Hemolytic complement test (CH50) demonstrates the function of classical complement components. The Nitroblue Tetrazolium test is used as a first-line test for suspected chronic granulomatous disease [35].

Circulating factors as disease markers or new therapeutic targets include soluble adhesion molecules, such as intercellular adhesion molecule 1 (ICAM-1); the antibacterial peptide LL-37; interleukins, such as IL-31 and IL-18; and numerous chemokines, such as mucosa-associated epithelial chemokine (CCL28), thymus and activation-regulated chemokine (CCL17), and cutaneous T cell–attracting chemokine (CCL27) in multiple allergic disorders, inflammatory disorders, or both. Although potentially suitable for disease monitoring, they are less likely to offer the specificity required for predictive biomarkers [36].

MicroRNAs (miRNAs) are approximately 22 nucleotides in length, small fragments of RNA that play a role in the post-transcriptional regulation of protein-coding genes. These small molecules play important roles in many biological processes. miRNAs play a regulatory role in many important cellular pathways such as inflammation, and changes in expression levels are directly related to many diseases. Circulating miRNAs are attractive biomarkers for the prediction of disease outcomes and are stable in the serum over long time frames, even in harsh conditions [37]. miRNAs have been shown to be involved in the pathogenesis of many allergic diseases including asthma, eosinophilic esophagitis, allergic rhinitis and eczema [38 - 42]. Wang *et al.* [43] indicate that circulating miR-3162-3p and miR-1260a should be further evaluated as potential non-invasive biomarkers in diagnosis and treatment for childhood asthma. A novel relationship between miR-3162-3p and β-catenin and clarify their mechanistic role in asthma etiopathogenesis was found in another study [44]. hsa-miR-206 was significantly associated with subsequent asthma exacerbations. When combined, 3 miRNAs (hsa-miR-146b-5p, hsa-miR-206 and hsa-miR-720) by themselves provided comparable predictive power to an established clinical model of exacerbations [45].

In asthma, the use of biomarkers examined in serum, exhale air or induced sputum in asthma is limited in their routine use due to their application problems, lack of standardization and cost. The number of eosinophils in the sputum is important in the diagnosis of eosinophilic phenotype and specific IgE determinations are important in the diagnosis of allergic phenotype. Serum total IgE level is important in the diagnosis and follow-up of the *Aspergillosis* phenotype in asthma [46]. Biomarkers can be used in more specific follow-up studies.

Filaggrin (FLG, filament aggregating protein), one of the epidermal barrier proteins, is encoded by the FLG gene. Until now, the strongest relationship with AD has been detected in this gene [47]. Decreased FLG expression is also observed in AD patients without filaggrin mutation, which is caused by local production of Th2 cytokines, interleukin IL-4 and IL-13, and filaggrin from keratinocytes. It is thought to be related to the decrease in production [48].

Allergic Rhinitis

Allergic rhinitis (AR) is an allergic disease mediated by type I allergic mechanisms that are defined as the inflammation of the mucosa lining the nose [49]. For the diagnosis and treatment of AR, several consensus reports have been prepared and made available in recent years [50 - 52]. Among these, the most prominent guide in recent years and updated from time to time is "Allergic Rhinitis and its Impact on Asthma (ARIA)" [50].

Allergic inflammation: AR occurs with IgE-dependent allergy mechanisms that cause nasal inflammation in varying degrees. Allergy is usually due to excessive IgE production [53]. Serum total IgE level is measured by radioimmunoassay and enzyme immunoassay methods. It rises from birth to adolescence and then begins to decline. In various conditions, especially allergic and parasitic diseases are affected. Therefore, it is not used in the diagnosis of allergic diseases and in screening [52].

Serum specific IgE measurements are also useful. Tests involving single or multiple allergens are performed. Sp IgE measurements are not much affected by drugs or skin diseases. The threshold value is usually 0.35 KU/L and is less sensitive than skin tests. Only a few allergens can be investigated, as the price is high. Multiple screening tests are 85% useful in detecting whether the patient is allergic (specificity and sensitivity), if the result is positive, further investigations are required [54]. A positive skin test alone does not indicate clinical reactivity. Skin testing cannot be performed on patients with skin disease or dermographism [55].

Mediators released during an allergic reaction: It is the measurement of mediators such as acute phase proteins, histamine, PGD2, cysteinyl leukotrienes, tryptase and ECP during the comparison test or allergic reaction. It is done for research purposes [10, 56]. Nitric oxide (NO) measurement in exhaled air: It is a non-invasive measurement method. It is easy to perform and has been found useful in differential diagnosis [57]. Serum PTX3 levels could be a reliable marker for symptom severity in children with AR [58]. Elevated hs-CRP is associated with allergic sensitization in school-aged children suggesting systemic low-grade inflammation as a phenotypic characteristic of this early-onset noncommunicable disease [59]. Serum IL-1beta can also be a biomarker in children with severe persistent AR [60]. All these tests are done for research purposes.

Peripheral blood activation markers: When the basophils of allergic persons are stimulated with a specific allergen, they release various mediators as degranulated. Measuring mediators, microscopic examination of cells or measuring the activation of cells can be useful in diagnosis. In new basophil activation tests, CD63, CD45, CD203 expression is measured in response to an allergen or nonspecific stimuli [61].

Pediatric Autoimmune Diseases

The host's cellular and humoral response to its own antigens are defined as autoimmunity, and clinical pictures resulting from tissue damage caused by these reactions are defined as autoimmune diseases. Among the multiple autoimmune diseases (AIDs) in pediatric rheumatology are juvenile idiopathic arthritis (JIA), systemic lupus erythematosus (SLE), childhood scleroderma or pediatric systemic sclerosis (PSS), juvenile dermatomyositis (JDM), Sjögren syndrome (SS), antineutrophil cytoplasmic antibody (ANCA)-associated vasculitides, and seronegative spondyloarthropathy. As well as timely diagnosis of autoimmune diseases, whether the precursor markers will be detected in the early stage of the disease is a matter of debate. Most autoimmune diseases cannot be diagnosed with a single test [62].

Autoinflammatory diseases include a group of diseases characterized by the involvement of different systems such as recurrent fever, erythema, lymphadenopathy, serositis and musculoskeletal system. Immunodeficiency, autoimmune disease and rheumatic diseases overlap with each other. Autoinflammatory diseases develop secondary to the gene mutation responsible for the inflammatory response. Systemic episodes of inflammation are seen as a result of insufficient regulation of the innate immune system. The onset of autoinflammatory diseases is most often in the first hours or first decade of life. It may rarely start later. Autoinflammatory diseases with different clinical forms are very rare and therefore, diagnosis is usually delayed [63]. However, in many autoinflammatory diseases, fever is not periodic, and some autoinflammatory diseases may not have fever [64, 65]. The information leading to diagnosis in pediatric rheumatology is obtained from approximately 80-85% by a good history of the disease, 10% by detailed physical examination, and 5% by laboratory examinations.

Anti-Nuclear Antibodies (ANAs), rheumatoid factor (RF), and anti-cyclic citrullinated peptide antibodies (anti-CCP) are used for the definitive evaluation of autoimmune diseases.

Anti-Nuclear Antibodies (ANA)

ANA are various antibodies that develop against structures such as DNA, histone, and centromere in the cell nucleus of mammals. These antibodies are not a single antibody; they form a class of antibodies. They are antibodies seen in many autoimmune diseases. The ANA is usually the first test performed when symptoms indicate an autoimmune disease. But, positive ANA results must be interpreted together with the clinic. It is a frequently used laboratory test for autoimmune disease screening, especially in patients with musculoskeletal complaints or skin findings. However, without any symptoms, ANA can also be positive in healthy people. It can also be detected positively in malignant diseases and infectious diseases [66]. The indirect immunofluorescence (IIF) method using HEp-2 cells is recommended as the gold standard in ANA screening [67]. ANA positivity is most frequently associated with systemic lupus erythematosus (SLE), and ANA positivity is detected in 97-100% of these patients. In addition, ANA positivity has been reported with a rate of 81-97% in juvenile systemic scleroderma and 40-63% in juvenile dermatomyositis [68 - 73]. It is known that 39% of juvenile idiopathic arthritis (JIA) patients have ANA positivity [74]. While ANA positivity in low titer (1/40) may be 36% in healthy adults, it can be detected in 5% in titers of 1/160 and above. In healthy children have been reported to be 5-18% of ANA positivity was found in a study from Turkey is still 4% rate in healthy children [75]. If the ANA is positive or there is high suspicion for autoimmune diseases, sub-definitions should be made with advanced tests. Although ANA is a sensitive test for the diagnosis of SLE, its specificity is very low.

The ANA profile usually includes anti-Sm, anti-double-stranded DNA anti-dsDNA), anti-Scl-70, anti-RNP, anti-chromatin, anti-SS-A, and anti-SS-B antibodies. Anti-double helix DNA (Anti-dsDNA) and ENA (extractable nuclear antigen) antibodies are more specific markers in the diagnosis and follow-up of the disease [76]. Antibodies against ENA [anti-SM, anti-Ro / SSA, anti-La / SSB, and antiribonucleoprotein (anti-RNP)] have been associated with various organ involvement of the disease and are shown in Table **2**.

Table 2. Autoantibodies and their properties detected in systemic lupus erythematosus (SLE).

Autoantibody	Features
ANA	High sensitivity, low specificity
Anti-dsDNA	Specific, disease activity particularly associated with renal involvement
Anti-smith (Anti-Sm)	Specific for SLE
Anti SS-A / Ro **Anti-SS-B / La**	It is also seen in sjögren and sicca syndrome besides SLE. Reduced risk of renal involvement associated with congenital heart block, neonatal lupus and subacute cutaneous lupus.
Lupus anticoagulant **Anti-cardiolipin** **Anti-B2 glycoprotein 1**	It is seen in antiphospholipid syndrome, increased arterial and venous thrombosis, recurrent abortion
Anti-ribosomal P	Associated with neuropsychiatric involvement
Antihistone antibodies	Associated with drug-associated lupus

Antiphospholipid syndrome (APS) lupus may accompany the disease at the beginning or any stage of the disease. Antiphospholipid syndrome (AFS) was first described in patients with SLE (secondary AFS). Patients without major clinical and serological features of SLE are called "primary AFS". Since the antiphospholipid antibodies detected in AFS initially accompany SLE, it is referred to as "lupus anticoagulant". This syndrome is characterized by arterial and venous thromboses, recurrent miscarriages, and increased antiphospholipid antibodies detected in serum. Antiphospholipid antibodies are lupus anticoagulant (LA), anticardiolipin antibody (aCL) IgG and / or IgM and B2 glycoprotein 1 (aB2-GPI) IgG and / or IgM. It is significant for APS that these antibodies are positive twice with an interval of 12 weeks. Except for autoantibodies, erythrocyte sedimentation rate (ESR) is the most important marker in the diagnosis and monitoring of disease activity in lupus. Unlike other rheumatological diseases, C-reactive protein (CRP), another inflammatory marker, is usually normal in lupus and does not show disease activity. CRP elevation is generally in favor of infection and may increase in patients with serositis. IgG elevation is another finding indicating immune system activation in SLE. Activation in the classical complement pathway starting with C1q in lupus and consumption of complement factors in this pathway are observed. Accordingly, low C4 and C3 are seen. Complement levels are a good indicator in disease follow-up; C3, C4 levels decrease correlated with disease activity. Apart from these analyses, complete urine analysis, urine protein measurement, and urine microscopy for renal involvement should be assessed routinely during the disease follow-up. Proteinuria can be an indicator of glomerular damage. The gold standard in the diagnosis of lupus nephritis is still kidney biopsy [77].

In the case of a positive ANA but a negative ANA profile, thyroid antibodies (anti-thyroglobulin and anti-thyroperoxidase) should be assessed [78].

Rheumatoid Factor (RF)

Rheumatoid arthritis (RA) is a chronic autoimmune disease characterized by chronic inflammation and joint damage. It can sometimes be quite difficult to distinguish RA from other inflammatory arthritis. In addition, while the disease is mild in some of the RA patients, it may continue in moderate severity and become chronic in others. In addition, it is known that it shows a slow course and causes widespread joint damage and systemic symptoms [16, 79]. The most commonly used laboratory tests in the diagnosis of RA are ESR, CRP, rheumatoid factor (RF), and anti-CCP. Sedimentation and CRP are non-specific tests that show inflammation associated with disease severity but are used in diagnosis and monitoring disease activity in patients with RA. Therefore, laboratory follow-up is important in the early diagnosis of these patients and in predicting the prognosis. There is no specific diagnostic test for RA, but serological tests are used in the diagnosis and follow-up of these patients. Measurement of RF level is the most widely used test for the diagnosis of RA. Sensitivity 60 -90%; its specificity is 85% [80 - 82]. RFs are basically autoantibodies produced by B lymphocytes activated by antigen presentation and react with the Fc part of IgG molecules. It is thought that the purpose of their production is to eliminate immune complexes more effectively. The most common RF isotype is IgM RF, but IgG, IgA, IgE and IgD type RFs can also be detected [83]. RF and anti-CCP antibodies help differentiate RA from other rheumatic diseases. However, RF can also be found in blood besides RA, especially in infections and other rheumatic diseases. Another condition in which RF can be detected as high is systemic autoimmune diseases. RF is detected in high titers in SLE and Sjögren's syndrome [84, 85]. RF can also be detected in some pathological conditions (Epstein-Barr virus (EBV), cytomegalovirus, parvovirus B-19, and hepatitis B and C other than RA. RF elevation is temporary in infections, RF increase is necessary for clearing immune complexes against pathogens [84, 86].

Although arthritis is one of the most common chronic diseases in childhood and adolescence, it is one of the important causes of short and long-term sequelae. There are many causes and types of chronic arthritis, but the most common is arthritis grouped under the name Juvenile Idiopathic Arthritis (JIA) [87]. There is no marker that can be called the gold standard for the diagnosis of JIA. RF is a positive, persistent, and high concentration autoantibody in polyarticular type JIA, which is the childhood equivalent of adult RF positive RA. It is found in 85% of adults with RA and only 5-10% of children with JIA. RF is helpful in the classification and prognosis of JIA, but it should not be used as a screening test as arthritis is a clinical diagnosis [88].

The role of inflammatory cytokines in JIA has been discussed. IL-1 is a cytokine involved in systemic and joint inflammation. IL-1β and TNF-α initiate the release of prostaglandins and proteases as a result of the destructive process in the joint and also induce cartilage resorption. IL-6 also plays an important role in the initiation of immune and inflammatory processes. On the other hand, TNF-α, initiates the production of IL-1 and IL-6 and causes the acute phase response to increasing gradually [68, 67]. CRP, ESR, platelet count, and hemoglobin level were also used in JIA monitoring [89 - 91].

Anticyclic Citrullinated Peptide Antibodies (Anti-CCP)

Anti-CCP antibodies are IgG-type antibody belonging to the family of antifilaggrin autoantibodies such as anticeratin antibodies (AKA) and antiperinuclear factor (APF). APF test is more sensitive than the AKA test. These are antibodies against citrullinated peptides found in the inflamed synovium. Anti-CCP is more specific for RA, and the likelihood of RA is much higher if detected high. These antibodies can become positive years before the symptoms and signs of RA occur [92, 93]. Anti-CCP is highly specific for RA, but is also often positive in older children with polyarticular RF positive JIA. Anti-CCP indicates an increased risk of aggressive disease and progressive joint damage. The high specificity of anti-CCP antibodies, their ability to be detected in the early stage of the disease, and their positivity in 20 43% of RF negative cases suggest that they are strong diagnostic markers [94 - 97].

Autoantibodies in Sjögren Syndrome

Sjögren's syndrome is a systemic autoimmune disease that occurs with permanent dryness of the mouth and eyes, mainly due to functional defects in the salivary and tears glands, particularly in the exocrine glands. Sjögren's syndrome is seen 9 times more frequently in women. The disease can be seen at any age. Typically 4--6th days of life begin in the decade [98, 99]. The clinical spectrum of the disease is very wide and ranges from organ-specific autoimmune disease (autoimmune exocrinopathy) to systemic disease (musculoskeletal, lung, kidney, nervous system involvement). In the diagnosis of primary Sjögren's syndrome, pancytopenia, increased ESR, and hypergammaglobulinemia is observed in laboratory tests. Cytopenias are most commonly observed as normochromic anemia, leukopenia, and thrombocytopenia. Among the autoimmune markers in Sjögren's syndrome, ANA 59-85%, RF 36-74%, SS-A 33-74%, SS-B 23-52% are positive [100 - 102].

Various cytokines and molecules can be used for research purposes in the systemic circulation and biological materials to characterize the inflammatory response and guide therapy. Studies may become more common in the coming years. Fecal calprotectin can be used as an indicator of underlying gastrointestinal inflammation [103 - 106].

Cells attach to the extracellular matrix (ECM) via adhesive glycoproteins such as fibronectin, laminin, vitronectin, thrombospondin, tenascin, entactin, nephronectin, and fibrinogen. Tenascin-C (TnC) can either increase or inhibit cell binding. TnC has high interaction properties with chondroitin sulfate and proteoglycan and is synthesized by interstitial fibroblasts. Although TnC levels are low in adult tissues, they increase in wound healing, especially during the formation of granulation tissue. Serum TnC level can be used as a new biochemical marker for diagnosis and predicting the prognosis of rheumatic carditis during childhood [107].

Type 1 Diabetes Mellitus (T1DM)

T1DM is the most common autoimmune disease caused by the destruction of insulin-producing β cells in the pancreas in children and adolescents. This disease, formerly called "Juvenile Diabetes" or "Insulin Dependent Diabetes Mellitus", can begin at any age between 6 months and 40 years. However, the frequency of onset around the age of 8 and 15 is very high. 90% of all T1DM cases start under the age of 18. In the presence of a complaint, a randomized blood glucose level above 200 mg/dL is diagnostic. It must be accompanied by glucose and ketone excretion in the urine. There are genetic, autoimmune, and metabolic markers in the development of T1DM. Autoantibodies [Anti-glutamic acid decarboxylase (Anti-GAD), islet cell antibody (ICA) as insulin autoantibody (IAA), anti-tyrosine phosphatase (IA2)] are usually positive in type 1 DM [108].

Laboratory tests used in diagnosis; the glucose tolerance test serum acetone, pH, PCO_2, sodium, potassium, leukocyte, glycated hemoglobin (HbA1c), and urine glucose includes [109].

Diagnostic criteria by the American Diabetes Association (ADA) include the following [110]:

- A fasting plasma glucose (FPG) level ≥126 mg/dL (7.0 mmol/L), or
- A 2-hour plasma glucose level ≥200 mg/dL (11.1 mmol/L) during a 75-g oral glucose tolerance test (OGTT), or
- A random plasma glucose ≥200 mg/dL (11.1 mmol/L) in a patient with classic symptoms of hyperglycemia or hyperglycemic crisis

Hemoglobin A1c (HbA1c, Glycolyzed Hemoglobin) Test

Normal HbA1c value in non-diabetic children is 5.7-5.4%. If this value is over 8, it indicates excessive serum glucose level. The HbA1c test shows the average blood glucose level for the previous 3-4 months. Glycolyzed hemoglobin levels should be measured 4 times a year in children with insulin-dependent T1DM [111].

Fasting C-peptide and insulin-like growth factor-binding protein-1 (IGFBP-1), CO2, and ketonuria are parameters measured to help distinguish childhood type 1 diabetes (T1D) from type 2 diabetes (T2D) at diagnosis [112].

Fructosamine indicates glycated albumin in plasma. It reflects the previous 1-3 weeks of glucose control [113].

Pediatric Malignancies

Cancer is a rare disease in childhood, and 10-15 out of every 100 thousand children in the 0-14 age group are diagnosed with cancer annually. The risk of cancer increases in familial / genetic diseases such as Down syndrome, neurofibromatosis, familial polyposis, Li-Fraumeni syndrome, immunodeficiencies, cancer treatment, or organ transplantation. It is often difficult to recognize childhood cancers early than the signs and symptoms are relatively non-specific. They can mimic the signs and symptoms of other common childhood diseases. For example, the symptoms of leukemia, which is the most common cancer of childhood, can be seen in many ordinary childhood diseases. Early symptoms are often suspicious and do not help diagnosis much [114, 115].

Tumor markers are chemical substances, most of which are in protein or glycoprotein structure, detected in the blood of patients with cancer. They can also be found in urine or other body fluids and tissues, but their levels in the blood are often checked. In normal, healthy people, they can be found in the blood to a certain extent. These chemical substances can be made by cancer cells themselves or they can occur as a reaction of the organism. Tumor markers are never used to diagnose cancer.

Distribution of childhood cancers by age is shown in Table **3**. Clinical use of tumor markers in childhood malignancies is shown in Table **4** [116, 117]. CA15-3 is a surrogate marker at diagnosis of hematological malignancies other than MDS and AML. It can determine relapse/deterioration or remission in idiopathic thrombocytosis, MDS, NHL, Hodgkin's lymphomas, and CLL. CA125 is a surrogate marker at diagnosis of NHL and M.M. It can determine relapse/deterioration or remission in NHL and CLL. CEA can help detect relapse/

deterioration in M.M. CA19-9 has no value in diagnosing and monitoring of hematological malignancies. Further studies in this area will likely bring new knowledge about tumor markers and their utility in current practice [118].

Table 3. Distribution of childhood cancers by age.

Early Childhood (First 5 years)
Embryonal tumors Wilms Tumor Retinoblastoma Rhabdomyosarcoma Medulloblastoma Neuroblastoma
2-5 age
Acute Lymphoblastic Leukemia Non-Hodgkin malignant lymphoma Wilms Tumor Neuroblastoma
Adolescent
Hodgkin lymphoma Bone Tumors Germ Cell Tumor Thyroid Cancer Malignant melanoma

Table 4. Clinical use of tumor markers in childhood malignancies.

Marker	Primary Cancer	Other Associated Malignancies
Alpha Feto-Protein (AFP)	Germ cell tumors that include embryonal carcinomas, endodermal sinus tumor and yolk sac tumors. Hepatocellular carcinoma (HCC, affecting the liver)	Stomach, colon, pancreatic, lung cancer
Human chorionic gonadotropin (β-hCG)	Choriocarcinoma Embryonal carcinoma	Neuroendocrine, bladder, kidney, lung, head, neck, gastrointestinal, cervix, uterus and vulva tumors, lymphoma, and leukemia
Lactate dehydrogenase (LDH)	Dysgerminoma Lymphoma Melanoma Neuroblastoma (NB)	Small-cell lung cancer Ewing sarcoma Osteogenic sarcoma

(Table 4) cont.....

Marker	Primary Cancer	Other Associated Malignancies
β2-microglobulin (β2M)	Lymphoma	Breast, prostate, lung, renal, gastrointestinal, nasopharyngeal cancers, and multiple myeloma
Catecholamines ***Homovanillic acid (HVA), vanillylmandelic acid (VMA) in urine***	NB Pheochromocytoma (PCC)	Medullary thyroid cancer (MTC)
Chromogranin A (CgA)	NB PCC Neuroendocrine tumors	Prostate, lung, breast, gastric, and colon cancers
Neuropeptide Y (NPY)	Neuroblastomas	Leukemia
Neuron specific enolase (NSE)	NB PCC MTC Gastric or lung carcinoids	Wilms tumor Rhabdomyosarcoma
Renin	Renal (Wilms, clear cell carcinoma Mesoblastic nephroma)	Ovarian, lung, pancreatic, adrenal, and colon cancers
Tissue polypeptide-specific antigen	Epithelial tumors such as carcinomas of the breast and lung.	NB or Wilms' tumor
Carcinoembryonic antigen (CEA)	Colorectal and other gastrointestinal cancers	-
-	Non- Hodgkin's(NHL) lymphomas	-
Carbohydrate antigen 125 (CA 12-5)	Ovary Lymphoma	Cervical and endometrial cancers, malignant ascites, renal cancer, non–small cell lung, breast, and stomach cancers, primary peritoneal carcinoma
Carbohydrate antigen 19-9 (CA 19-9)	Immature teratoma	-

Biomarkers in Vasculitis

All of the inflammatory changes that occur in the blood vessel wall are defined as vasculitis. Classification criteria of vasculitis are made only according to clinical and laboratory findings. The classification criteria have been developed to organize the clinical definitions of different vasculitis in the best possible way, not to diagnose the disease. Most of these markers are most useful for the diagnosis of vasculitis. Antibodies to proteinase-3 (PR3) or myeloperoxidase (MPO) are highly specific for granulomatosis with polyangiitis (GPA), microscopic polyangiitis (MPA), and eosinophilic granulomatosis with polyangiitis (ChurgStrauss; EGPA). An elevated peripheral eosinophil count has a high sensitivity for EGPA, and a positive cryoglobulin test has a high sensitivity for cryogobulinemic vasculitis. Inflammatory markers such as the erythrocyte sedimentation rate (ESR) and C reactive protein (CRP) have high, but by no means perfect, sensitivity for active vasculitis in untreated patients at first presentation, and thus are useful in ruling out vasculitis, particularly giant cell arteritis (GCA) [119].

Kawasaki disease (CD) is one of the most common vasculitides of childhood. There is no diagnostic laboratory test. Normochrome normocytic anemia, pyuria, ALT and GGT elevation, and hyponatremia can be detected. In KD, cytokines such as interleukin (IL) -1, IL2, IL-6, and tumor necrosis factor-α (TNF-α) are "up-regulated". There is a great deal of literature on inflammatory markers associated with prolonged fever or aneurysm development in the course of KD. However, to date, there is no marker that can differentiate CD and other inflammatory processes and be diagnostic [120]. In atypical CD, the risk of coronary artery involvement is higher than in classical CD. NT-proBNP can be used as an early marker of coronary artery involvement, especially in cases where ECHO cannot be performed or when the diagnosis is suspected [121, 122].

Takayasu's Arteritis (TA) is a rare chronic granulomatous large vessel vasculitis of unknown cause. In childhood, primary large vessel vasculitis is limited to TA. There are no laboratory tests specific to TA. The most useful tests in the diagnosis of TA and treatment follow-up are non-specific inflammatory markers C-reactive protein (CRP) and erythrocyte sedimentation rate (ESR> 20 mm/h). Research-oriented studies have been carried out on many molecules in TA [123].

Immunoglobulin A vasculitis/Henoch-Schönlein purpura (IgAV / HSP) is small vessel vasculitis and is generally characterized by non-thrombocytopenic purpura, arthritis, arthralgia, abdominal pain, gastrointestinal bleeding, and glomerulonephritis. There is no specific laboratory finding to diagnose the disease. IgAV / HSP is the most common primary vasculitis in childhood. A specific genetic factor causing the disease is not clearly known [124]. Laboratory findings are unspecific and include (usually) mildly elevated inflammatory parameters (ESR and CRP), normal or slightly reduced serum C3 and/or C4, the absence of high-titer autoantibodies (particularly ANCA and ANA) and, in 50% of all cases, elevated serum IgA and/or IgM. Fecal occult blood tests and urinary specimens are recommended at diagnosis. Since glomerulonephritis can manifest after the initial phase, regular urinary analysis at least for the first 6 months after disease-onset is recommended [125].

Polyarteritis nodosa is a necrotizing vasculitis involving medium-sized vessels. Laboratory evaluation shows leukocytosis, thrombocytosis, high acute phase reactants. Although rare in children with polyarteritis nodosa, positive hepatitis serology may be present. Antineutrophilic cytoplasmic antibodies are expected to be negative. Positive ANCA supports microscopic polyangiitis (MPA) with typical renal and possible lung involvement. Traditional angiogram is the gold standard for the diagnosis of polyarteritis nodosa (PAN). Magnetic resonance angiography may be insufficient to detect microaneurysms. Biopsy of the affected organs (kidney, intestine, nerve, muscle, skin) shows necrotizing arteritis and is important for a definitive diagnosis [126].

Adenosine deaminase 2 deficiency (DADA2, deficiency of ADA2, Sneddon syndrome) is a vasculitis syndrome, an autosomal recessive systemic disease associated with loss of function mutations in the ADA2 gene. Adenosine deaminase 2 deficiency was also known as childhood-onset polyarteritis nodosa. Sneddon syndrome and humoral immunodeficiency are the most common clinical presentations [127, 128].

Behçet's disease (BD) is a systemic vasculitis of unknown cause. Behçet's disease, which progresses with recurrent oral and genital ulcers as well as eye and skin findings, is an autoinflammatory disease that can affect many organ systems such as the central nervous system, musculoskeletal system, and gastrointestinal system. In 4-26% of Behçet's cases, the onset of the disease is before the age of 16. There is no specific laboratory finding for Behçet's disease. Behçet's disease, whose etiopathology is not known clearly, is mostly related to the HLA-B51 genetic marker which is linked to the more severe form of the disease. Pathergy skin test is positive in 60-70% of the patients [129].

INFECTIOUS DISEASES

Biomarkers in Inflammatory Diseases

Infectious diseases are in the first place among the reasons why patients in the pediatric age group consult a physician. The definitive diagnosis of infectious diseases is to reveal the causative microorganism, for this, microbiological and serological methods are used. The erythrocyte sedimentation rate (ESR), C-reactive protein (CRP), changes in the number and distribution of shaped elements in peripheral blood are all part of the acute phase response and are frequently used as indirect diagnostic methods in the diagnosis of infectious diseases [130].

The ideal biomarker for infections should distinguish patients with and without infection and distinguish between etiology (bacterial or viral). A biomarker ought to be independent of the duration of febrile illness and of comorbidities and to be a predictor of the severity of the disease. Importantly, any promising new biomarker will need a diagnostic test that is readily affordable, that can be measured in samples obtained with minimally invasive procedures, and which has a rapid turnaround time [131]. The most important characteristics of the ideal biomarker for infections are summarized in Table **5** [131].

Table 5. The most important characteristics of an ideal biomarker for infections.

• Has a positive test result in infected patient.
• Has a negative test result in patients without an infection
• Distinguishes etiology
• Is independent of comorbidities
• Is a predictor of severity
• Is a predictor of outcome
• Is a quick and easy test with small variation coefficient
• Is affordable

C-reactive Protein (CRP)

CRP is an acute-phase protein that was first discovered in the serum of patients with S. pneumoniae pneumonia by Tillet and Francis in 1930 and named as "C fraction protein". It was defined as an infection biomarker in the early stages, as it is higher in advanced infection periods than at baseline. CRP is the acute phase protein of the short pentraxin family produced in response to inflammatory signals in the liver, in particular, interleukin-6 (IL-6) enhanced synergistically by IL-1β [132]. CRP is found in very low levels (<1 mg/dL) in healthy individuals. Serum levels begin to rise 6-8 hours after the onset of inflammation, reach a peak within about 48 hours, and the half-life varies between 4-9 hours [133]. Unlike the ESR, CRP increases faster in the inflammatory process and decreases to normal earlier; CRP levels are not affected by conditions that affect ESR (anemia, polycythemia, erythrocyte shape, plasma protein level, age, gender). However, its synthesis is stimulated in all kinds of tissue destruction events such as trauma, burns, ischemia and infarction. Since its half-life is short (<24 hours), it can be used as a guide about the effectiveness of the treatment. In cases where the CRP is excessively increased, it may be falsely negative due to the prozone phenomenon. While CRP=150-300 mg/L in severe bacterial infections, it is generally below 40 mg/L in viral infections. While CRP increases approximately 10 times in bacterial infections, it increases 1-4 times in virus infections. Rarely, in some viral infections such as adenovirus, influenza, measles, mumps, the CRP value can be measured above 100 mg/L [134].

In the diagnosis of neonatal sepsis, CRP and immature neutrophil count may help the diagnosis [135 - 138]. CRP may be slightly higher (15 mg/L) in the first 4 days postnatal. CRP may increase some non-infectious perinatal conditions such as meconium aspiration syndrome, cephalhematoma, chest tube insertion [139].

The best studied acute phase reactant in neonatal sepsis is CRP. Although the main factor that increases the serum CRP level in newborns is an infection, some factors such as maternal fever, EMR, fetal distress, difficult birth, vacuum delivery and perinatal asphyxia may cause an increase in CRP level without systemic infection, and therefore, the specificity of CRP for early sepsis is low. The most appropriate acute phase reactant in the evaluation of infection in newborns is CRP. The negative predictive value of CRP for neonatal sepsis is over 98% [140 - 143].

Because of its rapid destruction, CRP measurement is useful in the follow-up of invasive diseases [144, 145]; however, CRP is not an alone indicator. It can also be used in the follow-up of non-infectious diseases such as Crohn's disease, rheumatological diseases, autoimmune vasculitis, Kawasaki disease and pancreatitis. CRP may also increase in liver neoplasia, lymphoma, post-surgical intervention, allograft rejection, and graft-versus-host disease [146, 147].

Inflammatory bowel diseases (IBD) are idiopathic, chronic disease that develops on a genetic basis and is characterized by activation and remission periods. IBD consists of two major clinical forms, Crohn's disease (CD) and ulcerative colitis (UC), and 'indeterminant colitis (IC)' seen in 10% of patients. Clinical symptoms and treatment of the disease are the same as in adults. The differentiation of these three diseases is made according to the differences in clinical, endoscopic, radiological, and histopathological findings [148]. The course of inflammatory bowel disease and its response to treatment vary in relation to the involvement site of the disease and the level of inflammation. In some intestinal infections, the complaints of the patients may be confused with the symptoms of IBD. For this reason, it should be investigated whether there is an infectious disease with stool and blood examinations. Many clinical and laboratory markers are used to evaluate the activation that is effective in the regulation of the treatment of the disease. Clinic studies have shown that CRP, pentraxin-3 (PTX-3), ESR, calprotectin, erythrocyte distribution volume (RDW), platelet count, and mean thrombocyte volume (MPV) levels vary significantly in individuals with IBD [149 - 152]. Together with the clinical findings of IBD, the activation of the disease should be determined by evaluating the hemogram parameters, and further examination and treatment should be provided in a short time.

The MEVF gene encoding pyrin, which is responsible for the regulation of neutrophil activity, is the gene responsible for familial Mediterranean fever (FMF). Since both epidemiological and clinical data suggest that MEFV may be a potential IBD gene, FMF and IBD have very similar clinical and biological characteristics, and FMF is frequently accompanied by IBD. It should be considered that other inflammatory diseases, especially FMF, may accompany IBD in cases where the disease does not progress as expected and the disease is resistant to treatment after being followed up with the diagnosis of IBD [153 - 155].

CRP plays an important role in many diseases, especially in follow-up. It is a powerful biomarker for inflammatory processes used for diagnostic and monitoring purposes. Reference percentiles of CRP have an important role in clinical practice for the diagnosis and monitoring of inflammatory diseases. The study (in Europe based on the IDEFICS study population) by Schlenz *et al.* [156] is the first to provide pan-European age-specific reference values of CRP for children that may guide paediatric practice and provide a basis for future clinical and epidemiological research on inflammation-related disorders in children. In the following years, various studies have been conducted on CRP based on the IDEFICS study population [157 - 159].

Conditions in which serum CRP levels may increase during the course of various pediatric diseases that indicate inflammatory and tissue damage are shown in Table **6**.

Table 6. Clinical use of CRP in pediatric diseases.

Evaluation of activity in inflammatory diseases
Juvenile rheumatoid arthritis Ankylosing spondylitis Reiter's disease Psoriatic arthropathy Vasculitis (Behçet's syndrome, Wegener Granulomatosis, Polyarteritis Nodosa, Polymyalgia Rheumatica) Chron’s disease Rheumatic fever Familial familial Mediterranean fever (FMF) Acute pancreatitis
Diagnosis and treatment of infectious diseases
Bacterial endocarditis Neonatal sepsis and meningitis Intercurrent infections (SLE and leukemias) Postoperative complications (infection, thromboembolism)
Differential diagnosis and classification of inflammatory diseases

(Table 6) cont.....

Evaluation of activity in inflammatory diseases
Rheumatoid arthritis with SLE Ulcerative colitis with Chron's disease

Serum Amyloid P (SAP) Component

SAP is a glycoprotein from the short pentraxin family that is secreted only in hepatocytes and plays key roles in immunity and inflammation. It has a half-life of 24 hours. The serum concentration of SAP in humans is □30-50 μg/mL. SAP and CRP have 66% of the same structure. Unlike CRP, serum concentrations are within normal limits in acute or chronic inflammations [160]. Tissue amyloid deposits due to systemic amyloidosis ubiquitously express SAP bound to amyloid fibrils and hence present themselves more abundantly compared to SAA [161].

In humans, SAP contributes to host defense, either via opsonins or through complement activation. In a calcium-dependent way, SAP binds to several lipoprotein ligands, which suggests that this process could have significant inferences in amyloidosis and atherosclerosis in humans. Moreover, many studies support the fact that SAP has a significant role in inflammatory regulation [162]. Cardiovascular disease (CVD), SAP is considered a valuable biomarker, as it contributes to CVD pathogenesis by modulating innate immunity and inflammation [163]. In a calcium-dependent manner, SAP binds to many different lipoprotein ligands, and this can have a significant contribution in the progression of amyloidosis [164].

Pentraxin 3 (PTX-3)

PTX-3 is the prototype of long pentraxins (molecular weight, 440 kDa) and is produced in response to inflammatory signals in various organs and tissues, including vascular endothelial cells, smooth muscle cells, macrophages, adipose tissue, and liver [165]. Protein PTX-3 may directly reflect disease activity [160]. Studies have reported that the correlation between CRP and PTX-3 is weak and may not be significant [166, 167].

PTX-3 levels in critically ill patients in intensive care units are associated with the severity of infection and disease. PTX-3 could be a better parameter than CRP in determining the degree of tissue damage caused by infection and inflammation [168]. Under normal conditions, blood PTX-3 level is approximately 25 μg/L in mice and <2 μg/L in humans. 6-8 hours after stimulation, it increases rapidly in proportion to the stimulus and the severity of the disease [169].

In the clinic, it is used in the follow-up and prognosis of rheumatic diseases, autoimmune diseases (especially systemic sclerosis), vasculitis involving small vessels, early membrane rupture and intraamniotic infections, myocardial diseases, atherosclerosis and infectious diseases in the follow-up and treatment of many diseases such as hypoxic ischemic brain damage and obesity [169 - 175]. PTX-3 is also very useful in the diagnosis of appendicitis, lower respiratory tract infection, sepsis, viral respiratory tract infections, allergic rhinitis, and pediatric epilepsy in children [176 - 182].

Erythrocyte Sedimentation Rate (ESR)

ESR is used to evaluate the course of a disease and its response to treatment. ESR elevation may increase systemic and localized inflammatory and infectious diseases, tissue damage, tissue ischemia, and trauma in malignant neoplasms, but it tends to increase more in infections than in other causes. Clinical use of ESR in pediatric diseases is shown in Table 7. The ESH half-life is 4-6 days. While ESR increases in most bacterial infections and mycobacterial infections; It is usually normal or slightly elevated in viral, rickettsia and Ehrlichia infections [183].

Table 7. Clinical use of ESR in pediatric diseases.

Increased ESR	Decreased ESR
•In chronic infections (such as osteomyelitis, septic arthritis and sometimes tuberculosis). •Autoimmune diseases (especially polymyalgia rheumatica, temporal arteritis, JRA, SLE, *etc.*). •High ESR (usually> 100 mm / hr) in solid tumors indicates metastasis. •Rheumatic fever •Anemia •Macrocytosis	•Hypofibrinogenemia, hypogammaglobulinemia and hyperviscosity causing dysproteinemia. •Polycythemia •Sickle cell anemia •Erythrocyte abnormalities such as leukocytosis, polycytosis, spherocytosis, acanthocytosis, microcytosis

ESR may also be high in non-infectious inflammatory conditions. If the ESR is below 50 mm/hour, the diagnosis of Kawasaki disease is excluded. ESR is not affected by fever and exercise [184].

CRP is generally a better indicator of inflammation than ESR. Elevation of ESR is typically due to the height of two types of proteins (globulins and/or fibrinogen). It may be helpful to look at fibrinogen and / or globulin (such as serum protein electrophoresis) to distinguish between them. ESR can rise without inflammation, as well as infection states. If the CRP is not high, only the increase in ESR indicates the high ESR level other than infection and other inflammatory factors [185, 186].

Macrophage activation syndrome (MAS) is a life□threatening complication of systemic juvenile idiopathic arthritis (sJIA). Ferritin/ESR ratio (with a cut point of 21.5 or greater) may be a biomarker for distinguishing sJIA□MAS from active sJIA without MAS [187].

ESR is also still commonly included in clinical algorithms for febrile children with a limp at risk for septic arthritis and osteomyelitis [188].

Procalcitonin (PCT)

The precursor of PCT calcitonin is a peptide (prohormone) that is released from the thyroid gland parafollicular cells. <72 hours there may be fluctuations in newborns; therefore the reference values are not clear,> 72 hours and normal levels in adults are ≤0.15 ng / mL.

It begins to increase within hours in response to infection and has a half-life of 25-30 hours [189].

It has been reported that serum PCT levels increase in bacterial infections such as septic shock and bacterial meningitis but remain low in viral infections [190 - 192]. Different studies have been conducted on whether serum PCT levels can be used in the distinction between bacterial and viral pneumonia, and different opinions have been reported on this issue [192 - 196].

PCT performs better than leukocyte count and CRP for detecting serious bacterial infection among children with fever without a source [197].

PCT has received substantial interest as a potential marker of infection to assess the presence, clearance, and eradication of infection; predict mortality, and guide antibiotic management [198]. PCT is a biomarker generally elevated in bacterial infections but not viral infections. PCT guidance may aid physicians in modestly decreasing antibiotic use in critically ill patients. However, the impact in settings with low baseline antibiotic use may be muted, and the effect on antibiotic resistance is unclear. PCT requires extensive observational and interventional studies to best determine its role [198].

White Blood Cell Count (WBC), Absolute Neutrophil Count (ANC)

WBC count, ANC, CRP, and PCT are generally not useful to discriminate nonsevere from severe disease in children with community-acquired pneumonia (CAP), although CRP and PCT may have some utility in predicting the most severe outcomes [199].

Kuppermann *et al.* [200] derived and validated an accurate prediction rule to identify febrile infants 60 days and younger at low risk for serious bacterial infections (SBIs) using urinalysis, ANC, and PCT levels.

WBC count and ANC had a better diagnostic performance with increasing age. Age-adjusted values of WBC count or ANC should be considered in diagnostic strategies for suspected pediatric appendicitis [201].

CRP performed better at longer fever duration. Both WBC and ANC were not reliable indicators of SBI at fever duration ≥12 h. In addition, the present study supports other recent data that CRP is a better predictor of SBI than either WBC or ANC regardless of fever duration. CRP has better sensitivity and specifi city than either WBC or ANC regardless of fever duration and is a useful laboratory screening tool in young patients with fever and no source of infection [202]. Combining WBC or ANC with other biomarkers did not improve diagnostic accuracy [203].

Coronavirus Disease 2019

The COVID-19 virus has been named SARS-CoV-2 due to its very close genetic similarity to SARS-CoV. While COVID-19 is seen in Wuhan South China seafood market employees, it soon spread around the world, and COVID-19 was recognized as a pandemic by WHO on March 11, 2020 [204]. In a study of 66 patients in which laboratory values were compiled in children, the total leukocyte count was found within normal limits in 69.6% of the patients, while an increase in the leukocyte count was observed in 15.2%, lymphopenia was found in only two patients [205]. Henry *et al.* [205] recommend clinicians monitor lymphocyte count and CRP as signs for severe infection while monitoring PCT for potential bacterial co-infection. IL-6 should be investigated as a potential prognostic indicator in severe COVID-19.

Multisystem inflammatory syndrome in children (MIS-C), also known as pediatric inflammatory, multisystem syndrome, is a new dangerous childhood disease that is temporally associated with coronavirus disease 2019 (COVID-19). Most patients (171 [92%]) had four or more laboratory biomarkers indicating inflammation. The majority had an elevated ESR or CRP level, lymphocytopenia, neutrophilia, elevated ferritin level, hypoalbuminemia, elevated ALT level, anemia, thrombocytopenia, and an elevated d-dimer level, prolonged international normalized ratio, or elevated fibrinogen level [206].

CONCLUSION

Biomarkers are ubiquitously incorporated into clinical practice as surrogate endpoints of physiological and pathological processes that cannot be directly measured. Pediatric biomarkers are more orphan markers than adults. Unlike adults, pediatric biomarkers should be developed specifically for children, and well-known developmental changes in children should be considered. In clinical practice, validated biomarkers are valuable tools in helping to understand the etiology, diagnosis, progression, regression, and treatment response of disease. Within the field of pediatrics, biomarker discovery remains dominated by targeted approaches. Further investigation and validation of emerging biomarkers to identify inflammatory or non-inflammatory conditions in childhood diseases is required. Because of the very low sensitivity and specificity, the available biomarkers have limited utility in the diagnosis and screening of pediatric diseases. Since validated pediatric biomarkers are vital in the care of the pediatric patient, collaborative efforts are required in the development of appropriate child-specific biomarkers.

CONSENT FOR PUBLICATION

Not applicable.

CONFLICT OF INTEREST

The author declares no conflict of interest, financial or otherwise

ACKNOWLEDGEMENT

Declared none.

REFERENCES

[1] Atkinson AJ, Colburn WA, DeGruttola VG, *et al.* Biomarkers and surrogate endpoints: preferred definitions and conceptual framework. Clin Pharmacol Ther 2001; 69(3): 89-95. [http://dx.doi.org/10.1067/mcp.2001.113989] [PMID: 11240971]

[2] Aronson JK. Biomarkers and surrogate endpoints. Br J Clin Pharmacol 2005; 59(5): 491-4. [http://dx.doi.org/10.1111/j.1365-2125.2005.02435.x] [PMID: 15842546]

[3] Goldman J, Becker ML, Jones B, Clements M, Leeder JS. Development of biomarkers to optimize pediatric patient management: what makes children different? Biomarkers Med 2011; 5(6): 781-94. [http://dx.doi.org/10.2217/bmm.11.96] [PMID: 22103612]

[4] Noncommunicable diseases progress monitor 2020. Geneva: World Health Organization 2020.

[5] Hoskins G, McCowan C, Neville RG, Thomas GE, Smith B, Silverman S. Risk factors and costs associated with an asthma attack. Thorax 2000; 55(1): 19-24. [http://dx.doi.org/10.1136/thorax.55.1.19] [PMID: 10607797]

[6] Martinez FD, Vercelli D. Asthma. Lancet 2013; 382(9901): 1360-72. [http://dx.doi.org/10.1016/S0140-6736(13)61536-6] [PMID: 24041942]

[7] Martinez FD. Maturation of immune responses at the beginning of asthma. J Allergy Clin Immunol 1999; 103(3 Pt 1): 355-61. [http://dx.doi.org/10.1016/S0091-6749(99)70456-2] [PMID: 10069865]

[8] Sears MR, Greene JM, Willan AR, *et al.* A longitudinal, population-based, cohort study of childhood asthma followed to adulthood. N Engl J Med 2003; 349(15): 1414-22. [http://dx.doi.org/10.1056/NEJMoa022363] [PMID: 14534334]

[9] DeVries A, Vercelli D. Early predictors of asthma and allergy in children: the role of epigenetics. Curr Opin Allergy Clin Immunol 2015; 15(5): 435-9. [http://dx.doi.org/10.1097/ACI.0000000000000201] [PMID: 26258923]

[10] Knipping K, Knippels LM, Dupont C, Garssen J. Serum biomarkers for allergy in children. Pediatr Allergy Immunol 2017; 28(2): 114-23. [http://dx.doi.org/10.1111/pai.12649] [PMID: 27590735]

[11] Schoos AM, Chawes BL, Følsgaard NV, Samandari N, Bønnelykke K, Bisgaard H. Disagreement between skin prick test and specific IgE in young children. Allergy 2015; 70(1): 41-8. [http://dx.doi.org/10.1111/all.12523] [PMID: 25224528]

[12] Kim J, Kim H, Lim D, Lee YK, Kim JH. Effects of Indoor Air Pollutants on Atopic Dermatitis. Int J Environ Res Public Health 2016; 13(12): E1220. [http://dx.doi.org/10.3390/ijerph13121220] [PMID: 27941696]

[13] Dupont C, Bradatan E, Soulaines P, Nocerino R, Berni-Canani R. Tolerance and growth in children with cow's milk allergy fed a thickened extensively hydrolyzed casein-based formula. BMC Pediatr 2016; 16(1): 96. [http://dx.doi.org/10.1186/s12887-016-0637-3] [PMID: 27430981]

[14] Gupta RS, Walkner MM, Greenhawt M, *et al.* Food Allergy Sensitization and Presentation in Siblings of Food Allergic Children. J Allergy Clin Immunol Pract 2016; 4(5): 956-62. [http://dx.doi.org/10.1016/j.jaip.2016.04.009] [PMID: 27421900]

[15] Kwon J, Kim J, Cho S, Noh G, Lee SS. Characterization of food allergies in patients with atopic dermatitis. Nutr Res Pract 2013; 7(2): 115-21. [http://dx.doi.org/10.4162/nrp.2013.7.2.115] [PMID: 23610604]

[16] Simon D, Braathen LR, Simon HU. Eosinophils and atopic dermatitis. Allergy 2004; 59(6): 561-70. [http://dx.doi.org/10.1111/j.1398-9995.2004.00476.x] [PMID: 15147438]

[17] Czech W, Krutmann J, Schöpf E, Kapp A. Serum eosinophil cationic protein (ECP) is a sensitive measure for disease activity in atopic dermatitis. Br J Dermatol 1992; 126(4): 351-5. [http://dx.doi.org/10.1111/j.1365-2133.1992.tb00677.x] [PMID: 1571256]

[18] Kim TY, Park HJ, Kim CW. Eosinophil cationic protein (ECP) level and its correlation with eosinophil number or IgE level of peripheral blood in patients with various skin diseases. J Dermatol Sci 1997; 15(2): 89-94. [http://dx.doi.org/10.1016/S0923-1811(97)00614-2] [PMID: 9279689]

[19] Taniuchi S, Chihara J, Kojima T, Yamamoto A, Sasai M, Kobayashi Y. Serum eosinophil derived neurotoxin may reflect more strongly disease severity in childhood atopic dermatitis than eosinophil cationic protein. J Dermatol Sci 2001; 26(1): 79-82. [http://dx.doi.org/10.1016/S0923-1811(00)00151-1] [PMID: 11323224]

[20] Wassom DL, Loegering DA, Solley GO, *et al.* Elevated serum levels of the eosinophil granule major basic protein in patients with eosinophilia. J Clin Invest 1981; 67(3): 651-61. [http://dx.doi.org/10.1172/JCI110080] [PMID: 7204553]

[21] Tischendorf FW, Brattig NW, Lintzel M, *et al.* Eosinophil granule proteins in serum and urine of patients with helminth infections and atopic dermatitis. Trop Med Int Health 2000; 5(12): 898-905. [http://dx.doi.org/10.1046/j.1365-3156.2000.00649.x] [PMID: 11169280]

[22] Breuer K, Kapp A, Werfel T. Urine eosinophil protein X (EPX) is an in vitro parameter of inflammation in atopic dermatitis of the adult age. Allergy 2001; 56(8): 780-4. [http://dx.doi.org/10.1034/j.1398-9995.2001.056008780.x] [PMID: 11488674]

[23] Oymar K, Bjerknes R. Urinary eosinophil protein X in children with atopic dermatitis: relation to atopy and disease activity. Allergy 2000; 55(10): 964-8. [http://dx.doi.org/10.1034/j.1398-9995.2000.00682.x] [PMID: 11030379]

[24] Najam FI, Giasuddin AS, Shembesh AH. Immunoglobulin isotypes in childhood asthma. Indian J Pediatr 1999; 66(3): 337-44. [http://dx.doi.org/10.1007/BF02845519] [PMID: 10798080]

[25] Kukkonen AK, Savilahti EM, Haahtela T, Savilahti E, Kuitunen M. Ovalbumin-specific immunoglobulins A and G levels at age 2 years are associated with the occurrence of atopic disorders. Clin Exp Allergy 2011; 41(10): 1414-21. [http://dx.doi.org/10.1111/j.1365-2222.2011.03821.x] [PMID: 21771118]

[26] Woof JM, Kerr MA. The function of immunoglobulin A in immunity. J Pathol 2006; 208(2): 270-82. [http://dx.doi.org/10.1002/path.1877] [PMID: 16362985]

[27] Stiehm RE. The four most common pediatric immunodeficiencies. Adv Exp Med Biol 2007; 601: 15-26. [http://dx.doi.org/10.1007/978-0-387-72005-0_2] [PMID: 17712988]

[28] Pesonen M, Kallio MJ, Siimes MA, Savilahti E, Ranki A. Serum immunoglobulin A concentration in infancy, but not human milk immunoglobulin A, is associated with subsequent atopic manifestations in children and adolescents: a 20-year prospective follow-up study. Clin Exp Allergy 2011; 41(5): 688-96. [http://dx.doi.org/10.1111/j.1365-2222.2011.03707.x] [PMID: 21418342]

[29] Hansen IS, Baeten DLP, den Dunnen J. The inflammatory function of human IgA. Cell Mol Life Sci 2019; 76(6): 1041-55. [http://dx.doi.org/10.1007/s00018-018-2976-8] [PMID: 30498997]

[30] Peng Z, Fisher R, Adkinson NF Jr. Total serum IgD is increased in atopic subjects. Allergy 1991; 46(6): 436-44. [http://dx.doi.org/10.1111/j.1398-9995.1991.tb04222.x] [PMID: 1720285]

[31] Salpietro DC, Masaracchio A, Turiaco A, Di Bella MR, Toscano V, Merlino MV. Serum IgD levels in children with atopic asthma. A longitudinal study. Minerva Pediatr 2001; 53(1): 1-5. [PMID: 11309536]

[32] Geha RS, Notarangelo LD, Casanova JL, *et al.* Primary immunodeficiency diseases: an update from the International Union of Immunological Societies Primary Immunodeficiency Diseases Classification Committee. J Allergy Clin Immunol 2007; 120(4): 776-94. [http://dx.doi.org/10.1016/j.jaci.2007.08.053] [PMID: 17952897]

[33] Notarangelo LD. Primary immunodeficiencies. J Allergy Clin Immunol 2010; 125(2) (Suppl. 2): S182-94. [http://dx.doi.org/10.1016/j.jaci.2009.07.053] [PMID: 20042228]

[34] Cantoni N, Recher M. [Primary and secondary immunodeficiencies]. Ther Umsch 2014; 71(1): 31-43. [http://dx.doi.org/10.1024/0040-5930/a000479] [PMID: 24394207]

[35] Turvey SE, Bonilla FA, Junker AK. Primary immunodeficiency diseases: a practical guide for clinicians. Postgrad Med J 2009; 85(1010): 660-6. [http://dx.doi.org/10.1136/pgmj.2009.080630] [PMID: 20075404]

[36] Skevaki C, Van den Berg J, Jones N, *et al.* Immune biomarkers in the spectrum of childhood noncommunicable diseases. J Allergy Clin Immunol 2016; 137(5): 1302-16. [http://dx.doi.org/10.1016/j.jaci.2016.03.012] [PMID: 27155027]

[37] Weiland M, Gao XH, Zhou L, Mi QS. Small RNAs have a large impact: circulating microRNAs as biomarkers for human diseases. RNA Biol 2012; 9(6): 850-9. [http://dx.doi.org/10.4161/rna.20378] [PMID: 22699556]

[38] Rebane A. microRNA: medical evidence from molecular biology to clinical practice foreword, Chapter 17: microRNA and Allergy. Microrna: Medical Evidence: From Molecular Biology to Clinical Practice 2015; 888: 331-52.

[39] Dissanayake E, Inoue Y. MicroRNAs in Allergic Disease. Curr Allergy Asthma Rep 2016; 16(9): 67. [http://dx.doi.org/10.1007/s11882-016-0648-z] [PMID: 27585977]

[40] Lu TX, Rothenberg ME. Diagnostic, functional, and therapeutic roles of microRNA in allergic diseases. J Allergy Clin Immunol 2013; 132(1): 3-13. [http://dx.doi.org/10.1016/j.jaci.2013.04.039] [PMID: 23735656]

[41] Liu Z, Zhang XH, Callejas-Díaz B, Mullol J. MicroRNA in United Airway Diseases. Int J Mol Sci 2016; 17(5): 17. [http://dx.doi.org/10.3390/ijms17050716] [PMID: 27187364]

[42] Zhang XH, Zhang YN, Liu Z. MicroRNA in chronic rhinosinusitis and allergic rhinitis. Curr Allergy Asthma Rep 2014; 14(2): 415. [http://dx.doi.org/10.1007/s11882-013-0415-3] [PMID: 24408538]

[43] Wang Y, Yang L, Li P, *et al.* Circulating microRNA signatures associated with childhood asthma. Clin Lab 2015; 61(5-6): 467-74. [http://dx.doi.org/10.7754/Clin.Lab.2014.141020] [PMID: 26118177]

[44] Fang C, Lu W, Li C, *et al.* MiR-3162-3p Is a Novel MicroRNA That Exacerbates Asthma by Regulating β-Catenin. PLoS One 2016; 11(3): e0149257. [http://dx.doi.org/10.1371/journal.pone.0149257] [PMID: 26959414]

[45] Kho AT, McGeachie MJ, Moore KG, Sylvia JM, Weiss ST, Tantisira KG. Circulating microRNAs and prediction of asthma exacerbation in childhood asthma. Respir Res 2018; 19(1): 128. [http://dx.doi.org/10.1186/s12931-018-0828-6] [PMID: 29940952]

[46] Soubani AO, Chandrasekar PH. The clinical spectrum of pulmonary aspergillosis. Chest 2002; 121(6): 1988-99. [http://dx.doi.org/10.1378/chest.121.6.1988] [PMID: 12065367]

[47] Boguniewicz M, Leung DY. Atopic dermatitis: a disease of altered skin barrier and immune dysregulation. Immunol Rev 2011; 242(1): 233-46. [http://dx.doi.org/10.1111/j.1600-065X.2011.01027.x] [PMID: 21682749]

[48] Howell MD, Kim BE, Gao P, *et al.* Cytokine modulation of atopic dermatitis filaggrin skin expression. J Allergy Clin Immunol 2007; 120(1): 150-5. [http://dx.doi.org/10.1016/j.jaci.2007.04.031] [PMID: 17512043]

[49] Bousquet J, Van Cauwenberge P, Khaltaev N. Allergic rhinitis and its impact on asthma. J Allergy Clin Immunol 2001; 108(5) (Suppl.): S147-334. [http://dx.doi.org/10.1067/mai.2001.118891] [PMID: 11707753]

[50] Dykewicz MS, Fineman S. Executive Summary of Joint Task Force Practice Parameters on Diagnosis and Management of Rhinitis. Ann Allergy Asthma Immunol 1998; 81(5 Pt 2): 463-8. [http://dx.doi.org/10.1016/S1081-1206(10)63152-3] [PMID: 9860024]

[51] van Cauwenberge P, Bachert C, Passalacqua G, *et al.* Consensus statement on the treatment of allergic rhinitis. Allergy 2000; 55(2): 116-34. [http://dx.doi.org/10.1034/j.1398-9995.2000.00526.x] [PMID: 10726726]

[52] Bousquet J, Khaltaev N, Cruz AA, *et al.* Allergic Rhinitis and its Impact on Asthma (ARIA) 2008 update (in collaboration with the World Health Organization, GA(2)LEN and AllerGen). Allergy 2008; 63(63) (Suppl. 86): 8-160. [http://dx.doi.org/10.1111/j.1398-9995.2007.01620.x] [PMID: 18331513]

[53] Pawankar R, Mullol J. ARIA update in collaboration with GA(2)LEN mechanisms of allergic rhinitis. Allergy 2008.

[54] Carr TF, Saltoun CA. Chapter 2: Skin testing in allergy. Allergy Asthma Proc 2012; 33(3) (Suppl. 1): 6-8.
[http://dx.doi.org/10.2500/aap.2012.33.3532] [PMID: 22794675]

[55] Incorvaia C, Cavaliere C, Frati F, Masieri S. Allergic rhinitis. J Biol Regul Homeost Agents 2018; 32(1) (Suppl. 1): 61-6.
[PMID: 29552875]

[56] Steiner I, Sobieska M, Pucher B, Grzegorowski M, Samborski W. [Examination of acute phase proteins concentrations in children with allergic rhinitis]. Ann Acad Med Stetin 2006; 52(2): 33-7.
[PMID: 17633124]

[57] Andersson M, Greiff L, Svensson C, Persson C. Various methods for testing nasal responses in vivo: a critical review. Acta Otolaryngol 1995; 115(6): 705-13.
[http://dx.doi.org/10.3109/00016489509139391] [PMID: 8749189]

[58] Marseglia GL, De Amici M, Marseglia A, *et al.* Pentraxin 3 serum levels in children with atopic dermatitis. J Biol Regul Homeost Agents 2012; 26(1) (Suppl.): S105-8.
[PMID: 22691256]

[59] Chawes BL, Stokholm J, Schoos AM, Fink NR, Brix S, Bisgaard H. Allergic sensitization at school age is a systemic low-grade inflammatory disorder. Allergy 2017; 72(7): 1073-80.
[http://dx.doi.org/10.1111/all.13108] [PMID: 27992959]

[60] Han MW, Kim SH, Oh I, Kim YH, Lee J. Serum IL-1β can be a biomarker in children with severe persistent allergic rhinitis. Allergy Asthma Clin Immunol 2019; 15(1): 58.
[http://dx.doi.org/10.1186/s13223-019-0368-8] [PMID: 31548841]

[61] Chinoy B, Yee E, Bahna SL. Skin testing versus radioallergosorbent testing for indoor allergens. Clin Mol Allergy 2005; 3(1): 4.
[http://dx.doi.org/10.1186/1476-7961-3-4] [PMID: 15833110]

[62] Dalrymple AM, Moore TL. Laboratory Evaluation in Pediatric Autoimmune Diseases. Pediatr Rev 2015; 36(11): 496-501.
[http://dx.doi.org/10.1542/pir.36.11.496] [PMID: 26527629]

[63] Zeft AS, Spalding SJ. Autoinflammatory syndromes: fever is not always a sign of infection. Cleve Clin J Med 2012; 79(8): 569-81.
[http://dx.doi.org/10.3949/ccjm.79a.11184] [PMID: 22854436]

[64] Hashkes PJ, Toker O. Autoinflammatory syndromes. Pediatr Clin North Am 2012; 59(2): 447-70.
[http://dx.doi.org/10.1016/j.pcl.2012.03.005] [PMID: 22560579]

[65] Cush JJ. Autoinflammatory syndromes. Dermatol Clin 2013; 31(3): 471-80.
[http://dx.doi.org/10.1016/j.det.2013.05.001] [PMID: 23827249]

[66] Fritzler MJ. Choosing wisely: Review and commentary on anti-nuclear antibody (ANA) testing. Autoimmun Rev 2016; 15(3): 272-80.
[http://dx.doi.org/10.1016/j.autrev.2015.12.002] [PMID: 26687321]

[67] Meroni PL, Schur PH. ANA screening: an old test with new recommendations. Ann Rheum Dis 2010; 69(8): 1420-2.
[http://dx.doi.org/10.1136/ard.2009.127100] [PMID: 20511607]

[68] Martini G, Foeldvari I, Russo R, *et al.* Systemic sclerosis in childhood: clinical and immunologic features of 153 patients in an international database. Arthritis Rheum 2006; 54(12): 3971-8.
[http://dx.doi.org/10.1002/art.22207] [PMID: 17133611]

[69] Scalapino K, Arkachaisri T, Lucas M, *et al.* Childhood onset systemic sclerosis: classification, clinical and serologic features, and survival in comparison with adult onset disease. J Rheumatol 2006; 33(5): 1004-13.
[PMID: 16583463]

[70] Kishi T, Miyamae T, Hara R, *et al.* Clinical analysis of 50 children with juvenile dermatomyositis. Mod Rheum 2013; 23(2): 311-7. [http://dx.doi.org/10.3109/s10165-012-0647-4] [PMID: 22526832]

[71] Gowdie PJ, Allen RC, Kornberg AJ, Akikusa JD. Clinical features and disease course of patients with juvenile dermatomyositis. Int J Rheum Dis 2013; 16(5): 561-7. [http://dx.doi.org/10.1111/1756-185X.12107] [PMID: 24164844]

[72] Bader-Meunier B, Armengaud JB, Haddad E, *et al.* Initial presentation of childhood-onset systemic lupus erythematosus: a French multicenter study. J Pediatr 2005; 146(5): 648-53. [http://dx.doi.org/10.1016/j.jpeds.2004.12.045] [PMID: 15870669]

[73] Hiraki LT, Benseler SM, Tyrrell PN, Hebert D, Harvey E, Silverman ED. Clinical and laboratory characteristics and long-term outcome of pediatric systemic lupus erythematosus: a longitudinal study. J Pediatr 2008; 152(4): 550-6. [http://dx.doi.org/10.1016/j.jpeds.2007.09.019] [PMID: 18346514]

[74] Berntson L, Andersson Gäre B, Fasth A, *et al.* Incidence of juvenile idiopathic arthritis in the Nordic countries. A population based study with special reference to the validity of the ILAR and EULAR criteria. J Rheumatol 2003; 30(10): 2275-82. [PMID: 14528529]

[75] Kasapçopur O, Ozbakir F, Arisoy N, Ingöl H, Yazici H, Ozdoğan H. Frequency of antinuclear antibodies and rheumatoid factor in healthy Turkish children. Turk J Pediatr 1999; 41(1): 67-71. [PMID: 10770678]

[76] El Sherbini HM, El Garf AK, El Din Mahmoud SS. Human leukocyte antigen and autoantibodies association with juvenile systemic lupus erythematosus. Egypt J Immunol 2009; 16(2): 107-14. [PMID: 22059358]

[77] Avcin T, O'Neil KM. Antiphospholipid syndrome.Textbook of Pediatric Rheumatology. Philadelphia, PA: Saunders 2011; pp. 344-60. [http://dx.doi.org/10.1016/B978-1-4160-6581-4.10022-6]

[78] Sperotto F, Cuffaro G, Brachi S, Seguso M, Zulian F. Prevalence of antinuclear antibodies in schoolchildren during puberty and possible relationship with musculoskeletal pain: a longitudinal study. J Rheumatol 2014; 41(7): 1405-8. [http://dx.doi.org/10.3899/jrheum.130948] [PMID: 24737914]

[79] Ingegnoli F, Castelli R, Gualtierotti R. Rheumatoid factors: clinical applications. Dis Markers 2013; 35(6): 727-34. [http://dx.doi.org/10.1155/2013/726598] [PMID: 24324289]

[80] Aletaha D, Alasti F, Smolen JS. Rheumatoid factor determines structural progression of rheumatoid arthritis dependent and independent of disease activity. Ann Rheum Dis 2013; 72(6): 875-80. [http://dx.doi.org/10.1136/annrheumdis-2012-201517] [PMID: 22798565]

[81] Nell VP, Machold KP, Stamm TA, *et al.* Autoantibody profiling as early diagnostic and prognostic tool for rheumatoid arthritis. Ann Rheum Dis 2005; 64(12): 1731-6. [http://dx.doi.org/10.1136/ard.2005.035691] [PMID: 15878904]

[82] Swedler W, Wallman J, Froelich CJ, Teodorescu M. Routine measurement of IgM, IgG, and IgA rheumatoid factors: high sensitivity, specificity, and predictive value for rheumatoid arthritis. J Rheumatol 1997; 24(6): 1037-44. [PMID: 9195506]

[83] Sweet RA, Cullen JL, Shlomchik MJ. Rheumatoid factor B cell memory leads to rapid, switched antibody-forming cell responses. J Immunol 2013; 190(5): 1974-81. [http://dx.doi.org/10.4049/jimmunol.1202816] [PMID: 23365079]

[84] Newkirk MM. Rheumatoid factors: host resistance or autoimmunity? Clin Immunol 2002; 104(1): 1-13.

[http://dx.doi.org/10.1006/clim.2002.5210] [PMID: 12139942]

[85] Díaz-López C, Geli C, Corominas H, *et al.* Are there clinical or serological differences between male and female patients with primary Sjögren's syndrome? J Rheumatol 2004; 31(7): 1352-5. [PMID: 15229956]

[86] Dresser DW, Popham AM. Induction of an IgM anti-(bovine)-IgG response in mice by bacterial lipopolysaccharide. Nature 1976; 264(5586): 552-4. [http://dx.doi.org/10.1038/264552a0] [PMID: 63917]

[87] Ravelli A, Martini A. Juvenile idiopathic arthritis. Lancet 2007; 369(9563): 767-78. [http://dx.doi.org/10.1016/S0140-6736(07)60363-8] [PMID: 17336654]

[88] Petty RE, Laxer RM, Wedderburn LR. Juvenile idiopathic arthritis.Textbook of pediatric rheumatology. 7th ed. Philadelphia: Elsevier 2016; pp. 188-204. [http://dx.doi.org/10.1016/B978-0-323-24145-8.00015-6]

[89] Spîrchez M, Samaşca G, Iancu M, Bolba C, Miu N. Relation of interleukin-6, TNF-alpha and interleukin-1alpha with disease activity and severity in juvenile idiopathic arthritis patients. Clin Lab 2012; 58(3-4): 253-60. [PMID: 22582498]

[90] Rooney M, David J, Symons J, Di Giovine F, Varsani H, Woo P. Inflammatory cytokine responses in juvenile chronic arthritis. Br J Rheumatol 1995; 34(5): 454-60. [http://dx.doi.org/10.1093/rheumatology/34.5.454] [PMID: 7788176]

[91] Mangge H, Kenzian H, Gallistl S, *et al.* Serum cytokines in juvenile rheumatoid arthritis. Correlation with conventional inflammation parameters and clinical subtypes. Arthritis Rheum 1995; 38(2): 211-20. [http://dx.doi.org/10.1002/art.1780380209] [PMID: 7848311]

[92] Schellekens GA, de Jong BAW, van den Hoogen FHJ, van de Putte LBA, van Venrooij WJ. Citrulline is an essential constituent of antigenic determinants recognized by rheumatoid arthritis-specific autoantibodies. J Clin Invest 1998; 101(1): 273-81. [http://dx.doi.org/10.1172/JCI1316] [PMID: 9421490]

[93] Sebbag M, Simon M, Vincent C, *et al.* The antiperinuclear factor and the so-called antikeratin antibodies are the same rheumatoid arthritis-specific autoantibodies. J Clin Invest 1995; 95(6): 2672-9. [http://dx.doi.org/10.1172/JCI117969] [PMID: 7539459]

[94] Schellekens GA, Visser H, de Jong BA, *et al.* The diagnostic properties of rheumatoid arthritis antibodies recognizing a cyclic citrullinated peptide. Arthritis Rheum 2000; 43(1): 155-63. [http://dx.doi.org/10.1002/1529-0131(200001)43:1<155::AID-ANR20>3.0.CO;2-3] [PMID: 10643712]

[95] Zendman AJ, van Venrooij WJ, Pruijn GJ. Use and significance of anti-CCP autoantibodies in rheumatoid arthritis. Rheumatology (Oxford) 2006; 45(1): 20-5. [http://dx.doi.org/10.1093/rheumatology/kei111] [PMID: 16188946]

[96] Visser H, le Cessie S, Vos K, Breedveld FC, Hazes JMW. How to diagnose rheumatoid arthritis early: a prediction model for persistent (erosive) arthritis. Arthritis Rheum 2002; 46(2): 357-65. [http://dx.doi.org/10.1002/art.10117] [PMID: 11840437]

[97] van Gaalen FA, Linn-Rasker SP, van Venrooij WJ, *et al.* Autoantibodies to cyclic citrullinated peptides predict progression to rheumatoid arthritis in patients with undifferentiated arthritis: a prospective cohort study. Arthritis Rheum 2004; 50(3): 709-15. [http://dx.doi.org/10.1002/art.20044] [PMID: 15022309]

[98] Ramos-Casals M, Tzioufas AG, Font J. Primary Sjögren's syndrome: new clinical and therapeutic concepts. Ann Rheum Dis 2005; 64(3): 347-54. [http://dx.doi.org/10.1136/ard.2004.025676] [PMID: 15498797]

[99] Pertovaara M, Korpela M, Uusitalo H, *et al.* Clinical follow up study of 87 patients with sicca

symptoms (dryness of eyes or mouth, or both). Ann Rheum Dis 1999; 58(7): 423-7. [http://dx.doi.org/10.1136/ard.58.7.423] [PMID: 10381486]

[100] Bowman SJ. Primary Sjögren's syndrome. Lupus 2018; 27(1_suppl) (Suppl.): 32-5. [http://dx.doi.org/10.1177/0961203318801673] [PMID: 30452329]

[101] Bournia V-K, Vlachoyiannopoulos PG. Subgroups of Sjögren syndrome patients according to serological profiles. J Autoimmun 2012; 39(1-2): 15-26. [http://dx.doi.org/10.1016/j.jaut.2012.03.001] [PMID: 22575069]

[102] Harley JB, Alexander EL, Bias WB, *et al.* Anti-Ro (SS-A) and anti-La (SS-B) in patients with Sjögren's syndrome. Arthritis Rheum 1986; 29(2): 196-206. [http://dx.doi.org/10.1002/art.1780290207] [PMID: 3485431]

[103] Aalto K, Lahdenne P, Kolho KL. Fecal calprotectin in juvenile idiopathic arthritis patients related to drug use. Pediatr Rheumatol Online J 2017; 15(1): 9. [http://dx.doi.org/10.1186/s12969-016-0132-2] [PMID: 28143478]

[104] Stoll ML, Punaro M, Patel AS. Fecal calprotectin in children with the enthesitis-related arthritis subtype of juvenile idiopathic arthritis. J Rheumatol 2011; 38(10): 2274-5. [http://dx.doi.org/10.3899/jrheum.110508] [PMID: 21965700]

[105] Biskou O, Gardner-Medwin J, Mackinder M, *et al.* Faecal Calprotectin in Treated and Untreated Children With Coeliac Disease and Juvenile Idiopathic Arthritis. J Pediatr Gastroenterol Nutr 2016; 63(5): e112-5. [http://dx.doi.org/10.1097/MPG.0000000000001384] [PMID: 27540707]

[106] Pikkarainen S, Martelius T, Ristimäki A, Siitonen S, Seppänen MRJ, Färkkilä M. A High Prevalence of Gastrointestinal Manifestations in Common Variable Immunodeficiency. Am J Gastroenterol 2019; 114(4): 648-55. [http://dx.doi.org/10.14309/ajg.0000000000000140] [PMID: 30747770]

[107] Karatas Z, Baysal T, Alp H, Toker A. Serum tenascin-C: a novel biomarker for diagnosis and predicting prognosis of rheumatic carditis? J Trop Pediatr 2013; 59(6): 476-82. [http://dx.doi.org/10.1093/tropej/fmt058] [PMID: 23868575]

[108] Orban T, Sosenko JM, Cuthbertson D, *et al.* Pancreatic islet autoantibodies as predictors of type 1 diabetes in the Diabetes Prevention Trial-Type 1. Diabetes Care 2009; 32(12): 2269-74. [http://dx.doi.org/10.2337/dc09-0934] [PMID: 19741189]

[109] Sacks DB, Arnold M, Bakris GL, *et al.* Guidelines and recommendations for laboratory analysis in the diagnosis and management of diabetes mellitus. Clin Chem 2011; 57(6): e1-e47. [http://dx.doi.org/10.1373/clinchem.2010.161596] [PMID: 21617152]

[110] American Diabetes Association. Diagnosis and classification of diabetes mellitus. Diabetes Care 2010; 33 (Suppl. 1): S62-9. [PMID: 20042775]

[111] Leow MK. Glycated Hemoglobin (HbA1c): Clinical Applications of a Mathematical Concept. Acta Inform Med 2016; 24(4): 233-8. [http://dx.doi.org/10.5455/aim.2016.24.233-238] [PMID: 27708483]

[112] Katz LE, Jawad AF, Ganesh J, Abraham M, Murphy K, Lipman TH. Fasting c-peptide and insulin-like growth factor-binding protein-1 levels help to distinguish childhood type 1 and type 2 diabetes at diagnosis. Pediatr Diabetes 2007; 8(2): 53-9. [http://dx.doi.org/10.1111/j.1399-5448.2007.00236.x] [PMID: 17448127]

[113] Ribeiro RT, Macedo MP, Raposo JF. HbA1c, fructosamine, and glycated albumin in the detection of dysglycaemic conditions. Curr Diabetes Rev 2016; 12(1): 14-9. [http://dx.doi.org/10.2174/1573399811666150701143112] [PMID: 26126638]

[114] Kilburn LB, Siegel SE, Steuber CP. Clinical assesment and differantial diagnosis of the child with suspected cancer.Principles and practice of pediatric oncology. 6th ed. Philadelphia, USA: Lippincott

Williams and Wolters 2011; pp. 123-35.

[115] Stiller CA, Marcos-Gragera R, Ardanaz E, *et al.* Geographical patterns of childhood cancer incidence in Europe, 1988-1997. Report from the Automated Childhood Cancer Information System project. Eur J Cancer 2006; 42(13): 1952-60. [http://dx.doi.org/10.1016/j.ejca.2006.05.017] [PMID: 16919763]

[116] Labdenne P, Heikinheimo M. Clinical use of tumor markers in childhood malignancies. Ann Med 2002; 34(5): 316-23. [http://dx.doi.org/10.1080/078538902320772070] [PMID: 12452476]

[117] Ricafort R. Tumor markers in infancy and childhood. Pediatr Rev 2011; 32(7): 306-8. [http://dx.doi.org/10.1542/pir.32.7.306] [PMID: 21724907]

[118] Kontoninas Z, Kaiafa GD, Saouli Z, *et al.* Tumor Markers in Diagnosis and Follow up of Haematologic Malignancies. Blood 2016; 128(22): 5337-7. [http://dx.doi.org/10.1182/blood.V128.22.5337.5337]

[119] Monach PA. Biomarkers in vasculitis. Curr Opin Rheumatol 2014; 26(1): 24-30. [http://dx.doi.org/10.1097/BOR.0000000000000009] [PMID: 24257367]

[120] De Ferranti SD, Newburger JW. Kawasaki disease (Mucocutaneous lymph node syndrome).Moss and Adams heart disease in infants, children, and adolescents: including the fetus and young adult. 8th ed., Philadelphia: Wolters Kluwer Health/Lippincott Williams & Wilkins 2013.

[121] Dahdah N, Siles A, Fournier A, *et al.* Natriuretic peptide as an adjunctive diagnostic test in the acute phase of Kawasaki disease. Pediatr Cardiol 2009; 30(6): 810-7. [http://dx.doi.org/10.1007/s00246-009-9441-2] [PMID: 19365652]

[122] McNeal-Davidson A, Fournier A, Spigelblatt L, *et al.* Value of amino-terminal pro B-natriuretic peptide in diagnosing Kawasaki disease. Pediatr Int 2012; 54(5): 627-33. [http://dx.doi.org/10.1111/j.1442-200X.2012.03609.x] [PMID: 22414326]

[123] O'Connor TE, Carpenter HE, Bidari S, Waters MF, Hedna VS. Role of inflammatory markers in Takayasu arteritis disease monitoring. BMC Neurol 2014; 14(1): 62. [http://dx.doi.org/10.1186/1471-2377-14-62] [PMID: 24678735]

[124] Gardner-Medwin JM, Dolezalova P, Cummins C, Southwood TR. Incidence of Henoch-Schönlein purpura, Kawasaki disease, and rare vasculitides in children of different ethnic origins. Lancet 2002; 360(9341): 1197-202. [http://dx.doi.org/10.1016/S0140-6736(02)11279-7] [PMID: 12401245]

[125] Kawasaki Y, Ono A, Ohara S, *et al.* Henoch-Schönlein purpura nephritis in childhood: pathogenesis, prognostic factors and treatment. Fukushima J Med Sci 2013; 59(1): 15-26. [http://dx.doi.org/10.5387/fms.59.15] [PMID: 23842510]

[126] Eleftheriou D, Batu ED, Ozen S, Brogan PA. Vasculitis in children. Nephrol Dial Transplant 2015; 30 (Suppl. 1): i94-i103. [PMID: 25550447]

[127] Navon Elkan P, Pierce SB, Segel R, *et al.* Mutant adenosine deaminase 2 in a polyarteritis nodosa vasculopathy. N Engl J Med 2014; 370(10): 921-31. [http://dx.doi.org/10.1056/NEJMoa1307362] [PMID: 24552285]

[128] Zhou Q, Yang D, Ombrello AK, *et al.* Early-onset stroke and vasculopathy associated with mutations in ADA2. N Engl J Med 2014; 370(10): 911-20. [http://dx.doi.org/10.1056/NEJMoa1307361] [PMID: 24552284]

[129] Koné-Paut I, Darce-Bello M, Shahram F, *et al.* Registries in rheumatological and musculoskeletal conditions. Paediatric Behçet's disease: an international cohort study of 110 patients. One-year follow-up data. Rheumatology (Oxford) 2011; 50(1): 184-8. [http://dx.doi.org/10.1093/rheumatology/keq324] [PMID: 21036877]

[130] Sáez-Llorens X, Lagrutta F. The acute phase host reaction during bacterial infection and its clinical impact in children. Pediatr Infect Dis J 1993; 12(1): 83-7. [http://dx.doi.org/10.1097/00006454-199301000-00017] [PMID: 8417431]

[131] Stol K, Nijman RG, van Herk W, van Rossum AMC. Biomarkers for infection in children: current clinical practice and future perspectives. Pediatr Infect Dis J 2019; 38(6S Suppl 1): S7-S13.

[132] Tillett WS, Francis T Jr. Serologıcal reactıons ın pneumonıa wıth a non-proteın somatıc fractıon of pneumococcus. J Exp Med 1930; 52(4): 561-71. [http://dx.doi.org/10.1084/jem.52.4.561] [PMID: 19869788]

[133] Clyne B, Olshaker JS. The C-reactive protein. J Emerg Med 1999; 17(6): 1019-25. [http://dx.doi.org/10.1016/S0736-4679(99)00135-3] [PMID: 10595891]

[134] Jaye DL, Waites KB. Clinical applications of C-reactive protein in pediatrics. Pediatr Infect Dis J 1997; 16(8): 735-46. [http://dx.doi.org/10.1097/00006454-199708000-00003] [PMID: 9271034]

[135] Da Silva O, Ohlsson A, Kenyon C. Accuracy of leukocyte indices and C-reactive protein for diagnosis of neonatal sepsis: a critical review. Pediatr Infect Dis J 1995; 14(5): 362-6. [http://dx.doi.org/10.1097/00006454-199505000-00005] [PMID: 7638010]

[136] Saboohi E, Saeed F, Khan RN, Khan MA. Immature to total neutrophil ratio as an early indicator of early neonatal sepsis. Pak J Med Sci 2019; 35(1): 241-6. [http://dx.doi.org/10.12669/pjms.35.1.99] [PMID: 30881431]

[137] Goldfinch CD, Korman T, Kotsanas D, Burgner DP, Tan K. C-reactive protein and immature-to-total neutrophil ratio have no utility in guiding lumbar puncture in suspected neonatal sepsis. J Paediatr Child Health 2018; 54(8): 848-54. [http://dx.doi.org/10.1111/jpc.13890] [PMID: 29602260]

[138] Khair KB, Rahman MA, Sultana T, Roy CK, Rahman MQ, Ahmed AN. Early diagnosis of neonatal septicemia by hematologic scoring system, C-reactive protein and serum haptoglobin. Mymensingh Med J 2012; 21(1): 85-92. [PMID: 22314460]

[139] Hofer N, Müller W, Resch B. Non-infectious conditions and gestational age influence C-reactive protein values in newborns during the first 3 days of life. Clin Chem Lab Med 2011; 49(2): 297-302. [http://dx.doi.org/10.1515/CCLM.2011.048] [PMID: 21126207]

[140] Edwards MS, Baker CJ. Sepsis in the newborn.Krugman's Infectious Diseases of Children. 11th ed. Philadelphia: Mosby 2004; pp. 545-61.

[141] Polin RA, Parravicini E, Regan JA, Taeusch HW. Bacterial sepsis and meningitis.Avery's Diseases of the Newborn. 8th ed. Philadelphia: Elsevier Inc. 2005; pp. 551-77. [http://dx.doi.org/10.1016/B978-072169347-7.50041-X]

[142] Gerdes JS. Diagnosis and management of bacterial infections in the neonate. Pediatr Clin North Am 2004; 51(4): 939-959, viii-ix. [http://dx.doi.org/10.1016/j.pcl.2004.03.009] [PMID: 15275982]

[143] Arnon S, Litmanovitz I. Diagnostic tests in neonatal sepsis. Curr Opin Infect Dis 2008; 21(3): 223-7. [http://dx.doi.org/10.1097/QCO.0b013e3282fa15dd] [PMID: 18448965]

[144] Kasapçopur O, Ozdoğan H, Yazici H. Obesity and erythrocyte sedimentation rate in children. J Pediatr 1991; 119(5): 773-5. [http://dx.doi.org/10.1016/S0022-3476(05)80300-2] [PMID: 1941385]

[145] Peltola H, Jaakkola M. C-reactive protein in early detection of bacteremic versus viral infections in immunocompetent and compromised children. J Pediatr 1988; 113(4): 641-6. [http://dx.doi.org/10.1016/S0022-3476(88)80372-X] [PMID: 3262735]

[146] Mortensen RF, Zhong W. Regulation of phagocytic leukocyte activities by C-reactive protein. J

Leukoc Biol 2000; 67(4): 495-500.
[http://dx.doi.org/10.1002/jlb.67.4.495] [PMID: 10770281]

[147] Unkila-Kallio L, Kallio MJT, Eskola J, Peltola H. Serum C-reactive protein, erythrocyte sedimentation rate, and white blood cell count in acute hematogenous osteomyelitis of children. Pediatrics 1994; 93(1): 59-62.
[http://dx.doi.org/10.1542/peds.93.1.59] [PMID: 8265325]

[148] Bamias G, Nyce MR, De La Rue SA, Cominelli F. New concepts in the pathophysiology of inflammatory bowel disease. Ann Intern Med 2005; 143(12): 895-904.
[http://dx.doi.org/10.7326/0003-4819-143-12-200512200-00007] [PMID: 16365470]

[149] Matowicka-Karna J. Markers of inflammation, activation of blood platelets and coagulation disorders in inflammatory bowel diseases. Postepy Hig Med Dosw 2016; 70: 305-12.
[http://dx.doi.org/10.5604/17322693.1199305] [PMID: 27117106]

[150] Kato S, Ochiai M, Sakurada T, *et al.* Increased expression of long pentraxin PTX3 in inflammatory bowel diseases. Dig Dis Sci 2008; 53(7): 1910-6.
[http://dx.doi.org/10.1007/s10620-007-0075-z] [PMID: 17990107]

[151] Vermeire S, Van Assche G, Rutgeerts P. C-reactive protein as a marker for inflammatory bowel disease. Inflamm Bowel Dis 2004; 10(5): 661-5.
[http://dx.doi.org/10.1097/00054725-200409000-00026] [PMID: 15472532]

[152] Ma C, Battat R, Parker CE, Khanna R, Jairath V, Feagan BG. Update on C-reactive protein and fecal calprotectin: are they accurate measures of disease activity in Crohn's disease? Expert Rev Gastroenterol Hepatol 2019; 13(4): 319-30.
[http://dx.doi.org/10.1080/17474124.2019.1563481] [PMID: 30791776]

[153] Uslu N, Yüce A, Demir H, *et al.* The association of inflammatory bowel disease and Mediterranean fever gene (MEFV) mutations in Turkish children. Dig Dis Sci 2010; 55(12): 3488-94.
[http://dx.doi.org/10.1007/s10620-010-1178-5] [PMID: 20306331]

[154] Beşer OF, Kasapçopur O, Cokuğraş FC, Kutlu T, Arsoy N, Erkan T. Association of inflammatory bowel disease with familial Mediterranean fever in Turkish children. J Pediatr Gastroenterol Nutr 2013; 56(5): 498-502.
[http://dx.doi.org/10.1097/MPG.0b013e31827dd763] [PMID: 23164758]

[155] Beşer ÖF, Çokuğraş FÇ, Kutlu T, *et al.* Association of familial Mediterranean fever in Turkish children with inflammatory bowel disease. Turk Pediatri Ars 2014; 49(3): 198-202.
[http://dx.doi.org/10.5152/tpa.2014.1998] [PMID: 26078663]

[156] Schlenz H, Intemann T, Wolters M, *et al.* C-reactive protein reference percentiles among pre-adolescent children in Europe based on the IDEFICS study population. Int J Obes 2014; 38(S2) (Suppl. 2): S26-31.
[http://dx.doi.org/10.1038/ijo.2014.132] [PMID: 25376218]

[157] González-Gil EM, Santabárbara J, Russo P, *et al.* Food intake and inflammation in European children: the IDEFICS study. Eur J Nutr 2016; 55(8): 2459-68.
[http://dx.doi.org/10.1007/s00394-015-1054-3] [PMID: 26419585]

[158] González-Gil EM, Santabárbara J, Siani A, *et al.* Whole-blood fatty acids and inflammation in European children: the IDEFICS Study. Eur J Clin Nutr 2016; 70(7): 819-23.
[http://dx.doi.org/10.1038/ejcn.2015.219] [PMID: 26757836]

[159] González-Gil EM, Tognon G, Lissner L, *et al.* Prospective associations between dietary patterns and high sensitivity C-reactive protein in European children: the IDEFICS study. Eur J Nutr 2018; 57(4): 1397-407.
[http://dx.doi.org/10.1007/s00394-017-1419-x] [PMID: 28315941]

[160] Pepys MB, Booth DR, Hutchinson WL, Gallimore JR, Collins IM, Hohenester E. Collins PM. Hohenester E. Amyloid P component. A critical review. Amyloid 1997; 4(4): 274-95.

[http://dx.doi.org/10.3109/13506129709003838]

[161] Blank N, Hegenbart U, Dietrich S, *et al.* Obesity is a significant susceptibility factor for idiopathic AA amyloidosis. Amyloid 2018; 25(1): 37-45. [http://dx.doi.org/10.1080/13506129.2018.1429391] [PMID: 29364741]

[162] Lu J, Marjon KD, Mold C, Du Clos TW, Sun PD. Pentraxins and Fc receptors. Immunol Rev 2012; 250(1): 230-8. [http://dx.doi.org/10.1111/j.1600-065X.2012.01162.x] [PMID: 23046133]

[163] Maekawa Y, Nagai T, Anzai A. Pentraxins: CRP and PTX3 and cardiovascular disease. Inflamm Allergy Drug Targets 2011; 10(4): 229-35. [http://dx.doi.org/10.2174/187152811796117744] [PMID: 21539513]

[164] Li XA, Yutani C, Shimokado K. Serum amyloid P component associates with high density lipoprotein as well as very low density lipoprotein but not with low density lipoprotein. Biochem Biophys Res Commun 1998; 244(1): 249-52. [http://dx.doi.org/10.1006/bbrc.1998.8248] [PMID: 9514915]

[165] Mantovani A, Garlanda C, Bottazzi B, *et al.* The long pentraxin PTX3 in vascular pathology. Vascul Pharmacol 2006; 45(5): 326-30. [http://dx.doi.org/10.1016/j.vph.2006.08.011] [PMID: 17023219]

[166] Boehme M, Kaehne F, Kuehne A, *et al.* Pentraxin 3 is elevated in haemodialysis patients and is associated with cardiovascular disease. Nephrol Dial Transplant 2007; 22(8): 2224-9. [http://dx.doi.org/10.1093/ndt/gfl747] [PMID: 17496115]

[167] Muller B, Peri G, Doni A, *et al.* Circulating levels of the long pentraxin PTX3 correlate with severity of infection in critically ill patients. Crit Care Med 2001; 29(7): 1404-7. [http://dx.doi.org/10.1097/00003246-200107000-00017] [PMID: 11445697]

[168] Kovacs A, Tornvall P, Nilsson R, Tegnér J, Hamsten A, Björkegren J. Human C-reactive protein slows atherosclerosis development in a mouse model with human-like hypercholesterolemia. Proc Natl Acad Sci USA 2007; 104(34): 13768-73. [http://dx.doi.org/10.1073/pnas.0706027104] [PMID: 17702862]

[169] Kacerovsky M, Tosner J, Drahosova M, Hornychova H, Andrys C. Pentraxin 3 in amniotic fluid as a marker of intra-amniotic inflammation in women with preterm premature rupture of membranes. Int J Gynaecol Obstet 2010; 108(3): 203-6. [http://dx.doi.org/10.1016/j.ijgo.2009.10.018] [PMID: 20022320]

[170] Cruciani L, Romero R, Vaisbuch E, *et al.* Pentraxin 3 in amniotic fluid: a novel association with intra-amniotic infection and inflammation. J Perinat Med 2010; 38(2): 161-71. [http://dx.doi.org/10.1515/jpm.2009.141] [PMID: 19792835]

[171] Kaess BM, Vasan RS. Heart failure: pentraxin 3—a marker of diastolic dysfunction and HF? Nat Rev Cardiol 2011; 8(5): 246-8. [http://dx.doi.org/10.1038/nrcardio.2011.50] [PMID: 21448138]

[172] Assi F, Fruscio R, Bonardi C, *et al.* Pentraxin 3 in plasma and vaginal fluid in women with preterm delivery. BJOG 2007; 114(2): 143-7. [http://dx.doi.org/10.1111/j.1471-0528.2006.01225.x] [PMID: 17305891]

[173] Mantovani A, Garlanda C, Doni A, Bottazzi B. Pentraxins in innate immunity: from C-reactive protein to the long pentraxin PTX3. J Clin Immunol 2008; 28(1): 1-13. [http://dx.doi.org/10.1007/s10875-007-9126-7] [PMID: 17828584]

[174] Ortega-Hernandez OD, Bassi N, Shoenfeld Y, Anaya JM. The long pentraxin 3 and its role in autoimmunity. Semin Arthritis Rheum 2009; 39(1): 38-54. [http://dx.doi.org/10.1016/j.semarthrit.2008.03.006] [PMID: 18614204]

[175] Manfredi AA, Rovere-Querini P, Bottazzi B, Garlanda C, Mantovani A. Pentraxins, humoral innate immunity and tissue injury. Curr Opin Immunol 2008; 20(5): 538-44.

[http://dx.doi.org/10.1016/j.coi.2008.05.004] [PMID: 18579363]

[176] Oztan MO, Aksoy Gokmen A, Ozdemir T, Müderris T, Kaya S, Koyluoglu G. Pentraxin-3: A strong novel biochemical marker for appendicitis in children. Am J Emerg Med 2019; 37(10): 1912-6. [http://dx.doi.org/10.1016/j.ajem.2019.01.010] [PMID: 30658876]

[177] Kim HS, Won S, Lee EK, *et al.* Pentraxin 3 as a clinical marker in children with lower respiratory tract infection. Pediatr Pulmonol 2016; 51(1): 42-8. [http://dx.doi.org/10.1002/ppul.23199] [PMID: 25832310]

[178] Kang X, Zhu Y, Zhang X. [Clinical study of Pentraxin 3 in diagnosing the severity and cardiovascular function of the children with sepsis]. Zhonghua Er Ke Za Zhi 2015; 53(8): 592-8. [PMID: 26717657]

[179] Ahout IML, Brand KH, Zomer A, *et al.* Prospective observational study in two Dutch hospitals to assess the performance of inflammatory plasma markers to determine disease severity of viral respiratory tract infections in children. BMJ Open 2017; 7(6): e014596. [http://dx.doi.org/10.1136/bmjopen-2016-014596] [PMID: 28667205]

[180] Marseglia GL, De Amici M, Leonardi S, *et al.* Pentraxin 3 in children suffering from allergic rhinitis. J Biol Regul Homeost Agents 2012; 26(1) (Suppl.): S109-12. [PMID: 22691257]

[181] Ishikawa N, Kobayashi Y, Fujii Y, Kobayashi M. Increased interleukin-6 and high-sensitivity C-reactive protein levels in pediatric epilepsy patients with frequent, refractory generalized motor seizures. Seizure 2015; 25: 136-40. [http://dx.doi.org/10.1016/j.seizure.2014.10.007] [PMID: 25455727]

[182] Ates U, Bahadir K, Ergun E, *et al.* Determination of pentraxin 3 levels in diagnosis of appendicitis in children. Pediatr Int 2020; 62(5): 624-8. [http://dx.doi.org/10.1111/ped.14131] [PMID: 31886585]

[183] Pizzo PA, Lovejoy FH, Smith DH. Prolonged fever in children: review of 100 cases. Pediatrics 1975; 55: 468-71. [http://dx.doi.org/10.1542/peds.55.4.468]

[184] Burns JC, Mason WH, Glodé MP, *et al.* Clinical and epidemiologic characteristics of patients referred for evaluation of possible Kawasaki disease. J Pediatr 1991; 118(5): 680-6. [http://dx.doi.org/10.1016/S0022-3476(05)80026-5] [PMID: 2019921]

[185] Nizet V, Klein JO. Bacterial sepsis and meningitis.Infectious diseases of the fetus and newborn Infant. 7th ed. Philadelphia: Elsevier Saunders 2011; pp. 222-76. [http://dx.doi.org/10.1016/B978-1-4160-6400-8.00006-7]

[186] Osei-Bimpong A, Meek JH, Lewis SM. ESR or CRP? A comparison of their clinical utility. Hematology 2007; 12(4): 353-7. [http://dx.doi.org/10.1080/10245330701340734] [PMID: 17654065]

[187] Eloseily EMA, Minoia F, Crayne CB, Beukelman T, Ravelli A, Cron RQ. Ferritin to Erythrocyte Sedimentation Rate Ratio: Simple Measure to Identify Macrophage Activation Syndrome in Systemic Juvenile Idiopathic Arthritis. ACR Open Rheumatol 2019; 1(6): 345-9. [http://dx.doi.org/10.1002/acr2.11048] [PMID: 31777812]

[188] Kocher MS, Zurakowski D, Kasser JR. Differentiating between septic arthritis and transient synovitis of the hip in children: an evidence-based clinical prediction algorithm. J Bone Joint Surg Am 1999; 81(12): 1662-70. [http://dx.doi.org/10.2106/00004623-199912000-00002] [PMID: 10608376]

[189] Assicot M, Gendrel D, Carsin H, Raymond J, Guilbaud J, Bohuon C. High serum procalcitonin concentrations in patients with sepsis and infection. Lancet 1993; 341(8844): 515-8. [http://dx.doi.org/10.1016/0140-6736(93)90277-N] [PMID: 8094770]

[190] Carrol ED, Thomson AP, Hart CA. Procalcitonin as a marker of sepsis. Int J Antimicrob Agents 2002;

20(1): 1-9.
[http://dx.doi.org/10.1016/S0924-8579(02)00047-X] [PMID: 12127705]

[191] Gendrel D, Raymond J, Assicot M, *et al.* Measurement of procalcitonin levels in children with bacterial or viral meningitis. Clin Infect Dis 1997; 24(6): 1240-2.
[http://dx.doi.org/10.1086/513633] [PMID: 9195090]

[192] Hedlund J, Hansson LO. Procalcitonin and C-reactive protein levels in community-acquired pneumonia: correlation with etiology and prognosis. Infection 2000; 28(2): 68-73.
[http://dx.doi.org/10.1007/s150100050049] [PMID: 10782390]

[193] Moulin F, Raymond J, Lorrot M, *et al.* Procalcitonin in children admitted to hospital with community acquired pneumonia. Arch Dis Child 2001; 84(4): 332-6.
[http://dx.doi.org/10.1136/adc.84.4.332] [PMID: 11259234]

[194] Toikka P, Irjala K, Juvén T, *et al.* Serum procalcitonin, C-reactive protein and interleukin-6 for distinguishing bacterial and viral pneumonia in children. Pediatr Infect Dis J 2000; 19(7): 598-602.
[http://dx.doi.org/10.1097/00006454-200007000-00003] [PMID: 10917215]

[195] Korppi M, Remes S. Serum procalcitonin in pneumococcal pneumonia in children. Eur Respir J 2001; 17(4): 623-7.
[http://dx.doi.org/10.1183/09031936.01.17406230] [PMID: 11401055]

[196] Korppi M, Remes S, Heiskanen-Kosma T. Serum procalcitonin concentrations in bacterial pneumonia in children: a negative result in primary healthcare settings. Pediatr Pulmonol 2003; 35(1): 56-61.
[http://dx.doi.org/10.1002/ppul.10201] [PMID: 12461740]

[197] Yo CH, Hsieh PS, Lee SH, *et al.* Comparison of the test characteristics of procalcitonin to C-reactive protein and leukocytosis for the detection of serious bacterial infections in children presenting with fever without source: a systematic review and meta-analysis. Ann Emerg Med 2012; 60(5): 591-600.
[http://dx.doi.org/10.1016/j.annemergmed.2012.05.027] [PMID: 22921165]

[198] Hamade B, Huang DT. Procalcitonin Where are we now? 2020; 36(1): 23-40.
[http://dx.doi.org/10.1016/j.ccc.2019.08.003] [PMID: 31733680]

[199] Florin TA, Ambroggio L, Brokamp C, *et al.* Biomarkers and Disease Severity in Children With Community-Acquired Pneumonia. Pediatrics 2020; 145(6): e20193728.
[http://dx.doi.org/10.1542/peds.2019-3728] [PMID: 32404432]

[200] Kuppermann N, Dayan PS, Levine DA, *et al.* Febrile Infant Working Group of the Pediatric Emergency Care Applied Research Network (PECARN). A clinical prediction rule to identify febrile infants 60 days and younger at low risk for serious bacterial infections. JAMA Pediatr 2019; 173(4): 342-51.
[http://dx.doi.org/10.1001/jamapediatrics.2018.5501] [PMID: 30776077]

[201] Bachur RG, Dayan PS, Dudley NC, *et al.* The influence of age on the diagnostic performance of white blood cell count and absolute neutrophil count in suspected pediatric appendicitis. Acad Emerg Med 2016; 23(11): 1235-42.
[http://dx.doi.org/10.1111/acem.13018] [PMID: 27251399]

[202] Pratt A, Attia MW. Duration of fever and markers of serious bacterial infection in young febrile children. Pediatr Int 2007; 49(1): 31-5.
[http://dx.doi.org/10.1111/j.1442-200X.2007.02316.x] [PMID: 17250502]

[203] Van den Bruel A, Thompson MJ, Haj-Hassan T, *et al.* Diagnostic value of laboratory tests in identifying serious infections in febrile children: systematic review. BMJ 2011; 342(jun08 2): d3082.
[http://dx.doi.org/10.1136/bmj.d3082] [PMID: 21653621]

[204] World Health Organization. https://www.who.int/china/news/detail/09-01-2020.

[205] Henry BM, Lippi G, Plebani M. Laboratory abnormalities in children with novel coronavirus disease 2019. Clin Chem Lab Med 2020; 58(7): 1135-8.
[http://dx.doi.org/10.1515/cclm-2020-0272] [PMID: 32172227]

[206] Feldstein LR, Rose EB, Horwitz SM, *et al.* Multisystem Inflammatory Syndrome in U.S. Children and Adolescents. N Engl J Med 2020; 383(4): 334-46.
[http://dx.doi.org/10.1056/NEJMoa2021680] [PMID: 32598831]

CHAPTER 18

Biomarkers in Liver Disease

Abdullah Sonsuz[1,*] and **Oguz Kagan Bakkaloglu**[1]

[1] *Department of Internal Medicine-Gastroenterology, Cerrahpasa Medical Faculty, Istanbul University – Cerrahpasa, Istanbul, Turkey*

Abstract: Symptoms and signs of liver diseases are highly variable depending on the etiology, disease stage, and type of liver involvement. There are different types of liver diseases; causes of liver diseases may be viral, toxic, metabolic, or autoimmune. However, in some cases, liver disease can develop as a result of diseases of other organs or systems. It is almost impossible to differentiate all of these solely on the basis of clinical symptoms and findings. Furthermore, the early stages of liver disease may be completely asymptomatic, or in some cases, the disease may progress with only subtle and non-specific symptoms. Therefore, biomarkers have a critical role in screening, diagnosis, staging, and evaluation of therapeutic response to treatment in liver diseases.

Keywords: 5'-Nucleotidase, Alanine aminotransferase, Albumin, Alkaline phosphatase, Alpha-fetoprotein, Aspartate aminotransferase, AST isoenzymes, Biomarkers, CA 19-9, Cholangiocarcinoma, Cholestasis, Fibrosis, Gamma-glutamyl transferase, Hepatocellular carcinoma, Hepatocellular injury, Liver disease, MELD-Na score, Prognosis, Prothrombin time, Staging of liver disease.

INTRODUCTION

Biomarkers are important tools for screening, diagnosis, assessment of treatment response, and evaluating the prognosis of diseases or other medical conditions. There are various definitions for biomarkers in the medical literature, and a more broadened definition of biomarkers may as well include the measurable clinical signs, radiologic or histopathologic findings [1].

National Institutes of Health Biomarkers Definitions Working Group defines a biomarker as "a characteristic that is objectively measured and evaluated as an indicator of normal biological processes, pathogenic processes, or pharmacologic responses to therapeutic interventions" [2].

* **Corresponding author Abdullah Sonsuz:** Istanbul University – Department of Internal Medicine-Gastroenterology, Cerrahpasa Medical Faculty, Istanbul University – Cerrahpasa, Istanbul, Turkey; Tel: +90 (532) 255 6 991; E-mail: abdullahsonsuz@yahoo.com

The burden of liver diseases has been underestimated for many years. Recent data show that liver diseases account for approximately 2 million deaths per year worldwide [3]. Although WHO aims to reduce the incidence of HBV and HCV infections and the associated mortality substantially by 2030, changes in the epidemiology and etiology of liver diseases indicate that nonalcoholic fatty liver disease and related other diseases will be much more important public health problems in the future [4, 5]. Rational usage of biomarkers is not important only for the diagnosis of liver diseases but also for the prediction and prevention of life-threatening complications and other clinical consequences. Unfortunately, most of the biomarkers used for liver diseases in daily clinical practice have some limitations.

The liver may be affected directly and specifically by viral, toxic, or hereditary disease, but also some systemic diseases may cause liver dysfunction or even overt liver disease that is indistinguishable from the liver-specific diseases. Clinical symptoms such as nausea, vomiting, right upper quadrant abdominal pain, fatigue, weakness, weight loss, and jaundice are not specific to liver diseases and may also be seen in many other diseases or medical conditions. Furthermore, these symptoms may not occur until late-stage liver disease. Diagnosis of liver diseases is more difficult than the diagnosis of many other systems' diseases and requires not only a high level of knowledge about liver diseases but also advanced clinical experience.

The intervals between the beginning of the disease, the appearance of the clinical signs and symptoms, and detectable laboratory findings of the liver disease are highly variable; thus, there may be different timeline models for the course of the liver disease (Fig. **1**).

One typical example for model-A is Bud-Chiari syndrome. Immediately after the occlusion of the hepatic veins with thrombus (cause of the liver disease), radiological findings of the venous obstruction can be demonstrated at the early phase of the disease; painful hepatomegaly and ascites (clinical findings) occur within a short time period, followed by other laboratory and clinical findings (Fig. **1A**).

The clinical course of some genetic liver diseases may be described as model-B. Although some of the laboratory findings are present from the beginning or earlier in the course of the disease, clinical findings may be delayed for a long time (Fig. **1B**). Wilson's disease is caused by a mutation in the ATP7B gene; low levels of serum ceruloplasmin and increased urinary copper excretion may be demonstrated many years before overt clinical signs and symptoms.

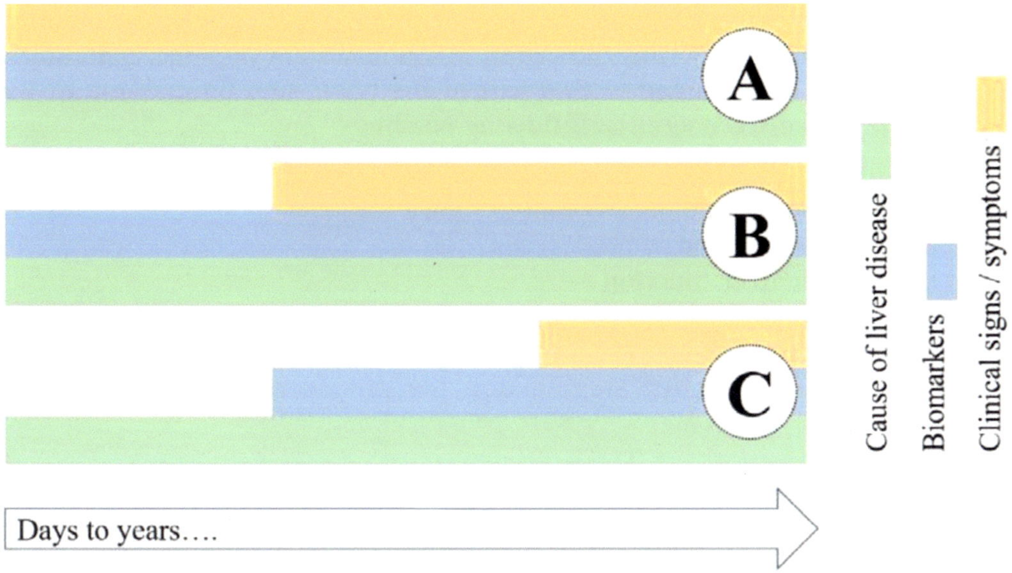

Fig. (1). Different timeline models depicting the manifestation sequence of "The cause", "The biomarker positivity" and "Clinically apparent signs and symptoms" in liver diseases.

Clinical models of some other liver diseases may have a long-time-interval between laboratory and clinical signs of the disease; the cause of the liver disease may persist for many years without any significant alteration of laboratory tests and before clinical symptoms or signs are evident (Fig. **1C**). The course of the immune tolerant phase of chronic hepatitis B virus infection is typical for this model. Despite the initial HBsAg positivity and the presence of high levels of HBV DNA, neither the biochemical findings of liver damage nor the clinical signs of liver disease are present. Aminotransferase elevation will manifest in the next phase (immune active phase), but many years later, clinical signs and symptoms of liver disease will be evident in some of these patients.

Biomarkers are of critical importance for the diagnosis and clinical evaluation of most of the liver diseases in any of the three clinical course models.

Biochemical findings used for the diagnosis of liver diseases are classified into four groups traditionally.

1. Tests used to evaluate the hepatocellular injury
2. Enzymes for detection of cholestasis
3. Tests of liver synthetic function
4. Quantitative liver function tests

Today, the term biomarker has been expanded to include the entire range of the measurable biological properties; however, imaging studies and histopathological examinations will be excluded in this part of the book, and biomarkers of liver disease will be classified under the following headings.

1. Tests used to evaluate the hepatocellular injury
2. Enzymes for the detection of cholestasis
3. Tests of synthetic liver function
4. Fibrosis markers
5. Tumor markers for primary liver malignancy
6. Scores and biomarkers for staging and the prognostic evaluation of liver diseases

Tests Used for the Detection of Liver Injury

Some degree of liver cell injury, with or without clinical symptoms or signs, is a common consequence of liver disease. When liver injury and hepatocyte damage are present without any clinical manifestation, hepatocellular injury biomarkers are the only tool for diagnosing liver disease.

There are several biomarkers related to hepatocellular injury, but some of them only have historical importance and are rarely used in daily clinical practice today (Table **1**).

Table 1. Biomarkers for hepatocellular injury.

ALT: Alanine aminotransferase
AST: Aspartate aminotransferase
ICD: Isocitrate dehydrogenase
SDH: Sorbitol dehydrogenase
LDH: Lactate dehydrogenase

Currently, ALT and AST are the main biomarkers used to detect liver cell injury. Both, ALT and AST are termed as aminotransferases or transaminases. Although LDH is an important biomarker for various diseases, it has limited value in the assessment of liver cell injury. However, LDH may be useful in the diagnosis of liver diseases accompanying hemolysis [6, 7]. Sorbitol dehydrogenase and isocitrate dehydrogenase were rarely used in the past and are currently not superior to other biomarkers in the diagnosis of liver cell injury in clinical practice [8, 9].

The article "Determination of aspartic-glutamic transaminase in tissue homogenates," published in 1946, showed the activity of aminotransferases in various tissues. In this study, "aspartic-glutamic transaminase" activity was reported in muscle, liver, kidney, and lung tissues [10]. Aminotransferase elevation provides us with a simple and easily available tool for detecting tissue damage in heart, skeletal muscle, and liver diseases [10].

Despite the widespread use of aminotransferases as screening tools for the detection of liver diseases, there are some limitations for both ALT and AST. First of all, normal levels of ALT and/or AST does not rule out any liver disease. Some patients with liver disease may have normal ALT and AST levels, even if their disease is at a very advanced stage. Elevated aminotransferase levels are not specific to liver disease. Aminotransferases originating from muscle tissue can also be detected at very high levels; thus, this muscle damage may be confused with liver diseases.

Aspartate Aminotransferase (AST)

Aspartate Aminotransferase; AST (Formerly serum glutamic-oxaloacetic transaminase, SGOT) is found in the liver, heart, skeletal muscle, kidneys, brain, and red blood cells [11]. Aspartate transaminase catalyzes the interconversion of aspartate and α-ketoglutarate to oxaloacetate and glutamate [12].

$$\text{L-Aspartate} + \alpha\text{-ketoglutarate} \rightleftharpoons \text{oxaloacetate} + \text{L-glutamate}$$

Aspartate aminotransferase is found in human tissues as two separate isoenzymes, one located in the cytoplasm (c-AST) and the other in mitochondria (m-AST), encoded by the GOT1 and GOT2 genes, respectively [13].

Alanine Aminotransferase (ALT)

Alanine transaminase; ALT, (Formerly serum glutamate-pyruvate transaminase, SGPT) catalyzes the transfer of an amino group from L-alanine to α-ketoglutarate [11, 14].

$$\text{L-alanine} + \alpha\text{-ketoglutarate} \rightleftharpoons \text{pyruvate} + \text{L-glutamate}$$

ALT is mainly found in the liver, heart and skeletal muscle. Unlike AST, ALT is found only in the cytoplasm [15].

The relative activities of transaminases in various human tissue are highly variable and should be considered in the diagnostic evaluation of aminotransferase elevation (Table **2**).

Table 2. Relative activities(*) of aminotransferases various human tissues [16].

-	AST	ALT
Heart	8000	500
Liver	7000	3000
Skeletal muscle	5000	300
Kidney	4500	1200
Spleen	700	150
Red blood cells	15	7

(*) Relative amounts of enzyme activity: Activity of the enzyme in the tissue (in IU/kg of tissue)/Upper reference of plasma activity of the enzyme (in IU/L).

Half-life of ALT and AST may also be important when evaluating the aminotransferase levels and disease activity. Half-life of ALT and total AST is 47 ± 10 hours and 17.5 ± 5 hours respectively, while half-life of mitochondrial AST is more than 80 hours [17]. AST isoenzymes are indistinguishable from standard AST assays; therefore, AST activity means total AST in routine clinical practice. Some physiological conditions or technical factors may affect the measurement of AST and ALT activity not related to liver injury, but meals have no effect [18].

Approach to Patients With Aminotransferase Elevation

Elevated ALT and AST are frequently found in liver disease; however, high levels are not unusual in those examined for other system diseases, or even in healthy individuals. A population-based survey in the United States conducted between 1999 and 2002 estimated that an abnormal ALT was present in 8.9 percent of respondents [19].

Although the term "liver function tests" (LFTs) is used commonly, it is imprecise and potentially misleading since many of the tests reflecting the health of the liver are not direct measures of its function. Furthermore, these commonly used tests can be abnormal even in patients with healthy livers. The pattern of aminotransferase elevation is important for differential diagnosis.

- The increment level of aminotransferase (how many times of the upper limit of normal (ULN) it has increased)
- Which aminotransferase, ALT or AST increased predominantly
- Duration of the aminotransferase elevation (a few days to months or years)

Toxic, acute viral hepatitis or ischemic hepatitis causes higher aminotransferase levels compared to chronic liver diseases (Fig. **2**). These cases usually have an increment that exceeds the ULN by 10 times or more. The increment level of aminotransferases in chronic liver diseases is usually between the ULN and 5 times of ULN. In some patients with chronic liver diseases, aminotransferase levels may be in between normal reference limits.

Elevated ALT and AST often coexist in patients with liver disease, and in most cases, ALT levels are higher than AST. However, in some diseases or specific clinical conditions, AST may be higher than ALT. The importance of the ratio "AST to ALT" was first described by De Ritis and today AST/ALT ratio is called the "De Ritis ratio" [20].

The AST/ALT ratio is usually higher in women than men. Alcoholic liver disease and liver cirrhosis also pose a De Ritis ratio greater than 1 (Table **3**).

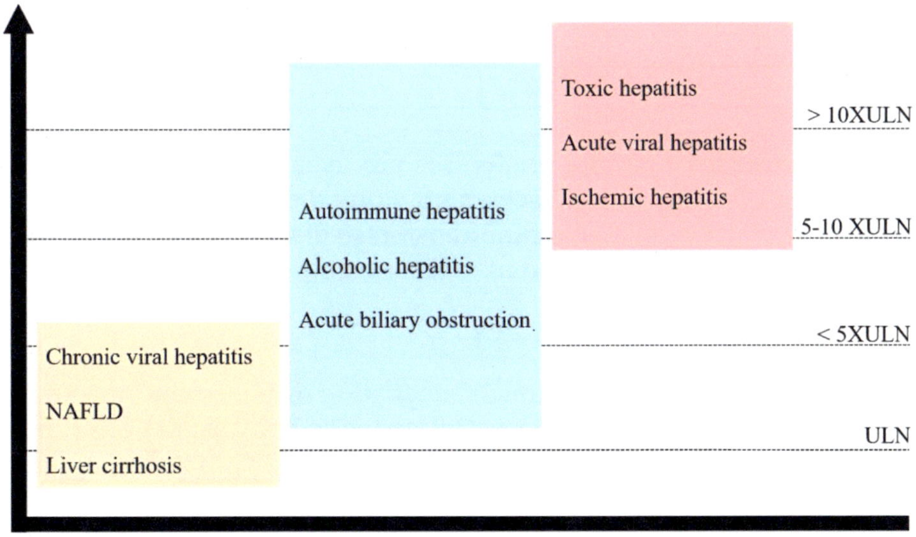

Fig. (2). Aminotransferase elevation levels in some liver diseases (ULN: Upper limit of normal, NAFLD: Nonalcoholic fatty liver disease).

Table 3. AST/ALT ratio in liver disease [21, 22].

-	% AST/ALT>1	% AST/ALT>2
Alcoholic liver disease	92	70
Liver cirrhosis	70	26
Viral hepatitis	12	4
Biliary obstructions	18	-

NAFLD often has a De Ritis ratio < 1, unless cirrhosis develops. The "De Ritis ratio" reported for various liver diseases is shown in the Table **4**.

Table 4. De Ritis ratio in liver diseases.

-	Parmar K, *et al* [23].	Rekha M, *et al* [24].	Karim S, *et al* [25].	Ferrari R, *et al* [26].	Kumar Shreevastva N, *et al* [27].	Khatroth S, *et al* [7].
Control	1.09 ± 0.14	1.6±1.0	-	-	-	-
Alcoholic Liver Disease	2.2 ±0.85	-	-	-	1.91±0.44	2.15±0.88
Viral hepatitis	0.8±0.15	1.6±1.4	-	0.6± 0.2	-	-
Chronic hepatitis	-	-	0.72±0.2	-	-	-
NAFLD	1.2 ± 0.18	-	-	-	-	-
Autoimmune hepatitis	-	-	-	1.20± 0.55	-	-
Cholestasis	1.14 ± 0.18	-	-	-	-	-
Liver cirrhosis	-	2.6±1.9	-	-	-	-
Alcoholic cirrhosis	-	2.9±2.7	-	-	-	-

Transient increments of aminotransferases are usually secondary to self-limited liver diseases or other systemic diseases affecting the liver. Chronic liver diseases are more likely when persistent aminotransferase elevations are encountered. The term chronicity is generally defined as a disease duration of more than 6 months.

Enzymes for Detection of Cholestasis

Cholestasis refers to severe or complete stagnation of bile secretion and flow. The blockage can occur either in the liver (intrahepatic cholestasis) or at the level of the extrahepatic bile ducts (extrahepatic cholestasis) [28]. Intrahepatic cholestasis is further subdivided into intra-lobular or extra-lobular cholestasis. Cholestasis may be transient or long lasting. Stagnation of bile secretion leads to retention of bilirubin and bile acids.

The clinical approach to diagnosing a patient with cholestasis necessitates a detailed medical history, careful physical examination, and rational use of laboratory tests. Identifying the cause of the liver disease is important and, determining the exact location and the nature of the obstructive pathology must be intended. Cholestasis is clinically manifested by generalized pruritis, jaundice, dark urine and light-colored stools. The main symptoms, itching and jaundice, are secondary to bile acid and bilirubin retention, respectively.

Characteristic laboratory findings of the cholestasis are elevation of serum alkaline phosphatase and other cholestatic enzymes; with or without elevated bilirubin levels, and in most cases, some degree of aminotransferases elevations.

Serum enzymes associated with cholestasis [29, 30]:

- Alkaline Phosphatase
- Gamma Glutamyl Transferase (previously γ-glutamyltranspeptidase)
- Leucine aminopeptidase
- 5' Nucleotidase

Alkaline Phosphatase

Alkaline phosphatase was first descripted by Suzuki and colleagues in 1907 [31]. Alkaline phosphatases (ALP) are a group of isoenzymes, found in almost all living organisms. ALP is located on the cell membrane [32]. In the liver, ALP is located on the canalicular membrane of hepatocytes.

ALP catalyzes the hydrolysis of phosphate monoesters at basic pH [33]. ALP is present in the liver, kidney, ileal mucosa and placenta [34]. Bone-specific alkaline phosphatase (BAP) is synthesized by the osteoblasts [19]. Liver, bone and kidney alkaline (L/B/K) phosphatases are tissue nonspecific but intestinal-ALP (ILAP), placental-ALP (PLAP) and Germ cell-ALP(GCALP) are tissue specific alkaline phosphatases [31]. The ILALP is slightly more thermostable than liver, bone and kidney ALP [35]. The majority of ALP in serum (more than 80%) is released from the liver and bone [33]. Little is known regarding the physiological function of ALP. Bone isoenzyme has a role in normal skeletal mineralization and an increase in the bone isoenzyme reflects increased osteoblastic activity [36]. The other functions of L/B/K remain obscure both in physiological or neoplastic conditions.

–Osteoblastic activity → Bone ALP activity ↑
–Bile flow obstruction → De novo synthesis of ALP at the canalicular surface of the hepatocyte and regurgitation of bile constituents into the systemic circulation. → Liver ALP activity ↑
–Placental tissue → ALP increases during pregnancy
–Intestinal ALP → High levels are found in the lymph

Normal reference values for ALP are closely related to age, with childhood starting to decline at 10-12 years of age and reaching adult levels at 14-18 years of age (Table **5**).

Table 5. Age- and gender-specific reference limits for ALP [37].

Age Group	Sex	Lower Limit (90% CI)	Upper Limit (90% CI)
2-3	F+M	160 [153-169]	376 [355-390]
5-9	F+M	143 [134-153]	362 [354-370]
10-12	F+M	116 [99 - 136]	426 [354-471]
13-14	F	48 [42 - 58]	232 [149-243]
13-14	M	111 [92-148]	486 [338-512]

M male, F female, CI Confidence interval.

An elevated ALP may be the result of a systemic disease or a physiological condition unrelated to any disease (Table **6**).

Table 6. Diseases and clinical conditions with increased alkaline phosphatase levels [38 - 40].

Liver Disease	Bone Disease	Other Diseases /Conditions
Viral hepatitis Alcoholic liver disease BRIC PFIC DILI PBC PSC Granulomatous liver disease - Sarcoidosis - Infections -Parasitic Infiltrative: - Amyloidosis - Lymphoma Autoimmune Cholangiopathy Prolonged TPN Postoperative state Malignancy - Hepatocellular - Metastatic Obstructive: cholestasis -Malign obstruction -Cholangiocarcinoma Choledocolithiasis Biliary strictures	Healing Fracture Paget's Disease Osteomalacia Vitamin D insufficiency Rickets Malignancy: osteogenic sarcoma, metastatic tumors	Renal Failure Heart Failure Hyperthyroidism Hyperparathyroidism Lymphoma Leukemia Renal cell carcinoma Multiple endocrine neoplasias (MEN) II <u>Physiological</u> -Pregnancy -Adolescence -Following a fatty meal in subjects -with blood group O or B

BRIC: Benign recurrent intrahepatic cholestasis, PFIC: Progressive familial intrahepatic cholestasis, DILI: Drug induced liver injury, PBC: Primary biliary cholangitis, PSC: Primary sclerosing cholangitis.

Although it is a less common condition, ALP may be below normal levels in some diseases or clinical conditions (Table 7).

Table 7. Low alkaline phosphatase [41, 42].

Malnutrition
Wilson's Disease
Hypothyroidism
Zinc deficiency
Vitamin C deficiency
Low phosphorus level
Pernicious anemia

Approach to a Patient with Elevated Serum Alkaline Phosphatase

Concomitant elevation of ALP, GGT, aminotransferases and other liver biomarkers are highly suggestive of hepatobiliary origin. Whether ALP originates from bone or liver should be distinguished before further diagnostic evaluation (Fig. **3**).

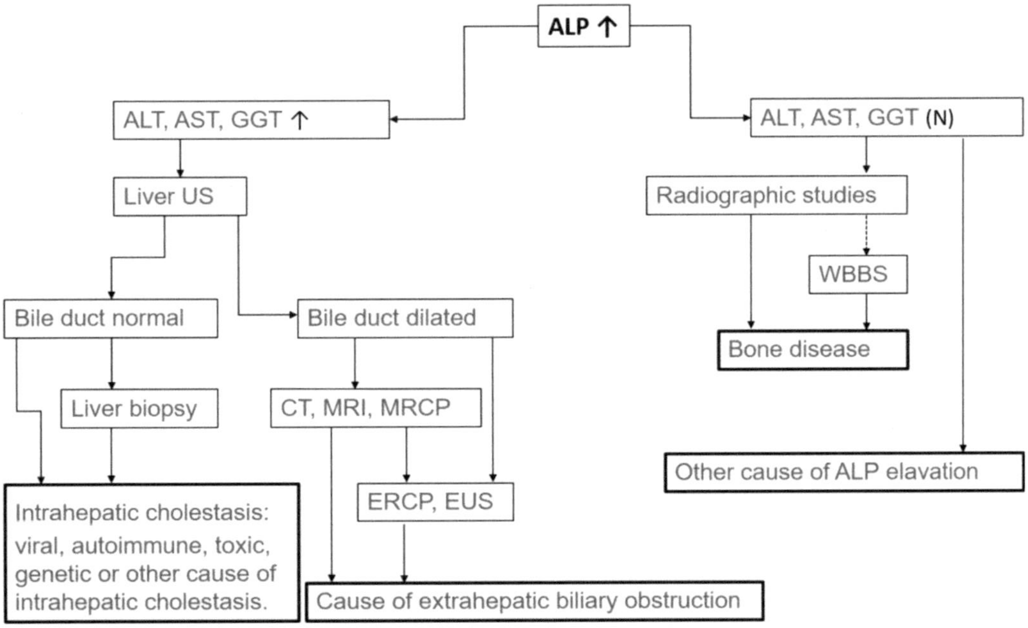

Fig. (3). Approach to a patient with elevated ALP. US: Ultrasonography, MRI: Magnetic Resonance Imaging, CT: Computerized Tomography, ERCP: Endoscopic retrograde cholangiopancreatography, EUS: Endoscopic ultrasound, MRCP: Magnetic resonance cholangiopancreatography, WBBS: Whole body bone scintigraphy

Gamma Glutamyl Transferase

Gamma Glutamyl Transferase (Gamma-glutamyl transpeptidase, GGT) is a transferase enzyme and catalyzes the transfer of gamma-glutamyl functional groups from molecules such as glutathione to acceptor molecules such as an amino acid or a peptide [43]. GGT's are found in the renal, prostatic, pancreatic, hepatobiliary tissue, spleen, heart, and brain. While there is electrophoretic heterogeneity, true isoenzyme forms have not yet been identified [44].

GGT activity is increased in all forms of hepatobiliary diseases. Intrahepatic or extrahepatic biliary obstruction can cause higher serum GGT levels than parenchymal liver diseases. Since bone diseases are not associated with increased serum GGT activity, dual elevations of GGT and ALP are highly suggestive of hepatobiliary disease. Elevated serum levels of GGT are also found in alcoholism and blood levels of GGT may be used as a marker of recent heavy alcohol use [45]. Patients receiving certain drugs, may have elevated GGT levels up to 2-5 times of normal, some of these drugs are listed below (Table **8**). This-elevation is probably a result of microsomal induction of the enzyme activity.

Table 8. Drugs that cause GGT elevation [38, 46 - 48].

Acetaminophen	Isotretinoin
Carbamazepine	Methotrexate
Cimetidine	Phenobarbital
Coumadin	Testosterone
Furosemide	Tricyclic antidepressants
Heparin	Valproic acid

High GGT activity is also seen in patients with acute or chronic pancreatitis. In addition, prostatic adenocarcinoma may be associated with increased serum GGT levels. Therefore, an elevated GGT activity should also not be considered as a specific indicator of hepatobiliary disease [44].

GGT is one of the indirect markers of liver fibrosis and may reflect alterations in hepatic functions like bilirubin, AST, ALT, GGT and platelet count [49].

Diagnostic approach to elevated GGT requires a detailed history, careful physical examination and use of appropriate imaging studies, liver function tests, tests specific for liver or other system diseases (Fig. **4**).

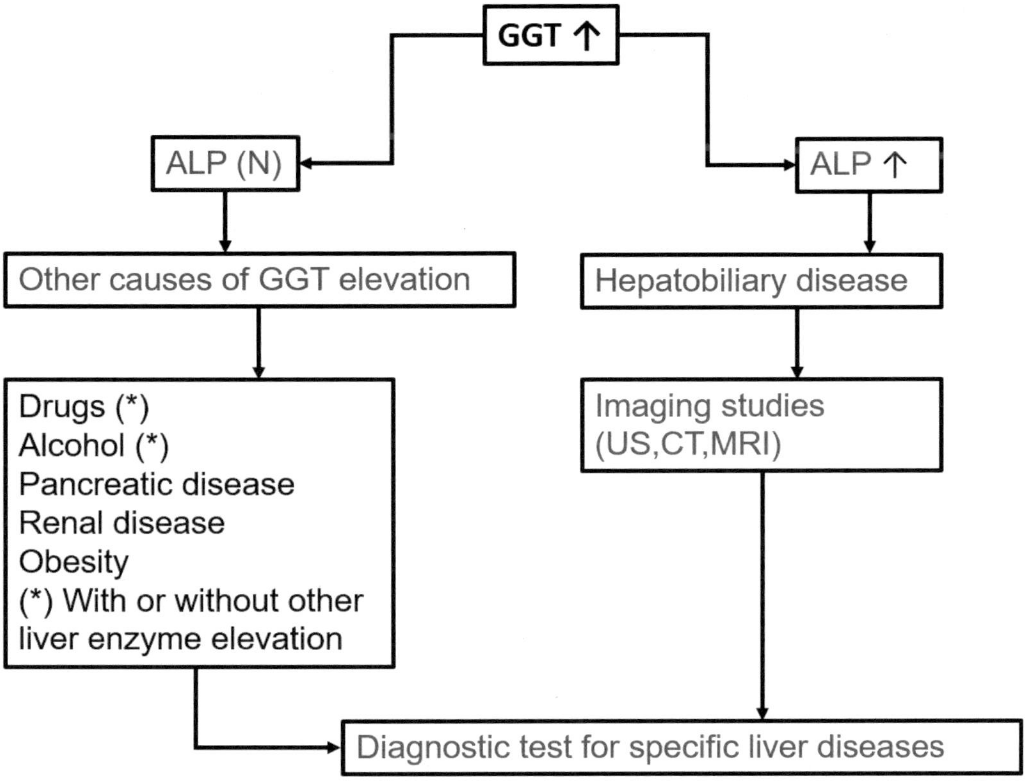

Fig. (4). Diagnostic approach for elevated GGT. US: Ultrasonography, MRI: Magnetic Resonance Imaging, CT: Computerized Tomography

5'- Nucleotidase

The 5'-nucleotidases are a family of enzymes that catalyze the dephosphorylation of nucleoside monophosphates [49]. 5′-Nucleotidase is found in the intestines, brain, heart, blood vessels, pancreas and liver [50, 51]. Although 5'-nucleotidase activity is present in a variety of tissues, elevated enzyme level is generally associated with hepatobiliary disease.

5′-nucleotidase activity is primarily located to the canalicular and sinusoidal cell membranes in hepatic tissue. Increased 5′-nucleotidase activity in a patient with elevated ALP level may indicate the hepatic origin of ALP, whereas normal 5′-nucleotidase level does not rule out liver disease.

Tissue Distribution of Cholestasis Biomarkers

Liver disease may cause any isolated or simultaneous increase in cholestasis biomarkers. None of these markers are specific to liver diseases. Each marker has several other tissue origins (Table **9**).

Table 9. Tissue origin of cholestasis biomarkers [52].

-	ALP	GGT	5'-Nucleotidase
Liver	+	+	+
Bone	+	NA	NA
Placenta	+	NA	NA
Pancreas	NA	+	+
Heart	NA	NA	+
Brain	NA	NA	+
Blood vessel	NA	NA	+
Kidney	+	+	NA
Increase by enzyme induction	NA	+ [53]	NA
Macro-enzyme	+ [54]	+ [55]	?

The approach to a patient with concomitant aminotransferase and cholestatic enzyme elevations requires determining whether the predominant component of liver disease is hepatocellular injury or cholestasis. The distinction between cholestatic, hepatocellular, or mixed injury models is particularly important for identifying the possible causative agent when drug-induced liver injury is suspected in the setting of multidrug treatment.

"R value" defined by ALT and ALP elevation ratio [56].

R= (ALT/ALT-ULN)/(ALP/ALP-ULN)

ULN: Upper Limits of Normal

R value >5.0: Hepatocellular injury

R <2.0: Cholestatic injury

R between 2.0 to 5.0: Mixed injury (hepatocellular- cholestatic)

Test of Liver Synthetic Function

The liver plays a central role in many metabolic processes, including lipid, carbohydrate and protein metabolism. Synthetic functions are also part of the metabolic functions of the liver and most of the plasma proteins are synthesized by the liver [57 - 59]. Synthetic functions of the liver may be evaluated with serum albumin level and prothrombin time.

Serum Albumin

Albumin, synthesized by the liver, has many important biological functions in both health and disease states. Maintenance of the oncotic pressure in the vascular bed is its main function; albumin also has other physiologically important functions such as immunomodulation, endothelial stabilization, antioxidant effects and binding of various drugs [60].

Albumin is synthesized exclusively by the liver, therefore serum albumin levels reflect the synthetic capacity of the liver; however, some other factors such as urinary or gastrointestinal albumin loss, malnutrition status, negative acute phase response or volume of distribution may also affect the serum albumin levels [61 - 66].

The rate of albumin synthesis in healthy adults is about 150 mg/kg/day and the half-life of the plasma albumin is approximately 20 days [67]. Albumin levels correlate with Child-Turcotte-Pugh stage of cirrhotic patients [68]. Albumin synthesis rate is also correlated with the galactose elimination capacity and the aminopyrine breath test [69]. Although albumin has a long half-life, its level is a reliable predictor of acute liver failure (ALF) in patients with acute hepatitis [67].

Prothrombin Time

Coagulation and clot formation are complex enzymatic reactions of several coagulation factors. Most of these coagulation factors are synthesized by the liver. The liver is the main site for the synthesis of factor I (fibrinogen), factor II (prothrombin), factor V, VII, IX, X and XII [70, 71]. Coagulation abnormalities seen in liver diseases may be secondary to the impaired synthetic capacity of the liver, inadequate absorption of fat-soluble vitamin K, or both.

The plasma level of each coagulation factor can be measured, but prothrombin time (PT) is a simple and useful index of coagulation capacity and liver synthetic function. PT can be expressed in seconds or in the international normalized ratio (INR). INR is often used in the management of patients receiving warfarin treatment. Most of the prognostic indices developed for liver diseases prefer to use the amount of PT prolongation (in seconds) rather than of INR. Normal value of PT is between 9 to 11 s.

A prolonged PT is not specific for liver diseases; some congenital or acquired diseases or consumption coagulopathy may also cause prolongation of PT. Vitamin K deficiency due to inadequate intake or prolonged cholestasis, malabsorption, or poor absorption due to antibiotic use, should also be excluded. In such cases, prothrombin times return to normal values after parenteral administration vitamin K but will remain prolonged in case of synthetic liver dysfunction [72].

Fibrosis Biomarkers

Liver fibrosis is the result of sustained liver injury. Liver fibrosis has been evaluated only by liver biopsy for many years and it is still the gold standard modality. However, diagnostic accuracy of biopsy may be influenced by sampling or inter-observer variability. Liver fibrosis is the result of a dynamic process; spontaneous or treatment-induced regression of the fibrosis stage is not uncommon [73, 74].

Liver biopsy is an invasive procedure with rare yet serious potential complications and side effects. Furthermore, it is not an easily reproducible procedure within a short time interval. Evaluation of the fibrosis stage, like the other findings of the liver disease, such as the level of hepatocellular injury, cholestasis or synthetic capacity, is also important for the clinical decision. Therefore, in this context, in new tools are needed in daily practice, apart from liver biopsy and histopathological examination.

Transient elastography and other elastography methods (Shear-wave elastography, MR elastography) are emerging modalities, but they are not readily available yet. A non-invasive fibrosis marker should be simple, easily accessible, reproducible and should have an acceptable correlation with liver biopsy. Biomarkers may be promising approaches for the assessment of liver fibrosis [75]. Unfortunately, most serum biomarkers have limited diagnostic utility and it is therefore reasonable to use multiple serum biomarkers or combined fibrosis scores for a more reliable prediction of liver fibrosis.

Fibrosis biomarkers may be classified as (A) direct biomarkers or (B) indirect biomarkers [76]. While direct biomarkers measure levels of collagen matrix components, their degradation products or growth factors; indirect fibrosis biomarkers give us an estimation of fibrosis using a group of liver tests and other laboratory findings that are unrelated to fibrogenesis or collagen metabolism.

Direct Biomarkers

Direct biomarkers (Class I Biomarkers) are further subdivided into four groups [77, 78].

I. Extracellular matrix (ECM) molecules: Procollagen N-terminal peptide, hyaluronic acid, type IV collagen, laminin, fibronectin, YKL-40, TIMP-1, TIMP-2, MMP-2, MMP-9.

II. Cytokines: TGF-β, TNF-α, angiotensin-II.

TGF-β signaling plays a role in liver injury, inflammation, fibrosis and even carcinogenesis in chronic liver diseases [79 - 81]. TGF-β, induces activation of stellate cells and transformation into myofibroblasts, a critical step for liver fibrogenesis [82].

Tumor necrosis factor-α (TNFα) is an inflammatory cytokine. Prolonged inflammatory processes eventually lead to liver fibrosis, but the role of TNFα in liver fibrosis is not fully understood [83]. TNFα may also have a role in hepatic stellate cell activation [84].

Several studies demonstrated elevated RAS components in liver inflammation and fibrosis [85, 86]. RAS components act in two different pathways: classical and alternative pathways. Classical pathway antagonists (ACE-i, ARB and AAA) have some promising results for prevention of liver fibrosis [84].

III. Proteomics marker: Proteomics is the analysis of the entire protein component of a cell, tissue, or organism under a specific, defined set of conditions [87]. Galectina-3 binding protein (G3BP), microfibril-associated protein 4 (MFAP-4], tropomyosin are proteomic markers of liver fibrosis [88 - 90].

IV. Genetic biomarkers of liver fibrosis. The progression of liver disease is a multifactorial process. Disease activity comorbidities such as obesity, poor nutrition or alcohol consumption all have an important impact on the natural course of liver disease. There is increasing evidence that genetic factors are associated with disease activity, fibrosis and progression rate of the liver disease. Several studies have been published on genetic disposition and nucleotide polymorphisms. Genetic mutations or polymorphisms may affect the fibrosis rate bi-directionally; either increasing or decreasing [91].

Indirect Biomarkers

Indirect biomarkers (Class II Biomarkers) are scores produced by statistical analyses and mathematical models that use a combination of various biochemical tests, including liver function tests, platelet count and other biochemical tests.

Currently, there are more than 20 scores to predict liver fibrosis, but most of them have not been validated for most liver diseases (Table **10**). There are only a few fibrosis scores that can be used for all liver diseases, regardless of etiology.

Table 10. Noninvasive (biomarker based) scores for liver fibrosis.

Index/Score/Test	Biomarkers ± Clinical Findings
PGAA [92]	PT, GGT, Apo-A1,α2-macroglobulin
FibroTest [93]	GGT, Total bilirubin, Alpha-2-macroglobulin, Apolipoprotein A1, Haptoglobin
GPR [94]	GGT, platelet count
Forns index [95]	Platelet, GGT, age, cholesterol
GAPRI score [96]	GGT, Platelet count
Hepascore [97]	Bilirubin, γGT, hyaluronan, α2-macroglobulin, age, gender
Fibroindex [93]	AST, gamma globulin, platelet
FIB-4 [98]	Platelet count, AST, ALT, age
APRI score [99]	AST, platelet count
NAFLD Fibrosis score [100]	Age (years), BMI (kg/m2], IGF/diabetes (Yes/No) AST, ALT, Platelets, Albumin
Hepamet Fibrosis Score [101]	Gender, Age, Diabetes, Glucose (mg/dL), Insulin (μU/mL) AST, ALT, Albumin (g/dL), Platelets (x 10^9]
Fortunato score [102]	Fibronectin, prothrombin, pseudocholinesterase, alanine aminotransferase, manganese superoxide dismutase, and N-acetyl-beta-glucosaminidase
Hui score [103]	Body mass index, bilirubin, albumin, and platelet count)
BARD score [94]	BMI, AST, ALT, presence of diabetes/hyperglycemia
Zeng score [104]	Age, α2-macroglobulin, hyaluronate, and GGT
Fibro Q [105]	Age, AST, ALT, INR, platelet count
PAHA score [106]	Platelet count, AST, haptoglobin, and Apo-A1
SHASTA index [107]	ALT, AST, AST to platelet ratio (APRI), albumin, total bilirubin, hyaluronic acid (HA) and YKL-40.
Fibroindex [93]	Platelet count, AST, and gamma globulin
King's Score [108]	Age, AST, INR, platelets
Lok index [108]	Platelet count, AST/ALT ratio, INR
Fibrometer [109]	Age, gender, ALT, AST, GGT, platelet count, INR, urea

Tumor Markers in Liver Disease

Primary liver carcinoma (PLC) can be categorized into three major histological subtypes: hepatocellular carcinoma (HCC), intrahepatic cholangiocarcinoma (ICC), and combined HCC-ICC [110]. HCC is the fifth most common cancer in men and 9th in women, but it is the second or third leading cause of the cancer-related deaths [111 - 113]. HCC is the most common PLC and 75-90% of PLC is HCC [114].

More than 80% of HCC arising in cirrhotic livers [115]. Chronic liver disease caused by viral or other etiologies also carries a high risk for developing HCC.

Although alpha- fetoprotein (AFP) is the most widely used biomarker for HCC, other biomarkers have been available for many years but are rarely used in clinical practice. These may be referred to as old/traditional biomarkers for HCC. Novel biomarkers are based on genes, DNA, RNA or other new molecules.

Old Biomarkers for HCC [116]

- *AFP (alpha-fetoprotein)*
- *AFP-L3 (AFP-L3: lens culinaris agglutinin-reactive fraction of alpha-fetoprotein*
- *DCP (des-γ-carboxyprothrombin)*
- *PVKA II (Protein Induced by Vitamin K Absence or Antagonist-II)*

***New biomarkers for HCC* [117]**

- CK19
- GP73
- OPN
- SCCA
- Annexin A2
- miRNA
- lncRNAs
- CTCs
- cfDNA
- Proteomic analysis and serum metabolite biomarkers
- Core-fucosylated (CF) proteins

Biomarkers can be used for diagnosis, treatment allocation, screening, and surveillance of HCC. AFP is the most widely used biomarker for HCC but as a biomarker, it has a suboptimal performance for screening and surveillance of HCC [118, 119].

Imaging modalities are the mainstay of screening and surveillance. The preferred imaging method is ultrasonography every 6 months, also MR imaging can be used in cases such as obesity that are not suitable for ultrasonographic examination.

It is recommended that AFP be combined with an imaging study for HCC screening and surveillance, but AFP alone shouldn't be used for HCC screening. Although AFP can improve the effectiveness of screening, there are some conflicts regarding the combined interpretation of imaging study and AFP (Table **11**).

Table 11. Imaging study and AFP in screening for HCC.

Imaging study (US)	AFP elevation	HCC possibility
+	+	Yes
+	-	Yes
-	+	?
-	-	No

Although a low level elevation in AFP is common in patients with chronic liver disease, very high elevations of AFP (>200 ng/ml) have also been observed in patients without HCC in rare cases [120, 121]. Anti-viral treatment may, in some cases, reduce AFP elevation not associated with HCC. In a small retrospective study, 10 of 17 patients with a negative initial MRI developed HCC after a mean of 4 months [122]. Imaging with an alternative second modality and close follow-up (3 months apart) may be reasonable.

AFP (Alpha-fetoprotein)

AFP was first identified as a major glycoprotein in human fetal serum in 1957 [123]. AFP plays two important screening roles, elevated AFP in a pregnant woman can indicate neural tube defects or other developmental abnormalities. Elevated AFP may be a sign of HCC in chronic liver disease. Other malignant tumors such as germ cell tumors or yolk sac tumors and non-malign conditions can also cause high AFP [119].

The sensitivity and specificity of AFP for HCC vary with the accepted cut-off level (Table **12**). The high diagnostic performance of AFP requires a higher cut-off, but a higher cut-off also reduces the sensitivity, so the screening cut-offs should be lower than the diagnostic cut-offs.

Table 12. Various cut-off values of AFP and their performances.

AFP (ng/ml)	Sensitivity (%)	Specificity (%)	Patients characteristics/Reference
> 20 > 100 > 200	60 31.2 22.4	90.6 98.8 99.4	Chronic hepatitis B,C [124]
> 20	64.3	91.4	Hepatitis B [125]
> 20 > 200 > 500	82.6 47.7 38.1	70.4 97.1 100	Chronic hepatitis B, C and others [126]
> 5 > 10 > 20 > 100 > 200 > 400	75.1 61.5 50.6 37.7 30.1 24.0	65.5 72.6 80.4 90.2 91.7 93.6	Chronic hepatitis B, C, alcohol and others [127]
> 10	82.6	71.2	Hepatitis B, C [128]
≥ 6 ≥ 20	80.7 38.6	80.4 98.9	Hepatitis B on Entecavir treatment [129]
> 20 > 40 > 100 > 200 > 24.6 *	57.7 50.1 44.2 39.6 56.1	85.2 90.3 93.8 96.0 88.1	Hepatitis B, C and others [130] (*) Best AUROC value: 0.775

CA 19-9 and HCC

CA 19-9 is not a biomarker for HCC, but a slight increase may be seen in cirrhotic patients with HCC. Serum levels of CA19-9 may be elevated in benign conditions and cancers of the pancreas, biliary system, stomach, esophagus, or colon [131]. Elevated CA19-9 may be an indicator of poor survival after liver transplantation or surgical liver resection [132, 133].

CA 19-9 and Cholangiocarcinoma (CCA)

Cholangiocarcinoma (CCA) is a malignant tumor of biliary epithelial cells and it is the second most common primary tumor of the liver. Although the prevalence is higher in Eastern countries, epidemiologic studies suggest that the incidence is also increasing in Western countries [134, 135]. A Systematic review and Meta-Analysis published in 2015 showed that although the sensitivity is not high, CA19-9 still has a good specificity in diagnosing CCA [134].

CEA and HCC

Carcinoembryonic antigen (CEA) is used as an indicator of tumor progression in various carcinomas. Some patients with HCC may have elevated levels of CEA, and this finding predicts a poor prognosis and a more aggressive clinical course [136, 137].

Liver Scores and Biomarkers for Staging and Prognostic Evaluation

Scores created by combining different biomarkers and some clinical features could be used for diagnostic or prognostic evaluation, staging, or risk analysis (Table **13**). We can also define it as biomarker-based scores.

Table 13. Biomarkers based clinical scores (diagnostic - prognostic scores).

Score/ Index	Biomarkers	Clinical finding	Clinical use		
			Diagnosis	Staging/Prognosis	Risk analysis
Child–Turcott	Albumin, Bilirubin	Ascites, encephalopathy, nutritional status	-	•	-
Child–Pugh	Albumin, Bilirubin, PT	Ascites, encephalopathy	-	•	-
MELD-Na	Creatinine, bilirubin, INR, Serum sodium level	-	-	•	-
UKLED score	Creatinine, bilirubin, INR, Serum sodium level	-	-	•	-
Maddrey's Discriminant Function	PT prolongation, bilirubin level	-	-	•	-
Fatty liver index(FLI) [138]	Triglycerides, GGT	BMI, waist circumference	•	-	-
Hepatic steatosis index (HSI) [139]	ALT/AST ratio	BMI, diabetes	•	-	-
Individual prediction model (IPM) [140]	HCV, HBV infection AFP, ALT level	Gender, Age Heavy alcoholics Chronic hepatitis Cirrhosis	-	-	•
GAG-HCC risk score [141]	HBV DNA level CP mutations	Age Cirrhosis	-	-	•

(Table 13) cont.....

Score/ Index	Biomarkers	Clinical finding	Clinical use		
			Diagnosis	Staging/Prognosis	Risk analysis
CUHK clinical scoring system [142]	Albumin Bilirubin HBV DNA level	Cirrhosis	-	-	•

LIMITATIONS AND FUTURE DIRECTIONS

The challenges in diagnosing liver diseases were mentioned in the introduction. Current and still being developed biomarkers will be important tools to address these challenges in the future. Prognostic biomarkers, by which we evaluate the clinical course of the disease, are as important as the diagnostic biomarkers. The development of reliable screening and diagnostic biomarkers for liver cancers, which have a high mortality, is particularly important. Unfortunately, it cannot be said that current biomarkers fully meet these needs.

CONSENT FOR PUBLICATION

Not applicable.

CONFLICT OF INTEREST

The author declares no conflict of interest, financial or otherwise

ACKNOWLEDGEMENT

Declared none.

REFERENCES

[1] Mayeux R. Biomarkers: potential uses and limitations. NeuroRx 2004; 1(2): 182-8. [http://dx.doi.org/10.1602/neurorx.1.2.182] [PMID: 15717018]

[2] Strimbu K, Tavel JA. What are biomarkers? Curr Opin HIV AIDS 2010; 5(6): 463-6. [http://dx.doi.org/10.1097/COH.0b013e32833ed177] [PMID: 20978388]

[3] Blachier M, Leleu H, Peck-Radosavljevic M, Valla D-C, Roudot-Thoraval F. The burden of liver disease in Europe: a review of available epidemiological data. J Hepatol 2013; 58(3): 593-608. [http://dx.doi.org/10.1016/j.jhep.2012.12.005] [PMID: 23419824]

[4] Bellentani S. The epidemiology of non-alcoholic fatty liver disease. Liver Int 2017; 37 (Suppl. 1): 81-4. [http://dx.doi.org/10.1111/liv.13299] [PMID: 28052624]

[5] Wong MCS, Huang JLW, George J, *et al.* The changing epidemiology of liver diseases in the Asia-Pacific region. Nat Rev Gastroenterol Hepatol 2019; 16(1): 57-73. [http://dx.doi.org/10.1038/s41575-018-0055-0] [PMID: 30158570]

[6] Kato GJ, McGowan V, Machado RF, *et al.* Lactate dehydrogenase as a biomarker of hemolysis-associated nitric oxide resistance, priapism, leg ulceration, pulmonary hypertension, and death in

patients with sickle cell disease. Blood 2006; 107(6): 2279-85.
[http://dx.doi.org/10.1182/blood-2005-06-2373] [PMID: 16291595]

[7] Khatroth S. Study of clinical and biochemical profile of acute alcoholic liver disease in a teaching hospital in Telangana. Int J Adv Med 2018; 5(4): 804.
[http://dx.doi.org/10.18203/2349-3933.ijam20182389]

[8] Chung YH, Jung SA, Song BC, *et al.* Plasma isocitrate dehydrogenase as a marker of centrilobular hepatic necrosis in patients with hyperthyroidism. J Clin Gastroenterol 2001; 33(2): 118-22.
[http://dx.doi.org/10.1097/00004836-200108000-00005] [PMID: 11468437]

[9] Chung YH, Kim JA, Song BC, *et al.* Isocitrate dehydrogenase as a marker of centrilobular hepatic necrosis in the experimental model of rats. J Gastroenterol Hepatol 2001; 16(3): 328-32.
[http://dx.doi.org/10.1046/j.1440-1746.2001.02447.x] [PMID: 11339426]

[10] Ames SR, Elvehjem CA. Determination of aspartic-glutamic transaminase in tissue homogenates. J Biol Chem 1946; 166(1): 81-6.
[http://dx.doi.org/10.1016/S0021-9258(17)34987-6] [PMID: 20273677]

[11] Aulbach CJ. Amuzie Biomarkers in Nonclinical Drug Development. In: Ali Said Faqi, Ed. A Comprehensive Guide to Toxicology in Nonclinical Drug Development. Academic Press 2017; pp. 447-71.

[12] Ellinger JJ, Lewis IA, Markley JL. Role of aminotransferases in glutamate metabolism of human erythrocytes. J Biomol NMR 2011; 49(3-4): 221-9.
[http://dx.doi.org/10.1007/s10858-011-9481-9] [PMID: 21380856]

[13] Panteghini M. Aspartate aminotransferase isoenzymes. Clin Biochem 1990; 23(4): 311-9.
[http://dx.doi.org/10.1016/0009-9120(90)80062-N] [PMID: 2225456]

[14] McGill MR. The past and present of serum aminotransferases and the future of liver injury biomarkers. EXCLI J 2016; 15: 817-28.
[PMID: 28337112]

[15] Rabkin SW, Desjardins P. Mitochondrial and cytoplasmic isoenzymes of aspartate aminotransferase in sera of patients after myocardial infarction. Clin Chim Acta 1984; 138(3): 245-57.
[http://dx.doi.org/10.1016/0009-8981(84)90131-1]

[16] Botros M, Sikaris KA. The de ritis ratio: the test of time. Clin Biochem Rev 2013; 34(3): 117-30.
[PMID: 24353357]

[17] Dufour DR, Lott JA, Nolte FS, Gretch DR, Koff RS, Seeff LB. Diagnosis and monitoring of hepatic injury. I. Performance characteristics of laboratory tests. Clin Chem 2000; 46(12): 2027-49.
[http://dx.doi.org/10.1093/clinchem/46.12.2027] [PMID: 11106349]

[18] Giannini EG, Testa R, Savarino V. Liver enzyme alteration: a guide for clinicians. CMAJ 2005; 172(3): 367-79.
[http://dx.doi.org/10.1503/cmaj.1040752] [PMID: 15684121]

[19] Ioannou GN, Boyko EJ, Lee SP. The prevalence and predictors of elevated serum aminotransferase activity in the United States in 1999-2002. Am J Gastroenterol 2006; 101(1): 76-82.
[http://dx.doi.org/10.1111/j.1572-0241.2005.00341.x] [PMID: 16405537]

[20] De Ritis F, Coltorti M, Giusti G. An enzymic test for the diagnosis of viral hepatitis: the transaminase serum activities. 1957. Clin Chim Acta 2006; 369(2): 148-52.
[http://dx.doi.org/10.1016/j.cca.2006.05.001] [PMID: 16781697]

[21] Cohen JA, Kaplan MM. The SGOT/SGPT ratio--an indicator of alcoholic liver disease. Dig Dis Sci 1979; 24(11): 835-8.
[http://dx.doi.org/10.1007/BF01324898] [PMID: 520102]

[22] Mera JR, Dickson B, Feldman M. Influence of gender on the ratio of serum aspartate aminotransferase (AST) to alanine aminotransferase (ALT) in patients with and without hyperbilirubinemia. Dig Dis Sci

2008; 53(3): 799-802.
[http://dx.doi.org/10.1007/s10620-007-9924-z] [PMID: 17717745]

[23] Parmar K, Singh G, Gupta G, Pathak T, Nayak S. Evaluation of De Ritis ratio in liver-associated diseases. Int J Med Sci Public Health 2016; 5(9): 1-6.
[http://dx.doi.org/10.5455/ijmsph.2016.24122015322]

[24] Rekha M, Murthy DS, Poornima RT, Dattatreya K. Evaluation of Deritis ratio, alkaline phosphatase and Bilirubin in liver diseases. J Pharm Sci Res 2011; 3: 1096-102.

[25] Karim S, Rahman M, Shermin S, Sultana R. Correlation between Aminotransferase Ratio (AST/ALT) and Other Biochemical Parameters in Chronic Liver Disease of Viral Origin. Delta Med Col J 2015; 3(1): 13-7.
[http://dx.doi.org/10.3329/dmcj.v3i1.22234]

[26] Ferrari R, Pappas G, Agostinelli D, *et al.* Type 1 autoimmune hepatitis: patterns of clinical presentation and differential diagnosis of the 'acute' type. QJM 2004; 97(7): 407-12.
[http://dx.doi.org/10.1093/qjmed/hch072] [PMID: 15208428]

[27] Kumar Shreevastva N, Pandeya A, Kumar Mishra D. A study of AST: ALT ratio in alcoholic and nonalcoholic liver diseases. Saudi J Med Pharm Sci 2017; (3): 1047-50.

[28] Desmet VJ, Bullens AM, De Groote J. A clinical and histochemical study of cholestasis. Gut 1970; 11(6): 516-23.
[http://dx.doi.org/10.1136/gut.11.6.516] [PMID: 5430380]

[29] Fukano M, Amano S, Hazama F, Hosoda S. 5′-Nucleotidase activities in sera and liver tissues of viral hepatitis patients. Gastroenterol Jpn 1990; 25(2): 199-205.
[http://dx.doi.org/10.1007/BF02776816] [PMID: 2347473]

[30] Sotil EU, Jensen DM. Serum enzymes associated with cholestasis. Clin Liver Dis 2004; 8(1): 41-54, vi.
[http://dx.doi.org/10.1016/S1089-3261(03)00136-3] [PMID: 15062192]

[31] Sharma U, Pal D, Prasad R. Alkaline phosphatase: an overview. Indian J Clin Biochem 2014; 29(3): 269-78.
[http://dx.doi.org/10.1007/s12291-013-0408-y] [PMID: 24966474]

[32] Štefková K, Procházková J, Pacherník J. Alkaline phosphatase in stem cells. Stem Cells Int 2015; 2015: 628368.
[http://dx.doi.org/10.1155/2015/628368] [PMID: 25767512]

[33] Millán JL. Alkaline Phosphatases : Structure, substrate specificity and functional relatedness to other members of a large superfamily of enzymes. Purinergic Signal 2006; 2(2): 335-41.
[http://dx.doi.org/10.1007/s11302-005-5435-6] [PMID: 18404473]

[34] Bourne G, MacKinnon M. The distribution of alkaline phosphatase in various tissues. Q J Exp Physiol Cogn Med Sci 1943; 32(1): 1-20.
[http://dx.doi.org/10.1113/expphysiol.1943.sp000870]

[35] Harris H. The human alkaline phosphatases: what we know and what we don't know. Clin Chim Acta 1990; 186(2): 133-50.
[http://dx.doi.org/10.1016/0009-8981(90)90031-M] [PMID: 2178806]

[36] Seibel MJ. Biochemical markers of bone turnover: part I: biochemistry and variability. Clin Biochem Rev 2005; 26(4): 97-122.
[PMID: 16648882]

[37] Li X, Wang D, Yang C, *et al.* Establishment of age- and gender-specific pediatric reference intervals for liver function tests in healthy Han children. World J Pediatr 2018; 14(2): 151-9.
[http://dx.doi.org/10.1007/s12519-018-0126-x] [PMID: 29546581]

[38] Mukaiyama K, Kamimura M, Uchiyama S, Ikegami S, Nakamura Y, Kato H. Elevation of serum

alkaline phosphatase (ALP) level in postmenopausal women is caused by high bone turnover. Aging Clin Exp Res 2015; 27(4): 413-8.
[http://dx.doi.org/10.1007/s40520-014-0296-x] [PMID: 25534961]

[39] Shipman KE, Holt AD, Gama R. Interpreting an isolated raised serum alkaline phosphatase level in an asymptomatic patient. BMJ 2013; 346(apr03 2): f976.
[http://dx.doi.org/10.1136/bmj.f976] [PMID: 23553977]

[40] Siddique A, Kowdley KV. Approach to a patient with elevated serum alkaline phosphatase. Clin Liver Dis 2012; 16(2): 199-229.
[http://dx.doi.org/10.1016/j.cld.2012.03.012] [PMID: 22541695]

[41] Saraff V, Narayanan VK, Lawson AJ, Shaw NJ, Preece MA, Högler W. A diagnostic algorithm for children with low alkaline phosphatase activities: Lessons learned from laboratory screening for hypophosphatasia. J Pediatr 2016; 172: 181-186.e1.
[http://dx.doi.org/10.1016/j.jpeds.2016.01.045] [PMID: 26896157]

[42] Riancho-Zarrabeitia L, García-Unzueta M, Tenorio JA, *et al.* Clinical, biochemical and genetic spectrum of low alkaline phosphatase levels in adults. Eur J Intern Med 2016; 29: 40-5.
[http://dx.doi.org/10.1016/j.ejim.2015.12.019] [PMID: 26783040]

[43] Whitfield JB. Gamma glutamyl transferase. Crit Rev Clin Lab Sci 200; 38(4): 263-355.
[http://dx.doi.org/10.1080/20014091084227]

[44] Vroon DH, Israili Z. Alkaline Phosphatase and Gamma Glutamyltransferase.Clinical Methods: The History, Physical, and Laboratory Examinations. 3rd ed., Boston: Butterworths 1990.

[45] van Beek JHDA, de Moor MHM, Geels LM, *et al.* The association of alcohol intake with γ-glutamyl transferase (GGT) levels: evidence for correlated genetic effects. Drug Alcohol Depend 2014; 134(1): 99-105.
[http://dx.doi.org/10.1016/j.drugalcdep.2013.09.016] [PMID: 24120856]

[46] Dumortier G, Cabaret W, Stamatiadis L, *et al.* [Hepatic tolerance of atypical antipsychotic drugs]. Encephale 2002; 28(6 Pt 1): 542-51.
[PMID: 12506267]

[47] Bell H, Steensland H. Serum activity of gamma-glutamyltranspeptidase (GGT) in relation to estimated alcohol consumption and questionnaires in alcohol dependence syndrome. Br J Addict 1987; 82(9): 1021-648.

[48] De Castro JJ, Nobre EL, Garcia e Costa J, Galvão-Teles A. [Asymptomatic hepatitis induced by propylthiouracil]. Acta Med Port 2001; 14(5-6): 523-7.
[PMID: 11878166]

[49] Joseph J. Serum Marker Panels for Predicting Liver Fibrosis - An Update. Clin Biochem Rev 2020; 41(2): 67-73.
[PMID: 32518428]

[50] Goldberg DM. 5'nucleotidase: recent advances in cell biology, methodology and clinical significance. Digestion 1973; 8(1): 87-99.
[http://dx.doi.org/10.1159/000197303] [PMID: 4359891]

[51] Gondal B, Aronsohn A. A Systematic Approach to Patients with Jaundice. Semin Intervent Radiol 2016; 33(4): 253-8.
[http://dx.doi.org/10.1055/s-0036-1592331] [PMID: 27904243]

[52] Green RM, Flamm S. AGA technical review on the evaluation of liver chemistry tests. Gastroenterology 2002; 123(4): 1367-84.
[http://dx.doi.org/10.1053/gast.2002.36061] [PMID: 12360498]

[53] Sano J, Kawada H, Yamaguchi N, Kawakita M, Kobayashi K. Effects of phenytoin on serum gamma-glutamyl transpeptidase activity. Epilepsia 1981; 22(3): 331-8.
[http://dx.doi.org/10.1111/j.1528-1157.1981.tb04117.x] [PMID: 6113134]

[54] McTaggart MP, Rawson C, Lawrence D, *et al.* Identification of a macro-alkaline phosphatase complex in a patient with inflammatory bowel disease. Ann Clin Biochem 2012; 49(Pt 4): 405-7. [http://dx.doi.org/10.1258/acb.2011.011224] [PMID: 22454544]

[55] Turecky L. Macroenzymes and their clinical significance. Bratisl Lek Listy 2004; 105(7-8): 260-3. [PMID: 15543847]

[56] Robles-Diaz M, Lucena MI, Kaplowitz N, *et al.* Use of Hy's law and a new composite algorithm to predict acute liver failure in patients with drug-induced liver injury. Gastroenterology 2014; 147(1): 109-118.e5. [http://dx.doi.org/10.1053/j.gastro.2014.03.050] [PMID: 24704526]

[57] Astegiano M, Sapone N, Demarchi B, Rossetti S, Bonardi R, Rizzetto M. Laboratory evaluation of the patient with liver disease. Eur Rev Med Pharmacol Sci 2004; 8(1): 3-9. [PMID: 15209149]

[58] Schreiber G. The synthesis and secretion of plasma proteins in the liver. Pathology 1978; 10(4): 394. [http://dx.doi.org/10.1016/S0031-3025(16)39817-8]

[59] Miller LL, Bly CG, Watson ML, Bale WF. The dominant role of the liver in plasma protein synthesis; a direct study of the isolated perfused rat liver with the aid of lysine-epsilon-C14. J Exp Med 1951; 94(5): 431-53. [http://dx.doi.org/10.1084/jem.94.5.431] [PMID: 14888824]

[60] Spinella R, Sawhney R, Jalan R. Albumin in chronic liver disease: structure, functions and therapeutic implications. Hepatol Int 2016; 10(1): 124-32. [http://dx.doi.org/10.1007/s12072-015-9665-6] [PMID: 26420218]

[61] Rothschild MA, Oratz M, Schreiber SS. Regulation of albumin metabolism. Annu Rev Med 1975; 26(1): 91-104. [http://dx.doi.org/10.1146/annurev.me.26.020175.000515] [PMID: 1096784]

[62] Rothschild MA, Oratz M, Schreiber SS. Albumin synthesis (second of two parts). N Engl J Med 1972; 286(15): 816-21. [http://dx.doi.org/10.1056/NEJM197204132861505] [PMID: 4551862]

[63] Rothschild MA, Oratz M, Schreiber SS. Albumin metabolism. Gastroenterology 1973; 64(2): 324-37. [http://dx.doi.org/10.1016/S0016-5085(73)80046-0] [PMID: 4568596]

[64] Sitar ME, Aydin S, Cakatay U. Human serum albumin and its relation with oxidative stress. Clin Lab 2013; 59(9-10): 945-52. [http://dx.doi.org/10.7754/Clin.Lab.2012.121115] [PMID: 24273915]

[65] Kim S, McClave SA, Martindale RG, Miller KR, Hurt RT. Hypoalbuminemia and clinical outcomes: What is the mechanism behind the relationship? Am Surg 2017; 83(11): 1220-7. [http://dx.doi.org/10.1177/000313481708301123] [PMID: 29183523]

[66] Tsirpanlis G, Bagos P, Ioannou D, *et al.* Serum albumin: a late-reacting negative acute-phase protein in clinically evident inflammation in dialysis patients. Nephrol Dial Transplant 2005; 20(3): 658-9.

[67] Jiang AA, Greenwald HS, Sheikh L, *et al.* Predictors of Acute Liver Failure in Patients With Acute Hepatitis A: An Analysis of the 2016-2018 San Diego County Hepatitis A Outbreak. Open Forum Infect Dis 2019; 6(11): ofz467. [http://dx.doi.org/10.1093/ofid/ofz467] [PMID: 31777757]

[68] Ballmer PE, Walshe D, McNurlan MA, Watson H, Brunt PW, Garlick PJ. Albumin synthesis rates in cirrhosis: correlation with Child-Turcotte classification. Hepatology 1993; 18(2): 292-7. [http://dx.doi.org/10.1002/hep.1840180211] [PMID: 8340057]

[69] Ballmer PE, Reichen J, McNurlan MA, Sterchi AB, Anderson SE, Garlick PJ. Albumin but not fibrinogen synthesis correlates with galactose elimination capacity in patients with cirrhosis of the liver. Hepatology 1996; 24(1): 53-9.

[http://dx.doi.org/10.1002/hep.510240111] [PMID: 8707282]

[70] Tripodi A. Liver Disease and Hemostatic (Dys)function. Semin Thromb Hemost 2015; 41(5): 462-7. [http://dx.doi.org/10.1055/s-0035-1550440] [PMID: 26080306]

[71] Monroe DM, Hoffman M. The coagulation cascade in cirrhosis. Clin Liver Dis 2009; 13(1): 1-9. [http://dx.doi.org/10.1016/j.cld.2008.09.014] [PMID: 19150304]

[72] Kamal AH, Tefferi A, Pruthi RK. How to interpret and pursue an abnormal prothrombin time, activated partial thromboplastin time, and bleeding time in adults. Mayo Clin Proc 2007; 82(7): 864-73. [http://dx.doi.org/10.4065/82.7.864] [PMID: 17605969]

[73] Regev A, Berho M, Jeffers LJ, *et al.* Sampling error and intraobserver variation in liver biopsy in patients with chronic HCV infection. Am J Gastroenterol 2002; 97(10): 2614-8. [http://dx.doi.org/10.1111/j.1572-0241.2002.06038.x] [PMID: 12385448]

[74] Ratziu V, Charlotte F, Heurtier A, *et al.* Sampling variability of liver biopsy in nonalcoholic fatty liver disease. Gastroenterology 2005; 128(7): 1898-906. [http://dx.doi.org/10.1053/j.gastro.2005.03.084] [PMID: 15940625]

[75] Patel K, Rockey DC. Clinical Utility of Biomarkers of Liver Fibrosis. Gastroenterol Hepatol (N Y) 2006; 2(1): 48-57. [PMID: 28210197]

[76] Czul F, Bhamidimarri KR. Noninvasive Markers to Assess Liver Fibrosis. J Clin Gastroenterol 2016; 50(6): 445-57. [http://dx.doi.org/10.1097/MCG.0000000000000534] [PMID: 27105176]

[77] Gressner OA, Weiskirchen R, Gressner AM. Biomarkers of hepatic fibrosis, fibrogenesis and genetic pre-disposition pending between fiction and reality. J Cell Mol Med 2007; 11(5): 1031-51. [http://dx.doi.org/10.1111/j.1582-4934.2007.00092.x] [PMID: 17979881]

[78] Cequera A, García de León Méndez MC. [Biomarkers for liver fibrosis: advances, advantages and disadvantages]. Rev Gastroenterol Mex 2014; 79(3): 187-99. [http://dx.doi.org/10.1016/j.rgmx.2014.05.003] [PMID: 24954541]

[79] Fabregat I, Moreno-Càceres J, Sánchez A, *et al.* TGF-β signalling and liver disease. FEBS J 2016; 283(12): 2219-32. [http://dx.doi.org/10.1111/febs.13665] [PMID: 26807763]

[80] Dooley S, ten Dijke P. TGF-β in progression of liver disease. Cell Tissue Res 2012; 347(1): 245-56. [http://dx.doi.org/10.1007/s00441-011-1246-y] [PMID: 22006249]

[81] Meindl-Beinker NM, Matsuzaki K, Dooley S. TGF-β signaling in onset and progression of hepatocellular carcinoma. Dig Dis 2012; 30(5): 514-23. [http://dx.doi.org/10.1159/000341704] [PMID: 23108308]

[82] Dewidar B, Meyer C, Dooley S, Meindl-Beinker AN. TGF-β in Hepatic Stellate Cell Activation and Liver Fibrogenesis-Updated 2019. Cells 2019; 8(11): 1419. [http://dx.doi.org/10.3390/cells8111419] [PMID: 31718044]

[83] Yang YM, Seki E. TNFα in liver fibrosis. Curr Pathobiol Rep 2015; 3(4): 253-61. [http://dx.doi.org/10.1007/s40139-015-0093-z] [PMID: 26726307]

[84] Osawa Y, Kojika E, Hayashi Y, *et al.* Tumor necrosis factor-α-mediated hepatocyte apoptosis stimulates fibrosis in the steatotic liver in mice. Hepatol Commun 2018; 2(4): 407-20. [http://dx.doi.org/10.1002/hep4.1158] [PMID: 29619419]

[85] Lubel JS, Warner FJ, Angus PW. The Role of the Renin-Angiotensin System in Hepatic Fibrosis. Frontiers in Research of the Renin-Angiotensin System on Human Disease 2007; Vol. 7: pp. 113-34. [http://dx.doi.org/10.1007/978-1-4020-6372-5_6]

[86] Shim KY, Eom YW, Kim MY, Kang SH, Baik SK. Role of the renin-angiotensin system in hepatic

fibrosis and portal hypertension. Korean J Intern Med (Korean Assoc Intern Med) 2018; 33(3): 453-61.
[http://dx.doi.org/10.3904/kjim.2017.317] [PMID: 29462546]

[87] Graves PR, Haystead TAJ. Molecular biologist's guide to proteomics. Microbiol Mol Biol Rev 2002; 66(1): 39-63.
[http://dx.doi.org/10.1128/MMBR.66.1.39-63.2002] [PMID: 11875127]

[88] Hara A, Niwa M, Noguchi K, *et al.* Galectin-3 as a Next-Generation Biomarker for Detecting Early Stage of Various Diseases. Iomolecules 2020; 10(3).

[89] Mölleken C, Sitek B, Henkel C, *et al.* Detection of novel biomarkers of liver cirrhosis by proteomic analysis. Hepatology 2009; 49(4): 1257-66.
[http://dx.doi.org/10.1002/hep.22764] [PMID: 19177598]

[90] Lurie Y, Webb M, Cytter-Kuint R, Shteingart S, Lederkremer GZ. Non-invasive diagnosis of liver fibrosis and cirrhosis. World J Gastroenterol 2015; 21(41): 11567-83.
[http://dx.doi.org/10.3748/wjg.v21.i41.11567] [PMID: 26556987]

[91] Bataller R, North KE, Brenner DA. Genetic polymorphisms and the progression of liver fibrosis: a critical appraisal. Hepatology 2003; 37(3): 493-503.
[http://dx.doi.org/10.1053/jhep.2003.50127] [PMID: 12601343]

[92] Lu LG, Zeng MD, Mao YM, *et al.* Relationship between clinical and pathologic findings in patients with chronic liver diseases. World J Gastroenterol 2003; 9(12): 2796-800.
[http://dx.doi.org/10.3748/wjg.v9.i12.2796] [PMID: 14669336]

[93] Koda M, Matunaga Y, Kawakami M, Kishimoto Y, Suou T, Murawaki Y. FibroIndex, a practical index for predicting significant fibrosis in patients with chronic hepatitis C. Hepatology 2007; 45(2): 297-306.
[http://dx.doi.org/10.1002/hep.21520] [PMID: 17256741]

[94] Cichoż-Lach H, Celiński K, Prozorow-Król B, Swatek J, Słomka M, Lach T. The BARD score and the NAFLD fibrosis score in the assessment of advanced liver fibrosis in nonalcoholic fatty liver disease. Med Sci Monit 2012; 18(12): CR735-40.
[http://dx.doi.org/10.12659/MSM.883601] [PMID: 23197236]

[95] Güzelbulut F, Çetınkaya ZA, Sezıklı M, Yaşar B, Özkara S, Övünç AO. AST-platelet ratio index, Forns index and FIB-4 in the prediction of significant fibrosis and cirrhosis in patients with chronic hepatitis C. Turk J Gastroenterol 2011; 22(3): 279-85.
[http://dx.doi.org/10.4318/tjg.2011.0213] [PMID: 21805418]

[96] Chrostek L, Przekop D, Gruszewska E, Gudowska-Sawczuk M, Cylwik B. Noninvasive indirect markers of liver fibrosis in alcoholics. In: Caputo F, Przekop D, Gruszewska E, Gudowska-Sawczuk M, Cylwik B, Eds. Biomed Res Int. 2019; p. 3646975.
[http://dx.doi.org/10.1155/2019/3646975]

[97] Adams LA, Bulsara M, Rossi E, *et al.* Hepascore: an accurate validated predictor of liver fibrosis in chronic hepatitis C infection. Clin Chem 2005; 51(10): 1867-73.
[http://dx.doi.org/10.1373/clinchem.2005.048389] [PMID: 16055434]

[98] Sterling RK, Lissen E, Clumeck N, *et al.* Development of a simple noninvasive index to predict significant fibrosis in patients with HIV/HCV coinfection. Hepatology 2006; 43(6): 1317-25.
[http://dx.doi.org/10.1002/hep.21178] [PMID: 16729309]

[99] Wai C-T, Greenson JK, Fontana RJ, *et al.* A simple noninvasive index can predict both significant fibrosis and cirrhosis in patients with chronic hepatitis C. Hepatology 2003; 38(2): 518-26.
[http://dx.doi.org/10.1053/jhep.2003.50346] [PMID: 12883497]

[100] Treeprasertsuk S, Björnsson E, Enders F, Suwanwalaikorn S, Lindor KD. NAFLD fibrosis score: a prognostic predictor for mortality and liver complications among NAFLD patients. World J Gastroenterol 2013; 19(8): 1219-29.

[http://dx.doi.org/10.3748/wjg.v19.i8.1219] [PMID: 23482703]

[101] Zambrano-Huailla R, Guedes L, Stefano JT, *et al.* Diagnostic performance of three non-invasive fibrosis scores (Hepamet, FIB-4, NAFLD fibrosis score) in NAFLD patients from a mixed Latin American population. Ann Hepatol 2020; 19(6): 622-6.
[http://dx.doi.org/10.1016/j.aohep.2020.08.066] [PMID: 32919087]

[102] Fortunato G, Castaldo G, Oriani G, *et al.* Multivariate discriminant function based on six biochemical markers in blood can predict the cirrhotic evolution of chronic hepatitis. Clin Chem 2001; 47(9): 1696-700.
[http://dx.doi.org/10.1093/clinchem/47.9.1696] [PMID: 11514405]

[103] Hui AY, Chan HLY, Wong VWS, *et al.* Identification of chronic hepatitis B patients without significant liver fibrosis by a simple noninvasive predictive model. Am J Gastroenterol 2005; 100(3): 616-23.
[http://dx.doi.org/10.1111/j.1572-0241.2005.41289.x] [PMID: 15743360]

[104] Zeng MD, Lu LG, Mao YM, *et al.* Prediction of significant fibrosis in HBeAg-positive patients with chronic hepatitis B by a noninvasive model. Hepatology 2005; 42(6): 1437-45.
[http://dx.doi.org/10.1002/hep.20960] [PMID: 16317674]

[105] Hsieh Y-Y, Tung S-Y, Lee I-L, *et al.* FibroQ: an easy and useful noninvasive test for predicting liver fibrosis in patients with chronic viral hepatitis. Chang Gung Med J 2009; 32(6): 614-22.
[PMID: 20035640]

[106] Lee KG, Seo YS, An H, *et al.* Usefulness of non-invasive markers for predicting liver cirrhosis in patients with chronic hepatitis B. J Gastroenterol Hepatol 2010; 25(1): 94-100.
[http://dx.doi.org/10.1111/j.1440-1746.2009.05953.x] [PMID: 19793171]

[107] Kelleher TB, Mehta SH, Bhaskar R, *et al.* Prediction of hepatic fibrosis in HIV/HCV co-infected patients using serum fibrosis markers: the SHASTA index. J Hepatol 2005; 43(1): 78-84.
[http://dx.doi.org/10.1016/j.jhep.2005.02.025] [PMID: 15894397]

[108] Deng H, Qi X, Guo X. Diagnostic Accuracy of APRI, AAR, FIB-4, FI, King, Lok, Forns, and FibroIndex Scores in Predicting the Presence of Esophageal Varices in Liver Cirrhosis: A Systematic Review and Meta-Analysis. Medicine (Baltimore) 2015; 94(42): e1795.
[http://dx.doi.org/10.1097/MD.0000000000001795] [PMID: 26496312]

[109] Calès P, Boursier J, Oberti F, *et al.* FibroMeters: a family of blood tests for liver fibrosis. Gastroenterol Clin Biol 2008; 32(6) (Suppl. 1): 40-51.
[http://dx.doi.org/10.1016/S0399-8320(08)73992-7] [PMID: 18973845]

[110] Feng M, Pan Y, Kong R, Shu S. Therapy of Primary Liver Cancer 1(2): 100032.

[111] Bray F, Ferlay J, Soerjomataram I, Siegel RL, Torre LA, Jemal A. Global cancer statistics 2018: GLOBOCAN estimates of incidence and mortality worldwide for 36 cancers in 185 countries. CA Cancer J Clin 2018; 68(6): 394-424.
[http://dx.doi.org/10.3322/caac.21492] [PMID: 30207593]

[112] McGlynn KA, Petrick JL, El-Serag HB. Epidemiology of hepatocellular carcinoma. Hepatology 2020; hep.31288.

[113] Zhu RX, Seto W-K, Lai C-L, Yuen M-F. Epidemiology of Hepatocellular Carcinoma in the Asia-Pacific Region. Gut Liver 2016; 10(3): 332-9.
[http://dx.doi.org/10.5009/gnl15257] [PMID: 27114433]

[114] Wong MCS, Jiang JY, Goggins WB, *et al.* International incidence and mortality trends of liver cancer: a global profile. Sci Rep 2017; 7(1): 45846.
[http://dx.doi.org/10.1038/srep45846] [PMID: 28361988]

[115] van Meer S, van Erpecum KJ, Sprengers D, *et al.* Hepatocellular carcinoma in cirrhotic *versus* noncirrhotic livers: results from a large cohort in the Netherlands. Eur J Gastroenterol Hepatol 2016; 28(3): 352-9.

[http://dx.doi.org/10.1097/MEG.0000000000000527] [PMID: 26629867]

[116] Lou J, Zhang L, Lv S, Zhang C, Jiang S. Biomarkers for Hepatocellular Carcinoma. Biomark Cancer 2017; 9: 1-9.
[http://dx.doi.org/10.1177/1179299X16684640] [PMID: 28469485]

[117] Zacharakis G, Aleid A, Aldossari KK. New and old biomarkers of hepatocellular carcinoma. Hepatoma Res 2018; 4(10): 65.
[http://dx.doi.org/10.20517/2394-5079.2018.76]

[118] Fujiwara N, Friedman SL, Goossens N, Hoshida Y. Risk factors and prevention of hepatocellular carcinoma in the era of precision medicine. J Hepatol 2018; 68(3): 526-49.
[http://dx.doi.org/10.1016/j.jhep.2017.09.016] [PMID: 28989095]

[119] Adigun OO, Yarrarapu SNS, Khetarpal S. Alpha Fetoprotein. Treasure Island (FL): StatPearls Publishing 2020.

[120] Kim KA, Lee JS, Jung ES, *et al.* [Usefulness of serum alpha-fetoprotein (AFP) as a marker for hepatocellular carcinoma (HCC) in hepatitis C virus related cirrhosis: analysis of the factors influencing AFP elevation without HCC development]. Korean J Gastroenterol 2006; 48(5): 321-6. [Internet].
[PMID: 17132920]

[121] Cheema AW, Hirschtritt T, Van Thiel DH. Markedly elevated alpha-fetoprotein levels without hepatocellular carcinoma. Hepatogastroenterology 2004; 51(60): 1676-8.
[PMID: 15532803]

[122] Kim BS, Hayashi PH, Kim SH, *et al.* Outcomes of patients with elevated α-fetoprotein level and initial negative findings at MR imaging. Radiology 2013; 268(1): 109-19.
[http://dx.doi.org/10.1148/radiol.13121314] [PMID: 23392426]

[123] Pucci P, Siciliano R, Malorni A, *et al.* Human α-fetoprotein primary structure: a mass spectrometric study. Biochemistry 1991; 30(20): 5061-6.
[http://dx.doi.org/10.1021/bi00234a032] [PMID: 1709810]

[124] Trevisani F, D'Intino PE, Morselli-Labate AM, *et al.* Serum alpha-fetoprotein for diagnosis of hepatocellular carcinoma in patients with chronic liver disease: influence of HBsAg and anti-HCV status. J Hepatol 2001; 34(4): 570-5.
[http://dx.doi.org/10.1016/S0168-8278(00)00053-2] [PMID: 11394657]

[125] Sherman M, Peltekian KM, Lee C. Screening for hepatocellular carcinoma in chronic carriers of hepatitis B virus: incidence and prevalence of hepatocellular carcinoma in a North American urban population. Hepatology 1995; 22(2): 432-8.
[PMID: 7543434]

[126] Chan SL, Mo F, Johnson PJ, *et al.* Performance of serum α-fetoprotein levels in the diagnosis of hepatocellular carcinoma in patients with a hepatic mass. HPB (Oxford) 2014; 16(4): 366-72.
[http://dx.doi.org/10.1111/hpb.12146] [PMID: 23980880]

[127] Ahn DG, Kim HJ, Kang H, *et al.* Feasibility of α-fetoprotein as a diagnostic tool for hepatocellular carcinoma in Korea. Korean J Intern Med (Korean Assoc Intern Med) 2016; 31(1): 46-53.
[http://dx.doi.org/10.3904/kjim.2016.31.1.46] [PMID: 26767857]

[128] Jasirwan COM, Fahira A, Siregar L, Loho I. The alpha-fetoprotein serum is still reliable as a biomarker for the surveillance of hepatocellular carcinoma in Indonesia. BMC Gastroenterol 2020; 20(1): 215.
[http://dx.doi.org/10.1186/s12876-020-01365-1] [PMID: 32646378]

[129] Wong GLH, Chan HLY, Tse YK, *et al.* On-treatment alpha-fetoprotein is a specific tumor marker for hepatocellular carcinoma in patients with chronic hepatitis B receiving entecavir. Hepatology 2014; 59(3): 986-95.
[http://dx.doi.org/10.1002/hep.26739] [PMID: 24123097]

[130] Hu J, Wang N, Yang Y, *et al.* Diagnostic value of alpha-fetoprotein combined with neutrophil-t--lymphocyte ratio for hepatocellular carcinoma. BMC Gastroenterol 2018; 18(1): 186.https://bmcgastroenterol.biomedcentral.com/articles/10.1186/s12876-018-0908-6 [http://dx.doi.org/10.1186/s12876-018-0908-6] [PMID: 30545306]

[131] Edoo MIA, Chutturghoon VK, Wusu-Ansah GK, *et al.* Serum biomarkers AFP, CEA and CA19-9 combined detection for early diagnosis of hepatocellular carcinoma. Iran J Public Health 2019; 48(2): 314-22. [PMID: 31205886]

[132] Wan P, Zhang J, Long X, *et al.* Serum levels of preoperative α-fetoprotein and CA19-9 predict survival of hepatic carcinoma patients after liver transplantation. Eur J Gastroenterol Hepatol 2014; 26(5): 553-61. [http://dx.doi.org/10.1097/MEG.0000000000000070] [PMID: 24589829]

[133] Kiliç M, Göçmen E, Tez M, Ertan T, Keskek M, Koç M. Value of preoperative serum CA 19-9 levels in predicting resectability for pancreatic cancer. Can J Surg 2006; 49(4): 241-4. [PMID: 16948881]

[134] Liang B, Zhong L, He Q, *et al.* Diagnostic Accuracy of Serum CA19-9 in Patients with Cholangiocarcinoma: A Systematic Review and Meta-Analysis. Med Sci Monit 2015; 21: 3555-63. [http://dx.doi.org/10.12659/MSM.895040] [PMID: 26576628]

[135] Blechacz B, Gores GJ. Cholangiocarcinoma: advances in pathogenesis, diagnosis, and treatment. Hepatology 2008; 48(1): 308-21. [http://dx.doi.org/10.1002/hep.22310] [PMID: 18536057]

[136] Yoshikawa M, Morine Y, Ikemoto T, *et al.* Elevated preoperative serum CEA level is associated with poor prognosis in patients with hepatocellular carcinoma through the epithelial-mesenchymal transition. Anticancer Res 2017; 37(3): 1169-75. [http://dx.doi.org/10.21873/anticanres.11430] [PMID: 28314278]

[137] Sawabu N, Wakabayashi T, Ozaki K, *et al.* Serum tumor markers in patients with hepatocellular carcinoma: diagnosis of α-fetoprotein -low or -negative patients. Gastroenterol Jpn 1985; 20(3): 209-15. [http://dx.doi.org/10.1007/BF02774706] [PMID: 2412927]

[138] Bedogni G, Bellentani S, Miglioli L, *et al.* The Fatty Liver Index: a simple and accurate predictor of hepatic steatosis in the general population. BMC Gastroenterol 2006; 6(1): 33. [http://dx.doi.org/10.1186/1471-230X-6-33] [PMID: 17081293]

[139] Lee JH, Kim D, Kim HJ, *et al.* Hepatic steatosis index: a simple screening tool reflecting nonalcoholic fatty liver disease. Dig Liver Dis 2010; 42(7): 503-8. [http://dx.doi.org/10.1016/j.dld.2009.08.002] [PMID: 19766548]

[140] Han K-H, Ahn SH. How to predict HCC development in patients with chronic B viral liver disease? Intervirology 2005; 48(1): 23-8. [http://dx.doi.org/10.1159/000082091] [PMID: 15785086]

[141] Yuen M-F, Tanaka Y, Fong DY-T, *et al.* Independent risk factors and predictive score for the development of hepatocellular carcinoma in chronic hepatitis B. J Hepatol 2009; 50(1): 80-8. [http://dx.doi.org/10.1016/j.jhep.2008.07.023] [PMID: 18977053]

[142] Wong VWS, Chan SL, Mo F, *et al.* Clinical scoring system to predict hepatocellular carcinoma in chronic hepatitis B carriers. J Clin Oncol 2010; 28(10): 1660-5. [http://dx.doi.org/10.1200/JCO.2009.26.2675] [PMID: 20194845]

CHAPTER 19

Application of New Acute Kidney Injury Biomarkers

Nurhan Seyahi[1,*] and **Seyda Gul Ozcan**[2]

[1] *Department of Nephrology, Cerrahpasa Medical Faculty, Istanbul University - Cerrahpasa, Istanbul, Turkey*

[2] *Department of Internal Medicine, Cerrahpasa Medical Faculty, Istanbul University - Cerrahpasa, Istanbul, Turkey*

Abstract: Kidney-related biomarkers can provide structural and functional information about different parts of the nephron. These biomarkers can be used to evaluate glomerular, tubular, or interstitial injury, inflammation, or repair, and glomerular or tubular function. Furthermore, biomarkers can improve the acute kidney injury diagnosis in various clinical conditions, including acute interstitial nephritis, acute tubular injury, hepatorenal and cardiorenal syndrome, ischemic and nephrotoxic acute kidney injury, and drug-induced acute kidney injury. Biomarkers might be used as an additional precision medicine tool in managing patients with acute kidney injury; they can help with clinical decision-making and impact patient outcomes. In this chapter, we reviewed the utility of biomarkers used in acute kidney injury.

Keywords: Acute kidney injury, Biomarkers, Nephrotoxicity, Tubular injury.

INTRODUCTION

Acute Kidney Injury, Biomarkers, And Definitions

Acute kidney injury (AKI) is the rapid loss of kidney function and is a serious problem, particularly in patients hospitalized in intensive care units (ICUs) [1, 2]. However, until about two decades ago, the condition lacked a uniform definition. The Acute Dialysis Quality Initiative first introduced the international consensus criteria, which was subsequently modified by the AKI Network and finally by the Kidney Disease Improving Global Outcomes (KDIGO) workgroup [1]. Although a standardization in the definition of AKI has been achieved using the serum creatinine and occasionally urine output, there is still a need for more robust tools to define the etiology, type, and prognosis of AKI [3].

* **Corresponding author Nurhan Seyahi:** Department of Nephrology, Cerrahpasa Medical Faculty, Istanbul University - Cerrahpasa, Istanbul, Turkey; Tel: +90 (532) 296 32 03; E-mail: nseyahi@yahoo.com

Pınar Atukeren and Hafize Uzun (Eds.)

The KDIGO guidelines emphasize early diagnosis and treatment in AKI, and the preferred diagnostic biomarker of the condition remains the measurement of the serum creatinine level because of its convenience and low cost. However, the utility of this marker is limited due to many reasons; for example, renal hypoperfusion may cause prerenal kidney injury, and creatinine levels may increase without any renal parenchymal damage [4]. Additionally, in patients with renal parenchymal injury, especially in chronic kidney disease (CKD) patients, renal compensation may delay the creatinine increase, and creatinine levels might not increase until a GFR loss of more than 50% occurs [5, 6]. Therefore, more sensitive and specific new biomarkers will likely improve the early diagnosis of AKI [7].

Biomarkers in kidney injury are new tools for risk assessment that can help determine which therapy to use. [8]. The Food and Drug Administration (FDA) has defined a "biomarker" as a characteristic that is objectively measured and evaluated as an indicator of normal biologic processes, pathogenic processes, or pharmacologic responses to therapeutic interventions [3].

Classification of AKI Biomarkers

Different schemes, usually based on the site of origin or pathophysiological process on the nephron (Table **1**) [9, 10], might be used to classify the AKI biomarkers. Biomarkers can help evaluate glomerular or tubular function, glomerular, tubular, or interstitial injury, inflammation, or repair. Selected biomarkers of AKI according to the site of origin on the nephron are shown in Fig. (**1**).

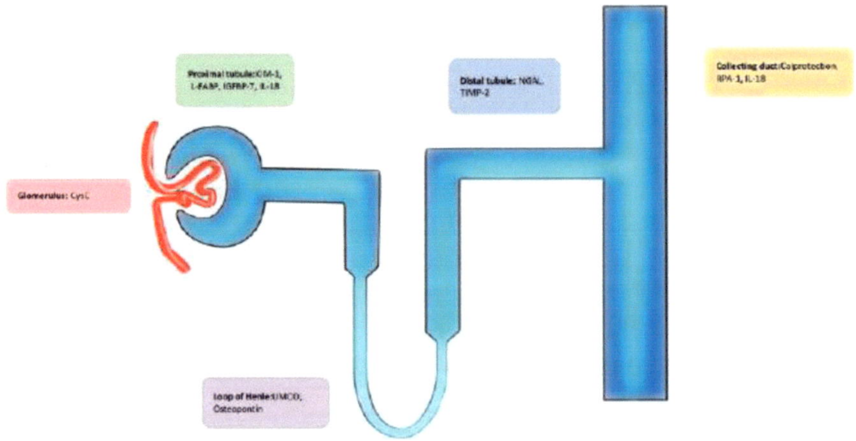

Fig. (1). Biomarkers and their predictive area of injury in the nephron. The illustration was prepared using the following references [10, 75, 94, 98].

Additionally, tissue inhibitors of metalloproteinases-2 (TIMP-2) and insulin-like growth factor binding protein-7 (IGFBP-7) might be classified as cell cycle arrest biomarkers [11].

Table 1. Acute kidney injury biomarkers classified according to the site of origin and pathophysiological process on the nephron [9, 10, 45].

Biomarkers of glomerular health	Cystatin-C, galectin-3, proenkephalin
Biomarkers of tubular health (function and injury)	KIM-1, L-FABP, IGFBP7, TIMP-2, UMOD, NGAL
Biomarkers of inflammation	TNF-alpha, IL-9, IL-18, IL-6, IL-10, TNFR1, TNFR2
Biomarkers of repair and fibrosis	MCP-1, CHI3L1 (YKL 40), PIIINP, EGF

KIM-1: kidney injury molecule-1, L-FABP: liver-type fatty acid-binding protein, TIMP-2: tissue inhibitor of metalloproteinases 2, IGFBP-7: insulin-like growth factor-binding protein 7, UMOD: uromodulin, NGAL: neutrophil gelatinase-associated lipocalin, TNF-alpha: tumor necrosis factor alpha, IL: interleukin, TNFR1: tumor necrosis factor receptor 1, TNFR2: tumor necrosis factor receptor 2, MCP-1: monocyte chemoattractant protein 1, PIIINP: procollagen type III N-terminal propeptide, EGF: epidermal growth factor.

Biomarkers of Glomerular Health

Cystatin C: Classically, creatinine is used as a biomarker of glomerular health, while cystatin C is a new biomarker that depicts the glomerular function. In contrast to creatinine, serum cystatin C is not correlated with age, sex, and muscle mass, and in certain situations, such as in patients with marginal muscle mass, it might be used as a more appropriate marker of glomerular function [12]. Cystatin C is eliminated entirely by glomerular filtration. Following filtration, the protein is almost completely catabolized in the proximal tubule. Therefore, in physiological conditions, it scarcely appears in the urine. An impairment in the proximal tubular reabsorption causes an increase in the urinary cystatin C levels [13].

Biomarkers of Tubular Health

Alpha-1-Microglobulin (α1M): Alpha-1-microglobulin is a biomarker of proximal tubular function and is mainly produced by the liver. The protein might be in free form, or it might form a complex with immunoglobulin A (IgA) [14]. Following free glomerular filtration and proximal tubular reabsorption, α1M is completely catabolized. Increased urinary concentration of this biomarker is associated with proximal tubular injury or dysfunction. Urinary α1M extraction is increased in patients who have a renal tubular disease [14].

Liver-type Fatty acid-binding Protein (L-FABP): There are several types of FABPs expressed in various organs, including the kidneys and the liver. Following glomerular filtration, the circulating L-FABP is reabsorbed by the proximal tubule. Acute ischemic injury in humans causes an increased expression of L-FABP in the proximal tubules [15]. Various studies performed in adults and children have shown that an elevated L-FABP level is a sensitive and specific biomarker of AKI [16, 17]. According to previous clinical studies, increased serum L-FABP levels do not significantly impact urinary L-FABP levels; additionally, in patients with hepatic disease, the urinary L-FABP concentrations are not different than that of the healthy subjects [17, 18].

Uromodulin (UMOD): Uromodulin, also known as Tamm-Horsfall protein, is produced by the cells of the thick ascending limb of the loop of Henle [19]. The physiologic function of UMOD is not clearly defined. However, according to animal models and clinical studies, UMOD might be used as a biomarker of tubular health. UMOD level is associated with the quantity of intact tubular cells of the ascending limb of the loop of Henle. Therefore, it is a marker for the remaining functional tubules [20].

Neutrophil Gelatinase-associated Lipocalin (NGAL): Plasma and urinary NGALs (pNGALs and uNGALs) are important biomarkers of kidney injury [21]. Urinary NGAL is produced by the kidney in response to injury, and it is more specific. NGAL is an early marker of kidney injury. According to studies performed on rodent models following tubular injury, NGAL genes are upregulated, especially in the distal nephron [22, 23]. Postoperative AKI in children who underwent cardiac surgery is associated with the elevation of NGAL levels within a time period of as early as two hours [5].

Kidney Injury Molecule-1 (KIM-1): This protein was previously named T-cell immunoglobulin and mucin domain-containing protein 1 (TIM-1) and hepatitis A virus cellular receptor 1 (HAVcr-1). KIM-1 expression is at low levels in healthy kidneys; however, following ischemic injury, the expression increases significantly, especially in proximal tubular cells [24, 25]. Therefore, KIM-1 is mainly a marker of proximal tubular injury. Ischemic and nephrotoxic kidney injury studies performed on animal models revealed that the molecule is a sensitive and specific kidney injury marker [26, 27]. KIM-1 is upregulated during the recovery of AKI. Thus, it may be associated with the healing of the kidney [28].

Tissue Inhibitor of Metalloproteinases-2 (TIMP-2] and Insulin-like Growth Factor-binding Protein 7 (IGFBP-7): Both IGFBP-7 (through p53 and p21] and TIMP-2 (through p27) block the effect of cyclin-dependent protein kinase complexes and cause short periods of G1 cell cycle arrest [29, 30]. Both molecules are produced and secreted from damaged tubular cells. The molecules inhibit endothelial cell proliferation by kinase activation and affect the bioavailability of insulin growth factors that contribute to tumor suppression and cell aging.

The FDA has approved these biomarkers for the evaluation of intensive care unit (ICU) patients with acute cardio-pulmonary compromise [31].

Biomarkers of Inflammation

Interleukin-18 (IL-18): Urine IL-18 levels are elevated in AKI [32]. It has been shown that urine IL-18 may serve as an early diagnostic marker of AKI and predict mortality across various clinical settings, including patients in the ICU and following cardiac surgery. IL-18 is both a mediator and a biomarker of AKI; IL-18 levels are augmented in several inflammatory processes [33, 34]. Its level starts increasing around six hours after ischemic injury, 24 to 48 hours earlier than the AKI diagnosis, and peaks about at the 12th hour [35].

Monocyte Chemoattractant Protein-1 (MCP-1): The MCP-1 protein has a significant role in the pathogenesis of ischemic and toxic kidney injury [9]. Its use as a potential biomarker for AKI was first described in urine samples in a mouse model and validated in critically ill patients [36]. Elevated levels of urineMCP-1 were associated with AKI and death following cardiac surgery [37].

Tumor Necrosis Factor Receptor 1 (TNF-r1): This membrane receptor has been associated with progression to end-stage renal disease and mortality in patients with Type 1 and Type 2 diabetes mellitus beyond other established risk factors such as proteinuria and was implicated in other kidney diseases, including various types of glomerulonephritis, obstructive kidney injury, and kidney transplant rejection [38, 39].

Biomarkers of Repair

YKL-40: It is also known as chitinase 3-like 1 protein (CHI3L1). This is a gene product that causes elevated serum and urine levels in the recipients post-transplantation and predicts the need for subsequent dialysis [40]. The protein is associated with the limitation of apoptosis based on ischemia/reperfusion injury is, therefore, considered an indicator of peritransplantation ischemia/reperfusion injury [9].

Procollagen Type III N-terminal Propeptide (PIIINP): Urinary PIIINP levels are associated with extracellular matrix deposition. Accumulation of extracellular matrix is increased in fibrosis; therefore, this biomarker might be useful for quantifying fibrosis during the early stages of CKD [10].

Other Biomarkers

Urine miRNA: Endogenous noncoding RNA molecules, called miRNAs, regulate gene expression by inhibiting protein translation. Both urine and plasma miR-21s are shown to be regulators of apoptosis in renal tubular epithelial cells in cardiovascular surgery patients and induce cellular proliferation after renal ischemia-reperfusion injury, which might be useful to detect AKI [41]. Urine and plasma miR-21s might be used as a biomarker to predict AKI development [42]. According to another pilot study, other sets of miRNAs were altered several days before the increase in serum creatinine level. Therefore, they have the potential to be predictive AKI biomarkers among ICU patients [43]. Lorenzen *et al.* revealed that in AKI patients who required renal replacement therapy, elevated levels of plasma miR-210 were an independent predictor of mortality [44].

Galectin-3: Plasma galectin-3 levels are increased with impaired renal function. Galectin-3 is linked to incident renal disease and may play a role in interstitial fibrosis and progressive kidney disease. Studies have shown that galectin-3 is associated with the combined endpoint of cardiovascular events in patients with impaired kidney function [45].

Proenkephalin: Proenkephalin is a predictor of the worsening renal function in acute heart failure and the short- and long-term prognosis independent of renal function [46]. Also, proenkephalin levels are elevated in sepsis-induced AKI [46].

Glutathione S-Transferase (GST): This enzyme family consists of four classes that play a role in detoxification. The alpha and pi classes are the most studied ones in AKI. When renal tubular epithelial cells are damaged, GSTs become detectable in the urine. Shu *et al.* pointed out that urinary pi-GST levels might be used to predict advanced AKI or hospital mortality in patients who underwent cardiovascular surgery [47].

Calprotectin: Calprotectin is an antimicrobial protein in the cytoplasm of neutrophil granulocytes [48]. According to the unilateral ureteral obstruction model of AKI, the monomers of calprotectin S100A8 and S100A9 are produced by the epithelial cells of the renal collecting duct [48]. Following ischemic reperfusion injury in mice, S100A8 and S100A9 are produced by neutrophils that infiltrate the injured kidney [48].

Urinary Angiotensinogen (UAGT): Urinary angiotensinogen is converted to angiotensin I by renin and converted to angiotensin II by the angiotensin-converting enzyme [49]. According to studies performed on animal models, activation of the renin-angiotensin system has a role in the pathogenesis of AKI. Increased levels of urine AGT is considered as a biomarker of AKI progression in decompensated heart failure patients [49, 50]. Yang *et al.* pointed out that urine AGT is a better predictor of AKI rather than NGAL [50].

Clinical Application of Biomarkers

There is a plethora of current and potentially promising applications of biomarkers in the setting of AKI. Biomarkers might be used to predict the development of AKI, to differentiate different types of AKI -such as ischemic and nephrotoxic AKI-, to differentiate the diagnosis of acute tubular injury (ATI) from prerenal azotemia and hepatorenal syndrome, to guide the management decisions in cardiorenal syndrome, and to define sepsis-related and drug-induced AKI [3]. Additionally, the outcome and the prognosis of AKI might be defined and renal recovery might be predicted using biomarkers.

Differentiating between Prerenal Azotemia and Acute Tubular Injury

Prerenal azotemia (PRA) is a clinical condition in which renal functions decrease, but no structural disorder occurs in the kidney. The condition can be treated with intravenous fluid resuscitation. Unlike PRA, ATI is a condition where the kidneys have structural disorders. The damaged tubules in ATI cause inflammatory responses. Subsequently, determining the proteins specific to the injured tubule or the markers of the kidney-specific inflammation in the urine would help differentiate ATI from PRA. IL-18 is a mediator of tubular apoptosis and its urine concentration, and is significantly higher in patients with ATI than patients with PRA, transient AKI, or urinary tract infections [32, 51]. In addition to IL-18, uNGAL is released from the renal tubules in response to injury. Compared to creatinine, uNGAL is a more specific marker of renal tubular injury because it is not affected by the use of diuretics and volume depletion. Moreover, uNGAL levels increase before creatinine [5]. It has also been shown that NGAL measured in the emergency department reliably distinguishes ATI from PRA [52]. Using adjudicated cases, NGAL levels of 104 μg/l were able to differentiate ATI from PRA with a likelihood ratio of 5.97 [53].

Likewise, it has been pointed out that the KIM-1 protein is broadly expressed in the kidney biopsies of patients with ATI and that KIM-1 concentration in the urine can distinguish ATI from other causes of AKI [3]. The use of IL-18, NGAL, LFABP, and KIM-1 would likely be able to differentiate between PRA and ATI and eventually help patient management [3].

Seibert *et al.* performed a multicenter study to evaluate the diagnostic accuracy of calprotectin to differentiate prerenal and intrinsic acute allograft kidney injury and showed that urine calprotectin levels were higher in the patients with intrinsic AKI [54].

Early Detection and Prognostic Stratification of AKI in Cardiac Surgery Patients

Cardiac surgery is one of the major causes of ischemic AKI in hospitalized patients and is associated with increased mortality. Serum creatinine levels increase with a delay of 48 to 72 hours following kidney injury. Therefore, its effectiveness in early detection and intervention of AKI is scant. The use of biomarkers can improve identifying the events of kidney injury and risk stratification [55]. It has been shown that the AKI prediction *via* renal tubular injury biomarkers *i.e.*, IL-18, NGAL, and KIM-1 is earlier than monitoring the serum creatinine increase in the first six hours of post-cardiac surgery. The higher values of serum IL-18 within six hours following surgery were associated with an increased risk of AKI, compared to the lower levels [56]. Comparable findings were reported for plasma NGAL (pNGAL) and urinary KIM-1 [56].

According to several preclinical studies, following ischemic kidney injury, creatinine levels might regress to normal values, however, inflammation, renal fibrosis, and abnormal kidney gene expression profiles might remain [57]. In line with this data, tubular injury biomarkers were also associated with the severity and progression of AKI in post-cardiac surgery. High values of urinary IL-18 were used to predict the patients who advanced from the early stage of AKI to more severe stages and were associated with a longer (more than one week) duration of AKI [58, 59]. Similarly, high values of pNGAL were related to progressive AKI [59], while high values of urinary KIM-1 were associated with a longer duration of AKI [59]. Similar associations were reported for L-FABP [59].

Long-term mortality following cardiac surgery might also be predicted with the use of these biomarkers. For example, the highest tertile of perioperative urinary KIM-1 was associated with increased mortality over a median follow-up time of 3.0 years, both in those who did and did not develop AKI. Parallel findings were also described for IL-18 [60]. These studies indicate that the regression of the serum creatinine values to baseline levels may have overlooked residual injury and fibrosis of the kidney. Moreover, residual injury and subclinical fibrosis might be associated with increased morbidity and mortality. The Health, Aging, and Body Composition (Health ABC) Study indicates that in elderly subjects with a normal baseline kidney function, tubular damage biomarkers such as α1M, PIIINP, and NGAL are associated with cardiovascular disease and mortality [61].

Differential Diagnosis of AKI in Patients with Cardiorenal and Hepatorenal Syndromes

Cardiorenal and hepatorenal syndromes (CRS and HRS) cause kidney injury through several different mechanisms. First, renal perfusion might be decreased because of cardiac dysfunction or cirrhosis; both decreased forward flow and increased venous congestion may be present in those patients. Additionally, comorbidities and treatments might also cause an increase in creatinine levels [10]. Biomarkers of tubular injury might help to differentiate these distinct pathological processes.

In patients with chronic heart failure, even with preserved GFR, uNGAL levels are increased compared to healthy controls, and urinary KIM-1 and N-acetyl--D-glucosaminidase levels are elevated [62]. This finding might indicate the possibility of tubular damage beyond that captured by creatinine. Therefore, low levels of NGAL, N-acetyl-beta-D-glucosaminidase, and KIM-1 might help to make the decision to continue with diuretic treatment in patients with increasing creatinine levels.

In patients with CRS, aggressive diuretic therapy is the mainstay of treatment in reducing congestion and restoring cardiovascular hemodynamics; however, this therapy might be associated with the elevations in serum creatinine level [63]. Yet, it should be noted that a rise in the serum creatinine level is not always related to intrinsic kidney damage and adverse outcomes. For instance, Dupont *et al.* showed that uNGAL levels did not rise significantly in patients with acute decompensated heart failure who developed AKI after diuretic therapy [64]. In another study that was performed with 283 patients who were receiving aggressive diuretic therapy, increases in creatinine levels were not associated with the elevations in the biomarkers of tubular injuries, such as NGAL and KIM-1, nor with adverse outcomes [65]. These findings suggest that an increase in creatinine level may reflect the clinically benign alterations in infiltration and thus may not require discontinuation of valuable therapies [65].

Likewise, biomarkers might be used to distinguish HRS and ATI in patients with cirrhosis. The differential diagnosis of HRS and intrinsic tubular injury is vital since their managements differ significantly [66].

Biomarkers of tubular health may differentiate the causes of AKI and permit faster diagnoses and correct treatments. The levels of uNGAL, IL-18, and L-FABP were found to differentiate between ATI and other causes of kidney impairment in cirrhosis patients, with the highest levels to be observed in patients with ATI, followed by HRS and PRA, respectively [67 - 69]. A combination of numerous biomarkers was also assessed in identifying the type of AKI in patients with cirrhosis. The likelihood of being diagnosed with ATI increases in a stepwise fashion by the number of biomarkers above the cut-off value [70]. Biomarker levels also led to an increase in the post-test probability of having ATI; for instance, having an NGAL level of 0.365 ng/ml raised the pre-test probability of 40% to a post-test probability of 76% [69].

Defining Prognosis of AKI in Deceased Donor Kidney Transplantation

Acute kidney injury due to reperfusion injury is frequently observed following deceased donor kidney transplantation and is associated with higher rejection rates [10]. Because of the scarcity of kidneys for transplantation, the use of all suitable grafts is vital. Therefore, kidney injury and repair biomarkers are studied in the deceased donor's urine to define which donor AKI kidneys will have poor or favorable outcomes.

According to a study performed in deceased donors with AKI, elevated urinary levels of NGAL, KIM-1, IL-18, and L-FABP were significantly associated with donor AKI but not with delayed graft function [70]. Additionally, it was demonstrated in the same study that in recipients with delayed graft function six months after transplantation, the levels of kidney injury biomarkers were correlated with eGFRs [70]. Ischemic preconditioning, associated with the upregulation of anti-ischemic mediators, might be the underlying mechanism for the protective effects of the intrinsic kidney injury from delayed graft function. In conclusion, tubular injury biomarkers provided a moderate incremental value in predicting the worse recipient eGFR six months postoperatively.

Chitinase-3-like protein 1 (YKL 40) is a glycoprotein associated with adaptive repair following kidney injury. The protein might demonstrate cytoprotective effects in the setting of kidney transplantation [72]. Higher levels of this biomarker were associated with fewer delayed graft functions and a higher long-term eGFR. According to a study performed by Reese *et al.*, deceased kidney donors who had the highest YKL-40 levels had higher eGFRs at the sixth postoperative month when compared to recipients whose donors had the lowest YKL-40 levels. The predictive value of YKL-40 holds following the correction for donor and recipient factors and the degree of kidney injury. Additionally, kidney recipients with the highest donor YKL-40 levels also had a 50% lower hazard of graft loss [71]. Recovery markers such as YKL-40 might be helpful for a better assessment of deceased donor kidneys. By this approach, discard rates might decrease and more kidneys might be provided to waitlist patients [3].

Drug-induced AKI

Drug-induced kidney injury is a common cause of AKI; patients in the ICU possess more risk for this condition. Urinary biomarkers of renal tubular dysfunction may detect nephrotoxicity and predict AKI in the early period [72]. Recently, Da *et al.* serially measured a panel of urinary protein biomarkers to predict drug-induced AKI in patients using aminoglycosides, vancomycin, amphotericin, or calcineurin inhibitors. Their biomarker panel consisted of clusterin, beta-2-microglobulin, KIM-1, MCP-1, cystatin C, trefoil factor 3, NGAL, IL-18, GST-Pi, calbindin, and osteopontin [72]. The authors showed that the levels of clusterin, beta-2-microglobulin, KIM-1, MCP-1, and cystatin C biomarkers were raised 1 to 3 days before developing AKI. The average accuracy for predicting AKI was the highest for the average value for clusterin [72].

Moreover, minimizing its side effects remains one of the main difficulties in introducing a new drug to market. In the last few decades, many groups have worked on this problem and performed extensive omics studies to identify specific biomarkers for nephrotoxicity and discover the underlying mechanism of different kidney injury mechanisms [73]. Omics technologies are now accepted to be an essential methodological part of systems or integrative biology, by which it is possible to observe global changes of transcripts (genomics), proteins (proteomics), or metabolites (metabolomics) in cellular systems or tissues. Initially, rats were used as a model system, which led to the use of a large variety of biomarkers in preclinical studies [74].

These new developments led to identifying urinary protein biomarkers, including KIM-1, clusterin, osteopontin or RPA-1, and other transcriptional biomarkers that enable early detection of AKI and deliver further information about the extent of the nephron damage or the underlying mechanism [75].

Sepsis-induced AKI

Sepsis is the primary cause of AKI in critically ill patients. More than half of the AKIs in ICU patients are due to sepsis and are associated with high in-hospital mortality [76]. Recent studies showed that several biomarkers might be used to predict sepsis-associated AKI, including a soluble form of the triggering receptor expressed on myeloid cells 1(sTREM-1), NGAL, cell cycle arrest proteins, KIM-1, and netrin-1 [7]. sTREM-1 is a promising biomarker of sepsis-induced AKI. In 2011, Su *et al.* described the role of urine sTREM-1 in predicting sepsis-associated renal injury for the first time [77]. In 2015, a clinical trial by Dai *et al.* suggested that both serum and urine sTREM-1 levels have diagnostic usefulness for sepsis-induced AKI [78].

A meta-analysis that included 1,263 sepsis patients concluded that the sensitivity and specificity of uNGAL for diagnosing sepsis-associated AKI were both 0.80 [79]. A prospective study suggested that uNGAL was a better predictor than pNGAL in patients with sepsis [80]. A multicenter prospective study of 232 sepsis patients hospitalized at ICUs revealed that the urine test of TIMP-2 and IGFBP7 might be used to predict the development of AKI in septic patients, with an AUC of 0.84 in a receiver operating characteristic analysis [81]. The association of TIMP-2 and IGFBP7 with AKI development was confirmed by the results of two clinical trials that examined more than 500 critically ill patients. A prospective clinical study of 150 patients with sepsis showed that the early diagnosis of sepsis-induced AKI using the urine KIM-1 level measured 24 hours after admission had an AUC of 0.912 and that the AUC of urinary netrin-1 was as high as 0.858 [82].

Prognostication of AKI Outcomes

Risk stratification for the need for dialysis or death in severe AKI cases can be made using biomarkers. Additionally, some biomarkers might be used to predict AKI progression regardless of GFR changes [7]. For instance, high urine IL-18 level >100 pg/mL has been associated with the development of AKI within the first 24 hours and also with increased mortality [55]. Another study revealed that urine KIM-1 levels were correlated with AKI progression in hospitalized patients [83].

In patients who had Stage 1 AKI after cardiac surgery, a pNGAL level >323 ng/mL and a urine IL-18 level >185 pg/mL showed an odds ratio (OR) of 7.72 and 3.0, respectively, in predicting the progression of AKI [55]. Alge *et al.* showed that a high urine AGT-to-creatinine ratio had been associated with the progression of AKI and increased mortality [84]. Likewise, increased levels of IL-6, IL-8, IL-10, IL-18, TNFR1, macrophage migration inhibitory factor, and death receptor 5 were associated with higher mortality rates [85].

TIMP-2 and IGFBP7, two FDA-approved biomarkers, are useful in predicting the prognosis in patients in IUC. According to Joyner *et al.*, urine, TIMP-2, and IGFBP7 levels at ICU admission were associated with combined long-term outcome of death or RRT requirement during the next nine months in patients who developed AKI, where a [TIMP-2] × [IGFBP7] level >2.0 was associated with a HR of 2.16 [86]. Other studies have also examined the usefulness of TIMP-2 and IGFBP7 in cardiac and major surgery patients [87, 88] nevertheless, there is still no robust data on the performance of TIMP-2 and IGFBP7 in diverse patient groups excluding the ICU or perioperative setting.

Combining the clinical models with the biomarkers increase the predictive ability [89]. Pike *et al.* combined plasma IL-8 with a four-variable clinical model that consisted of age, mean arterial blood pressure, mechanical ventilation, and bilirubin, and showed that the combination model had a better prediction for recovery of kidney function and death compared to the clinical model [89].

Chen *et al.* examined the predictive performance of urinary AGT, NGAL, IL-18, and KIM-1 levels in AKI progression and suggested that a higher urine AGT level was associated with a greater risk of AKI progression and that urine AGT was a better predictor than the other three biomarkers for AKI progression with death [49].

Biomarkers of Recovery Following AKI

Several studies examined the utility of biomarkers as a tool for predicting recovery from AKI. Srisawat *et al.* assessed the pNGAL and interleukin-6 (IL-6) as biomarkers for predicting recovery from AKI in a cohort of 181 patients and noticed that 93 patients experienced renal recovery and that the recovery group had a lower pNGAL concentration compared to the non-recovery groups. However, the two groups were similar in terms of plasma IL-6 levels [90]. Among 138 patients that survived, pNGAL predicted non-recovery with an AUC of 0.71 [90]. In addition, a cut-off level of 257 ng/mL was able to predict non-recovery with a sensitivity of 68% and a specificity of 75% [90].

Moon *et al.* [91] measured the uNGAL and cystatin C levels every two days for eight days in 66 AKI patients which were classified as recovery and non-recovery groups, with a similar numbers subjects in each group. The uNGAL levels were significantly different between the two groups from Day 0 until Day 8. The authors concluded that uNGAL was a superior marker of recovery from AKI than plasma creatinine and that uNGAL at Day 0 was a valuable predictor of renal recovery (AUC=0.781). The authors also reported that for a cut-off value of 348.2 ng/ml, the sensitivity and specificity were 0.84 and 0.687, respectively.

Through the Biological Markers of Recovery for the Kidney (BioMaRK) study, different biomarkers including uNGAL, urinary hepatocyte growth factor (uHGF), urinary cystatin C (cystatin C), urinary IL-18 (uIL-18], uNGAL/matrix metalloproteinase protein 9, and urinary creatinine were assessed for their predictive value of renal recovery. The investigators collected urine samples on Days 1, 7, and 14 from 76 patients who developed AKI and required RRT. The patients who recovered had higher cystatin C levels on Day 1 and lower uHGF levels on Days 7 and 14. In the prediction of recovery, the AUC for uHGF on Day 14 was 0.74 while the AUC for uNGAL on Day 14 was 0.66 [90].

Another study examined urinary biomarkers that could predict AKI recovery by comparing 24 critically ill patients with early AKI recovery (<1 week) patients with late recovery (>1 week) or no recovery [92]. The eight candidate urinary biomarkers investigated in the study were: α1M, α-1 antitrypsin, apolipoprotein D, calreticulin, cathepsin D, CD59, IGFBP7, and NGAL. Of these eight biomarkers, only urinary IGFBP7 (uIGFBP-7) and uNGAL adequately predicted renal recovery (uIGFBP-7 AUC: 0.74; uNGAL AUC: 0.70) and were associated with the duration of AKI [92]. IGFBP7 was also found to be a more accurate predictor of renal outcome than uNGAL [92]. In line with this data, in a large multicenter study of cell cycle arrest with 728 participants with high uIGFBP-7 and metallopeptidase tissue inhibitor 2 (TIMP-2) levels, the multiplication of the two markers (urinary IGFBP*TIMP-2) doubled the risk of major adverse kidney events (death, RRT, or the persistence of renal dysfunction); all suggesting a role for this combination of biomarkers as predictors of recovery. Additionally, in patients undergoing cardiac surgery, the decline in the TIMP-2*IGFBP7 level between four and 24 hours after surgery was an accurate marker of renal recovery with an AUC of 0.79 [87].

LIMITATIONS AND FUTURE DIRECTIONS

Multiple AKI biomarkers, including uNGAL, KIM-1, IL-18, L-FABP, calprotectin, AGT, miRNAs, and the combination of TIMP-2 and IGFBP7 being studied and returned, encouraging results [48]. However, each of these biomarkers has its own advantages and limitations. Although no new biomarker has been unanimously accepted for routine use in clinical practice, some of the biomarkers are locally available (e.g., NGAL in Europe, L-FABP in Japan, [TIMP-2] × [IGFBP7] in the USA) [93]. Moreover, FDA approved KIM-1 for preclinical drug development [74].

There is still progress to be made before these biomarkers can be accepted for routine clinical use. None of the reported biomarkers are totally specific for AKI, and this is a critical limitation. For instance, NGAL, IL-18, and calprotectin levels are elevated in urinary tract infections and sepsis, regardless of AKI. Additionally, NGAL, KIM-1, and IL-18 levels are also elevated in patients with CKD [94]. TIMP-2 and IGFBP7 were mainly examined in critically ill patients with diabetes who were at ICU [95]. Following AKI development, elevated biomarker levels persist for some time; therefore, the timing of the AKI assessment is rather challenging [58].

Another weakness lies in comparing the biomarker performance to serum creatinine performance when serum creatinine defines the AKI event. To decide whether an increase in the biomarkers alone should allow for an intervention, functional and damage biomarkers should be independently compared with hard endpoints. The appropriate biomarker cut-offs could thus be identified [45]. Pickering and Endre proposed using the sensitivity of creatinine for dialysis requirement and the incidence of death to determine the sensitivity needed to validate a particular cut-off point for novel damage biomarkers of AKI [96]. However, such an approach might be used in studies with a high event rate or a large number of subjects.

Additionally, in clinical trials, the use of biomarkers to enroll patients in a study may lead to a higher screening failure rate and oblige screening with a greater number of subjects. Furthermore, patient recruitment based on biomarker levels may limit the generalizability of the findings [45].

Despite all those limitations, biomarkers have a great potential to improve patient management in nephrology *via* precision medicine. Specifically, in patients with AKI, there is a need for individualization of management approaches [10]. Since AKI is a syndrome, an individual set of clinical decisions and a personalized biomarker panel for each setting is needed. The Kidney Precision Medicine Project is an initiative started by the National Institutes of Health. This project aims to obtain and assess kidney biopsy specimens to create kidney tissue atlas and thus to create tissue-based definitions for renal injury and disease instead of a creatinine-based diagnosis for the disease [97]. Following the expectant transformation from creatinine-defined to tissue-defined renal injury, the field of nephrology will likely develop a deeper understanding of kidney injury biomarkers. This way, physicians might include kidney injury biomarkers into clinical practice more vigorously, which will help them better personalize patient care [98].

CONSENT FOR PUBLICATION

Not applicable.

CONFLICT OF INTEREST

The authors declare no conflicts of interest, financial or otherwise.

ACKNOWLEDGEMENT

Declared none.

REFERENCES

[1] Bellomo R, Kellum JA, Ronco C. Acute kidney injury. Lancet 2012; 380(9843): 756-66. [http://dx.doi.org/10.1016/S0140-6736(11)61454-2] [PMID: 22617274]

[2] Outcomes KDIG. KDIGO clinical practice guideline for acute kidney injury. Section 2: AKI Definition. Kidney Int Suppl 2012; 2: 19-36.

[3] Parikh CR, Mansour SG. Perspective on clinical application of biomarkers in AKI. J Am Soc Nephrol 2017; 28(6): 1677-85. [http://dx.doi.org/10.1681/ASN.2016101127] [PMID: 28220028]

[4] Ichai C, Vinsonneau C, Souweine B, *et al.* Acute kidney injury in the perioperative period and in intensive care units (excluding renal replacement therapies). Ann Intensive Care 2016; 6(1): 48. [http://dx.doi.org/10.1186/s13613-016-0145-5] [PMID: 27230984]

[5] Mishra J, Dent C, Tarabishi R, *et al.* Neutrophil gelatinase-associated lipocalin (NGAL) as a biomarker for acute renal injury after cardiac surgery. Lancet 2005; 365(9466): 1231-8. [http://dx.doi.org/10.1016/S0140-6736(05)74811-X] [PMID: 15811456]

[6] Bellomo R, Kellum JA, Ronco C. Defining acute renal failure: physiological principles. Intensive Care Med 2004; 30(1): 33-7. [http://dx.doi.org/10.1007/s00134-003-2078-3] [PMID: 14618231]

[7] Wang K, Xie S, Xiao K, Yan P, He W, Xie L. Biomarkers of sepsis-induced acute kidney injury. Biomed Res Int 2018; 6937947. [http://dx.doi.org/10.1155/2018/6937947]

[8] Ronco C, Rizo-Topete L, Serrano-Soto M, Kashani K. Pro: Prevention of acute kidney injury: time for teamwork and new biomarkers. Nephrol Dial Transplant 2017; 32(3): 408-13. [http://dx.doi.org/10.1093/ndt/gfx016] [PMID: 28375502]

[9] Menez S, Parikh CR. Assessing the health of the nephron in acute kidney injury: biomarkers of kidney function and injury. Curr Opin Nephrol Hypertens 2019; 28(6): 560-6. [http://dx.doi.org/10.1097/MNH.0000000000000538] [PMID: 31369422]

[10] Zhang WR, Parikh CR. Biomarkers of acute and chronic kidney disease. Annu Rev Physiol 2019; 81(1): 309-33. [http://dx.doi.org/10.1146/annurev-physiol-020518-114605] [PMID: 30742783]

[11] Ortega LM, Heung M. The use of cell cycle arrest biomarkers in the early detection of acute kidney injury. Is this the new renal troponin? Nefrologia (Engl Ed) 2018; 38(4): 361-7.

[12] Onopiuk A, Tokarzewicz A, Gorodkiewicz E. Cystatin C: a kidney function biomarker Advances in clinical chemistry 68. Elsevier 2015; pp. 57-69. [http://dx.doi.org/10.1016/bs.acc.2014.11.007]

[13] Togashi Y, Sakaguchi Y, Miyamoto M, Miyamoto Y. Urinary cystatin C as a biomarker for acute kidney injury and its immunohistochemical localization in kidney in the CDDP-treated rats. Exp Toxicol Pathol 2012; 64(7-8): 797-805. [http://dx.doi.org/10.1016/j.etp.2011.01.018] [PMID: 21377848]

[14] Itoh Y, Kawai T. Human α 1-microglobulin: its measurement and clinical significance. J Clin Lab Anal 1990; 4(5): 376-84. [http://dx.doi.org/10.1002/jcla.1860040511] [PMID: 1700091]

[15] Yamamoto T, Noiri E, Ono Y, *et al.* Renal L-type fatty acid--binding protein in acute ischemic injury. J Am Soc Nephrol 2007; 18(11): 2894-902. [http://dx.doi.org/10.1681/ASN.2007010097] [PMID: 17942962]

[16] Portilla D, Dent C, Sugaya T, *et al.* Liver fatty acid-binding protein as a biomarker of acute kidney injury after cardiac surgery. Kidney Int 2008; 73(4): 465-72. [http://dx.doi.org/10.1038/sj.ki.5002721] [PMID: 18094680]

[17] Parikh CR, Thiessen-Philbrook H, Garg AX, *et al.* Performance of kidney injury molecule-1 and liver fatty acid-binding protein and combined biomarkers of AKI after cardiac surgery. Clin J Am Soc Nephrol 2013; 8(7): 1079-88. [http://dx.doi.org/10.2215/CJN.10971012] [PMID: 23599408]

[18] Kamijo A, Sugaya T, Hikawa A, *et al.* Urinary liver-type fatty acid binding protein as a useful biomarker in chronic kidney disease. Mol Cell Biochem 2006; 284(1-2): 175-82. [http://dx.doi.org/10.1007/s11010-005-9047-9] [PMID: 16532260]

[19] Devuyst O, Olinger E, Rampoldi L. Uromodulin: from physiology to rare and complex kidney disorders. Nat Rev Nephrol 2017; 13(9): 525-44. [http://dx.doi.org/10.1038/nrneph.2017.101] [PMID: 28781372]

[20] Steubl D, Block M, Herbst V, *et al.* Plasma uromodulin correlates with kidney function and identifies

early stages in chronic kidney disease patients. Medicine (Baltimore) 2016; 95(10): e3011. [http://dx.doi.org/10.1097/MD.0000000000003011] [PMID: 26962815]

[21] Wheeler DS, Devarajan P, Ma Q, *et al.* Serum neutrophil gelatinase-associated lipocalin (NGAL) as a marker of acute kidney injury in critically ill children with septic shock. Crit Care Med 2008; 36(4): 1297-303. [http://dx.doi.org/10.1097/CCM.0b013e318169245a] [PMID: 18379258]

[22] Supavekin S, Zhang W, Kucherlapati R, Kaskel FJ, Moore LC, Devarajan P. Differential gene expression following early renal ischemia/reperfusion. Kidney Int 2003; 63(5): 1714-24. [http://dx.doi.org/10.1046/j.1523-1755.2003.00928.x] [PMID: 12675847]

[23] Yuen PS, Jo S-K, Holly MK, Hu X, Star RA. Ischemic and nephrotoxic acute renal failure are distinguished by their broad transcriptomic responses. Physiol Genomics 2006; 25(3): 375-86. [http://dx.doi.org/10.1152/physiolgenomics.00223.2005] [PMID: 16507785]

[24] Ichimura T, Bonventre JV, Bailly V, *et al.* Kidney injury molecule-1 (KIM-1), a putative epithelial cell adhesion molecule containing a novel immunoglobulin domain, is up-regulated in renal cells after injury. J Biol Chem 1998; 273(7): 4135-42. [http://dx.doi.org/10.1074/jbc.273.7.4135] [PMID: 9461608]

[25] Hubank M, Schatz DG. Identifying differences in mRNA expression by representational difference analysis of cDNA. Nucleic Acids Res 1994; 22(25): 5640-8. [http://dx.doi.org/10.1093/nar/22.25.5640] [PMID: 7838717]

[26] Han WK, Bailly V, Abichandani R, Thadhani R, Bonventre JV. Kidney Injury Molecule-1 (KIM-1): a novel biomarker for human renal proximal tubule injury. Kidney Int 2002; 62(1): 237-44. [http://dx.doi.org/10.1046/j.1523-1755.2002.00433.x] [PMID: 12081583]

[27] Vaidya VS, Ramirez V, Ichimura T, Bobadilla NA, Bonventre JV. Urinary kidney injury molecule-1: a sensitive quantitative biomarker for early detection of kidney tubular injury. Am J Physiol Renal Physiol 2006; 290(2): F517-29. [http://dx.doi.org/10.1152/ajprenal.00291.2005] [PMID: 16174863]

[28] Ko GJ, Grigoryev DN, Linfert D, *et al.* Transcriptional analysis of kidneys during repair from AKI reveals possible roles for NGAL and KIM-1 as biomarkers of AKI-to-CKD transition. Am J Physiol Renal Physiol 2010; 298(6): F1472-83. [http://dx.doi.org/10.1152/ajprenal.00619.2009] [PMID: 20181666]

[29] Price PM, Safirstein RL, Megyesi J. The cell cycle and acute kidney injury. Kidney Int 2009; 76(6): 604-13. [http://dx.doi.org/10.1038/ki.2009.224] [PMID: 19536080]

[30] Boonstra J, Post JA. Molecular events associated with reactive oxygen species and cell cycle progression in mammalian cells. Gene 2004; 337: 1-13. [http://dx.doi.org/10.1016/j.gene.2004.04.032] [PMID: 15276197]

[31] Kashani K, Al-Khafaji A, Ardiles T, *et al.* Discovery and validation of cell cycle arrest biomarkers in human acute kidney injury. Crit Care 2013; 17(1): R25. [http://dx.doi.org/10.1186/cc12503] [PMID: 23388612]

[32] Parikh CR, Jani A, Melnikov VY, Faubel S, Edelstein CL. Urinary interleukin-18 is a marker of human acute tubular necrosis. Am J Kidney Dis 2004; 43(3): 405-14. [http://dx.doi.org/10.1053/j.ajkd.2003.10.040] [PMID: 14981598]

[33] Lochner M, Förster I. Anti-interleukin-18 therapy in murine models of inflammatory bowel disease. Pathobiology 2002-2003; 70(3): 164-9. [http://dx.doi.org/10.1159/000068149] [PMID: 12571421]

[34] Wu H, Craft ML, Wang P, *et al.* IL-18 contributes to renal damage after ischemia-reperfusion. J Am Soc Nephrol 2008; 19(12): 2331-41. [http://dx.doi.org/10.1681/ASN.2008020170] [PMID: 18815244]

[35] Gonzalez F, Vincent F. Biomarkers for acute kidney injury in critically ill patients. Minerva Anestesiol 2012; 78(12): 1394-403.
[PMID: 23032924]

[36] Munshi R, Johnson A, Siew ED, *et al.* MCP-1 gene activation marks acute kidney injury. J Am Soc Nephrol 2011; 22(1): 165-75.
[http://dx.doi.org/10.1681/ASN.2010060641] [PMID: 21071523]

[37] Moledina DG, Isguven S, McArthur E, *et al.* Plasma monocyte chemotactic protein-1 is associated with acute kidney injury and death after cardiac operations. Ann Thorac Surg 2017; 104(2): 613-20.
[http://dx.doi.org/10.1016/j.athoracsur.2016.11.036] [PMID: 28223055]

[38] Speeckaert MM, Speeckaert R, Laute M, Vanholder R, Delanghe JR. Tumor necrosis factor receptors: biology and therapeutic potential in kidney diseases. Am J Nephrol 2012; 36(3): 261-70.
[http://dx.doi.org/10.1159/000342333] [PMID: 22965073]

[39] Saulnier P-J, Gand E, Ragot S, *et al.* Association of serum concentration of TNFR1 with all-cause mortality in patients with type 2 diabetes and chronic kidney disease: follow-up of the SURDIAGENE Cohort. Diabetes Care 2014; 37(5): 1425-31.
[http://dx.doi.org/10.2337/dc13-2580] [PMID: 24623026]

[40] Schmidt IM, Hall IE, Kale S, *et al.* Chitinase-like protein Brp-39/YKL-40 modulates the renal response to ischemic injury and predicts delayed allograft function. J Am Soc Nephrol 2013; 24(2): 309-19.
[http://dx.doi.org/10.1681/ASN.2012060579] [PMID: 23291472]

[41] Godwin JG, Ge X, Stephan K, Jurisch A, Tullius SG, Iacomini J. Identification of a microRNA signature of renal ischemia reperfusion injury. Proc Natl Acad Sci USA 2010; 107(32): 14339-44.
[http://dx.doi.org/10.1073/pnas.0912701107] [PMID: 20651252]

[42] Du J, Cao X, Zou L, *et al.* MicroRNA-21 and risk of severe acute kidney injury and poor outcomes after adult cardiac surgery. PLoS One 2013; 8(5): e63390.
[http://dx.doi.org/10.1371/journal.pone.0063390] [PMID: 23717419]

[43] Aguado-Fraile E, Ramos E, Conde E, *et al.* A pilot study identifying a set of microRNAs as precise diagnostic biomarkers of acute kidney injury. PLoS One 2015; 10(6): e0127175.
[http://dx.doi.org/10.1371/journal.pone.0127175] [PMID: 26079930]

[44] Lorenzen JM, Kielstein JT, Hafer C, *et al.* Circulating miR-210 predicts survival in critically ill patients with acute kidney injury. Clin J Am Soc Nephrol 2011; 6(7): 1540-6.
[http://dx.doi.org/10.2215/CJN.00430111] [PMID: 21700819]

[45] Teo SH, Endre ZH. Biomarkers in acute kidney injury (AKI). Best Pract Res Clin Anaesthesiol 2017; 31(3): 331-44.
[http://dx.doi.org/10.1016/j.bpa.2017.10.003] [PMID: 29248140]

[46] Ng LL, Squire IB, Jones DJL, *et al.* Proenkephalin, renal dysfunction, and prognosis in patients with acute heart failure: a GREAT network study. J Am Coll Cardiol 2017; 69(1): 56-69.
[http://dx.doi.org/10.1016/j.jacc.2016.10.038] [PMID: 28057251]

[47] Shu K-H, Wang C-H, Wu C-H, *et al.* Urinary π-glutathione S-transferase predicts advanced acute kidney injury following cardiovascular surgery. Sci Rep 2016; 6(1): 26335.
[http://dx.doi.org/10.1038/srep26335] [PMID: 27527370]

[48] Kashani K, Cheungpasitporn W, Ronco C. Biomarkers of acute kidney injury: the pathway from discovery to clinical adoption. Clin Chem Lab Med 2017; 55(8): 1074-89.
[http://dx.doi.org/10.1515/cclm-2016-0973] [PMID: 28076311]

[49] Chen C, Yang X, Lei Y, *et al.* Urinary biomarkers at the time of AKI diagnosis as predictors of progression of AKI among patients with acute cardiorenal syndrome. Clin J Am Soc Nephrol 2016; 11(9): 1536-44.
[http://dx.doi.org/10.2215/CJN.00910116] [PMID: 27538426]

[50] Yang X, Chen C, Tian J, *et al.* Urinary angiotensinogen level predicts AKI in acute decompensated heart failure: a prospective, two-stage study. J Am Soc Nephrol 2015; 26(8): 2032-41. [http://dx.doi.org/10.1681/ASN.2014040408] [PMID: 25722365]

[51] Nejat M, Pickering JW, Devarajan P, *et al.* Some biomarkers of acute kidney injury are increased in pre-renal acute injury. Kidney Int 2012; 81(12): 1254-62. [http://dx.doi.org/10.1038/ki.2012.23] [PMID: 22418979]

[52] Parikh CR, Butrymowicz I, Yu A, *et al.* Urine stability studies for novel biomarkers of acute kidney injury. Am J Kidney Dis 2014; 63(4): 567-72. [http://dx.doi.org/10.1053/j.ajkd.2013.09.013] [PMID: 24200462]

[53] Singer E, Elger A, Elitok S, *et al.* Urinary neutrophil gelatinase-associated lipocalin distinguishes pre-renal from intrinsic renal failure and predicts outcomes. Kidney Int 2011; 80(4): 405-14. [http://dx.doi.org/10.1038/ki.2011.41] [PMID: 21412214]

[54] Seibert FS, Rosenberger C, Mathia S, *et al.* Urinary calprotectin differentiates between prerenal and intrinsic acute renal allograft failure. Transplantation 2017; 101(2): 387-94. [http://dx.doi.org/10.1097/TP.0000000000001124] [PMID: 26901081]

[55] Koyner JL, Parikh CR. Clinical utility of biomarkers of AKI in cardiac surgery and critical illness. Clin J Am Soc Nephrol 2013; 8(6): 1034-42. [http://dx.doi.org/10.2215/CJN.05150512] [PMID: 23471130]

[56] Parikh CR, Coca SG, Thiessen-Philbrook H, *et al.* Postoperative biomarkers predict acute kidney injury and poor outcomes after adult cardiac surgery. J Am Soc Nephrol 2011; 22(9): 1748-57. [http://dx.doi.org/10.1681/ASN.2010121302] [PMID: 21836143]

[57] Basile DP, Donohoe D, Roethe K, Osborn JL. Renal ischemic injury results in permanent damage to peritubular capillaries and influences long-term function. Am J Physiol Renal Physiol 2001; 281(5): F887-99. [http://dx.doi.org/10.1152/ajprenal.00050.2001] [PMID: 11592947]

[58] Koyner JL, Garg AX, Coca SG, *et al.* Biomarkers predict progression of acute kidney injury after cardiac surgery. J Am Soc Nephrol 2012; 23(5): 905-14. [http://dx.doi.org/10.1681/ASN.2011090907] [PMID: 22383693]

[59] Coca SG, Nadkarni GN, Garg AX, *et al.* First post-operative urinary kidney injury biomarkers and association with the duration of AKI in the TRIBE-AKI cohort. PLoS One 2016; 11(8): e0161098. [http://dx.doi.org/10.1371/journal.pone.0161098] [PMID: 27537050]

[60] Coca SG, Garg AX, Thiessen-Philbrook H, *et al.* Urinary biomarkers of AKI and mortality 3 years after cardiac surgery. J Am Soc Nephrol 2014; 25(5): 1063-71. [http://dx.doi.org/10.1681/ASN.2013070742] [PMID: 24357673]

[61] Jotwani V, Katz R, Ix JH, *et al.* Urinary biomarkers of kidney tubular damage and risk of cardiovascular disease and mortality in elders. Am J Kidney Dis 2018; 72(2): 205-13. [http://dx.doi.org/10.1053/j.ajkd.2017.12.013] [PMID: 29602632]

[62] Damman K, van Veldhuisen DJ, Navis G, Voors AA, Hillege HL. Urinary neutrophil gelatinase associated lipocalin (NGAL), a marker of tubular damage, is increased in patients with chronic heart failure. Eur J Heart Fail 2008; 10(10): 997-1000. [http://dx.doi.org/10.1016/j.ejheart.2008.07.001] [PMID: 18804416]

[63] Weinfeld MS, Chertow GM, Stevenson LW. Aggravated renal dysfunction during intensive therapy for advanced chronic heart failure. Am Heart J 1999; 138(2 Pt 1): 285-90. [http://dx.doi.org/10.1016/S0002-8703(99)70113-4] [PMID: 10426840]

[64] Dupont M, Shrestha K, Singh D, *et al.* Lack of significant renal tubular injury despite acute kidney injury in acute decompensated heart failure. Eur J Heart Fail 2012; 14(6): 597-604. [http://dx.doi.org/10.1093/eurjhf/hfs039] [PMID: 22505396]

[65] Ahmad T, Jackson K, Rao VS, *et al.* Worsening renal function in patients with acute heart failure undergoing aggressive diuresis is not associated with tubular injury. Circulation 2018; 137(19): 2016-28.
[http://dx.doi.org/10.1161/CIRCULATIONAHA.117.030112] [PMID: 29352071]

[66] Velez JCQ, Kadian M, Taburyanskaya M, *et al.* Hepatorenal acute kidney injury and the importance of raising mean arterial pressure. Nephron 2015; 131(3): 191-201.
[http://dx.doi.org/10.1159/000441151] [PMID: 26485256]

[67] Puthumana J, Ariza X, Belcher JM, Graupera I, Ginès P, Parikh CR. Urine interleukin 18 and lipocalin 2 are biomarkers of acute tubular necrosis in patients with cirrhosis: a systematic review and meta-analysis. Clin Gastroenterol Hepatol 2017; 15(7): 1003-3.
[http://dx.doi.org/10.1016/j.cgh.2016.11.035]

[68] Dobre M, Demirjian S, Sehgal AR, Navaneethan SD. Terlipressin in hepatorenal syndrome: a systematic review and meta-analysis. Int Urol Nephrol 2011; 43(1): 175-84.
[http://dx.doi.org/10.1007/s11255-010-9725-8] [PMID: 20306131]

[69] Belcher JM, Sanyal AJ, Peixoto AJ, *et al.* Kidney biomarkers and differential diagnosis of patients with cirrhosis and acute kidney injury. Hepatology 2014; 60(2): 622-32.
[http://dx.doi.org/10.1002/hep.26980] [PMID: 24375576]

[70] Reese PP, Hall IE, Weng FL, *et al.* Associations between deceased-donor urine injury biomarkers and kidney transplant outcomes. J Am Soc Nephrol 2016; 27(5): 1534-43.
[http://dx.doi.org/10.1681/ASN.2015040345] [PMID: 26374609]

[71] Puthumana J, Hall IE, Reese PP, *et al.* YKL-40 associates with renal recovery in deceased donor kidney transplantation. J Am Soc Nephrol 2017; 28(2): 661-70.
[http://dx.doi.org/10.1681/ASN.2016010091] [PMID: 27451287]

[72] Da Y, Akalya K, Murali T, *et al.* Serial Quantification of urinary protein biomarkers to predict drug-induced acute kidney injury. Curr Drug Metab 2019; 20(8): 656-64.
[http://dx.doi.org/10.2174/1389200220666190711114504] [PMID: 31296157]

[73] Blank M, Thompson A, Hausner E, Rouse R. Biomarkers of drug-induced acute kidney injury: a regulatory perspective. Expert Opin Drug Metab Toxicol 2018; 14(9): 929-36.
[http://dx.doi.org/10.1080/17425255.2018.1511701] [PMID: 30099912]

[74] Dieterle F, Sistare F, Goodsaid F, *et al.* Renal biomarker qualification submission: a dialog between the FDA-EMEA and Predictive Safety Testing Consortium. Nat Biotechnol 2010; 28(5): 455-62.
[http://dx.doi.org/10.1038/nbt.1625] [PMID: 20458315]

[75] Fuchs TC, Hewitt P. Biomarkers for drug-induced renal damage and nephrotoxicity-an overview for applied toxicology. AAPS J 2011; 13(4): 615-31.
[http://dx.doi.org/10.1208/s12248-011-9301-x] [PMID: 21969220]

[76] Uchino S, Kellum JA, Bellomo R, *et al.* Acute renal failure in critically ill patients: a multinational, multicenter study. JAMA 2005; 294(7): 813-8.
[http://dx.doi.org/10.1001/jama.294.7.813] [PMID: 16106006]

[77] Su LX, Feng L, Zhang J, *et al.* Diagnostic value of urine sTREM-1 for sepsis and relevant acute kidney injuries: a prospective study. Crit Care 2011; 15(5): R250.
[http://dx.doi.org/10.1186/cc10508] [PMID: 22023777]

[78] Dai X, Zeng Z, Fu C, Zhang S, Cai Y, Chen Z. Diagnostic value of neutrophil gelatinase-associated lipocalin, cystatin C, and soluble triggering receptor expressed on myeloid cells-1 in critically ill patients with sepsis-associated acute kidney injury. Crit Care 2015; 19(1): 223.
[http://dx.doi.org/10.1186/s13054-015-0941-6] [PMID: 25944130]

[79] Böttiger BW, Bernhard M, Knapp J, Nagele P. Influence of EMS-physician presence on survival after out-of-hospital cardiopulmonary resuscitation: systematic review and meta-analysis. Crit Care 2016; 20(1): 4.

[http://dx.doi.org/10.1186/s13054-015-1156-6] [PMID: 26747085]

[80] Mårtensson J, Bell M, Oldner A, Xu S, Venge P, Martling C-R. Neutrophil gelatinase-associated lipocalin in adult septic patients with and without acute kidney injury. Intensive Care Med 2010; 36(8): 1333-40.
[http://dx.doi.org/10.1007/s00134-010-1887-4] [PMID: 20397003]

[81] Honore PM, Nguyen HB, Gong M, *et al.* Urinary tissue inhibitor of metalloproteinase-2 and insulin-like growth factor-binding protein 7 for risk stratification of acute kidney injury in patients with sepsis. Crit Care Med 2016; 44(10): 1851-60.
[http://dx.doi.org/10.1097/CCM.0000000000001827] [PMID: 27355527]

[82] Tu Y, Wang H, Sun R, *et al.* Urinary netrin-1 and KIM-1 as early biomarkers for septic acute kidney injury. Ren Fail 2014; 36(10): 1559-63.
[http://dx.doi.org/10.3109/0886022X.2014.949764] [PMID: 25154466]

[83] Xie Y, Wang Q, Wang C, Qi C, Ni Z, Mou S. High urinary excretion of kidney injury molecule-1 predicts adverse outcomes in acute kidney injury: a case control study. Crit Care 2016; 20(1): 286.
[http://dx.doi.org/10.1186/s13054-016-1455-6] [PMID: 27613644]

[84] Alge JL, Karakala N, Neely BA, *et al.* Urinary angiotensinogen and risk of severe AKI. Clin J Am Soc Nephrol 2013; 8(2): 184-93.
[http://dx.doi.org/10.2215/CJN.06280612] [PMID: 23143504]

[85] Murugan R, Wen X, Shah N, *et al.* Plasma inflammatory and apoptosis markers are associated with dialysis dependence and death among critically ill patients receiving renal replacement therapy. Nephrol Dial Transplant 2014; 29(10): 1854-64.
[http://dx.doi.org/10.1093/ndt/gfu051] [PMID: 24619058]

[86] Koyner JL, Shaw AD, Chawla LS, *et al.* Tissue inhibitor metalloproteinase-2 (TIMP-2)□ IGF-binding protein-7 (IGFBP7) levels are associated with adverse long-term outcomes in patients with AKI. J Am Soc Nephrol 2015; 26(7): 1747-54.
[http://dx.doi.org/10.1681/ASN.2014060556] [PMID: 25535301]

[87] Meersch M, Schmidt C, Van Aken H, *et al.* Urinary TIMP-2 and IGFBP7 as early biomarkers of acute kidney injury and renal recovery following cardiac surgery. PLoS One 2014; 9(3): e93460.
[http://dx.doi.org/10.1371/journal.pone.0093460] [PMID: 24675717]

[88] Gunnerson KJ, Shaw AD, Chawla LS, *et al.* TIMP2•IGFBP7 biomarker panel accurately predicts acute kidney injury in high-risk surgical patients. J Trauma Acute Care Surg 2016; 80(2): 243-9.
[http://dx.doi.org/10.1097/TA.0000000000000912] [PMID: 26816218]

[89] Pike F, Murugan R, Keener C, *et al.* Biomarker enhanced risk prediction for adverse outcomes in critically ill patients receiving RRT. Clin J Am Soc Nephrol 2015; 10(8): 1332-9.
[http://dx.doi.org/10.2215/CJN.09911014] [PMID: 26048891]

[90] Srisawat N, Murugan R, Lee M, *et al.* Plasma neutrophil gelatinase-associated lipocalin predicts recovery from acute kidney injury following community-acquired pneumonia. Kidney Int 2011; 80(5): 545-52.
[http://dx.doi.org/10.1038/ki.2011.160] [PMID: 21677634]

[91] Moon SJ, Park HB, Yoon SY, Lee SC. Urinary biomarkers for early detection of recovery in patients with acute kidney injury. J Korean Med Sci 2013; 28(8): 1181-6.
[http://dx.doi.org/10.3346/jkms.2013.28.8.1181] [PMID: 23960445]

[92] Aregger F, Uehlinger DE, Witowski J, *et al.* Identification of IGFBP-7 by urinary proteomics as a novel prognostic marker in early acute kidney injury. Kidney Int 2014; 85(4): 909-19.
[http://dx.doi.org/10.1038/ki.2013.363] [PMID: 24067438]

[93] Pickering JW, Endre ZH. Bench to bedside: the next steps for biomarkers in acute kidney injury. Am J Physiol Renal Physiol 2016; 311(4): F717-21.
[http://dx.doi.org/10.1152/ajprenal.00268.2016] [PMID: 27465990]

[94] Schrezenmeier EV, Barasch J, Budde K, Westhoff T, Schmidt-Ott KM. Biomarkers in acute kidney injury - pathophysiological basis and clinical performance. Acta Physiol (Oxf) 2017; 219(3): 554-72. [http://dx.doi.org/10.1111/apha.12764] [PMID: 27474473]

[95] Bell M, Larsson A, Venge P, Bellomo R, Mårtensson J. Assessment of cell-cycle arrest biomarkers to predict early and delayed acute kidney injury. Dis Markers 2015; 2015: 158658. [http://dx.doi.org/10.1155/2015/158658]

[96] Pickering JW, Endre ZH. Linking injury to outcome in acute kidney injury: a matter of sensitivity. PLoS One 2013; 8(4): e62691. [http://dx.doi.org/10.1371/journal.pone.0062691] [PMID: 23626850]

[97] Basile DP, Fredrich K, Alausa M, *et al.* Identification of persistently altered gene expression in the kidney after functional recovery from ischemic acute renal failure. Am J Physiol Renal Physiol 2005; 288(5): F953-63. [http://dx.doi.org/10.1152/ajprenal.00329.2004] [PMID: 15632414]

[98] Emlet DR, Pastor-Soler N, Marciszyn A, *et al.* Insulin-like growth factor binding protein 7 and tissue inhibitor of metalloproteinases-2: differential expression and secretion in human kidney tubule cells. Am J Physiol Renal Physiol 2017; 312(2): F284-96. [http://dx.doi.org/10.1152/ajprenal.00271.2016] [PMID: 28003188]

CHAPTER 20

Biomarkers to Predict Sudden Cardiac Death

Ali Uğur Soysal[1,*] and **Kivanc Yalin**[1]

[1] *Department of Cardiology, Cerrahpaşa Faculty of Medicine, Istanbul University-Cerrahpaşa, Istanbul, Turkey*

Abstract: Sudden cardiac death (SCD) is a common disorder and an unsolved issue for health care providers. Despite several risk factors for SCD, some cases experience SCD as an initial presentation of cardiovascular disease. Prediction of individuals at increased risk for SCD is important for implementing community-based approaches and individual-based therapies with high costs, such as implantable defibrillators. This chapter discusses the potential role of biomarkers in predicting SCD in different cardiovascular diseases.

Keywords: Biomarkers, Cardiovascular disease, Health care providers, Implantable defibrillators, Risk factors, Sudden cardiac death (SCD).

INTRODUCTION

Sudden cardiac death (SCD) is defined as the unexpected death of a person from the loss of cardiac function either within one hour of medical observation or 24 hours without medical observation. Cardiovascular diseases are responsible for 17 million deaths every year, and 25% of these are SCD. Males are currently at higher risk in terms of cardiovascular disease. SCD affects 6.68 out of every 100,000 males and 1.4 out of every 100,000 females. The highest prevalence of SCD is amongst individuals with a history of cardiac arrest or myocardial infarction, and this SCD risk is at the highest level between the 6th and 18th months after the initial cardiac event. There are also racial differences observed among the SCD cases; the black race is at more risk than the white race. SCD occurs mostly outside of the hospital or in the emergency rooms. There are different mechanisms identified that lead to SCD. Physical fragility of the atherosclerotic plaque, thrombus formation due to this, and unstable electrical conductivity are amongst the reasons for SCD in individuals with coronary atherosclerosis. Arrhythmic events can also be accelerated by hypokalemia and

* **Corresponding author Ali Uğur Soysal:** Department of Cardiology, Cerrahpaşa Faculty of Medicine, Istanbul University-Cerrahpaşa, Istanbul, Turkey; Tel: +90 (212) 414 30 00; E-mail: aliugursoysal993@gmail.com

Pınar Atukeren and Hafize Uzun (Eds.)

certain drugs that can alter QT time. Genetic causes of arrhythmogenic cardiac death, such as channelopathies, are amongst the common reasons behind SCD in relatively young population [1 - 5]. This review aims to discuss possible risk factors, especially biomarkers, among different cardiovascular substrates.

SCD in the Setting of an Acute Coronary Syndrome, Ischemic Cardiomyopathy

Sudden death is mostly due to cardiac causes and is mainly seen in patients with heart disease secondary to ischemia. Sudden death due to cardiac causes constitutes approximately 50% of all cardiovascular deaths. STEMIs that lead to SCD are associated with more proximal acute coronary artery occlusion than STEMIs without SCD. Compared to patients with STEMI, NSTEMI patients who develop SCD have been shown to be elderly patients with diffuse and multiple coronary artery disease and have more cardiovascular risk factors [4].

The risk of SCD is higher, especially in the first 30 days after MI. Although its predictive value for myocardial scar formation is weak, an echocardiogram is one of the imaging methods that can predict SCD due to ventricular tachycardia or ventricular fibrillation. As ejection fraction of the left ventricle is an indirect measure of an LV scar and can be affected by hemodynamic-autonomic factors, direct visualization of the scar and its architecture may increase risk stratification accuracy. Cardiac MR imaging with late gadolinium enhancement is superior to echocardiography in determining total scar tissue [6]. The relationship between certain ECG changes has been identified with SCD, either with or without the coronary syndrome. QRS-T angle between depolarization and repolarization, QRS-T area angle, QRS area, QT prolongation, QT dispersion, QRS fragmentation, QRS amplitude, and T wave elevation are independent predictors of SCD on an ECG reading. The angle of the QRS-T area below 112 degrees was associated with the low rate of SCD in 30-month follow-up cases [7]. The presence of microvolt T-wave alternation in patients with MI was shown to be risky in terms of arrhythmic events [8]. It is also shown that the presence of early repolarization pattern in the ECG, especially in the inferolateral leads, in patients who have no previous history of cardiac disease, who have had the first acute coronary event, creates the environment for fatal arrhythmias due to which the risk of SCD is doubled [9]. QRS wave fragmentation seen on the ECG has been associated with heterogeneous activation of ventricles due to myocardial scar in individuals with known or suspected coronary artery disease. The presence of fragmented QRS waves in patients with acute coronary syndrome triggers arrhythmogenic mechanisms and can shorten the process of fatal events [10].

The autonomic nervous system plays a key role in the development of ventricular arrhythmias, especially in patients with a history of MI. Sympathetic stimulation in cardiac disease may accelerate the development of VT or VF. Therefore, autonomic tonus measurements are included in the risk classification. Heart rate variability may occur as a decrease in spontaneous variation secondary to the loss of vagal tone after MI. The association of decreased heart rate variability with both SCD and non-sudden death has been documented. Heart rate turbulence, on the other hand, measures the short-term change in heart rate after ventricular premature beats and is closely related to baroreceptor sensitivity. Unlike baroreceptor sensitivity, heart rate turbulence can be obtained from ECG recordings alone. In addition, the REFINE study revealed heart rate turbulence association with fatal or non-fatal cardiac arrest. Compared to heart rate variability, heart rate turbulence may be more specific for SCD due to VT or VF. Advances in scintigraphical imaging (SPECT / PET) may allow visualization of the cardiac sympathetic function of the left ventricle using norepinephrine analogs (C-meta-hydroxyepinephrine). This may reveal both arrhythmic and non arrhythmic outcomes [6]. In patients presenting acute coronary syndrome symptoms, several biomarkers have been evaluated for endpoints. In a study conducted with NT-proBNP, GDF-15, CRP, cystatin-C, troponin I, and troponin T values, it was shown that basal levels of NT-proBNP and GDF-15 are associated with arrhythmia and SCD [11]. Natriuretic peptides are associated with an increased risk of SCD in both men and women. There is evidence that effective HF treatments reduce BNP values, as in the MADIT-CRT study, and transform into a better prognosis for ventricular arrhythmias and SCD. Patients with decreased BNP levels by more than a third of baseline have a lower risk of VT/VF [12]. Cardiac troponins play an important role in the evaluation of patients with the acute coronary syndrome. Detectable levels of hs-Tn in the circulation reflect cardiac myocyte damage. Basal hs-Tn levels have been shown to be associated with SCD, and it is postulated that detectable levels of hs-Tn may provide an anatomical substrate for scar-related re-entry into ventricular tachycardia. The fact that troponin levels are specific for cardiac myocytes and can be measured more easily than other lethal rhythm predictive biomarkers provide clear advantages to the clinician [13]. Sst-2 is a member of the interleukin-1 family and is associated with myocardial dysfunction and fibrosis. It plays a role in the pathophysiological processes of different ischemic events, including the progression of coronary atherosclerosis. Although serum levels of Sst-2 increase with ischemic damage, they may remain high even after MI. As a result of the analysis obtained from the MUSIC records, Pascual-Dual *et al.* were able to show that high Sst-2 levels are associated with SCD. The authors suggested that the combination of natriuretic peptide and Sst-2 may be valuable in predicting SCD [12]. Cystatin-c, which is one of the indicators of renal dysfunction and plays a role in measuring the degree of dysfunction, increases the risk of SCD. Neurohormonal dysregulation involving renin-angiotensin-aldosterone upregulation is associated with atherosclerosis and heart failure. In the LURIC study, increased renin and aldosterone levels in elderly men undergoing coronary angiography were associated with adverse cardiovascular events, including SCD. In addition, there is an increased risk of cardiac arrest with increased levels of aldosterone in a group of post-ST segment elevation MI [5]. In a study conducted by Ying-Hsin Chen et al., it has been shown that patients with ACE DD genotype are in the risk group for developing acute MI, and patients with this genotype have increased ACS severity and increased risk of cardiac death [14]. Although more proof is needed for the relationship between plasma fibrinogen levels and SCD, available data suggest that fibrinogen levels are associated with the risk of SCD [15].

The relationship between the development of coronary artery disease and blood lipid levels has been proven. However, its relationship with the risk of SCD has not been clearly demonstrated as in studies with homocysteine. In a prospective study by Wannamethee *et al.* in middle-aged British men, blood lipid levels were found to be associated with an increased risk of SCD. It has been reported that unesterified free fatty acids are pro-arrhythmic, either directly or with the effect of modifying potassium and calcium channels, and these free fatty acids are potentially a risk factor for SCD in people with ischemic heart disease. Although the vast majority of SCD cases consist of patients with coronary artery disease, plaque rupture plays the most important role. From a physiological standpoint, an increase in tissue inflammation increases the sensitivity of plaque. Therefore, biomarkers reflecting inflammatory activity have increasingly been the focus for SCD. Analyzes have shown that one of these biomarkers, CRP level, stands out as a risk factor for SCD in men after correcting potential confounding parameters. Blangy *et al.* reported an association between CRP levels and ventricular tachycardia in ischemic heart patients with ICD. Another inflammatory marker associated with coronary artery disease is IL-6. In the PRIME study conducted with approximately 10 thousand asymptomatic middle-aged men, there was no significant relationship between hs-CRP and SCD, but increased risk of SCD was found to be associated with IL-6. Further evaluation is required to establish the relationship between SCD and markers such as myeloperoxidase and IL-18. Magnesium is a membrane stabilizer and has an anti-arrhythmic effect. Studies have shown that people with high magnesium levels have a 40% reduction in risk of SCD compared to those with low levels of magnesium [16]. Cytokines have a direct effect on ion channels. They also have a direct effect on the QT interval that leads to a change action potential and thus determines the arrhythmia threshold. Prolongation of the QT period is an important indicator of SCD risk. It would be promising to investigate the relationship between cytokines and SCD as they affect QTc independently. In a study of patients with end-stage renal disease, there is the first line of evidence indicating IL-6 levels are linked to SCD. These results were also supported by showing the strong association of IL-6 levels with SCD in a study conducted in European men aged 50-59 years. Another cytokine associated with SCD is IL-18. In the Helsinki Sudden Death Study, it was thought that the IL-18 genotype was independently associated with SCD, but this effect was directly related to subclinical atherosclerosis rather than its proarrhythmic feature. Animal models with MI have shown that TNF-alpha expression after early infarction is associated with ventricular arrhythmia. In humans, TNF-alpha levels in the first 96 hours after acute coronary syndrome were not found to be accurate in predicting ventricular arrhythmias. In the study conducted on 186 AMI patients, it was reported that measured levels of IL-6 in the first 72 hours of the initial event were effective in predicting ventricular arrhythmias. Similarly, IL-6 levels of the patients during the first 24 hours after percutaneous coronary intervention predicted the risk of developing ventricular arrhythmia. More studies are needed to fully explain these relationships. However, preliminary results suggest that cytokines are an attractive target for predicting the occurrence, prognosis, and treatment of ventricular arrhythmias [17].

SCD in the Setting of Nonischemic Cardiomyopathy

Dilated cardiomyopathy is defined as the systolic dysfunction along with the left ventricular dilation in the absence of any epicardial coronary diseases or other situations causing increased workload to the myocardium. NICM is a heterogeneous group that includes idiopathic dilated cardiomyopathy, valvular cardiomyopathies, genetically related cardiomyopathy syndromes such as laminopathies, chemotherapy-induced cardiomyopathies, hypertrophic cardiomyopathy, arrhythmogenic cardiomyopathy, *etc*. In this review, we focused on idiopathic dilated cardiomyopathy, and the term NICM refers to this entity. Hypertrophic and arrhythmogenic cardiomyopathies are discussed under separate subtitles. NICM is more frequently observed in males than in females. Death usually occurs as a result of pump failure or SCD. The left ventricular remodeling and the accompanying myocardial fibrosis are the common sequelae of the diseases causing mechanical dysfunction. Multivariate risk stratification models incorporating multiple cardiovascular variables and laboratory data have been established, along with not being specific to dilated cardiomyopathy. Even though the direct relationship between ejection fraction measured by echocardiography to the myocardial scar load has not been determined, it is known to be an important prognostic factor [18]. A study performed by Haugaa *et al.* on 94 NICMP patients reported that global longitudinal strain (GLS) on echocardiography was superior to LVEF in predicting the arrhythmic events [19]. Cardiac MRI can assess the density of myocardial fibrosis. Approximately 30% of the patients with NICM show myocardial fibrosis on MRI [20]. A relationship between late gadolinium enhancement and major cardiac arrhythmias in the cardiac MRI, which is the gold standard, has been established [21]. Similar to its use in ischemic cardiac diseases, ECG is also an easy and important tool for risk classification in NICM. Even though there is a relationship between the ECG parameters and cardiovascular diseases, there is no specific analysis for non-ischemic cardiac diseases. Cohort studies in NICM failed to show a relationship between SCD and a preexisting branch block. On the contrary, the QRS wave fragmentation in the absence of QRS widening and QT interval dispersion is related to increased arrhythmic events in patients with NICM. A meta-analysis has shown that the T wave alternation (TWA) is related to increased arrhythmic events. Holter analysis, similar to its use in ischemic diseases, is valuable for determining NSVT and measuring autonomic activities in NICM. The MACAS study has shown that the presence of NSVT alone is not correlated with major arrhythmic events; however, its combination with the decrease in LVEF increases the arrhythmic events risk by 8.2 times. Even though decreased heart rate variation (HRV) is related to the progression of heart failure, the heart rate turbulence (HRT) in NICM patients is not related to SCD. The role of electrophysiologic studies in NICM patients was also investigated. In light of these studies, the routine use of EPS for risk classification in NICM patients is not proposed in recent guidelines [20]. Besides other risk stratification tools, biochemical evaluation of the patient gives additional information for estimating SCD risk. Parameters using a daily practice such as proBNP and CRP are useful. The ProBNP and NT ProBNP values were found to be related to death in many studies [18]. Simon *et al.* have shown that the risk of major ventricular arrhythmias increases in NICM patients with increased proBNP levels. This relationship is called the mechano-electric feedback phenomenon. The natriuretic peptides released secondary to the mechanical stretch delay the conduction and increase the length of the action potential. CRP, an inflammation marker, is frequently related to coronary artery diseases; nevertheless, the basal increase in CRP in the male population without coronary artery diseases increases the SCD rate. Biomarkers outside of daily practice may also give clues to identify high-risk patients in whom classical risk stratification is inclusive or in the mid range. The extracellular matrix, similar to its role in acute coronary events, plays a role in ventricular remodeling in cardiomyopathies as well. The unbalanced activity between matrix metalloproteinases (MMP) and its tissue inhibitor (TIMO) leads to the damage of the myocardial fibrillary network and accelerates myocardial fibrosis and myocardial dilation. High levels of myocardial fibrosis are a substrate for ventricular arrhythmias leading to SCD. Kanoupakis *et al.* have shown that TIMP and MMP-1 levels in low EF NICMP patients who were implanted with primarily preventive ICDs are a predictor for malignant ventricular arrhythmias [22]. Osteopontin is an extracellular matrix glycoprotein playing a role in the extracellular matrix turnover that is synthesized from the macrophages and fibroblasts. Even though the collagen turnover is not specific to the myocardial tissue, osteopontin is related to the ventricular arrhythmias observed in heart failure patients with ICDs due to its role in profibrotic remodeling [23]. Francia *et al.* reported that galectin-3 and osteopontin levels in ICD implanted patients are predictors of continuous VT/VF incidences; they are promising biomarkers that need to be investigated more [12]. sST-2 is a member of the interleukin family and has a role in the pathophysiology of ischemic events. It is related to the reverse remodeling and fibrosis of the left ventricle. It is released from the cardiac cells and fibroblasts during heart failure [18]. The sST-2 levels are related to poor prognosis, along with its increased levels being related to increased cardiovascular death and death due to all causes [24]. A case-control study performed in 2009 reported that the sST-2 levels in patients with advanced systolic dysfunction are prognostically additive to the natriuretic peptides [25]. Galectin-3 beta-galactosidase is a binding protein that is released from the myofibroblasts due to cardiac stress. Even though some studies showed that it is a predictor of all-cause deaths, similar to sST-2, this relationship has not yet been fully elucidated. The genetic basis of dilated cardiomyopathy is complex; more than 50 genes have been identified. There is evidence for the increased probability of receiving adequate ICD shock in people carrying the titin mutation variants [18]. Lamin A mutation is an important inherited cause of dilated cardiomyopathy, and SCD risk due to ventricular arrhythmias is increased in these patients [26]. The risk of ventricular arrhythmias is increased in patients with Flamin-C mutation as well [18].

Hypertrophic Cardiomyopathy (HCM) is one of the most common inherited cardiomyopathies. It presents left ventricular hypertrophy without any known secondary cause. The genetic basis of HCM relies mainly upon the sarcomeric protein mutations [27]. The disease is diagnosed in adult patients by measuring the left ventricular wall width equal to or greater than 15 mm by any diagnostic method. The cardiovascular death incidence in adult patients with HCM is 1-2%; SCD, heart failure, and thromboembolism are the major causes of cardiac death. Although the annual SCD rate is around 1% in the whole HCM cohort, some clinical characteristics have been related to the increased risk of SCD. Age, NSVT attacks, maximal left ventricular wall thickness, first-degree relatives with SCD before the age of 40, syncope, left atrial diameter, left ventricular outflow obstruction, and exercise-induced blood pressure response are some of the clinical risk factors related to SCD risk. A model has been developed to assess the risk of SCD in the following 5 years, and recommendation for ICD implantation is generally determined with this model. The risk calculation is performed annually or in two years, or in the case of clinical changes [28]. The natriuretic peptide and troponin levels are correlated with the HCM progression; however, BNP is a more sensitive marker in patients with left ventricular hypertrophy. Moreover, natriuretic peptides are strong markers of hemodynamic parameters. These biomarkers are of use when the traditional SCD risk markers are inconclusive or show midlevel risk. Increased hs-TN is related to late gadolinium enhancement in MRI, which is a marker for increased SCD risk. Even though the combination of Hs-Tn>0.007 ng/ml or BNP>70 pg/ml has high sensitivity and specificity, more studies and data are required to support this approach [29].

Arrhythmogenic Right Ventricular Cardiomyopathy (ARVC) is a cause of malignant ventricular arrhythmia in young patients. Mortality is usually caused due to SCD, which has a 2-3% incidence per year and is commonly the first manifestation of the disease. Previous literature has too little evidence regarding the sudden arrhythmic death risk factors. Right ventricular dilation with angiography or echocardiography, T wave inversion in precordial derivations in ECG, ST elevation in right-handed derivations, QRS wave dispersion, left precordial JT interval lengthening, and complete completion right bundle branch block are characteristics related to increased sudden arrhythmic death [30]. Characterization of the disease is progressive and includes an inflammatory process with fibrofatty infiltration of the right ventricular myocardium. In terms of biomarkers in ARVC, a study comparing patients with ARVC to RVOT VT showed that patients with ARVC had significantly higher levels of CRP. Bonny *et al.* also reported that the CRP levels are increased in patients with ARVD within the 24 hours following VT. The authors noticed that the CRP increase corresponded to the lengthening of the QT interval; thus, the cytokines of inflammation may interfere with the function of ion channels and change the QT length. A small study of 29 controls and 24 ARVC patients revealed that the galectin-3 levels are increased in ARVC patients, and it is predictive for the initiation of both VT and VF [17].

FUTURE DIRECTIONS

Biomarkers like troponin and BNP derivatives have dramatically changed cardiology practice in acute coronary syndromes and heart failure management. Despite biomarkers being a key component of modern medicine since they have a major role in the early diagnosis and the treatment of patients by categorising the biologic markers [31], currently available biomarkers for predicting SCD are not sensitive and specific enough to assist patient management. Therefore, it is unclear for a clinician how to manage a patient with a high biomarker level associated with SCD. The development of biomarkers that enable the early diagnosis of patients carrying the risk of SCD is an important cornerstone. Further studies are needed in order to show the value of pre-existing specific biomarkers in determining the risk of SCD in specific patient groups. Large community-based research combining pre-existing and novel biomarkers should be conducted, with results evaluated using machine learning and artificial intelligence systems to improve both sensitivity and specificity in risk assessment. Despite the fact that the multi-marker strategy is thought to be more important, evidence shows that it is only of medium strength in risk stratification. Newer biomarkers will be determined with the ongoing genome and proteome studies. The evolution of the potential SCD biomarkers will necessitate the use of systematic strategy in the evaluation of these biomarkers [16].

CONSENT FOR PUBLICATION

Not applicable.

CONFLICT OF INTEREST

The authors declare no conflict of interest, financial or otherwise.

ACKNOWLEDGEMENT

Declared none.

REFERENCES

[1] Priori SG, Blomstrom-Lundqvist C, Mazzanti A, *et al.* 2015 ESC Guidelines for the management of patients with ventricular arrhythmias and the prevention of SCD the Task Force for the Management of Patients with Ventricular Arrhythmias and the Prevention of SCD of the Europea. Eur Heart J 2015; 36(41): 2793-2867l.
[http://dx.doi.org/10.1093/eurheartj/ehv316] [PMID: 26320108]

[2] Virmani R, Burke AP, Farb A. Sudden cardiac death. Cardiovasc Pathol 2001; 10(5): 211-8.
[http://dx.doi.org/10.1016/S1054-8807(01)00091-6] [PMID: 11673058]

[3] Inegbenebor U. Sudden Cardiac Death: Pathophysiology and Evidence Based Intervention Strategy. Int J Bas Appl Innov Res 2013; 2(4): 73-7.

[4] Carrizo S, Peinado RP, Sanchez-Recalde A, *et al.* Clinical and angiographic characteristics of patients with acute coronary syndrome associated with sudden cardiac death. Hell J Cardiol 2015; 56(2): 136-41.
[PMID: 25854442]

[5] Dhindsa DS, Khambhati J, Sandesara PB, Eapen DJ, Quyyumi AA. Biomarkers to Predict Cardiovascular Death. Card Electrophysiol Clin 2017; 9(4): 651-64.
[http://dx.doi.org/10.1016/j.ccep.2017.07.014] [PMID: 29173408]

[6] Deyell MW, Krahn AD, Goldberger JJ. Sudden cardiac death risk stratification. Circ Res 2015; 116(12): 1907-18.
[http://dx.doi.org/10.1161/CIRCRESAHA.116.304493] [PMID: 26044247]

[7] Lingman M, Hartford M, Karlsson T, *et al.* Value of the QRS-T area angle in improving the prediction of sudden cardiac death after acute coronary syndromes. Int J Cardiol 2016; 218: 1-11.
[http://dx.doi.org/10.1016/j.ijcard.2016.05.005] [PMID: 27203188]

[8] Ikeda T, Yoshino H, Sugi K, *et al.* Predictive value of microvolt T-wave alternans for sudden cardiac death in patients with preserved cardiac function after acute myocardial infarction: results of a collaborative cohort study. J Am Coll Cardiol 2006; 48(11): 2268-74.
[http://dx.doi.org/10.1016/j.jacc.2006.06.075] [PMID: 17161258]

[9] Tikkanen JT, Wichmann V, Junttila MJ, *et al.* Association of early repolarization and sudden cardiac death during an acute coronary event. Circ Arrhythm Electrophysiol 2012; 5(4): 714-8. [http://dx.doi.org/10.1161/CIRCEP.112.970863] [PMID: 22730409]

[10] Das MK, El Masry H. Fragmented QRS and other depolarization abnormalities as a predictor of mortality and sudden cardiac death. Curr Opin Cardiol 2010; 25(1): 59-64. [http://dx.doi.org/10.1097/HCO.0b013e328333d35d] [PMID: 19881337]

[11] Lindholm D, James SK, Gabrysch K, *et al.* Association of Multiple Biomarkers With Risk of All-Cause and Cause-Specific Mortality After Acute Coronary Syndromes: A Secondary Analysis of the PLATO Biomarker Study. JAMA Cardiol 2018; 3(12): 1160-6. [http://dx.doi.org/10.1001/jamacardio.2018.3811] [PMID: 30427997]

[12] Shomanova Z, Ohnewein B, Schernthaner C, *et al.* Classic and Novel Biomarkers as Potential Predictors of Ventricular Arrhythmias and Sudden Cardiac Death. J Clin Med 2020; 9(2): 578. [http://dx.doi.org/10.3390/jcm9020578] [PMID: 32093244]

[13] McCarthy CP, Yousuf O, Alonso A, Selvin E, Calkins H, McEvoy JW. High-Sensitivity Troponin as a Biomarker in Heart Rhythm Disease. Am J Cardiol 2017; 119(9): 1407-13. [http://dx.doi.org/10.1016/j.amjcard.2017.01.032] [PMID: 28256250]

[14] Chen YH, Liu JM, Hsu RJ, *et al.* Angiotensin converting enzyme DD genotype is associated with acute coronary syndrome severity and SCD in Taiwan: A case-control emergency room study. BMC Cardiovasc Disord 2012; 12: 6.

[15] Kunutsor SK, Kurl S, Zaccardi F, Laukkanen JA. Baseline and long-term fibrinogen levels and risk of sudden cardiac death: A new prospective study and meta-analysis. Atherosclerosis 2016; 245: 171-80. [http://dx.doi.org/10.1016/j.atherosclerosis.2015.12.020] [PMID: 26724527]

[16] Havmöller R, Chugh SS. Plasma biomarkers for prediction of sudden cardiac death: another piece of the risk stratification puzzle? Circ Arrhythm Electrophysiol 2012; 5(1): 237-43. [http://dx.doi.org/10.1161/CIRCEP.111.968057] [PMID: 22334431]

[17] Kouvas N, Kontogiannis C, Georgiopoulos G, *et al.* The complex crosstalk between inflammatory cytokines and ventricular arrhythmias. Cytokine 2018; 111(May): 171-7. [http://dx.doi.org/10.1016/j.cyto.2018.08.007] [PMID: 30172113]

[18] Marrow BA, Cook SA, Prasad SK, McCann GP. Emerging Techniques for Risk Stratification in Nonischemic Dilated Cardiomyopathy: JACC Review Topic of the Week. J Am Coll Cardiol 2020; 75(10): 1196-207. [http://dx.doi.org/10.1016/j.jacc.2019.12.058] [PMID: 32164893]

[19] Travin MI, Feng D, Taub CC. Novel Imaging Approaches for Predicting Arrhythmic Risk. Circ Cardiovasc Imaging 2015; 8(11)e003019 [http://dx.doi.org/10.1161/CIRCIMAGING.115.003019] [PMID: 26565012]

[20] Pimentel M, Rohde LE, Zimerman A, Zimerman LI. Sudden cardiac death markers in non-ischemic cardiomyopathy. J Electrocardiol 2016; 49(3): 446-51. [http://dx.doi.org/10.1016/j.jelectrocard.2016.03.012] [PMID: 27016256]

[21] Becker MAJ, Cornel JH, van de Ven PM, van Rossum AC, Allaart CP, Germans T. The prognostic value of Late gadolinium-enhanced cardiac magnetic resonance imaging in nonischemic dilated cardiomyopathy: a review and meta-analysis. JACC Cardiovasc Imaging 2018; 11(9): 1274-84. [http://dx.doi.org/10.1016/j.jcmg.2018.03.006] [PMID: 29680351]

[22] Balla C, Vitali F, Brieda A, *et al.* Predictors of cardiac arrhythmic events in non coronary artery disease patients. BMC Cardiovasc Disord 2019; 19(1): 104. [http://dx.doi.org/10.1186/s12872-019-1083-6] [PMID: 31046686]

[23] Francia P, Adduci C, Semprini L, *et al.* Osteopontin and galectin-3 predict the risk of ventricular tachycardia and fibrillation in heart failure patients with implantable defibrillators. J Cardiovasc Electrophysiol 2014; 25(6): 609-16.

[http://dx.doi.org/10.1111/jce.12364] [PMID: 24400815]

[24] Aimo A, Vergaro G, Passino C, *et al.* Prognostic Value of Soluble Suppression of Tumorigenicity-2 in Chronic Heart Failure: A Meta-Analysis. JACC Heart Fail 2017; 5(4): 280-6. [http://dx.doi.org/10.1016/j.jchf.2016.09.010] [PMID: 27816512]

[25] Pascual-Figal DA, Lax A, Perez-Martinez MT, del Carmen Asensio-Lopez M, Sanchez-Mas J. Clinical relevance of sST2 in cardiac diseases. Clin Chem Lab Med 2016; 54(1): 29-35. [http://dx.doi.org/10.1515/cclm-2015-0074] [PMID: 26544104]

[26] Wahbi K, Yaou RB, Gandjbakhch E, *et al.* Development and validation of a new risk prediction score for life- threatening ventricular tachyarrhythmias in laminopathies. Circulation 2019; 140(4): 293-302.

[27] Geske JB, Ommen SR, Gersh BJ. Hypertrophic Cardiomyopathy: Clinical Update. JACC Heart Fail 2018; 6(5): 364-75. [http://dx.doi.org/10.1016/j.jchf.2018.02.010] [PMID: 29655825]

[28] Elliott PM, Anastasakis A, Borger MA, *et al.* 2014 ESC guidelines on diagnosis and management of hypertrophic cardiomyopathy: The task force for the diagnosis and management of hypertrophic cardiomyopathy of the European Society of Cardiology (ESC). Eur Heart J 2014; 35(39): 2733-79. [http://dx.doi.org/10.1093/eurheartj/ehu284] [PMID: 25173338]

[29] Kehl DW, Buttan A, Siegel RJ, Rader F. Clinical utility of natriuretic peptides and troponins in hypertrophic cardiomyopathy. Int J Cardiol 2016; 218: 252-8. [http://dx.doi.org/10.1016/j.ijcard.2016.05.031] [PMID: 27236124]

[30] Peters S, Peters H, Thierfelder L. Risk stratification of sudden cardiac death and malignant ventricular arrhythmias in right ventricular dysplasia-cardiomyopathy. Int J Cardiol 1999; 71(3): 243-50. [http://dx.doi.org/10.1016/S0167-5273(99)00142-4] [PMID: 10636530]

[31] Coats CJ, Heywood WE, Mills K, Elliott PM. Current applications of biomarkers in cardiomyopathies. Expert Rev Cardiovasc Ther 2015; 13(7): 825-37. [http://dx.doi.org/10.1586/14779072.2015.1053873] [PMID: 26106935]

CHAPTER 21

Oxidative Stress Biomarkers in the Diagnosis and Prognosis

Pınar Atukeren[1,2,*]

[1] *Department of Medical Biochemistry, Cerrahpasa Medical Faculty, Istanbul University-Cerrahpasa, Istanbul, Turkey*

[2] *Faculty of Pharmacy, Istanbul University-Cerrahpasa, Istanbul, Turkey*

Abstract: Oxidative stress describes the state of a cell where there is an imbalance between free radical formation and antioxidants due to either excess formation of reactive oxygen species (ROS) or inadequate antioxidant defence. It is very well known that oxidative stress plays an important role in the pathophysiology of various diseases through impaired intracellular redox homeostasis. To evaluate and imply the excess production of ROS, various biomarkers are used and suggested, yet it is also known that there is a lack of standardization and validation for these methods. It is almost very difficult to measure ROS directly because of their short half-life, yet it is still possible with a suitable technique. The most frequently used biomarkers are represented by oxidized macromolecules such as lipids, proteins, and nucleic acids, which are modified *via* ROS, and also the amounts or activities of antioxidant molecules and enzymes, respectively. There are also various genetic biomarkers measuring the susceptibility of modification due to oxidative stress. However, the preferred biomarker would be dependent on the aim of the study and the clinical relevance.

Keywords: Antioxidants, Oxidative biomarkers, Oxidative stress, Reactive oxygen species.

INTRODUCTION

Oxidative Stress

In normal cells, there is balanced redox homeostasis which is regulated by free radicals' production and the antioxidant defense systems [1]. Reactive oxygen species (ROS), so-called free radical molecules, are produced not only during normal cellular metabolism but also there are various exogenous sources of ROS (Tables **1** and **2**). In both experimental and clinical medicine , oxidative stress,

* **Corresponding author Pınar Atukeren**: Department of Medical Biochemistry, Cerrahpaşa Medical Faculty, Istanbul University-Cerrahpasa Istanbul, Turkey and Faculty of Pharmacy, Istanbul University-Cerrahpasa, Istanbul, Turkey; Tel: +90 (532) 513 93 10; E-mail: p_atukeren@yahoo.com

Pınar Atukeren and Hafize Uzun (Eds.)

which is mediated by ROS due to either an excess of ROS or a reduction in the amount of activity of antioxidant defense mechanisms, is important [2]. ROS are known to have a dual role as they might also have beneficial effects [3]. The intracellular redox imbalance is induced by oxidative stress, which triggers a number of factors such as cancer stimulation, increased metabolic activity, and mitochondrial dysfunction [4, 5]. The cell's redox status includes physiological and pathological effects related to the endogenous antioxidant status, which has a vital role in the prevention against oxidative stress due to the imbalanced defense mechanism of antioxidants, and the healthy living system will be exposed to serious damage [6].

Table 1. Reactive oxygen species.

$O_2^{\cdot-}$ (Superoxide radical)
H_2O_2 (Hydrogen peroxide)
$HO^{\cdot}$ (Hydroxyl radical)
HOCl (Hypochlorus acid)
Singlet O_2
$R^{\cdot}$ (Alkyl radical)
$ROO^{\cdot}$ (Peroxyl radical)
$RCOO^{\cdot}$ (Organic peroxyl radical)
$HO_2^{\cdot}$ (Peroxyl radical)
$RO^{\cdot}$ (Alkoxyl radical)

Table 2. Sources of reactive oxygen species.

Endogenous sources	**Exogenous Sources**
Mitochondrial electron transport chain	Radiation (UV lights, X-rays, Gamma-rays)
Autooxidation reactions	Chemotherapeutics, Chemicals, Drugs
Redox cycles	Environmental toxins
Cytochrome p450 reactions	Diet; Charcohalbroiled foods
Xanthine oxidase	Smoke
NADPH oxidase, Lipoxygenase, Prostaglandin synthase	
Inflammation; Oxidative reactions in phagocytic cells (Myeloperoxidase)	
Transition metals	

Indeed biological oxidation is the natural process of aerobic metabolism, but during this process, these destructive reactive molecules are produced constantly and express deleterious effects on human cells and damage [7].

Antioxidants

When the antioxidant system is unable to neutralise the overly generated ROS, oxidative stress develops as a result of the impaired redox balance [8]. There are many antioxidant defense systems in the human body to cope with the oxidative damaging effects of ROS [9]. The antioxidants might be either endogenous or exogenous too. They might be water or lipid-soluble due to their molecular structure. Some antioxidants are vitamins such as vitamin C and vitamin E; some antioxidants are enzymes such as catalase (CAT), superoxide dismutase (SOD), glutathione peroxidase (GPx), and glutathione reductase (GR). SOD is the first defending enzyme dismutating O_2^- . Also, there are some endogenous metabolites having antioxidant activities (Table **3**).

Table 3. Antioxidants.

Enzymatic Antioxidants	Superoxide dismutase (SOD) Catalase (CAT) Glutathione peroxidase (GPx) Glutathione reductase (GR) Glutathione-S-transferase Thioredoxine system (Trx/TrxR)
Non-enzymatic Antioxidants	Glutathione (GSH), Ubiquinone, Selenium Vitamin A, Vitamin C, Vitamin E, α-Lipoic acid, Flavonoids Albumin, Haptoglobin, Ceruloplasmin, Lactoferrin Uric acid, Bilirubin, Melatonin

Cells have various antioxidant mechanisms that balance the concentrations of ROS. SOD, CAT and GPx are the most effective [10]. There are non-enzymatic antioxidants such as vitamin C, vitamin E, carotenoids and glutathione, thioredoxin, lipoic acid, which are thiol antioxidants [11]. Glutathione (GSH) is a key thiol-containing antioxidant that detoxifies intracellular H2O2 to H2O and O2 through GPx and recycles through (GR), which is critical for maintaining cellular redox equilibrium [12, 13].

Oxidative Stress Biomarkers

Oxidative stress status is evaluated by investigating the cell's oxidant and antioxidant status as various biomolecules will be damaged. Oxidative stress biomarkers are related to the status of the disease and the effects of antioxidant molecules. The frequently measured oxidative stress related biomarkers and their properties are shown in Table **4**. These biomarkers can be measured in biological samples, mostly blood and tissue samples and also in other biological fluids such as urine and saliva.

Table 4. Oxidative stress related biomarkers.

Redox Homeostasis Status	Biomarker
Lipid oxidation	Malondialdehyde (MDA); Thiobarbituric acid reacting substances (TBARS) Lipid hydroperoxides (LHPs) F2-isoprotane 4-hyroxynonenal (4-HNE)
Protein oxidation	Protein carbonyl oxidation (PCO) Advanced oxidised protein products (AOPP)
Glyco oxidation	Advanced glycated end products (AGEs)
DNA oxidation	8-hydroxydeoxyguanosine (8-OHdG)
Tyrosine-derived oxidation	Dityrosine (DT) 3-Nitrotyrosine (3-NTyr)
Triptophan-derived oxidation	Kynurenine (KYN) N- formylkynurenine (N-FKYN)
Antioxidant status	Superoxide dismutase (SOD) Catalase (CAT) Glutathione (GSH) Glutathione peroxidase (GPx) Glutathione reductase (GR) Total thiol (T-SH) Total antioxidant status (TAS) Ferric reducing antioxidant power (FRAP)

Measurement of the oxidative stress biomarkers has a causal role in the diagnosis and prognosis of diseases. Not only oxidative stress causes the disease but also any disease might cause oxidative stress too, or sometimes a third factor such as inflammation might be the reason [14, 15]. ROS has a very short half-life and it is usually very hard to measure ROS directly in biological samples, but there are various imaging techniques that have been developed and this might make this direct measurement possible in the near future [16, 17].

Lipid Oxidation Products

Polyunsaturated fatty acids (PUFA) in cellular membranes are the major targets for ROS, triggering the lipid peroxidation reaction chain with a number of reaction steps. In each step, various byproducts are produced, which might be used as a biomarker. During the first three reactions, lipid peroxyl radicals and lipid hydroperoxides are formed. In the following reactions, the degradation products such as malondialdehyde (MDA), 4-hydroxy nonenal (4-HNE) are produced as in the presence of transition metals, lipid hydroperoxides or lipid alkoxyl radicals are degraded into these reactive aldehydes which can also react with proteins [18]. MDA reacts with thiobarbituric acid, so the biomarker, namely thiobarbituric acid reacting substances (TBARS), also involves this. During the ROS catalyzed oxidation of arachidonic acid, which is a membrane PUFA, isoprostanes (IsoPs) are formed [19]. IsoPs are chemically stable and sensitive against changes in oxidative status so they are the most reliable marker [20]. There are various qualitative and quantitative methods for measuring these byproducts and aldehydes. 4-HNE and MDA or TBARS are among the most investigated biomarkers [21, 22].

Protein Oxidation Products

Protein molecules are modified irreversibly through various mechanisms such as carbonylation, nitrosylation, breaking of the histidine and tryptophan rings [23]. Protein oxidation biomarkers, being chemically stabil, have good clinical relevance. Carbonyl groups might be formed *via* ROS attacks, especially HO• radicals, oxidative cleavage of the protein structure, or oxidative deamination [24]. The widely used biomarker for protein oxidation is measuring carbonyl levels [23, 25]. It has some advantages compared with the other oxidation products as they are formed relatively early and more stabile.

Advanced glycation end products (AGEs) are heterogeneous molecules with carboxymethyl valine and pentosidine, which are protein products, and carboxymethyl lysine, which is a product of lipid peroxidation and glycoxidation [26 - 28]. AGEs are not only oxidative stress biomarkers but also they have a role in the pathophysiology of inflammatory diseases.

Proteins also react with hypochlorous acid forming 3-chloro-tyrosine (3-ClTyr) and 3,5-dichloro-tyrosine. as main products, which are advanced oxidation protein products (AOPP) and involve protein aggregates *via* disulphide bridges and tyrosine cross links. 3-nitro-tyrosine (3-NTyr) is formed *via* tyrosine oxidation. These 3-ClTyr and 3-NTyr might be formed reacting with hipochlorous acid [29]. Kynurenine (KYN) and N- formylkynurenine (N-FKYN) are tryptophan metabolism metabolite, and they can be used biomarkers, too.

Oxidative DNA Damage

ROS also can damage DNA in the cells so the generated oxidatively modified nucleobases might also be measured as biomarkers in DNA [30 - 32]. Frequently 8-hydroxydeoxyguanosine (8OHdG) is measured as an oxidative DNA damage biomarker [30, 33, 34]. DNA might also undergo nitration and deamination of purines and damage. There are also methods evaluating DNA base nitration and deamination, but they should also be validated before routine use [30].

Glutathione and Protein Thiols

GSH is the main nonprotein thiol. Measuring GSH levels in the cell, if any elevation, will be an important indicator of antioxidant defense systems to scavenge ROS [35]. Protein cysteine residues might be formed through oxidation. Glutathionylation of the protein molecules is suggested as biomarkers reflecting oxidative stress [36, 37].

Nuclear Factor (erythroid-derived2)-like2 (Nrf2) and the Antioxidant Response as a Biomarker

When oxidative stress is induced, an adaptive response mediated by Nrf2 activation through the oxidation of its redox sensor, Kelch-like ECH-associated protein (Keap1) will occur, and the activities of antioxidant expressing genes will be enhanced so the activities of the antioxidant enzymes will be elevated [38]. Nrf2 induces the expression of antioxidant enzymes when there is oxidative stress, and Keap1 liberates Nrf2, as both were connected as a complex so that it will move from the cytoplasm to the nucleus. In the nucleus, Nrf2 binds to the antioxidant response element (ARE) located on the DNA molecule, and this binding will activate the endogenous antioxidant coding genes' expression [39]. So, measuring the biological response would also be remarkable as the biomarkers of Keap1-Nrf2 activation will indicate not only oxidative stress but also will imply the protective antioxidant status, which detoxifies ROS [40 - 42] and will correlate with the severity of the diseases.

MiRNA Biomarkers

The main source of genetic damage is ROS but on the other hand, it is also possible for the enhanced levels of ROS to work against cancer cells [43]. Epigenetic mechanisms are also involved in oxidative stress status as it is very well known that in various diseases related to ROS, transcriptional and epigenetic dysregulation are seen in the signaling pathways. These mechanisms also involve micro-RNA (miRNA) regulation which are heritable changes in gene expression with no change in the sequence of DNA [44, 45]. Due to the relationship between oxidative stress and regulation of miRNAs, new diagnostic and prognostic biomarkers are also suggested [46 - 49]. The diagnostic importance of miRNAs are progressing each day and is implied in various studies, and also a number of miRNA panels are already suggested to be used as biomarkers as the dysregulation of physiological miRNAs might trigger oxidative stress, modulate redox signaling pathways, and have a role in the pathophysiology of the disease regarding the relationship between redox homeostasis and specific miRNAs.

CONCLUSION

Oxidative stress involves various cellular processes with many pathways so assessment of oxidative stress status is challenging. None of the biomarkers are more reliable than another, so there is not a gold standard oxidative stress biomarker yet. Main points should be considered when evaluating oxidative stress status, such as selecting the biomarker and method, selecting the appropriate population, sample collection, transport and storage, testing, interpretation of the results, antioxidant supplemantation; if there are any. The preanalytical, analytical and postanalytic aspects should be investigated in further studies for each of these biomarkers for the validation of the methods. These biomarkers will be attractive in the future, especially when the clinicians will follow patients if benefiting from antioxidant supplements as using antioxidants in the treatment of many diseases would be very beneficial.

CONSENT FOR PUBLICATION

Not applicable.

CONFLICT OF INTEREST

The authors declare no conflict of interest, financial or otherwise.

ACKNOWLEDGEMENT

Declared none.

REFERENCES

[1] Devi GS, Prasad MH, Saraswathi I, Raghu D, Rao DN, Reddy PP. Free radicals antioxidant enzymes and lipid peroxidation in different types of leukemias. Clin Chim Acta 2000; 293(1-2): 53-62. [http://dx.doi.org/10.1016/S0009-8981(99)00222-3] [PMID: 10699422]

[2] Halliwell B, Gutteridge JMC. Free Radicals in Biology and Medicine. 3rd ed., Oxford, UK: Oxford University Press 1999.

[3] Valko M, Izakovic M, Mazur M, Rhodes CJ, Telser J. Role of oxygen radicals in DNA damage and cancer incidence. Mol Cell Biochem 2004; 266(1-2): 37-56. [http://dx.doi.org/10.1023/B:MCBI.0000049134.69131.89] [PMID: 15646026]

[4] Pelicano H, Carney D, Huang P. ROS stress in cancer cells and therapeutic implications. Drug Resist Updat 2004; 7(2): 97-110. [http://dx.doi.org/10.1016/j.drup.2004.01.004] [PMID: 15158766]

[5] Valko M, Rhodes CJ, Moncol J, Izakovic M, Mazur M. Free radicals, metals and antioxidants in

oxidative stress-induced cancer. Chem Biol Interact 2006; 160(1): 1-40.
[http://dx.doi.org/10.1016/j.cbi.2005.12.009] [PMID: 16430879]

[6] Halliwell B. Role of free radicals in the neurodegenerative diseases: therapeutic implications for antioxidant treatment. Drugs Aging 2001; 18(9): 685-716.
[http://dx.doi.org/10.2165/00002512-200118090-00004] [PMID: 11599635]

[7] Castro L, Freeman BA. Reactive oxygen species in human health and disease. Nutrition 2001; 17(2): 161-165, 163-165.
[http://dx.doi.org/10.1016/S0899-9007(00)00570-0] [PMID: 11240347]

[8] Pervaiz S. Pro-oxidant milieu blunts scissors: insight into tumor progression, drug resistance, and novel druggable targets. Curr Pharm Chem 2006; 12(34): 4469-77.
[http://dx.doi.org/10.2174/138161206779010503] [PMID: 17168754]

[9] Aoyama K, Watabe M, Nakaki T. Regulation of neuronal glutathione synthesis. J Pharmacol Sci 2008; 108(3): 227-38.
[http://dx.doi.org/10.1254/jphs.08R01CR] [PMID: 19008644]

[10] Matés JM, Pérez-Gómez C, Núñez de Castro I. Antioxidant enzymes and human diseases. Clin Biochem 1999; 32(8): 595-603.
[http://dx.doi.org/10.1016/S0009-9120(99)00075-2] [PMID: 10638941]

[11] McCall MR, Frei B. Can antioxidant vitamins materially reduce oxidative damage in humans? Free Radic Biol Med 1999; 26(7-8): 1034-53.
[http://dx.doi.org/10.1016/S0891-5849(98)00302-5] [PMID: 10232849]

[12] Meister A. Glutathione biosynthesis and its inhibition. Methods Enzymol 1995; 252: 26-30.
[http://dx.doi.org/10.1016/0076-6879(95)52005-8] [PMID: 7476360]

[13] Smith CV, Jones DP, Guenthner TM, Lash LH, Lauterburg BH. Compartmentation of glutathione: implications for the study of toxicity and disease. Toxicol Appl Pharmacol 1996; 140(1): 1-12.
[http://dx.doi.org/10.1006/taap.1996.0191] [PMID: 8806864]

[14] Ghezzi P, Ghiara V, Davies K. Epistemological Challenges of the Oxidative Stress Theory of Disease and the Problem of Biomarkers, Oxidative Stress. London: Academic Press 2020; pp. 13-27.

[15] Ghezzi P, Jaquet V, Marcucci F, Schmidt HHHW. The oxidative stress theory of disease: levels of evidence and epistemological aspects. Br J Pharmacol 2017; 174(12): 1784-96.
[http://dx.doi.org/10.1111/bph.13544] [PMID: 27425643]

[16] Berkowitz BA. Oxidative stress measured in vivo without an exogenous contrast agent using QUEST MRI. J Magn Reson 2018; 291: 94-100.
[http://dx.doi.org/10.1016/j.jmr.2018.01.013] [PMID: 29705036]

[17] Kühl A, Dixon A, Hali M, *et al.* Novel QUEST MRI in vivo measurement of noiseinduced oxidative stress in the cochlea. Sci Rep 2019; 9(1): 16265.
[http://dx.doi.org/10.1038/s41598-019-52439-4] [PMID: 31700007]

[18] Esterbauer H, Schaur RJ, Zollner H. Chemistry and biochemistry of 4-hydroxynonenal, malonaldehyde and related aldehydes. Free Radic Biol Med 1991; 11(1): 81-128.
[http://dx.doi.org/10.1016/0891-5849(91)90192-6] [PMID: 1937131]

[19] Morrow JD, Hill KE, Burk RF, Nammour TM, Badr KF, Roberts LJ II. A series of prostaglandin F2-like compounds are produced in vivo in humans by a non-cyclooxygenase, free radical-catalyzed mechanism. Proc Natl Acad Sci USA 1990; 87(23): 9383-7.
[http://dx.doi.org/10.1073/pnas.87.23.9383] [PMID: 2123555]

[20] Kadiiska MB, Gladen BC, Baird DD, *et al.* Biomarkers of oxidative stress study II: are oxidation products of lipids, proteins, and DNA markers of CCl4 poisoning? Free Radic Biol Med 2005; 38(6): 698-710.
[http://dx.doi.org/10.1016/j.freeradbiomed.2004.09.017] [PMID: 15721980]

[21] Negre-Salvayre A, Auge N, Ayala V, *et al.* Pathological aspects of lipid peroxidation. Free Radic Res 2010; 44(10): 1125-71. [http://dx.doi.org/10.3109/10715762.2010.498478] [PMID: 20836660]

[22] Zarkovic N. 4-hydroxynonenal as a bioactive marker of pathophysiological processes. Mol Aspects Med 2003; 24(4-5): 281-91. [http://dx.doi.org/10.1016/S0098-2997(03)00023-2] [PMID: 12893006]

[23] Berlett BS, Stadtman ER. Protein oxidation in aging, disease, and oxidative stress. J Biol Chem 1997; 272(33): 20313-6. [http://dx.doi.org/10.1074/jbc.272.33.20313] [PMID: 9252331]

[24] Stadtman ER, Levine RL. Free radical-mediated oxidation of free amino acids and amino acid residues in proteins. Amino Acids 2003; 25(3-4): 207-18. [http://dx.doi.org/10.1007/s00726-003-0011-2] [PMID: 14661084]

[25] Chevion M, Berenshtein E, Stadtman ER. Human studies related to protein oxidation: protein carbonyl content as a marker of damage. Free Radic Res 2000; 33 (Suppl.): S99-S108. [PMID: 11191280]

[26] Anguizola J, Matsuda R, Barnaby OS, *et al.* Review: Glycation of human serum albumin. Clin Chim Acta 2013; 425: 64-76. [http://dx.doi.org/10.1016/j.cca.2013.07.013] [PMID: 23891854]

[27] Monnier VM, Sell DR. Prevention and repair of protein damage by the Maillard reaction in vivo. Rejuvenation Res 2006; 9(2): 264-73. [http://dx.doi.org/10.1089/rej.2006.9.264] [PMID: 16706654]

[28] Fu MX, Requena JR, Jenkins AJ, Lyons TJ, Baynes JW, Thorpe SR. The advanced glycation end product, Nepsilon-(carboxymethyl)lysine, is a product of both lipid peroxidation and glycoxidation reactions. J Biol Chem 1996; 271(17): 9982-6. [http://dx.doi.org/10.1074/jbc.271.17.9982] [PMID: 8626637]

[29] Dalle-Donne I, Giustarini D, Colombo R, Rossi R, Milzani A. Protein carbonylation in human diseases. Trends Mol Med 2003; 9(4): 169-76. [http://dx.doi.org/10.1016/S1471-4914(03)00031-5] [PMID: 12727143]

[30] Halliwell B, Whiteman M. Measuring reactive species and oxidative damage in vivo and in cell culture: how should you do it and what do the results mean? Br J Pharmacol 2004; 142(2): 231-55. [http://dx.doi.org/10.1038/sj.bjp.0705776] [PMID: 15155533]

[31] Dizdaroglu M, Jaruga P, Birincioglu M, Rodriguez H. Free radical-induced damage to DNA: mechanisms and measurement. Free Radic Biol Med 2002; 32(11): 1102-15. [http://dx.doi.org/10.1016/S0891-5849(02)00826-2] [PMID: 12031895]

[32] Halliwell B. Effect of diet on cancer development: is oxidative DNA damage a biomarker? Free Radic Biol Med 2002; 32(10): 968-74. [http://dx.doi.org/10.1016/S0891-5849(02)00808-0] [PMID: 12008112]

[33] Griffiths HR, Møller L, Bartosz G, *et al.* Biomarkers. Mol Aspects Med 2002; 23(1-3): 101-208. [http://dx.doi.org/10.1016/S0098-2997(02)00017-1] [PMID: 12079771]

[34] Collins AR, Cadet J, Möller L, Poulsen HE, Viña J. Are we sure we know how to measure 8-oxo-7-8-dihydroguanine in DNA from human cells? Arch Biochem Biophys 2004; 423(1): 57-65. [http://dx.doi.org/10.1016/j.abb.2003.12.022] [PMID: 14989265]

[35] Atkuri KR, Mantovani JJ, Herzenberg LA, Herzenberg LA. N-Acetylcysteine--a safe antidote for cysteine/glutathione deficiency. Curr Opin Pharmacol 2007; 7(4): 355-9. [http://dx.doi.org/10.1016/j.coph.2007.04.005] [PMID: 17602868]

[36] Ghezzi P. Protein glutathionylation in health and disease. Biochim Biophys Acta 2013; 1830(5): 3165-72.

[http://dx.doi.org/10.1016/j.bbagen.2013.02.009] [PMID: 23416063]

[37] Sies H. Glutathione and its role in cellular functions. Free Radic Biol Med 1999; 27(9-10): 916-21. [http://dx.doi.org/10.1016/S0891-5849(99)00177-X] [PMID: 10569624]

[38] Cuadrado A, Manda G, Hassan A, *et al.* Lopez, B. Oliva, M. Pajares, A.I. Rojo, N. Robledinos-Anton, A.M. Valverde, E. Guney, H. Schmidt, Transcription factor NRF2 as a therapeutic target for chronic diseases: a systems medicine approach. Pharmacol Rev 2018; 70(2): 348-83. [http://dx.doi.org/10.1124/pr.117.014753] [PMID: 29507103]

[39] Barančík M, Grešová L, Barteková M, Dovinová I. Nrf2 as a key player of redox regulation in cardiovascular diseases. Physiol Res 2016; 65 (Suppl. 1): S1-S10. [http://dx.doi.org/10.33549/physiolres.933403] [PMID: 27643930]

[40] Yin J, Wang AP, Li WF, Shi R, Jin HT, Wei JF. Sensitive biomarkers identification for differentiating Cd and Pb induced toxicity on zebrafish embryos. Environ Toxicol Pharmacol 2017; 56: 340-9. [http://dx.doi.org/10.1016/j.etap.2017.10.010] [PMID: 29102874]

[41] Yadetie F, Karlsen OA, Lanzén A, *et al.* Global transcriptome analysis of Atlantic cod (Gadus morhua) liver after in vivo methylmercury exposure suggests effects on energy metabolism pathways. Aquat Toxicol 2013; 126: 314-25. [http://dx.doi.org/10.1016/j.aquatox.2012.09.013] [PMID: 23103053]

[42] Mondal NK, Saha H, Mukherjee B, Tyagi N, Ray MR. Inflammation, oxidative stress, and higher expression levels of Nrf2 and NQO1 proteins in the airways of women chronically exposed to biomass fuel smoke. Mol Cell Biochem 2018; 447(1-2): 63-76. [http://dx.doi.org/10.1007/s11010-018-3293-0] [PMID: 29363060]

[43] Perillo B, Di Donato M, Pezone A, *et al.* ROS in cancer therapy: the bright side of the moon. Exp Mol Med 2020; 52(2): 192-203. [http://dx.doi.org/10.1038/s12276-020-0384-2] [PMID: 32060354]

[44] Kebede AF, Schneider R, Daujat S. Novel types and sites of histone modifications emerge as players in the transcriptional regulation contest. FEBS J 2015; 282(9): 1658-74. [http://dx.doi.org/10.1111/febs.13047] [PMID: 25220185]

[45] Moore LD, Le T, Fan G. DNA methylation and its basic function. Neuropsychopharmacology 2013; 38(1): 23-38. [http://dx.doi.org/10.1038/npp.2012.112] [PMID: 22781841]

[46] Baker JR, Vuppusetty C, Colley T, *et al.* Oxidative stress dependent microRNA-34a activation via PI3Kα reduces the expression of sirtuin-1 and sirtuin-6 in epithelial cells. Sci Rep 2016; 6(1): 35871. [http://dx.doi.org/10.1038/srep35871] [PMID: 27767101]

[47] Büssing I, Slack FJ, Grosshans H. let-7 microRNAs in development, stem cells and cancer. Trends Mol Med 2008; 14(9): 400-9. [http://dx.doi.org/10.1016/j.molmed.2008.07.001] [PMID: 18674967]

[48] Zhang WC. microRNAs Tune Oxidative Stress in Cancer Therapeutic Tolerance and Resistance. Int J Mol Sci 2019; 20(23): 6094. [http://dx.doi.org/10.3390/ijms20236094] [PMID: 31816897]

[49] He J, Jiang BH. Interplay between Reactive oxygen Species and MicroRNAs in Cancer. Curr Pharmacol Rep 2016; 2(2): 82-90. [http://dx.doi.org/10.1007/s40495-016-0051-4] [PMID: 27284501]

CHAPTER 22

Current Biomarkers for Endometrial Receptivity

Ozlem Erten[1,*], **Aysun Tekeli Taskomur**[2] and **Mustafa Albayrak**[3]

[1] *Department of Obstetrics and Gynecology, School of Medicine, Kutahya Health Sciences University, Kutahya, Turkey*

[2] *Department of Obstetrics and Gynecology, Faculty of Medicine, Amasya University, Amasya, Turkey*

[3] *Department of Obstetrics and Gynecology, Sisli Florence Nightingale Hospital, Istanbul, Turkey*

Abstract: Implantation and the continuation of pregnancy occur through a complicated and sophisticated dialogue, called "cross-talk," which starts between the embryo and the endometrium in the early stages of oocyte maturation. This dialogue provides synchronization of the journey of the embryo to be implanted with the receptive endometrium. Understanding the activity and function of the hormones and factors involved in this dialogue will provide an understanding of endometrial receptivity, which plays a key role in implantation, and the determination of biomarkers specific for this period. As a result of the development of omics technology, it has become possible to identify biomarkers specific to endometrial receptivity by performing genomic, proteomic, and lipidomic analyses of these hormones and factors. The determination of these biomarkers, their optimization, and making them usable in the clinic will allow increased success in ART.

Keywords: Biomarkers, Endometrial receptivity, Genomic analysis, Implantation, Lipidomic analysis, Omics, Proteomic analysis.

INTRODUCTION

Endometrial Receptivity

A successful pregnancy is achieved only with successful implantation, with two important factors forming the basis of successful implantation. The first is to obtain a good-quality embryo, and our chance to obtain quality embryos has increased with the development of assisted reproduction techniques. The second is the receptive endometrium [1]. The allowance of the endometrium of implantation of the blastocyst. The receptive endometrium has a very important physiological role in implantation and also in the maintenance of pregnancy. However, the

* **Corresponding author Özlem Erten:** Department of Obstetrics and Gynecology, School of Medicine, Kutahya Health Sciences University, Kutahya, Turkey; Tel: +90 530 563 0015; E-mail: ozlem erten@ksbu.edu.tr

Pınar Atukeren and Hafize Uzun (Eds.)

endometrium is not always receptive through the luteal phase of the menstrual cycle, and its time is limited. This short period lasting about 4 days at most in the mid-secretory phase is called the implantation window, which usually corresponds to the 20th-24th day of the cycle [2, 3]. Other than this window, the endometrium is non-receptive. In order to achieve successful implantation, the journey of the embryo and the receptive period of the endometrium must be synchronized with each other [4, 5].

This happens through "cross-talk" between the embryo and the adhesive endometrium by means of endocrine, paracrine, and autocrine factors [6].

The mechanism for the non-receptive endometrium to become receptive is not fully understood. Studies have determined that endometrial receptivity occurs due to the development of changes such as adhesion, invasion, growth, differentiation, decidualization, and immunomodulation. The factors arising during these developing changes play an important role in the preparation process for implantation [7, 8].

Various studies have been conducted to determine these synthesized factors in order to better understand endometrial receptivity. These studies have been conducted with a wide range of endometrial sampling for the use of omics technology [6, 9].

Determining the activity and function of these factors in the dialogue between the embryo and the endometrium allows us to use these factors as endometrial receptivity biomarkers in the clinic. Considering that two-thirds of recurrent implantation failure (RIF, *i.e.*, failure of three IVF cycles in which one or two high-quality embryos are transferred to the patient per cycle) is due to endometrial receptivity; it is very important to determine potential biomarkers and evaluate their significance in order to provide utility in the clinic and increase the success of implantation [1, 10].

Markers Brought by New Technology

The complete sequencing of the human genome has opened a new era of system biology, called omics technologies. The term omics is expressed as a comprehensive analysis of biological systems. The omics technologies driving these new fields of research include DNA and protein microarrays, mass spectrometry, and many other tools that enable high throughput analysis. Due to these developments, it is possible to examine the genome, transcriptome, proteome, and metabolome of cells, tissues, and organisms and to investigate the interactions between them [11, 12].

The histology of the endometrium has been studied since the 1950s to evaluate the developmental cycle and receptivity of the endometrium [9]. When the endometrial surface was examined with electron microscopy, it was determined that new projections were formed on the endometrial surface where the embryo first came into contact. Pinopodes, which were first defined by Psychoyos, were determined by Niklas in a certain period of the cycle, namely in the implantation window, and were named as endometrial receptivity markers [13, 14]. The development of omics technology, genomic, proteomic, and lipidomic studies have enabled the search for new biomarkers that can evaluate endometrial receptivity and provide utility in the clinic. Diagnostic markers of the implantation stage can be found by examining the microarray of many genes expressed in the endometrium at the stages of the menstruation cycle. Many proteins related to reproduction can be discovered by proteomic analysis. Thus, the detection of biomarkers specific to the receptive and non-receptive period of the endometrium could be identified [15].

Genomic Study of the Endometrium

In recent years, many transcriptomic studies in which important information has been obtained in determining the receptive endometrium have been conducted. These studies have found 107-2878 genes that are expressed differently in the endometrium by up and down-regulation during the transformation of the endometrium from the pre-receptive (LH+1/5] period to the receptive period (LH+7/9] [5, 16, 17]. However, the results have not shown a correlation with each other due to the different microarrays used for the analysis and the different sampling methods and conditions in these studies. For this reason, a common consensus could not be reached in the transcript studies. Another issue that causes confusion in these studies is whether the evaluation of the endometrium should be based on the natural cycle of fertile women or the cycle of infertile women. The studies conducted to solve this problem are classified into three groups: 1) the comparison of fertile women with the cycles of infertile women, 2) studies on the natural cycles of fertile women [16, 18 - 20] or fertile donors [21,22], and 3) the comparison of natural cycles with stimulated cycles [21]. Having such a classification in studies will facilitate the determination of biomarkers of the non-receptive endometrial period (proliferative or early secretory) and the receptive endometrium (mid-secretory), and will enable the determination of common biomarkers from the results of the studies on the cycles, increasing the specificity of the biomarkers.

In 2011, Diaz-Gimeno *et al.* identified 238 genes with expression specific to early and mid-secretory phases, 134 of which carry transcriptomic signatures of the receptive phase. These genes were customized to be in a single sequence and turned into a genomic diagnostic test that detects pathologies of endometrial origin [5]. This test was modified many times and is currently used under the name of NGS ERA predictor. This test has high specificity (0.8857) and sensitivity (0.9976) to determine the endometrial date. However, the relationship between molecular data and clinical results is still being investigated [22].

Proteomics Study of the Endometrium

Proteomics means the analysis of proteins in any sample. Since there are many regulatory steps between the transcriptome and functional proteins, proteomics provides more information on physiology compared to genomics [23, 24]. The proteomic analyses to date have mainly used two-dimensional differential in-gel electrophoresis (2D-DiGE) followed by matrix-assisted desorption/ionization tandem time-of-flight mass spectrometry (MALDI-TOF MS) [24, 25] to define protein exchanges between proliferative and secretory phases or between pre-receptive and receptive endometrium [17, 26]. The presence of these proteins has been confirmed by immunohistochemistry, and their cellular localization has been determined in the endometrium [26].

Proliferative, early secretory, and mid-secretory proteins have been determined by examining the uterine fluids of fertile and infertile women with proteomic analysis, and protein differences in fertile-infertile cycles are understood [27, 28]. The proteins confirmed to be associated with endometrial receptivity discovered by proteomic studies from endometrial tissues of women with proven fertility include chloride intracellular channel protein 1, Rho GDP-dissociation inhibitor 1, membrane-associated progesterone receptor component 1, Ezrin, stathmin 1, annexin A2, annexin IV, calreticulin and a1-antitrypsin [15]. The five genes associated with endometrial receptivity are progesterone receptor-membrane component 1, stathmin 1, apolipoprotein L2, aldehyde dehydrogenase 1 family member A3, and S100-calcium-binding protein A10 (S100A10). Their presence has been confirmed by both transcriptomic analysis and proteomic study. However, their functions have not yet been fully determined [5, 18, 29, 30].

Lipidomic Analysis of the Endometrium

Though less used than genomic and proteomic analyses, another method applied in studies evaluating the endometrium is lipidomic analysis [31]. The endometrium has very important lipid components, and lipid mediators have roles in important functions, including signal transmission, proliferation, apoptosis, and in the cell membrane [32]. Therefore, prostaglandins, thromboxanes, leukotrienes, endocannabinoids, and sphingolipids from the eicosanoid family, which are important lipid mediators secreted in the endometrium, have been studied and shown to have effects in reproduction. It has been confirmed in animal experiments conducted with these agents that lipids play an important role in implantation [33, 34]. Prostaglandins, the most studied lipids, have been shown to have an important role in implantation in both animal [35] and human [36] studies.

Most of the information was obtained from animal experiments since the studies conducted to evaluate the roles of these agents in implantation are difficult to apply in humans. More studies are needed to elucidate the functions of lipids in implantation.

Potential Markers of Endometrial Receptivity

There are three stages of embryo implantation in the endometrium: apposition, adhesion and invasion [37]. In the apposition stage, the blastocyst adheres to the epithelium of the receptive endometrium with trophoblastic cells. This adhesion occurs when the pinopodes formed in the endometrium and mucin-1 (MUC-1] secreted from the surface epithelium carry the blastocyst to the appropriate area and direct it. In this process, the endometrium is under the influence of progesterone. This hormone provides for the formation of pinopodes and the secretion of MUC-1. In addition, the formation of pinopodes takes place via HOXA-10 gene transcription [38]. There is an intense interaction between the blastocyst carried to the appropriate endometrial area and the endometrium. In this interaction, many molecules along with growth factors and cytokines help signal transmission between the embryo and the endometrium, allowing adhesion to occur [39]. Intercellular and extracellular matrix interactions occur through cell surface adhesion molecules. The best known of these surface adhesion molecules are integrin, cadherin, selectin and immunoglobulins [40]. Invasion occurs when trophoblasts penetrate into the endometrial stroma by passing through the luminal epithelium and basal lamina. Thus, the formation of the maternal vascular bed and placental system required for the embryo to maintain its vitality is provided [41].

Implantation is controlled by this complex and sophisticated interaction between the embryo and the endometrium that begins in the early stages of oocyte maturation. Synchronization between the endometrium and the embryo is provided by this interaction. Endometrial hormones and factors provide this interaction (Table **1**) [42].

By understanding the activity and functions of these hormones and factors, it may be possible to use them in clinical practice as predictive biomarkers of endometrial receptivity or embryo quality. Many studies have been conducted for this purpose. Especially with the development of omics technology, it has become possible to carry out proteomic, genomic and lipidomic analyses of these hormones and factors [43]. In order to identify biomarkers specific to endometrial receptivity, active and functional mediators are examined, especially in the mid-secretory phase. A considerable amount of data has been obtained regarding this subject in the studies conducted so far. However, the overlap between the results of the different studies is relatively small [44]. This is due to differences in study design, different time and conditions of collection of samples, different patient selection criteria, use of different transcriptome sequence/sequencing platforms and genome annotation for data analysis, and lack of consistent standards for data presentation [6, 21, 45 - 47].

Table 1. Endometrial factors that act as mediators in all stages of the implantation process [42].

Localization	Factors
Trophoblasts	Adhesion molecules, cytokine and chemokine receptors, proteases
Luminal epithelium	Glycocalyx components, adhesion molecules, cytokines and chemokines, proteases, calcium regulators
Glands	Cytokines and chemokines, growth factors, nutrients
Extracellular matrix	Hyaluronan, collagen, degradative enzymes and inhibitors
Decidualizing stroma	Prostaglandins, cytokines and chemokines, growth factors, proteases

In order to obtain clearer and more consistent results from these studies, application of certain standardizations may be required in the studies. In addition, although massive omics data have been created from the studies conducted so far, there are only three silico data-mining studies (studies using gene expression data published to date) that have used gene expression data published so far [48 - 50]. Bhagwat *et al.* created a database of human gene expression endometrial receptivity (HGEx-ERdb) consisting of 19,285 genes expressed in the endometrium, identifying 179 of these genes that were specific to the receptive endometrium [50]. Zhang *et al.* [48] identified potential biomarkers of 148 receptive endometrium genes by analyzing raw data from three microarray studies, while Tapia *et al.* [49] identified 61 potential biomarkers specific to the receptive endometrium from seven microarray studies. As a result of these three in silico analysis studies, only nine genes were found in common. In addition, these studies have emphasized that there are great differences in the designs of transcriptomic studies, analysis methods and data processing. In a meta-analysis study, it was attempted to determine meta-signature biomarkers with high-assumed endometrial receptivity to reduce the differences in endometrial transcriptomic analysis by using the robust rank aggregation (RRA) method [51] followed by enrichment analysis. As a result of this meta-analysis, they identified 57 different gene expression differences between pre-receptive and mid-secretory endometrial phases. Of these genes, 52 were up-regulated, 5 were down-regulated and carrying meta-signatures. Up-regulated and highly expressed transcripts in the receptive endometrium were PAEP, SPP1, GPX3, MAOA and GADD45A, while the 5 transcripts that were identified as down-regulated were SFRP4, EDN3, OLFM1, CRABP2 and MMP7 [44]. As in this study, joint evaluation of the data obtained from the previous study results and the implementation of certain standardizations in future studies will provide an increase in specificity in determining biomarkers of endometrial receptivity.

Molecules with Determined Functions in Endometrial Receptivity

Endometrial receptivity can be considered as a barrier that includes the activation of adhesion ligands and inactivation of inhibitor components and provides the adhesion of the embryo [52]. With the functional evaluation of endometrial receptivity, a large number of molecular mediators, including adhesion molecules, cytokines, growth factors and lipids have been identified to date [53]. The discovery of these molecules, which are of high importance for successful embryo implantation, provided valuable information on endometrial receptivity. However, the mechanisms of these molecules, which are functional during the implantation process, are still not fully understood [10]. The molecules whose efficiency during the implantation process was determined are given in Table **2**.

Table 2. Molecules whose effectiveness during the implantation process has been determined.

Cell Adhesion Molecules	
1- Integrins	
α1β1, α4β1 and αVβ3	- Histologically, expression of these proteins increases in the endometrium on the 20-24th days of the cycle. Therefore, they are recommended as markers for the implantation window [54].
β3	- Integrin β3 expression can be used to predict success in the IVF cycle [55].
αVβ3	- This represents a promising clinical and research marker of the human implantation process [56]. - In the endometrium, expression of αVda3 integrin is regulated by both positive (e.g. epidermal growth factor (EGF), heparin-binding EGF (HB-EGF)) and negative factors (e.g. 17β-estradiol (E2]) [56]. - The association of abnormal αVβ3 integrin expression with unexplained infertility [57], endometriosis [58], hydrosalpinx [59], luteal phase defect [60] and polycystic ovary syndrome [61] has been determined.
α1β1 and α5β1	- These function in embryonic adhesion, being found on both the uterine epithelial surface and the trophoblast [62].

(Table 2) cont.....

Cell Adhesion Molecules	
2- Selectins	
L-selectin	- This enables the blastocyst to find the appropriate localization in the uterine cavity and adhere to the endometrial epithelium for successful implantation [63, 64]. - Strong L-selectin staining was observed on the blastocyst side compared to the entire embryo surface [65]. - On the maternal side, expression of selectin oligosaccharide-based ligands such as MECA-79 or HECA-452 is up-regulated during the implantation window [65].
3- Cadherins	
E-cadherin	- In studies on mouse embryo implantation, mutations in the E-cadherin gene have resulted in defective preimplantation development [66]. Its role in human embryo implantation is unknown. However, looking at the expression patterns, it is thought to be important for this process [10]. - E-cadherin mRNA levels have been shown to be significantly higher during the luteal phase [67].
4- Immunoglobulins	
ICAM-1 or CD54	- ICAM-1 expression in endometrial cells has been demonstrated in the luteal phase [68]. - The effect of ICAM-1 on the interaction between the endometrium and the blastocyst could not be demonstrated. However, ICAM is thought to play a role in the process due to its close relationship with the immune system [30].
Mucins	
MUC-1	- Endometrial MUC-1 is up-regulated during the peri-implantation process [69]. - It is the first molecule that the embryo encounters on its attachment path. - It guides the blastocyst until it finds the right time and place for implantation [66]. - Mucins are found on the epithelial cell surface and protect the cell surface against pathological processes [70]. High progesterone levels decrease MUC-1 expression. Thus, the effect of CAMs on the endometrial surface is removed, and the embryo-epithelial interaction is facilitated [71]. Therefore, MUC-1 inhibits implantation; its down-regulation contributes to the formation of endometrial receptivity [72].
Cytokines	
LIF	- It has been shown that the production of LIF gradually increases in uterine washes taken from fertile patients from LH+7th day to LH+12 [73]. - LIF levels have been found to be lower in infertile women compared to fertile women, particularly in those with recurrent implantation failure [73]. - The main role of LIF in implantation has been established on the basis of abnormal LIF levels in infertile patients and particularly in recurrent implantation failures [74].
IL-6	- Endometrial IL-6 mRNA expression steadily increases in the mid- to late secretory phase and decreases in the late secretory phase [75]. - Strong immunoreactivity is detected for IL-6 during the implantation window process [76].
IL-1	- Expression of IL-1 antagonist has been found to decrease during the implantation window [77].

(Table 2) cont.....

Cell Adhesion Molecules	
PG	- PG's have a role in the timing of the implantation window [34]. - In the delayed timing of blastocyst implantation, the embryo placement near the cervix in mice has a fluctuating effect, manifested as abnormal placentation and fetal resorption [78].
Endometrial Growth Factors	
HBEGF	- This is expressed at the time of maximal endometrial receptivity. - It has a role in both adhesion and embryo development [79].
IGFBP-1	- This secretory protein is the major product of the endometrium and deciduan. It inhibits the effect of IGFBP-1 on target cells. Its role in endometrial receptivity requires further research [79].
Endometrial Immune Markers	
CD8+, CD56+, CD4+	- Significantly lower CD8+ (T suppressor/cytotoxic) and CD56+ levels and higher CD4+ (T helper/inducer) levels were reported in women with unexplained infertility compared to the fertile control group [80].
Matrix proteins	
Laminin, fibronectin, collagen IV	- These are present in the secretory endometrium. They are absent in the endometrium of patients with unexplained infertility. This indicates that these matrix proteins are probably necessary proteins for implantation [81].
Glycodelin	- This is thought to have an immunomodulatory role in the implantation period [82].

PG: Prostaglandin, HBEGF: Heparin Binding-Epidermal Growth Factor, IGFBP-1: Insulin-like growth factor binding protein-1.

Omics and why is the Measurement of Endometrial Receptivity Important

The evaluation of endometrial receptivity is particularly important in three applications. The first of these is that it will be useful in evaluating endometrial receptivity for those who experience recurrent implantation failure, since no matter how well the embryo transfer is performed, the transfer of the embryo to an endometrium whose receptivity is defective will result in failure. For this reason, having tests that can evaluate endometrial receptivity will enable evaluating the success of IVF by standardizing the receptivity window in patients. Women who experience recurrent implantation failure will particularly benefit from this [15]. The second important reason for evaluating endometrial receptivity is to obtain information about the receptivity of the endometrium before IVF. With the stimulation of ovaries performed to increase the chances of success in IVF, the increased levels of ovarian steroids, namely estrogen and progesterone, affect the morphology and function of the endometrium and disrupt endometrial receptivity. Therefore, tests evaluating endometrial receptivity allow the transfer to be in the same cycle in cases with good receptivity and to freeze the embryo for a more appropriate cycle in cases where the receptivity is not good [83, 84].

The third important reason for the application of tests that evaluate endometrial receptivity is that if the tests are reliable and inexpensive, they will be performed routinely in the evaluation of infertility and will ensure the detection of endometrial receptivity disorders before treatment, thus reducing the treatment process and cost.

With omics technologies, it may be possible to detect new biomarkers related to specific biomarkers for endometrial receptivity. Nowadays, they have been customized as a single sequence and turned into diagnostic tests to evaluate endometrial receptivity. However, the reliability of these tests has not yet been proven [85]. In order to benefit from these tests, they must first be optimized.

CONCLUSION

The determination of biomarkers of endometrial receptivity will make an important contribution to increasing IVF success, particularly for patients with recurrent implantation failure. Some potential biomarkers have been identified, particularly with the studies conducted on this subject as a result of the development of omics technology. However, most are still unclear and controversial. For this reason, it would be appropriate to carry out further detailed studies on this subject and to increase the sensitivity of biomarkers determined from the data to be obtained by providing standardization in carrying out these studies.

CONSENT FOR PUBLICATION

Not applicable.

CONFLICT OF INTEREST

The authors declare no conflict of interest, financial or otherwise.

ACKNOWLEDGMENT

We gratefully thank Prof. Dr. Hafize UZUN for giving us the opportunity to be a part of this book, “Biomarkers in Medicine”.

REFERENCES

[1] Haouzi D, Dechaud H, Assou S, De Vos J, Hamamah S. Insights into human endometrial receptivity from transcriptomic and proteomic data. Reprod Biomed Online 2012; 24(1): 23-34. [http://dx.doi.org/10.1016/j.rbmo.2011.09.009] [PMID: 22119322]

[2] Croxatto HB, Ortiz ME, Díaz S, Hess R, Balmaceda J, Croxatto HD. Studies on the duration of egg transport by the human oviduct. II. Ovum location at various intervals following luteinizing hormone peak. Am J Obstet Gynecol 1978; 132(6): 629-34. [http://dx.doi.org/10.1016/0002-9378(78)90854-2] [PMID: 717467]

[3] Bergh PA, Navot D. The impact of embryonic development and endometrial maturity on the timing of implantation. Fertil Steril 1992; 58(3): 537-42. [http://dx.doi.org/10.1016/S0015-0282(16)55259-5] [PMID: 1521649]

[4] Lessey BA, Young SL. What exactly is endometrial receptivity? Fertil Steril 2019; 111(4): 611-7. [http://dx.doi.org/10.1016/j.fertnstert.2019.02.009] [PMID: 30929718]

[5] A genomic diagnostic tool for human endometrial receptivity based on the transcriptomic signature. Fertil Steril, 2011. 95(1): p. 50-60.

[6] Aghajanova L, Hamilton AE, Giudice LC. Uterine receptivity to human embryonic implantation: histology, biomarkers, and transcriptomics. Semin Cell Dev Biol 2008; 19(2): 204-11. [http://dx.doi.org/10.1016/j.semcdb.2007.10.008] [PMID: 18035563]

[7] Carson DD, Bagchi I, Dey SK, *et al.* Embryo implantation. Dev Biol 2000; 223(2): 217-37. [http://dx.doi.org/10.1006/dbio.2000.9767] [PMID: 10882512]

[8] Giudice LC. Elucidating endometrial function in the post-genomic era. Hum Reprod Update 2003; 9(3): 223-35. [http://dx.doi.org/10.1093/humupd/dmg019] [PMID: 12859044]

[9] Noyes RW, Hertig AT, Rock J. Dating the endometrial biopsy. Am J Obstet Gynecol 1975; 122(2): 262-3. [http://dx.doi.org/10.1016/S0002-9378(16)33500-1] [PMID: 1155504]

[10] Achache H, Revel A. Endometrial receptivity markers, the journey to successful embryo implantation. Hum Reprod Update 2006; 12(6): 731-46. [http://dx.doi.org/10.1093/humupd/dml004] [PMID: 16982667]

[11] Baugher JL, Klaenhammer TR. Invited review: Application of omics tools to understanding probiotic functionality. J Dairy Sci 2011; 94(10): 4753-65. [http://dx.doi.org/10.3168/jds.2011-4384] [PMID: 21943727]

[12] Fukushima A, Kusano M, Redestig H, Arita M, Saito K. Integrated omics approaches in plant systems biology. Curr Opin Chem Biol 2009; 13(5-6): 532-8. [http://dx.doi.org/10.1016/j.cbpa.2009.09.022] [PMID: 19837627]

[13] Nikas G, Drakakis P, Loutradis D, *et al.* Uterine pinopodes as markers of the 'nidation window' in cycling women receiving exogenous oestradiol and progesterone. Hum Reprod 1995; 10(5): 1208-13. [http://dx.doi.org/10.1093/oxfordjournals.humrep.a136120] [PMID: 7657767]

[14] Mikołajczyk M, Skrzypczak J, Wirstlein P. No correlation between pinopode formation and LIF and MMP2 expression in endometrium during implantation window. Folia Histochem Cytobiol 2011; 49(4): 615-21. [http://dx.doi.org/10.5603/FHC.2011.0084] [PMID: 22252755]

[15] Edgell TA, Rombauts LJ, Salamonsen LA. Assessing receptivity in the endometrium: the need for a rapid, non-invasive test. Reprod Biomed Online 2013; 27(5): 486-96. [http://dx.doi.org/10.1016/j.rbmo.2013.05.014] [PMID: 23933033]

[16] Carson DD, Lagow E, Thathiah A, *et al.* Changes in gene expression during the early to mid-luteal (receptive phase) transition in human endometrium detected by high-density microarray screening. Mol Hum Reprod 2002; 8(9): 871-9. [http://dx.doi.org/10.1093/molehr/8.9.871] [PMID: 12200466]

[17] Domínguez F, Garrido-Gómez T, López JA, *et al.* Proteomic analysis of the human receptive versus non-receptive endometrium using differential in-gel electrophoresis and MALDI-MS unveils stathmin 1 and annexin A2 as differentially regulated. Hum Reprod, 2009. 24(10): p. 2607-17..

[18] Haouzi D, Assou S, Mahmoud K, *et al.* Gene expression profile of human endometrial receptivity: comparison between natural and stimulated cycles for the same patients. Hum Reprod 2009; 24(6): 1436-45.

[http://dx.doi.org/10.1093/humrep/dep039] [PMID: 19246470]

[19] Kao LC, Tulac S, Lobo S, *et al.* Global gene profiling in human endometrium during the window of implantation. Endocrinology 2002; 143(6): 2119-38. [http://dx.doi.org/10.1210/endo.143.6.8885] [PMID: 12021176]

[20] Kuokkanen S, Chen B, Ojalvo L, Benard L, Santoro N, Pollard JW. Genomic profiling of microRNAs and messenger RNAs reveals hormonal regulation in microRNA expression in human endometrium. Biol Reprod 2010; 82(4): 791-801. [http://dx.doi.org/10.1095/biolreprod.109.081059] [PMID: 19864316]

[21] Ruiz-Alonso M, Blesa D, Simón C. The genomics of the human endometrium. Biochim Biophys Acta 2012; 1822(12): 1931-42. [http://dx.doi.org/10.1016/j.bbadis.2012.05.004] [PMID: 22634130]

[22] Clemente-Ciscar M. R.-A.M., Blesa D, Jimenez-Almazan J, and B.M.e.a. Bahceci M. Endometrial receptivity analysis (ERA) using a next generation sequencing (NGS) predictor improves reproductive outcome in recurrent implantation failure (RIF) patients when compared to ERA arrays. 34th Annual Meeting of the European Society of Human Reproduction and Embryology. i8.

[23] Burney RO, Talbi S, Hamilton AE, *et al.* Gene expression analysis of endometrium reveals progesterone resistance and candidate susceptibility genes in women with endometriosis. Endocrinology 2007; 148(8): 3814-26. [http://dx.doi.org/10.1210/en.2006-1692] [PMID: 17510236]

[24] Chen JI, Hannan NJ, Mak Y, *et al.* Proteomic characterization of midproliferative and midsecretory human endometrium. J Proteome Res 2009; 8(4): 2032-44. [http://dx.doi.org/10.1021/pr801024g] [PMID: 19714818]

[25] DeSouza L, Diehl G, Yang EC, *et al.* Proteomic analysis of the proliferative and secretory phases of the human endometrium: protein identification and differential protein expression. Proteomics 2005; 5(1): 270-81. [http://dx.doi.org/10.1002/pmic.200400920] [PMID: 15602768]

[26] Li J, Tan Z, Li MT, *et al.* Study of altered expression of annexin IV and human endometrial receptivity. Zhonghua Fu Chan Ke Za Zhi 2006; 41(12): 803-5. [PMID: 17327107]

[27] Casado-Vela J, Rodriguez-Suarez E, Iloro I, *et al.* Comprehensive proteomic analysis of human endometrial fluid aspirate. J Proteome Res 2009; 8(10): 4622-32. [http://dx.doi.org/10.1021/pr9004426] [PMID: 19670903]

[28] Hannan NJ, Stephens AN, Rainczuk A, Hincks C, Rombauts LJ, Salamonsen LA. 2D-DiGE analysis of the human endometrial secretome reveals differences between receptive and nonreceptive states in fertile and infertile women. J Proteome Res 2010; 9(12): 6256-64. [http://dx.doi.org/10.1021/pr1004828] [PMID: 20925431]

[29] Mirkin S, Arslan M, Churikov D, *et al.* In search of candidate genes critically expressed in the human endometrium during the window of implantation. Hum Reprod 2005; 20(8): 2104-17. [http://dx.doi.org/10.1093/humrep/dei051] [PMID: 15878921]

[30] Talbi S, Hamilton AE, Vo KC, *et al.* Molecular phenotyping of human endometrium distinguishes menstrual cycle phases and underlying biological processes in normo-ovulatory women. Endocrinology 2006; 147(3): 1097-121. [http://dx.doi.org/10.1210/en.2005-1076] [PMID: 16306079]

[31] Wenk MR. The emerging field of lipidomics. Nat Rev Drug Discov 2005; 4(7): 594-610. [http://dx.doi.org/10.1038/nrd1776] [PMID: 16052242]

[32] Loizides-Mangold U. On the future of mass-spectrometry-based lipidomics. FEBS J 2013; 280(12): 2817-29. [http://dx.doi.org/10.1111/febs.12202] [PMID: 23432956]

[33] Wang H, Dey SK. Lipid signaling in embryo implantation. Prostaglandins Other Lipid Mediat 2005; 77(1-4): 84-102.
[http://dx.doi.org/10.1016/j.prostaglandins.2004.09.013] [PMID: 16099394]

[34] Ye X, Hama K, Contos JJ, *et al.* LPA3-mediated lysophosphatidic acid signalling in embryo implantation and spacing. Nature 2005; 435(7038): 104-8.
[http://dx.doi.org/10.1038/nature03505] [PMID: 15875025]

[35] Arosh JA, Banu SK, McCracken JA. Novel concepts on the role of prostaglandins on luteal maintenance and maternal recognition and establishment of pregnancy in ruminants. J Dairy Sci 2016; 99(7): 5926-40.
[http://dx.doi.org/10.3168/jds.2015-10335] [PMID: 27179861]

[36] Vilella F, Ramirez L, Berlanga O, *et al.* PGE2 and PGF2α concentrations in human endometrial fluid as biomarkers for embryonic implantation. J Clin Endocrinol Metab 2013; 98(10): 4123-32.
[http://dx.doi.org/10.1210/jc.2013-2205] [PMID: 23979956]

[37] Enders AC, Schlafke S. Cytological aspects of trophoblast-uterine interaction in early implantation. Am J Anat 1969; 125(1): 1-29.
[http://dx.doi.org/10.1002/aja.1001250102] [PMID: 5770161]

[38] Du H, Taylor HS. The Role of Hox Genes in Female Reproductive Tract Development, Adult Function, and Fertility. Cold Spring Harb Perspect Med 2015; 6(1): a023002.
[http://dx.doi.org/10.1101/cshperspect.a023002] [PMID: 26552702]

[39] Navot D, Scott RT, Droesch K, Veeck LL, Liu HC, Rosenwaks Z. The window of embryo transfer and the efficiency of human conception in vitro. Fertil Steril 1991; 55(1): 114-8.
[http://dx.doi.org/10.1016/S0015-0282(16)54069-2] [PMID: 1986951]

[40] Carson DD, Tang JP, Julian J. Heparan sulfate proteoglycan (perlecan) expression by mouse embryos during acquisition of attachment competence. Dev Biol 1993; 155(1): 97-106.
[http://dx.doi.org/10.1006/dbio.1993.1010] [PMID: 8416848]

[41] Guzeloglu-Kayisli O, Basar M, Arici A. Basic aspects of implantation. Reprod Biomed Online 2007; 15(6): 728-39.
[http://dx.doi.org/10.1016/S1472-6483(10)60541-X] [PMID: 18062873]

[42] Diedrich K, Fauser BC, Devroey P, Griesinger G. The role of the endometrium and embryo in human implantation. Hum Reprod Update 2007; 13(4): 365-77.
[http://dx.doi.org/10.1093/humupd/dmm011] [PMID: 17548368]

[43] Demiral İ, Doğan M, Baştu E, Buyru F. Genomic, proteomic and lipidomic evaluation of endometrial receptivity. Turk J Obstet Gynecol 2015; 12(4): 237-43.
[http://dx.doi.org/10.4274/tjod.98475] [PMID: 28913076]

[44] Altmäe S, Koel M, Võsa U, *et al.* Meta-signature of human endometrial receptivity: a meta-analysis and validation study of transcriptomic biomarkers. Sci Rep 2017; 7(1): 017-10098.

[45] Horcajadas JA, Pellicer A, Simón C. Wide genomic analysis of human endometrial receptivity: new times, new opportunities. Hum Reprod Update 2007; 13(1): 77-86.
[http://dx.doi.org/10.1093/humupd/dml046] [PMID: 16960016]

[46] Altmäe S, Martínez-Conejero JA, Salumets A, Simón C, Horcajadas JA, Stavreus-Evers A. Endometrial gene expression analysis at the time of embryo implantation in women with unexplained infertility. Mol Hum Reprod 2010; 16(3): 178-87.
[http://dx.doi.org/10.1093/molehr/gap102] [PMID: 19933690]

[47] Ulbrich SE, Groebner AE, Bauersachs S. Transcriptional profiling to address molecular determinants of endometrial receptivity--lessons from studies in livestock species. Methods 2013; 59(1): 108-15.
[http://dx.doi.org/10.1016/j.ymeth.2012.10.013] [PMID: 23178633]

[48] Zhang D, Sun C, Ma C, Dai H, Zhang W. Data mining of spatial-temporal expression of genes in the

human endometrium during the window of implantation. Reprod Sci 2012; 19(10): 1085-98. [http://dx.doi.org/10.1177/1933719112442248] [PMID: 22915151]

[49] Tapia A, Vilos C, Marín JC, Croxatto HB, Devoto L. Bioinformatic detection of E47, E2F1 and SREBP1 transcription factors as potential regulators of genes associated to acquisition of endometrial receptivity. Reprod Biol Endocrinol 2011; 9(14): 14. [http://dx.doi.org/10.1186/1477-7827-9-14] [PMID: 21272326]

[50] Bhagwat SR, Chandrashekar DS, Kakar R, *et al.* Endometrial receptivity: a revisit to functional genomics studies on human endometrium and creation of HGEx-ERdb. PLoS One 2013; 8(3): e58419. [http://dx.doi.org/10.1371/journal.pone.0058419] [PMID: 23555582]

[51] Kolde R, Laur S, Adler P, Vilo J. Robust rank aggregation for gene list integration and meta-analysis. Bioinformatics 2012; 28(4): 573-80. [http://dx.doi.org/10.1093/bioinformatics/btr709] [PMID: 22247279]

[52] Aplin JD. The cell biological basis of human implantation. Best Pract Res Clin Obstet Gynaecol 2000; 14(5): 757-64. [http://dx.doi.org/10.1053/beog.2000.0116] [PMID: 11023797]

[53] Simón C, Martín JC, Pellicer A. Paracrine regulators of implantation. Best Pract Res Clin Obstet Gynaecol 2000; 14(5): 815-26. [http://dx.doi.org/10.1053/beog.2000.0121] [PMID: 11023802]

[54] Lessey BA, Castelbaum AJ, Wolf L, *et al.* Use of integrins to date the endometrium. Fertil Steril 2000; 73(4): 779-87. [http://dx.doi.org/10.1016/S0015-0282(99)00604-4] [PMID: 10731541]

[55] Thomas K, Thomson A, Wood S, Kingsland C, Vince G, Lewis-Jones I. Endometrial integrin expression in women undergoing in vitro fertilization and the association with subsequent treatment outcome. Fertil Steril 2003; 80(3): 502-7. [http://dx.doi.org/10.1016/S0015-0282(03)00792-1] [PMID: 12969689]

[56] Lessey BA. Two pathways of progesterone action in the human endometrium: implications for implantation and contraception. Steroids 2003; 68(10-13): 809-15. [http://dx.doi.org/10.1016/j.steroids.2003.09.004] [PMID: 14667972]

[57] Tei C, Maruyama T, Kuji N, Miyazaki T, Mikami M, Yoshimura Y. Reduced expression of alphavbeta3 integrin in the endometrium of unexplained infertility patients with recurrent IVF-ET failures: improvement by danazol treatment. J Assist Reprod Genet 2003; 20(1): 13-20. [http://dx.doi.org/10.1023/A:1021254620888] [PMID: 12645863]

[58] Lessey BA, Castelbaum AJ, Sawin SW, *et al.* Aberrant integrin expression in the endometrium of women with endometriosis. J Clin Endocrinol Metab 1994; 79(2): 643-9. [PMID: 7519194]

[59] Meyer WR, Castelbaum AJ, Somkuti S, *et al.* Hydrosalpinges adversely affect markers of endometrial receptivity. Hum Reprod 1997; 12(7): 1393-8. [http://dx.doi.org/10.1093/humrep/12.7.1393] [PMID: 9262264]

[60] Lessey BA, Damjanovich L, Coutifaris C, Castelbaum A, Albelda SM, Buck CA. Integrin adhesion molecules in the human endometrium. Correlation with the normal and abnormal menstrual cycle. J Clin Invest 1992; 90(1): 188-95. [http://dx.doi.org/10.1172/JCI115835] [PMID: 1378853]

[61] Apparao KB, Lovely LP, Gui Y, Lininger RA, Lessey BA. Elevated endometrial androgen receptor expression in women with polycystic ovarian syndrome. Biol Reprod 2002; 66(2): 297-304. [http://dx.doi.org/10.1095/biolreprod66.2.297] [PMID: 11804942]

[62] Damsky CH, Librach C, Lim KH, *et al.* Integrin switching regulates normal trophoblast invasion. Development 1994; 120(12): 3657-66. [http://dx.doi.org/10.1242/dev.120.12.3657] [PMID: 7529679]

[63] Smalley DM, Ley K. L-selectin: mechanisms and physiological significance of ectodomain cleavage. J Cell Mol Med 2005; 9(2): 255-66. [http://dx.doi.org/10.1111/j.1582-4934.2005.tb00354.x] [PMID: 15963248]

[64] Dominguez F, Yáñez-Mó M, Sanchez-Madrid F, Simón C. Embryonic implantation and leukocyte transendothelial migration: different processes with similar players? FASEB J 2005; 19(9): 1056-60. [http://dx.doi.org/10.1096/fj.05-3781hyp] [PMID: 15985528]

[65] Genbacev OD, Prakobphol A, Foulk RA, *et al.* Trophoblast L-selectin-mediated adhesion at the maternal-fetal interface. Science 2003; 299(5605): 405-8. [http://dx.doi.org/10.1126/science.1079546] [PMID: 12532021]

[66] Riethmacher D, Brinkmann V, Birchmeier C. A targeted mutation in the mouse E-cadherin gene results in defective preimplantation development. Proc Natl Acad Sci USA 1995; 92(3): 855-9. [http://dx.doi.org/10.1073/pnas.92.3.855] [PMID: 7846066]

[67] Fujimoto J, Ichigo S, Hori M, Tamaya T. Alteration of E-cadherin, alpha- and beta-catenin mRNA expression in human uterine endometrium during the menstrual cycle. Gynecol Endocrinol 1996; 10(3): 187-91. [http://dx.doi.org/10.3109/09513599609027987] [PMID: 8862494]

[68] Gaffuri B, Airoldi L, Di Blasio AM, *et al.* Unexplained habitual abortion is associated with a reduced endometrial release of soluble intercellular adhesion molecule-1 in the luteal phase of the cycle. Eur J Endocrinol 2000; 142(5): 477-80. [http://dx.doi.org/10.1530/eje.0.1420477] [PMID: 10802525]

[69] Aplin JD, Hey NA, Graham RA. Human endometrial MUC1 carries keratan sulfate: characteristic glycoforms in the luminal epithelium at receptivity. Glycobiology 1998; 8(3): 269-76. [http://dx.doi.org/10.1093/glycob/8.3.269] [PMID: 9451036]

[70] Strous GJ, Dekker J. Mucin-type glycoproteins. Crit Rev Biochem Mol Biol 1992; 27(1-2): 57-92. [http://dx.doi.org/10.3109/10409239209082559] [PMID: 1727693]

[71] Surveyor GA, Gendler SJ, Pemberton L, *et al.* Expression and steroid hormonal control of Muc-1 in the mouse uterus. Endocrinology 1995; 136(8): 3639-47. [http://dx.doi.org/10.1210/endo.136.8.7628404] [PMID: 7628404]

[72] DeSouza MM, Mani SK, Julian J, Carson DD. Reduction of mucin-1 expression during the receptive phase in the rat uterus. Biol Reprod 1998; 58(6): 1503-7. [http://dx.doi.org/10.1095/biolreprod58.6.1503] [PMID: 9623612]

[73] Laird SM, Tuckerman EM, Dalton CF, Dunphy BC, Li TC, Zhang X. The production of leukaemia inhibitory factor by human endometrium: presence in uterine flushings and production by cells in culture. Hum Reprod 1997; 12(3): 569-74. [http://dx.doi.org/10.1093/humrep/12.3.569] [PMID: 9130761]

[74] Steck T, Giess R, Suetterlin MW, *et al.* Leukaemia inhibitory factor (LIF) gene mutations in women with unexplained infertility and recurrent failure of implantation after IVF and embryo transfer. Eur J Obstet Gynecol Reprod Biol 2004; 112(1): 69-73. [http://dx.doi.org/10.1016/S0301-2115(03)00315-4] [PMID: 14687743]

[75] von Wolff M, Stieger S, Lumpp K, Bücking J, Strowitzki T, Thaler CJ. Endometrial interleukin-6 in vitro is not regulated directly by female steroid hormones, but by pro-inflammatory cytokines and hypoxia. Mol Hum Reprod 2002; 8(12): 1096-102. [http://dx.doi.org/10.1093/molehr/8.12.1096] [PMID: 12468642]

[76] Tabibzadeh S, Kong QF, Babaknia A, May LT. Progressive rise in the expression of interleukin-6 in human endometrium during menstrual cycle is initiated during the implantation window. Hum Reprod 1995; 10(10): 2793-9. [http://dx.doi.org/10.1093/oxfordjournals.humrep.a135793] [PMID: 8567815]

[77] Boucher A, Kharfi A, Al-Akoum M, Bossù P, Akoum A. Cycle-dependent expression of interleukin-1

receptor type II in the human endometrium. Biol Reprod 2001; 65(3): 890-8.
[http://dx.doi.org/10.1095/biolreprod65.3.890] [PMID: 11514355]

[78] Song H, Lim H, Paria BC, *et al.* Cytosolic phospholipase A2alpha is crucial [correction of A2alpha deficiency is crucial] for 'on-time' embryo implantation that directs subsequent development. Development 2002; 129(12): 2879-89.
[http://dx.doi.org/10.1242/dev.129.12.2879] [PMID: 12050136]

[79] Tamada H, Higashiyama C, Takano H, Kawate N, Inaba T, Sawada T. The effects of heparin-binding epidermal growth factor-like growth factor on preimplantation-embryo development and implantation in the rat. Life Sci 1999; 64(22): 1967-73.
[http://dx.doi.org/10.1016/S0024-3205(99)00128-9] [PMID: 10374922]

[80] Klentzeris LD, Bulmer JN, Warren MA, Morrison L, Li TC, Cooke ID. Lymphoid tissue in the endometrium of women with unexplained infertility: morphometric and immunohistochemical aspects. Hum Reprod 1994; 9(4): 646-52.
[http://dx.doi.org/10.1093/oxfordjournals.humrep.a138564] [PMID: 7519197]

[81] Bilalis DA, Klentzeris LD, Fleming S. Immunohistochemical localization of extracellular matrix proteins in luteal phase endometrium of fertile and infertile patients. Hum Reprod 1996; 11(12): 2713-8.
[http://dx.doi.org/10.1093/oxfordjournals.humrep.a019196] [PMID: 9021377]

[82] Mueller MD, Vigne JL, Vaisse C, Taylor RN. Glycodelin: a pane in the implantation window. Semin Reprod Med 2000; 18(3): 289-98.
[http://dx.doi.org/10.1055/s-2000-12566] [PMID: 11299967]

[83] Bourgain C, Devroey P. The endometrium in stimulated cycles for IVF. Hum Reprod Update 2003; 9(6): 515-22.
[http://dx.doi.org/10.1093/humupd/dmg045] [PMID: 14714588]

[84] Kolibianakis E, Bourgain C, Albano C, *et al.* Effect of ovarian stimulation with recombinant follicle-stimulating hormone, gonadotropin releasing hormone antagonists, and human chorionic gonadotropin on endometrial maturation on the day of oocyte pick-up. Fertil Steril 2002; 78(5): 1025-9.
[http://dx.doi.org/10.1016/S0015-0282(02)03323-X] [PMID: 12413988]

[85] Mikołajczyk M, Skrzypczak J. Endometrial receptivity-- can it be diagnosed and controlled? And why does it matter? Ginekol Pol 2014; 85(2): 149-53.
[http://dx.doi.org/10.17772/gp/1706] [PMID: 24745162]

CHAPTER 23

Biomarkers for Preterm Delivery

Ismail BIYIK[1,*] and **Mustafa ALBAYRAK**[2]

[1] *Department of Obstetrics and Gynecology, Kutahya Health Sciences University, Kutahya, Turkey*

[2] *Department of Gynecologic Oncology, Istanbul Faculty of Medicine, Istanbul University, Istanbul, Turkey*

Abstract: Preterm birth occurring before the thirty-seventh gestational week complicates 4.5%-18% of pregnancies worldwide. The pathogenesis of spontaneous preterm delivery is not fully understood. Among the factors held to be responsible for its pathogenesis, the most emphasized is the inflammatory process. Studies in terms of the prediction of preterm delivery are basically divided into 3 categories: 1) Prediction in pregnant women who are asymptomatic and without risk factors, 2) Prediction in pregnant women who are asymptomatic and have risk factors, 3) Prediction in symptomatic pregnant women who have threatened preterm labour. In this chapter, the topic of biomarkers in relation to preterm delivery is discussed. The most commonly used markers in published studies are fetal fibronectin, cervical pIGFBP-1 and cervical length measurement by transvaginal ultrasound. For prediction in symptomatic pregnant women applying to the hospital with threatened preterm labour, the markers used are fetal fibronection, insulin-like growth factors (IGFs) and inflammatory markers. Preterm labour prediction with markers checked in the first and second trimesters are fetal fibronection, insulin-like growth factors (IGFs), micro RNAs, progesterone, circulating microparticles (CMPs), inflammatory markers, matrix metalloproteinases, aneuploidy syndrome screening test parameters and other hormones.

Keywords: Aneuploidy syndrome screening test parameters, Circulating microparticles, Corticotropin-releasing factor, Corticotropin-releasing hormone, C-reactive protein, Fetal fibronectin, Hsa-let-7a-5p, Hsa-miR-15b-5p, Hsa-mi--19b-3p, Hsa-miR-23a-3p, Hsa-miR-93-5p, Hsa-miR-150-5p, Hsa-miR-185-5p, Hsa-miR-191-5p, Hsa-miR-374a-5p, IGFs, IGFBPs, Inflammatory markers, Insulin-like growth factor-binding protein 4, Interleukin 6, Lipopolysaccharide-binding protein, Macrophage inflammatory protein, Matrix metalloproteinases, MicroRNAs, PAPP-A, Placental alpha microglobulin-1, Prediction, Preterm birth, Preterm delivery, Toll-like receptors.

* **Corresponding author Ismail BIYIK:** Department of Obstetrics and Gynecology, Kutahya Health Sciences University, Kutahya, Turkey; Tel: +90 (532) 786 70 31; E-mail: dribiyik@hotmail.com

Pınar Atukeren and Hafize Uzun (Eds.)

INTRODUCTION

Preterm Delivery

Preterm birth occurring before the thirty-seventh gestational week complicates 4.5%-18% of pregnancies worldwide [1 - 3]. Preterm birth can occur in three ways: 1) Spontaneous preterm delivery without preterm premature rupture of membranes, 2) preterm delivery after preterm premature rupture of membranes, 3) Preterm labour induced by maternal or fetal indications. This chapter aims to cover spontaneous preterm labour without preterm premature rupture of membranes. Preterm delivery occurs as a result of 40-50% spontaneous preterm labour, 25-30% after preterm prelabour membrane rupture and 30-35% medical or obstetric reasons [4]. Spontaneous preterm delivery is one of the most important causes of neonatal mortality and morbidity [5].

The pathogenesis of spontaneous preterm delivery is not fully understood. Among the factors held to be responsible for its pathogenesis, the most emphasized is the inflammatory process [6 - 8]. Spontaneous preterm delivery occurs as a result of the onset of labour after pathological signal transmission [8]. The onset of labour begins and progresses with similar mechanisms in both term and preterm pregnant women. All mechanisms cause cervical maturation and initiation of uterine contractions [9]. Cervical maturation is associated with an increase in prostaglandins, chemokines and cytokines, and infiltration of inflammatory cells and increased matrix metalloproteinase activity [10]. Spontaneous delivery occurs from proinflammatory cytokines in the myometrium, amniotic membranes and cervix, and through increased uterine contractility and cervical dilatation due to the increase in prostaglandins [11].

The onset of uterine contractions before the 37th gestational week is called threatened preterm labour (TPL). 9% -24% of pregnant women are hospitalized due to TPL [12 - 15]. However, spontaneous preterm delivery occurs in approximately 50% of pregnant women with complaints related to TPL [15].

Prediction of which women in preterm labour will have a preterm delivery would enable us to apply tocolysis and antenatal steroids to the right patients. Also, this would make it possible for us to refer the mother to a tertiary maternity centre, for the mother to give birth and the baby to be born under safer conditions [16]. In addition, pre-determination of women who will have preterm delivery, and referral of risky pregnancies to perinatal centres and corticosteroid therapy may be possible [17]. There are still no markers with high sensitivity and specificity that can be used for this purpose [18 - 20]. Women who had a preterm birth in their previous pregnancy are in the high-risk group in terms of preterm delivery in their next pregnancy. However, most women with preterm birth do not have a history of preterm birth [21]. Therefore, there is a need for markers to predict preterm birth, except for those with a history of preterm birth.

Studies in terms of the prediction of preterm delivery are basically divided into 3 categories: 1) Prediction for pregnant women who are asymptomatic and without risk factors, 2) Prediction for pregnant women who are asymptomatic and have risk factors, 3) Prediction for symptomatic pregnant women who have threatened preterm labour. These 3 categories will be examined under separate subheadings.

Prediction in Symptomatic Pregnant Women Applying to the Hospital with Threatened Preterm Labour

The most commonly used markers in published studies are fetal fibronectin, cervical pIGFBP-1 and cervical length measurement by transvaginal ultrasound [22].

a). Fetal Fibronectin

Fetal fibronectin (FFN) is found in maternal choroidal tissues, and with the inflammatory process, it is separated from the tissues and mixed with cervicovaginal tissue [23]. The quantitative fetal fibronectin (qFFN) measured in the cervicovaginal fluids of pregnant women who went to the hospital due to preterm labour had a high negative predictive value in predicting preterm labour, but its positive predictive value was lower [24 - 26]. Iams *et al.* reported that fetal fibronectin (in cervicovaginal fluid) may be useful in determining the birth that will develop within 7 days for those who go to the hospital with threatened preterm labour [27]. In their large series prospective study, Esplin *et al.* reported that quantitative vaginal fetal fibronectin and serial transvaginal ultrasound cervical length measurement have low predictive value in predicting spontaneous preterm birth in patients with a nulliparous singleton pregnancy, and therefore cannot be recommended for routine use [28]. In the conducted studies, the sensitivity and positive predictive value of cervicovaginal fetal fibronectin measurement in predicting preterm labour were found to be low [29 - 32].

b). IGFBPs

Insulin-like growth factors (IGFs) play a role in fetal and placental development [33]. In circulation, IGFs bind to IGFBP-3. The clearance of IGFs slows down through this binding [34]. Eroglu *et al.* were able to predict preterm deliveries before 35 weeks of gestation by examining insulin-like growth factor binding protein-1 (phIGFBP-1) in cervicovaginal secretions in 90 pregnant women going to the hospital with threatened preterm labour [35]. IGF binding protein-1 (IGFBP-1) is produced in the decidua. When labour begins and fetal membranes separate from the decidua parietalis, phIGFBP-1 reaches cervical secretions. There are studies showing that phIGFBP-1 provides prediction of preterm delivery [36, 37]. In the study conducted by Fuchs *et al.*, the predictive value of preterm labour was low because the phIGFBP-1 test was used alone or with cervical length [38]. The advantage of phIGFBP-1 over fetal fibronectin is that it is not affected by urine and seminal plasma remaining in the vagina after coitus [39]. Also, phIGFBP-1 is less expensive to measure than fetal fibronectin. Furthermore, bleeding in the cervical region negatively affects the results of both fetal fibronectin and phIGFBP-1 [35].

c). Inflammatory Markers

In a prospective study conducted by Ducarme *et al.*, in which they tried to predict preterm labour using serum procalcitonin levels of pregnant women between 24-36 weeks of gestation who were hospitalized for TPL, they showed that there was no difference in serum procalcitonin levels in terms of sPTB and admission-t--delivery intervals of ≤ 7 and ≤ 14 days between preterm and term deliveries [40]. In the study carried out by Elovitz *et al.*, they looked at 5640 types of miRNA from maternal blood they received at the time of admission from pregnant women hospitalized with a diagnosis of TPL, and they stated that miRNAs do not predict preterm labour [41].

In their study including pregnant women with preterm labour symptoms at 28 - 37 weeks, Shahshahan and Iravani reported that maternal serum CRP (C-reactive protein), ALK-P (alkaline phosphatase) and cervical length are suitable biomarkers to predict preterm labour and response to tocolytic treatment [42]. There are also different studies showing that maternal CRP increase is related to preterm labour [43 - 46]. In addition, there are studies linking high CRP with tocolysis failure [47, 48]. CRP is an increasing marker in cases of infection and inflammation. These findings suggest that preterm labour is the result of increased inflammation and intraamniotic infection.

$CD14^+CD16^+$ cells are the cells associated with the development of inflammation. Toll-like receptors (TLRs) are mostly expressed by monocytes and regulate the inflammatory response [49]. TLR4 is the receptor that is activated early in response to bacterial lipopolysaccharide [50]. In the study of Kim *et al.* comparing pregnant women with preterm labour, term labour and non-labour, they reported that increased TLR expression was measured with high CD16 monocytes and increased CD14 expression in peripheral blood during preterm labour. They also argued that monocyte TLR4 levels could predict preterm labour

[51]. In the study conducted by Pawelczyk *et al.*, increased $TLR4^+/CD14^+$ monocytes and enhanced TLR4 expression were found in the blood of pregnant women in preterm labour [52].

Interleukin 6 (IL-6) is an inflammation-related cytokine. Production of prostaglandins that provide cervical ripening and uterine contractions is increased by IL-6 [53]. There are studies reporting a positive correlation between IL-6 level in amniotic and cervical fluid and subsequent preterm delivery [54, 55]. Brik *et al* showed that cervical interleukin-6 was predictive for delivery at 48 h, 7 days, <32 and <34 weeks in a study involving 100 pregnant women subjects between 24-34 weeks of gestation who were hospitalized due to threatened preterm delivery [56].

Urocortin is a neuropeptide belonging to the corticotropin-releasing factor (CRF) family, which binds to CRF type 1 and type 2 receptors. It is expressed in amnion, chorion, decidua, trophoblast and myometrium. Urocortin enhances the contractile response of the myometrium to prostaglandins [57]. Maternal plasma urocortin levels increase in preterm births [58]. In the study of Florio *et al*, in which they evaluated 85 cases hospitalized due to TPL at 28-34 gestational weeks, they showed that an increase in maternal plasma urocortin level was associated with preterm labour [59].

In a retrospective case-control study, Hong *et al* examined maternal plasma lipopolysaccharide-binding protein (LBP), endostatin and macrophage inflammatory protein (MIP) -1α levels of women who had preterm and term deliveries who came to the hospital because of PTL. Women who gave birth within 48 hours of their admission to the hospital had higher maternal plasma LBP, endostatin levels and lower MIP-1α compared to women who gave birth at term. Prediction rates were found to be high when plasma endostatin, LBP levels, nulliparity and cervical length were combined in determining delivery within 48 hours after admission (area under the curve = 0.920) [60]. LBP and MIP-1α are inflammatory molecules with angiogenic properties. This suggests that SPTD is related to the increased inflammatory response [60].

Preterm Labour Prediction with Markers Checked in First and Second Trimesters

Fetal Fibronectin

In the second trimester, those with a positive cervical fFN test have double the risk of developing PTD [61].

IGFBPs

Those with a high phIGFBP-1 level in cervical fluid in the first trimester have an increased risk of spontaneous PTD. However, phIGFBP-1 can mix with cervical and vaginal secretions since decidua capsularis and decidua parietalis have not yet combined in the first trimester. Therefore, the sensitivity and specificity of phIGFBP-1 taken from cervical fluid in the first trimester in determining PTD are low [62]. However, in symptomatic cases, phIGFBP-1 has a high negative predictive value [34, 63 - 66]. On the other hand, there are also studies reporting that IGFBP-1, which is examined in vaginal swab in the second trimester, cannot predict PTD in high-risk asymptomatic women [67]. Data on the prediction of IGFs and IGFBPs in early mid-trimester for preterm delivery are limited. In placental ischemia, when placentation and fetal development are impaired, there are studies arguing that IGFs mediate the development of preterm labour and delivery [68]. In some studies, it has been shown that phosphorylated IGFBP-1 in vaginal secretions is associated with preterm delivery [69]. Wang *et al.* reported that IGF-1 or IGFBP-1 could not predict preterm birth [70]. In another study by Wang *et al*, maternal IGFBP-3 levels were found to be low in those who gave birth before the 32nd gestational week [71]. In the study conducted by Shin *et al.*, in which they evaluated cervical length (CL) and blood insulin-like growth factor binding protein 1 and 3 (IGFBP-1 and IGFBP-3) levels in asymptomatic pregnant women at 11-18 weeks, they reported that CL and IGFBP-3 predicted preterm labour, but IGFBP-1 did not [72].

Micro RNAs

Cook *et al* showed that first-trimester plasma miRNAs could predict preterm delivery and cervical shortening in their prospective studies in which they followed up pregnant women with high risk for preterm labour. Subsequently, it was shown that miRNAs (hsa-let-7a-5p, hsa-miR-374a-5p, hsa-miR-15b-5p, hsa-miR-19b-3p, hsa-miR-23a-3p, hsa-miR-93-5p, hsa-miR-150-5p, hsa-miR-185-5p and hsa-miR-191-5p) levels of women who had preterm birth or cervical shortening were expressed differently compared to those who gave birth at term. Among miRNAs, Hsa-miR-150-5p was reported to be the best predictor of preterm labour (AUC = 0.8725) and cervical shortening (AUC = 0.8514) [73]. Gray *et al* compared the micro-RNA levels in the plasma taken at the 20th gestational week from asymptomatic pregnant women who had preterm delivery and term delivery. Subsequently, miR302b and miR1253 expression were low in the plasma of women who had a preterm birth, whereas miR223 expression was high [74]. MiRNAs are RNA particles with 19-22 nucleotides that do not code a protein but are involved in the regulation of gene expression. There are studies showing that they play a role in various pregnancy complications such as preeclampsia [75]. Also, there are studies showing that some miRNAs can initiate birth by affecting progesterone receptor activity [76]. In their prospective studies, Sanders *et al.* followed pregnant women from the 16-19th gestational week to delivery and found high plasma miR223 levels in preterm births [77].

Progesterone

In various studies, important information that progesterone withdrawal plays an important role in the initiation of labour was found [78]. Priya *et al* tried to determine the risk of preterm delivery by cervical length and salivary progesterone levels in the study in which asymptomatic pregnant women were followed up starting from 24-28 weeks of gestation at 3-4 week intervals. They stated that low saliva progesterone concentration could be used to predict preterm labour [79]. In different studies, it was shown that measuring free progesterone in saliva could predict spontaneous preterm labour in risky pregnancies [80].

Circulating Microparticles

Circulating microparticles (CMPs) are extracellular vesicles that are nanosized and contain lipids, proteins and metabolites. These provide specific intercellular communication [81]. They are released into the maternal circulation from the early first trimester [82]. In the study in which McElrath *et al.* looked at CMPs in maternal plasma at 10-12 weeks of gestation, CMPs were successful in predicting preterm labour [83].

Inflammatory Markers

Manning *et al.* examined cytokines from samples taken from the cervical fluid at the 22-24th week of pregnancy in asymptomatic women with a history of spontaneous preterm birth or cervical surgery. Subsequently, low levels of IL-8 and IL-1β were found in pregnant women with preterm birth [84]. They claimed that low cervical cytokine levels did not prevent increasing infections due to low immune response. In their study, Bastek *et al.* followed 104 pregnant women between 20 weeks and 23 6/7 weeks of gestation in the risk group for preterm delivery, and they showed that preterm labour history and cervical length predicted preterm birth, but fetal fibronectin and elafin, soluble E-cadherin and IL-6 did not. In the same study, they reported that combining biomarkers did not provide a prediction [85]. In the study conducted by Amabebe *et al.*, they evaluated asymptomatic high-risk pregnant women at 20^{+0}–22^{+6} weeks of gestation and investigated cervicovaginal fluid cytokine levels. RANTES (regulated on activation, normal T cell expressed and secreted) was successful in predicting IL-1β preterm labour. In the study, it was emphasized that sPTB is associated with ongoing choriodecidual inflammation [86].

Thioredoxin (TXN) is a molecule that protects cells from reactive oxygen species and inhibits apoptosis [87]. TXN indirectly regulates the production of proinflammatory cytokines, prostaglandin and COX-2 [88]. In high-risk asymptomatic women, TXN levels in the cervicovaginal fluid were shown to decrease during labour [89]. IL1ß provides induction of prostaglandin synthesis in human amnion tissue [90]. The proinflammatory effects of IL1 are regulated by IL1RN. IL1RN was found to be decreased in the cervicovaginal fluid of women who gave spontaneous term delivery [89]. In the study of Liong *et al.*, it was shown that in samples taken from the cervicovaginal fluid of high-risk symptomatic women, TXN and IL1RN levels decreased 90 days before spontaneous preterm labour developed [91]. IL-10 regulates the immune response by inhibiting IL-1 synthesis [92]. Ruiz *et al.* indicated that the risk of PTB was increased with increased IL-1Ra and decreased IL-10 levels [93]. Blanco-Quiros *et al* [94] found that PTD was associated with increased IL-10, and Makhseed *et al.* [95] reported that it was associated with low IL-10 levels. Despite the findings of numerous studies, none of the inflammatory cytokines were found to be suitable for clinical use in predicting preterm birth [96 - 98].

Neopterin is a cytokine that is activated by activated macrophages and monocytes and is associated with cell-mediated immune activation [99]. In the conducted studies, it was shown that inflammatory changes in the amnion and decidua initiate preterm labour, and these changes occur before preterm labour occurs [100, 101]. Navolan *et al.* showed that serum neopterin level, which is increased in early pregnancy in asymptomatic pregnant women, was successful in predicting preterm labour [102]. There are studies reporting that serum neopterin levels predict preterm delivery in pregnant women hospitalized due to symptomatic threatened preterm labour [103].

Matrix Metalloproteinases

In various studies, it was claimed that cytokine-mediated stimulation of prostaglandins and matrix metalloproteinases (MMP) weakens the cervical and amniotic membrane barriers, and in this way bacteria reach the amniotic cavity, resulting in myometrial activation and membrane rupture [23, 104 - 106].

Vitamin D binding protein (VDBP) is a protein found on inflammatory cells and fibroblasts and has a role in the extracellular remodelling of the cervix [107, 108].

In the study conducted by Yoo *et al.*, they found high levels of VDBP and tissue inhibitor of metalloproteinases (TIMP) -1 in the cervicovaginal fluids of asymptomatic women who would later give preterm birth [109].

Aneuploidy Syndrome Screening Test Parameters

There are also studies trying to predict pregnancy complications with aneuploidy syndrome screening test parameters. PP-13 is produced by syncytiotrophoblasts and is involved in the vascular remodelling process [110]. Pregnancy-associated plasma protein-A (PAPP-A) is a metalloproteinase used in first-trimester aneuploidy screening. Low PP-13 and PAPP-A levels in the first trimester were shown to be associated with increased uterine artery resistance and fetal growth retardation and preeclampsia [111 - 114]. There are also studies arguing that fetal growth retardation and placental dysfunction similar to preeclampsia are found in the pathogenesis of preterm labour [115, 116]. In a retrospective study of Huang *et al.*, including 141,698 women, it was reported that low second-trimester unconjugated oestriol (uE3] and first-trimester maternal serum PAPP-A, increased second-trimester serum alpha-fetoprotein (AFP) or increased second trimester total human chorionic gonadotrophin (hCG) were associated with pregnancy complications in preterm delivery [117]. In the study of Stout and *et al*, it was shown that low serum PP-13 and PAPP-A levels in the first trimester and increased uterine artery pulsatility index are associated with preterm birth [118]. Beta *et al.* reported that low PP-13 and PAPP-A levels predicted delivery below 34 weeks [119]. Spencer *et al.* showed that low PAPP-A levels are associated with preterm labour [120].

When labour begins, placental alpha microglobulin-1 (PAMG-1] reaches the vagina through the amniotic fluid. Nikolova *et al.* reported that a positive PAMG-1 test was successful in predicting spontaneous preterm birth in a study in which they followed up pregnant women who were admitted to the hospital for preterm labour [121]. In other studies it was shown that PAMG could predict preterm birth [122].

Insulin-like growth factor-binding protein 4 (IBP4] negatively regulates IGF1 and IGF2 [123]. IBP4 is expressed by syncytiotrophoblasts. In pregnant women complicated by preeclampsia characterized by placental dysfunction and fetal growth restriction, IBP4 levels increase compared to healthy pregnancies [124]. Saade *et al.* found a high rate of maternal IBP4/sex hormone-binding globulin (SHBG) in asymptomatic pregnancies with subsequent preterm birth [125].

Other Hormones

Some argue that corticotropin-releasing hormone (CRH) plays a role in the onset of labour [126 - 128]. However, Sibai *et al.* reported that CRH is not a good marker in predicting preterm labour in their study involving high-risk pregnancies at 16-20 weeks [129].

In summary, there is no inexpensive marker that has been validated for clinical use for preterm labour prediction with high sensitivity and specificity.

CONCLUSION

In summary, there is no inexpensive marker that has been validated for clinical use for preterm labour prediction with high sensitivity and specificity.

CONSENT FOR PUBLICATION

Not applicable.

CONFLICT OF INTEREST

The authors declare no conflict of interest, financial or otherwise.

ACKNOWLEDGEMENTS

I gratefully thank Prof. Dr. Pınar ATUKEREN and Prof. Dr. Hafize UZUN for giving us the opportunity to be a part of this successful book "Biomarkers in Medicine".

REFERENCES

[1] Lawn JE, Gravett MG, Nunes TM, Rubens CE, Stanton C, Group GR. Global report on preterm birth and stillbirth (1 of 7): definitions, description of the burden and opportunities to improve data. BMC Pregnancy Childbirth 2010; 10(S1) (Suppl. 1): S1. [http://dx.doi.org/10.1186/1471-2393-10-S1-S1] [PMID: 20233382]

[2] Gyamfi-Bannerman C, Ananth CV. Trends in spontaneous and indicated preterm delivery among singleton gestations in the United States, 2005-2012. Obstet Gynecol 2014; 124(6): 1069-74. [http://dx.doi.org/10.1097/AOG.0000000000000546] [PMID: 25415156]

[3] Frey HA, Klebanoff MA. The epidemiology, etiology, and costs of preterm birth. Semin Fetal Neonatal Med 2016; 21(2): 68-73. [http://dx.doi.org/10.1016/j.siny.2015.12.011] [PMID: 26794420]

[4] Ananth CV, Vintzileos AM. Epidemiology of preterm birth and its clinical subtypes. J Matern Fetal Neonatal Med 2006; 19(12): 773-82. [http://dx.doi.org/10.1080/14767050600965882] [PMID: 17190687]

[5] Han S, Crowther CA, Moore V. Magnesium maintenance therapy for preventing preterm birth after threatened preterm labour. Cochrane Database Syst Rev 2013; 5(5): CD000940. [http://dx.doi.org/10.1002/14651858.CD000940.pub3] [PMID: 23728634]

[6] Dudley DJ. Pre-term labor: an intra-uterine inflammatory response syndrome? J Reprod Immunol 1997; 36: 93-109.12.

[7] Gonzalez JM, Romero R, Girardi G. Comparison of the mechanisms responsible for cervical remodeling in preterm and term labor. J Reprod Immunol 2013; 97: 112-9.

[8] Romero R, Espinoza J, Kusanovic JP, *et al.* The preterm parturition syndrome. BJOG 2006; 113 (Suppl. 3): 17-42.
[http://dx.doi.org/10.1111/j.1471-0528.2006.01120.x] [PMID: 17206962]

[9] Timmons B, Akins M, Mahendroo M. Cervical remodeling during pregnancy and parturition. Trends Endocrinol Metab 2010; 21(6): 353-61.
[http://dx.doi.org/10.1016/j.tem.2010.01.011] [PMID: 20172738]

[10] Terzidou V. Preterm labour. Biochemical and endocrinological preparation for parturition. Best Pract Res Clin Obstet Gynaecol 2007; 21(5): 729-56.
[http://dx.doi.org/10.1016/j.bpobgyn.2007.05.001] [PMID: 17616441]

[11] Romero R, Dey SK, Fisher SJ. Preterm labor: one syndrome, many causes. Science 2014; 345(6198): 760-5.
[http://dx.doi.org/10.1126/science.1251816] [PMID: 25124429]

[12] Bennett TA, Kotelchuck M, Cox CE, Tucker MJ, Nadeau DA. Pregnancy-associated hospitalizations in the United States in 1991 and 1992: a comprehensive view of maternal morbidity. Am J Obstet Gynecol 1998; 178(2): 346-54.
[http://dx.doi.org/10.1016/S0002-9378(98)80024-0] [PMID: 9500498]

[13] Romero R, Erez O, Maymon E, Pacora P. Is an episode of suspected preterm labor that subsequently leads to a term delivery benign? Am J Obstet Gynecol 2017; 216(2): 89-94.
[http://dx.doi.org/10.1016/j.ajog.2016.12.030] [PMID: 28148450]

[14] Gazmararian JA, Petersen R, Jamieson DJ, *et al.* Hospitalizations during pregnancy among managed care enrollees. Obstet Gynecol 2002; 100: 94-100.

[15] McPheeters ML, Miller WC, Hartmann KE, *et al.* The epidemiology of threatened preterm labor: a prospective cohort study. Am J Obstet Gynecol 2005; 192(4): 1325-9.
[http://dx.doi.org/10.1016/j.ajog.2004.12.055] [PMID: 15846230]

[16] Chandiramani M, Shennan A. Preterm labour: update on prediction and prevention strategies. Curr Opin Obstet Gynecol 2006; 18(6): 618-24.
[http://dx.doi.org/10.1097/GCO.0b013e3280106228] [PMID: 17099332]

[17] Bittar RE, Yamasaki AA, Sasaki S, Zugaib M. Cervical fetal fibronectin in patients at increased risk for preterm delivery. Am J Obstet Gynecol 1996; 175(1): 178-81.
[http://dx.doi.org/10.1016/S0002-9378(96)70271-5] [PMID: 8694047]

[18] Goldenberg RL, Goepfert AR, Ramsey PS. Biochemical markers for the prediction of preterm birth. Am J Obstet Gynecol 2005; 192(5) (Suppl.): S36-46.
[http://dx.doi.org/10.1016/j.ajog.2005.02.015] [PMID: 15891711]

[19] Agarwal N, Suneja A, Arora S, Tandon OP, Sircar S. Role of uterine artery velocimetry using color-flow Doppler and electromyography of uterus in prediction of preterm labor. J Obstet Gynaecol Res 2004; 30(6): 402-8.
[http://dx.doi.org/10.1111/j.1447-0756.2004.00222.x] [PMID: 15566453]

[20] Dyson DC, Danbe KH, Bamber JA, *et al.* Monitoring women at risk for preterm labor. N Engl J Med 1998; 338(1): 15-9.
[http://dx.doi.org/10.1056/NEJM199801013380103] [PMID: 9414326]

[21] Slattery MM, Morrison JJ. Preterm delivery. Lancet 2002; 360(9344): 1489-97.
[http://dx.doi.org/10.1016/S0140-6736(02)11476-0] [PMID: 12433531]

[22] Leitich H. Controversies in diagnosis of preterm labour. BJOG 2005; 112 (Suppl. 1): 61-3.

[http://dx.doi.org/10.1111/j.1471-0528.2005.00587.x] [PMID: 15715597]

[23] Agrawal V, Hirsch E. Intrauterine infection and preterm labor. Semin Fetal Neonatal Med 2012; 17(1): 12-9.
[http://dx.doi.org/10.1016/j.siny.2011.09.001] [PMID: 21944863]

[24] Abbott DS, Hezelgrave NL, Seed PT, *et al.* Quantitative fetal fibronectin to predict preterm birth in asymptomatic women at high risk. Obstet Gynecol 2015; 125(5): 1168-76.
[http://dx.doi.org/10.1097/AOG.0000000000000754] [PMID: 25932845]

[25] Heng YJ, Liong S, Permezel M, Rice GE, Di Quinzio MK, Georgiou HM. Human cervicovaginal fluid biomarkers to predict term and preterm labor. Front Physiol 2015; 6: 151.
[http://dx.doi.org/10.3389/fphys.2015.00151] [PMID: 26029118]

[26] Stock SJ, Morris RK, Chandiramani M, Shennan AH, Norman JE. Variation in management of women with threatened preterm labour. Arch Dis Child Fetal Neonatal Ed 2015; 100(3): F276.
[http://dx.doi.org/10.1136/archdischild-2014-307806] [PMID: 25526911]

[27] Iams JD, Casal D, McGregor JA, *et al.* Fetal fibronectin improves the accuracy of diagnosis of preterm labor. Am J Obstet Gynecol 1995; 173(1): 141-5.
[http://dx.doi.org/10.1016/0002-9378(95)90182-5] [PMID: 7631671]

[28] Esplin MS, Elovitz MA, Iams JD, *et al.* Predictive Accuracy of Serial Transvaginal Cervical Lengths and Quantitative Vaginal Fetal Fibronectin Levels for Spontaneous Preterm Birth Among Nulliparous Women. JAMA 2017; 317(10): 1047-56.
[http://dx.doi.org/10.1001/jama.2017.1373] [PMID: 28291893]

[29] Deshpande SN, van Asselt AD, Tomini F, *et al.* Rapid fetal fibronectin testing to predict preterm birth in women with symptoms of premature labour: a systematic review and cost analysis. Health Technol Assess 2013; 17(40): 1-138.
[http://dx.doi.org/10.3310/hta17400] [PMID: 24060096]

[30] Hellemans P, Gerris J, Verdonk P. Fetal fibronectin detection for prediction of preterm birth in low risk women. Br J Obstet Gynaecol 1995; 102(3): 207-12.
[http://dx.doi.org/10.1111/j.1471-0528.1995.tb09095.x] [PMID: 7794844]

[31] Faron G, Boulvain M, Lescrainier JP, Vokaer A. A single cervical fetal fibronectin screening test in a population at low risk for preterm delivery: an improvement on clinical indicators? Br J Obstet Gynaecol 1997; 104(6): 697-701.
[http://dx.doi.org/10.1111/j.1471-0528.1997.tb11980.x] [PMID: 9197873]

[32] Greenhagen JB, Van Wagoner J, Dudley D, *et al.* Value of fetal fibronectin as a predictor of preterm delivery for a low-risk population. Am J Obstet Gynecol 1996; 175(4 Pt 1): 1054-6.
[http://dx.doi.org/10.1016/S0002-9378(96)80052-4] [PMID: 8885775]

[33] Cooley SM, Donnelly JC, Collins C, Geary MP, Rodeck CH, Hindmarsh PC. The relationship between maternal insulin-like growth factors 1 and 2 (IGF-1, IGF- 2) and IGFBP-3 to gestational age and preterm delivery. J Perinat Med 2010; 38: 255e9.

[34] Francis GL, McMurtry JP, Johnson RJ, Ballard FJ. Plasma clearance of chicken and human insulin-like growth factor-I and their association with circulating binding proteins in chickens. J Endocrinol 1990; 124: 361e70.

[35] Eroglu D, Yanik F, Oktem M, Zeyneloglu HB, Kuscu E. Prediction of preterm delivery among women with threatened preterm labor. Gynecol Obstet Invest 2007; 64(2): 109-16.
[http://dx.doi.org/10.1159/000100120] [PMID: 17339775]

[36] Lembet A, Eroglu D, Ergin T, *et al.* New rapid bed-side test to predict preterm delivery: phosphorylated insulin-like growth factor binding protein-1 in cervical secretions. Acta Obstet Gynecol Scand 2002; 81(8): 706-12.
[http://dx.doi.org/10.1034/j.1600-0412.2002.810804.x] [PMID: 12174153]

[37] Akercan F, Kazandi M, Sendag F, *et al.* Value of cervical phosphorylated insulinlike growth factor

binding protein-1 in the prediction of preterm labor. J Reprod Med 2004; 49(5): 368-72.
[PMID: 15214710]

[38] Fuchs F, Houllier M, Leparco S, Guyot A, Senat MV, Fernandez H. Performance of cervical phIGFBP-1 test alone or combined with short cervical length to predict spontaneous preterm birth in symptomatic women. Sci Rep 2017; 7(1): 10856.
[http://dx.doi.org/10.1038/s41598-017-11447-y] [PMID: 28883617]

[39] Rutanen EM. Insulin-like growth factors in obstetrics. Curr Opin Obstet Gynecol 2000; 12(3): 163-8.
[http://dx.doi.org/10.1097/00001703-200006000-00001] [PMID: 10873115]

[40] Ducarme G, Desroys du Roure F, Le Thuaut A, Grange J, Vital M, Dimet J. Efficacy of serum procalcitonin to predict spontaneous preterm birth in women with threatened preterm labour: a prospective observational study. BMC Pregnancy Childbirth 2018; 18(1): 65.
[http://dx.doi.org/10.1186/s12884-018-1696-2] [PMID: 29514604]

[41] Elovitz MA, Anton L, Bastek J, Brown AG. Can microRNA profiling in maternal blood identify women at risk for preterm birth? Am J Obstet Gynecol 2015; 212(6): 782.e1-5.
[http://dx.doi.org/10.1016/j.ajog.2015.01.023] [PMID: 25617732]

[42] Shahshahan Z, Iravani H. Comparison of CRP and ALK-P serum levels in prediction of preterm delivery. Adv Biomed Res 2016; 5(1): 17.
[http://dx.doi.org/10.4103/2277-9175.175903] [PMID: 26962519]

[43] Sorokin Y, Romero R, Mele L, *et al.* Maternal serum interleukin-6, C-reactive protein, and matrix metalloproteinase-9 concentrations as risk factors for preterm birth <32 weeks and adverse neonatal outcomes. Am J Perinatol 2010; 27(8): 631-40.
[http://dx.doi.org/10.1055/s-0030-1249366] [PMID: 20195952]

[44] Shahshahan Z, Rasouli O. The use of maternal C-reactive protein in the predicting of preterm labor and tocolytic therapy in preterm labor women. Adv Biomed Res 2014; 3(1): 154.
[http://dx.doi.org/10.4103/2277-9175.137864] [PMID: 25221757]

[45] Malek A, Bersinger NA, Di Santo S, *et al.* C-reactive protein production in term human placental tissue. Placenta 2006; 27(6-7): 619-25.
[http://dx.doi.org/10.1016/j.placenta.2005.05.009] [PMID: 16026834]

[46] Huras H, Ossowski P, Jach R, Reron A. Usefulness of marking alkaline phosphatase and C-reactive protein in monitoring the risk of preterm delivery. Med Sci Monit 2011; 17(11): CR657-62.
[http://dx.doi.org/10.12659/MSM.882052] [PMID: 22037746]

[47] Michalowicz BS, Novak MJ, Hodges JS, *et al.* Serum inflammatory mediators in pregnancy: changes after periodontal treatment and association with pregnancy outcomes. J Periodontol 2009; 80(11): 1731-41.
[http://dx.doi.org/10.1902/jop.2009.090236] [PMID: 19905943]

[48] Škrablin S, Lovrić H, Banović V, Kralik S, Dijaković A, Kalafatić D. Maternal plasma interleukin-6, interleukin-1β and C-reactive protein as indicators of tocolysis failure and neonatal outcome after preterm delivery. J Matern Fetal Neonatal Med 2007; 20(4): 335-41.
[http://dx.doi.org/10.1080/14767050701227877] [PMID: 17437242]

[49] Sabroe I, Jones EC, Usher LR, Whyte MK, Dower SK. Toll-like receptor (TLR)2 and TLR4 in human peripheral blood granulocytes: a critical role for monocytes in leukocyte lipopolysaccharide responses. J Immunol. 2002;168(9):4701-4710.; Kopp E, Medzhitov R. Recognition of microbial infection by Toll-like receptors. Curr Opin Immunol 2003; 15(4): 396-401.
[PMID: 12900270]

[50] Kopp E, Medzhitov R. Recognition of microbial infection by Toll-like receptors. Curr Opin Immunol 2003; 15(4): 396-401.
[http://dx.doi.org/10.1016/S0952-7915(03)00080-3] [PMID: 12900270]

[51] Kim J, Ko Y, Kwon K, *et al.* Analysis of monocyte subsets and toll-like receptor 4 expression in

peripheral blood monocytes of women in preterm labor. J Reprod Immunol 2012; 94(2): 190-5. [http://dx.doi.org/10.1016/j.jri.2012.02.002] [PMID: 22440522]

[52] Pawelczyk E, Nowicki BJ, Izban MG, *et al.* Spontaneous preterm labor is associated with an increase in the proinflammatory signal transducer TLR4 receptor on maternal blood monocytes. BMC Pregnancy Childbirth 2010; 10(1): 66. [http://dx.doi.org/10.1186/1471-2393-10-66] [PMID: 20964862]

[53] Mitchell MD, Romero RJ, Edwin SS, Trautman MS. Prostaglandins and parturition. Reprod Fertil Dev 1995; 7(3): 623-32. [http://dx.doi.org/10.1071/RD9950623] [PMID: 8606975]

[54] El-Bastawissi AY, Williams MA, Riley DE, Hitti J, Krieger JN. Amniotic fluid interleukin-6 and preterm delivery: a review. Obstet Gynecol 2000; 95(6 Pt 2) (Suppl.): 1056-64. [http://dx.doi.org/10.1097/00006250-200006001-00032] [PMID: 10808034]

[55] Rizzo G, Capponi A, Vlachopoulou A, Angelini E, Grassi C, Romanini C. Interleukin-6 concentrations in cervical secretions in the prediction of intrauterine infection in preterm premature rupture of the membranes. Gynecol Obstet Invest 1998; 46(2): 91-5. [http://dx.doi.org/10.1159/000010009] [PMID: 9701687]

[56] Brik M, Antonio P, Perales-Puchalt A, Diago V, Perales A. Cervical interleukin-6 as a predictive test for preterm delivery in symptomatic women: preliminary results. Eur J Obstet Gynecol Reprod Biol 2011; 155(1): 14-8. [http://dx.doi.org/10.1016/j.ejogrb.2010.11.001] [PMID: 21130555]

[57] Petraglia F, Florio P, Benedetto C, *et al.* Urocortin stimulates placental adrenocorticotropin and prostaglandin release and myometrial contractility *in vitro.* J Clin Endocrinol Metab 1999; 84(4): 1420-3. [http://dx.doi.org/10.1210/jcem.84.4.5585] [PMID: 10199789]

[58] Florio P, Torricelli M, Galleri L, *et al.* High fetal urocortin levels at term and preterm labor. J Clin Endocrinol Metab 2005; 90(9): 5361-5. [http://dx.doi.org/10.1210/jc.2005-0109] [PMID: 15956084]

[59] Florio P, Linton EA, Torricelli M, *et al.* Prediction of preterm delivery based on maternal plasma urocortin. J Clin Endocrinol Metab 2007; 92(12): 4734-7. [http://dx.doi.org/10.1210/jc.2007-0457] [PMID: 17986644]

[60] Hong S, Park KH, Kim YM, Lee YE, Park Y, Lee JE. A Protein Microarray Analysis of Plasma Proteins for the Prediction of Spontaneous Preterm Delivery in Women with Preterm Labor. Reprod Sci 2020; 27(5): 1187-96. [http://dx.doi.org/10.1007/s43032-019-00114-4] [PMID: 32046457]

[61] Goldenberg RL, Klebanoff M, Carey JC, *et al.* Vaginal fetal fibronectin measurements from 8 to 22 weeks' gestation and subsequent spontaneous preterm birth. Am J Obstet Gynecol 2000; 183(2): 469-75. [http://dx.doi.org/10.1067/mob.2000.106073] [PMID: 10942489]

[62] Rahkonen L, Rutanen EM, Nuutila M, Sainio S, Saisto T, Paavonen J. Elevated levels of decidual insulin-like growth factor binding protein-1 in cervical fluid in early and mid-pregnancy are associated with an increased risk of spontaneous preterm delivery. BJOG 2010; 117(6): 701-10. [http://dx.doi.org/10.1111/j.1471-0528.2010.02551.x] [PMID: 20374609]

[63] Ting HS, Chin PS, Yeo GS, Kwek K. Comparison of bedside test kits for prediction of preterm delivery: phosphorylated insulin-like growth factor binding protein-1 (pIGFBP-1) test and fetal fibronectin test. Ann Acad Med Singap 2007; 36(6): 399-402. [PMID: 17597963]

[64] Rahkonen L, Unkila-Kallio L, Nuutila M, *et al.* Cervical length measurement and cervical phosphorylated insulin-like growth factor binding protein-1 testing in prediction of preterm birth in patients reporting uterine contractions. Acta Obstet Gynecol Scand 2009; 88(8): 901-8.

[http://dx.doi.org/10.1080/00016340903104281] [PMID: 19575312]

[65] Riboni F, Vitulo A, Dell'avanzo M, Plebani M, Battagliarin G, Paternoster D. Biochemical markers predicting pre-term delivery in symptomatic patients: phosphorylated insulin-like growth factor binding protein-1 and fetal fibronectin. Arch Gynecol Obstet 2011; 284(6): 1325-9. [http://dx.doi.org/10.1007/s00404-011-1839-4] [PMID: 21274721]

[66] Altinkaya O, Gungor T, Ozat M, Danisman N, Mollamahmutoglu L. Cervical phosphorylated insulin-like growth factor binding protein-1 in prediction of preterm delivery. Arch Gynecol Obstet 2009; 279(3): 279-83. [http://dx.doi.org/10.1007/s00404-008-0703-7] [PMID: 18551300]

[67] Khambay H, Bolt LA, Chandiramani M, De Greeff A, Filmer JE, Shennan AH. The Actim Partus test to predict pre-term birth in asymptomatic high-risk women. J Obstet Gynaecol 2012; 32(2): 132-4. [http://dx.doi.org/10.3109/01443615.2011.637649] [PMID: 22296421]

[68] Goetzinger KR, Cahill AG, Kemna J, Odibo L, Macones GA, Odibo AO. First-trimester prediction of preterm birth using ADAM12, PAPP-A, uterine artery Doppler, and maternal characteristics. Prenat Diagn 2012; 32

[69] Kallioniemi H, Rahkonen L, Heikinheimo O, Paavonen J. Early pregnancy vaginal fluid phosphorylated insulin-like growth factor binding protein-1 predicts preterm delivery. Prenat Diagn 2013; 33(4): 378-e83. [http://dx.doi.org/10.1002/pd.4072]

[70] Wang HS, Lee CL, Chard T. Levels of insulin-like growth factor-I and insulin- like growth factor-binding protein-1 in pregnancy with preterm delivery. Br J Obstet Gynaecol 1993; 100: 472e5.

[71] Wang HS, Perry LA, Kanisius J, Iles RK, Holly JM, Chard T. Purification and assay of insulin-like growth factor-binding protein-1: measurement of circulating levels throughout pregnancy. J Endocrinol 1991; 128: 161e8.

[72] Shin JE, Shin JC, Kim SJ, Lee Y, Park IY, Lee S. Early midtrimester serum insulin-like factors and cervical length to predict preterm delivery. Taiwan J Obstet Gynecol 2016; 55(1): 45-9. [http://dx.doi.org/10.1016/j.tjog.2015.03.007] [PMID: 26927247]

[73] Cook J, Bennett PR, Kim SH, *et al.* First Trimester Circulating MicroRNA Biomarkers Predictive of Subsequent Preterm Delivery and Cervical Shortening. Sci Rep 2019; 9(1): 5861. [http://dx.doi.org/10.1038/s41598-019-42166-1] [PMID: 30971726]

[74] Gray C, McCowan LM, Patel R, Taylor RS, Vickers MH. Maternal plasma miRNAs as biomarkers during mid-pregnancy to predict later spontaneous preterm birth: a pilot study. Sci Rep 2017; 7(1): 815. [http://dx.doi.org/10.1038/s41598-017-00713-8] [PMID: 28400603]

[75] Li H, Ge Q, Guo L, Lu Z. Maternal plasma miRNAs expression in preeclamptic pregnancies. BioMed Res Int 2013; 2013: 970265. [http://dx.doi.org/10.1155/2013/970265] [PMID: 24195082]

[76] Williams KC, Renthal NE, Condon JC, Gerard RD, Mendelson CR. MicroRNA-200a serves a key role in the decline of progesterone receptor function leading to term and preterm labor. Proc Natl Acad Sci USA 2012; 109(19): 7529-34. [http://dx.doi.org/10.1073/pnas.1200650109] [PMID: 22529366]

[77] Sanders AP, Burris HH, Just AC, *et al.* microRNA expression in the cervix during pregnancy is associated with length of gestation. Epigenetics 2015; 10(3): 221-8. [http://dx.doi.org/10.1080/15592294.2015.1006498] [PMID: 25611922]

[78] Csapo AI. Progesterone "block". Am J Anat 1956;98:273–92.; Mesiano S. Myometrial progesterone responsiveness and the control of human parturition. J Soc Gynecol Investig 2004; 11: 193-202.

[79] Priya B, Mustafa MD, Guleria K, Vaid NB, Banerjee BD, Ahmed RS. Salivary progesterone as a biochemical marker to predict early preterm birth in asymptomatic high-risk women. BJOG 2013;

120(8): 1003-11.
[http://dx.doi.org/10.1111/1471-0528.12217] [PMID: 23551599]

[80] Lachelin GC, McGarrigle HH, Seed PT, Briley A, Shennan AH, Poston L. Low saliva progesterone concentrations are associated with spontaneous early preterm labour in women at increased risk of preterm delivery. BJOG 2009; 116: 1515-9.
[http://dx.doi.org/10.1111/j.1471-0528.2009.02293.x] [PMID: 19614937]

[81] Colombo M, Raposo G, Théry C. Biogenesis, secretion, and intercellular interactions of exosomes and other extracellular vesicles. Annu Rev Cell Dev Biol 2014; 30(1): 255-89.
[http://dx.doi.org/10.1146/annurev-cellbio-101512-122326] [PMID: 25288114]

[82] Sarker S, Scholz-Romero K, Perez A, *et al.* Placenta-derived exosomes continuously increase in maternal circulation over the first trimester of pregnancy. J Transl Med 2014; 12(1): 204.
[http://dx.doi.org/10.1186/1479-5876-12-204] [PMID: 25104112]

[83] McElrath TF, Cantonwine DE, Jeyabalan A, *et al.* Circulating microparticle proteins obtained in the late first trimester predict spontaneous preterm birth at less than 35 weeks' gestation: a panel validation with specific characterization by parity. Am J Obstet Gynecol 2019; 220(5): 488.e1-488.e11.
[http://dx.doi.org/10.1016/j.ajog.2019.01.220] [PMID: 30690014]

[84] Manning R, James CP, Smith MC, *et al.* Predictive value of cervical cytokine, antimicrobial and microflora levels for pre-term birth in high-risk women. Sci Rep 2019; 9(1): 11246.
[http://dx.doi.org/10.1038/s41598-019-47756-7] [PMID: 31375740]

[85] Bastek JA, Hirshberg A, Chandrasekaran S, *et al.* Biomarkers and cervical length to predict spontaneous preterm birth in asymptomatic high-risk women. Obstet Gynecol 2013; 122(2 Pt 1): 283-9.
[http://dx.doi.org/10.1097/AOG.0b013e31829ab714] [PMID: 23969796]

[86] Amabebe E, Chapman DR, Stern VL, Stafford G, Anumba DOC. Mid-gestational changes in cervicovaginal fluid cytokine levels in asymptomatic pregnant women are predictive markers of inflammation-associated spontaneous preterm birth. J Reprod Immunol 2018; 126: 1-10.
[http://dx.doi.org/10.1016/j.jri.2018.01.001] [PMID: 29367099]

[87] Saitoh M, Nishitoh H, Fujii M, *et al.* Mammalian thioredoxin is a direct inhibitor of apoptosis signal-regulating kinase (ASK) 1. EMBO J 1998; 17(9): 2596-606.
[http://dx.doi.org/10.1093/emboj/17.9.2596] [PMID: 9564042]

[88] Hirota K, Matsui M, Murata M, *et al.* Nucleoredoxin, glutaredoxin, and thioredoxin differentially regulate NF-kappaB, AP-1, and CREB activation in HEK293 cells (published correction appears in Biochem Biophys Res Commun 2000 Aug 18;275(1):247. Junji, Y (corrected to Yodoi, J)). Biochem Biophys Res Commun 2000; 274(1): 177-82.
[http://dx.doi.org/10.1006/bbrc.2000.3106] [PMID: 10903915]

[89] Heng YJ, Di Quinzio MK, Permezel M, Rice GE, Georgiou HM. Temporal expression of antioxidants in human cervicovaginal fluid associated with spontaneous labor. Antioxid Redox Signal 2010; 13(7): 951-7.
[http://dx.doi.org/10.1089/ars.2010.3122] [PMID: 20446766]

[90] Romero R, Durum S, Dinarello CA, Oyarzun E, Hobbins JC, Mitchell MD. Interleukin-1 stimulates prostaglandin biosynthesis by human amnion. Prostaglandins 1989; 37(1): 13-22.
[http://dx.doi.org/10.1016/0090-6980(89)90028-2] [PMID: 2785698]

[91] Liong S, Di Quinzio MK, Fleming G, Permezel M, Rice GE, Georgiou HM. Prediction of spontaneous preterm labour in at-risk pregnant women. Reproduction 2013; 146(4): 335-45.
[http://dx.doi.org/10.1530/REP-13-0175] [PMID: 23858477]

[92] Corwin EJ. Understanding cytokines. Part I: Physiology and mechanism of action. Biol Res Nurs 2000; 2(1): 30-40.
[http://dx.doi.org/10.1177/109980040000200104] [PMID: 11232509]

[93] Ruiz RJ, Jallo N, Murphey C, Marti CN, Godbold E, Pickler RH. Second trimester maternal plasma levels of cytokines IL-1Ra, Il-6 and IL-10 and preterm birth. J Perinatol 2012; 32(7): 483-90. [http://dx.doi.org/10.1038/jp.2011.193] [PMID: 22460542]

[94] Blanco-Quirós A, Arranz E, Solis G, Villar A, Ramos A, Coto D. Cord blood interleukin-10 levels are increased in preterm newborns. Eur J Pediatr 2000; 159(6): 420-3. [http://dx.doi.org/10.1007/s004310051299] [PMID: 10867846]

[95] Makhseed M, Raghupathy R, El-Shazly S, Azizieh F, Al-Harmi JA, Al-Azemi MMK. Pro-inflammatory maternal cytokine profile in preterm delivery. Am J Reprod Immunol 2003; 49(5): 308-18. [http://dx.doi.org/10.1034/j.1600-0897.2003.00038.x] [PMID: 12854735]

[96] Conde-Agudelo A, Papageorghiou AT, Kennedy SH, Villar J. Novel biomarkers for the prediction of the spontaneous preterm birth phenotype: a systematic review and meta-analysis. BJOG 2011; 118(9): 1042-54. [http://dx.doi.org/10.1111/j.1471-0528.2011.02923.x] [PMID: 21401853]

[97] Gervasi MT, Romero R, Bracalente G, *et al.* Midtrimester amniotic fluid concentrations of interleukin-6 and interferon-gamma-inducible protein-10: evidence for heterogeneity of intra-amniotic inflammation and associations with spontaneous early (<32 weeks) and late (>32 weeks) preterm delivery. J Perinat Med 2012; 40(4): 329-43. [http://dx.doi.org/10.1515/jpm-2012-0034] [PMID: 22752762]

[98] Hee L. Likelihood ratios for the prediction of preterm delivery with biomarkers. Acta Obstet Gynecol Scand 2011; 90(11): 1189-99. [http://dx.doi.org/10.1111/j.1600-0412.2011.01187.x] [PMID: 21568946]

[99] Berdowska A, Zwirska-Korczala K. Neopterin measurement in clinical diagnosis. J Clin Pharm Ther 2001; 26(5): 319-29. [http://dx.doi.org/10.1046/j.1365-2710.2001.00358.x] [PMID: 11679022]

[100] Vrachnis N, Vitoratos N, Iliodromiti Z, Sifakis S, Deligeoroglou E, Creatsas G. Intrauterine inflammation and preterm delivery. Ann N Y Acad Sci 2010; 1205(1): 118-22. [http://dx.doi.org/10.1111/j.1749-6632.2010.05684.x] [PMID: 20840262]

[101] Hamilton S, Oomomian Y, Stephen G, *et al.* Macrophages infiltrate the human and rat decidua during term and preterm labor: evidence that decidual inflammation precedes labor. Biol Reprod 2012; 86(2): 39. [http://dx.doi.org/10.1095/biolreprod.111.095505] [PMID: 22011391]

[102] Navolan DB, Vladareanu S, Lahdou I, *et al.* Early pregnancy serum neopterin concentrations predict spontaneous preterm birth in asymptomatic pregnant women. J Perinat Med 2016; 44(5): 517-22. [http://dx.doi.org/10.1515/jpm-2015-0081] [PMID: 25918916]

[103] Keskin U, Ulubay M, Kurt YG, Fidan U, Koçyiğit YK, Honca T, *et al.* Increased neopterin level and chitotriosidase activity in pregnant women with threatened preterm labor. J Matern Fetal Neonatal Med 2014; 1-5. [PMID: 25005858]

[104] Kalinka J, Sobala W, Wasiela M, Brzezińska-Błaszczyk E. Decreased proinflammatory cytokines in cervicovaginal fluid, as measured in midgestation, are associated with preterm delivery. Am J Reprod Immunol 2005; 54(2): 70-6. [http://dx.doi.org/10.1111/j.1600-0897.2005.00289.x] [PMID: 16105098]

[105] Oh KJ, Kim SM, Hong JS, *et al.* Twenty-four percent of patients with clinical chorioamnionitis in preterm gestations have no evidence of either culture-proven intraamniotic infection or intraamniotic inflammation. Am J Obstet Gynecol 2017; 216(6): 604.e1-604.e11. [http://dx.doi.org/10.1016/j.ajog.2017.02.035] [PMID: 28257964]

[106] Park JY, Romero R, Lee J, Chaemsaithong P, Chaiyasit N, Yoon BH. An elevated amniotic fluid

prostaglandin F2α concentration is associated with intra-amniotic inflammation/infection, and clinical and histologic chorioamnionitis, as well as impending preterm delivery in patients with preterm labor and intact membranes. J Matern Fetal Neonatal Med 2016; 29(16): 2563-72.
[PMID: 26669519]

[107] DiMartino SJ, Kew RR. Initial characterization of the vitamin D binding protein (Gc-globulin) binding site on the neutrophil plasma membrane: evidence for a chondroitin sulfate proteoglycan. J Immunol 1999; 163(4): 2135-42.
[PMID: 10438954]

[108] Emerson DL, Werner PA, Cheng MH, Galbraith RM. Presence of Gc (vitamin D-binding protein) and interactions with actin in human placental tissue. Am J Reprod Immunol Microbiol 1985; 7(1): 15-21.
[http://dx.doi.org/10.1111/j.1600-0897.1985.tb00257.x] [PMID: 4039112]

[109] Yoo HN, Park KH, Jung EY, Kim YM, Kook SY, Jeon SJ. Non-invasive prediction of preterm birth in women with cervical insufficiency or an asymptomatic short cervix (≤25 mm) by measurement of biomarkers in the cervicovaginal fluid. PLoS One 2017; 12(7): e0180878.
[http://dx.doi.org/10.1371/journal.pone.0180878] [PMID: 28700733]

[110] Than NG, Pick E, Bellyei S, *et al.* Functional analyses of placental protein 13/galectin-13. Eur J Biochem 2004; 271(6): 1065-78.
[http://dx.doi.org/10.1111/j.1432-1033.2004.04004.x] [PMID: 15009185]

[111] Goetzinger KR, Singla A, Gerkowicz S, Dicke JM, Gray DL, Odibo AO. Predicting the risk of pre-eclampsia between 11 and 13 weeks' gestation by combining maternal characteristics and serum analytes, PAPP-A and free β-hCG. Prenat Diagn 2010; 30(12-13): 1138-42.
[http://dx.doi.org/10.1002/pd.2627] [PMID: 20936638]

[112] Goetzinger KR, Singla A, Gerkowicz S, Dicke JM, Gray DL, Odibo AO. The efficiency of first-trimester serum analytes and maternal characteristics in predicting fetal growth disorders. Am J Obstet Gynecol 2009; 201(4): 412. e1-6.

[113] Poon LC, Kametas NA, Maiz N, Akolekar R, Nicolaides KH. First-trimester prediction of hypertensive disorders in pregnancy. Hypertension 2009; 53(5): 812-8.
[http://dx.doi.org/10.1161/HYPERTENSIONAHA.108.127977] [PMID: 19273739]

[114] Ong CY, Liao AW, Spencer K, Munim S, Nicolaides KH. First trimester maternal serum free beta human chorionic gonadotrophin and pregnancy associated plasma protein A as predictors of pregnancy complications. BJOG 2000; 107(10): 1265-70.
[http://dx.doi.org/10.1111/j.1471-0528.2000.tb11618.x] [PMID: 11028579]

[115] Kim YM, Bujold E, Chaiworapongsa T, *et al.* Failure of physiologic transformation of the spiral arteries in patients with preterm labor and intact membranes. Am J Obstet Gynecol 2003; 189(4): 1063-9.
[http://dx.doi.org/10.1067/S0002-9378(03)00838-X] [PMID: 14586356]

[116] Kim YM, Chaiworapongsa T, Gomez R, *et al.* Failure of physiologic transformation of the spiral arteries in the placental bed in preterm premature rupture of membranes. Am J Obstet Gynecol 2002; 187(5): 1137-42.
[http://dx.doi.org/10.1067/mob.2002.127720] [PMID: 12439491]

[117] Huang T, Hoffman B, Meschino W, Kingdom J, Okun N. Prediction of adverse pregnancy outcomes by combinations of first and second trimester biochemistry markers used in the routine prenatal screening of Down syndrome. Prenat Diagn 2010; 30(5): 471-7.
[http://dx.doi.org/10.1002/pd.4090] [PMID: 20440736]

[118] Stout MJ, Goetzinger KR, Tuuli MG, Cahill AG, Macones GA, Odibo AO. First trimester serum analytes, maternal characteristics and ultrasound markers to predict pregnancies at risk for preterm birth. Placenta 2013; 34(1): 14-9.
[http://dx.doi.org/10.1016/j.placenta.2012.10.013] [PMID: 23199792]

[119] Beta J, Akolekar R, Ventura W, Syngelaki A, Nicolaides KH. Prediction of spontaneous preterm

delivery from maternal factors, obstetric history and placental perfusion and function at 11-13 weeks. Prenat Diagn 2011; 31(1): 75-83.
[http://dx.doi.org/10.1002/pd.2662] [PMID: 21210482]

[120] Spencer K, Cowans NJ, Molina F, Kagan KO, Nicolaides KH. First-trimester ultrasound and biochemical markers of aneuploidy and the prediction of preterm or early preterm delivery. Ultrasound Obstet Gynecol 2008; 31(2): 147-52.
[http://dx.doi.org/10.1002/uog.5163] [PMID: 17992705]

[121] Nikolova T, Bayev O, Nikolova N, Di Renzo GC. Evaluation of a novel placental alpha microglobulin-1 (PAMG-1) test to predict spontaneous preterm delivery. J Perinat Med 2014; 42(4): 473-7.
[http://dx.doi.org/10.1515/jpm-2013-0234] [PMID: 24334429]

[122] Lee SE, Park JS, Norwitz ER, Kim KW, Park HS, Jun JK. Measurement of placental alpha-microglobulin-1 in cervicovaginal discharge to diagnose rupture of membranes. Obstet Gynecol 2007; 109(3): 634-40.
[http://dx.doi.org/10.1097/01.AOG.0000252706.46734.0a] [PMID: 17329514]

[123] Forbes K, Westwood M, Baker PN, Aplin JD. Insulin-like growth factor I and II regulate the life cycle of trophoblast in the developing human placenta. Am J Physiol Cell Physiol 2008; 294(6): C1313-22.
[http://dx.doi.org/10.1152/ajpcell.00035.2008] [PMID: 18400990]

[124] Qiu Q, Bell M, Lu X, *et al.* Significance of IGFBP-4 in the development of fetal growth restriction. J Clin Endocrinol Metab 2012; 97(8): E1429-39.
[http://dx.doi.org/10.1210/jc.2011-2511] [PMID: 22689691]

[125] Saade GR, Boggess KA, Sullivan SA, *et al.* Development and validation of a spontaneous preterm delivery predictor in asymptomatic women. Am J Obstet Gynecol 2016; 214(5): 633.e1-633.e24.
[http://dx.doi.org/10.1016/j.ajog.2016.02.001] [PMID: 26874297]

[126] Goffinet F, Maillard F, Fulla Y, Cabrol D. Biochemical markers (without markers of infection) of the risk of preterm delivery. Implications for clinical practice. Eur J Obstet Gynecol Reprod Biol 2001; 94(1): 59-68.
[http://dx.doi.org/10.1016/S0301-2115(00)00317-1] [PMID: 11134827]

[127] Reis FM, Fadalti M, Florio P, Petraglia F. Putative role of placental corticotropin-releasing factor in the mechanisms of human parturition. J Soc Gynecol Investig 1999; 6(3): 109-19.
[http://dx.doi.org/10.1016/S1071-5576(99)00009-X] [PMID: 10376266]

[128] Wadhwa PD, Porto M, Chicz-DeMet A, Sandman CA. Ma- ternal CRH levels in early third trimester predict length of gestation in human pregnancy. Am J Obstet Gynecol 1998; 179: 1079-85.
[http://dx.doi.org/10.1016/S0002-9378(98)70219-4] [PMID: 9790402]

[129] Sibai B, Meis PJ, Klebanoff M, *et al.* Maternal Fetal Medicine Units Network of the National Institute of Child Health and Human Development. Plasma CRH measurement at 16 to 20 weeks' gestation does not predict preterm delivery in women at high-risk for preterm delivery. Am J Obstet Gynecol 2005; 193(3 Pt 2): 1181-6.
[http://dx.doi.org/10.1016/j.ajog.2005.06.036] [PMID: 16157134]

CHAPTER 24

Biomarkers in Obesity and Clinical Applications to Surgical Practice: From Pharmacogenomics to Surgigenomics

Halit Eren Taskin[1,*]

[1] *Department of General Surgery, Cerrahpasa Medical Faculty, Istanbul University-Cerrahpasa, Istanbul, Turkey*

Abstract: Personalized medicine is an evolving field and helps in detecting putative targets for pharmacological treatment modalities in obesity treatment. Obesity has become a pandemic and major threat, increasing all causes of mortality and bringing a heavy cost burden on health budgets. Although bariatric/metabolic surgery justified its role as the most effective treatment regimen for obesity and related comorbidities in the long-term, some patients require revisional operations, do not respond to this treatment option, or experience a relapse of their co-morbidities. For this reason, detection of possible non-responders and prediction of patients at high risk for developing cardiovascular or metabolic adverse events in long-term follow-up is of utmost importance. Therefore, the individualization of surgical treatment by the use of novel biomarkers has come into prominence recently. Many biomarkers and gene loci had been identified related to adiposity, insulin resistance, and oxidative stress. In this brief chapter, the recent advances and clinical implications of these biomarkers will be elaborated.

Keywords: Bariatric surgery, Biomarkers, Obesity, Oxidative stress, Inflammation, Personalized medicine, Proteomics.

INTRODUCTION

Obesity has become a global health challenge, affecting almost 27.5% of adults and 47.1% of children worldwide [1]. Obesity increases the development of many chronic diseases such as Type II diabetes (T2DM), hypertension, coronary heart disease, and certain types of cancer, decreasing the quality and life span of an individual. Moreover, obesity has shown to be a multifactorial, heritable disease for which no gold-standard treatment or preventive strategy has been developed [2]. World Health Organization defined obesity as an abnormal or excessive

* **Corresponding author Halit Eren Taskin:** Department of General Surgery, Cerrahpasa Medical Faculty, Istanbul University-Cerrahpasa, Istanbul, Turkey; Tel: +90 (505) 669 39 05; E-mail: eren.taskin@istanbul.edu.tr

Pınar Atukeren and Hafize Uzun (Eds.)

accumulation of fat that impairs the health status of an individual. Diagnosis of obesity has shown variation methodology, and certain indexes, most commonly body mass index and anthropometric measurements, are currently being used as simple methods for the classification of obesity [3].

Bariatric surgery proved its role to be the most effective treatment modality for the long-term management of overweight, obesity, and related comorbidities. Currently, a variety of operations have been approved by international organizations for the treatment of obesity and metabolic disorders; however, there have been no classification criteria developed on which operation should be used for specific patients. It is unfortunate that even today, the decision-making process for the type of operation for the treatment of obese patients depends on the expertise of the surgeon and the patient's choice. It has been well known that certain types of operations yield better results in patients with T2DM and dyslipidemia. There are many comparative retrospectives and prospective studies in the literature regarding the potency and the long-term results of the bariatric procedures; however, randomized trials with concrete evidence are scarce. Also, hundreds of biomarkers have been defined in literature as a result of animal and human studies, but only a couple of these are used routinely in our daily practice.

Identification of predictors of diabetes remission and the markers that reveal the early start of serious complications such as endothelial injury and oxidative stress resulting from diabetes is immensely important for the decision of treatment strategy in metabolically ill patients. Currently, the ability to do whole genome sequencing allows us to detect major genetic polymorphisms that may leave the patients vulnerable to metabolic disorders. Also detecting putative genetic targets for pharmacological therapy in obese and diabetic patients may help the development of drugs that can show a similar mechanism of highly effective bariatric surgical procedures [4].

Genetic variations among patients treated with bariatric surgery might provide clues for differences in outcomes. Patients with single-gene mutations that lead to early-onset monogenic obesity were recently found to be more resistant to weight loss induced after sleeve gastrectomy. This fact might explain the presence of non-responders after revisional surgical procedures, that who cannot accomplish satisfactory weight loss. Thus, in this chapter, we tried to make a thorough evaluation of the biomarkers that can be used for the decision-making process for the treatment of obesity and co-morbidities [5].

Importance of Personalized Medicine

Up-to-date evidence implies that patients may show variable responses to medical and surgical treatment of obesity and metabolic disorders. Yet every patient has a unique pharmacogenetic profile. This genetic variability determines the outcomes of both primary and revisional bariatric surgical procedures. It has been evident that pharmacotherapy alone cannot accomplish optimal results compared to surgery. However, it is known that even the most effective methods of bariatric surgical practice may fail in the long-term resulting in weight recidivism and relapse of co-morbid disorders. This is why detecting genetic mutations that may lead to this failure before the operation may help decide on the best procedure tailored for the individual patient [6, 7].

Personalized health risk assessment and a greater understanding of the specific obesity and co-morbid conditions of an individual may help providers to make more rational and specific medical decisions and may prescribe personalized interventions that will best prevent obesity and associated diseases, delay their onset, or reduce their negative impact on the individual. This fact also is amenable to surgical decision-making strategy. This is why we created the word surgigenomics which means the use of the genetic makeup of an individual to decide on the optimal surgical treatment method for the optimal care of an illness. Also, studying biomarkers in obesity and bariatric surgical practice may increase the number of drugs that can be used for obesity treatment. The best example of this fact is the novel use of liraglutide, which is a synthetic form of glucagon-like peptide-1 (GLP-1). Thus, both the gastric bypass procedure and the use of liraglutide reduce gastric emptying, increase satiety, and decrease food intake. In long-term follow-up, patients with weight regain on this medication showed promising results; personalized medicine can predict patients who might respond to surgery and/or medical treatment modalities and thus provide cost-effective strategies [8].

The Impact of Genes on Obesity Treatment

The impact of genes on obesity is complex, and since obesity is a multifactorial disease, the predictive ability to genotype in bariatric surgery can only be of value with clinical correlation with outcomes. Swedish Obese Subjects study (SOS) is the most prominent non-randomized, prospective trial studying intentional weight reduction in severely obese health effects. As a continuation of this study on patients in the surgical intervention group during 6 years of follow-up with full genomic and bodyweight data, single nucleotide polymorphisms (SNAP) has been studied on 11 obesity candidate genes. Twelve SNPs in the ADIPOQ, FTO, LEP, LEPR, MC4R, PPARGC1A, and TNF gene loci were found to be nominally associated with maximum weight loss after bariatric surgery. However, the correctional analysis showed that only the FTO SNP rs16945088 remained statistically associated with maximum weight loss. Although more than 20 culprit genes are associated with obesity, less than 5% of obese subjects carry these

SNP's. Moreover, certain biomarkers found to be beneficial for predicting favourable outcomes are not procedure-specific, and none can be elucidated for easy and efficient clinical use [9 - 11].

Proteomics and Lipidomic Profiling

In recent years, proteomic and lipidomic profiling of the patients before and after surgery can help us understand the weight loss independent effects of bariatric and metabolic surgery on the cardiovascular system, mental status, and immune regulation [11]. Global protein expression (proteomics) is an emerging scientific advancement successfully applied to certain fields of medicine as cancer and transplantation. This technique has a great potential for use in the field of bariatric and metabolic surgery. The prominence of this technique is that post-translational modification of protein-protein interactions can be determined, thus providing a better understanding of the effects of surgery on the cellular level. It is an immense opportunity to detect novel biomarkers and find putative targets for emerging pharmaceuticals and nano-nutrients [12].

Lipids play a significant role in the cardiovascular health and metabolic stability of an individual. Obesity usually alters metabolism, resulting in dyslipidemia and type-2 diabetes that increase cardiovascular mortality. Current clinically used biomarkers such as triglycerides, total cholesterol, HDL-C, and LDL-C provide a very narrow glimpse of the dynamic processes of lipid metabolism.

Novel Biomarkers Associated with Bariatric Surgery and Eating Control

Recently, with the plasma proteome profiling technique, hundreds of plasma proteins ranging from apolipoproteins to other transporters and functional proteins of the complement system can be studied in individuals. The biological impact of losing weight on humane proteome has been studied on individuals with calorie restriction basically on 800 kcal/day diet of 8-week range and 12 months range. This study also shows that a significant decrease of adipocyte-secreted SERINF1 in contrast to the known weight loss marker of sex hormone-binding globulin (SHBG) makes it possible to be used in a clinical context used as a weight maintenance marker [13].

In addition to these emerging findings, bariatric surgery also altered clinically significant circulating plasma proteins. This change in plasma proteins may explain the independent weight improvement in increased hepatic and peripheral insulin sensitivity. Thus, it is found that the plasma proteome dynamically changes after Roux-en-Y gastric bypass surgery. It is seen that lipoproteins, inflammatory proteins, and also proteins responsible for the cross-talk of organ systems have been altered after surgery. Also, there is a notable relationship between inflammation-related protein alterations and increased insulin sensitivity [14].

A recent study revealed that proteins also varied among individuals with different eating behaviours, and also there is a change before and after surgical interventions. Subjects with food cravings and loss of control scores (LC) according to the State and Trait Food Cravings Questionnaire Trait scale (FCQ-T) seem to have higher optical densities of clusterine values on mass spectrophotometer analysis. Clusterine showed a significant correlation with eating behaviour in animal studies, showing opposite effects compared to ghrelin and leptin on the activity of hypothalamic neurons and the homeostatic control of appetite. This biomarker is promising to be used for the pre-operative identification of non-responders for bariatric surgical procedures in the future [15].

Novel Biomarkers for Inflammation and Endothelial Dysfunction Associated to Bariatric Surgery

Many studies justified the role of low-grade inflammation for the cause of serious co-morbidities such as cardiovascular disease and possibly cancers in obese individuals. Recently, our group studied the association between peroxisome proliferator-activated receptor-γ (PPAR-γ), Zn-alpha 2 glycoprotein (ZAG), nuclear factor kappa-light-chain-enhancer of activated B cell (NF-κB) and obesity in patients who underwent sleeve gastrectomy procedure. It is found that although the preoperative peroxisome proliferator-activated receptor-γ (PPAR-γ) was lower in-patient group compared to the control group, in the follow-up period of 1st to 12th month, there was a significant increase in the patient group with the effect of weight loss. Also, post-operative values of ZAG increased, and NF-κB decreased significantly over the post-operative period in patients' group (postoperative month 1 and month 12 by $P < 0.05$, $P < 0.01$). The importance of this finding is due to the knowledge that ZAG plays a major role in fat metabolism, weight loss, and reduction in muscle mass and fat tissue, especially in cachexia. Furthermore, many studies revealed that PPAR-γ is the main regulator of adipocyte differentiation (adipogenesis). Also, NF-κB contributes to the regulation of apoptosis, cell growth, angiogenesis, metastasis, and cell life cycle and plays a prominent role in the development of insulin resistance [16].

Our recent study showed clearly that there was a significant negative correlation among EWL% and NF-κB, HOMA-IR, fasting plasma glucose, and insulin parameters, whereas the maximum correlation value was calculated in the relationship was between EWL% and fasting plasma glucose level. Moreover, there was a significant positive correlation with ZAG protein (Fig. **1**). The results of the study suggest that weight loss achieved by sleeve gastrectomy ameliorates upregulating PPAR-γ expression, thereby inhibiting NF-κB-mediated inflammation [16].

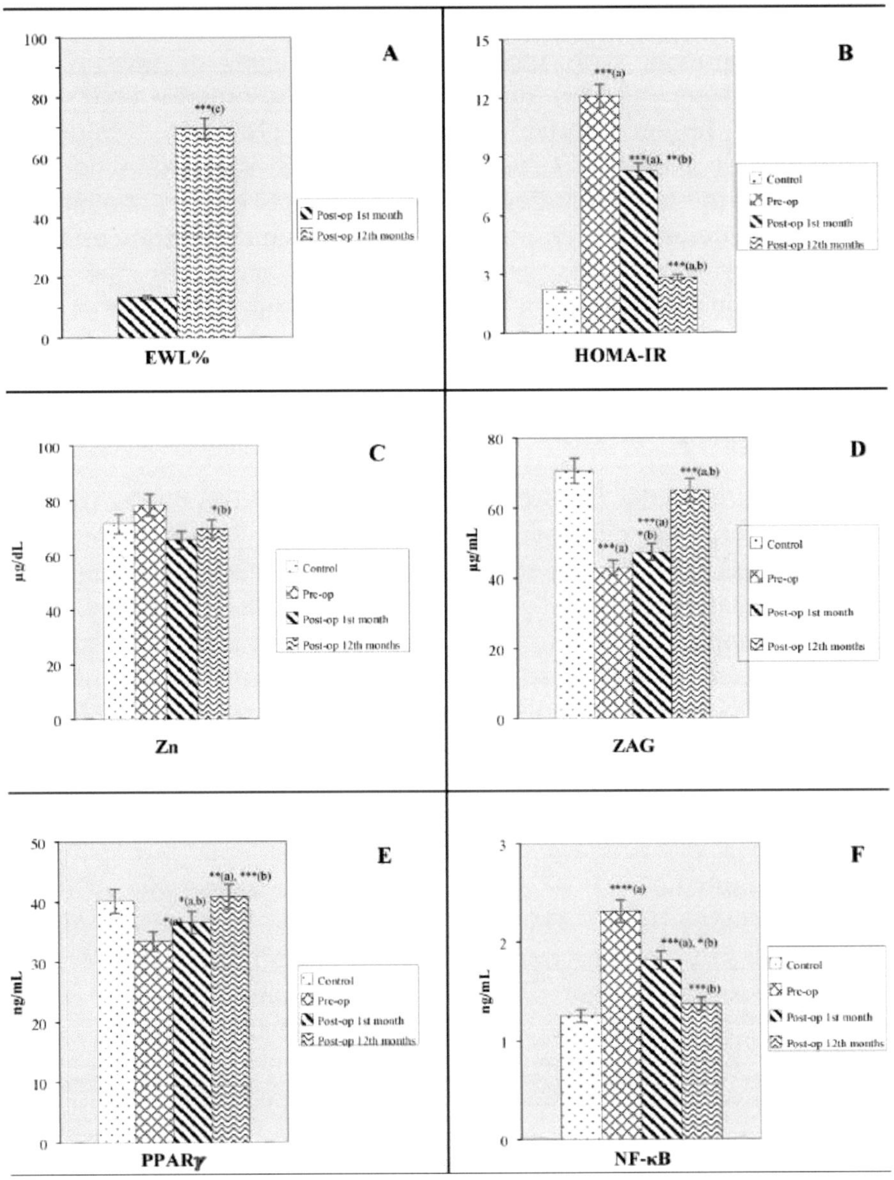

Fig. (1). A- EWL%, B- HOMA-IR, C- Zn, D- ZAG, E- PPAR-γ, and F- NF-κB levels of the study groups (Reprinted with permission from the corresponding author).

Other biomarkers we have studied related to the relation of inflammation and morbid obesity were procalcitonin (PCT), high-sensitivity CRP (hsCRP), interleukin 33 (IL-33), and the IL-33 receptor sST2. The results of the study suggested that morbidly obese patients at baseline (pre-surgery) had markedly higher levels of the 4 studied pro-inflammatory biomarkers relative to healthy non-obese patients. At both 1 and 6 months after undergoing SG, morbidly obese patients' serum pro-inflammatory biomarker levels dropped significantly, trending toward those of non-obese patients (Table **1**). Post-SG, insulin resistance (assessed by change in insulin and HOMA-IR) and endothelial dysfunction (assessed by change in hsCRP and sST2) were seen to improve with weight loss. The current findings are in agreement with those of several studies that demonstrate that SG and other bariatric procedures (*i.e.*, Roux-en-Y gastric bypass, biliopancreatic diversion/duodenal switch, and adjustable gastric banding) appear to effectively reduce systemic inflammation in obese patients [17, 18].

Oxidative stress (OS) has been linked to the pathogenesis of obesity and its comorbid diseases. Biomarkers of OS ease the assessment of obesity-related disease activity and the health-enhancing effects of therapeutic interventions. Ischemia-modified albumin (IMA) has been found as a novel blood biomarker of OS and ischemia, and, recently, obesity has been discovered as an independent cause of IMA [19, 20]. Morbidly obese patients are at particularly high risk of ischemic injury and oxidative damage. Satisfactory significant weight loss has been associated with reduced OS and increased antioxidant protection [21]. Our recent unpublished study showed the role of adjustable gastric band applications (AGB) in the regulation of serum IMA, and additional OS parameters for 6 months post-operative follow-up. There was a significant reduction (65%) of IMA levels and additional parameters of the advanced oxidation protein product (AOPP) and pro-oxidant antioxidant balance (PAB) from baseline at 6 months after the surgery even though the mean BMI of operated patients remained significantly higher than controls in this period. Also, AGB-induced changes in body mass index (BMI) and HOMA-IR were found to be directly and significantly correlated with changes in OS parameters: BMI/IMA, $r = 0.83$, $p < 0.001$; BMI/AOPP, $r = 0.87$, $p < 0.001$; HOMA-IR/IMA, $r = 0.84$, $p < 0.001$; HOMA-IR/AOPP, $r = 0.86$, $p < 0.001$ (Table **2**). These findings are consequential since OS may trigger obesity by stimulating white adipose tissue deposition and increasing preadipocyte proliferation, adipocyte size, and differentiation and interestingly, obesity is associated with overproduction of free radicals, and thus, may induce systemic OS. Also, OS may play a promoting role in the pathogenesis and catenation of obesity-related comorbidities. For instance, in type 2 diabetes, OS compromises glucose uptake in fat and muscle and encumbers pancreatic β cells from secreting insulin [21 - 23]. In the essence of these prominent findings, these novel biomarkers may play a critical role in predicting adverse comorbidities and major cardiovascular mortality-related factors in patients who underwent bariatric-metabolic surgery.

Table 1. Pre- and post-sleeve gastrectomy inflammatory and metabolic measurements in morbidly obese patients.

-	Baseline (n=20)	1 month after Operation* (n=20)	6 months after Operation† (n=20)
Body weight, kg	136.55±22.46	130.00±22.29	102.70±18.55
BMI, kg/m^2	49.12±6.49	46.75±6.43	36.94±5.66
Leukocytes, mm^3	10.73±4.56	8.43±4.55	8.57±5.78
Glucose, mg/dL	162.85±51.96	150.35±36.95	122.85±26.44
Insulin, µIU/mL	17.28±7.73	14.45±6.30	7.91±2.95
PCT, pg/mL	27.46±10.29	21.97±6.06	13.45±5.33
IL-33, ng/mL	804.94±513.03	489.39±233.96	254.10±97.1
sST2, ng/mL	2.96±0.24	2.22±0.19	1.77±0.15
hsCRP, mg/dL	6.67±4.89	4.34±1.88	2.78±1.03
HOMA-IR	7.47±5.62	5.58±3.43	2.44±1.21

Values presented as mean ± SD. BMI, body mass index; PCT, procalcitonin; IL-33, interleukin 33; sST2, soluble suppression of tumorigenicity 2; hsCRP, high-sensitivity C-reactive protein; HOMA-IR, homeostatic model assessment–insulin-resistant index. * One month post SG vs baseline (paired t-test; $p < 0.05$) † Six months post SG vs baseline (paired t-test; $p < 0.001$).

Table 2. Metabolic parameters, ischemia, and oxidative stress parameters in morbidly obese patients at baseline (preoperative) and 1 and 6 months after laparoscopic adjustable gastric banding.

-	Baseline (n=20)	1 month after operation* (n=20)	6 months after operation† (n=20)
Body weight, kg	136.55±22.46	130.00±22.29	102.70±18.55
BMI, kg/m^2	49.12±6.49	46.75±6.43	36.94±5.66
Glucose, mg/dL	162.85±51.96	150.35±36.95	122.85±26.44
Insulin, µIU/mL	17.28±7.73	14.45±6.30	7.91±2.95
HOMA-IR	7.47±5.62	5.58±3.43	2.44±1.21
IMA, mg/dL	6.67±4.89	4.12±1.45	2.33±1.11
AOPP, µM chloramin T	32.89±17.39	21.97±6.06	13.45±5.33
TAC, mM uric acid	0.35±0.12	0.46±0.91	0.58±0.17
PAB, U/ml	1.39±0.16	1.24±0.19	0.97±0.11

BMI, body mass index; HOMA-IR, homeostatic model assessment–insulin resistant; NS, not significant; IMA, ischemia-modified albumin; AOPP, advanced oxidation protein product; TAC, total antioxidant capacity; PAB, pro-oxidant antioxidant balance. * One month post AGB vs baseline (paired t-test; $p < 0.05$) † Six months post AGB vs baseline (paired t-test; $p < 0.001$).

THE FUTURE OF PERSONALIZED MEDICINE IN THE BARIATRIC-METABOLIC SURGICAL PRACTICE AND CONCLUSION

Bariatric/metabolic surgery is highly effective not only by sustaining weight loss but also resolution of co-morbidities by complex hormonal, nutritional and biochemical pathways. However, there are no adequate sources to offer multidisciplinary treatment modalities to all obese patients. Therefore, patient selection and individualization of the optimal method are immensely important. Weight regain, and relapse of co-morbidities is another cumbersome issue in the long-term follow-up of operated patients. Effective use of novel biomarkers and personalized medicine may help us develop new pre-operative selection criteria and algorithms that can predict non-responders for surgical treatment in long term. Also, these biomarkers may help the follow-up process of the patients by easing the prediction of adverse mortality-related factors. Although personalized medicine is still an evolving field, its initiation into clinical, surgical practice will lead to its practical and cost-efficient use in the future. Nevertheless, currently, its capability to designate putative targets for future pharmacologic treatment methods makes it a frontier element of clinical translational research.

CONSENT FOR PUBLICATION

Not applicable.

CONFLICT OF INTEREST

The authors declare no conflict of interest, financial or otherwise.

ACKNOWLEDGEMENT

Declared none.

REFERENCES

[1] Lajunen H-R, Kaprio J, Keski-Rahkonen A, *et al.* Genetic and environmental effects on body mass index during adolescence: a prospective study among Finnish twins. Int J Obes 2009; 33(5): 559-67. [http://dx.doi.org/10.1038/ijo.2009.51] [PMID: 19337205]

[2] Nimptsch K, Konigorski S, Pischon T. Diagnosis of obesity and use of obesity biomarkers in science and clinical medicine. Metabolism 2019; 92: 61-70. [http://dx.doi.org/10.1016/j.metabol.2018.12.006] [PMID: 30586573]

[3] Zhukouskaya VV, Rothenbuhler A, Colao A, *et al.* Increased prevalence of overweight and obesity in children with X-linked hypophosphatemia. Endocr Connect 2020; 9(2): 144-53. [http://dx.doi.org/10.1530/EC-19-0481] [PMID: 31910157]

[4] Andrew AB, Robert W O'R. Bariatric surgery in the era of personalized medicine. Enterology 2013; 144(3): 497-500. [http://dx.doi.org/10.1053/j.gastro.2013.01.027]

[5] Li Y, Zhang H, Tu Y, *et al.* Monogenic Obesity Mutations Lead to Less Weight Loss After Bariatric

Surgery: a 6-Year Follow-Up Study. Obes Surg 2019; 29(4): 1169-73.
[http://dx.doi.org/10.1007/s11695-018-03623-4]

[6] Awad H, Ahmed A, Urman RD, Stoicea N, Bergese SD. Potential role of pharmacogenomics testing in the setting of enhanced recovery pathways after surgery. Pharmgenomics Pers Med 2019; 26(12): 145-54.
[http://dx.doi.org/10.2147/PGPM.S198224]

[7] Schauer PR, Bhatt DL, Kirwan JP, *et al.* Bariatric Surgery versus Intensive Medical Therapy for Diabetes - 5-Year Outcomes. N Engl J Med 2017; 376(7): 641-51.
[http://dx.doi.org/10.1056/NEJMoa1600869] [PMID: 28199805]

[8] Wharton S, Kuk JL, Luszczynski M, Kamran E, Christensen RAG. Liraglutide 3.0 mg for the management of insufficient weight loss or excessive weight regain post-bariatric surgery. Clin Obes 2019; 9(4): e12323.
[http://dx.doi.org/10.1111/cob.12323] [PMID: 31183988]

[9] Sarzynski MA, Jacobson P, Rankinen T, *et al.* Associations of markers in 11 obesity candidate genes with maximal weight loss and weight regain in the SOS bariatric surgery cases. Int J Obes 2011; 35(5): 676-83.
[http://dx.doi.org/10.1038/ijo.2010.166] [PMID: 20733583]

[10] Neff KJ, Olbers T, le Roux CW. Bariatric surgery: the challenges with candidate selection, individualizing treatment and clinical outcomes. BMC Med 2013; 11: 8.
[http://dx.doi.org/10.1186/1741-7015-11-8]

[11] Graessler J, Bornstein TD, Goel D, *et al.* Lipidomic profiling before and after Roux-en-Y gastric bypass in obese patients with diabetes. Pharmacogenomics J 2014; 14(3): 201-7.
[http://dx.doi.org/10.1038/tpj.2013.42] [PMID: 24365785]

[12] Brandacher G, Golderer G, Kienzl K, Werner ER, Margreiter R, Weiss HG. Potential applications of global protein expression analysis (proteomics) in morbid obesity and bariatric surgery. Obes Surg 2008; 18(7): 905-10.
[http://dx.doi.org/10.1007/s11695-007-9337-2] [PMID: 18330661]

[13] Geyer PE, Wewer Albrechtsen NJ, Tyanova S, *et al.* Proteomics reveals the effects of sustained weight loss on the human plasma proteome. Mol Syst Biol 2016; 12(12): 901.
[http://dx.doi.org/10.15252/msb.20167357] [PMID: 28007936]

[14] Wewer Albrechtsen NJ, Geyer PE, Doll S, *et al.* Plasma proteome profiling reveals dynamics of inflammatory and lipid homeostasis markers after roux-en-y gastric bypass surgery. Cell Syst 2018; 7(6): 601-612.e3.
[http://dx.doi.org/10.1016/j.cels.2018.10.012] [PMID: 30528273]

[15] Rodríguez-Rivera C, Pérez-García C, Muñoz-Rodríguez JR, *et al.* Proteomic identification of biomarkers associated with eating control and bariatric surgery outcomes in patients with morbid obesity. World J Surg 2019; 43(3): 744-50.
[http://dx.doi.org/10.1007/s00268-018-4851-z] [PMID: 30426189]

[16] Ergun S, Ergun DD, Akinci O, Taskin HE, *et al.* The role of laparoscopic sleeve gastrectomy on inflammatory parameters in morbidly obese patients. J Visc Surg 2020; S1878-7886(20): 30303-9.
[http://dx.doi.org/10.1016/j.jviscsurg.2020.11.009]

[17] Askarpour M, Khani D, Sheikhi A, Ghaedi E, Alizadeh S. Effect of bariatric surgery on serum inflammatory factors of obese patients: A systematic review and meta-analysis. Obes Surg 2019; 29(8): 2631-47.
[http://dx.doi.org/10.1007/s11695-019-03926-0] [PMID: 31093862]

[18] Miller GD, Nicklas BJ, Fernandez A. Serial changes in inflammatory biomarkers after Roux-en-Y gastric bypass surgery. Surg Obes Relat Dis 2011; 7(5): 618-24.
[http://dx.doi.org/10.1016/j.soard.2011.03.006] [PMID: 21546319]

[19] Marrocco I, Altieri F, Peluso I. Measurement and clinical significance of biomarkers of oxidative stress in humans. Oxid Med Cell Longev 2017; 2017: 6501046. [http://dx.doi.org/10.1155/2017/6501046] [PMID: 28698768]

[20] Chawla R, Loomba R, Guru D, Loomba V. Ischemia modified albumin (IMA)-A marker of glycaemic control and vascular complications in type 2 diabetes mellitus. J Clin Diagn Res 2016; 10(3): BC13-6. [http://dx.doi.org/10.7860/JCDR/2016/15282.7432] [PMID: 27134857]

[21] Mehmetoğlu I, Kurban S, Yerlikaya FH, Polat H. Obesity is an independent determinant of ischemia-modified albumin. Obes Facts 2012; 5(5): 700-9. [http://dx.doi.org/10.1159/000343954] [PMID: 23108391]

[22] Marseglia L, Manti S, D'Angelo G, *et al.* Oxidative stress in obesity: a critical component in human diseases. Int J Mol Sci 2014; 16(1): 378-400. [http://dx.doi.org/10.3390/ijms16010378] [PMID: 25548896]

[23] Rudich A, Tirosh A, Potashnik R, Hemi R, Kanety H, Bashan N. Prolonged oxidative stress impairs insulin-induced GLUT4 translocation in 3T3-L1 adipocytes. Diabetes 1998; 47(10): 1562-9. [http://dx.doi.org/10.2337/diabetes.47.10.1562] [PMID: 9753293]

CHAPTER 25

Clinical Application of New Possible Biomarkers in the Assessment and Monitoring of Nutritional Status

Burcu Yesilkaya[1,*]

[1] *Departments of Nutrition and Dietetics, İstanbul Okan University, İstanbul, Turkey*

Abstract: Nutrition is directly related to human health. It is very critical to determine the nutritional status to prevent or diagnose diseases and create the right treatment plans. The determination of the nutritional status provides an early diagnosis of growth and development retardation such as malnutrition. It also plays a major role in preventing diseases that may be caused by vitamin and mineral deficiencies. It helps in the surveillance of one of the world's most serious health problems, namely "obesity." Different ways can be used to assess nutritional status. One of the best ways to assess the nutritional and health status is to use biomarkers. A biomarker is a substance whose detection indicates a specific disease state or a response to a therapeutic intervention. Biomarkers are used to detect nutrient consumption and deficiencies as early as possible, enabling early intervention for metabolic problems. Biomarkers also allow the visualization of diseases that a person might develop or potentially have with a sample, such as blood, tissue, and urine, from the person. Health interventions such as nutritional advice will preserve health or promote rapid recovery. In this chapter, the topic of biomarkers related to nutrition and nutrient deficiencies is discussed. The existence of new possible biomarkers is also reviewed.

Keywords: Biomarker, Dietary biomarkers, Food Exposure Markers, Macronutrient, Metabolomics, Micronutrient, Nutrient, Nutrigenomics, Nutrition, Nutritional Assessment, Omics technologies.

INTRODUCTION

Nutrition and Health

According to World Health Organization (WHO), health is a state of complete physical, mental, and social well-being [1]. In a healthy life, nutrition is an important key factor. Nutrition is also a part of healthy growth and development.

* **Corresponding author Burcu Yesilkaya:** Departments of Nutrition and Dietetics, İstanbul Okan University, İstanbul, Turkey; E-mail: Tel: +90(543)5278676; burcuyavunc@hotmail.com

Pınar Atukeren and Hafize Uzun (Eds.)

The body, as a living organism, needs nutrients to perform mental and physical activities [2].

Nutrition is a compulsory human behavior for survival. Food consumption and hormonal control mechanisms such as hunger and satiety involve many physiological processes. However, this needs to be a conscious behavior to protect and improve one's health and the quality of life. Healthy nutrition is defined as adequate and balanced nutrient consumption, with macronutrients and micronutrients consumed according to personal requirements. These recommended amounts can change according to age, gender, preferences, and physical activity status [3, 4].

Nutrients provide the required "energy" to the body. This energy comes from carbohydrates, proteins, and fats, which are called macronutrients. Micronutrients such as vitamins and minerals do not provide energy but play major roles in the processes of energy metabolism. These include chemical reactions that maintain tissues, electrical conduction of nerves, mechanical workings of muscles, and heat generation to keep body temperature stable [5]. Healthy nutrition is also based on a variety of nutrients. Daily requirements of energy and nutrients are taken into the body with foods. Foods are divided into five groups according to the similarity of nutrients they contain. These are vegetables, fruits, grains, proteins, and dairy products that need to be consumed daily [3].

The nutritional status is stable when a person's nutrient intake meets the requirements. Determination of the nutritional status may prevent diseases and promote health. Most chronic diseases show up due to poor nutrition. Assessing the nutritional status will help determine the lack of nutrients. Early diagnosis of diseases also prevents severe deficiencies and functional disorders. As an intervention program, nutritional support and counseling help improve insufficient nutrition and poor nutritional status [6]. A sufficient and balanced diet, including macronutrients and micronutrients, is beneficial for complete health and, ultimately, the quality of life, especially for patients with chronic diseases/wounds [7].

Malnutrition, such as undernutrition, overnutrition, or particular nutrient deficiencies, are identified during the nutritional assessment. These problems are usually caused by poor or overnutrition. Lack of essential nutrients causes various diseases. Malnutrition problems such as obesity, malnourishment, or diseases caused by nutrient deficiencies are examples of an unhealthy nutritional status [8].

Assessment of Nutritional Status and Methods

The nutritional statuses of people are evaluated by dietitians who are nutrition experts. However, it requires a multidisciplinary screening to determine the nutritional status. As a result of the evaluation, the requirements and deficiencies are seen, and the possible risks are revealed [9].

The nutrition care process has 4 steps: 1) assessment of nutritional status, 2) identifying the nutritional diagnosis, 3. interventions such as education, supports, counseling, *etc.* 4. monitoring and evaluation of the effectiveness of the interventions [6]. To determine the nutritional status, many methods may be used. Dietary indicators, anthropometric indicators, laboratory indicators, and clinical indicators help to define health problems.

The most harmful nutritional status is malnutrition. It refers to deficiencies, excesses, or imbalances in energy and/or nutrient intake. Early diagnosis of malnutrition can both prevent the long-term treatment of the person and contribute greatly to the national economy. Nutritional assessment methods (such as anthropometric measurements) can easily identify malnutrition. The early detection of malnutrition can save the infant's life [10, 11].

In addition to continuous and regular follow-up, a good nutrition plan is required in the treatment plan. This plan must be constantly evaluated and its usefulness measured. Generally, the ABCD nutrition assessment method is used. It includes *anthropometric (A)* indices based on body measurements, *biochemical* indicators (B), an examination of *clinical* status (C), and *dietary* assessment (D) [12, 13].

Anthropometric

Anthropometry investigates the measurements of the human body. Anthropometric measurements include measurement methods used to evaluate body components such as bone, muscle, and fat tissue [14, 15]. Anthropometry has an important role in the nutritional evaluation and is used in general nutritional screening, surveillance, and follow-up, as well as clinical use. A single measure may not provide a comprehensive overview of the patients' condition and the other measurements combinations are required for certain assessments [16, 17].

Anthropometric measurement components vary according to the age, gender, ethnicity, status (pregnancy, athletes, *etc.*) of individuals. The basic elements of anthropometry are height (length for newborns), weight, body circumferences (waist, hip, and limbs), and skinfold thickness. The cut-off points and the classifications of these measurements depend on the person's status [15, 18].

Body mass index (BMI), a person's weight (in kilograms) divided by the square of his/her height (in meters). It is kg/m^2. WHO classification for BMI is generally used all over the world [19]. With the increase in BMI, the risks of chronic diseases such as cardiovascular diseases or diabetes increase. BMI alone is not sufficient to determine the nutritional status, but it provides information about the condition. The assessment of nutritional status and the growth & development levels are an essential part of clinical evaluation in pediatric patients. Growth charts such as Percentiles and Z- score tables (known as gold standard) are used for screening the growth and development period of a child's nutritional assessment [1]. BMI provides a good warning to raise awareness of what can happen if not intervened. Body circumferences can complement BMI estimates. Such as waist circumference, hips circumference and their ratio (waist/hips) are good indicators for a nutritional assessment [20]. Also, waist/ height ratio is a newly used predictor of metabolic syndrome if it is evaluated with BMI and Waist/hip ratio [21].

Biochemical

Nutritional status has a wide range starting from insufficiency to toxicity. Both may cause death. Biochemical tests are objective methods in the assessment of nutritional status. The accuracy and precision depend on the method used. A single biochemical indicator may not be differential for a diagnosis in some situations. Biochemical data, physical attributes, and functional changes should be tested and observed for a medical or nutritional diagnosis. Acute or chronic diseases or interventions such as surgery or injuries cause changes in biochemical tests parameters. Biochemical tests are performed on tissues or biologic fluids such as blood (plasma, serum), blood cells, urine, feces, saliva, breath tests, sweat, liver, bone, hair, and nail. Blood proteins (albumin, transferrin, prealbumin), blood lipids (total cholesterol, HDL, LDL, VLDL, triglycerides), hemoglobin and hematocrit levels, blood and urine vitamin and mineral levels determine the nutritional status of a person. Specimen types are selected according to the aim of the test. Complete blood tests, stool tests, and urinalysis are group tests. These are routine medical laboratory tests that interpret nutrition [22].

Body hydration status is vital because it affects other balances such as electrolyte balance. Dehydration or edema is the result of imbalanced hydration. Serum sodium, serum osmolarity, blood urea nitrogen, and urine-specific gravity are the biomarkers of hydration [22, 23].

Inflammation markers include positive acute phase reactants like C- reactive protein (CRP), which increases because of an infection or negative acute phase reactants like albumin, which is also related to inflammation [22]. Albumin is also an indicator of nutritional status. Inflammation is thought to be a risk factor for malnutrition. Other negative acute phase reactants like prealbumin, transferrin, and retinol-binding protein are also indicators for malnutrition [24 - 26].

Markers of anemia depend on the classification of the disease. There are three types; Microcytic anemia which is mostly caused by iron deficiency, macrocyte anemia, which is mostly caused by folate or B_{12} deficiency, and normocyte anemia which is associated with anemia of chronic and inflammatory disease (ACD) is caused by chronic diseases. While the microcyte and macrocyte anemia respond to mineral supplementation, normocyte anemia does not [22]. Serum ferritin, hematocrit (Hct), hemoglobin (Hb), serum iron, total iron-binding capacity (TIBC), and transferrin saturation are the laboratory indicator tests for microcyte anemia. Serum homocysteine, folate, vitamin $B_{12,}$ and methylmalonic acid assessment are the macrocyte anemia indicators [27, 28]. Also, zinc and copper analysis with iron, folate, and B_{12} are useful for the diagnosis of anemia [29].

Micronutrient (vitamins and minerals) deficiencies are huge and negatively impact morbidity, mortality, and life quality in the world [22]. Micronutrients are essential for S-adenosylmethionine synthesis, which is the donor of the methyl group for methylation, which is required by all types of RNAs and biochemical reactions. So, micronutrient deficiencies can affect RNA methylation. Vitamin deficiencies can cause serious medical conditions. Mostly inadequate and unbalanced nutrition or bariatric surgeries may cause micronutrient deficiencies. Food intolerance may occur in postop periods of bariatric surgeries. Vitamin B12, folate, and vitamin D deficiencies are mostly seen in patients with bariatric surgeries. Also, fat malabsorption induces fat-soluble vitamin (Vit A, D, E, and K) deficiencies. Fat-soluble vitamins are required for most physiological functions. Deficiencies of these vitamins have been associated with a high risk of cancer, type 2 diabetes mellitus (DM), and several immune system disorders. Vitamin A has a great role in many biologic functions such as vision, the structure of epithelial surfaces, immunity, reproduction, and in utero growth and development. Weight loss is mostly seen during malnutrition, after gastric bypass surgeries, and biliopancreatic surgeries [34]. Serum retinol concentrations (Retinoic acid receptor (RBP), Dried blood spots (DBS)), are the major biomarkers for Vitamin A deficiency. Breastmilk retinol concentration, Relative-dose-response tests (RDR), modified relative dose-response test (MRDR), isotope dilution, serum retinyl esters are the biomarkers to detect vitamin A deficiency. In addition to these biomarkers, some analytical tests, such as diagnosis-carotene tests, can also be parameters in diagnosing vitamin A deficiency [30 - 35].

Vitamin D can be provided through the endogenous synthesis in the skin with sunbathing or can be consumed through food or supplements [36]. The vitamin D status can be defined according to the plasma 25-hydroxy vitamin D (25-[OH]D_3

or Calciferol), which is the most frequently referred biomarker for epidemiologic studies [37]. The active form of vitamin D is 1,25-dihydroxy vitamin D3 (1,25-$(OH)_2D_3$). The deficiency of vitamin D causes many chronic diseases such as cardiovascular diseases, cancer, and malabsorption of calcium. Calcium deficiency causes chronic renal problems and renal failure. Renal hydroxylation is required active vitamin D, which promotes the synthesis of calcium-binding protein by intestinal absorptive cells [22, 32, 38, 39].

Vitamin E is a fat-soluble vitamin that is an antioxidant metabolite and an essential dietary factor. It is related to cancer, aging, cell destruction, and protection of cell structure. Vitamin E has also been associated with suppressing the synthesis of TNF, IL-1, IL-6, and IL-8. Vitamin E status can be identified with serum α-tocopherol or the ratio of serum α-tocopherol to serum lipids. New evidence has shown that β- tocopherol and γ- tocopherol is more important for human, so new biomarkers to determine β- tocopherol and γ- tocopherol need to be researched [22, 32].

Vitamin K has two primary forms as K1(phylloquinone) and K2 (menaquinones). Vitamin K status is about both nutrition and intestinal bacterial counts. Vitamin K is also responsible for blood clotting. There is not a certain biomarker to detect vitamin K status. Recent studies have shown that Uncarboxylated (UC) protein forms of vitamin K dependent on proteins (VKDPs) can determine the vitamin K status in blood circulation. The reduced vitamin K is related to reduced osteocalcin. It may explain the connection between vitamin K deficiency and osteoporosis. Desphospho-uncarboxylated matrix gla protein (dp-ucMGP) is used to determine the vitamin K status in cardiovascular diseases. The status of vitamin K is also measurable in urine excretion. These methods are new possible biomarkers for vitamin K. There is no gold standard biomarker for vitamin K. More studies and trials are needed for definitive results [22, 40 - 42].

Ascorbic acid (vitamin C) is an essential and water-soluble vitamin and antioxidant. It cannot be synthesized or stored by the human body. It protects cells from oxidation. It is also the cofactor of metalloenzymes which is responsible for vasopressors noradrenaline and vasopressin has a vital role in inflammation/sepsis. Concentrations of plasma vitamin C reflect the vitamin C status of individuals [22, 43]. In lymphocytes, vitamin C levels are 80 - 100 times higher than in plasma concentration; moreover, transient dietary modifications do not affect it. In a current study on some chronic diseases like diabetes mellitus, low lymphocyte vitamin C levels are linked with low oxidative stress. It has been seen that it is a potentially beneficial biomarker of Parkinson's disease progression [44].

Vitamin B_1 (Thiamine) is a water-soluble vitamin that has a great role in decarboxylation during energy production. Thiamine deficiency can cause many chronic diseases like cardiovascular, nervous, and immune system diseases. The most seen deficiency problem is called Beriberi or Wernicke-Korsakoff syndrome. Metabolic acidosis can also occur due to the cumulation of lactate. The thiamine pyrophosphate (TPP) which activates the transketolase is measured to diagnose the B_1 deficiency [45]. Vitamin B_{12} (cobalamin) is a dietary essential nutrient. Cyanocobalamin is another name of cobalamin (B_{12}-active corrinoid) which is a cobalt (Co) atom with a cyanide ion (CN–) at the β-position in the molecular structure. The main functions of B_{12} are red blood cell synthesis and prevention of megaloblastic anemia, neurologic dysfunction, cognitive dysfunction, prevention of dementia, and hyperhomocysteinemia [46]. The biomarkers of B_{12} can be categorized into two, direct measurement of vitamin B_{12} in blood and, those that measure metabolites that stimulate with inadequate amounts of vitamin B_{12}. Serum B_{12} and newly used holotranscobalamin (holoTC) measure circulating vitamin B_{12} concentrations. The measurements of vitamin B_{12} methylmalonic acid (MMA) and total homocysteine (tHcy), accumulate when concentrations of vitamin B_{12} are inadequate; tHcy also accumulates with insufficient vitamin B_2 B_6 and B_9 [47].

The magnesium (Mg) content of the body is important for vitamin D activation. It is also related to many chronic diseases and health. Magnesium deficiency causes electrolyte imbalance, acute phase deficiency, and inflammation [48, 49]. Zinc has a direct effect on the immune system, growth, and development of children. Usually, plasma zinc is the best indicator for zinc deficiency. Zinc is a good biomarker to detect growth and development retardation in children, hair loss, and geophagy [50].

Iron is required by humans for oxygen transportation, cellular functions, energy production, and pathogen blocking. Iron status biomarkers are hemoglobin (Hb), Serum Ferritin (SF), Iron Binding Capacity (IBC), Soluble Transferrin Receptor (sTfR), and zinc protoporphyrin (ZPP). The first three parameters are the most reliable parameters to define iron deficiency. The cerebrospinal fluid may show brain iron deficiency, but as a potential biomarker, it has not been examined yet. Hepcidin and novel red blood cell indexes are yet to be investigated. But it is a future clinical tool to define the iron status [51].

Iodine: Urine iodine is a worldwide parameter to assess the body iodine because 90% of dietary iodine is excreted in the urine. But it reflects the last 3-4 days iodine intake. The serum thyroglobulin (Tg) level is a helper biomarker for iodine nutrition status, and it reflects abnormal thyroid function [50, 52].

Selenium (Se) is an essential element where it is an integral part of selenoproteins. Selenium is a key factor for immune function, oxidative stress, human development (especially cognitive development), and health. Se status can be defined with blood, nails, hair, and urine samples. Casual urine Se concentration is a possible biomarker to plasma selenium concentration and selenium status [53].

Clinical

Clinical assessments include checking of clinical findings and symptoms that might indicate a potential specific nutrient deficiency, disease, or imbalance of body fluids in addition to the anthropometric, dietary, and biochemical assessments. Clinical symptoms are mostly caused by imbalanced body electrolytes, acid-base concentrations, and water content of the body. The lack of stabilization on fluid, electrolyte, and acid-base balance affects homeostasis, especially for hospitalized people. The composition, volume, and distribution of body fluids directly affect the cell structure and function. It may cause both acute and chronic problems in the body [23].

Dietary

Diet has a direct relationship with health and disease. All epidemiological and medical studies show that a balanced and adequate diet can improve health and quality of life. Changing dietary habits may also prevent most chronic diseases such as cancer and diabetes. Dietary assessment is an indicator of the nutritional status of a person. Dietary assessment methods should be selected according to the situation. Food consumption can be performed by subjective and objective methods. Subjective assessment methods are general surveys. The open-ended surveys such as dietary recalls or records (DR), or closed-ended surveys like food frequency questionnaires (FFQ) [54]. Self-reported dietary intake may be assessed with prospective (ex: food diaries) or retrospective (ex: dietary histories, FFQ, and 24H) methods. 24-hour recall (24H), FFQ, dietary record (DR), and dietary history are the most common methods to assess the dietary assessment. As an objective assessment, nutritional biomarkers are valid for dietary intake in other assessment methods (like clinical or anthropometric). The FFQs, assess diet and disease relationships, and the 24-H DRs, usually used in nutrition surveillance studies. These surveys monitor the population's intake generally or the part of the population. It may help to identify if a dietary guide is needed [54 - 57].

Biomarkers of Dietary Patterns

A biomarker is a chemical, a metabolite, or a product of interactivity between a chemical and a target metabolite that is measured in body fluids and tissue to realize a clinical assessment. It also helps monitor the status of disease or health status [56, 58]. Biomarkers can be the biochemical, functional, or clinical status index of any food/nutrient. It is not easy to record everything that people eat to assess their nutritional status. Even the best-known methods have possible errors that cannot be prevented. Subjective methods are based on one's self-reports. Assessment of adequate or insufficient food consumption is important for tracking the diet and health status. This is important for patients especially in serious clinical conditions [55]. FFQ, dietary history, and 24-H recall methods are valid to assess nutritional status. The validation studies are done using specific biomarkers. If needed, biomarkers are used to emerge as a complementary approach to subjective methods. Biomarkers are independent markers of food consumption. They respond to the intake dose of dependent nutrients [56, 59].

The General Classification of Biomarkers

Recovery biomarkers are based on the physiological balance between food consumption and use in the body. It measures the dose-response relationship over a period. These biomarkers are directly related to food consumption. A doubly labeled water (DLW), which measures the total energy expenditure (TEE) over 14 days is an example of recovery biomarkers. Another example is measuring the urinary nitrogen, sodium, and potassium, to estimate the intake of total daily protein, sodium, and potassium [55, 60].

Concentration biomarkers are related to food consumption, but personal metabolism or characteristics (such as age, gender, smoking, or sports) directly affect them. Therefore, they can neither be a measure of absolute intake or nutrient nor cannot be used for error correction of self-reported dietary questionnaires. Tissue concentration and disease association are the areas of this kind of biomarkers. Adipose tissue fatty acid concentrations or vitamin concentrations in blood are examples of concentration biomarkers [55, 60, 61].

Replacement biomarkers refer to compounds with restricted, unsatisfactory, or unavailable information in nutrient composition databases or a metabolic response. Examples of these biomarkers are sodium, potassium, and chloride [55, 60].

Predictive biomarkers are new biomarkers that are similar to recovery biomarkers. They are responsive to intake for a dose-response manner, time depended, but their overall recovery is reduced. Predictive biomarkers can be used to identify errors such as 24-hour urinary sucrose and fructose [55].

Nutritional classification of dietary biomarkers is: Biomarkers of dietary exposure assess food consumption (all nutrients) and non-nutritive components.

Biomarkers of nutritional status assess the nutrient or nutrients' food intake and metabolic function. Generally, it is not only one nutrient but also interactions of several nutrients [60].

Time-related classification of biomarkers is Short-term biomarkers represent the metabolites in hours or days and are measured in body liquids. *Medium-term* biomarkers represent the metabolites over weeks or longer and are generally measured in adipose tissue or red blood cells. *Long-term* biomarkers represent metabolites over months or longer and are usually measured from teeth, hair, or nail [60].

Omics

The human genome project exposed the effects of the genome and 'gene-environment' interaction on health and disease. The progress of this project made it possible to examine the molecular effects of nutrients on metabolism. Nutrition and nutrition-related diseases have been reported to be linked to a large number of genes [62]. In genetic diversity, it has been determined that the dietary response varies from person to person. Understanding how genetic variation affects gene expression and identifying risk factors for nutritional disorders is part of the nutrigenomics discipline. Nutrigenomics investigates the nutrient-gene interaction in three parts. 1. nutrients can act as a transcription factor that can bind to DNA by interacting with receptors and changing gene expression. 2. nutrients can create epigenetic interactions (such as DNA methylation and chromatin remodeling) that affect gene expression. 3. the reaction to dietary response may change according to the individuals due to genetic variations (single nucleotide polymorphism) [63].

The -omics terms (phenomics, transcriptomics, metabolomics, epigenomics, lipidomics, *etc.*) have enabled the identification of biological markers of disease (64). Omics technologies, it may be possible to detect new biomarkers related to specific nutrient compounds and dietary factors, it may be possible to learn at an early stage of disease whether the person has any disease risk with genomic data. It may also be possible to determine the relationship between DNA and chronic diseases such as cancer, cardiovascular diseases, obesity, diabetes mellitus [63 - 65].

CONCLUSION

Biomarkers are widely used in healthcare. Biomarkers are being developed in the field of nutrition day by day. Monitoring of nutritional status is very important for children. Growth and development screening. The issue of determining the nutritional status is directly related to the early diagnosis and often provides early intervention to the situation. More experimental studies are needed to improve the use of biomarkers in nutritional science. There is a need for more investigation about -omics. It is thought that a new era will begin in the concepts of health and disease from the dissolution of -omics and human genome projects.

CONSENT FOR PUBLICATION

Not applicable.

CONFLICT OF INTEREST

The authors declare no conflict of interest, financial or otherwise.

ACKNOWLEDGEMENTS

I gratefully thank Prof. Dr. Pınar ATUKEREN and Prof. Dr. Hafize UZUN for allowing me to be a part of this successful book "Biomarkers in Medicine". I am sure that this book will be a guide for everybody working in this field.

REFERENCES

[1] World Health Organization (WHO) [homepage on the Internet]. Constitution 2020. [updated: 2020; cited: 5th April 2020]. Available from: https://www.who.int/about/who-we-are/constitution

[2] U.S. Department of Health & Human Services (HHS). Importance of Good Nutrition 2020. https://www.hhs.gov/fitness/eat-healthy/importance-of-good-nutrition/index.html

[3] Ministry of health of turkish republic, healthy eating and active: turkey nutrition guide (TUBER). Ankara: General Directorate of Public Health, Ankara. 2015.

[4] Reber E, Gomes F, Vasiloglou MF, Schuetz P, Stanga Z. Nutritional risk screening and assessment. J Clin Med 2019; 8(7): 1-19. [http://dx.doi.org/10.3390/jcm8071065] [PMID: 31330781]

[5] Ireton-Jones I, Intake CS. Energy.Krause's food & the nutrition care process. 14th ed. Canada: Elsevier 2017; pp. 2-27.

[6] Raymond JL, Mahan LK. Intake: Analysis of the diet.Raymond JL, Mahan LK Krause's food & the nutrition care process. 14th ed. Canada: Elsevier 2017; pp. 52-62.

[7] Herberger K, Müller K, Protz K, Zyriax BC, Augustin M, Hagenström K. Nutritional status and quality of nutrition in chronic wound patients. Int Wound J 2020; 17(5): 1246-54. [http://dx.doi.org/10.1111/iwj.13378] [PMID: 32378317]

[8] World Health Organization & Food and Agriculture Organization of the United Nations. The nutrition challenge: food system solutions 2018. https://apps.who.int/iris/handle/10665/277440

[9] Ferrie S. What is nutritional assessment? A quick guide for critical care clinicians. Aust Crit Care 2020; 33(3): 295-9. [http://dx.doi.org/10.1016/j.aucc.2020.02.005] [PMID: 32303438]

[10] Langley-Evans S, King CR. Editorial: Assessment of nutritional status in clinical settings. J Clin Nurs 2014; 23(3-4): 299-300. [http://dx.doi.org/10.1111/jocn.12533] [PMID: 24467587]

[11] https://www.who.int/features/qa/malnutrition/en/

[12] Earthman CP. Evaluation of nutrition assessment parameters in the presence of human immunodeficiency virus infection. Nutr Clin Pract 2004; 19(4): 330-9. [http://dx.doi.org/10.1177/0115426504019004330] [PMID: 16215124]

[13] Srinivasan B, Lee S, Erickson D, Mehta S. Precision nutrition - review of methods for point-of-care assessment of nutritional status. Curr Opin Biotechnol 2017; 44(1): 103-8. [http://dx.doi.org/10.1016/j.copbio.2016.12.001] [PMID: 28043002]

[14] https://wwwn.cdc.gov/nchs/nhanes/Nhanes3/Default.aspx

[15] Casadei K, Kiel J. Anthropometric Measurement. StatPearls. Treasure Island, FL: StatPearls Publishing 2020. https://www.ncbi.nlm.nih.gov/books/NBK537315/ Updated 2020 Apr 28

[16] Ulijaszek SJ, Kerr DA. Anthropometric measurement error and the assessment of nutritional status. Br J Nutr 1999; 82(3): 165-77. [http://dx.doi.org/10.1017/S0007114599001348] [PMID: 10655963]

[17] https://www.bapen.org.uk/nutrition-support/assessment-and-planning/nutritio-al-assessment?showall=1

[18] Leong C, Aspalilah C, Razif A, *et al.* Application of anthropometric measurements analysis for stature in the human vertebral column: A systematic review. Forensic Imaging 2020; 20: 1-9.

[19] https://www.who.int/gho/ncd/risk_factors/bmi_text/en/

[20] World Health Organization (WHO). 2008.

[21] Rajput R, Rajput M, Bairwa M, Singh J, Saini O, Shankar V. Waist height ratio: A universal screening tool for prediction of metabolic syndrome in urban and rural population of Haryana. Indian J Endocrinol Metab 2014; 18(3): 394-9. [http://dx.doi.org/10.4103/2230-8210.131201] [PMID: 24944937]

[22] Litchford MD. Clinical: biochemical, physical, and functional assessment.Raymond JL, Mahan LK Krause's food & the nutrition care process. 14th ed. Canada: Elsevier 2017; pp. 98-121.

[23] Corrigan ML. Clinical: Water, electrolytes, and acid-base balance.Raymond JL, Mahan LK Krause's food & the nutrition care process. 14th ed. Canada: Elsevier 2017; pp. 85-91.

[24] Bharadwaj S, Ginoya S, Tandon P, *et al.* Malnutrition: laboratory markers vs nutritional assessment. Gastroenterol Rep (Oxf) 2016; 4(4): 272-80. [PMID: 27174435]

[25] Keller U. Nutritional laboratory markers in malnutrition. J Clin Med 2019; 8(6): 775. [http://dx.doi.org/10.3390/jcm8060775] [PMID: 31159248]

[26] Tan CS, Read JA, Phan VH, Beale PJ, Peat JK, Clarke SJ. The relationship between nutritional status, inflammatory markers and survival in patients with advanced cancer: a prospective cohort study. Support Care Cancer 2015; 23(2): 385-91. [http://dx.doi.org/10.1007/s00520-014-2385-y] [PMID: 25112562]

[27] Ali ET, Jabbar AS, Mohammed AN. A comparative study of interleukin 6, inflammatory markers, ferritin, and hematological profile in rheumatoid arthritis patients with anemia of chronic disease and iron deficiency anemia. Anemia 2019; 2019: 3457347. [http://dx.doi.org/10.1155/2019/3457347] [PMID: 31057960]

[28] Vanarsa K, Ye Y, Han J, Xie C, Mohan C, Wu T. Inflammation associated anemia and ferritin as disease markers in SLE. Arthritis Res Ther 2012; 14(4): R182. [http://dx.doi.org/10.1186/ar4012] [PMID: 22871034]

[29] Picó C, Serra F, Rodríguez AM, Keijer J, Palou A. Biomarkers of nutrition and health: new tools for new approaches. Nutrients 2019; 11(5): 1092. [http://dx.doi.org/10.3390/nu11051092] [PMID: 31100942]

[30] Tanumihardjo SA, Russell RM, Stephensen CB, *et al.* Biomarkers of nutrition for development

(BOND)-vitamin A review. J Nutr 2016; 146(9): 1816S-48S.
[http://dx.doi.org/10.3945/jn.115.229708] [PMID: 27511929]

[31] Tanumihardjo SA. Vitamin A: biomarkers of nutrition for development. Am J Clin Nutr 2011; 94(2): 658S-65S.
[http://dx.doi.org/10.3945/ajcn.110.005777] [PMID: 21715511]

[32] Albahrani AA, Greaves RF. Fat-soluble vitamins: clinical indications and current challenges for chromatographic measurement. Clin Biochem Rev 2016; 37(1): 27-47.
[PMID: 27057076]

[33] Lee S, Srinivasan B, Vemulapati S, Mehta S, Erickson D. Personalized nutrition diagnostics at the point-of-need. Lab Chip 2016; 16(13): 2408-17.
[http://dx.doi.org/10.1039/C6LC00393A] [PMID: 27272753]

[34] Agbaba N, Ates Ozcan B. Food intolerance after bariatric surgery. Journal of Health and Life Sciences 2020; 2(1): 16-21.

[35] Mosca P, Leheup B, Dreumont N. Nutrigenomics and RNA methylation: Role of micronutrients. Biochimie 2019; 164: 53-9.
[http://dx.doi.org/10.1016/j.biochi.2019.07.008] [PMID: 31302162]

[36] Jacobs ET, Martínez ME, Jurutka PW. Vitamin D: marker or mechanism of action? Cancer Epidemiol Biomarkers Prev 2011; 20(4): 585-90.
[http://dx.doi.org/10.1158/1055-9965.EPI-10-1257] [PMID: 21454423]

[37] Jones G. Metabolism and biomarkers of vitamin D. Scand J Clin Lab Invest Suppl 2012; 243: 7-13.
[PMID: 22536757]

[38] Zerwekh JE. Blood biomarkers of vitamin D status. Am J Clin Nutr 2008; 87(4): 1087S-91S.
[http://dx.doi.org/10.1093/ajcn/87.4.1087S] [PMID: 18400739]

[39] Faridi KF, Lupton JR, Martin SS, *et al.* Vitamin D deficiency and non-lipid biomarkers of cardiovascular risk. Arch Med Sci 2017; 13(4): 732-7.
[http://dx.doi.org/10.5114/aoms.2017.68237] [PMID: 28721139]

[40] Theuwissen E, Smit E, Vermeer C. The role of vitamin K in soft-tissue calcification. Adv Nutr 2012; 3(2): 166-73.
[http://dx.doi.org/10.3945/an.111.001628] [PMID: 22516724]

[41] Simes DC, Viegas CSB, Araújo N, Marreiros C. Vitamin K as a Powerful Micronutrient in Aging and Age-Related Diseases: Pros and Cons from Clinical Studies. Int J Mol Sci 2019; 20(17): 4150.
[http://dx.doi.org/10.3390/ijms20174150] [PMID: 31450694]

[42] Shearer MJ, Fu X, Booth SL. Vitamin K nutrition, metabolism, and requirements: current concepts and future research. Adv Nutr 2012; 3(2): 182-95.
[http://dx.doi.org/10.3945/an.111.001800] [PMID: 22516726]

[43] Carr AC, Rosengrave PC, Bayer S, Chambers S, Mehrtens J, Shaw GM. Hypovitaminosis C and vitamin C deficiency in critically ill patients despite recommended enteral and parenteral intakes. Crit Care 2017; 21(1): 300.
[http://dx.doi.org/10.1186/s13054-017-1891-y] [PMID: 29228951]

[44] Ide K, Yamada H, Umegaki K, *et al.* Lymphocyte vitamin C levels as potential biomarker for progression of Parkinson's disease. Nutrition 2015; 31(2): 406-8.
[http://dx.doi.org/10.1016/j.nut.2014.08.001] [PMID: 25592020]

[45] Wiley KD, Gupta M. Vitamin B1 Thiamine Deficiency (Beriberi). StatPearls. Treasure Island (FL): StatPearls Publishing 2020.
[PMID: 30725889]

[46] Allen LH, Miller JW, de Groot L, *et al.* Biomarkers of nutrition for development (BOND): vitamin B-12 review. J Nutr 2018; 148 (Suppl. 4): 1995S-2027S.

[http://dx.doi.org/10.1093/jn/nxy201] [PMID: 30500928]

[47] Yetley EA, Pfeiffer CM, Phinney KW, *et al.* Biomarkers of vitamin B-12 status in NHANES: a roundtable summary. Am J Clin Nutr 2011; 94(1): 313S-21S. [http://dx.doi.org/10.3945/ajcn.111.013243] [PMID: 21593512]

[48] Franz KB. A functional biological marker is needed for diagnosing magnesium deficiency. J Am Coll Nutr 2004; 23(6): 738S-41S. [http://dx.doi.org/10.1080/07315724.2004.10719418] [PMID: 15637224]

[49] Erem S, Atfi A, Razzaque MS. Anabolic effects of vitamin D and magnesium in aging bone. J Steroid Biochem Mol Biol 2019; 193: 105400. [http://dx.doi.org/10.1016/j.jsbmb.2019.105400] [PMID: 31175968]

[50] Hambidge M. Biomarkers of trace mineral intake and status. J Nutr 2003; 133(3): 948S-55S. [http://dx.doi.org/10.1093/jn/133.3.948S] [PMID: 12612181]

[51] Thomas DW, Hinchliffe RF, Briggs C, Macdougall IC, Littlewood T, Cavill I. British Committee for Standards in Haematology. Guideline for the laboratory diagnosis of functional iron deficiency. Br J Haematol 2013; 161(5): 639-48. [http://dx.doi.org/10.1111/bjh.12311] [PMID: 23573815]

[52] Du Y, Gao YH, Feng ZY, Meng FG, Fan LJ, Sun DJ. Serum Thyroglobulin-A Sensitive Biomarker of Iodine Nutrition Status and Affected by Thyroid Abnormalities and Disease in Adult Populations. Biomed Environ Sci 2017; 30(7): 508-16. [PMID: 28756810]

[53] Phiri FP, Ander EL, Lark RM, *et al.* Urine selenium concentration is a useful biomarker for assessing population level selenium status. Environ Int 2020; 134: 105218. [http://dx.doi.org/10.1016/j.envint.2019.105218] [PMID: 31715489]

[54] Shim JS, Oh K, Kim HC. Dietary assessment methods in epidemiologic studies. Epidemiol Health 2014; 36: e2014009. [http://dx.doi.org/10.4178/epih/e2014009] [PMID: 25078382]

[55] Naska A, Lagiou A, Lagiou P. Dietary assessment methods in epidemiological research: current state of the art and future prospects. F1000 Res 2017; 6(6): 926. [http://dx.doi.org/10.12688/f1000research.10703.1] [PMID: 28690835]

[56] Dietary Assessment FAO. A resource guide to method selection and application in low resource settings 2018. http://www.fao.org/3/i9940en/I9940EN.pdf

[57] Walton J. Dietary assessment methodology for nutritional assessment. Topics Clin Nutr 2015; 30(1): 33-46. [http://dx.doi.org/10.1097/TIN.0000000000000018]

[58] Noland D, Raj S. Academy of Nutrition and Dietetics: revised 2019 standards of practice and standards of professional performance for registered dietitian nutritionists (competent, proficient, and expert) in nutrition in integrative and functional medicine. J Acad Nutr Diet 2019; 119(6): 1019-1036.e47. [http://dx.doi.org/10.1016/j.jand.2019.02.010] [PMID: 31130171]

[59] Gibbons H, Brennan L. Metabolomics as a tool in the identification of dietary biomarkers. Proc Nutr Soc 2017; 76(1): 42-53. [http://dx.doi.org/10.1017/S002966511600032X] [PMID: 27221515]

[60] National Institute for Health Research (NIHR). https://www.dapa-toolkit.mrc.ac.uk/diet/objectiv--methods/biomarkers

[61] https://www.dietassessmentprimer.cancer.gov/learn/biomarkers.html

[62] de Toro-Martín J, Arsenault BJ, Després JP, Vohl MC. Precision Nutrition: a review of personalized nutritional approaches for the prevention and management of the metabolic syndrome. Nutrients 2017;

9(8): 913.
[http://dx.doi.org/10.3390/nu9080913] [PMID: 28829397]

[63] Gülşah K. The Impact of Nutrigenomics from Genotype to Phenotype. Journal of Medical Clinics 2018; 1(1): 79-92.

[64] Olivier M, Asmis R, Hawkins GA, Howard TD, Cox LA. The need for multi-omics biomarker signatures in precision medicine. Int J Mol Sci 2019; 20(19): 4781.
[http://dx.doi.org/10.3390/ijms20194781] [PMID: 31561483]

[65] Hasin Y, Seldin M, Lusis A. Multi-omics approaches to disease. Genome Biol 2017; 18(1): 83.
[http://dx.doi.org/10.1186/s13059-017-1215-1] [PMID: 28476144]

CHAPTER 26

A Physiological Approach to Inflammatory Markers in Obesity

Sila H. Bozdogan Polat[1,*] and **Nuran Dariyerli**[1]

[1] *Cerrahpaşa Faculty of Medicine, Istanbul University-Cerrahpaşa, Istanbul, Turkey*

Abstract: Obesity is one of the most critical health problems all over the world; it is associated with metabolic dysfunction and overnutrition. Changes in the physiological function of adipose tissue, leading to altered secretion of adipocytokines, inflammatory mediators release, and chronic low-grade inflammation, are seen in obesity. Macrophages, neutrophils, CD4+ and CD8+ T cells, B cells, natural killer T (NKT) cells, eosinophils, mast cells, and adipocytes are involved in the inflammatory response that occurs during obesity. Various inflammatory markers are released from these cells. In this chapter, we will mention inflammatory mechanisms and markers of obesity.

Keywords: Adipocytes, Adiponectin, Apelin, Cardiotrophin-1, Ceramide, CfDNA, Fetuin-A, Galectin-3, GlycA, HIF-1α, Interleukin-1, Interleukin-6, Leptin, MCP-1, Omentin, TNF- α, TWEAK, Vaspin.

INTRODUCTION

Obesity is abnormal or excessive fat accumulation. As a result of the excess of macronutrients in adipose tissues, inflammatory mediators such as leptin, interleukin (IL-6), tumor necrosis factor-α (TNF-α), monocyte chemoattractant protein-1 (MCP-1), resistin are released, and adiponectin production is reduced. In this way, a state sensitive to the inflammatory process occurs [1]. It is unclarified how obesity-induced inflammation in adipose tissue (AT) is triggered. However, potential mechanisms include dysregulation of fatty acids homeostasis, increased adipose cell size and death, local hypoxia, mitochondrial dysfunction, ER stress, and mechanical stress [2]. These are thought to link chronic caloric excess and AT inflammation or factors that may perpetuate chronic tissue inflammation (Fig. **1**) [3].

[*] **Corresponding author Sila H. Bozdogan Polat:** Cerrahpaşa Faculty of Medicine, Istanbul University-Cerrahpaşa, Istanbul, Turkey; Tel: +90 (534) 844 83 94; E-mail: sila.polat@istanbul.edu.tr

Pınar Atukeren and Hafize Uzun (Eds.)

Numerous immune cell types are involved in the inflammatory response during obesity. These are macrophages, neutrophils, CD4+ and CD8+ T cells, B cells, natural killer T (NKT) cells, eosinophils, and mast cells. They can regulate inflammation either positively or negatively [4].

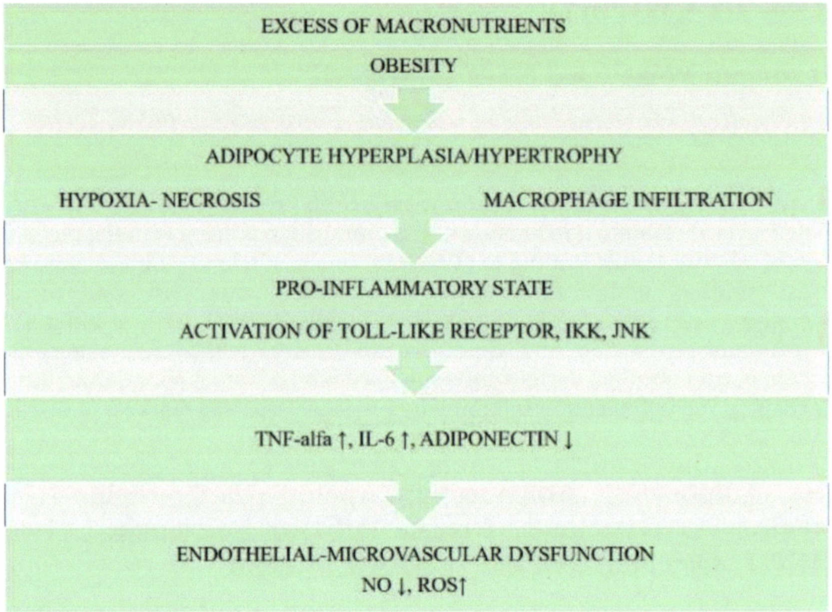

Fig. (1). Pathogenesis of obesity.

Macrophages

Tissue macrophage populations comprise a heterogeneous group of cells, and they have a variety of phenotypes and functions. Mantovani *et al.* defined polarized macrophages in two classes; M1 and M2. M1 macrophages (integrin CD11c +), are proinflammatory, whereas M2 macrophages, which do not express CD11c, are anti-inflammatory [5]. In the obese WAT (white adipose tissue), proinflammatory M1 macrophages form crown-like structures around dead adipocytes, contributing to the obesity-induced, low-grade inflammation. The elevated M1 macrophages in the obese WAT are the main sources of TNFα and IL-6 [6].

Neutrophils

Within a week of starting a high-fat diet (HFD), an increase in neutrophil uptake into adipose tissue occurs, suggesting that neutrophils may play a role in initiating the inflammatory cascade in response to excess nutrients and energy [7].

B Cells

Shortly after initiating an HFD, B cells accumulate in adipose tissues [8]. In addition, the transfer of IgG antibodies from obese WT mice to B cell-deficient mice enhances TNFα production and proinflammatory macrophage polarization [9].

$CD8^+$T Cells

$CD8^+$ T cells are recruited to obese adipose tissue, which begins within two weeks of HFD feeding [10]. Obesity is associated with an increase in the number of CD8+ T cells in AT, and these cells promote macrophage differentiation and chemotaxis [11].

CD4+ T Cells: T-Helper and T-regulatory Cell Subsets

CD4+ T cells can be separated into distinct subsets with diverse phenotypes and functions referred to as T-helper (Th1, Th2, and Th17) cells and T-regulatory (Treg) cells, and each of these subsets is present in AT [12].

Treg and Th2 cells predominate in lean AT, where they prevent the onset of inflammation [13].

Natural Killer T Cells

NKT cells are primarily responsible for target cell killing and cytotoxic activity, but they are also a source of pro-inflammatory cytokines and chemokines. They are present in significant proportions in lean WAT and the liver, although their numbers are depleted after the onset of obesity [14, 15].

Eosinophils

Eosinophils accumulate in adipose tissue by an integrin-dependent process. This promotes M2 macrophage polarization by the secretion of IL-4- or IL-13 [16].

Mast Cells

Mast cells are mainly associated with allergic response [17]. Increased numbers of mast cells are found in obese adipose tissues of mice and humans compared with their lean counterparts, and this is accompanied by increased circulating levels of the mast cell protease tryptase. Genetic depletion of mast cells or pharmacologic inhibition of mast cell function reduces body weight gain, reduces the levels of inflammatory cytokines, chemokines, and proteases in serum and adipose tissue, increases energy expenditure, and improves glucose homeostasis [18].

Proinflammatory Cytokines

The inflammatory state of obese adipose tissue leads to increased local cytokine secretion. Macrophage is the major cell type that releases proinflammatory cytokines. TNFα is the most well studied.

During inflammatory responses, a variety of interleukins are released. The two most prominent proinflammatory interleukins upregulated in obesity are IL-1β and IL-6 [19, 20].

Tumor Necrosis Factor-alpha

Tumor necrosis factor-alpha (TNF-α), a pluripotent cytokine, is produced mainly from macrophages [21]. Its expression was shown to be increased in animal models of obesity [22].

TNF-α suppresses the expression of most adipose-specific genes in murine adipocytes, including the enzymes involved in lipogenesis, *in vitro* [23].

TNF- α also induces the formation of reactive oxygen species (ROS) [24].

ROS production increased in adipose tissue of obese mice and caused irregular production of adipocytokines, such as adiponectin, plasminogen activator inhibitor-1, interleukin-6 (IL-6), and monocyte chemotactic protein-1 [25].

TNF-Like Weak Inducer of Apoptosis

TNF-like weak inducer of apoptosis (TWEAK) is a member of the TNF superfamily. Expression levels of membrane-bound TWEAK (mTWEAK) and Fn14, its receptor, are increased during obesity, thus increasing proinflammatory activity caused by TNF-α [26].

It has proinflammatory action in adipocytes mediated by the ERK and nuclear factor-κB (NFκB) (Fig. **2**) [27].

Interleukin-1

Interleukin-1 (IL-1) is a cytokine that has effects similar to those of TNF-α and is also secreted by stimulated macrophages. IL-1 gene's common polymorphisms are associated with fat mass in humans [28].

Pro-inflammatory pathways in adipose tissue are directly activated by free fatty acids (FFA). In turn, the inflammatory status of macrophages is linked to body fat content. In lean mice, macrophages in WAT are in their active M2 state and produce immunosuppressive factors. However, in obese mice, macrophages are in a pro-inflammatory M1 state (F4/80+, CD11b+, CD11c+), highly responsive to the pro-inflammatory effect of FFA that binds the Toll-Like Receptors (TLRs) [29] (Fig. **2**). IL-1 also increases hepatic triglyceride secretion and serum triglyceride levels [30].

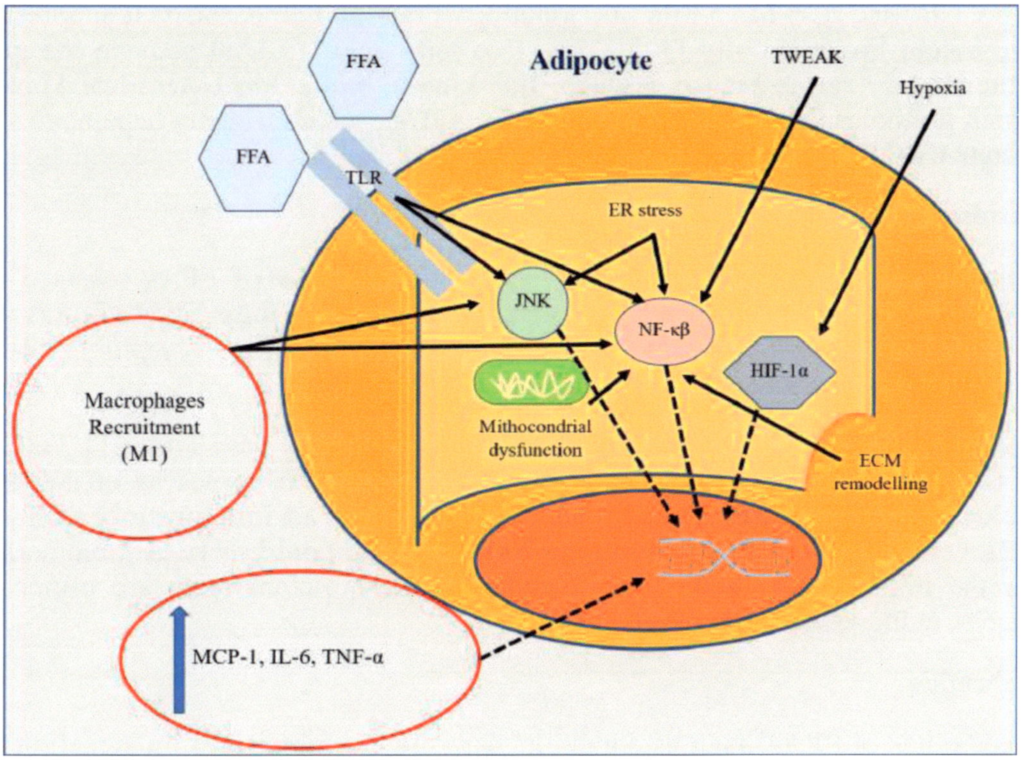

Fig. (2). Inflammation mechanisms in obesity.

Interleukin-6

Circulating IL-6 has been hypothesized to play a substantial role in developing several diseases [31]. IL-6 is an essential cytokine in the acute-phase inflammatory response that releases white blood cells and platelets from bone marrow stimulates CRP and fibrinogen production in the liver [32].

Positive associations have been shown between plasma IL-6 levels and different measures of obesity [33]. It has been described that one-third of total circulating concentrations of IL-6 comes from adipose tissue [34].

Many different cells, including immune cells and adipose tissue, produce IL-6 [35]. IL-6 is called the endocrine cytokine because, unlike other cytokines, it is unusual in that its major effects take place at sites distinct from its origin and are consequent upon its circulating concentrations [36].

The IL-6 receptor is expressed in several regions of the brain, such as the hypothalamus, in which it controls appetite and energy intake. In the hypothalamus, IL-6 has a role in regulating of energy homeostasis. It suppresses lipoprotein lipase activity [37]. Also, free fatty acids (FFAs) activate the pro-inflammatory serine kinase cascade, IkB kinase, and c-Jun N-terminal kinase, which promotes the release of adipose tissue IL-6, which triggers hepatocytes to secrete CRP [38].

Fetuin-A (FetA)

Immune cells are exposed to high local concentrations of FFAs released by adipocyte lipolysis in obesity. The effects of saturated fatty acids (SFAs) are mediated in part through activation of the pattern recognition receptors (PRRs) (Toll-like receptor 4 (Tlr4) and/or Tlr2) [29, 39, 40].

Recent studies indicated that SFAs do not interact directly with Tlr4 [41]. Fetuin-A (FetA) is a liver-derived circulating glycoprotein. It serves as an adaptor molecule presenting FFAs to Tlr4 and activating the Tlr4 inflammatory pathway [42]. FetA levels are elevated in type 2 diabetes and could serve as a biomarker for the inflammatory state [43]. Besides, fetuin-deficient mice are protected against aging-associated obesity [44].

Ceramide

Metabolites of SFAs, such as ceramide, can directly inhibit insulin signaling by inhibiting Akt. Virtually all stress stimuli increase rates of ceramide synthesis, and numerous studies have demonstrated a strong association between intracellular ceramide levels and the development of insulin resistance [44, 45]. Ceramide accumulation is a hallmark in the manifestation of numerous obesity-related diseases, such as type 2 diabetes mellitus and atherosclerosis. Until the early 2000s, ceramides were viewed as a homogenous class of sphingolipids. However, it has now become clear that ceramides exert fundamentally different effects depending on the specific fatty acyl chain lengths, which are integrated into ceramides by a group of enzymes known as dihydroceramide synthases. In addition, alterations in ceramide synthesis, trafficking, and metabolism in specific cellular compartments exert distinct consequences on metabolic homeostasis. The synthesis of ceramides with varying acyl chain lengths is regulated by six ceramide synthases (CerS1–CerS6), and their expression profiles differ throughout the body. Genetic manipulation of individual ceramide synthases has identified a highly regulated specificity of distinct acyl chain ceramides in the pathogenesis of obesity. Targeting ceramides with different acyl chains within different intracellular locations could help to provide new therapeutics for a range of obesity-related diseases [46].

ER Stress

ER stress increases the degradation of proteins from the ER, inhibits protein synthesis, and enhances the level of chaperone proteins to assist in protein folding. It activates the unfolded protein response (UPR) to restore ER homeostasis [47]. When these mechanisms are not enough to restore ER homeostasis, the cell goes under programmed cell death [48]. PKR-like eukaryotic initiation factor 2α kinase (PERK), inositol- requiring enzyme 1 (IRE-1), and activating transcription factor 6 (ATF-6) are downstream signaling through regulators of the UPR. They can activate both JNK and IKK. In this way, these molecules lead to the expression of inflammatory cytokines [49, 50]. ER stress response genes not only play a role in protein folding, but they are also active in lipid metabolism. It is shown that abnormal lipid processing occurs in the absence of these genes [51, 52].

There are various folding enzymes and ER chaperones like glucose-regulated protein (GRP) 78, GRP94, protein disulfide isomerase (PDI), calnexin (CNX), and calreticulin (CRT). They occur in protein folding and prevent aggregation of misfolded or unfolded proteins [53]. Increased ER stress in fatty liver tissues from obese mice has been shown in numerous studies. Obese or diabetic mice treatment with overexpressing GRP78 improves hepatic steatosis. X-box-binding protein-1 (XBP-1), a transcription factor, deficient mice have an increased UPR response, get insulin resistance and glucose intolerance. So it can be suggested that ER stress could be a factor in initiating metabolic inflammation [54, 55].

Hypoxia-Inducible Factor (HIF)

It is shown that fat depots in obese subjects exist in a hypoxic state and contain distinct regions of microvascularity [56, 57]. Hypoxia-inducible factor (HIF) is related to this hypoxic adipose tissue. Evidence suggests that HIF -1α in VAT is critical for developing glucose intolerance and insulin resistance in HFD mice, and HIF-1α can activate inflammatory pathways [58]. Hypoxia leading to stabilization of HIF-1α serves as an early upstream initiator for adipose tissue (AT) dysfunction. Monocyte-derived macrophage infiltration in AT contributes to inflammation, fibrosis, and obesity-related metabolic dysfunction. It was previously reported that myeloid cell-specific deletion of HIF-1α protected against high-fat diet (HFD)-induced AT dysfunction [59].

C-reactive Protein (CRP)

C-reactive protein is a sensitive marker of systemic inflammation. It is synthesized by the liver. C-reactive protein is a nonspecific acute-phase reactant, and it has traditionally been used to detect acute injury infection and inflammation [60]. Obesity is characterized by low-grade chronic inflammation. As an acute-phase reactant to inflammation and infection, C-reactive protein (CRP) is the strongest factor associated with obesity [61]. According to the meta-analysis of Choi *et al.*, each degree of obesity is related directly to CRP, regardless of ethnic characteristics and sex [62].

The association between obesity and elevated serum level of CRP has been well clarified by the pathophysiological mechanism. The liver has a central role in inflammation as it drains free fatty acids and circulating triacylglycerol, promoting the release of cytokines (IL-6) by adipose tissue, which triggers hepatocyte expression and release of CRP [61]. The induction of IL-6 could be inhibited by the farnesoid X receptor (FXR; NR1H4), which is a member of the nuclear hormone receptor superfamily that functions as a ligand-activated transcription factor and is highly expressed in the liver, intestine, kidney, and adrenal glands [63]. FXR regulates many genes about bile acid synthesis and lipid and lipoprotein metabolism [64].

Subcutaneous abdominal fat thickness (SCFT) is important for predisposition to metabolic and cardiovascular diseases. In pregnant women between 24 and 28 weeks of gestation, Köşüş *et al.* [65] evaluated maternal SCFT and metabolic changes. The results indicated a significant association between SCFT and glycated haemoglobin (HbA1c). At the same time, higher levels of CRP were found in 47.9% of cases with SCFT over 15 mm.

In a recent study, Gaillard *et al.* [66] concluded that a higher second-trimester maternal CRP level was associated with higher mid-childhood overall and central adiposity.

On the other hand, many factors such as age, gender, smoking, chronic illnesses, and sedentary lifestyle have correlated closely with CRP and systemic cytokine concentrations [67 - 69].

GlycA

GlycA is a marker of systemic inflammation with lower intra-individual variability than high sensitivity C-reactive protein. GlycA is elevated in obesity [70].

Levine *et al.* showed that GlycA was associated with elevated lipolysis, independent of adiposity, in adults with obesity. Their findings suggest that GlycA and hsCRP have distinct inflammation-mediated metabolic effects, with GlycA greater association with adipose tissue dysfunction [71].

Cell-free DNA (cfDNA)

Cell-free DNA (cfDNA) has received much attention as a potential biomarker for monitoring physiological and pathological conditions. Physiologically low levels of cfDNA are detected in the plasma of healthy individuals [72]. However, the increase in cfDNA occurs due to various acute and chronic pathological conditions such as cancer, autoimmune diseases, sepsis, stroke, myocardial infarction, and trauma, as reported in the literature [72, 73]. cfDNA is released into plasma by several cells, including neutrophils, monocytes, eosinophils, and tumor cells, and may originate from apoptosis and necrosis (thereby reflecting the degree of cellular damage) or be deliberately released [72]. An association between cfDNA levels in obese individuals and BMI and visceral fat has been suggested in the literature [74]. The process of systemic inflammation associated with metabolic dysfunction in obese subjects may contribute to the release of cfDNA into the bloodstream [74].

Galectin-3

The differentiation and proliferation of adipocytes are modulated by a series of factors or molecules, which include galectin-3 (Gal-3), a member of the family of Gals that bind carbohydrates or glycoproteins. Gal-3 is expressed in various cell types and is secreted into the extracellular fluid. Gal-3 has a variety of biological activities which depend on different subcellular localization. For example, extracellular Gal-3 modulates interactions between cells; Gal-3 in the cytoplasm is related to cell survival and death, while Gal-3 in the nucleus may influence gene expression [75, 76].

Gal-3 is positively associated with obesity. In obese adults or children, the serum level of Gal-3 (which may reflect its secretion from cells) was found to be elevated, and this elevation was related to an increase in body fat [77, 78].

In mice with diet-induced obesity (DIO), Gal-3 was found to be increased in serum, white adipose tissue (WAT), and liver [79].

As compared to wild-type mice, the Gal-3 gene knockout mice had reduced body weight and WAT and were protected from DIO.

Gal-3 stimulates the differentiation and proliferation of adipocytes. During the differentiation and maturation of adipocytes, the mRNA and protein expression levels of Gal-3 increased in a time-dependent manner. Furthermore, knockout or knockdown of Gal-3 gene blocked the differentiation of adipocytes, as indicated by the reduction of cell size and lipid accumulation, as well as the down-regulation of key factors involved in adipogenesis, such as peroxisome proliferator-activated receptor γ (PPARγ) and CCAAT/enhancer-binding protein α (C/EBPα) [80]. In addition, the mRNA level of Gal-3 was up-regulated in growing but not quiescent adipocytes, and administration of exogenous Gal-3 protein promoted DNA synthesis and increased cell number of adipocytes *in vitro* [81].

Adiponectin

Adiponectin is a 247-amino acid peptide discovered in 1995. It circulates at 2–20 μg/ml serum concentrations [82]. Adiponectin is derived from adipocytes.

In the liver, adiponectin improves insulin sensitivity, decreases uptake of non-esterified fatty acids, reduces gluconeogenesis, and increases the oxidation process [83]. Centrally, it has high potency in regulating weight gain and increasing energy expenditure (Fig. **3**) [84].

However, circulating adiponectin level is inversely related to body weight, especially visceral fat accumulation. The mechanism of this paradoxical relation remains obscure. Low circulating adiponectin concentrations (hypo-adiponectinemia; <4 mg/mL) are associated with a variety of diseases, including dysmetabolism (type 2 diabetes, insulin resistance, hypertension, dyslipidemia, metabolic syndrome, hyperuricemia), atherosclerosis (coronary artery disease, stroke, peripheral artery disease), sleep apnea, non-alcoholic fatty liver disease, gastritis, and gastroesophageal reflux disease, inflammatory bowel diseases, pancreatitis, osteoporosis, and cancer (endometrial cancer, postmenopausal breast cancer, leukemia, colon cancer, gastric cancer, prostate cancer). On the other hand, hyperadiponectinemia is associated with cardiac, renal, and pulmonary diseases.

Adiponectin increases local nitric oxide production, which protects against endothelial dysfunction, and inhibits plaque initiation and thrombosis [85]. Thus, it acts as a vasoprotective factor [86] and reduces oxidative stress [87].

Adiponectin plays a critical autocrine or paracrine role, having local effects. It influences adipose tissue functions by overexpressing the Adipoq gene which increases fat tissue mass *via* increased adipocytes [88].

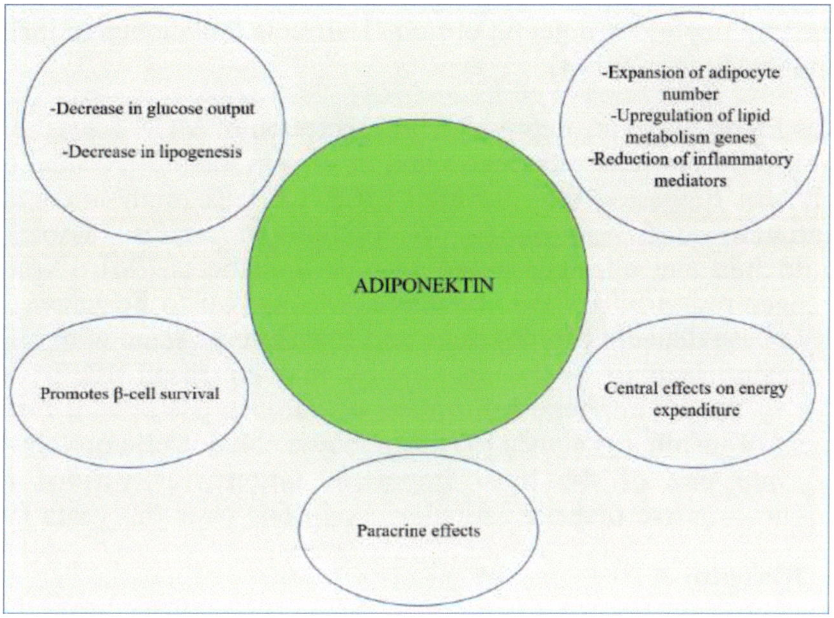

Fig. (3). Effects of adiponectin.

Various studies showed that serum level of adiponectin is reduced significantly with weight gain and obesity [89]. In the study of Dekker *et al.* [90], Adiponectin levels were divided into quartiles; the lowest quartile was associated with higher BMI, and the highest quartile was associated with lower BMI insignificant trends in both men and women. Jaleel *et al.* [91] compared the adiponectin level in normal weight and obese post-menopause women. The level of adiponectin was significantly reduced in obese subjects compared with normal-weight ones.

Snijder *et al.* [92] showed that higher adiponectin is associated with decreased peripheral arterial stiffness, and also adiponectin is negatively associated with abdominal and total fat.

Eglit *et al.* [93] found that Adiponectin is negatively associated with obesity in both males and females, as it correlated positively with high-density lipoprotein cholesterol (HDL-c) and negatively with fasting blood glucose (FBG) and total cholesterol.

Monocyte Chemotactic Protein-1

Monocyte chemotactic protein-1(MCP-1) represents the first discovered and most extensively studied human CC chemokine and is also known as CCL2 (Chemokine (C-C motif) ligand 2). CC chemokines are characterized by four cysteine residues responsible for protein stabilization [95]. The pro-inflammatory chemokine monocyte chemotactic protein-1 attracts leukocytes to inflamed sites and is regulated by NFκB [94].

Insulin was found to induce expression and secretion of MCP-1 substantially both *in vitro* in insulin-resistant adipocytes and in vivo in insulin-resistant obese mice (ob/ob). It was suggested that elevated MCP-1 levels might induce adipocyte dedifferentiation and contribute to pathologic states associated with hyperinsulinemia and obesity, including type 2 diabetes [96]. Expression and plasma concentration of MCP-1, however, was shown to be increased both in genetically obese diabetic (db/db) mice and in wild-type mice with high-fat diet-induced obesity, leading to the assumption that increased MCP-1 expression contributes to the macrophage infiltration of adipose tissue and, finally, to the development of insulin resistance [97]. Monocyte chemotactic protein-1 has been developed into one of the most important targets for various therapeutic approaches to improve diabetic vascular conditions over the years [98].

Toll-Like Receptor-4

Toll-like receptors are membrane-spanning, non-catalytic receptors that respond to different microbial antigens, representing an essential factor of innate immunity [99].

Toll-like Receptor-4 (TLR-4) is another important factor in fatty acid-induced insulin resistance. Scherer and coworkers were the first ones that found it expressed on 3T3-L1 adipocytes and activated by lipopolysaccharides (LPS) [100]. Characterization of TLR-4 as the main endogenous sensor for LPS in adipocytes supports the relevance of fat tissue in immune processes [101]. Additionally, TLR-4 was recently shown to be directly activated by dietary saturated fatty acids, thereby promoting inflammatory aspects of metabolic syndrome and atherosclerosis [102].

In addition, stimulation of TLR-4 with activation of the Erk pathway was shown to upregulate IL-6 and MCP-1 release in adipose tissue. Therefore, it can be suggested that activation of TLR-4 in adipocytes induces inflammation and, consequently, promotes the progression towards diabetes.

Leptin

Leptin was the first adipokine discovered to influence body fat mass. It is predominantly secreted from white adipose tissue and exerts its main function by repressing food intake and promoting energy expenditure through sites of action in the central nervous system [103].

The leptin receptor is expressed in the arcuate, ventromedial, dorsomedial, and lateral hypothalamic nuclei, which regulate food intake [104]. Mutation of both the leptin gene (ob) as well as the leptin receptor gene (db) leads to severe obesity, hyperphagia, and insulin resistance in mice [105].

The ob mutation was first hypothesized in 1950, when animal caretakers of the Jackson Laboratory observed the spontaneous occurrence of an obese phenotype in a mouse, but was not described as a non-sense mutation in the leptin gene until more than 40 years later [106].

The expression and secretion of leptin are correlated with the amount of body fat and adipocyte size [107]. Humans with mutations in both alleles of either leptin or the leptin receptor are obese, but these homozygous mutations are extremely rare [108]. On the contrary, most obese individuals display high plasma leptin levels in proportion to their increased body fat. Consequently, attempts to treat obesity by leptin administration have been mostly unsuccessful due to an apparent leptin resistance of these patients [109].

Nevertheless, leptin improves insulin sensitivity by several mechanisms. In the liver and skeletal muscle, leptin enhances glucose homeostasis by decreasing intracellular lipid accumulation and, in skeletal muscle, by direct activation of AMP-activated protein kinase (AMPK) [110, 111].

In addition, leptin can inhibit insulin secretion by both a direct effect on pancreatic β-cells and an indirect mechanism *via* activation of the SNS (sympathetic nervous system) by the CNS (central nervous system) [112 - 116].

Visfatin

Visfatin, found in the visceral adipose tissue, also known as nicotinamide phosphoribosyltransferase (Nampt) and pre-B-cell colony-enhancing factor (PBEF), is a multifaceted molecule with a triple action: a cytokine, a growth factor, and an enzyme. It exerts a pivotal role in a multitude of metabolic and stress responses and cellular bioenergetics, specifically nicotinamide adenine dinucleotide (NAD) synthesis. Visfatin/Nampt exhibits antiapoptotic, prolifera tive, pro-inflammatory, pro-angiogenic, and metastatic properties [117].

As an adipokine, the protein had also been found in the bloodstream, which has been shown to exert insulin-like functions. In mice, administration of visfatin was shown to lower blood glucose levels, whereas mice with a mutation in visfatin had increased levels of circulating glucose [118]. However, subsequent studies have produced conflicting results regarding the association between visceral fat mass and plasma visfatin in humans, and the initial study was, in part, retracted

[119 - 121].

Despite these inconsistencies, a positive correlation between visfatin gene expression in visceral adipose tissue and BMI was seen in some human studies, as well as a negative correlation between BMI and visfatin gene expression in subcutaneous fat [122].

In summary, the provided evidence of a direct link between visfatin action and human type 2 diabetes mellitus is still weak, and its role in obesity and insulin resistance remains to be elucidated.

Vaspin

Visceral adipose tissue-derived serpin or serpinA12 (Vaspin) was originally identified as an adipokine predominantly secreted from visceral adipose tissue. Notably, several different animal models have recently implicated vaspin as an adipokine that promotes insulin sensitivity during metabolic stress. In a diabetic mouse model, administration of recombinant vaspin increased insulin sensitivity and glucose tolerance whilst reducing the expression of pro-inflammatory adipokines, including TNF-α and resistin, and increasing the expression of adiponectin [123].

Similar increases in insulin sensitivity have been reported in both db/db and C57BL6 mice following administration with recombinant vaspin [124], whilst transgenic mice overexpressing vaspin display improved glucose tolerance and are protected from obesity when challenged with a high-fat diet [125].

Omentin-1

Omentin is a 38–40 kDa adipokine that was identified from a cDNA library in visceral omental adipose tissue [126]. Omentin is predominantly produced by the stromal-vascular fraction of visceral adipose tissue [127].

There are two omentin genes located adjacent to each other in the 1q22–q23 chromosomal region, which produce omentin-1 and omentin-2 [128]. Both isoforms show a different pattern of tissue expression. In humans, omentin-1 is the predominant isoform in plasma and adipose tissue [129].

Physiological concentrations of omentin-1 in humans are in the range of 100–800 ng/mL [130]. It is known that its expression and production are modified in several pathological situations, such as obesity and insulin resistance. Omentin-1 expression is also altered in inflammatory states [131].

Circulating levels of omentin are reduced in the obese state and patients with T2DM. The beneficial effects of omentin are thought to be caused by vasodilatation of blood vessels and attenuation of C-reactive protein-induced angiogenesis, potentially *via* the nuclear factor B signaling pathway, a potent pro-inflammatory signaling pathway [132]. In addition, omentin has been shown to block TNFα-induced JNK and NF-κB activation [133].

As with adiponectin, circulating levels of omentin are lower in obesity and inversely correlated with measures of insulin resistance (HOMA-IR), and lower serum omentin concentrations were found in individuals with impaired glucose tolerance and type 2 diabetes compared to healthy individuals [134].

However, it has to be elucidated whether the association of circulating omentin levels with the risk of T2DM is independent of BMI or can entirely be explained by obesity. In addition, further studies are needed to distinguish entirely between adiponectin and omentin action and whether omentin also shows a counter-regulatory increase in pro-inflammatory conditions [135].

Apelin

The peptide Apelin, as a member of the adipose tissue-derived peptides, might contribute to obesity-related disorders. Nevertheless, its role remains unclear, and experimental findings have been inconsistent. In the studies examining obese young patients, the apelin-12 plasma level was decreased relative to healthy patients, which could be associated with the severity of insulin resistance and adiposity [136, 137].

However, another study found that its plasma level was significantly increased in obese female children compared to non-obese children [138]. Another study reported increased apelin-12 serum concentration in patients with obesity and obesity-related insulin resistance, which could be caused by impaired insulin sensitivity [139].

The expression changes of the apelin/APJ system under obesity are shown in Table **1**.

Moreover, obese and hyperinsulinemic patients demonstrate elevated plasma levels of apelin. However, apelin plasma levels depend on several factors, including blood glucose levels and plasma triglyceride concentration [140].

Currently, apelin is being considered as a biomarker and drug target, but its role in the development of both obesity and type 2 diabetes needs to be clarified with convincing clinical studies [141, 142].

Table 1. Expression of apelin/APJ in obesity.

Disease	Patient/tissue/cell line	Protein	References
Obesity	Human plasma	Apelin-12 ↓	Tapan *et al.*, 2010; Krist *et al.*, 2013; Kotanidou *et al.*, 2015 [137 - 139]
-	-	Apelin-12 ↑	Ba *et al.*, 2014 [140]

Cardiotrophin-1

Cardiotrophin-1 (Ctf1) is expressed in different tissues and secreted as an adipokine. Targeted disruption of cardiotrophin-1 in mice leads to obesity and insulin resistance [143].

However, studies in humans have yielded contradictory results regarding cardiotrophin-1 levels and their association with obesity [144]. The role of cardiothrophin-1 in regulating metabolic circadian rhythms is the focus of current research [145].

Fatty Acid Esters of Hydroxy Fatty Acids

Recently, a novel family of lipids, the so-called fatty acid esters of hydroxy fatty acids (FAHFAs), has been identified. These branched fatty acid esters can be found in various tissues, with the highest amounts in adipose tissues. A specific group of FAHFAs, the PAHSAs (Palmitic acid esters of hydroxy-stearic acids), have been shown to have beneficial metabolic effects. Circulating PAHSA levels are reduced in insulin-resistant people, and serum levels correlate highly with insulin sensitivity. Moreover, treating obese mice with PAHSAs leads to improved glucose tolerance and increased insulin secretion. In adipocytes, PAHSAs signal through the omega-3 fatty acid receptor GPR120 to enhance insulin-stimulated glucose uptake [146].

The production of FAHFAs in adipose tissue is tightly linked to the abundance of the insulin-responsive glucose transporter 4 (GLUT4) and the ability of adipocytes to transport glucose into the cell. Increased glucose uptake activates the nuclear transcription factor carbohydrate response element-binding protein (CREBP), thereby enhancing lipogenesis and synthesizing FAHFA's [147].

In addition, PAHSAs have been demonstrated to exert anti-inflammatory effects by repressing macrophage-induced tissue inflammation [148].

Micro RNA (MiRNA)- Containing Exosomes

Micro RNAs are small, noncoding sequences of RNAs that control a multiplicity of gene expression processes in diverse organs. In adipose tissue, miRNAs are important regulators of cellular metabolisms, such as cell differentiation and lipid storage. So it is expected that miRNAs may be a new layer of regulation of the different functions of the adipose tissue in obesity [149].

Many miRNAs are differentially regulated in white adipose tissue of obese subjects compared to non-obese human Individuals [150 - 153].

Several studies have illustrated a correlation between the expression of miRNAs in adipose tissue and different metabolic parameters (BMI, adipogenesis, glycemia, leptinemia) in humans [154].

In accordance with this observation, Heneghan *et al.* discovered that the expression of miR-17-5p and miR-132 differed significantly between obese and non-obese omental fat. Interestingly, the expression of these two miRNAs in omental fat and blood from obese patients correlated significantly with glycosylated hemoglobin, leptin, body mass index, and fasting blood glucose [155]. An increase in the expression of miR-21 was found in the white adipose tissue (WAT) of obese humans compared to lean controls, and it was positively correlated with BMI [156].

Endogenous miRNAs produced within adipocytes have been well-characterized. However, recently an increasing number of exogenous miRNAs have been proved to exist in different biological fluids, such as plasma, serum, urine, and saliva [157, 158]. Many fluid-based miRNAs are present in small extracellular vesicles, called exosomes, secreted by various cell types, including adipose cells. Adipose tissue can release exosomal miRNAs that act as signaling molecules [159].

The alteration in miRNA expression could induce changes in the pattern of genes controlling a range of biological processes, including adipogenesis and inflammation of the adipose tissue, lipid metabolism, and insulin resistance.

These different studies suggest that miRNAs can be used as biomarkers in the context of metabolic diseases, but they could also be targeted for pharmaceutical approaches.

CONCLUSION

Obesity is a kind of epidemic that leads to impaired metabolic health. As we mentioned above, a low-grade inflammatory state occurs with obesity. Recent researches showed that inflammation was also a critical component of the pathway from obesity to its comorbidities. The underlying inflammatory molecular mechanisms of obesity have not been fully understood, but the primary source of inflammatory markers in obesity is adipose tissue. Biomarker research is a highly active field with significant progress to be expected in the coming years, and further comprehensive prospective studies need to be conducted to elucidate the inflammatory mechanisms in obesity.

CONSENT FOR PUBLICATION

Not applicable.

CONFLICT OF INTEREST

The authors declare no conflict of interest, financial or otherwise.

ACKNOWLEDGEMENT

Declared none.

REFERENCES

[1] Lafontan M. Fat cells: afferent and efferent messages define new approaches to treat obesity. Annu Rev Pharmacol Toxicol 2005; 45(1): 119-46. [http://dx.doi.org/10.1146/annurev.pharmtox.45.120403.095843] [PMID: 15822173]

[2] Reilly SM, Saltiel AR. Adapting to obesity with adipose tissue inflammation. Nat Rev Endocrinol 2017; 13(11): 633-43. [http://dx.doi.org/10.1038/nrendo.2017.90] [PMID: 28799554]

[3] Burhans MS, Hagman DK, Kuzma JN, Schmidt KA, Kratz M. Contribution of adipose tissue inflammation to the development of type 2 diabetes mellitus. Compr Physiol 2018; 9(1): 1-58. [http://dx.doi.org/10.1002/cphy.c170040] [PMID: 30549014]

[4] Gregor MF, Hotamisligil GS. Inflammatory mechanisms in obesity. Annu Rev Immunol 2011; 29(1): 415-45. [http://dx.doi.org/10.1146/annurev-immunol-031210-101322] [PMID: 21219177]

[5] Liu G, Yang H. Modulation of macrophage activation and programming in immunity. J Cell Physiol 2013; 228(3): 502-12. [http://dx.doi.org/10.1002/jcp.24157] [PMID: 22777800]

[6] Kern L, Mittenbühler MJ, Vesting AJ, Ostermann AL, Wunderlich CM, Wunderlich FT. Obesity-Induced TNFα and IL-6 Signaling: The Missing Link between Obesity and Inflammation—Driven Liver and Colorectal Cancers. Cancers (Basel) 2019; 11(1): 24. [http://dx.doi.org/10.3390/cancers11010024]

[7] Elgazar-Carmon V, Rudich A, Hadad N, Levy R. Neutrophils transiently infiltrate intra-abdominal fat early in the course of high-fat feeding. J Lipid Res 2008; 49(9): 1894-903. [http://dx.doi.org/10.1194/jlr.M800132-JLR200] [PMID: 18503031]

[8] Duffaut C, Galitzky J, Lafontan M, Bouloumié A. Unexpected trafficking of immune cells within the adipose tissue during the onset of obesity. Biochem Biophys Res Commun 2009; 384(4): 482-5. [http://dx.doi.org/10.1016/j.bbrc.2009.05.002] [PMID: 19422792]

[9] Winer DA, Winer S, Shen L, *et al.* B cells promote insulin resistance through modulation of T cells and production of pathogenic IgG antibodies. Nat Med 2011; 17(5): 610-7.

[http://dx.doi.org/10.1038/nm.2353] [PMID: 21499269]

[10] Nishimura S, Manabe I, Nagasaki M, *et al.* CD8+ effector T cells contribute to macrophage recruitment and adipose tissue inflammation in obesity. Nat Med 2009; 15(8): 914-20. [http://dx.doi.org/10.1038/nm.1964] [PMID: 19633658]

[11] Henao-Mejia J, Elinav E, Thaiss CA, Flavell RA. Inflammasomes and metabolic disease. Annu Rev Physiol 2014; 76(1): 57-78. [http://dx.doi.org/10.1146/annurev-physiol-021113-170324] [PMID: 24274736]

[12] Furuhashi M, Fucho R, Görgün CZ, Tuncman G, Cao H, Hotamisligil GS. Adipocyte/macrophage fatty acid-binding proteins contribute to metabolic deterioration through actions in both macrophages and adipocytes in mice. J Clin Invest 2008; 118(7): 2640-50. [http://dx.doi.org/10.1172/JCI34750] [PMID: 18551191]

[13] Feuerer M, Herrero L, Cipolletta D, *et al.* Lean, but not obese, fat is enriched for a unique population of regulatory T cells that affect metabolic parameters. Nat Med 2009; 15(8): 930-9. [http://dx.doi.org/10.1038/nm.2002] [PMID: 19633656]

[14] Fauriat C, Long EO, Ljunggren H-G, Bryceson YT. Regulation of human NK-cell cytokine and chemokine production by target cell recognition. Blood 2010; 115(11): 2167-76. [http://dx.doi.org/10.1182/blood-2009-08-238469] [PMID: 19965656]

[15] Schipper HS, Rakhshandehroo M, van de Graaf SF, *et al.* Natural killer T cells in adipose tissue prevent insulin resistance. J Clin Invest 2012; 122(9): 3343-54. [http://dx.doi.org/10.1172/JCI62739] [PMID: 22863618]

[16] Wu D, Molofsky AB, Liang H-E, *et al.* Eosinophils sustain adipose alternatively activated macrophages associated with glucose homeostasis. Science 2011; 332(6026): 243-7. [http://dx.doi.org/10.1126/science.1201475] [PMID: 21436399]

[17] Zhang J, Shi G-P. Mast cells and metabolic syndrome. Biochimica et Biophysica Acta (BBA)-. Biochim Biophys Acta Mol Basis Dis 2012; 1822(1): 14-20. [http://dx.doi.org/10.1016/j.bbadis.2010.12.012]

[18] Liu J, Divoux A, Sun J, *et al.* Genetic deficiency and pharmacological stabilization of mast cells reduce diet-induced obesity and diabetes in mice. Nat Med 2009; 15(8): 940-5. [http://dx.doi.org/10.1038/nm.1994] [PMID: 19633655]

[19] Tack CJ, Stienstra R, Joosten LA, Netea MG. Inflammation links excess fat to insulin resistance: the role of the interleukin-1 family. Immunol Rev 2012; 249(1): 239-52. [http://dx.doi.org/10.1111/j.1600-065X.2012.01145.x] [PMID: 22889226]

[20] Pradhan AD, Manson JE, Rifai N, Buring JE, Ridker PM. C-reactive protein, interleukin 6, and risk of developing type 2 diabetes mellitus. JAMA 2001; 286(3): 327-34. [http://dx.doi.org/10.1001/jama.286.3.327] [PMID: 11466099]

[21] Old LJ. Tumor necrosis factor (TNF). Science 1985; 230(4726): 630-2. [http://dx.doi.org/10.1126/science.2413547] [PMID: 2413547]

[22] Hotamisligil GS, Shargill NS, Spiegelman BM. Adipose expression of tumor necrosis factor-alpha: direct role in obesity-linked insulin resistance. Science 1993; 259(5091): 87-91. [http://dx.doi.org/10.1126/science.7678183] [PMID: 7678183]

[23] Torti FM, Dieckmann B, Beutler B, Cerami A, Ringold GM. A macrophage factor inhibits adipocyte gene expression: an *in vitro* model of cachexia. Science 1985; 229(4716): 867-9. [http://dx.doi.org/10.1126/science.3839597] [PMID: 3839597]

[24] Kamata H, Honda S, Maeda S, Chang L, Hirata H, Karin M. Reactive oxygen species promote TNFalpha-induced death and sustained JNK activation by inhibiting MAP kinase phosphatases. Cell 2005; 120(5): 649-61. [http://dx.doi.org/10.1016/j.cell.2004.12.041] [PMID: 15766528]

[25] Furukawa S, Fujita T, Shimabukuro M, *et al.* Increased oxidative stress in obesity and its impact on metabolic syndrome. J Clin Invest 2004; 114(12): 1752-61. [http://dx.doi.org/10.1172/JCI21625] [PMID: 15599400]

[26] Maymó-Masip E, Fernández-Veledo S, Garcia España A, *et al.* The rise of soluble TWEAK levels in severely obese subjects after bariatric surgery may affect adipocyte-cytokine production induced by TNFα. J Clin Endocrinol Metab 2013; 98(8): E1323-33. [http://dx.doi.org/10.1210/jc.2012-4177] [PMID: 23783093]

[27] Vendrell J, Maymó-Masip E, Tinahones F, *et al.* Tumor necrosis-like weak inducer of apoptosis as a proinflammatory cytokine in human adipocyte cells: up-regulation in severe obesity is mediated by inflammation but not hypoxia. J Clin Endocrinol Metab 2010; 95(6): 2983-92. [http://dx.doi.org/10.1210/jc.2009-2481] [PMID: 20382683]

[28] Strandberg L, Lorentzon M, Hellqvist A, *et al.* Interleukin-1 system gene polymorphisms are associated with fat mass in young men. J Clin Endocrinol Metab 2006; 91(7): 2749-54. [http://dx.doi.org/10.1210/jc.2005-2786] [PMID: 16636119]

[29] Nguyen MT, Favelyukis S, Nguyen A-K, *et al.* A subpopulation of macrophages infiltrates hypertrophic adipose tissue and is activated by free fatty acids *via* Toll-like receptors 2 and 4 and JNK-dependent pathways. J Biol Chem 2007; 282(48): 35279-92. [http://dx.doi.org/10.1074/jbc.M706762200] [PMID: 17916553]

[30] Feingold KR, Soued M, Adi S, *et al.* A subpopulation of macrophages infiltrates hypertrophic adipose tissue and is activated by free fatty acids via Toll-like receptors 2 and 4 and JNK-dependent pathways. Journal of biological chemistry 2007; 282(48): 35279-92. 1991.

[31] Danesh J, Kaptoge S, Mann AG, *et al.* Long-term interleukin-6 levels and subsequent risk of coronary heart disease: two new prospective studies and a systematic review. PLoS Med 2008; 5(4): e78. [http://dx.doi.org/10.1371/journal.pmed.0050078] [PMID: 18399716]

[32] Hansson GK. Inflammation, atherosclerosis, and coronary artery disease. N Engl J Med 2005; 352(16): 1685-95. [http://dx.doi.org/10.1056/NEJMra043430] [PMID: 15843671]

[33] Straub RH, Hense HW, Andus T, Schölmerich J, Riegger GA, Schunkert H. Hormone replacement therapy and interrelation between serum interleukin-6 and body mass index in postmenopausal women: a population-based study. J Clin Endocrinol Metab 2000; 85(3): 1340-4. [http://dx.doi.org/10.1210/jcem.85.3.6355] [PMID: 10720088]

[34] Fontana L, Eagon JC, Trujillo ME, Scherer PE, Klein S. Visceral fat adipokine secretion is associated with systemic inflammation in obese humans. Diabetes 2007; 56(4): 1010-3. [http://dx.doi.org/10.2337/db06-1656] [PMID: 17287468]

[35] Brichory FM, Misek DE, Yim A-M, *et al.* An immune response manifested by the common occurrence of annexins I and II autoantibodies and high circulating levels of IL-6 in lung cancer. Proc Natl Acad Sci USA 2001; 98(17): 9824-9. [http://dx.doi.org/10.1073/pnas.171320598] [PMID: 11504947]

[36] Papanicolaou DA, Vgontzas AN. Interleukin-6: the endocrine cytokine. J Clin Endocrinol Metab 2000; 85(3): 1331-3. [http://dx.doi.org/10.1210/jcem.85.3.6582] [PMID: 10720086]

[37] Stenlöf K, Wernstedt I, Fjällman T, Wallenius V, Wallenius K, Jansson J-O. Interleukin-6 levels in the central nervous system are negatively correlated with fat mass in overweight/obese subjects. J Clin Endocrinol Metab 2003; 88(9): 4379-83. [http://dx.doi.org/10.1210/jc.2002-021733] [PMID: 12970313]

[38] Rocha VZ, Libby P. Obesity, inflammation, and atherosclerosis. Nat Rev Cardiol 2009; 6(6): 399-409. [http://dx.doi.org/10.1038/nrcardio.2009.55] [PMID: 19399028]

[39] Lee J, Sohn KH, Rhee SH, Hwang D. Saturated fatty acids, but not unsaturated fatty acids, induce the

expression of cyclooxygenase-2 mediated through Toll-like receptor. 4: 16683-9.

[40] Senn JJ. Toll-like receptor-2 is essential for the development of palmitate-induced insulin resistance in myotubes. J Biol Chem 2006; 281(37): 26865-75. [http://dx.doi.org/10.1074/jbc.M513304200] [PMID: 16798732]

[41] Schaeffler A, Gross P, Buettner R, *et al.* Fatty acid-induced induction of Toll-like receptor-4/nuclear factor-kappaB pathway in adipocytes links nutritional signalling with innate immunity. Immunology 2009; 126(2): 233-45. [http://dx.doi.org/10.1111/j.1365-2567.2008.02892.x] [PMID: 18624726]

[42] Pal D, Dasgupta S, Kundu R, *et al.* Fetuin-A acts as an endogenous ligand of TLR4 to promote lipid-induced insulin resistance. Nat Med 2012; 18(8): 1279-85. [http://dx.doi.org/10.1038/nm.2851] [PMID: 22842477]

[43] Heinrichsdorff J, Olefsky JM. Fetuin-A: the missing link in lipid-induced inflammation. Nat Med 2012; 18(8): 1182-3. [http://dx.doi.org/10.1038/nm.2869] [PMID: 22869185]

[44] Holland WL, Bikman BT, Wang LP, *et al.* Lipid-induced insulin resistance mediated by the proinflammatory receptor TLR4 requires saturated fatty acid-induced ceramide biosynthesis in mice. J Clin Invest 2011; 121(5): 1858-70. [http://dx.doi.org/10.1172/JCI43378] [PMID: 21490391]

[45] Ussher JR, Koves TR, Cadete VJ, *et al.* Inhibition of de novo ceramide synthesis reverses diet-induced insulin resistance and enhances whole-body oxygen consumption. Diabetes 2010; 59(10): 2453-64. [http://dx.doi.org/10.2337/db09-1293] [PMID: 20522596]

[46] Turpin-Nolan SM, Brüning JC. The role of ceramides in metabolic disorders: when size and localization matters. Nat Rev Endocrinol 2020; 16(4): 224-33. [http://dx.doi.org/10.1038/s41574-020-0320-5] [PMID: 32060415]

[47] Hotamisligil GS. Endoplasmic reticulum stress and the inflammatory basis of metabolic disease. Cell 2010; 140(6): 900-17. [http://dx.doi.org/10.1016/j.cell.2010.02.034] [PMID: 20303879]

[48] Engin F, Hotamisligil GS. Restoring endoplasmic reticulum function by chemical chaperones: an emerging therapeutic approach for metabolic diseases. Diabetes Obes Metab 2010; 12 (Suppl. 2): 108-15. [http://dx.doi.org/10.1111/j.1463-1326.2010.01282.x] [PMID: 21029307]

[49] Hu P, Han Z, Couvillon AD, Kaufman RJ, Exton JH. Autocrine tumor necrosis factor alpha links endoplasmic reticulum stress to the membrane death receptor pathway through IRE1alpha-mediated NF-kappaB activation and down-regulation of TRAF2 expression. Mol Cell Biol 2006; 26(8): 3071-84. [http://dx.doi.org/10.1128/MCB.26.8.3071-3084.2006] [PMID: 16581782]

[50] Urano F, Wang X, Bertolotti A, *et al.* Coupling of stress in the ER to activation of JNK protein kinases by transmembrane protein kinase IRE1. Science 2000; 287(5453): 664-6. [http://dx.doi.org/10.1126/science.287.5453.664] [PMID: 10650002]

[51] Lee AH, Scapa EF, Cohen DE, Glimcher LH. Regulation of hepatic lipogenesis by the transcription factor XBP1. Science 2008; 320(5882): 1492-6. [http://dx.doi.org/10.1126/science.1158042] [PMID: 18556558]

[52] Zhang K, Wang S, Malhotra J, *et al.* The unfolded protein response transducer IRE1α prevents ER stress-induced hepatic steatosis. EMBO J 2011; 30(7): 1357-75. [http://dx.doi.org/10.1038/emboj.2011.52] [PMID: 21407177]

[53] Ni M, Lee AS. ER chaperones in mammalian development and human diseases. FEBS Lett 2007; 581(19): 3641-51. [http://dx.doi.org/10.1016/j.febslet.2007.04.045] [PMID: 17481612]

[54] Kammoun HL, Chabanon H, Hainault I, *et al.* GRP78 expression inhibits insulin and ER stress-induced SREBP-1c activation and reduces hepatic steatosis in mice. J Clin Invest 2009; 119(5): 1201-15.
[http://dx.doi.org/10.1172/JCI37007] [PMID: 19363290]

[55] Ozcan U, Cao Q, Yilmaz E, *et al.* Endoplasmic reticulum stress links obesity, insulin action, and type 2 diabetes. Science 2004; 306(5695): 457-61.
[http://dx.doi.org/10.1126/science.1103160] [PMID: 15486293]

[56] Ye J, Gao Z, Yin J, He Q. Hypoxia is a potential risk factor for chronic inflammation and adiponectin reduction in adipose tissue of ob/ob and dietary obese mice. Am J Physiol Endocrinol Metab 2007; 293(4): E1118-28.
[http://dx.doi.org/10.1152/ajpendo.00435.2007] [PMID: 17666485]

[57] Pasarica M, Sereda OR, Redman LM, *et al.* Reduced adipose tissue oxygenation in human obesity: evidence for rarefaction, macrophage chemotaxis, and inflammation without an angiogenic response. Diabetes 2009; 58(3): 718-25.
[http://dx.doi.org/10.2337/db08-1098] [PMID: 19074987]

[58] Krishnan J, Danzer C, Simka T, *et al.* Dietary obesity-associated Hif1α activation in adipocytes restricts fatty acid oxidation and energy expenditure *via* suppression of the Sirt2-NAD+ system. Genes Dev 2012; 26(3): 259-70.
[http://dx.doi.org/10.1101/gad.180406.111] [PMID: 22302938]

[59] Poblete JMS, Ballinger MN, Bao S, *et al.* Macrophage HIF-1α mediates obesity-related adipose tissue dysfunction *via* interleukin-1 receptor-associated kinase M. Am J Physiol Endocrinol Metab 2020; 318(5): E689-700.
[http://dx.doi.org/10.1152/ajpendo.00174.2019] [PMID: 32154744]

[60] Backes JM, Howard PA, Moriarty PM. Role of C-reactive protein in cardiovascular disease. Ann Pharmacother 2004; 38(1): 110-8.
[http://dx.doi.org/10.1345/aph.1D203] [PMID: 14742804]

[61] Brooks GC, Blaha MJ, Blumenthal RS. Relation of C-reactive protein to abdominal adiposity. Am J Cardiol 2010; 106(1): 56-61.
[http://dx.doi.org/10.1016/j.amjcard.2010.02.017] [PMID: 20609648]

[62] Choi J, Joseph L, Pilote L. Obesity and C-reactive protein in various populations: a systematic review and meta-analysis. Obes Rev 2013; 14(3): 232-44.
[http://dx.doi.org/10.1111/obr.12003] [PMID: 23171381]

[63] Forman BM, Goode E, Chen J, *et al.* Identification of a nuclear receptor that is activated by farnesol metabolites. Cell 1995; 81(5): 687-93.
[http://dx.doi.org/10.1016/0092-8674(95)90530-8] [PMID: 7774010]

[64] Rader DJ. Liver X receptor and farnesoid X receptor as therapeutic targets. Am J Cardiol. 2007; 100(11 a): n15-9.

[65] Köşüş N, Köşüş A, Turhan N. Relation between abdominal subcutaneous fat tissue thickness and inflammatory markers during pregnancy. Arch Med Sci 2014; 10(4): 739-45.
[http://dx.doi.org/10.5114/aoms.2014.44865] [PMID: 25276159]

[66] Gaillard R, Rifas-Shiman SL, Perng W, Oken E, Gillman MW. Maternal inflammation during pregnancy and childhood adiposity. Obesity (Silver Spring) 2016; 24(6): 1320-7.
[http://dx.doi.org/10.1002/oby.21484] [PMID: 27094573]

[67] Pearson TA, Mensah GA, Alexander RW, *et al.* Markers of inflammation and cardiovascular disease: application to clinical and public health practice: A statement for healthcare professionals from the Centers for Disease Control and Prevention and the American Heart Association. Circulation 2003; 107(3): 499-511.
[http://dx.doi.org/10.1161/01.CIR.0000052939.59093.45] [PMID: 12551878]

[68] Swiatkiewicz I, Kozinski M, Magielski P, *et al.* Usefulness of C-reactive protein as a marker of early post-infarct left ventricular systolic dysfunction. Inflamm Res 2012; 61(7): 725-34. [http://dx.doi.org/10.1007/s00011-012-0466-2] [PMID: 22446726]

[69] Kawamoto R, Kusunoki T, Abe M, Kohara K, Miki T. An association between body mass index and high-sensitivity C-reactive protein concentrations is influenced by age in community-dwelling persons. Ann Clin Biochem 2013; 50(Pt 5): 457-64. [http://dx.doi.org/10.1177/0004563212473445] [PMID: 23880621]

[70] Manmadhan A, Lin BX, Zhong J, *et al.* Elevated GlycA in severe obesity is normalized by bariatric surgery. Diabetes Obes Metab 2019; 21(1): 178-82. [http://dx.doi.org/10.1111/dom.13481] [PMID: 30047224]

[71] Levine JA, Han JM, Wolska A, *et al.* Associations of GlycA and high-sensitivity C-reactive protein with measures of lipolysis in adults with obesity. J Clin Lipidol 2020; 14(5): 667-74. [http://dx.doi.org/10.1016/j.jacl.2020.07.012] [PMID: 32863171]

[72] Rodriguez AL, Whitehurst M, Fico BG, *et al.* Acute high-intensity interval exercise induces greater levels of serum brain-derived neurotrophic factor in obese individuals. Exp Biol Med (Maywood) 2018; 243(14): 1153-60. [http://dx.doi.org/10.1177/1535370218812191] [PMID: 30453773]

[73] Xu Y, Song Y, Chang J, *et al.* High levels of circulating cell-free DNA are a biomarker of active SLE. Eur J Clin Invest 2018; 48(11): e13015. [http://dx.doi.org/10.1111/eci.13015] [PMID: 30079480]

[74] Camuzi Zovico PV, Gasparini Neto VH, Venâncio VA, *et al.* Cell-free DNA as an obesity biomarker. Physiol Res 2020; 69(3): 515-20.

[75] Pugliese G, Iacobini C, Pesce CM, Menini S. Galectin-3: an emerging all-out player in metabolic disorders and their complications. Glycobiology 2015; 25(2): 136-50. [http://dx.doi.org/10.1093/glycob/cwu111] [PMID: 25303959]

[76] Menini S, Iacobini C, Blasetti Fantauzzi C, Pesce CM, Pugliese G. Role of Galectin-3 in Obesity and Impaired Glucose Homeostasis. Oxid Med Cell Longev 2016; 2016: 9618092. [http://dx.doi.org/10.1155/2016/9618092] [PMID: 26770660]

[77] Pang J, Nguyen VT, Rhodes DH, Sullivan ME, Braunschweig C, Fantuzzi G. Relationship of galectin-3 with obesity, IL-6, and CRP in women. J Endocrinol Invest 2016; 39(12): 1435-43. [http://dx.doi.org/10.1007/s40618-016-0515-8] [PMID: 27444618]

[78] Dencker M, Arvidsson D, Karlsson MK, Wollmer P, Andersen LB, Thorsson O. Galectin-3 levels relate in children to total body fat, abdominal fat, body fat distribution, and cardiac size. Eur J Pediatr 2018; 177(3): 461-7. [http://dx.doi.org/10.1007/s00431-017-3079-5] [PMID: 29327139]

[79] Rhodes DH, Pini M, Castellanos KJ, *et al.* Adipose tissue-specific modulation of galectin expression in lean and obese mice: evidence for regulatory function. Obesity (Silver Spring) 2013; 21(2): 310-9. [http://dx.doi.org/10.1002/oby.20016] [PMID: 23401338]

[80] Baek JH, Kim SJ, Kang HG, *et al.* Galectin-3 activates PPARγ and supports white adipose tissue formation and high-fat diet-induced obesity. Endocrinology 2015; 156(1): 147-56. [http://dx.doi.org/10.1210/en.2014-1374] [PMID: 25343273]

[81] Kiwaki K, Novak CM, Hsu DK, Liu FT, Levine JA. Galectin-3 stimulates preadipocyte proliferation and is up-regulated in growing adipose tissue. Obesity (Silver Spring) 2007; 15(1): 32-9. [http://dx.doi.org/10.1038/oby.2007.526] [PMID: 17228029]

[82] Oh DK, Ciaraldi T, Henry RR. Adiponectin in health and disease. Diabetes Obes Metab 2007; 9(3): 282-9. [http://dx.doi.org/10.1111/j.1463-1326.2006.00610.x] [PMID: 17391153]

[83] Harmancey R, Wilson CR, Taegtmeyer H. Adaptation and maladaptation of the heart in obesity. Hypertension 2008; 52(2): 181-7. [http://dx.doi.org/10.1161/HYPERTENSIONAHA.108.110031] [PMID: 18574077]

[84] Park S, Kim DS, Kwon DY, Yang HJ. Long-term central infusion of adiponectin improves energy and glucose homeostasis by decreasing fat storage and suppressing hepatic gluconeogenesis without changing food intake. J Neuroendocrinol 2011; 23(8): 687-98. [http://dx.doi.org/10.1111/j.1365-2826.2011.02165.x] [PMID: 21599766]

[85] Szmitko PE, Teoh H, Stewart DJ, Verma S. Adiponectin and cardiovascular disease: state of the art? Am J Physiol Heart Circ Physiol 2007; 292(4): H1655-63. [http://dx.doi.org/10.1152/ajpheart.01072.2006] [PMID: 17142348]

[86] Nakamura Y, Shimada K, Fukuda D, *et al.* Implications of plasma concentrations of adiponectin in patients with coronary artery disease. Heart 2004; 90(5): 528-33. [http://dx.doi.org/10.1136/hrt.2003.011114] [PMID: 15084551]

[87] Ouedraogo R, Wu X, Xu SQ, *et al.* Adiponectin suppression of high-glucose-induced reactive oxygen species in vascular endothelial cells: evidence for involvement of a cAMP signaling pathway. Diabetes 2006; 55(6): 1840-6. [http://dx.doi.org/10.2337/db05-1174] [PMID: 16731851]

[88] Turer AT, Scherer PE. Adiponectin: mechanistic insights and clinical implications. Diabetologia 2012; 55(9): 2319-26. [http://dx.doi.org/10.1007/s00125-012-2598-x] [PMID: 22688349]

[89] Ricci R, Bevilacqua F. The potential role of leptin and adiponectin in obesity: a comparative review. Vet J 2012; 191(3): 292-8. [http://dx.doi.org/10.1016/j.tvjl.2011.04.009] [PMID: 21592831]

[90] Dekker JM, Funahashi T, Nijpels G, *et al.* Prognostic value of adiponectin for cardiovascular disease and mortality. J Clin Endocrinol Metab 2008; 93(4): 1489-96. [http://dx.doi.org/10.1210/jc.2007-1436] [PMID: 18211973]

[91] Jaleel F, Jaleel A, Rahman MA, Alam E. Comparison of adiponectin, leptin and blood lipid levels in normal and obese postmenopausal women. J Pak Med Assoc 2006; 56(9): 391-4. [PMID: 17091749]

[92] Snijder MB, Flyvbjerg A, Stehouwer CD, *et al.* Relationship of adiposity with arterial stiffness as mediated by adiponectin in older men and women: the Hoorn Study. Eur J Endocrinol 2009; 160(3): 387-95. [http://dx.doi.org/10.1530/EJE-08-0817] [PMID: 19095778]

[93] Eglit T, Ringmets I, Lember M. Obesity, high-molecular-weight (HMW) adiponectin, and metabolic risk factors: prevalence and gender-specific associations in Estonia. PLoS One 2013; 8(9): e73273. [http://dx.doi.org/10.1371/journal.pone.0073273] [PMID: 24039900]

[94] Aljada A, Ghanim H, Saadeh R, Dandona P. Insulin inhibits NFkappaB and MCP-1 expression in human aortic endothelial cells. J Clin Endocrinol Metab 2001; 86(1): 450-3. [PMID: 11232040]

[95] Van Coillie E, Van Damme J, Opdenakker G. The MCP/eotaxin subfamily of CC chemokines. Cytokine Growth Factor Rev 1999; 10(1): 61-86. [http://dx.doi.org/10.1016/S1359-6101(99)00005-2] [PMID: 10379912]

[96] Sartipy P, Loskutoff DJ. Monocyte chemoattractant protein 1 in obesity and insulin resistance. Proc Natl Acad Sci USA 2003; 100(12): 7265-70. [http://dx.doi.org/10.1073/pnas.1133870100] [PMID: 12756299]

[97] Kanda H, Tateya S, Tamori Y, *et al.* MCP-1 contributes to macrophage infiltration into adipose tissue, insulin resistance, and hepatic steatosis in obesity. J Clin Invest 2006; 116(6): 1494-505. [http://dx.doi.org/10.1172/JCI26498] [PMID: 16691291]

[98] Panee J. Monocyte Chemoattractant Protein 1 (MCP-1) in obesity and diabetes. Cytokine 2012; 60(1): 1-12.
[http://dx.doi.org/10.1016/j.cyto.2012.06.018] [PMID: 22766373]

[99] Netea MG, van der Graaf C, Van der Meer JW, Kullberg BJ. Toll-like receptors and the host defense against microbial pathogens: bringing specificity to the innate-immune system. J Leukoc Biol 2004; 75(5): 749-55.
[http://dx.doi.org/10.1189/jlb.1103543] [PMID: 15075354]

[100] Lin Y, Lee H, Berg AH, Lisanti MP, Shapiro L, Scherer PE. The lipopolysaccharide-activated toll-like receptor (TLR)-4 induces synthesis of the closely related receptor TLR-2 in adipocytes. J Biol Chem 2000; 275(32): 24255-63.
[http://dx.doi.org/10.1074/jbc.M002137200] [PMID: 10823826]

[101] Beutler B. Inferences, questions and possibilities in Toll-like receptor signalling. Nature 2004; 430(6996): 257-63.
[http://dx.doi.org/10.1038/nature02761] [PMID: 15241424]

[102] Fessler MB, Rudel LL, Brown JM. Toll-like receptor signaling links dietary fatty acids to the metabolic syndrome. Curr Opin Lipidol 2009; 20(5): 379-85.
[http://dx.doi.org/10.1097/MOL.0b013e32832fa5c4] [PMID: 19625959]

[103] Houseknecht KL, Baile CA, Matteri RL, Spurlock ME. The biology of leptin: a review. J Anim Sci 1998; 76(5): 1405-20.
[http://dx.doi.org/10.2527/1998.7651405x] [PMID: 9621947]

[104] Fei H, Okano HJ, Li C, *et al.* Anatomic localization of alternatively spliced leptin receptors (Ob-R) in mouse brain and other tissues. Proc Natl Acad Sci USA 1997; 94(13): 7001-5.
[http://dx.doi.org/10.1073/pnas.94.13.7001] [PMID: 9192681]

[105] Hamann A, Matthaei S. Regulation of energy balance by leptin. Exp Clin Endocrinol Diabetes 1996; 104(4): 293-300.
[http://dx.doi.org/10.1055/s-0029-1211457] [PMID: 8886745]

[106] Ingalls AM, Dickie MM, Snell GD. Obese, a new mutation in the house mouse. J Hered 1950; 41(12): 317-8.
[http://dx.doi.org/10.1093/oxfordjournals.jhered.a106073] [PMID: 14824537]

[107] Houseknecht KL, Flier SN, Frevert EU, Frederich RC, Flier JS, Kahn BB. Leptin secretion correlates with adipocyte size in genetic and dietary obesity. Diabetes 1996; 45: 144-4.

[108] Rosen ED, Spiegelman BM. Adipocytes as regulators of energy balance and glucose homeostasis. Nature 2006; 444(7121): 847-53.
[http://dx.doi.org/10.1038/nature05483] [PMID: 17167472]

[109] Minokoshi Y, Toda C, Okamoto S. Regulatory role of leptin in glucose and lipid metabolism in skeletal muscle. Indian J Endocrinol Metab 2012; 16(9) (Suppl. 3): S562-8.
[http://dx.doi.org/10.4103/2230-8210.105573] [PMID: 23565491]

[110] Heymsfield SB, Greenberg AS, Fujioka K, *et al.* Recombinant leptin for weight loss in obese and lean adults: a randomized, controlled, dose-escalation trial. JAMA 1999; 282(16): 1568-75.
[http://dx.doi.org/10.1001/jama.282.16.1568] [PMID: 10546697]

[111] Kamohara S, Burcelin R, Halaas JL, Friedman JM, Charron MJ. Acute stimulation of glucose metabolism in mice by leptin treatment. Nature 1997; 389(6649): 374-7.
[http://dx.doi.org/10.1038/38717] [PMID: 9311777]

[112] Minokoshi Y, Kim Y-B, Peroni OD, *et al.* Leptin stimulates fatty-acid oxidation by activating AMP-activated protein kinase. Nature 2002; 415(6869): 339-43.
[http://dx.doi.org/10.1038/415339a] [PMID: 11797013]

[113] Kulkarni RN, Wang Z-L, Wang R-M, *et al.* Leptin rapidly suppresses insulin release from insulinoma

cells, rat and human islets and, in vivo, in mice. J Clin Invest 1997; 100(11): 2729-36. [http://dx.doi.org/10.1172/JCI119818] [PMID: 9389736]

[114] Fan W, Dinulescu DM, Butler AA, Zhou J, Marks DL, Cone RD. The central melanocortin system can directly regulate serum insulin levels. Endocrinology 2000; 141(9): 3072-9. [http://dx.doi.org/10.1210/endo.141.9.7665] [PMID: 10965876]

[115] Meek TH, Morton GJ. Leptin, diabetes, and the brain. Indian J Endocrinol Metab 2012; 16(9) (Suppl. 3): S534-42. [http://dx.doi.org/10.4103/2230-8210.105568] [PMID: 23565487]

[116] Park S, Ahn IS, Kim DS. Central infusion of leptin improves insulin resistance and suppresses β-cell function, but not β-cell mass, primarily through the sympathetic nervous system in a type 2 diabetic rat model. Life Sci 2010; 86(23-24): 854-62. [http://dx.doi.org/10.1016/j.lfs.2010.03.021] [PMID: 20388519]

[117] Dalamaga M, Christodoulatos GS. Visfatin, obesity, and cancer Adipocytokines, energy balance, and cancer. Springer 2017; pp. 109-36. [http://dx.doi.org/10.1007/978-3-319-41677-9_6]

[118] Fukuhara A, Matsuda M, Nishizawa M, *et al.* Visfatin: a protein secreted by visceral fat that mimics the effects of insulin. Science 2005; 307(5708): 426-30. [http://dx.doi.org/10.1126/science.1097243] [PMID: 15604363]

[119] Berndt J, Klöting N, Kralisch S, *et al.* Plasma visfatin concentrations and fat depot-specific mRNA expression in humans. Diabetes 2005; 54(10): 2911-6. [http://dx.doi.org/10.2337/diabetes.54.10.2911] [PMID: 16186392]

[120] Arner P. Visfatin—a true or false trail to type 2 diabetes mellitus. Oxford University Press 2006. [http://dx.doi.org/10.1210/jc.2005-2391]

[121] Fukuhara A, Matsuda M, Nishizawa M, *et al.* Retraction. Science 2007; 318(5850): 565. [http://dx.doi.org/10.1126/science.318.5850.565b] [PMID: 17962537]

[122] Varma V, Yao-Borengasser A, Rasouli N, *et al.* Human visfatin expression: relationship to insulin sensitivity, intramyocellular lipids, and inflammation. J Clin Endocrinol Metab 2007; 92(2): 666-72. [http://dx.doi.org/10.1210/jc.2006-1303] [PMID: 17090638]

[123] Hida K, Wada J, Eguchi J, *et al.* Visceral adipose tissue-derived serine protease inhibitor: a unique insulin-sensitizing adipocytokine in obesity. Proc Natl Acad Sci USA 2005; 102(30): 10610-5. [http://dx.doi.org/10.1073/pnas.0504703102] [PMID: 16030142]

[124] Heiker JT, Klöting N, Kovacs P, *et al.* Vaspin inhibits kallikrein 7 by serpin mechanism. Cell Mol Life Sci 2013; 70(14): 2569-83. [http://dx.doi.org/10.1007/s00018-013-1258-8] [PMID: 23370777]

[125] Nakatsuka A, Wada J, Iseda I, *et al.* Vaspin is an adipokine ameliorating ER stress in obesity as a ligand for cell-surface GRP78/MTJ-1 complex. Diabetes 2012; 61(11): 2823-32. [http://dx.doi.org/10.2337/db12-0232] [PMID: 22837305]

[126] Yang R, Xu A, Pray J, *et al.* Cloning of omentin, a new adipocytokine from omental fat tissue in humans. Diabetes 2003; 52.

[127] Yang R-Z, Lee M-J, Hu H, *et al.* Identification of omentin as a novel depot-specific adipokine in human adipose tissue: possible role in modulating insulin action. Am J Physiol Endocrinol Metab 2006; 290(6): E1253-61. [http://dx.doi.org/10.1152/ajpendo.00572.2004] [PMID: 16531507]

[128] FU M. Systematic analysis of omentin 1 and omentin 2 on 1q23 as candidate genes for type 2 diabetes in the Old Order Amish. Diabetes 2004; 53.

[129] Lee JA, Park HS, Song YS, *et al.* Relationship between vaspin gene expression and abdominal fat distribution of Korean women. Endocr J 2011; 58(8): 639-46.

[http://dx.doi.org/10.1507/endocrj.K11E-073] [PMID: 21646731]

[130] Hernández-Díaz A, Arana-Martínez JC, Carbó R, Espinosa-Cervantes R, Sánchez-Muñoz F. Omentin: Role in insulin resistance, inflammation and cardiovascular protection. Arch Cardiol Mex 2016; 86(3): 233-43.
[PMID: 26778502]

[131] Kuperman DA, Lewis CC, Woodruff PG, *et al.* Dissecting asthma using focused transgenic modeling and functional genomics. J Allergy Clin Immunol 2005; 116(2): 305-11.
[http://dx.doi.org/10.1016/j.jaci.2005.03.024] [PMID: 16083784]

[132] Tan BK, Adya R, Randeva HS. Omentin: a novel link between inflammation, diabesity, and cardiovascular disease. Trends Cardiovasc Med 2010; 20(5): 143-8.
[http://dx.doi.org/10.1016/j.tcm.2010.12.002] [PMID: 21742269]

[133] Yamawaki H, Kuramoto J, Kameshima S, Usui T, Okada M, Hara Y. Omentin, a novel adipocytokine inhibits TNF-induced vascular inflammation in human endothelial cells. Biochem Biophys Res Commun 2011; 408(2): 339-43.
[http://dx.doi.org/10.1016/j.bbrc.2011.04.039] [PMID: 21514279]

[134] Yan P, Liu D, Long M, Ren Y, Pang J, Li R. Changes of serum omentin levels and relationship between omentin and adiponectin concentrations in type 2 diabetes mellitus. Exp Clin Endocrinol Diabetes 2011; 119(4): 257-63.
[http://dx.doi.org/10.1055/s-0030-1269912] [PMID: 21374544]

[135] Herder C, Carstensen M, Ouwens DM. Anti-inflammatory cytokines and risk of type 2 diabetes. Diabetes Obes Metab 2013; 15(s3) (Suppl. 3): 39-50.
[http://dx.doi.org/10.1111/dom.12155] [PMID: 24003920]

[136] Tapan S, Tascilar E, Abaci A, *et al.* Decreased plasma apelin levels in pubertal obese children. J Pediatr Endocrinol Metab 2010; 23(10): 1039-46.
[http://dx.doi.org/10.1515/jpem.2010.165] [PMID: 21158215]

[137] Kotanidou EP, Kalinderi K, Kyrgios I, *et al.* Apelin and G212A apelin receptor gene polymorphism in obese and diabese youth. Pediatr Obes 2015; 10(3): 213-9.
[http://dx.doi.org/10.1111/ijpo.251] [PMID: 25060841]

[138] Ba H-J, Chen H-S, Su Z, *et al.* Associations between serum apelin-12 levels and obesity-related markers in Chinese children. PLoS One 2014; 9(1): e86577.
[http://dx.doi.org/10.1371/journal.pone.0086577] [PMID: 24475149]

[139] Krist J, Wieder K, Klöting N, *et al.* Effects of weight loss and exercise on apelin serum concentrations and adipose tissue expression in human obesity. Obes Facts 2013; 6(1): 57-69.
[http://dx.doi.org/10.1159/000348667] [PMID: 23429279]

[140] Castan-Laurell I, Dray C, Attané C, Duparc T, Knauf C, Valet P. Apelin, diabetes, and obesity. Endocrine 2011; 40(1): 1-9.
[http://dx.doi.org/10.1007/s12020-011-9507-9] [PMID: 21725702]

[141] Bertrand C, Valet P, Castan-Laurell I. Apelin and energy metabolism. Front Physiol 2015; 6: 115.
[http://dx.doi.org/10.3389/fphys.2015.00115] [PMID: 25914650]

[142] Alipour FG, Ashoori MR, Pilehvar-Soltanahmadi Y, Zarghami N. An overview on biological functions and emerging therapeutic roles of apelin in diabetes mellitus. Diabetes Metab Syndr 2017; 11 (Suppl. 2): S919-23.
[http://dx.doi.org/10.1016/j.dsx.2017.07.016] [PMID: 28712823]

[143] Moreno-Aliaga MJ, Pérez-Echarri N, Marcos-Gómez B, *et al.* Cardiotrophin-1 is a key regulator of glucose and lipid metabolism. Cell Metab 2011; 14(2): 242-53.
[http://dx.doi.org/10.1016/j.cmet.2011.05.013] [PMID: 21803294]

[144] Escoté X, Gómez-Zorita S, López-Yoldi M, *et al.* Role of omentin, vaspin, cardiotrophin-1, TWEAK and NOV/CCN3 in obesity and diabetes development. Int J Mol Sci 2017; 18(8): 1770.

[http://dx.doi.org/10.3390/ijms18081770] [PMID: 28809783]

[145] López-Yoldi M, Stanhope KL, Garaulet M, *et al.* Role of cardiotrophin-1 in the regulation of metabolic circadian rhythms and adipose core clock genes in mice and characterization of 24-h circulating CT-1 profiles in normal-weight and overweight/obese subjects. FASEB J 2017; 31(4): 1639-49.
[http://dx.doi.org/10.1096/fj.201600396RR] [PMID: 28096235]

[146] Moraes-Vieira PM, Saghatelian A, Kahn BB. GLUT4 expression in adipocytes regulates de novo lipogenesis and levels of a novel class of lipids with antidiabetic and anti-inflammatory effects. Diabetes 2016; 65(7): 1808-15.
[http://dx.doi.org/10.2337/db16-0221] [PMID: 27288004]

[147] Yore MM, Syed I, Moraes-Vieira PM, *et al.* Discovery of a class of endogenous mammalian lipids with anti-diabetic and anti-inflammatory effects. Cell 2014; 159(2): 318-32.
[http://dx.doi.org/10.1016/j.cell.2014.09.035] [PMID: 25303528]

[148] Oh DY, Talukdar S, Bae EJ, *et al.* GPR120 is an omega-3 fatty acid receptor mediating potent anti-inflammatory and insulin-sensitizing effects. Cell 2010; 142(5): 687-98.
[http://dx.doi.org/10.1016/j.cell.2010.07.041] [PMID: 20813258]

[149] Vienberg S, Geiger J, Madsen S, Dalgaard LT. MicroRNAs in metabolism. Acta Physiol (Oxf) 2017; 219(2): 346-61.
[http://dx.doi.org/10.1111/apha.12681] [PMID: 27009502]

[150] Ortega FJ, Moreno-Navarrete JM, Pardo G, *et al.* MiRNA expression profile of human subcutaneous adipose and during adipocyte differentiation. PLoS One 2010; 5(2): e9022.
[http://dx.doi.org/10.1371/journal.pone.0009022] [PMID: 20126310]

[151] Martinelli R, Nardelli C, Pilone V, *et al.* miR-519d overexpression is associated with human obesity. Obesity (Silver Spring) 2010; 18(11): 2170-6.
[http://dx.doi.org/10.1038/oby.2009.474] [PMID: 20057369]

[152] Arner P, Kulyté A. MicroRNA regulatory networks in human adipose tissue and obesity. Nat Rev Endocrinol 2015; 11(5): 276-88.
[http://dx.doi.org/10.1038/nrendo.2015.25] [PMID: 25732520]

[153] Heneghan HM, Miller N, Kerin MJ. Role of microRNAs in obesity and the metabolic syndrome. Obes Rev 2010; 11(5): 354-61.
[http://dx.doi.org/10.1111/j.1467-789X.2009.00659.x] [PMID: 19793375]

[154] Klöting N, Berthold S, Kovacs P, *et al.* MicroRNA expression in human omental and subcutaneous adipose tissue. PLoS One 2009; 4(3): e4699.
[http://dx.doi.org/10.1371/journal.pone.0004699] [PMID: 19259271]

[155] Heneghan HM, Miller N, McAnena OJ, O'Brien T, Kerin MJ. Differential miRNA expression in omental adipose tissue and in the circulation of obese patients identifies novel metabolic biomarkers. J Clin Endocrinol Metab 2011; 96(5): E846-50.
[http://dx.doi.org/10.1210/jc.2010-2701] [PMID: 21367929]

[156] Keller P, Gburcik V, Petrovic N, *et al.* Gene-chip studies of adipogenesis-regulated microRNAs in mouse primary adipocytes and human obesity. BMC Endocr Disord 2011; 11(1): 7.
[http://dx.doi.org/10.1186/1472-6823-11-7] [PMID: 21426570]

[157] Turchinovich A, Weiz L, Langheinz A, Burwinkel B. Characterization of extracellular circulating microRNA. Nucleic Acids Res 2011; 39(16): 7223-33.
[http://dx.doi.org/10.1093/nar/gkr254] [PMID: 21609964]

[158] Cortez MA, Calin GA. MicroRNA identification in plasma and serum: a new tool to diagnose and monitor diseases. Expert Opin Biol Ther 2009; 9(6): 703-11.
[http://dx.doi.org/10.1517/14712590902932889] [PMID: 19426115]

[159] Guay C, Regazzi R. Exosomes as new players in metabolic organ cross-talk. Diabetes Obes Metab 2017; 19 (Suppl. 1): 137-46.
[http://dx.doi.org/10.1111/dom.13027] [PMID: 28880477]

CHAPTER 27

Application of Biomarkers in the Diagnostic Distinction of Bacterial and Viral Infections

Fatma Köksal Çakırlar[1,*]

[1] Department of Medical Microbiology, Faculty of Cerrahpaşa Medicine, University of İstanbul-Cerrahpaşa, Istanbul, Turkey

Abstract: Infectious diseases, which pose a great threat worldwide, have a significant impact on public health and the world economy. It contributes to increased healthcare costs, unnecessary drug-related side effects, and increased antimicrobial resistance. It is not always easy to distinguish the etiological differentiation of diseases that can develop with bacteria and viruses. Therefore, one of the biggest challenges in medicine is how to correctly distinguish between the different causes of these infections and how to manage the patient. Because bacterial and viral infections often present similar symptoms. The real decision is whether the infection is caused by bacteria or viruses and whether to treat the patient with antibiotics. There are many different methodological approaches to diagnosing infections. Biomarkers have been used in the diagnosis of diseases and other conditions for many years. Biomarkers are molecules found in blood and body fluids in measurable amounts, which can evaluate biological and pathological processes. These key indicators can provide vital information in determining disease prognosis, predicting response to treatments, adverse events and drug interactions, and identifying key risks. An effective biomarker is extremely important for the early diagnosis of various diseases. The explosion of interest in biomarker research is driving the development of new predictive, diagnostic, and prognostic products in modern medical practice. The purpose of this review is to demonstrate the use and diagnostic potential of current and investigational biomarkers in the distinction between bacterial and viral infections.

Keywords: Current and investigational biomarkers, Bacterial and viral infections.

INTRODUCTION

Infectious diseases are defined as clinical diseases that occur as a result of many inflammatory reactions in the host when an infectious agent such as bacteria, virus, parasite, and fungus enters the body and multiplies. It can be transmitted directly or indirectly from person to person under appropriate conditions, causing epidemics and pandemics [1].

[*] **Corresponding author Fatma Köksal Çakırlar:** Department of Medical Microbiology, Faculty of Cerrahpaşa Medicine, University of İstanbul-Cerrahpaşa, Istanbul, Turkey; Tel: +90 (534) 844 83 94; E-mail: fatma.koksal@istanbul.edu.tr

Pınar Atukeren and Hafize Uzun (Eds.)

The most of emergency infections are caused by pathogens that are already present in the environment or that gain a selective advantage as conditions change, allowing them to infect new host populations. In rare cases, a new variant may also develop and cause a new disease. Emergency infections are a concern, especially when they are spreading rapidly and death rates are high. Previously unidentified or unknown pathogens may be agents that affect new geographic regions or new populations or re-emergent agents with significantly reduced disease incidence in the past. The world is still struggling with various infectious diseases such as malaria, H1N1, Ebola, dengue Fever, Chagas disease, yellow fever, West Nile, Japanese encephalitis, and chikungunya [1].

Gram-negative bacteria caused by newly evolved organisms such as multidrug-resistant tuberculosis and plasmid-mediated colistin-resistant, newly emerging organisms in the human population, HIV-1 infection, respiratory syndrome (SARS) coronavirus in 2003 and MERS-CoV (Middle East respiratory disease) The current coronavirus disease-2019 (COVID-19) pandemic caused by the eastern respiratory syndrome caused by the syndrome coronavirus, and the new coronavirus SARS-CoV-2, which affects the whole world, is challenging the world [2]. Globally, as of October 17, 2020, the COVID-19 outbreak has spread to more than 200 countries and 39,023,292 confirmed cases of COVID-19, including 1,099,586 deaths, have been reported by WHO [3].

With the ongoing evolution of viral and microbial variants and the choice of drug resistance, it appears that new infections will continue to emerge, possibly increase, and will make human beings very challenging. Infectious diseases contribute to increased healthcare costs, unnecessary drug-related side effects, and increased antimicrobial resistance. These diseases have a significant impact on public health and the world economy. In cases where the focus of infection cannot be effectively eliminated or treated appropriately, sepsis, septic shock, and multi-organ failure may develop. In the presence of infection in any part of the body, the immune system begins to react and sepsis occurs due to the body's exaggerated response to this infection. If early treatment is not initiated, organ failure, shock, and death are inevitable. It is not always easy to distinguish the etiological differentiation of diseases that can develop with both bacteria and viruses. Bacterial and viral infections usually present with similar symptoms. In most cases, the decision is whether the infection is caused by bacteria or viruses and whether to treat the patient with antibiotics. This is primarily a challenge for clinicians and healthcare administrators. In addition, correct separation is important in reducing unnecessary antibiotic use [4].

There are many different methodological approaches to diagnosing infections. It is based on patient symptoms, laboratory tests, radiological examinations as well as clinical findings. These procedures take a long time and are limited. Therefore, additional tools are needed to increase diagnostic accuracy. Biomarkers have been used for many years in the diagnosis of diseases and other conditions. Biomarkers are structures found in the blood or other body fluids, with little or no normal state, that appear or increase in the disease state. Protein metabolites are often used as biomarkers. Nucleic acids and various analytical targets are also used as biomarkers. A better understanding of disease mechanisms is required to fully utilize biomarkers. Bacteria and viruses activate or block different signaling pathways in the cells they infect. Because biomarkers are molecules that can evaluate biological and pathological processes and can be found in measurable amounts in blood and body fluids, they are considered physical, functional or biochemical markers of disease processes. Biomarkers used in the diagnosis of sepsis may play a role in early diagnosis, risk stratification, evaluation, and prognosis prediction. These key indicators can provide vital information in determining disease prognosis, predicting response to treatments, adverse events and drug interactions, and identifying key risks. An effective biomarker is extremely important for the early diagnosis of various diseases [5]. However, the problem with existing biomarkers; It gives late results, low specificity, complicate clinical decisions, and interact with corticosteroid and immunosuppressive treatments. For this reason, new biomarkers are needed to make early and etiological diagnosis, decide on early, correct and adequate treatment, determine the severity of the infection, and evaluate the response to treatment. The explosion of interest in biomarker research will enable the development of new predictive, diagnostic, and prognostic products in modern medical practice. This could advance the use of personalized medicine in the treatment of infectious disease and will also raise important considerations for validating the use of biomarkers as diagnostic or prognostic markers. Many different classes of molecules have been studied in the search for new diagnostic biomarkers to overcome the problems in currently used methods that could help distinguish between bacterial and viral infections. The purpose of this review is to demonstrate the use and diagnostic potential of current and investigational biomarkers in the distinction between bacterial and viral infections.

STANDARD BİOMARKERS

White Blood Cell Counts

White blood cell counts, especially differences in white blood cell counts or neutrophil counts in the blood, are the most commonly used tests to diagnose acute infections. Counting can easily be done manually under the microscope or with automatic cell counters. However, at least 400 cells must be counted for differentials. Relying on white blood cell counts alone can be misleading because we cannot know if the dominant cell in a patient is a neutrophil, lymphocyte, or in some cases, eosinophils. In a study conducted on patients with acute infection, patients were classified by all available means for the agent causing the bacteria and the agent causing the virus. Neutrophil counts measured with an automatic cell counter in the blood before any treatment was given, it was found to be clearly increased in patients with bacterial infection, but not in most of those with a viral infection, and this selected group of patients, at the level of 7.3 × 109/L, negative predictive value was 79% at the optimal diagnostic accuracy. In patients with very severe infections, e.g., septicemia, low numbers of neutrophils are often seen as signs of increased consumption of cells in the process. Therefore, in these cases, the neutrophil count does not distinguish it from normal blood cell counts or those with an acute viral infection. Low or normal neutrophil counts are also common in patients with *Mycoplasma* infection [5].

POTENTİAL DİAGNOSTİC BİOMARKERS

Acute phase proteins, cytokines/chemokines, vasodilator biomarkers, biomarkers related to organ dysfunction, biomarkers related to vascular damage, cell surface biomarkers, receptor biomarkers, and other (lactate) biomarkers are in this group.

ACUTE PHASE PROTEİNS

Serum amyloid A, ceruloplasmin, C-reactive protein (CRP), ferritin, α1-acid glycoprotein, hepcidin, procalcitonin (PCT), Lipopolysaccharide binding protein (LBP), Pentraxin-3 are in this group.

C-Reactive Protein

C-reactive protein (CRP) is referred to as the capacity of *Streptococcus pneumoniae* to precipitate the somatic C polysaccharide. CRP, is the first acute phase protein described, and it is produced by the liver in response to a variety of cytokines, particularly interleukin (IL)-6 and tumor necrosis factor (TNF)-α. CRP binds the phospholipid components of microorganisms in acute inflammations and facilitates their removal by macrophages. Normal CRP concentrations are very low, <2 mg/L, but may increase in the elderly. In bacterial infections, these concentrations may increase more than 100 times. Its sensitivity and specificity for bacterial infections are 68-92% and 40-67%, respectively [6]. CRP has become one of the most commonly used blood tests in the clinic, partly because it is available as a bedside test and is also a very sensitive biomarker for detecting ongoing inflammation in the body [7, 8]. High CRP measurements in patients in ICUs can be helpful in estimating the risk of organ failure and mortality and in predicting risk in early sepsis. One disadvantage of CRP is the fact that it is present in high concentrations in many inflammatory conditions other than acute infections.

CRP has been examined the diagnostic accuracy of CRP in the distinction between acute viral and acute bacterial infections. The use of CRP as a rule-out test is limited because slightly increased concentrations of CRP, *i.e.*, above 5 mg/L, can often be seen even in participants without any obvious symptoms or other signs of illness. A disadvantage of CRP as a diagnostic tool of acute infections is its kinetics. The liver production of CRP in response to IL6 is very slow, and periodic concentrations in the circulation take 12–24 hours after the onset of clinical symptoms of acute infections. For this reason, CRP, may be normal only hours after the onset of symptoms in patients calling the doctor. This finding often contradicts with the number of neutrophils already increasing at baseline. Also, CRP production continues for 1-2 days after the infection subsides. This feature, in contrast, means that patients may be misclassified as an active infection persistent, and antibiotics may be prescribed unnecessarily [9]. Extremely high concentrations of CRP were detected 48-72 hours after the initiation of treatment, even if symptoms and signs indicate that the bacterial challenge has been eliminated [5].

Procalcitonin

Procalcitonin (PCT) is the precursor to calcitonin, a calcium regulating hormone secreted from the thyroid tissue in healthy individuals, showing host response. Released from almost all tissues, including lung, liver, kidney, pancreas, spleen, colon, and adipose tissues. Its production is induced by endotoxins and various cytokines. Procalcitonin determination in serum and plasma has become relatively common in many countries and has been shown to be useful in intensive care units in severe and complicated diseases [10].

PCT has been shown to correlate strongly with the extent and severity of bacterial infections and helps distinguish bacterial infections from viral diseases. In addition, the studies have found that PCT increases immediately within 6 to 12 hours after stimulation and circulating PCT levels are halved each day when the infection is controlled by the host immune system or antibiotic therapy, and PCT has been found to be associated with the bacterial load and the severity of the infection. It is used as a biomarker in the diagnosis of bacterial infections such as sepsis and septic shock [6, 11]. Serum level is below 0.1 ng/mL in healthy adults. The determined limit values are 2 ng/mL for sepsis and 11.6 ng/mL for septic shock [12]. Procalcitonin, which increases in the blood in severe bacterial infections such as sepsis, decreases with the successful treatment of patients [5]. It has been reported that knowledge of procalcitonin concentrations does not reduce antibiotic prescription because procalcitonin rises after surgical trauma without obvious signs of concomitant infections, and this precludes interpretation of moderately increased procalcitonin concentrations. Therefore, the researchers concluded that procalcitonin is not suitable as an exclusion analysis of acute bacterial infections [6].

Pentraxin-3

Pentraxin-3 (PTX3) is a member of the long pentraxine subfamily. It is released from monocytes, dendritic cells and macrophages, epithelial cells, vascular endothelial cells, and smooth muscle cells after stimulation by microbial parts such as toll-like receptor agonists, lipopolysaccharides or inflammatory cytokines. Circulating PTX-3 modulates complement activity, cell extravasation, and pathogen recognition by myeloid cells and contributes to resistance to microorganisms through the creation of an antimicrobial microenvironment and plays a variety of roles in different situations such as cardiovascular disease wound healing. Several studies have shown that PTX-3 has a good diagnostic value, significantly increased in the sepsis group compared to the healthy group, and PTX3 has been evaluated as a prognostic marker of sepsis [13, 14].

Hepcidin

Hepcidin has been shown to be a potential acute phase biomarker in inflammation and sepsis. It is a new peptide hormone of hepatic origin and has a very important role in iron metabolism that prevents microorganism access to iron. Hepcidin synthesis in the liver is induced by IL-6 in response to inflammation, and increased serum levels have been observed in different diseases, particularly sepsis [15].

There are studies showing that it is a reliable biological sign of early and late-onset sepsis in newborns. Tests have been developed to evaluate serum and urine hepcidin levels [16].

Lipopolysaccharide Binding Protein (LBP)

Lipopolysaccharide binding protein (LBP), which is mainly synthesized in the liver, is a polypeptide that binds lipopolysaccharide (LPS). It has 100% sensitivity and specificity with a cut-off value of 46.3 mg/L in febrile and neutropenic patients [17]. It has been stated that LBP can be used for diagnostic purposes as it starts to increase in the first 24 hours in sepsis, but it has been stated that the high LBP is non-specific in showing the inflammatory response. Therefore, its use in distinguishing between infectious and non-infectious etiologies is controversial. It has been stated that LPS and LBP levels are affected by the administration of antibiotics, but are not related to the clinical course of sepsis. It is not a good indicator for determining the severity of inflammation in LBP, because induction is slow and elimination is fast. For this reason, there are limited studies on clinical use. Its use may be advantageous when evaluated together with clinical scores and other biomarkers for the diagnosis of sepsis [16].

CYTOKİNES AND CHEMOKİNES

Cytokines function in both cellular and humoral immunity. They are stimulated in a short time and secreted from cells. They are glycoproteins that have great tasks such as regulating the immune response by intensifying and suppressing it, activating the cells involved in inflammation, recruiting and moving them in the reaction zone, and showing antiviral activity. Interleukin-1 (IL-1) and IL-6 are responsible for the innate immune response in tissue damage or infection. These proinflammatory cytokines are effective in inducing fever and acute phase response. However, their use in the diagnosis of sepsis is limited as they are not specific and cannot distinguish infection from inflammation. Chemokines have proinflammatory properties and play a role in chemotaxis [18].

Interleukin 8 (IL-8) indicates the severity of the non-specific IL-6 inflammatory response for bacterial infections; it is a proinflammatory chemokine with an increase in viral infection, organ transplant rejection, autoimmune diseases, and surgical intervention. IL-8 is a promising marker for many clinical situations and is used by various subspecialties of medicine as a predictor of rapid diagnosis and prognosis. However, the level of IL8 increases as a result of many inflammatory conditions, so careful interpretation of the IL-8 level is required to relate it to the diagnosis or prognosis of the desired clinical condition. In some studies have found a link between increased IL-8 levels in the first 48 hours and poor prognosis and increased mortality rate [23, 24]. Weitkamp *et al.* Investigated the correlation of IL-6 and IL-8 values during neonatal clinical infection and whether IL-8 would be a more useful infection marker, especially in the preterm neonatal intensive care unit (NICU). They found that plasma levels of IL-6 and IL-8 correlated well and did not differ significantly in sensitivity and specificity for clinical infection. They emphasized that IL-8 could be a more useful marker of infection, especially in preterm NICU patients, as detection of IL-8 is faster and requires less sample volume. A panel of multiple biomarkers, including IL-8, the diagnostic and prognostic approach for many clinical disorders, could be promising. Large-scale studies are needed to prove the accuracy and outcome of IL-8's usefulness and efficacy as a biomarker [19, 20].

High Mobility Group Box 1 (HMGB1)

The high mobility group Box 1 (HMGB1) is a nuclear protein. It acts as a DNA chaperone, and in the nucleus it participates in a number of activities, including transcription, replication, DNA repair, and nucleosome assembly. It binds to DNA and causes the transcription of various inflammatory markers. It also has some extracellular roles such as promoting migration and increasing the production of proinflammatory markers and cytokines such as Interleukin 6 (IL-6), Tumor necrosis factor (TNF) or Interferon gamma (IFN-γ). HMGB1 is secreted in a number of diseases, including chronic inflammatory disorders, autoimmune diseases, trauma, and cancer, and is elevated during especially bacterial and Influenza virus co-infection [21, 22].

In a study by Zhou *et al.*, Instead of measuring HMGB1 concentration in serum, changes in expression of the HMGB1 gene in peripheral monocytes were measured by PCR technique by comparing the density of HMGB1 proteins with an 18S ribosomal ribonucleotide acid and found that coinfection (virus and bacteria) can result when HMGB1 expression is greater than 1.0256. In addition, in this study, HMGB1 expression <1.0256 and WBC> 13 × 109/L were determined to have 92.3% positive predictive value for single bacterial pneumonia. Therefore HMGB1 seems to be a good token. However, the need to isolate specific blood cells (monocytes) and use PCR makes this method long-term and expensive. Also, RNA measurement may not be associated with functional serum HMGB1 protein [23]. It is known that HMGB-1 increases, especially in the late and lethal stages of sepsis. Therefore, it can be predicted that it will not be used for diagnostic purposes but rather have a prognostic value. In one study, HMGB-1 measurements on day 3 differentiated survivors with sensitivity and specificity of 66% and 67%, respectively, and mortality was reported to increase 5.5 times when HMGB1 level was> 4 ng/mL [24].

There is a need for new studies on the use of HMGB-1, which has limited information in the literature, as a biomarker during the course of bacterial infections [16].

VASODİLATOR BİOMARKERS

Angiopoietin, endoxane, ELAM-1, ICAM-1, Adrenomedullin, Copeptin, Laminin, Elastin are in this group.

Proadrenomedullin

Like PCT, proadrenomedullin (proADM) is a type of "hormokine" that expresses the cytokine-like behavior of hormones during inflammation and infections.

Adrenomedullin (ADM) is a 52 amino acid peptide produced by the adrenal medulla. ADM is produced during physiological stress and has various functions such as vasodilation, anti-inflammatory, and antimicrobial effects [25]. Plasma ADM concentration and ADM gene expression increase in patients with sepsis [26]. ADM is rapidly cleared from circulation, has a short half-life, and tends to attach to vessel walls and surfaces while in circulation. Therefore his measurements are not reliable. Determination of the serum amount of the mid-regional fragment of proADM instead of ADM has been studied. Recent clinical data have shown that circulating mid-regional proADM levels are significantly higher in patients with sepsis compared with patients with systemic inflammatory response syndrome (SIRS) [18, 27]. proADM has been reported to be responsible for severe sepsis-associated hypotension, which has been proposed as a good marker for risk assessment and prediction of sepsis prognosis [18, 28].

In future studies, in light of the data to be obtained about the predictive value of proADM, this biomarker may be useful both as a diagnostic and a prognostic marker in the early stages of infections [16].

BİOMARKERS RELATED TO ORGAN DYSFUNCTİON

Natriuretic peptides (ANP/BNP), troponin, surfactant protein, glial fibrillary acidic protein, endothelin-1, α-glutathione, pancreatitis-related protein-1, and Carbomyl phosphate synthetase (CPS) -1 are biomarkers in this group.

Endothelin-1

Endothelin-1 (ET-1) is the strongest known vasoconstrictor released by endothelial cells. In alveolar macrophages, leukocytes have high levels of ET-1. It is thought that ET-1 increases superoxide anion production and cytokine secretion and induces pro-inflammatory mechanisms. In endotoxemia, the precursor protein prepro-endothelin is released from the heart and lungs. Plasma level increases in systemic infections. The selective ET-1 receptor antagonist has been shown to be useful in the treatment of septic shock in animal experiments [29].

Clinical and experimental results show that ET-1 plays a role in the pathogenesis of sepsis, viral and bacterial pneumonia, *Rickettsia conorii* infections, Chagas disease and severe malaria. ET-1 levels increase in patients with sepsis and septic shock, and it is thought that there may be a relationship between this high level and the mortality and morbidity rate of the patients. However, there are limited studies [16].

BİOMARKERS RELATED TO VASCULAR ENDOTHELİAL DAMAGE

L-selectin, P-selectin, VEGF, Neopterin, VCAM-1, Endoksan, Angiopoietin (Ang 1-2) are this group of biomarkers.

Angiopoietins

Angiopoietin-1 (Ang-1) and Ang-2 are endothelium-derived vascular growth factors that play a reciprocal role during sepsis. Ang-1 stabilizes the endothelium, while Ang-2 facilitates loss of endothelial integrity and vascular leakage. Ang-1 or Ang-2 activates the transmembrane endothelial tyrosine kinase Tie2, which mediates the healthy state of blood vessels. Macrophages express the angiopoietin receptor, a tyrosine kinase known as Tie2. The interaction between Ang 2 and Tie2 promotes endothelial growth, cell survival, and angiogenesis. This interaction promotes pro-inflammatory and pro-thrombotic pathways, microvascular leakage, and angiogenic stimuli. Therefore Ang-2 plays a critical role in the induction of inflammation [30].

Increased Ang-2 levels in the systemic circulation have been associated with multiple organ dysfunction and sepsis, an indicator of impaired vascular endothelial integrity. A cohort study revealed that high Ang1 and low Ang-2 levels were observed in sepsis patients [31].

CELL SURFACE BİOMARKERS

In this group, CD 1b, CD 10, CD14 (s-CD14-ST = Presepsin), CD 18, CD 64, CD 25, CD 73 are the biomarkers.

CD64

CD64 represents a membrane glycoprotein that mediates endocytosis, phagocytosis, antibody dependent cellular toxicity (ADCC), cytokine release, and superoxide generation [32].

It is structurally expressed on monocytes and macrophages. However, it is expressed in low concentration in non-activated neutrophils, and it can be significantly upregulated at the onset of the sepsis process from neutrophils [33]. Neutrophil CD64 could be a valuable diagnostic, prognostic and follow-up biomarker for newborns with sepsis [34]. CD64 expression can indicate very early stages of infection and can aid in both early diagnosis and prediction of prognosis. Thus, the surface expression of CD64, *i.e.*, the FccyI receptor, appears to be the most promising biomarker [5]. The sensitivity of CD64 in the diagnosis of sepsis has been reported as 86% and its specificity as 87% [16]. In some studies, when comparing blood samples from children with bacterial infections, adults, and non-infected controls, it was found that the distinction between infected and non-infected groups was good. In another study, in Influenza, CD64 production was found to be increased and clearly different from healthy participants. However, in this study, it was found to be moderate compared to patients with bacterial infections. The disadvantage of CD64 analysis is the rather complex setup and requirement of flow cytometric techniques. A specific advantage of CD64 is that its kinetics are very slow, *i.e.* upregulated surface expression on neutrophils occurs within the first 12-24 hours after the onset of signs and symptoms of bacterial infection. Because neutrophil expression of CD64 is not seen in patients after surgery unless they resist an infection [35].

Presepsin

CD14 is a glycoprotein expressed on the surface of immune cells such as monocytes and macrophages, acting as a receptor for lipopolysaccharide (LPS). The physiological role of preseps is thought to be related to bacterial phagocytosis and lysosomal division of microorganisms, but is not yet fully understood. In the immune response to sepsis, serum presepsin levels rise before procalcitonin or IL-6, so it has been suggested as a potential biomarker for the diagnosis of sepsis. Significant differences in presepsin levels have been found between Gram-positive and Gram-negative bacterial infections [36]. Presepsin is one of the receptors for lipopolysaccharides found in the cell walls of Gram-negative bacteria, and a prospective study reported that the bacteremia status of SIRS patients admitted to the emergency room was associated with an increase in presepsin levels [14].

RECEPTOR BİOMARKERS

This group, TCR-2/4, suPAR (Soluble urokinase plasminogen activator receptor), IL-2R, s TREM-1 (Soluble triggering receptor expressed on myeloid cells), and CCR2/3 are receptor biomarkers.

TREM-1 (Trigger Receptor Released from Myeloid Cells)

Trigger receptor-1 (TREM-1) is a member of the immunoglobulin superfamily. It is a surface receptor of mature polymorphonuclear and monocyte cells expressed in myeloid cells in non-bacterial infections and contributes to innate immunity [37].

Immunoglobulin is in the G superfamily and has a glycopeptide structure. It is released from neutrophils, monocytes, and macrophages in acute inflammatory response. TREM-1 enhances the inflammatory response by increasing the production of pro-inflammatory cytokines while inhibiting IL-10 synthesis [38]. Since soluble TREM-1 (sTREM-1) levels increase in bacterial infections and sepsis, it is a promising biomarker in clinical practice. Especially in organ-specific infections such as pneumonia and meningitis, sTREM-1 levels in bronchoalveolar lavage fluid, pleural fluid and CSF can be detected higher than plasma, and therefore their diagnostic value may increase, but sTREM-1 values have been reported to be insufficient in predicting survival in septic patients or those with SIRS [39].

Soluble Urokinase Plasminogen Activator Receptor (suPAR)

The soluble urokinase plasminogen activator receptor (suPAR) has been shown to have prognostic value for many diseases involving inflammation, both infectious and non-infectious. suPAR; It is involved in a variety of immunological functions, including cell adhesion, migration, chemotaxis, proteolysis, immune activation, tissue remodeling, invasion, and signal transduction. Cerebrospinal fluid (CSF), urine and serum levels can be measured by commercial kits and ELISA method. In healthy adults, the mean value of suPAR is 1.5 ng/mL (range: 1.2-1.9 ng/mL, N = 44) or 2.6 ng/mL (range: 1.5 - 4.0 ng/mL, N = 31) [40, 41].

It has been reported to be associated with mortality in patients with malaria, tuberculosis and human immunodeficiency virus (HIV) infection [16].

NEW BİOMARKERS

Lactate

Lactate is a glucose metabolite produced by tissues in the body during oxygen deficiency. Lactate excretion is mostly 60-70% in the liver, then 25-30% in the kidneys, and less in other organs. Lactate levels in the blood increase as a result of liver blood supply or impaired circulation, which cannot be metabolized in the liver. In patients with hypoperfusion or septic shock, the lactate value is usually measured above 2 mmol/L. A lactate level above 4 mmol/L determines the urgent need for resuscitation and intensive care hospitalization. This situation has been associated with increased mortality in sepsis [42].

Measuring lactate levels in sepsis can provide useful information about disease progression and the effectiveness of treatment [43]. Therefore, lactate has been reported to be an important biomarker in the diagnosis, determination of treatment response and prediction of prognosis in sepsis [16].

In a study investigating the ability of cerebrospinal fluid (CSF) biomarkers to differentiate acute bacterial and viral meningitis, CSF lactate was found to have better operational properties than other biomarkers [44].

TRAIL AND IP-10

Interferon (IFN)-induced protein 10 kDa (IP-10) or chemokine 10 (CXCL10) with a C-X-C motif is a small cytokine belonging to the CXC family of chemokines. This family of signaling molecules is known to control various biological functions and also play important roles in the initiation and progression of the disease. CXCL10 binds to its unique cognate receptor CXCR3, critically regulating chemotaxis during various immune-inflammatory processes. Induce apoptosis, cell growth and angiostasis Controls chemotaxis during the inflammatory response resulting from allograft rejection after transplantation. It has been associated with inflammatory diseases, including infectious diseases, immune dysfunction, and tumor development. It is considered as a biomarker that predicts the severity of various diseases. However, its mechanism in infectious disease pathogenesis remains unclear [6].

IP-10 has been found to play a role in *Helicobacter pylori* and *Mycoplasma*, and it is associated with the severity of Legionnaires' disease and tuberculosis (TB) [45, 46]. Antimicrobial therapy has been found to significantly lower the IP-10 level in a murine model of *Mycoplasma pneumonia.* IP-10 has been shown to be a new biomarker for the severity of *Trypanosoma bruc*ei infection, *Plasmodium falciparum, Plasmodium vivax*, *Toxoplasma gondii* and *Leishmania major* [6, 47]. TNF-associated apoptosis-inducing ligand (TRAIL) and IP-10 have been reported to have higher plasma and serum concentrations in viral infections than in bacterial infections. TRAIL concentrations were even higher than normal in bacterial infections. This difference makes proteins interesting candidates for inclusion in diagnostic algorithms as complementary to biomarkers that primarily indicate bacterial infections. Taking advantage of this finding, an Israeli group added CRP to the algorithm, showing that the diagnostic performance of acute infections was significantly improved compared to both biomarkers [48]. The studies based on the scoring system of serum concentrations of 3 biomarkers achieved high sensitivities and specificity in the 90% range. In the Bio-X study conducted on adults, it was shown that TRAIL and IP-10 had ROC curves (AuROC) of approximately 0.80, with sensitivities varying between 65% and 82% and specificity ranging from 79% to 81%, respectively [5].

The biomarker signature composed of TRAIL, IP-10, and CRP is thought to be a promising new tool to assist emergency room doctors in determining the bacterial and viral etiology of infectious disease [49, 50].

Human Neutrophil Lipocalin (HNL)

HNL is stored as a preformed protein in secondary granules of human neutrophils and is readily released from neutrophils as a result of stimulation, and can also be produced by epithelial cells after the cells are exposed to cytokines such as TNF-α. It is a highly complex molecule as it exists as monomers, homodimers, and heterodimers. It is the dimeric form that should be targeted in the diagnosis of bacterial infection, as dimeric HNL concentrations only show the participation of neutrophils [51]. Serum and plasma measurements of HNL have shown promising diagnostic performances in patients with sepsis. The concentrations of HNL in serum have been found to be much higher in patients with acute infections caused by bacteria compared to viral infections [5, 52]. In newborns, concentrations have been found in the same range as adults with high concentrations in those diagnosed with or suspected sepsis [51]. It is predicted that it may be a valuable tool for monitoring HNL antibiotic therapy [9]. The neutrophil-specific dimeric form HNL test may mean that high serum concentrations seen in other diseases may be signs of neutrophil involvement. It can be seen in patients with chronic obstructive pulmonary disease or exacerbation of cystic fibrosis, except in patients with active rheumatoid arthritis. HNL measurements and bacterial infections can therefore be interesting tools to investigate the role of neutrophils in a variety of inflammatory conditions.

In a study by Venge, a whole blood assay with an assay time of <10 minutes in which blood neutrophils are activated to release HNL was developed to eliminate the need to produce serum, blood neutrophil counts were shown to be superior to tests for CRP, procalcitonin and CD64 expression. . Another approach to make the distinction more accurate is the combination of biomarkers. Thus, significant improvements were achieved by combining whole blood -HNL (B-HNL) with IP-10 or TRAIL. When comparing the 3-biomarker algorithm of TRAIL, IP-10, and CRP with HNL after whole blood activation, HNL appeared to be independently superior [53].

Calprotectin

Calprotectin is one of the most abundant proteins expressed in neutrophil granulocytes and found in the cytosol of neutrophil granulocytes. It makes up 40-50% of the total protein content. Calprotectin is released during the activation and conversion of neutrophils. As a result of neutrophil activation with bacterial infection, calprotectin is released into the bloodstream. As part of the body's defense against bacteria, neutrophil granulocytes are activated. Activation of neutrophils is one of the earliest responses to bacterial infection. At the site of infection, neutrophil granulocytes attack the invading bacteria by releasing their cytosol. In this process, a wide variety of proteins can be released into the bloodstream. These proteins, including calprotectin, can be used as biomarkers for the activation of neutrophil granulocytes. Because neutrophils are activated early in the course of an infection, there are studies showing that calprotectin has the potential to be an early biomarker for bacterial infections [54, 55].

Several studies have reported that plasma/serum calprotectin may be a useful early biomarker of bacterial infections in critically ill patients and in sepsis, with better predictive properties than WBC and PCT [56, 57].

Adenosin Deaminaz (ADA)

Adenosine deaminase (ADA) is an enzyme that catalyzes the deamination reaction from adenosine to inosine. It is an important enzyme of the purine catabolic pathway. ADA plays a role in the proliferation and differentiation of lymphocytes, especially T lymphocyte. It also plays a role in the maturation of monocytes, converting them into macrophages. ADA is an important indicator of active cellular immunity. In the course of infectious disease characterized by microorganisms that infect macrophages, it increases in biological fluids. ADA levels increase in tuberculosis (TB) due to the stimulation of T cells by mycobacterial antigens. ADA has been suggested to be a useful biomarker for TB, as it can be detected in body fluids such as pleural, pericardial, and peritoneal fluid. ADA is widely used for the diagnosis of tuberculosis due to its simplicity, low cost, and fast results. Several studies have confirmed the high sensitivity and specificity of ADA (sensitivity 92% and specificity 89%) for early diagnosis of EPTB such as tuberculous pleuritis, pericarditis, ascites and meningitis [58, 59]. The main reason for increased ADA levels in pleural effusion is the movement of T lymphocytes towards this area. When TB infection is brought under control, growth markers of lymphocytes are reduced. Leukocytes decrease in serum ADA levels simultaneously with the decrease in lymphocytes. Therefore, serum ADA level can be used as an index of treatment response. Mycobacterial culture, PCR etc. In addition, ADA tests can be performed in many health centers with limited diagnostic possibilities. It is also inexpensive and has good sensitivity. Culture is still the gold standard and essential for inflammatory diagnosis [59].

However, high levels of ADA have been reported in peritoneal, meningeal, pleural pericardial effusions in various non-tuberculous diseases such as liver cirrhosis, typhoid fever, infectious mononucleosis, brucellosis, and bronchogenic carcinoma, which includes simulation of cell-mediated immunity.

miRNA

MicroRNAs (miRNAs) represent a small, non-coding RNA class of 17-25 nucleotides, and their main role is to regulate the degradation of mRNA and also to regulate protein levels. In recent years, many studies have been conducted to identify the origin and full functions of miRNA, focusing on its potential uses in both research and clinical fields [6]. miRNA plays an important role in various biological processes, including cell proliferation, cell cycle, cell growth, tissue organization, apoptosis, and different metabolisms. In infectious diseases, miRNAs also play a comprehensive role in maintaining regulatory signaling at the cellular level, providing protection from disease or promoting its persistence by inducing protected or unprotected signaling pathways, respectively. In the case of disease, miRNAs are found in serum derived from infected epithelial cells and other types of immune cells. Changes in various miRNA levels in plasma, serum, urine, and saliva have been associated with various diseases. Circulating miRNAs are stable and protected from endogenous RNAse activity. Recent studies have established an important correlation between miRNA and the cause of diseases [60]. Therefore, today several miRNAs are considered as molecular biomarkers for the diagnosis of human infectious diseases such as tuberculosis, HIV/AIDS, Hepatitis, H. pylori. To date, miRNA-150 and miRNA-146b-5p in infectious diseases, such as human immunodeficiency virus (HIV); miRNA-122, miRNA-21 and miRNA-34a in hepatitis; miRNA-361-5p and miRNA-29c in tuberculosis; miRNA-16 and miRNA-451 in malaria and miRNA-181 in Helicobacter pylori infection, several miRNAs have been reported as molecular biomarkers [6, 61, 62]. MiRNAs as biomarkers are an impressive area of research, despite some limitations. It is anticipated that miRNAs will become a routine approach to the development of personalized patient profiles as current techniques improve.

Myxoma Resistance Protein (MxA1)

The MxA1 protein shows promise compared to other markers as a biomarker that tends to be significantly elevated during viral infection rather than a bacterial infection. Type I or III Interferon (IFN) can activate MxA1 but not type II IFN signaling pathway or direct interaction of bacteria or viruses. IFN is also increased in autoimmune conditions and some hematological cancers. Therefore, the value of IFN-induced MxA1 in this population may not be significant. In a study analyzing the role of MxA1, a cut-off value of 200 ng/mL was 96.4% sensitive and 66.7% specific for identifying patients with viral CAP (Community-acquired pneumonia). This study also assumed that when a bacterial infection is diagnosed, and a high level of MxA1 is detected, this is an indication that the bacterial organism precedes a viral cause [63]. This is because MxA1 remains high for about ten days after a viral damage compared to IFN, which has a very short half-life [64]. When the authors also establish a correlation with CRP. They stated that low CRP (<40mg/l) and MxA1> 200ng/ml might indicate a high rate of viral etiology, but further studies are needed to confirm the diagnostic value of MxA1 [65].

Lipocalin 2 (Lcn2)

Lcn2 protein is stored and released by neutrophils that disrupt iron transport within bacteria. This marker attracts great attention in the diagnosis of the etiological factor of Community-acquired pneumonia (CAP). One study concluded that Lcn2 is a poor predictor compared to CRP or WBC [66]. It has been reported that Lcn2 can be used to help differentiate bacterial and viral pneumonia [67].

FUTURE BİOMARKERS

Genome-Based Sepsis Biomarkers

Genome-based biomarkers are expected to advance rapidly in the near future as a result of advances in analysis techniques. Data collection and interrogation of DNA sequences for a large number of septic patients showed that there are many genes that change the outcome of sepsis [68]. Genes that regulate the immune system are affected during systemic inflammation and sepsis. Changes in gene transcription that affect cytokine synthesis, cytokine receptor expression, T-cell differentiation, protein synthesis regulation, and apoptosis regulators can be easily observed in severely septic patients. This has led to the development of gene expression assays in biomarkers for sepsis [69]. Genome-based sepsis biomarkers may be a new hope for differentiating sterile inflammation from sepsis.

Proteomics and Sepsis Markers

Proteomics is a biological study area in terms of revealing its importance for the whole organism for understanding the functions of proteins, their mutual communication with other proteins, where, for how long, and in what form they are in the cell. The purpose of examining proteins in sepsis is because secreted proteins are signal systems that convert genetic signals into enzymatic activity [68]. Advances in systems biology and technical aspects of proteomics will bring the understanding of sepsis to a new level in the future rather than measuring individual proteins, and early experience with proteomics and systems biology will reveal many interesting findings in experimental models and sepsis.

Intestinal Microbiota

The microbiome is known the gut microbiome play an important role in the development and maturation of the immune system and protection against pathogens. Therefore, intestinal dysbiosis is considered to be a potent biomarker for the development and progression of sepsis. Among patients with sepsis, the abundance of Enterococcus species has increased in non-survivors and

Enterococcus species have been suggested as a potential prognostic biomarker of sepsis [70].

Mitochondrial Markers

It has become increasingly clear that mitochondria are involved in the innate immune system. Mitochondria are associated with energy production through oxidative phosphorylation, but also play a role in numerous other functions such as ion homeostasis, cellular signaling, differentiation, and cell death/survival.

In studies on RNA viruses, viral proteins responsible for inhibiting mitochondrial pathways have been identified. Pathways that are inhibited in HIV and Hepatitis B infections have been demonstrated. As far as it is known from studies on SARS-CoV, changes in mitochondria occur in Coronavirus infections. Many pathogens, including RNA viruses such as CoVs, have developed targeted strategies to capture mitochondria, damage them or use them for purposes such as survival, proliferation or avoidance of host immunity. Some viruses use mitochondrial membranes for genome replication. It has been shown that in HIV infection, viral RNAs can also enter the mitochondria, compromising mitochondrial integrity. Recent study shows that SARS-CoV-2 RNA is preferentially localized to mitochondria [71]. Changes in mitochondria in Covid-19 sepsis have been shown in some studies. It is not known which of these are virus specific. Experimental data are not yet sufficient. In SARS-CoV-, the accumulation of RNA in the mitochondria can likewise impair mitochondrial function, potentially causing a decrease in energy and increased ROS production [72]. These data may pave the way for the development of mitochondrial biomarkers for response to SARS-CoV-2-infection, vulnerability and expected disease severity. If significant mitochondrial dysfunction occurs in COVID-19, it would be interesting to consider the use of accessible mitochondrial biomarkers such as oxidative in monitoring the susceptibility to infection and the severity of disease progression, for example in the blood. In addition, targeting common features of a wide array of viruses can be an important approach for future outbreaks, especially unless effective antibody treatments are available. A better understanding of mitochondrial involvement in immune functions will eventually contribute to our understanding of the role of infections and inflammation in chronic diseases and aging.

CONCLUSION

There is no biomarker effective in differentiating bacterial and viral infections. Their use in combination is beneficial. New biomarkers are strongly needed to support the next generation of diagnostic platforms. In recent years, research on how to use biomarkers more effectively has been increasing in the literature, and valuable information has been provided. More studies are needed in the future for the use of bacterial and viral infections in the diagnosis, treatment and prognosis follow-up of sepsis. In the near future, it is possible that biomarkers will help clinicians treat infections in a personalized way, reducing uncertainty and increasing patient safety.

CONSENT FOR PUBLICATION

Not applicable.

CONFLICT OF INTEREST

The author declares no conflict of interest, financial or otherwise.

ACKNOWLEDGEMENT

Declared none.

REFERENCES

[1] Hwang H, Hwang BY, Bueno J. Biomarkers in Infectious Diseases. Dis Markers 2018; 2018: 8509127.
[http://dx.doi.org/10.1155/2018/8509127]

[2] Zoumpourlis V, Goulielmaki M, Rizos E, Baliou S, Spandidos DA. The COVID-19 pandemic as a scientific and social challenge in the 21st century. Mol Med Rep 2020; 22(4): 3035-48.
[http://dx.doi.org/10.3892/mmr.2020.11393] [PMID: 32945405]

[3] https://covid19.who.int

[4] Levy MM, Fink MP, Marshall JC, *et al.* 2001 SCC/ESICM/ACCP/ATS/SIS International sepsis definition conference. Crit Care Med 2003; 31(4): 1250-6.
[http://dx.doi.org/10.1097/01.CCM.0000050454.01978.3B] [PMID: 12682500]

[5] Venge P, Xu S. Diagnosis and Monitoring of Acute Infections with Emphasis on the Novel Biomarker Human Neutrophil Lipocalin. J Appl Lab Med 2019; 3(4): 664-74.
[http://dx.doi.org/10.1373/jalm.2018.026369] [PMID: 31639734]

[6] Chakraborty A, Monica S. Host biomarkers for early diagnosis of infectious diseases: A comprehensive review. Int J Clin Microbiol Biochem Technol 2019; 2: 001-7.
[http://dx.doi.org/10.29328/journal.ijcmbt.1001005]

[7] Zhang X, Guo J, Fan S, *et al.* Screening and identification of six serum microRNAs as novel potential combination biomarkers for pulmonary tuberculosis diagnosis. PLoS One 2013; 8(12): e81076.
[http://dx.doi.org/10.1371/journal.pone.0081076] [PMID: 24349033]

[8] Weh J, Antoni C, Weiß C, Findeisen P, Ebert M, Böcker U. Discriminatory potential of C-reactive protein, cytokines, and fecal markers in infectious gastroenteritis in adults. Diagn Microbiol Infect Dis 2013; 77(1): 79-84.
[http://dx.doi.org/10.1016/j.diagmicrobio.2013.05.005] [PMID: 23773676]

[9] Fjaertoft G, Foucard T, Xu S, Venge P. Human neutrophil lipocalin (HNL) as a diagnostic tool in children with acute infections: a study of the kinetics. Acta Paediatr 2005; 94(6): 661-6.
[http://dx.doi.org/10.1111/j.1651-2227.2005.tb01961.x] [PMID: 16188765]

[10] Müller B, Becker KL. Procalcitonin: how a hormone became a marker and mediator of sepsis. Swiss

Med Wkly 2001; 131(41-42): 595-602.
[PMID: 11820070]

[11] Schuetz P, Albrich W, Mueller B. Procalcitonin for diagnosis of infection and guide to antibiotic decisions: past, present and future. BMC Med 2011; 9(1): 107.
[http://dx.doi.org/10.1186/1741-7015-9-107] [PMID: 21936959]

[12] Kopterides P, Siempos II, Tsangaris I, Tsantes A, Armaganidis A. Procalcitonin-guided algorithms of antibiotic therapy in the intensive care unit: a systematic review and meta-analysis of randomized controlled trials. Crit Care Med 2010; 38(11): 2229-41.
[http://dx.doi.org/10.1097/CCM.0b013e3181f17bf9] [PMID: 20729729]

[13] Tian R, Wang X, Pan T, *et al.* Plasma PTX3, MCP1 and Ang2 are early biomarkers to evaluate the severity of sepsis and septic shock. Scand J Immunol 2019; 90(6): e12823.
[http://dx.doi.org/10.1111/sji.12823] [PMID: 31489646]

[14] Kim MH, Choi JH. An Update on Sepsis Biomarkers. Infect Chemother 2020; 52(1): 1-18.
[http://dx.doi.org/10.3947/ic.2020.52.1.1] [PMID: 32239808]

[15] Yapakçi E, Tarcan A, Celik B, Özbek N, Gürakan B. Serum pro-hepcidin levels in term and preterm newborns with sepsis. Pediatr Int 2009; 51(2): 289-92.
[http://dx.doi.org/10.1111/j.1442-200X.2008.02688.x] [PMID: 19405934]

[16] Arabacı C. Current and future sepsis biomarkers.Current Microbiology Studies. 2nd ed. Ankara: Academician Bookstore press 2019; pp. 11-27.

[17] Jerala R. Structural biology of the LPS recognition. Int J Med Microbiol 2007; 297(5): 353-63.
[http://dx.doi.org/10.1016/j.ijmm.2007.04.001] [PMID: 17481951]

[18] Qiu P, Cui X, Barochia A, Li Y, Natanson C, Eichacker PQ. The evolving experience with therapeutic TNF inhibition in sepsis: considering the potential influence of risk of death. Expert Opin Investig Drugs 2011; 20(11): 1555-64.
[http://dx.doi.org/10.1517/13543784.2011.623125] [PMID: 21961576]

[19] Shahzad A, Knapp M, Lang I, Köhler G. Interleukin 8 (IL-8) - a universal biomarker? Int Arch Med 2010; 3(1): 11.
[http://dx.doi.org/10.1186/1755-7682-3-11] [PMID: 20550702]

[20] Niehues N. C-reactive protein and other biomarkers-the sense and non-sense of using inflammation biomarkers for the diagnosis of severe bacterial infection. LymphoSign Journal 2018; 5(2): 35-47.
[http://dx.doi.org/10.14785/lymphosign-2018-0001]

[21] Venereau E, De Leo F, Mezzapelle R, Careccia G, Musco G, Bianchi ME. HMGB1 as biomarker and drug target. Pharmacol Res 2016; 111: 534-44.
[http://dx.doi.org/10.1016/j.phrs.2016.06.031] [PMID: 27378565]

[22] Thomas J, Pociute A, Kevalas R, Malinauskas M, Jankauskaite L. Thomas j, Pociute A, Kevalas R, Malinauskas M, Jankauskaite L. Blood biomarkers differentiating viral *versus* bacterial pneumonia aetiology: a literature review. Ital J Pediatr 2020; 46(1): 4.
[http://dx.doi.org/10.1186/s13052-020-0770-3]

[23] Zhou WF, Chen Q, Jin MF, *et al.* The diagnostic accuracy of high-mobility group box 1 protein and twelve other markers in discriminating bacterial, viral and co-infected bronchial pneumonia in Han children. Microbiol Immunol 2011; 55(4): 279-88.
[http://dx.doi.org/10.1111/j.1348-0421.2011.00306.x] [PMID: 21223368]

[24] Gibot S, Massin F, Cravoisy A, *et al.* High-mobility group box 1 protein plasma concentrations during septic shock. Intensive Care Med 2007; 33(8): 1347-53.
[http://dx.doi.org/10.1007/s00134-007-0691-2] [PMID: 17525840]

[25] Linscheid P, Seboek D, Zulewski H, Keller U, Müller B. Autocrine/paracrine role of inflammation-mediated calcitonin gene-related peptide and adrenomedullin expression in human adipose tissue. Endocrinology 2005; 146(6): 2699-708.

[http://dx.doi.org/10.1210/en.2004-1424] [PMID: 15761041]

[26] Al Shuaibi M, Bahu RR, Chaftari AM, *et al.* Pro-adrenomedullin as a novel biomarker for predicting infections and response to antimicrobials in febrile patients with hematologic malignancies. Clin Infect Dis 2013; 56(7): 943-50.
[http://dx.doi.org/10.1093/cid/cis1029] [PMID: 23288950]

[27] Christ-Crain M, Jaccard-Stolz D, Bingisser R, *et al.* Effect of procalcitonin-guided treatment on antibiotic use and outcome in lower respiratory tract infections: cluster-randomised, single-blinded intervention trial. Lancet 2004; 363(9409): 600-7.
[http://dx.doi.org/10.1016/S0140-6736(04)15591-8] [PMID: 14987884]

[28] Hatada T, Wada H, Nobori T, *et al.* Plasma concentrations and importance of High Mobility Group Box protein in the prognosis of organ failure in patients with disseminated intravascular coagulation. Thromb Haemost 2005; 94(5): 975-9.
[http://dx.doi.org/10.1160/TH05-05-0316] [PMID: 16363239]

[29] Schuetz P, Christ-Crain M, Zimmerli W, Mueller B. Repeated measurements of endothelin-1 precursor peptides predict the outcome in community-acquired pneumonia. Intensive Care Med 2011; 37(6): 970-80.
[http://dx.doi.org/10.1007/s00134-011-2208-2] [PMID: 21394616]

[30] David S, Kümpers P, van Slyke P, Parikh SM. Mending leaky blood vessels: the angiopoietin-Tie2 pathway in sepsis. J Pharmacol Exp Ther 2013; 345(1): 2-6.
[http://dx.doi.org/10.1124/jpet.112.201061] [PMID: 23378191]

[31] Ricciuto DR, dos Santos CC, Hawkes M, *et al.* Angiopoietin-1 and angiopoietin-2 as clinically informative prognostic biomarkers of morbidity and mortality in severe sepsis. Crit Care Med 2011; 39(4): 702-10.
[http://dx.doi.org/10.1097/CCM.0b013e318206d285] [PMID: 21242795]

[32] Delanghe JR, Speeckaert MM. Translational research and biomarkers in neonatal sepsis. Clin Chim Acta 2015; 451(Pt A): 46-64.
[http://dx.doi.org/10.1016/j.cca.2015.01.031] [PMID: 25661089]

[33] Elkareem RMA, Ahmed HM, Meabed MH, Elias SS, Elmaraghy MA. Diagnostic value of CD64 in early detection of neonatal sepsis. Comp Clin Pathol 2020; 29(3): 639-43.
[http://dx.doi.org/10.1007/s00580-020-03100-4]

[34] Hashem HE, El Masry SA, Mokhtar AM, Ismail EA, Abdelaal NM. Valuable role of neutrophil CD64 and highly sensitive CRP biomarkers for diagnostic, monitoring, and prognostic evaluations of sepsis patients in neonatal ICUs. Hindawi BioMed Research International 2020; 1-12.

[35] Gros A, Roussel M, Sauvadet E, *et al.* The sensitivity of neutrophil CD64 expression as a biomarker of bacterial infection is low in critically ill patients. Intensive Care Med 2012; 38(3): 445-52.
[http://dx.doi.org/10.1007/s00134-012-2483-6] [PMID: 22310872]

[36] Lu B, Zhang Y, Li C, *et al.* The utility of presepsin in diagnosis and risk stratification for the emergency patients with sepsis. Am J Emerg Med 2018; 36(8): 1341-5.
[http://dx.doi.org/10.1016/j.ajem.2017.12.038] [PMID: 29276032]

[37] de Oliveira Matos A, Dos Santos Dantas PH, Figueira Marques Silva-Sales M, Sales-Campos H. The role of the triggering receptor expressed on myeloid cells-1 (TREM-1) in non-bacterial infections. Crit Rev Microbiol 2020; 46(3): 237-52.
[http://dx.doi.org/10.1080/1040841X.2020.1751060] [PMID: 32326783]

[38] Dupuy AM, Philippart F, Péan Y, *et al.* Role of biomarkers in the management of antibiotic therapy: an expert panel review: I - currently available biomarkers for clinical use in acute infections. Ann Intensive Care 2013; 3(1): 22.
[http://dx.doi.org/10.1186/2110-5820-3-22] [PMID: 23837559]

[39] Bopp C, Hofer S, Bouchon A, Zimmermann JB, Martin E, Weigand MA. Soluble TREM-1 is not

suitable for distinguishing between systemic inflammatory response syndrome and sepsis survivors and nonsurvivors in the early stage of acute inflammation. Eur J Anaesthesiol 2009; 26(6): 504-7. [http://dx.doi.org/10.1097/EJA.0b013e328329afca] [PMID: 19307971]

[40] Andaluz-Ojeda D, Bobillo F, Iglesias V, *et al.* A combined score of pro- and anti-inflammatory interleukins improves mortality prediction in severe sepsis. Cytokine 2012; 57(3): 332-6. [http://dx.doi.org/10.1016/j.cyto.2011.12.002] [PMID: 22197776]

[41] Gibot S, Béné MC, Noel R, *et al.* Combination biomarkers to diagnose sepsis in the critically ill patient. Am J Respir Crit Care Med 2012; 186(1): 65-71. [http://dx.doi.org/10.1164/rccm.201201-0037OC] [PMID: 22538802]

[42] Andersen LW, Mackenhauer J, Roberts JC, Berg KM, Cocchi MN, Donnino MW. Etiology and therapeutic approach to elevated lactate levels. Mayo Clin Proc 2013; 88(10): 1127-40. [http://dx.doi.org/10.1016/j.mayocp.2013.06.012] [PMID: 24079682]

[43] Dellinger RP, Levy MM, Rhodes A, *et al.* Surviving sepsis campaign: international guidelines for management of severe sepsis and septic shock: 2012. Crit Care Med 2013; 41(2): 580-637. [http://dx.doi.org/10.1097/CCM.0b013e31827e83af] [PMID: 23353941]

[44] de Almeida SM, Furlan SMP, Cretella AMM, *et al.* Comparison of cerebrospinal fluid biomarkers for differential diagnosis of acute bacterial and viral meningitis with atypical cerebrospinal fluid characteristics. Med Princ Pract 2020; 29(3): 244-54. [http://dx.doi.org/10.1159/000501925] [PMID: 31480054]

[45] Eck M, Schmausser B, Scheller K, *et al.* CXC chemokines Gro(alpha)/IL-8 and IP-10/MIG in Helicobacter pylori gastritis. Clin Exp Immunol 2000; 122(2): 192-9. [http://dx.doi.org/10.1046/j.1365-2249.2000.01374.x] [PMID: 11091274]

[46] Lettinga KD, Weijer S, Speelman P, Prins JM, Van Der Poll T, Verbon A. Reduced interferon-gamma release in patients recovered from Legionnaires' disease. Thorax 2003; 58(1): 63-7. [http://dx.doi.org/10.1136/thorax.58.1.63] [PMID: 12511724]

[47] Liu M, Guo S, Hibbert JM, *et al.* CXCL10/IP-10 in infectious diseases pathogenesis and potential therapeutic implications. Cytokine Growth Factor Rev 2011; 22(3): 121-30. [http://dx.doi.org/10.1016/j.cytogfr.2011.06.001] [PMID: 21802343]

[48] Oved K, Cohen A, Boico O, *et al.* A novel host-proteome signature for distinguishing between acute bacterial and viral infections. PLoS One 2015; 10(3): e0120012. [http://dx.doi.org/10.1371/journal.pone.0120012] [PMID: 25785720]

[49] van Houten CB, de Groot JAH, Klein A, *et al.* A host-protein based assay to differentiate between bacterial and viral infections in preschool children (OPPORTUNITY): a double-blind, multicentre, validation study. Lancet Infect Dis 2017; 17(4): 431-40. [http://dx.doi.org/10.1016/S1473-3099(16)30519-9] [PMID: 28012942]

[50] Eden E, Srugo I, Gottlieb T, *et al.* Diagnostic accuracy of a TRAIL, IP-10 and CRP combination for discriminating bacterial and viral etiologies at the Emergency Department. J Infect 2016; 73(2): 177-80. [http://dx.doi.org/10.1016/j.jinf.2016.05.002] [PMID: 27255416]

[51] Yu Z, Jing H, Hongtao P, *et al.* Distinction between bacterial and viral infections by serum measurement of human neutrophil lipocalin (HNL) and the impact of antibody selection. J Immunol Methods 2016; 432: 82-6. [http://dx.doi.org/10.1016/j.jim.2016.02.014] [PMID: 26899825]

[52] Mårtensson J, Bell M, Xu S, *et al.* Association of plasma neutrophil gelatinase-associated lipocalin (NGAL) with sepsis and acute kidney dysfunction. Biomarkers 2013; 18(4): 349-56. [http://dx.doi.org/10.3109/1354750X.2013.787460] [PMID: 23627612]

[53] Venge P. Human Neutrophil Lipocalin (HNL) in the Distinction between Bacterial and Viral Infections. J Microbiol Biotechnol 2018; 7(1): 7-14.

[54] Havelka A, Sejersen K, Venge P, Pauksens K, Larsson A. Calprotectin, a new biomarkerfor diagnosis of acute respiratory infections. Sci Rep 2020; 10: 4208.

[55] Lipcsey M, Hanslin K, Stålberg J, Smekal D, Larsson A. The time course of calprotectin liberation from human neutrophil granulocytes after *Escherichia coli* and endotoxin challenge. Innate Immun 2019; 25(6): 369-73. [http://dx.doi.org/10.1177/1753425919848476] [PMID: 31109223]

[56] Huang L, Li J, Han Y, *et al.* Serum Calprotectin Expression as a Diagnostic Marker for Sepsis in Postoperative Intensive Care Unit Patients. J Interferon Cytokine Res 2016; 36(10): 607-16. [http://dx.doi.org/10.1089/jir.2016.0037] [PMID: 27610929]

[57] Bartáková E, Štefan M, Stráníková A, *et al.* Calprotectin and calgranulin C serum levels in bacterial sepsis. Diagn Microbiol Infect Dis 2019; 93(3): 219-26. [http://dx.doi.org/10.1016/j.diagmicrobio.2018.10.006] [PMID: 30420210]

[58] Saleh MA, Hammad E, Ramadan MM, Abd El-Rahman A, Enein AF. Use of adenosine deaminase measurements and QuantiFERON in the rapid diagnosis of tuberculous peritonitis. J Med Microbiol 2012; 61(Pt 4): 514-9. [http://dx.doi.org/10.1099/jmm.0.035121-0] [PMID: 22174374]

[59] Barua R, Hossain M. Adenosine deaminase in diagnosis of tuberculosis: a review. Anwer Khan Modern Medical College Journal 2014; 5(2): 43-8. [http://dx.doi.org/10.3329/akmmcj.v5i2.21132]

[60] Pandey RK, Sundar S, Prajapati VK. Differential Expression of miRNA Regulates Tcell differentiation and plasticity during visceral leishmaniasis infection. Front Microbiol 2016; 7: 206. [http://dx.doi.org/10.3389/fmicb.2016.00206] [PMID: 26941729]

[61] Verma P, Pandey RK, Prajapati P, Prajapati VK. Circulating microRNAs: potential and emerging biomarkers for diagnosis of human infectious diseases. Front Microbiol 2016; 7: 1274. [http://dx.doi.org/10.3389/fmicb.2016.01274] [PMID: 27574520]

[62] Ojha R, Nandani R, Pandey RK, Mishra A, Prajapati VK. Emerging role of circulating microRNA in the diagnosis of human infectious diseases. J Cell Physiol 2019; 234(2): 1030-43. [http://dx.doi.org/10.1002/jcp.27127] [PMID: 30146762]

[63] Engelmann I, Dubos F, Lobert P-E, *et al.* Diagnosis of viral infections using myxovirus resistance protein A (MxA). Pediatrics 2015; 135(4): e985-93. [http://dx.doi.org/10.1542/peds.2014-1946] [PMID: 25802344]

[64] Haller O, Kochs G. Human MxA protein: an interferon-induced dynamin-like GTPase with broad antiviral activity. J Interferon Cytokine Res 2011; 31(1): 79-87. [http://dx.doi.org/10.1089/jir.2010.0076] [PMID: 21166595]

[65] Hoshina T, Nanishi E, Kanno S, Nishio H, Kusuhara K, Hara T. The utility of biomarkers in differentiating bacterial from non-bacterial lower respiratory tract infection in hospitalized children: difference of the diagnostic performance between acute pneumonia and bronchitis. J Infect Chemother 2014; 20(10): 616-20. [http://dx.doi.org/10.1016/j.jiac.2014.06.003] [PMID: 25027057]

[66] Esposito S, Bianchini S, Gambino M, *et al.* Measurement of lipocalin-2 and syndecan-4 levels to differentiate bacterial from viral infection in children with community-acquired pneumonia. BMC Pulm Med 2016; 16(1): 103. [http://dx.doi.org/10.1186/s12890-016-0267-4] [PMID: 27439403]

[67] Naydenova E, Tsanas A, Howie S, Casals-Pascual C, De Vos M. The power of data mining in diagnosis of childhood pneumonia. J R Soc Interface 2016; 13(120): 1-10. [http://dx.doi.org/10.1098/rsif.2016.0266] [PMID: 27466436]

[68] LaRosa SP, Opal SM. Biomarkers: the future. Crit Care Clin 2011; 27(2): 407-19. [http://dx.doi.org/10.1016/j.ccc.2010.12.012] [PMID: 21440209]

[69] Johnson SB, Lissauer M, Bochicchio GV, Moore R, Cross AS, Scalea TM. Gene expression profiles differentiate between sterile SIRS and early sepsis. Ann Surg 2007; 245(4): 611-21. [http://dx.doi.org/10.1097/01.sla.0000251619.10648.32] [PMID: 17414611]

[70] Agudelo-Ochoa GM, Valdés-Duque BE, Giraldo-Giraldo NA, *et al.* Gut microbiota profiles in critically ill patients, potential biomarkers and risk variables for sepsis. Gut Microbes 2020; 12(1): 1707610.
[http://dx.doi.org/10.1080/19490976.2019.1707610] [PMID: 31924126]

[71] Wu KE, Fazal FM, Parker KR, Zou J, Chang HY. RNA-GPS Predicts SARS-CoV-2 RNA Residency to Host Mitochondria and Nucleolus. Cell Syst 2020; 11(1): 102-108.e3. [http://dx.doi.org/10.1016/j.cels.2020.06.008] [PMID: 32673562]

[72] Burtscher J, Cappellano G, Omori A, Koshiba T, Millet GP. Mitochondria: In the Cross Fire of SARS-CoV-2 and Immunity. iScience 2020; 23(10): 101631. [http://dx.doi.org/10.1016/j.isci.2020.101631] [PMID: 33015593]

SUBJECT INDEX

A

Pınar Atukeren and Hafize Uzun (Eds.)

B

D

E

F

G

H

I

J

K

L

N

O

P

Q

R

S

T

U

V

W

X

Z

Made in the USA
Las Vegas, NV
03 November 2026

57424372R00440